Computer Imaging

Digital Image Analysis and Processing

Computer Imaging

Digital Image Analysis and Processing

Scott E Umbaugh

Southern Illinois University Edwardsville

Boca Raton London New York Singapore

A CRC title, part of the Taylor & Francis imprint, a member of the Taylor & Francis Group, the academic division of T&F Informa plc.

Library of Congress Cataloging-in-Publication Data

Umbaugh, Scott E.
Computer imaging : digital image analysis and processing / Scott E. Umbaugh.
p. cm.
Includes bibliographical references and index.
ISBN 0-8493-2919-1 (alk. paper)
1. Computer vision. 2. Image processing—Digital techniques. I. Title.

TA1634.U45 2005
006.4'2--dc22 2004058331

Direct all inquiries to CRC Press, 2000 N.W. Corporate Blvd., Boca Raton, Florida 33431.

Visit the CRC Press Web site at www.crcpress.com

International Standard Book Number 0-8493-2919-1
Library of Congress Card Number 2004058331
Printed in the United States of America 1 2 3 4 5 6 7 8 9 0
Printed on acid-free paper

To the survivors
Jeanie, Michael, Robin and David

Preface

Computer imaging is a rapidly expanding field, with applications in areas such as medicine, space exploration, and the entertainment industry. The diversity of applications is one of the driving forces that make it such an exciting topic to be involved in for the 21st century. *Computer imaging* can be defined as the acquisition and processing of visual information by computer. This book presents a unique approach to the practice of computer imaging, and will be of interest to those who want to learn about and use computer imaging techniques.

Computer imaging can be divided into two primary application areas, computer vision and image processing, with image analysis tying these two together. Although the book focuses on image analysis and processing, the image analysis part provides the reader with the tools necessary for developing computer vision applications. The automatic classification of abnormalities in medical images and the robotic control of a Mars rover are examples of computer vision applications. Image processing applications involve manipulation of image data for viewing by people. Examples include special effects imaging for motion pictures, and the restoration of satellite images distorted by atmospheric disturbance.

Why Write this Book?

This book takes an engineering approach to computer imaging and brings together image analysis and image processing into a unified framework that provides a useful paradigm for both computer vision and image processing applications. Additionally, the theoretical foundation is presented as needed in order to fully understand the material. Although theoretical-based textbooks are available, they do not really take what I consider an engineering approach. I felt that there was a need for an application-oriented book that would bring image analysis and processing together in a unified framework, and this book fills that gap.

The book's development was initiated by my experience using my first book in my imaging courses. The first book was also written to take an engineering approach, but was targeted more at working professionals. For the new book I wanted to reorganize, update, expand, and add more materials that make it more useful as a textbook. To meet those requirements I have reorganized the topics and added key aspect summaries and exercises. Additionally, the original CVIPtools software was UNIX-based and it reached the point where almost 95% of the users are using the Windows operating system. Although a Windows port of the UNIX version of CVIPtools is available, it seemed that the creation of a real Windows version of CVIPtools would be greatly welcomed by the majority of our users. The new CVIPtools is integrated throughout the book, which provides the readers with practical examples and creates an environment where they are encouraged to do more exploration on their own. In addition to CVIPtools providing a valuable environment for learning about computer imaging, it also contains a set of reusable tools for application development that can be used in projects and research.

Who Will Use this Book?

Computer Imaging: Digital Image Analysis and Processing is intended for use by the academic community in teaching and research, as well as working professionals doing research and development in the commercial sectors. This includes image analysis, computer vision and image processing academics and practicing engineers, consultants, and programmers, as well as those in the graphics fields, medical imaging professionals, multimedia specialists, and others. The book can be used for self study and is of interest to anyone involved with developing computer imaging applications, whether they are engineers, geographers, biologists, oceanographers, or astronomers. At the university it can be used as a text in standard computer vision and/or image processing senior-level or graduate courses, or may be used at any level in an application-oriented course. One essential component that is missing from standard theoretical textbooks is a conceptual presentation of the material, which is fundamental to gaining a solid understanding of these complex topics. Additionally, this book provides the theory necessary to understand the foundations of computer imaging, as well as that which is needed for new algorithm development.

The prerequisites for the book are an interest in the field, a basic background in computers, and a basic math background provided in an undergraduate science or engineering program. Knowledge of the C or C++ programming language will be necessary for those intending to develop algorithms at the programming level. Some background in signal and system theory is required for those intending to gain a deep understanding of the sections on transforms and compression. However, the book is written so that those without this background can learn to *use* the tools and achieve a conceptual understanding of the material.

Approach

To help motivate the reader I have taken an approach to learning that presents topics as needed. This approach stars by presenting a global model to help gain an understanding of the overall process, followed by a breakdown and explanation of each individual topic. Instead of presenting techniques or mathematical tools when they fit into a nice, neat *theoretical* framework, topics are presented as they become necessary for understanding the practical imaging model under study. This approach provides the reader with the motivation to learn about and use the tools and topics, because they see an immediate need for them. For example, the mathematical process of convolution is introduced when it is needed for an image zoom algorithm, and morphological operations are introduced when morphological filtering operations are needed after image segmentation. This approach also makes the book more useful as a reference, or for those who may not work through the book sequentially, but will reference a specific section as the need arises.

Organization of the Book

The book is divided into five major sections: I. Introduction to Computer Imaging, II. Digital Image Analysis, III. Digital Image Processing. IV. Programming with CVIPtools, and V. Appendices. The first section of the book contains all of the basic concepts and definitions necessary to understand computer imaging. The second section describes image analysis and provides the tools, concepts, and models required to analyze digital images and develop computer vision applications. Section III discusses topics and application areas for the processing of images for human consumption, so it starts with a chapter on visual perception. Each chapter includes numerous references and examples for the material presented. The material is presented in a conceptual and application-oriented manner, so that the reader will immediately understand how each topic fits into the overall framework of computer imaging applications development.

The programming section of the book, Section IV, provides all the necessary information required to use the CVIPtools environment for algorithm development. This section also includes information to assist with the implementation of the programming exercises included with each chapter. The appendices also contain reference material for use with CVIPtools, as well as other useful computer imaging related information.

Using the Book in Your Courses

The book is intended for use in both image processing and computer vision courses. Both types of courses will use the introductory chapters in Section I. After the introduction, computer vision courses will concentrate on Section II, Digital Image Analysis, where the introductory chapter presents a model of image analysis and concludes with the development of a pattern classification algorithm for geometric objects in images. This model provides a foundation for all the tools that are developed and discussed throughout Section II. Digital image processing courses will focus on Section III, Digital Image Processing, which contains an introductory chapter on human visual perception, followed by chapters on image enhancement, image restoration, and image compression. Most image processing courses will also want to cover Chapter 5 on image transforms. Both computer vision and image processing courses can use the programming parts of the book, depending on the desire of the instructor. I encourage all who use the book to explore the programming exercises because they provide a valuable tool to learn about computer imaging. There are also many tutorial exercises using CVIPtools included with each chapter, which provide hands-on experience and allow the user to gain insight into the various algorithms and parameters. The following table outlines using the book in your course.

SENIOR LEVEL/GRADUATE COURSE TITLES	REQUIRED CHAPTERS	ADDITIONAL/OPTIONAL CHAPTERS/SECTIONS	REFERENCE CHAPTERS
• **Image Analysis** • **Computer Vision** • **Machine Vision**	1, 2, 3, 4, 5.1, 5.2, 5.7, 6, 11	5.3, 5.4, 5.5, 5.6, 5.8	12, Appendices
• **Digital Image Processing** • **Digital Picture Processing** • **Image Processing**	1, 2, 3.1, 3.2, 5.1, 5.2, 5.7, 5.8, 7, 8 9, 10	5.3, 5.4, 5.5, 5.6, 11	12, Appendices

After the CVIPtools environment is installed from the CD, an image database will be in the default images directory, which contains the images used in the book. In addition to the CVIPtools environment, other material useful to instructor is contained on the CD. This includes all the book figures in linked html format, which can be readily used during lectures via a web browser or any other software tool that can display html pages. Additionally, the Word documents of all the original figures are available on the CD. The web site itself, *www.ee.siue.edu/CVIPtools*, is a resource which has many useful imaging examples, information and links to other imaging web sites.

The CVIPtools Software Development Environment

The software development environment includes an extensive set of standard C libraries, a skeleton program for using the C libraries called *CVIPlab*, a dynamically linked library (cviptools.dll) based on the common object module (COM) interface, and a GUI-based program for the exploration of computer imaging called *CVIPtools*. The CVIPlab program and all the standard libraries are all ANSI-C compatible. The new version of CVIPtools has been developed exclusively for the Windows operating system, but various UNIX versions are available at the web site (*www.ee.siue.edu/CVIPtools*). The CVIPtools software, the libraries, the CVIPlab program, images used in the textbook, and associated documentation are included on the CD-ROM.

The CVIPtools software has been used in projects funded by the National Institutes of Health, the United States Department of Defense, and numerous corporations in the commercial sector. CVIPtools has been used for applications in the medical, aerospace, printing, and manufacturing fields in applications such as the development of a helicopter simulation, automated classification of lumber, skin tumor evaluation and analysis, embedded image processing for print technology, and the automatic classification of defects in microdisplay chips. Since it is a university-sponsored project, it is continually being upgraded and expanded, and updates are available via the Internet (see Appendix B). This software allows the reader to learn about imaging topics in an interactive and exploratory manner, and to develop their own programming expertise with the CVIPlab program and the associated laboratory exercises. With the CVIPlab program they cab link any of the already defined CVIPtools functions, ranging from general purpose input/output and matrix functions to more advanced transform functions and complex imaging algorithms; some of these functions are state-of-the-art algorithms since CVIPtools is continually being improved at the Computer Vision and Image Processing Laboratory at Southern Illinois University Edwardsville (SIUE).

Author

Dr. Scott E Umbaugh is professor of electrical and computer engineering and a member of the graduate faculty at Southern Illinois University Edwardsville (SIUE). He is also the director of the Computer Vision and Image Processing (CVIP) Laboratory at SIUE. He has been teaching computer vision and image processing, as well as computer and electrical engineering design, for over 15 years. His professional interests include computer imaging education, computer imaging research, and engineering design education.

Prior to this academic career, Dr. Umbaugh spent 6 years as a computer design engineer and project manager in the avionics and telephone industries. He has been a computer imaging consultant since 1986 and has provided consulting services for the aerospace, medical, and manufacturing industries with projects ranging from helicopter simulation development to automatic identification of defects in microdisplay chips. He has performed research and development for projects funded by the National Institutes of Health, the National Science Foundation, and the U.S. Department of Defense.

Dr. Umbaugh served on the editorial board for the *IEEE Engineering in Medicine in Biology Magazine* for 8 years and is currently an associate editor for the *Pattern Recognition* journal. He served as a reviewer for a variety of *IEEE* journals and has evaluated research monographs and textbooks in the computer imaging field. He has written a previous book on computer vision and image processing, has authored numerous papers, and co-authored several book chapters.

Dr. Umbaugh received his B.S.E. degree with honors from Southern Illinois University Edwardsville in 1982, M.S.E.E. in 1987 and Ph.D. in 1990 from the University of Missouri-Rolla, where he was a Chancellor's Fellow. He is a senior member of the Institute of Electrical and Electronic Engineers (IEEE), a member of Sigma Xi and the International Society for Optical Engineering (SPIE). Dr. Umbaugh is also the primary developer of the CVIPtools software package used throughout this book.

Acknowledgments

I thank Southern Illinois University Edwardsville, specifically the School of Engineering and the Electrical and Computer Engineering Department, for their support of this endeavor. I also thank all the students who have taken my computer imaging courses and provided valuable feedback regarding the learning and teaching of computer imaging.

The initial version of the CVIPtools software was developed primarily by myself and a few graduate students: Gregory Hance, Arve Kjoelen, Kun Luo, Mark Zuke, and Yansheng Wei; without their hard work and dedication the foundation that was built upon for this new version would not be solid. Our new Windows version of CVIPtools was developed primarily by myself and Iris Cheng, Xiaohe Chen, Dejun Zhang, and Huashi Ding. Additional students who contributed were Husain Kagalwalla and Sushama Gouravaram.

Iris Cheng deserves special credit for helping extensively with the CVIPtools for Windows project. She was always available throughout the project to assist with any and every part of the project. She dedicated a major amount of her time to the development of the CVIPtools software, and helped us greatly in project organization and in solving many problems. Overall, Iris's contributions to this project have been substantial, and her extra efforts deserve special recognition.

In small but important parts of CVIPtools public domain software was used, and kudos to those who provided it: Jef Pokanzer's pbmplus, Sam Leffler's TIFF library, Paul Heckbert's Graphics Gems, the Independent JPEG Group's software, Yuval Fisher's fractal code, and the Texas Agricultural Experiment Station's code for texture features.

I'd like to thank those who contributed photographs and images: Mark Zuke, Mike Wilson, Tony Berke, George Dean, Sara Sawyer, Sue Eder, Jeff Zuke, Bill White, the National Oceanic and Atmospheric Administration, NASA, and MIT. H. Kipsang Choge deserves credit for helping out with the figures, especially for Chapters 2 and 5, and I thank him for this work. Thanks also to David, Robin, Michael, Angi, Tyler, Kayla, Jeanie, Chris, and Chad for letting me use their smiling faces in some of the figures.

I also thank the publisher, CRC Press, for having the foresight and good sense in publishing the book. Nora Konopka and Helena Redshaw and their staff have been very helpful, and special thanks go to them. Their encouragement and enthusiasm is much appreciated. Joette Lynch has done a wonderful job managing project details during the production phase and Dan Collacott of Keyword Publishing and his staff survived my many requests regarding the layout of the book. Both Joette and Dan deserve my thanks for making this book happen.

Finally I thank my family for all their contributions; without them this book would not have been possible. I thank my mom who instilled in me a love of learning and a sense of wonder about the world around me; my dad, who taught me how to think like an engineer and the importance of self-discipline and hard work. And I am especially grateful to Jeanie, Michael, Robin, and David, who lived through the ordeal and endured the many long hours I put into this book.

Contents

Section I Introduction to Computer Imaging

Section II Digital Image Analysis

Chapter 3 Introduction to Digital Image Analysis

Chapter 4 Segmentation and Edge/Line Detection

Section III Digital Image Processing

Chapter 7 Digital Image Processing and Visual Perception

Chapter 8 Image Enhancement

Chapter 12 CVIPtools C Function Libraries

Section V Appendices

Index

Section I

Introduction to Computer Imaging

1

Computer Imaging

1.1 Overview

Computer imaging is a field that continues to grow, with new applications being developed at an ever increasing pace. It is a fascinating and exciting area to be involved in today with application areas ranging from the entertainment industry to the space program. The Internet, with its ease of use via the World Wide Web browsers, combined with the advances in computer power and network bandwidth have brought the world into our offices and into our homes. One of the most interesting aspects of this information revolution is the ability to send and receive complex data that transcends ordinary written text. Visual information, transmitted in the form of digital images, has become a major method of communication for the 21st century.

Computer imaging can be defined as the acquisition and processing of visual information by computer. The importance of computer imaging is derived from the fact that our primary sense is our visual sense, and the information that can be conveyed in images has been known throughout the centuries to be extraordinary—one picture *is* worth a thousand words. Fortunately, this is the case, because the computer representation of an image requires the equivalent of many thousands of words of data, and without a corresponding amount of information the medium would be prohibitively inefficient. The massive amount of data required for images is a primary reason for the development of many subareas within the field of computer imaging, such as image segmentation and compression. Another important aspect of computer imaging involves the ultimate "receiver" of the visual information—in some cases the human visual system, in others the computer itself.

This distinction allows us to separate computer imaging into two primary application areas: (1) computer vision, and (2) image processing, with image analysis being a key component in the development and deployment of both (Figure 1.1.1). In *computer vision* applications the processed (output) images are for use by a computer, while in *image processing* applications the output images are for human consumption. The human visual system and the computer as a vision system have varying limitations and strengths, and the computer imaging specialist needs to be aware of the functionality of these two very different systems.

Historically, the field of image processing grew from electrical engineering as an extension of the signal processing branch, while the computer science discipline was largely responsible for developments in computer vision. Recently, these two primary groups have come together to create the modern field of computer imaging. At some universities these two are still separate and distinct, but the commonalities and the perceived needs have brought the two together in the past few years.

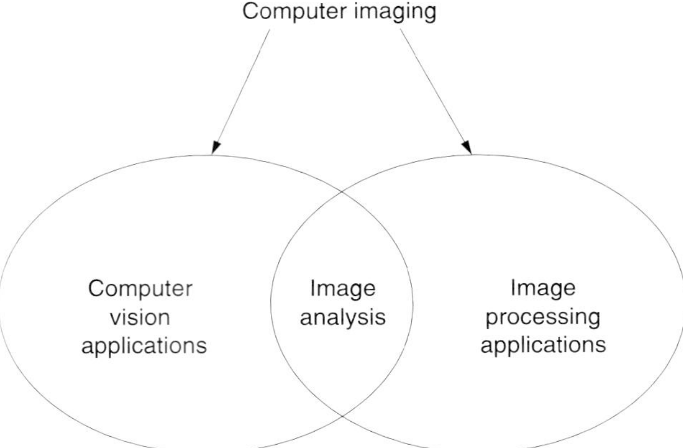

FIGURE 1.1.1
Computer imaging. Computer imaging can be separated into computer vision and image processing applications, with image analysis being part of both.

Image analysis involves the examination of the image data to facilitate solving an imaging problem. Image analysis methods comprise the major components of a computer vision system, where the system is to analyze images and have a computer act on the results. In one sense a computer vision application is simply a *deployed* image analysis system. In the development of an image processing algorithm or system, many images must be examined and tested so image analysis is necessary during the *development* of an image processing system.

This book focuses on image analysis and image processing, and, following this introductory part it is divided into three main parts: (1) Image Analysis, (2) Image Processing, and (3) Programming with CVIPtools. Chapters 1 and 2 provide an introduction to the basic concepts involved in computer imaging, and will provide the necessary background for those who are new to the field. This includes a discussion of image acquisition, imaging systems, and image representation. Chapters 3 through 6 comprise the image analysis part of the book, beginning with a system model for the image analysis process, and then describing each major part of this model in separate chapters. The image processing part of the book starts with an introductory chapter, Chapter 7, which discusses human visual perception. Following the introduction, Chapters 8, 9, and 10 examine different application areas by presenting a system model followed by representative algorithms within each area. Each of these chapters concludes with a *key points* section, followed by references and suggestions for further reading, and a series of exercises to help the learning process. The exercises include definitions, discussions, problems, computer exercises using the CVIPtools software, and programming exercises.

For the programming exercises the CVIPlab prototype program contained in the next part of the book can be used (Chapter 11), or the experienced programmer can use the platform of their choice. The programming environment provided with the CVIPtools software is a comprehensive environment for computer imaging education and application development. Chapter 12 has the C function library descriptions and function prototypes. Finally, the appendices contain information necessary for using the CD-ROM, installing CVIPtools, getting updates via the Internet, function quick reference lists, useful computer imaging resources, and a description of the CVIPtools software organization.

1.2 Image Analysis and Computer Vision

Image analysis involves investigation of the image data for a specific application. Typically, we have a set of images and want to look beyond the raw image data, to gain insight into what is happening with the images and determine how they can be used to extract the information we need. The image analysis process requires the use of tools such as image segmentation, image transforms, feature extraction, and pattern classification. *Image segmentation* is often one of the first steps in finding higher level objects from the raw image data. *Feature extraction* is the process of acquiring higher level image information, such as shape or color information, and may require the use of *image transforms* to find spatial frequency information. *Pattern classification* is the act of taking this higher level information and identifying objects within the image.

Image analysis methods comprise the major components of a computer vision system. Computer vision may be best understood by considering different types of applications. Many of these applications involve tasks that are either tedious for people to perform, require work in a hostile environment, require a high rate of processing, or require access and use of a large database of information. Computer vision systems are used in many and various types of environments—from manufacturing plants to hospital surgical suites to the surface of Mars. For example, in manufacturing systems, computer vision is often used for quality control. There, the computer vision system will scan manufactured items for defects and provide control signals to a robotic manipulator to automatically remove defective parts. To develop an application of this nature an image database consisting of sample images is first created. Next, image analysis is applied to develop the necessary algorithms to solve the problem. One interesting example of this type of system involves the automated inspection of microdisplay chips.

Microdisplay chips are used in digital cameras, projection systems, televisions, heads-up-displays, and any application that requires a small imaging device. Prior to the design of this system these chips were inspected manually—a process that is slow and prone to error. Once the market demand for these chips accelerated the manual inspection process was not practical. In Figure 1.2.1 we see the microdisplay chip inspection system along with two sample images. The original images were captured at 3:1 magnification, which means each picture element in the microdisplay chip corresponds to a 3×3 array in the image. The system automatically finds various types of defects in the chips; such as the pixel defects and the faint display defects shown here.

Another interesting computer vision application that required image analysis for algorithm development involved the automatic counting and grading of lumber. Before this system was implemented this task was done manually; which was a boring task, had an unacceptable error rate, and was inefficient. This application was challenging due to the variation in the lumber stack, such as gaps in between boards, variation in the wood color, cracks in the boards, or holes in the boards. Figure 1.2.2 shows the system in operation, and a sample input image and a processed image. The processed image is used by high level software to count and grade the lumber in stack. With the system in place the lumberyard can minimize errors, increase efficiency, and provide their workers with more rewarding tasks.

Image analysis is used in the development of many computer vision applications for the medical community, with the only certainty being that the types of applications will continue to grow. Current examples of medical systems being developed include: systems to automatically diagnosis skin tumors, systems to aid neurosurgeons during brain surgery, and systems to automatically perform clinical tests. Systems that automate

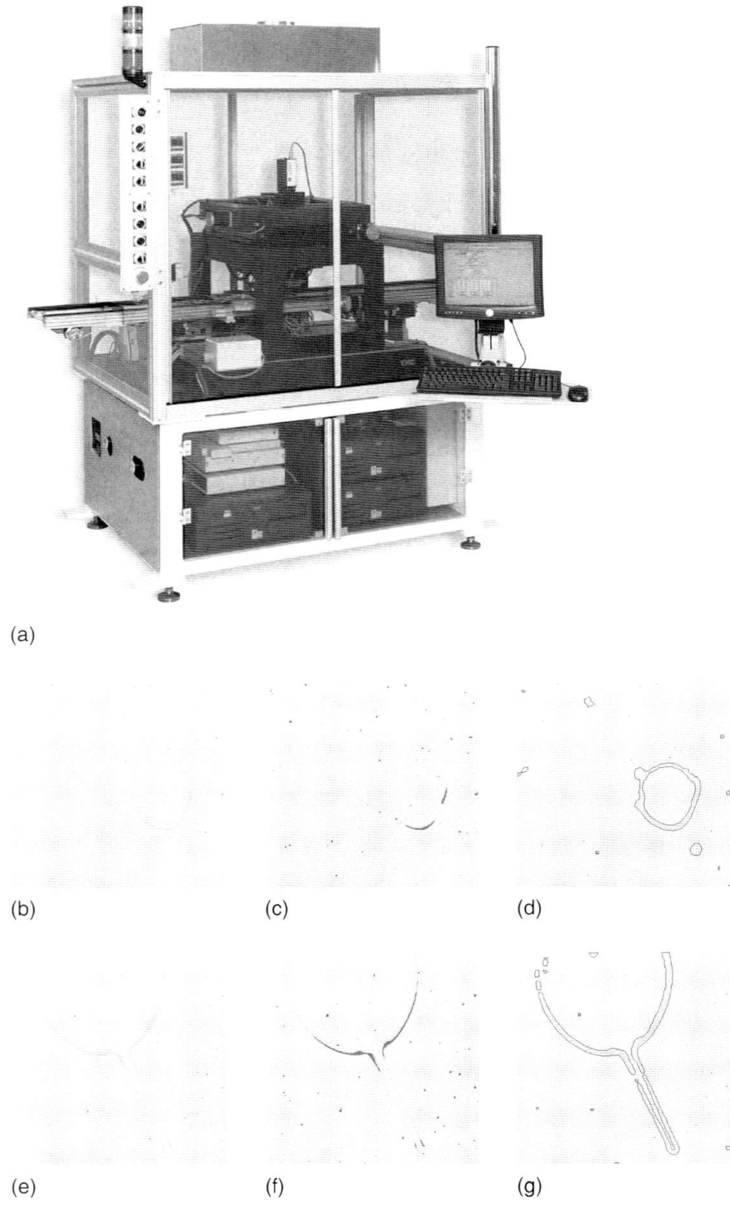

FIGURE 1.2.1
(a) Computer vision system for microdisplay chip inspection. (b) Microdisplay chip image1 at 3:1. (c) Image1 after pixel defect detection. (d) Image1 after blob analysis to find faint defects. (e) Microdisplay chip image2 at 3:1.(f) Image2 after pixel defect detection. (g) Image2 after blob analysis to find faint defects. (Photos courtesy of Mike Wilson and Iris Cheng, Westar Display Technologies Inc.)

the diagnostic process are being developed primarily to be used as tools by medical professionals where specialists are unavailable, or to act as consultants to the primary care givers, and may serve their most useful purpose in the training of medical professionals. Many of these types of systems are highly experimental, and it may be a long time before we actually see computers playing doctor like the holographic doctor

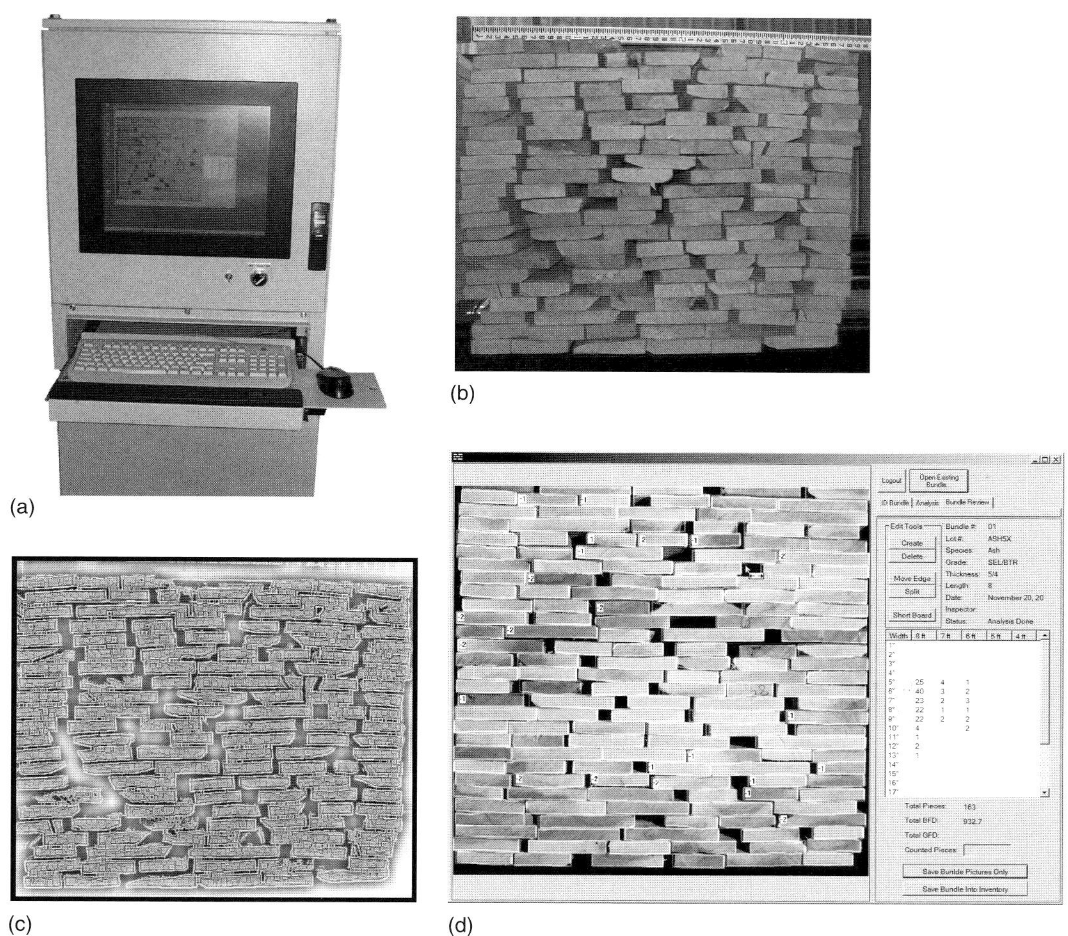

FIGURE 1.2.2
(a) Computer vision system for lumber counting and grading. (b) Image captured by the system. (c) Intermediate image after processing. (d) Example of system output after software analysis. (Photos courtesy of Tony Berke, River City Software.)

in the *Star Trek* series. Computer vision systems that are being used in the surgical suite have already been used to improve the surgeon's ability to "see" what is happening in the body during surgery, and consequently improve the quality of medical care available. Systems are also currently being used for tissue and cell analysis; for example, to automate applications that require the identification and counting of certain types of cells.

The field of law enforcement and security is an active area for image analysis research and development, with applications ranging from automatic identification of fingerprints to DNA analysis. Security systems to identify people by retinal scans, facial scans, and the veins in the hand have been developed. Reflected ultraviolet imaging systems are being used to find latent fingerprints, shoeprints, body fluids, and bite marks that are not visible to the human visual system. Infrared imaging to count bugs has been used at Disney World to help keep their greenery green. Currently, systems are in place to automatically check our highways for speeders, and in the future, computer vision systems may be used to fully automate our transportation systems to make travel safer. The U.S. space program and the Defense department, with their needs for robots with visual capabilities, are actively involved in image analysis research and development. Applications range from autonomous vehicles to target tracking and identification. Satellites orbiting the earth collect massive amounts of image data every day, and these images are automatically scanned to aid in making maps, predicting the weather, and helping us to understand the changes taking place on our home planet.

1.3 Image Processing

Image processing is computer imaging where the application involves a human being in the visual loop. In other words, the images are to be examined and acted upon by people. For these types of applications we require some understanding of how the human visual system operates. The major topics within the field of image processing include image restoration, image enhancement, and image compression. As was previously mentioned, image analysis is used in the development of image processing algorithms. In order to restore, enhance, or compress digital images in a meaningful way we need to examine the images and understand how the raw image data relates to human visual perception.

Image restoration is the process of taking an image with some known, or estimated, degradation, and restoring it to its original appearance. Image restoration is often used in the field of photography or publication where an image was somehow degraded, but needs to be improved before it can be printed. For this type of application we need to know something about the degradation process in order to develop a model for the distortion. Once we have a model for the degradation process, we can apply the inverse process to the image to restore it to its original form. This type of image restoration is often used in space exploration—for example to eliminate artifacts generated by mechanical jitter in a spacecraft (Figure 1.3.1) or to compensate for flaws in the optical system of a telescope. Restoration techniques can be used in noise removal, Figure 1.3.2, or in fixing geometric distortion, Figure 1.3.3.

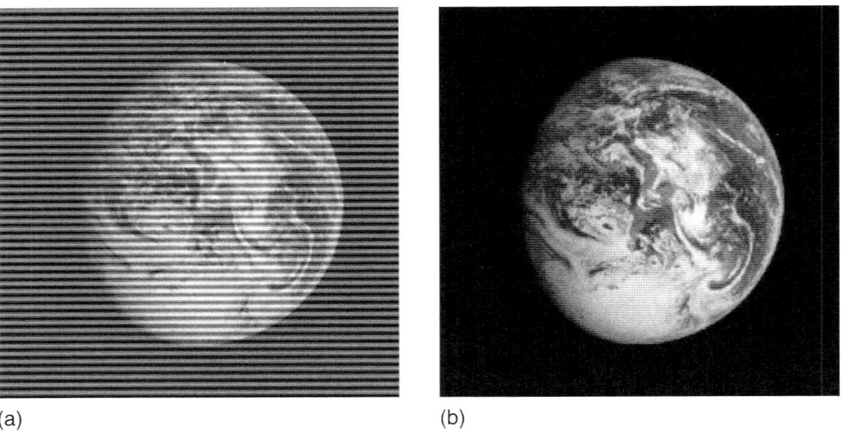

FIGURE 1.3.1
Image restoration. (a) Image with distortion. (b) Restored image.

(a)

(b)

FIGURE 1.3.2
(See color insert following page 362.)
Noise removal. (a) Noisy image. (b) Noise removed with image restoration.

Image enhancement involves taking an image and improving it visually, typically by taking advantage of the human visual system's response. One of the simplest and often dramatic enhancement techniques is to simply stretch the contrast of an image (Figure 1.3.4). Another common enhancement is image sharpening, shown in Figure 1.3.5. Enhancement methods tend to be problem specific. For example, a method that is used to enhance satellite images may not be suitable for enhancing medical images. Although enhancement and restoration are similar in aim, to make an image look better, they differ in how they approach the problem. Restoration methods attempt to model the distortion to the image and reverse this degradation, whereas enhancement methods use knowledge of the human visual system's response to improve an image visually.

Image compression involves reducing the typically massive amount of data needed to represent an image. This is done by eliminating data that is visually unnecessary, and by taking advantage of the redundancy that is inherent in most images. Although image compression is used in computer vision systems, it is included as an image processing topic because much of the work being done in the field is in areas where we want to compress images that are to be examined by people, so we want to understand exactly what part of the image data is important for human perception. By taking advantage of the physiological and psychological aspects of the human visual system,

(a) (b)

FIGURE 1.3.3
Geometric distortion correction. (a) Distorted image. (b) Restored image—note the process is not perfect.

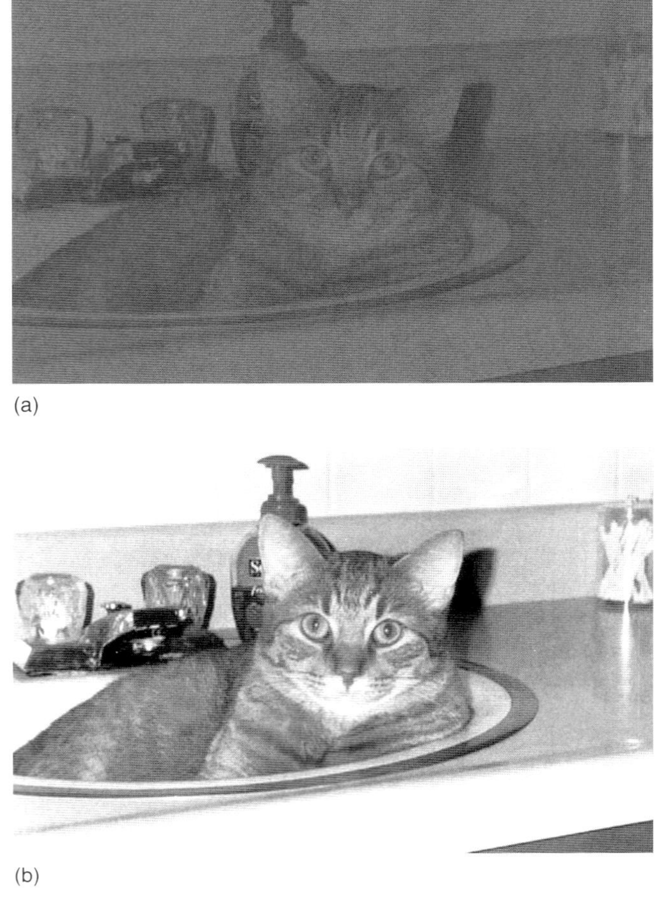

(a)

(b)

FIGURE 1.3.4
Contrast stretching. (a) Image with poor contrast. (b) Image enhanced by contrast stretching.

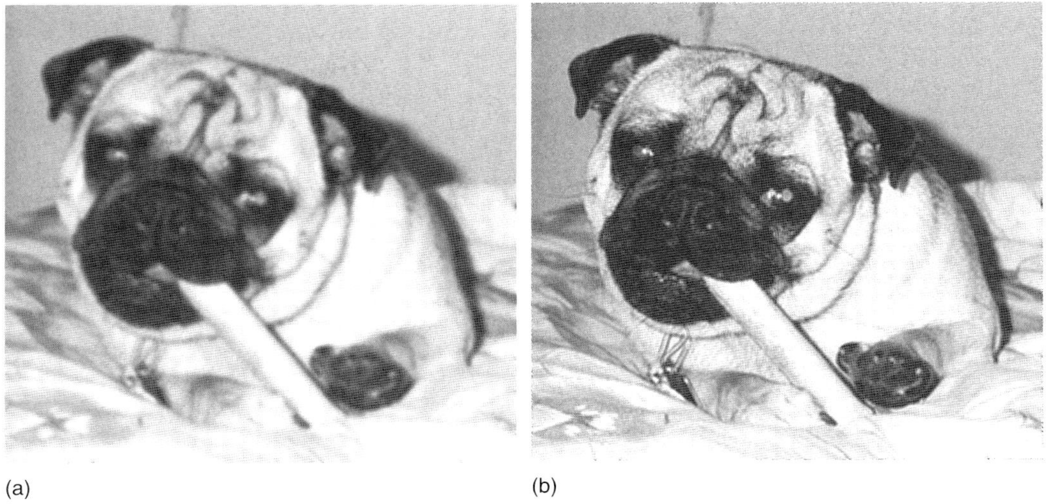

(a) (b)

FIGURE 1.3.5
Image sharpening. (a) Original image. (b) Sharpened image.

still image data can be reduced 10 to 50 times, and motion image data (video) can be reduced by factors of 100 or even 200. Figure 1.3.6 shows an image with various degrees of compression. It should be noted the amount of compression and the quality of the compressed image is highly image dependent and will vary widely.

The medical community has many important applications for image processing, often involving various types of diagnostic imaging. The beauty of the diagnostic imaging modalities, including PET, CT, and MRI scanning, is that they allow the medical professional to look into the human body without the need to cut it open. Image processing is also widely used in many different types of biological research, for example, to enhance microscopic images to bring out features that are otherwise indiscernible. The entertainment industry uses image processing for special effects, editing, creating artificial scenes and beings—computer animation, closely allied with the field of computer graphics. Image processing is being used to enable people to see what they would look like with a new haircut, a new pair of eyeglasses, or even a new nose. Computer aided design, which uses tools from image processing and computer graphics, allows the user to design a new building or spacecraft and explore it from the inside out. This type of capability can be used, for example, by people wanting to explore different modifications to their homes, from a new room to new carpeting, and will let them see the end result before the work has even begun. Virtual reality is one application that exemplifies future possibilities, where applications are without bound, and image processing techniques, combined with new developments in allied areas, will continue to affect our lives in ways we can scarcely imagine.

1.4 Key Points

Computer imaging—the acquisition and processing of visual information by computer. It can be divided into two main application areas: (1) computer vision, and (2) image processing, with image analysis being a key component of both.

FIGURE 1.3.6
Image compression. (a) Original image. (b) File compressed to 1/10 its original size. (c) File compressed to 1/20 its original size. (d) File compressed to 1/30 its original size.

Computer vision—imaging applications where the output images are for computer use

Image processing—imaging applications where the output images are for human consumption

- **Image restoration**—the process of taking an image with some known, or estimated, degradation, and restoring it to its original appearance
- **Image enhancement**—improving an image visually
- **Image compression**—reducing the amount of data needed to represent an image

Image analysis—the examination of image data to solve a computer imaging problem

- **Image segmentation**—used to find higher level objects from raw image data
- **Feature extraction**—acquiring higher-level information, such as shape or color of objects

- **Image transforms**—may be used in feature extraction to find spatial frequency information
- **Pattern classification**—used for identifying objects in an image

1.5 References and Further Reading

A comprehensive treatment of computer vision can be found in [Forsyth/Ponce 03], [Shapiro/Stockman 01], [Davies 97], [Ballard/Brown 82], [Haralick/Shapiro 92], [Horn 86], [Granland/Knutsson 95] and [Jain/Kasturi/Schnuck 95]. Comprehensive image processing texts include [Gonzalez/Woods 02], [Castleman 96], [Jain 89], [Pratt 91], [Bracewell 95] and [Rosenfeld/Kak 82].

Books that bring computer vision and image processing together include [Sonka,/ Hlavac/Boyle 99], [Schalkoff 89], [Granland/Knutsson 95] and [Banks 90]. One book that takes a more practical, lab oriented approach to computer vision and image processing is [Galbiati 90]. A good conceptual and practical approach to computer imaging is taken by [Baxes 94], and [Myler/Weeks 93]. Two books that are useful for practical algorithm implementation (including code!) are [Seul/O'Gorman/Sammon 00] and [Parker 97]. [Russ 99] provides a good handbook for the computer imaging specialist. Some of the applications discussed can be found in the trade magazines *Advanced Imaging, Biophotonics International, Design News, Photonics Spectra*, and the *Journal of Forensic Identification.*

Ballard, D.H., Brown, C.M., *Computer Vision*, Upper Saddle River, NJ: Prentice Hall, 1982.

Banks, S., *Signal Processing, Image Processing and Pattern Recognition*, Cambridge, UK: Prentice Hall International (UK) Ltd., 1990.

Baxes, G.A., *Digital Image Processing: Principles and Applications*, NY: Wiley, 1994.

Bracewell, R.N., *Two-Dimensional Imaging*, Upper Saddle River, NJ: Prentice Hall, 1995.

Castleman, K.R., *Digital Image Processing*, Upper Saddle River, NJ: Prentice Hall, 1996.

Davies, E.R., *Machine Vision*, San Diego, CA: Academic Press, 1997.

Forsyth, D.A., Ponce, J., *Computer Vision*, Upper Saddle River, NJ: Prentice Hall, 2003.

Galbiati, L.J., *Machine Vision and Digital Image Processing Fundamentals*, Upper Saddle River, NJ: Prentice Hall, 1990.

Gonzalez, R.C., Woods, R.E., *Digital Image Processing*, Upper Saddle River, NJ: Prentice Hall, 2002.

Granlund, G., Knutsson, H., *Signal Processing for Computer Vision*, Boston: Kluwer Academic Publishers, 1995.

Haralick, R.M., Shapiro, L.G., *Computer and Robot Vision*, Reading, MA: Addison-Wesley, 1992.

Horn, B.K.P., *Robot Vision*, Cambridge, MA: The MIT Press, 1986.

Jain, A.K., *Fundamentals of Digital Image Processing*, Upper Saddle River, NJ: Prentice Hall, 1989.

Jain, R., Kasturi, R., Schnuck, B.G., *Machine Vision*, NY: McGraw Hill, 1995.

Myler, H.R., Weeks, A.R., *Computer Imaging Recipes in C*, Upper Saddle River, NJ: Prentice Hall, 1993.

Parker, J.R., *Algorithms for Image Processing and Computer Vision*, NY: Wiley, 1997.

Pratt, W.K., *Digital Image Processing*, NY: Wiley, 1991.

Rosenfeld, A., Kak, A.C., *Digital Picture Processing*, San Diego, CA: Academic Press, 1982.

Russ, J.C., *The Image Processing Handbook*, Boca Raton, FL: CRC Press, 1999.

Schalkoff, R.J., *Digital Image Processing and Computer Vision*, NY: Wiley, 1989.

Seul, M., O'Gorman, L., Sammon, M.J., *Practical Algorithms for Image Analysis*, Cambridge, UK: Cambridge University Press, 2000.

Shapiro, L., Stockman, G., *Computer Vision*, Upper Saddle River, NJ: Prentice Hall, 2001.

Sonka, M., Hlavac, V., Boyle, R., *Image Processing, Analysis and Machine Vision*, Pacific Grove, CA: Brooks/Cole Publishing Company, 1999.

West, M., Barsley, et al., *Journal of Forensic Identification*, 40[5], 249–255, 1990.

1.6 Exercises

1. Discuss how computer imaging, image processing, image analysis, and computer vision are related.
2. Discuss two computer vision applications.
3. List and describe the tools used in image analysis.
4. What are the major topics in the field of image processing? Discuss two image processing applications.
5. Suppose we need to develop a new image compression algorithm. Discuss the factors that must be considered.
6. What is the difference between image enhancement and image restoration?

2

Computer Imaging Systems

2.1 Imaging Systems Overview

Computer imaging systems come in many different configurations, depending on the application. As technology advances these systems get smaller, faster and more sophisticated. In Chapter 1 we saw some imaging systems used for computer vision applications. In this chapter we will focus on the primary aspects of a generic imaging system. We will look at how images are sensed and transformed into computer files, how the CVIPtools software can be used for image analysis and processing, and how these computer files are used to represent image information.

Computer imaging systems are comprised of two primary component types, hardware and software. The hardware components, as seen in Figure 2.1.1 can be divided into the image acquisition subsystem, the computer itself, and the display devices. The software allows us to manipulate the image and perform any desired analysis or processing on the image data. Additionally, we may also use software to control the image acquisition and storage process.

The computer system may be a general purpose computer with an imaging device connected. A digital camera can be interfaced with the computer via USB (Universal Serial Bus), FireWire (IEEE 1394), Camera Link, or Gigabit Ethernet (IEEE 802.3). Specifications for these are shown in Table 2.1. A standard video camera requires a frame grabber, or image digitizer, to interface with the computer. The *frame grabber* is a special purpose piece of hardware that accepts a standard analog video signal, and outputs an image in the form that a computer can understand—a digital image. Video standards vary throughout the world, RS-170 or its equivalent NTSC-RS-330 is used in North America and Japan. Northern Europe uses CCIR or PAL, while France and Russia use SECAM, a CCIR equivalent.

The process of transforming a standard video signal into a digital image is called digitization. This transformation is necessary because the standard video signal is in analog (continuous) form, and the computer requires a digitized or sampled version of that continuous signal. A typical video signal contains frames of video information, where each *frame* corresponds to a full screen of visual information. Each frame may then be broken down into *fields*, and each field consists of alternating lines of video information. In Figure 2.1.2a, we see the typical image on a display device, where the solid lines represent one field of information and the dotted lines represent the other field. These two fields make up one frame of visual information. This two-fields-per-frame model is referred to as *interlaced* video. Some types of video signals, called *noninterlaced* video, have only one field per frame. Noninterlaced video is typically used in computer monitors.

In Figure 2.1.2b we see the electrical signal that corresponds to one line of video information. Note the *horizontal synch pulse* between each line of information, this

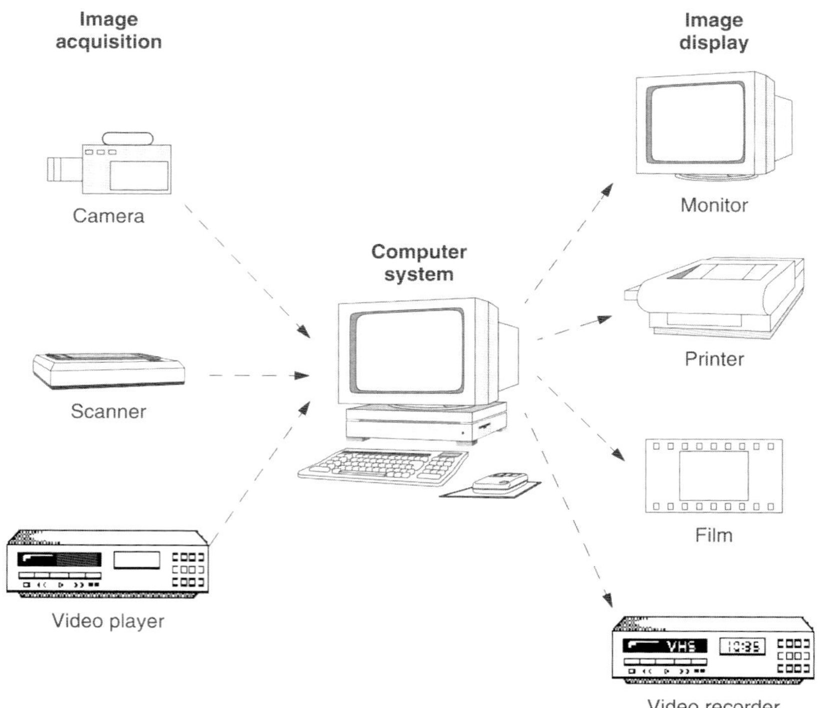

FIGURE 2.1.1
Computer imaging system hardware.

TABLE 2.1
Camera Interface Specifications

	Gigabyte Ethernet	Firewire	USB 2.0	Camera Link
Type of standard	Commercial	Public	Public	Commercial
Connection type	Point-to-point or Local Area Network (LAN)	Peer-to-peer, shared bus	Master–slave, shared bus	Point-to-point
Maximum bandwidth for images	~1000 Megabits/sec (Mbs)	~800 Mbs	~480 Mbs	~2000 to 7000 Mbs
Distance	~100 meters, no limit with switches or fiber	~4.5 meters, ~72 meters with switches, ~200 meters with fiber	~5 meters, ~30 meters with switches	~10 meters
PC interface	Network	PCI card	PCI card	PCI frame grabber
Wireless support	Yes	No	No	No
Max # of devices	Unlimited	63	127	1

synchronization pulse tells the display hardware to start a new line. After one frame has been displayed, a longer synchronization pulse, called the *vertical synch pulse*, tells the display hardware to start a new field or frame.

The analog video signal is turned into a digital image by sampling the continuous signal at a fixed rate. In Figure 2.1.3, we see one line of a video signal being sampled

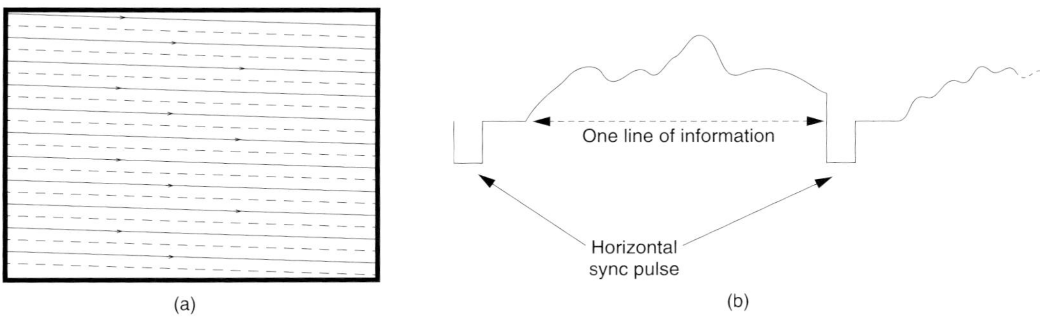

FIGURE 2.1.2
The video signal. (a) One frame, two fields. (b) The video signal.

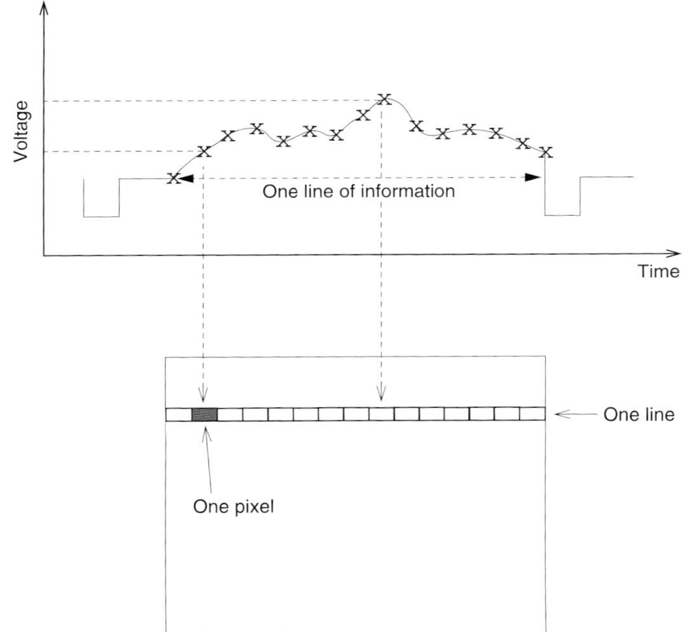

FIGURE 2.1.3
Digitizing (sampling) an analog video signal.

(digitized) by instantaneously measuring the voltage of the signal at fixed intervals in time. The value of the voltage at each instant is converted into a number that is stored, corresponding to the brightness of the image at that point. Note that the image brightness at a point depends on both the intrinsic properties of the object *and* the lighting conditions in the scene. Once this process has been completed for an entire frame of video information, we have "grabbed" a frame, and the computer can store it and process it as a digital image.

The image can now be accessed as a two-dimensional array of data, where each data point is referred to as a *pixel* (picture element). For digital images we will use the following notation:

$$I(r, c) = \text{the brightness of the image at the point } (r, c)$$

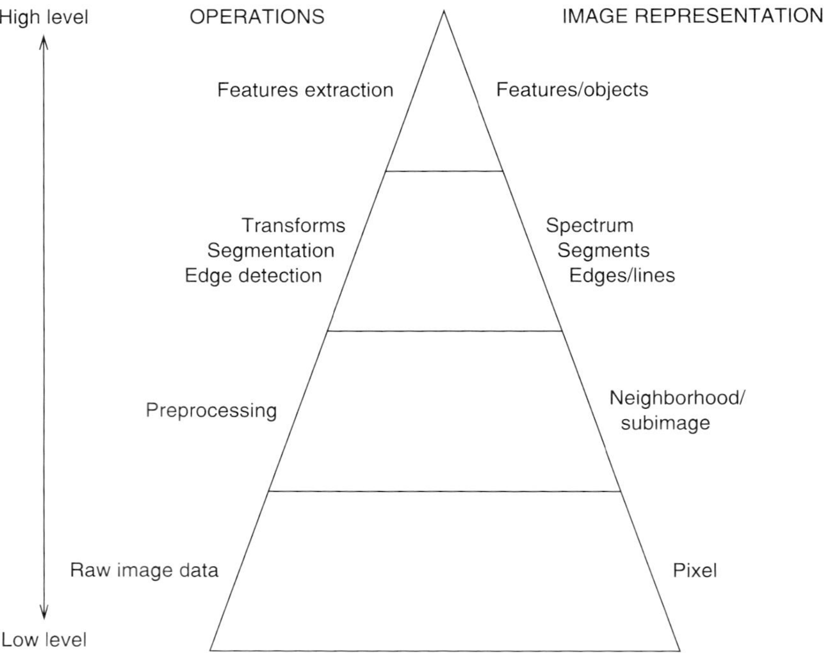

FIGURE 2.1.4
The hierarchical image pyramid.

where

$$r = \text{row} \quad \text{and} \quad c = \text{column}$$

Note that this notation, $I(r, c)$, is used throughout the book, to be consistent with the way in which matrices are defined in most programming languages. But also note that most imaging software tools (Photoshop, CVIPtools, Scion Image, etc.), and some other textbooks, list the column first and row coordinate second in the form $I(x, y)$. So, do not be confused, but look carefully at which coordinate is the row and which is the column.

Once we have the data in digital form we can use the software to process the data. This processing can be illustrated by the *hierarchical image pyramid*, as seen in Figure 2.1.4. At the very lowest level we deal with the individual pixels, where we may perform some low-level preprocessing. The next level up is the *neighborhood*, which typically consists of a single pixel and the surrounding pixels, and we may continue to perform some preprocessing operations at this level. As we continue to go up the pyramid, we get higher and higher level representations of the image, and consequently, a reduction in the amount of data. All of the types of operations and image representations in Figure 2.1.4 will be explored in the following chapters.

2.2 Image Formation and Sensing

Digital images are formed by energy interacting with a device that responds in a way that can be measured. These measurements are taken at various points across a

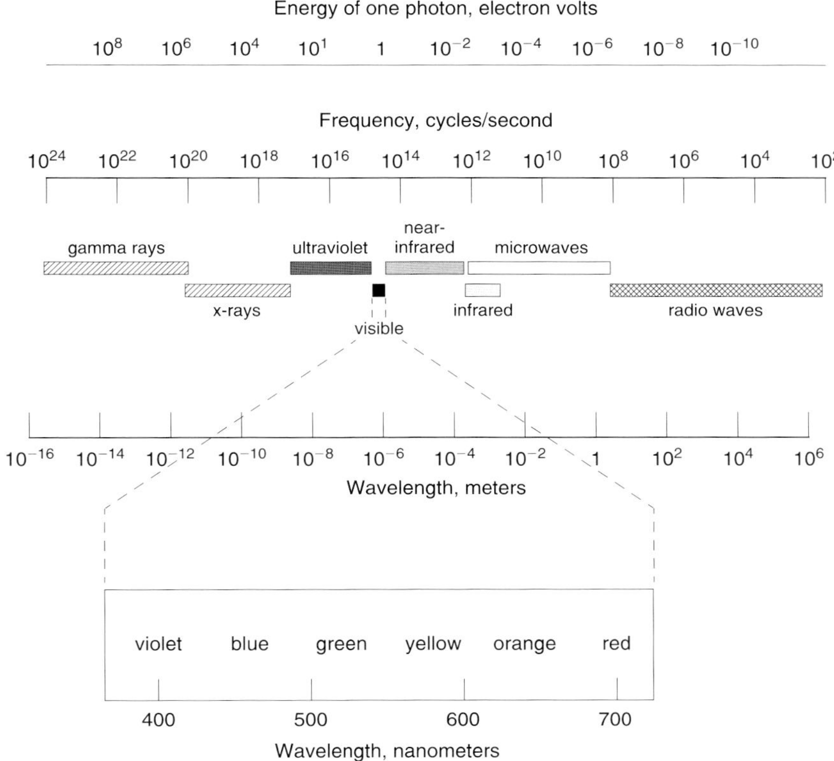

FIGURE 2.2.1
The electromagnetic spectrum.

two-dimensional grid in the world in order to create the image. These measuring devices are called *sensors*, and many different types are in use. Sensors may respond to various parts of the electromagnetic (EM) spectrum, acoustical (sound) energy, electron beams, lasers, or any other signal that can be measured.

The EM spectrum consists of visible light, infrared, ultraviolet, x-rays, microwaves, radio waves, or gamma waves (see Figure 2.2.1). Electromagnetic radiation consists of alternating (sinusoidal) electric and magnetic fields that are perpendicular to each other as well as to the direction of propagation. These waves all travel at the speed of light in free space, approximately 3×10^8 *meters/second* (m/s), and are classified by their frequency or wavelength. Figure 2.2.1 shows the various spectral bands and their associated names, wavelengths, frequencies, and energy. The various bands in the EM spectrum are named for historical reasons related to their discovery, or their application.

In addition to the wave model EM radiation can be modeled as a stream of massless particles called *photons*, where a photon corresponds to the minimum amount of energy, the quantum, which can be measured in the EM signal. The energy of a photon is measured in electron volts, a very small unit, which is the kinetic (motion) energy that an electron acquires in being accelerated through an electronic potential of one volt. In Figure 2.2.1 we see that as frequency increases, the energy contained in a photon increases. Radio waves have the smallest frequencies so we believe that it is safe to be immersed in them (they're everywhere!), whereas gamma rays have the highest energy which makes them very dangerous to biological systems.

Sensors may also respond to acoustical energy, as in ultrasound images. In some cases images are created to produce *range* images, which do not correspond to what we typically think of as images, but are measures of distance to objects, and may be created by radar (*ra*dio *d*etection *a*nd *r*anging), sound energy or lasers. In this section, and in this book, we will primarily focus on visible light images; however, we will briefly discuss other types of images. Once an image is acquired it can be analyzed or processed using all the tools discussed in this book, regardless of the type of acquisition process.

We will consider two key components of image formation:

- Where will the image point appear?
- What value will be assigned to that point?

The first question can be answered by investigating basic properties of lenses and the physics of light, the science of *optics;* and the second will require a look into sensor and electronic technology.

2.2.1 Visible Light Imaging

The basic model for visible light imaging is shown in Figure 2.2.2. Here the light source emits light that is reflected from the object, and focused by the lens onto the image sensor. The sensor responds to the light energy by converting it into electrical energy which is then measured. This measurement is proportional to the incident energy, and we describe it as the brightness of the image at that point. The way an object appears in an image is highly dependent on the way in which it reflects light, this is called the *reflectance function* of the object and is related to what we think of as color and texture. The color determines which wavelengths of light are absorbed and which are reflected, and the texture determines the angle at which the light is reflected. In Figure 2.2.3 objects of very different reflectance functions are shown.

In imaging two terms are necessary to define brightness. What is measured is called *irradiance*, while the light reflected from an object is referred to as *radiance*. Figure 2.2.4 illustrates the difference between these two terms. Irradiance is the amount of light falling

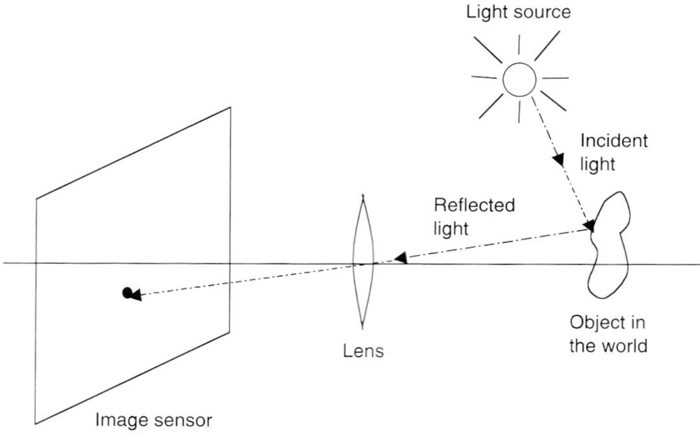

FIGURE 2.2.2
Model for visible light imaging. The light source emits light that is reflected from the object, and focused by the lens onto the image sensor.

FIGURE 2.2.3
(See color insert following page 362.) The reflectance function. Here we see that the way in which an object reflects the incident light, the reflectance function, has a major effect on how it appears in the resulting image. The reflectance function is an intrinsic property of the object and relates to both color and texture. (a) Monochrome image showing brightness only, the color determines how much light is reflected and the surface texture determines the angle at which the light is reflected. (b) Color image, the color determines which wavelengths are absorbed and which are reflected.

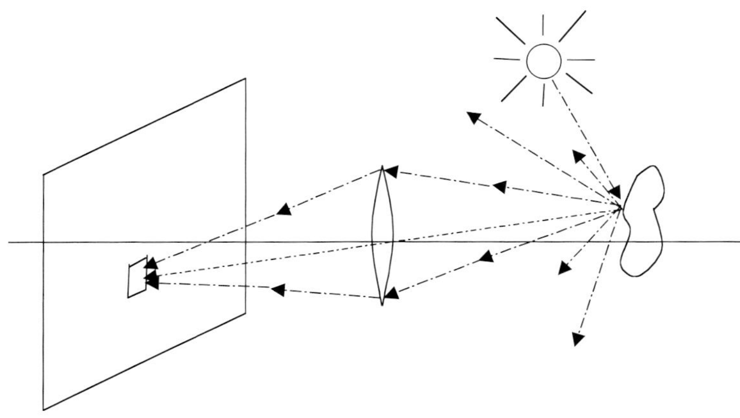

FIGURE 2.2.4
Irradiance and radiance. Irradiance is the measured light falling on the image plane. It is measured in power per unit area. Radiance is the power reflected or emitted per unit area into a directional cone having a unit of solid angle. Note that all the reflected light is not captured by the lens.

on a surface, such as an image sensor, while radiance is the amount of light emitted from a surface into a solid unit angle. So the units used for these two measures are different:

$$\text{Irradiance} \rightarrow \frac{\text{Power}}{\text{Area}}$$

$$\text{Radiance} \rightarrow \frac{\text{Power}}{(\text{Area})(\text{SolidAngle})}$$

The irradiance is the brightness of an image at a point, and is proportional to the scene radiance.

A lens is necessary to focus light in a real imaging system. In Figure 2.2.5 we see the relationship between points in the world and points in the image. The relationship of distance of the object in the world and the image plane is defined by the lens equation:

$$\frac{1}{a} + \frac{1}{b} = \frac{1}{f}$$

where f is the *focal length* of the lens and is an intrinsic property of the lens, and a and b are the two distances in question. In Figure 2.2.4 we see three rays of light shown; note that the one through the center of the lens goes straight through to the image plane, and, if the system is in focus, the other rays will meet at that point. If the object is moved closer to the lens, the single point will become a blur circle; the diameter of the circle is given by the *blur equation*:

$$c = \frac{d}{b'}|b - b'|$$

where c is the circle diameter, d is the diameter of the lens, and a' and b' are the distances shown in Figure 2.2.6. This equation can be derived by the property of similar triangles.

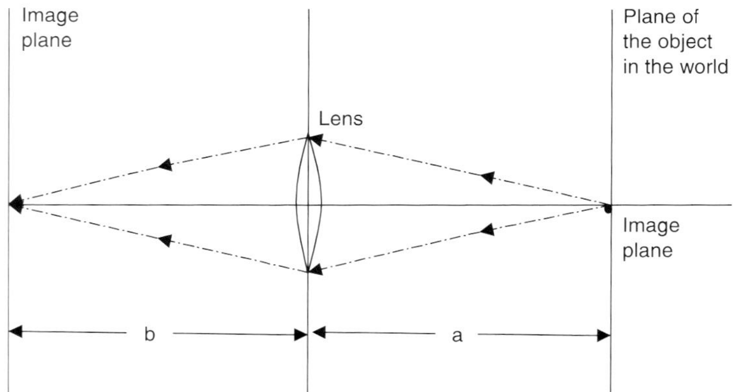

FIGURE 2.2.5
Relationship between points in the world and points in the image. A lens will focus an image of an object only at a specific distance given by the lens equation.

$$\text{Lens equation:} \quad \frac{1}{a} + \frac{1}{b} = \frac{1}{f}$$

where f is the *focal length* of the lens and is an intrinsic property of the lens, and a and b are the two distance shown.

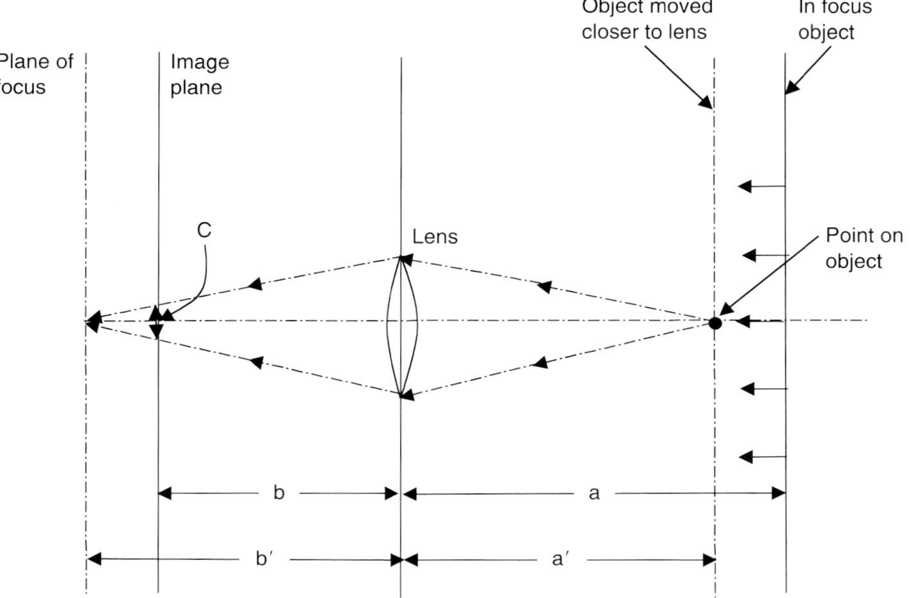

FIGURE 2.2.6
The blur circle from a poorly focused lens. As the object is moved closer to the lens, it gets blurry. Application of the lens equation shows the object is actually focused behind the image plane. The blur equation defines the amount of blur.

A real object typically does not appear in a single plane, so some blurring will occur. The question is—what are the conditions that will allow an object to be focused sufficiently well? This will be determined by the spatial resolution of the imaging device. If the blur circles are equal to, or smaller than the device resolution, the object will be focused sufficiently well. The range of distances over which objects are focused sufficiently well is called the *depth of field*. With many imaging devices, a diaphragm can be adjusted to allow some control over the depth of field (also called depth of focus). If the diaphragm is closed somewhat, not only does it let less light in, but it changes the effective diameter, $D_{\text{effective}}$, of the lens. The *f-number* (or *f-stop*) is defined as the ratio of the focal length to the lens diameter, and as the f-number increases the depth of field increases.

$$f\text{-number} = \frac{f}{D_{\text{effective}}}$$

Another important parameter of an imaging device is the field of view (FOV). The *field of view* is the amount of the scene that the imaging device actually "sees"; that is, the angle of the cone of directions from which the device will create the image. Note that the FOV depends not only on the focal length of the lens, but also on the size of the imaging sensor. In Figure 2.2.7, we can see that the field of view can be defined as:

$$\text{FOV} = 2\phi, \quad \text{where } \phi = \tan^{-1}\left(\frac{d/2}{f}\right)$$

with d being the diagonal size of the image sensor and f is the focal length of the lens. From this figure we can also see that for a fixed size image sensor, in order to get a wider

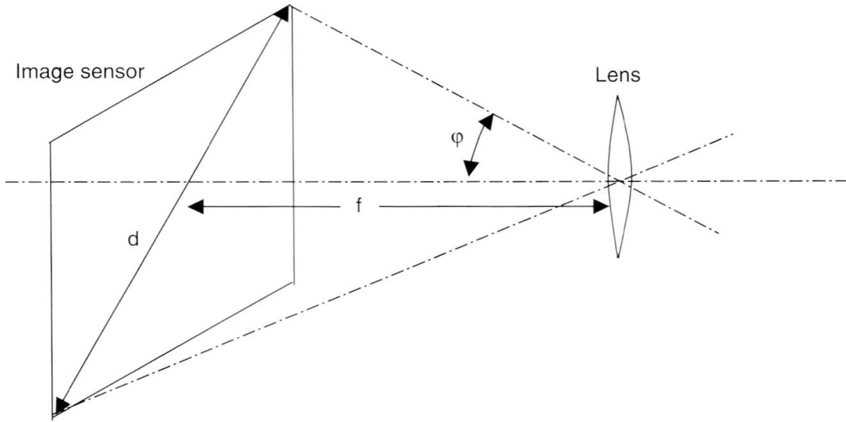

FIGURE 2.2.7
Field of view (FOV). The FOV for an imaging system depends on both focal length of the lens and the size of image sensor.

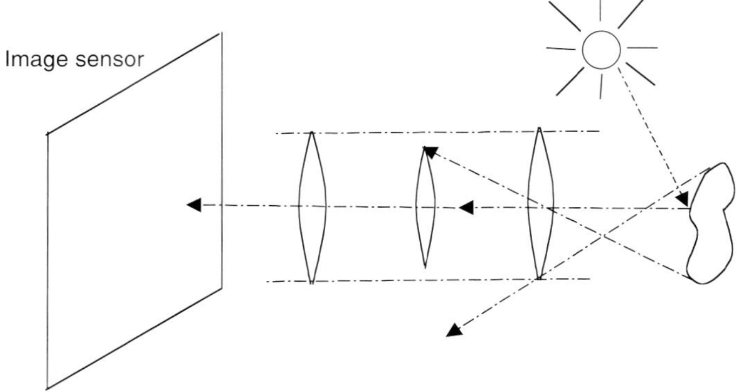

FIGURE 2.2.8
The vignetting effect. A compound lens causes less light on the edges of the image to get through to the image sensor. This has the effect of decreasing brightness as we move away from the center of the image.

field of view, we need a lens with a shorter focal length. A lens with a very short focal length compared to image sensor size is called a wide-angle lens. The three basic types of lenses are: (1) wide-angle–short focal length, FOV greater than 45°, (2) normal–medium focal length, FOV 25° to 45°, and (3) telephoto–long focal length, FOV less than 25°.

Real lenses do not typically consist of a single lens, but are multiple lenses aligned together. This is primarily due to the fact that a single lens will have various types of distortions, called aberrations. The effect of these aberrations can be mitigated by aligning multiple lenses of varying types and sizes to create a compound lens. One of the negative effects of a compound lens is the vignetting effect. This effect, shown in Figure 2.2.8, causes the amount of energy that actually makes it through the lens to the image plane to decrease as we move farther away from the center of the image. This effect can be avoided by only using the center portion of the lens. It is interesting to note that the human visual system is not sensitive to these types of slow spatial variations, which is explored more in Chapter 7.

We have briefly considered where the image will appear; now we will consider how bright the image will be. How will we sense the object and turn it into an image? How will we measure the energy? As mentioned before, sensors are used to convert the light energy into electrical energy; this is done by using a material that emits electrons when bombarded with photons. In Figure 2.2.9 generic imaging sensors are shown. The array sensor is the primary type used in digital cameras, and the sensing element is typically a charge-coupled device (CCD). These CCD arrays are packaged in arrays of up to 4000×4000 elements, and continue to get larger as technology advances. A newer type of image sensing element is the CMOS (complementary metal-oxide-semiconductor) device. Although these devices have been around for a while it has only recently become practical to use them in image sensor arrays. Currently, the CMOS image sensors are cheaper and require less power than the CCDs, but the image quality is not as good. This makes the CMOS sensors attractive for mass market applications where cost is a factor, low power is desired, and low-quality images are acceptable; such as in cell phones and toys.

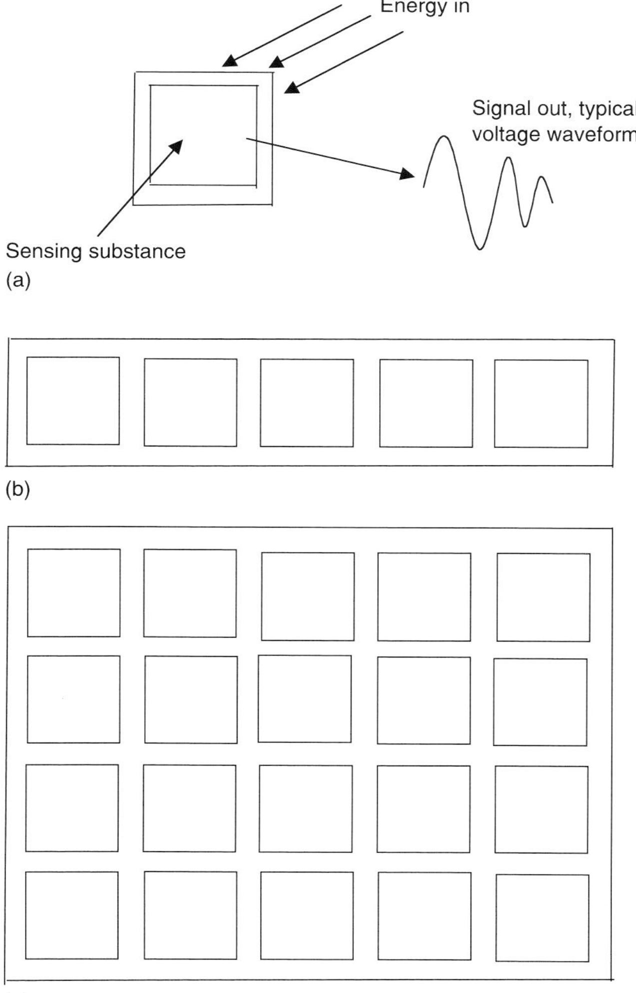

FIGURE 2.2.9
Generic imaging sensors. (a) Single imaging sensor. (b) linear or line sensor. (c) two-dimensional or array sensor.

When light energy (photonic energy) impinges upon the sensor, the sensing substance will output electrons to form electrical energy. We can approximate the number of electrons liberated in a sensor with the following *sensor equation*:

$$N = \delta A \delta t \int b(\lambda) q(\lambda) \mathrm{d}\lambda$$

where N is the approximate number of electrons liberated, δA is the area, δt is the time interval, $q(\lambda)$ is the quantum efficiency, and $b(\lambda)$ is the incident photon flux, and the integration takes place over the wavelengths of interest. The *quantum efficiency* of the material is the ratio of the electron flux produced to the incident photon flux; in other words, it is the amount of incoming light energy that is converted to electrical energy. Older tube technology devices had a quantum efficiency of about 5%, modern solid state devices may vary from about 60% to 95% efficiency.

The equation above tells us that we need to measure the light energy over a finite area and a finite time interval—these measurements cannot be performed instantaneously. This is because the devices measuring the output signal are not sensitive enough to count only a few electrons, and, even if they did, the signal would be overwhelmed by random noise which exists in electronic systems. One bonus of these requirements is that some of the noise will be averaged out by measuring over time and over a fixed area. In Chapters 4 and 9 we will explore methods to reduce this random noise even more.

2.2.2 Imaging Outside the Visible Range of the EM Spectrum

Imaging with gamma-rays is done by measuring the rays as they are emitted from the object. In nuclear medicine using positron emission tomography (PET), a patient is injected with a radioactive isotope and as it decays gamma rays are detected and measured. X-rays are used in medical diagnostics by using film that responds to x-ray energy which is placed between the energy source and the patient. X-rays are also used in computerized tomography (CT) where a ring of detectors encircles the patient and is rotated to obtain two-dimensional "slices" which can be assembled into a three-dimensional image. Fluorescence microscopy works by using dyes that emit visible light when ultraviolet light (UV) is beamed upon it. Examples of x-ray and UV images are shown in Figure 2.2.10.

Ultraviolet (UV) imaging is used in industrial applications, law enforcement, microscopy, and astronomy. Reflected UV imaging is used in forensics to find evidence that is invisible to the human visual system. For example, fingerprints, body fluids, bite marks, and even shoe prints on waxed floors have been found. Since these systems use UV illumination the background elements in the image are suppressed, which is a bonus for these types of applications. These systems use short UV, below 300 nanometer wavelengths, and have the added advantage of not requiring powders or chemicals on nonporous surfaces.

Infrared (IR) images are often used in satellite imaging (remote sensing), since features of interest, for example moisture content and mineral mapping, are found in the IR spectral bands (Figure 2.2.11). Infrared imaging is also used in law enforcement and fire detection, primarily in the middle and long wave ranges. Infrared images can be divided into four primary spectral ranges—near IR, 780 nm to 1.3 μm, middle wave IR, 3 to 5 μm, long wave IR, 7 to 14 μm, and very long wave IR, 30 μm and above. Recent advances in technology have dramatically reduced size, power consumption and cost of

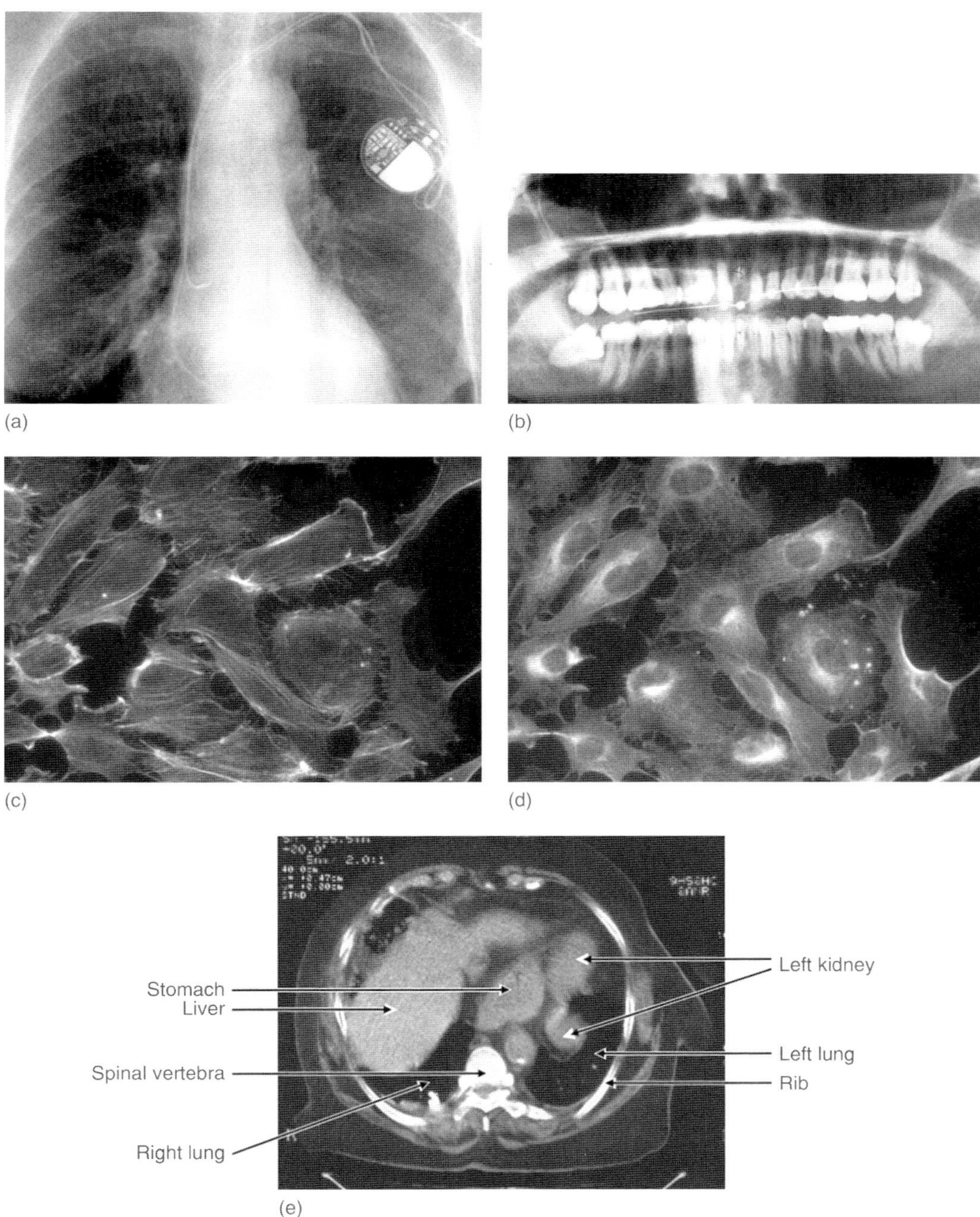

FIGURE 2.2.10
X-ray and UV images. (a) X-ray of a chest with an implanted electronic device to assist the heart. (Image courtesy of George Dean.) (b) Dental x-ray. (c) and (d) Fluorescence microscopy images of cells, generated by emitting visible light when illuminated by ultraviolet (UV) light. (Cell images courtesy of Sara Sawyer, SIUE.) (e) One "slice" of a computerized tomograpy (CT) image of a patient's abdomen, multiple 2-D image "slices" are taken at various angles and are then assembled to create a 3-D image. (Image courtesy of George Dean.)

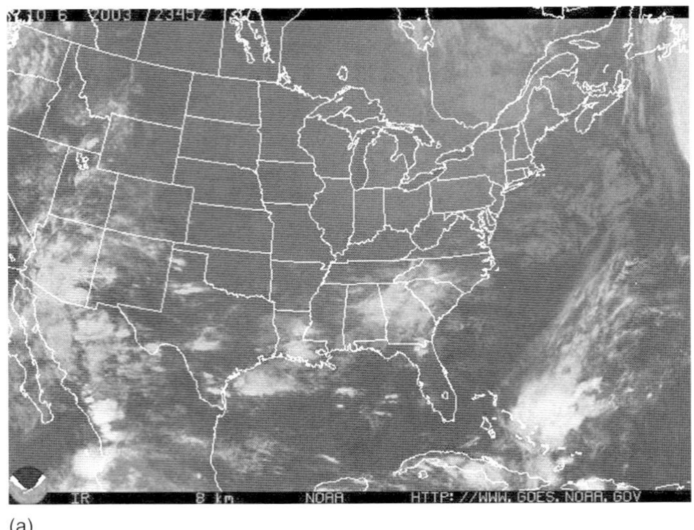

(a)

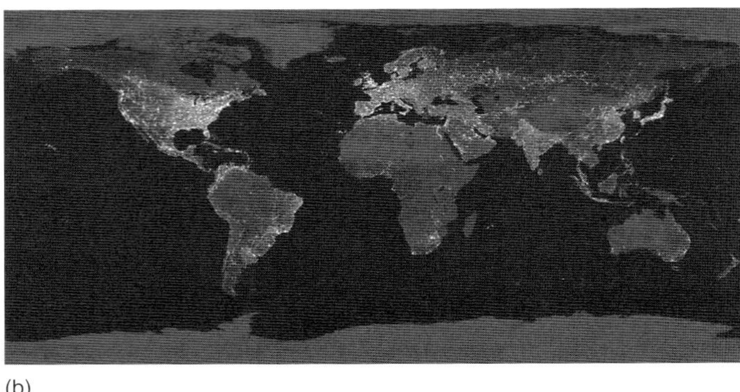

(b)

FIGURE 2.2.11
(a) Infrared satellite image showing water vapor. (b) Infrared satellite imagery in the near infrared band. (Images courtesy of National Oceanic and Atomspheric Administration (NOAA.)

these IR units; thereby making these devices much more widely available, more practical, and more cost effective.

Multispectral images, which include IR bands, are often used in weather analysis (Figure 2.2.12a). Microwave images are used most often in radar applications, where the primary requirement is the capability to acquire information even through clouds or other obstacles, regardless of lighting conditions. In the radio band of the EM spectrum, applications are primarily in astronomy and medicine. In astronomy, radio waves can be detected and measured in a manner similar to collecting visible light information, except that the sensor responds to radio wave energy.

In medicine, magnetic resonance imaging (MRI) works by sending radio waves through a patient in short pulses in the presence of a powerful magnetic field. The body responds to these pulses by emitting radio waves, which are measured to create an image of any part of the patient's body (Figure 2.2.12b). MRI systems use a special antenna (receiver coil) to detect these interactions between radio-frequency EM and atomic nuclei in the patient's body. The superconducting magnets used in MRI systems can generate fields with magnitudes from 0.1 to 3.0 Tesla (1,000 to 30,000 Gauss).

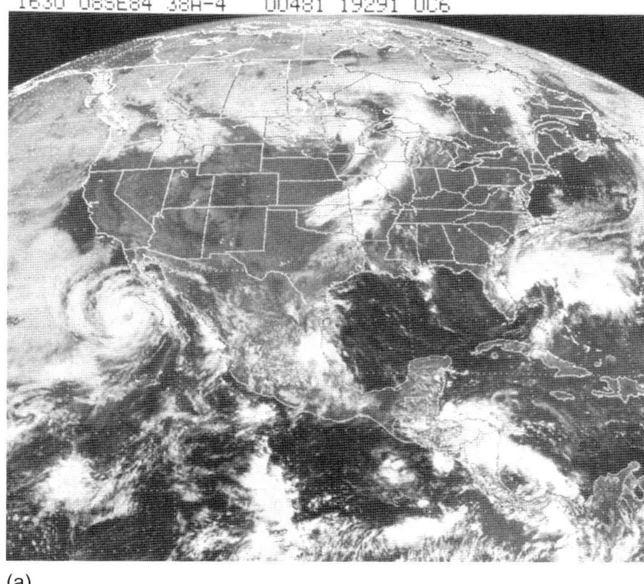

(a)

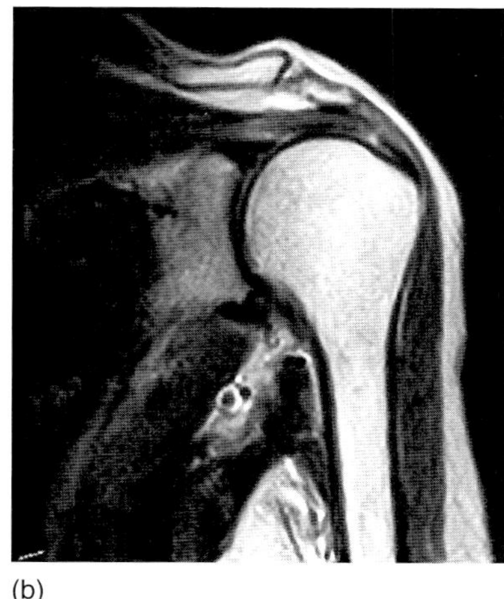

(b)

FIGURE 2.2.12
Multispectral and radio wave images. (a) Multispectral Geostationary Operational Environment Satellite (GOES) image of North America, showing a large tropical storm off Baja California, a frontal system over the Midwest, and tropical storm Diana off the east coast of Florida. (Courtesy of NOAA.) (b) Magnetic resonance image (MRI) of a patient's shoulder. MRI images are created using radio waves. This is a single 2-D "slice," multiple images are taken at different angles and assembled to create a 3-D image. (Image courtesy of George Dean.)

By comparison, the magnetic field of the Earth is 0.00005 Tesla (0.5 Gauss). MRI systems have excellent contrast resolution, which means they are much better at showing subtle differences among the soft tissues and organs of the body that are not easily viewed on conventional x-ray or CT films.

2.2.3 Acoustic Imaging

Acoustic imaging works by sending out pulses of sonic energy (sound) at various frequencies, and then measuring the reflected waves. The time it takes for the reflected signal to appear contains distance information, and the amount of energy reflected contains information about the object's density and material. The measured information is then used to create a two- or three-dimensional image. Acoustic imaging is used in biological systems; for example, bats use it to "see," and in man-made systems such as the sonar used in submarines.

The frequency of the acoustic signals depends on the application and the medium in which the signal is transmitted. Geological applications, for example oil and mineral exploration, typically use low-frequency sounds (around hundreds of Hertz). Ultrasonic, or high-frequency sound, imaging is often used in manufacturing and in medicine. The most common use in medicine is to follow the development of the unborn baby inside the womb. Here, at frequencies ranging from 1 to 5 megahertz, the health (and gender) of the baby can be determined (see Figure 2.2.13). Because ultrasonic imaging allows us to "see" inside opaque objects, it is also commonly used in manufacturing to detect defects in materials.

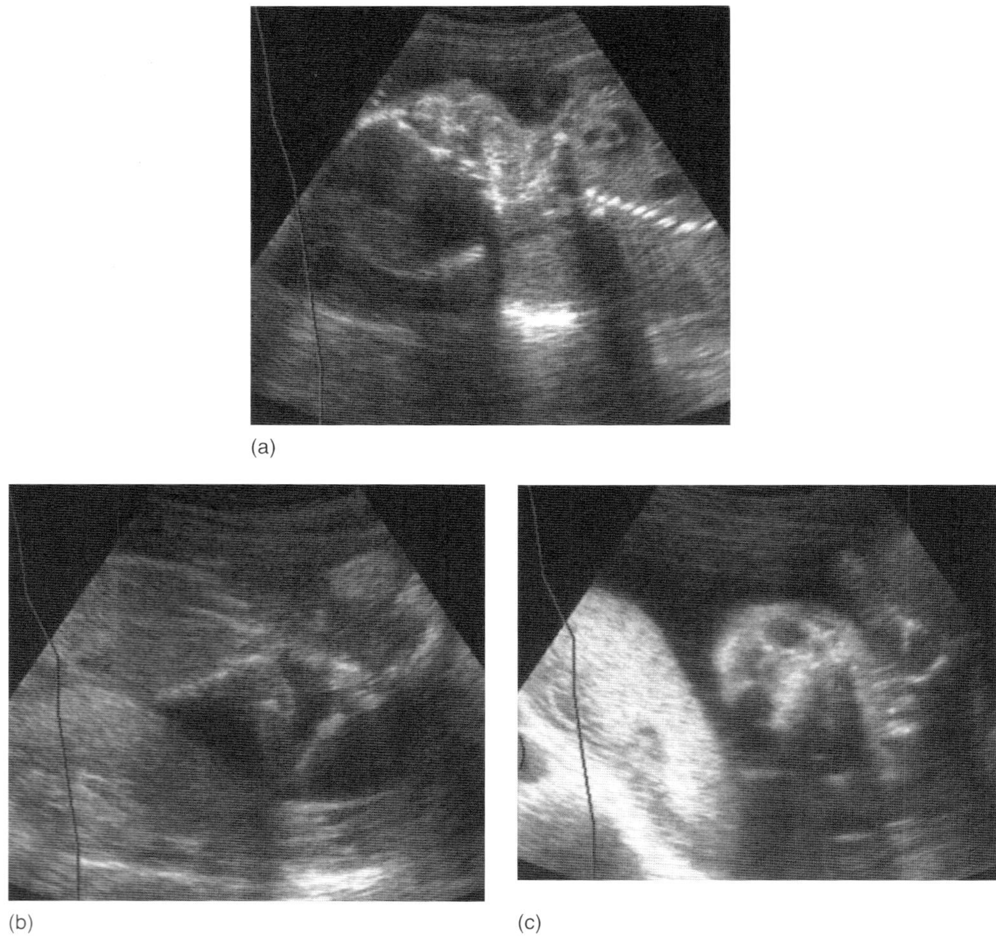

FIGURE 2.2.13
(a) Ultrasound image of a baby showing the head and spine. (b) Ultrasound image of a baby sucking its thumb. (c) Ultrasound image of a baby smiling (Isn't it amazing what parents can see!) (Images courtesy of Angi and Dave Lunning.)

2.2.4 Electron Imaging

Electron microscopes are used in applications which require extremely high magnification. Standard light microscopes can magnify one-thousand times, but electron microscopes can magnify up to 200 thousand times. These microscopes function by producing a focused beam of electrons, which is used to image a specimen like a light beam is used in a standard microscope. These microscopes come in two types: a transmission electron microscope (TEM), and a scanning electron microscope (SEM).

A TEM works by transmitting a beam of electrons through the specimen and then projecting the results onto a screen for viewing. A SEM, as the name implies, scans the electronic beam across the specimen and detects various signals generated by the electrons interacting with the specimen and uses these to produce an image. Figure 2.2.14 shows a SEM and sample images.

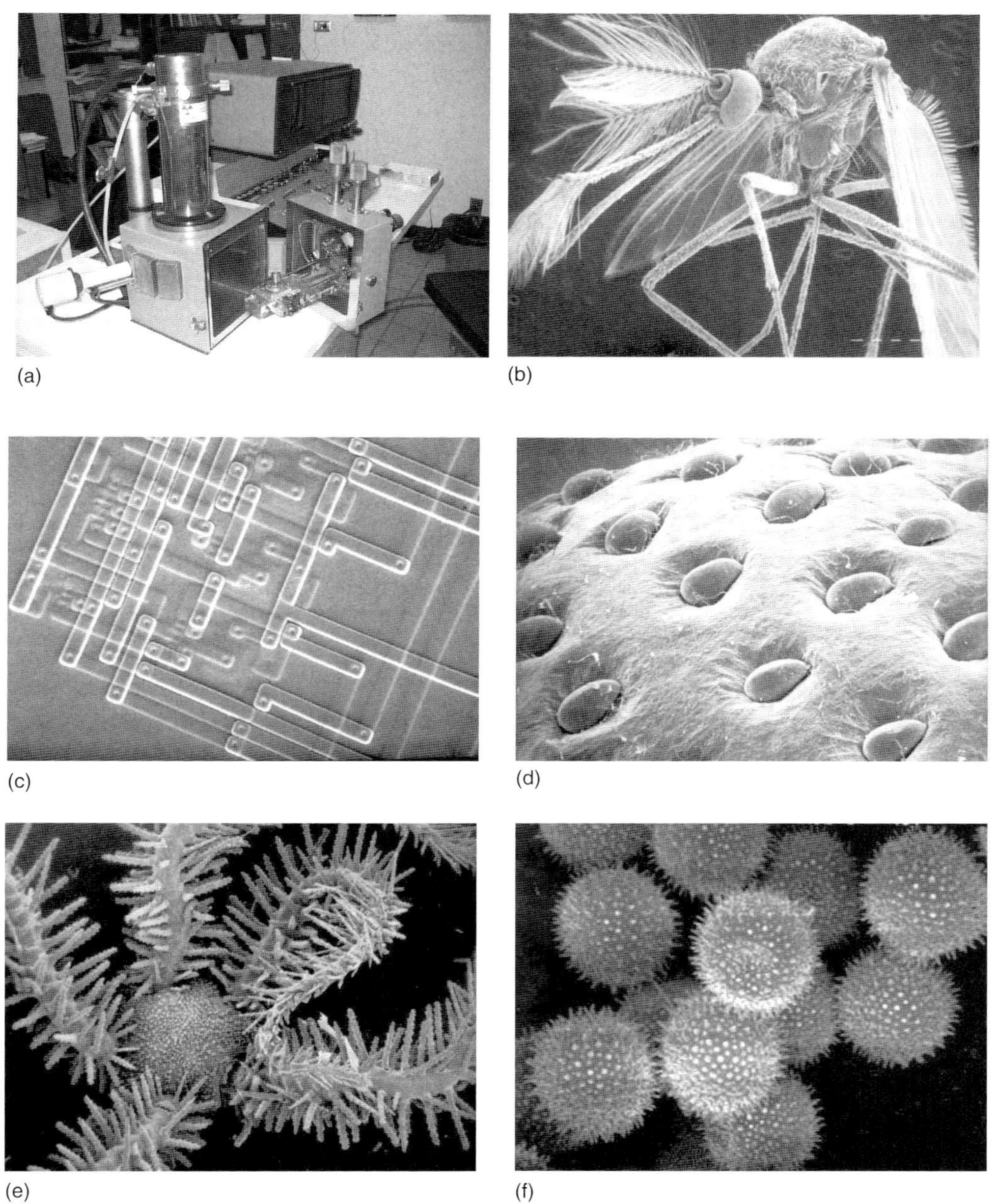

FIGURE 2.2.14
(a) Scanning electron microscope (SEM). (b) SEM image of a mosquito. (c) Logic gate in a microchip. (d) Strawberry. (e) Brittlestar. (f) Hollyhock pollen. (Photos courtesy of Sue Eder, Southern Illinois University Edwardsville.)

2.2.5 Laser Imaging

Lasers (*l*ight *a*mplification by *s*timulated *e*mission of *r*adiation) are specialized light sources that produce a narrow light beam in the visible, IR or UV range of the EM spectrum. In standard light sources, such as light bulbs, the atoms do not cooperate as they emit photons; they behave in a random or chaotic manner which produces *incoherent* light. Lasers are designed so that all the atoms cooperate, which produces a *coherent* light source that is highly intense and monochromatic (one color). Thus lasers, first developed in the 1960s, provide a method of controlling visible light energy in a manner similar to that in which radio waves and microwaves can be controlled.

Lasers are often used to create range images (also called *depth maps*) which contain information about the distance of a point in the world to the image sensor. One of the methods for this involves using structured lighting, which can be generated with a laser and two rotating mirrors. Special lenses of cylindrical shape can also be used to create a plane of light so that only one rotating mirror is needed. These techniques will not work for objects which are highly reflective, unless the sensor happens to be in the direction of the surface normal (perpendicular), since the light will not be reflected back to the sensor.

Another approach is to measure time-of-flight; that is, how long does it take a transmitted signal to be returned? As in radar imaging, a transmitter and receiver are required, and an electronic device (for example, a computer) measures the time it takes for a signal to be sent, reflected and received. Various types of signals are used, including pulses, amplitude-modulated phase shift (AM), and frequency-modulated (FM) beat signals.

2.2.6 Computer Generated Images

Images are not always generated by sensing real world objects; for example, computers can be used to create images for a myriad of applications. These include computer generated models for engineering, medicine and education, computer graphics for movies, art and games, and many other applications. In engineering, computer generated models are used in design and drafting; while in medicine they are used for surgical planning and training. Three-dimensional computer generated simulations are also created for training pilots in both military and commercial aviation. The quality of computer generated images has improved dramatically in the past several years as a result of advancements in technology and applied research in computer graphics.

Images are also created by computers as a result of applying image processing methods to images of real world objects. For example, the output of an image segmentation (see Chapter 4) process can be thought of as a computer generated image which uses the original real image as a model. Any image that has been remapped for display is, in a sense, a computer generated image since what you are looking at is not really the data itself but a representation of the underlying image data. Figure 2.2.15 shows examples of computer generated images.

2.3 The CVIPtools Software

The CVIPtools (Computer Vision and Image Processing tools) software was developed at Southern Illinois University Edwardsville, and contains functions to perform all the operations that are discussed in this book. These were originally written in ANSI-compatible C code and divided into libraries based on function category. For the new version of CVIPtools, a wrapper based on the Common Object Module (COM) interface was added to each function, and these COM functions are all contained in a dynamically linked library (dll) for use under the Windows operating system.

(a)

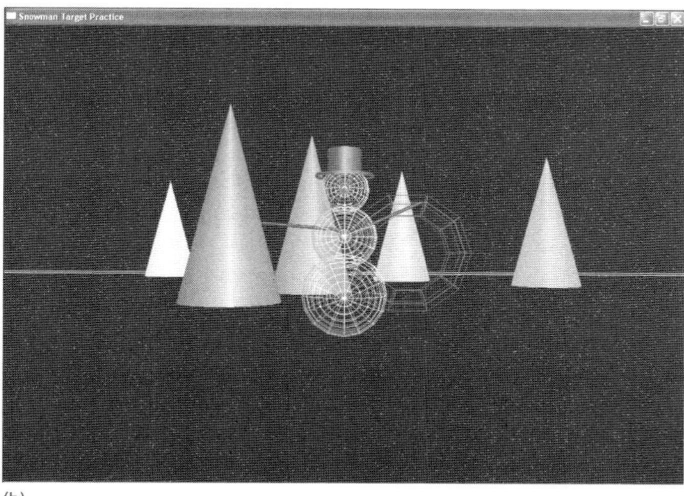

(b)

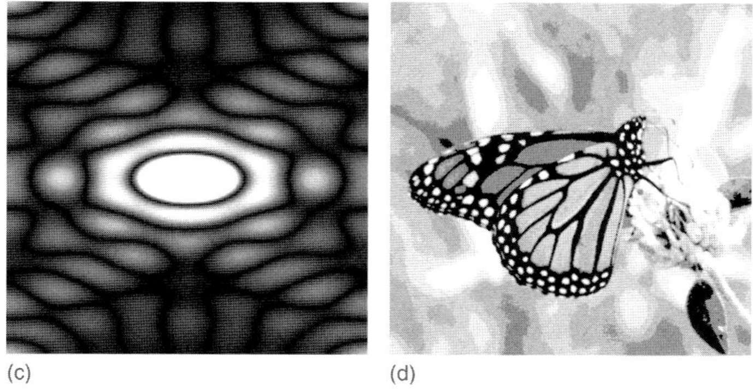

(c) (d)

FIGURE 2.2.15
Computer generated images. (a) Graphics image of a snowman scene. (b) The wireframe model of the snowman image. (c) An fft image of an ellipse (see Chapter 5). (d) A segmented image of a butterfly. (Snowman image courtesy of Dr Bill White, Southern Illinois University Edwardsville; original butterfly photo courtesy of Mark Zuke.)

These functions are explored in more detail in the fourth part of the book. A graphical user interface (GUI) was created for algorithm development and exploratory learning. The CVIPtools software is on CD-ROM (see Appendix A), and can also be accessed via the Internet (see Appendix B).

The only requirement for the new CVIPtools software, version 4.x, is a Windows operating system (2000/NT/ME/XP). Note that version 3.9 of CVIPtools is available for UNIX operating systems, including Sun Solaris, FreeBSD and Linux, and version 3.7c is available for these UNIX flavors and additionally SGI IRIX systems. The libraries, and the CVIPlab program which is used for all the programming exercises, are also available for all platforms.

The philosophy underlying the development of the CVIPtools is to allow the non-programmer to have access to a wide variety of computer imaging operations (not just the "standard" ones), and provide a platform for the exploration of these operations by allowing the user to vary all the parameters and observe the results in almost real time. This is especially facilitated by the CVIPlab program with the associated programming exercises and tutorials. Additionally, the function libraries allow those with programming skills to develop their own imaging applications with a minimum of coding.

The CVIPtools software will perform computer imaging operations from simple image editing to complex analysis, enhancement, restoration, or compression algorithms. One of the primary advantages of the software is that it is continually under development in a university environment, so as algorithms are developed they are made available for exploration and research. Another advantage is that it is being developed for educational purposes, not simply end-user results, so the focus is on *learning* about computer imaging. Because it is designed specifically for research and education, and the user has access to the many different parameters for the different algorithms, it is not constrained by what the market has deemed "works best." In some cases, the algorithms may not work very well (for commercial applications), but have educational and research value. Some of these same algorithms that "do not work very well" may be useful to researchers for specific applications, or may become part of a larger processing algorithm that does work well.

2.3.1 Main Window

When CVIPtools is first invoked the main window appears, as shown in Figure 2.3.1a. The main window contains the image queue, the image viewing area, the toolbar, the status bar, and access to all the windows and operations. The *image queue* is on the left of the main window and contains the names of all the images loaded, as well as any images that are created by CVIPtools. The image queue was implemented to facilitate fast processing—output images are automatically put into the queue and are not written to disk files unless you explicitly save them. Note that there is a checkbox at the top of the image queue labeled *Lock Input*. If it is checked it will retain (lock) the current image as input for each successive operation. This is useful when comparing various operations on the same image. When applying a sequence of operations to an image it may be desirable to have each sequential function operate on the output image, which happens when the *Lock Input* box is unchecked. Above the *Lock Input* checkbox are buttons to delete all the images from the queue or to delete one at a time.

Across the top of the window are the standard *File* and *View* selections, and the primary window selections for analysis and processing—*Analysis, Enhancement, Restoration, Compression, Utilities* and *Help*. Directly under these we see the toolbar which contains icons for opening, saving, printing and capturing image files as well as frequently used functions such as histogram display and RGB band extraction. To the right of these icons the column, row, red, green, and blue values are displayed for the current pixel position (see Figure 2.3.1b). The status bar at the bottom contains image specific information as determined by the image viewer.

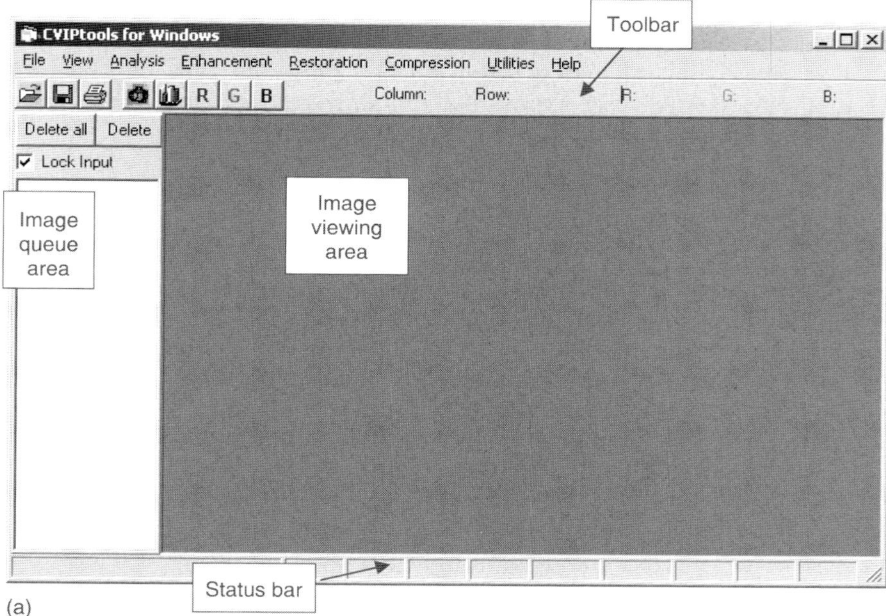

(a)

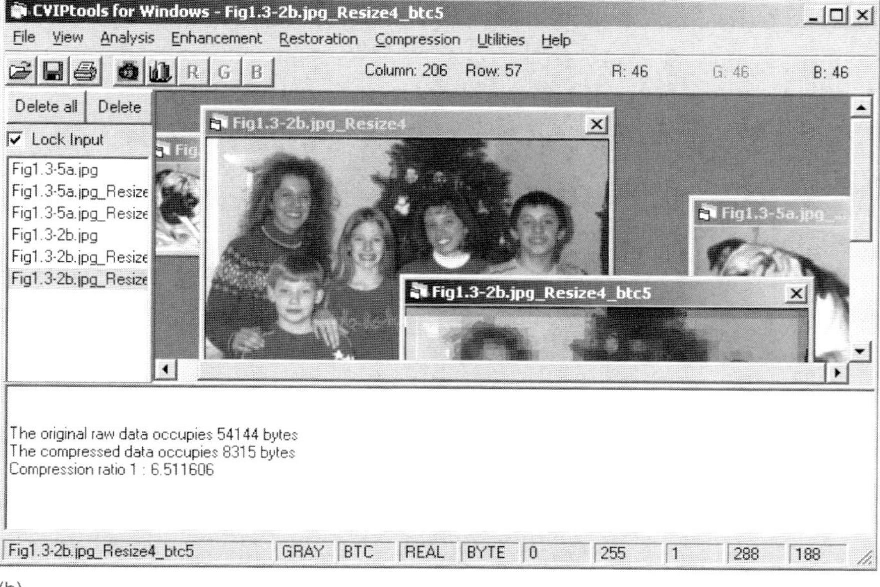

(b)

FIGURE 2.3.1
CVIPtools main window. (a) The main CVIPtools window when the program is first invoked. (b) Main window with images in the queue, and the *View* option *CVIPtools Function Information* at the bottom.

The items on the *View* menu provide useful options. Here the user can select the *Toolbar*, *Image Queue*, *CVIP Function Information*, and/or the *Status Bar* to appear (or not) on the main window. Removing any or all of these will free up screen space, if desired. The default setting, which appears when CVIPtools is first invoked, is to have the toolbar, image queue, and the status bar on the window. The *CVIP Function Information* option brings up a text window at the bottom of the main window, as shown in Figure 2.3.1b. This window displays information that is text output from the lower level functions and is often useful to gain insight into the inner workings of a specific function. Examples of information displayed include: convolution mask coefficients, type of data conversion, type of remapping, compression ratios of output images, and the number of clusters found for a segmentation algorithm.

2.3.2 Image Viewer

To load an image into CVIPtools you simply *open* it using the standard file open icon of a file folder opening in the upper left of the main window. When this is done the image is read into memory and its name will appear in the image queue, and the image will be displayed in the main window. Additionally, image information will appear in the status bar at the bottom of the main window (see Figure 2.3.1b). This information includes color format, image (file) format, data format, data type, data range (minimum and maximum), number of bands, and image width and height.

When an image is loaded it becomes the active image, and the active image can be changed at anytime by either clicking on the name in the queue, or clicking on the image itself. When this is done the image is brought to the front and as the mouse is rolled around the image the row and column coordinates, and the gray or color pixel values, will appear in the toolbar at the top of the main window. The active image can then be processed by selecting functions on the other windows.

The image viewer allows the user to perform standard image geometry operations, such as resizing, rotating, flipping, as well as image enhancement via histogram equalization. It is important to note that *these operations affect only the image that is displayed, not the image in the CVIPtools image queue*. They are for viewing convenience only, and any changes to the image itself (in the queue) can be accomplished by use of the standard CVIPtools windows. Even if the image is resized within the viewer, the row and column coordinates displayed will still correspond to the original image. Therefore, the image can be enlarged to ease the selection of small image features, or the image can be shrunk to minimize screen use. The keyboard and mouse can be used to perform the operations listed in Table 2.2. In addition to the keyboard commands, the user can stretch the image by grabbing the lower right corner of the image with the left mouse button and dragging it.

The CVIPtools image viewer allows the user to select a specific portion of an image by drawing a box with a press of the *Shift* key and drag of the left mouse button. This information is automatically passed back to the CVIPtools GUI for use in, for example, the image crop function. A new select box can be created at anytime and automatically destroys the first select box, or the middle mouse button can be used to remove the drawn box. Once a select box has been drawn, it retains its position throughout any image geometry operations. The viewer can also be used to draw borders on images by pressing the *Control* key and using the left mouse button, and the middle mouse button will remove it. Drawn borders are useful to extract features about specific objects, or to provide more control on an image crop function. Other functions are listed in Table 2.2.

TABLE 2.2

CVIPtools Image Viewer Keyboard and Mouse Commands

DRAW	Shift key–left mouse button	Draw a box on an image, used in crop, etc.
	Control key–left mouse button	Draw border for *Utilities->Create->Border Mask* and *Border Image* and *crop*, etc.
	Alt key–left mouse button	Mark mesh points for geometric transforms
	Right mouse button on image	Mesh display select box (followed by left button to select)
	Middle mouse button on image	Removes drawn boxes and borders
ROTATE	t	Turn 90 degrees clockwise
	T	Turn 90 degrees counter-clockwise
FLIP	h, H	Horizontal flip
	v, V	Vertical flip
OTHERS	N	Change back to original image, including size
	n	Change back to original image, without changing size
	q, Q	Quit—removes image from display but leaves in queue (clicking on the X in the upper right corner will remove the image from the queue)
	e, E	Histogram equalization
	Right mouse button in image viewing area (workspace)	Brings up *Utilities* menu

2.3.3 Analysis Window

When Analysis is first selected from the main window, the drop-down menu appears as shown in Figure 2.3.2a. Upon selection of one of the menu items the Analysis window appears with the tab corresponding to the menu item selected (Figure 2.3.2b). At any time the user can select another category of image analysis operations: *Geometry, Edge/Line Detection, Segmentation, Transforms*, and *Features*. When the user makes a selection, by clicking one of the file tabs with the left mouse button, the CVIPtools functions available under that selection will appear.

Most of the functions can be selected by the round buttons on the left of the window. These are called *option buttons*—only one can be active at a time. Once the operation has been selected, the necessary input parameters can be typed in the text boxes, or selected with the mouse using the arrows. Note that the text boxes will initially contain default values, which allow for immediate use of an operation by simply selecting it via the option button on the left, and the clicking on the *Apply* button (assuming an image has been selected). Any parameters that are not used with a particular function will be grayed out, or disappear from the screen, whenever that particular function is selected. The individual tabs and functions will be discussed in more detail in Chapters 3–6.

In addition to the *Apply* button at the bottom of the window, there are buttons for *Help*, *Cancel*, and *Reset*. The *Cancel* button will remove the window from the screen, but leave all the parameters unchanged. The *Reset* button will leave the window on the screen, but reset all the parameters in the current tab to the default values. Use of the X in the upper right corner will remove the window from the screen and automatically reset all the parameters in all the tabs in the window to their default values. These buttons, and their functionality, are also standard in the Enhancement, Restoration, Compression, and Utilities windows.

2.3.4 Enhancement Window

The *Enhancement* window is shown in Figure 2.3.3. Across the top of the window are file tabs that allow for selections that pertain to image enhancement: *Histograms, Pseudocolor,*

(a)

(b)

FIGURE 2.3.2
CVIPtools analysis window. (a) The drop-down menu for the Analysis window. (b) The Analysis window with the Edge/Line Detection tab selected.

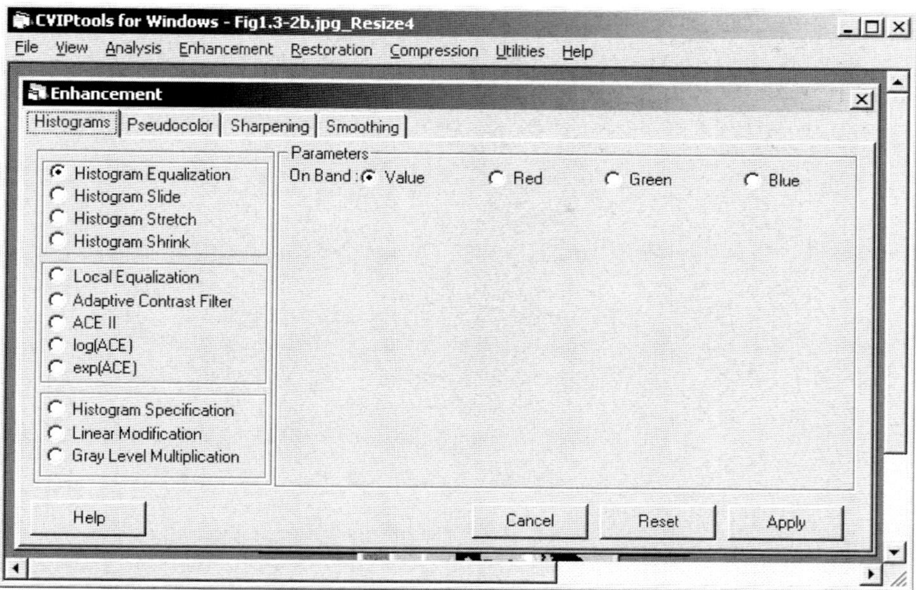

FIGURE 2.3.3
CVIPtools enhancement window. The Enhancement window with the Histograms tab selected.

Sharpening, and *Smoothing*. The image enhancement functions are used to make images more appealing to the human visual system, to mitigate noise effects, to improve image contrast or to bring out image information that is not readily visible. The histograms tab contains functions that are primarily used to improve contrast and brightness, and the pseudocolor tab has functions that will change a monochrome image into a color image. The sharpening and smoothing tabs have filter functions and algorithms that will perform these effects on images. The enhancement methods are discussed in more detail in Chapter 8.

2.3.5 Restoration Window

The *Restoration* window is shown in Figure 2.3.4. Across the top of the window are file tabs that allow for selections that pertain to image restoration: *Noise, Spatial Filters, Frequency Filters,* and *Geometric Transforms*. The image restoration functions are used to mitigate noise effects, to restore blurry images or to fix images that have been spatially distorted. The noise tab has functions to add noise to images or to create noise only images, which can be used in restoration algorithm development. The spatial filters tab contains functions that are used to mitigate noise effects, and the frequency filters tab has functions to restore noisy, degraded and blurred images. Geometric transforms are used to restore spatially distorted images. The restoration functions are discussed in more detail in Chapter 9.

2.3.6 Compression Window

The *Compression* window is shown in Figure 2.3.5. Across the top of the window are file tabs that allow for selections that pertain to image compression: *Preprocessing,*

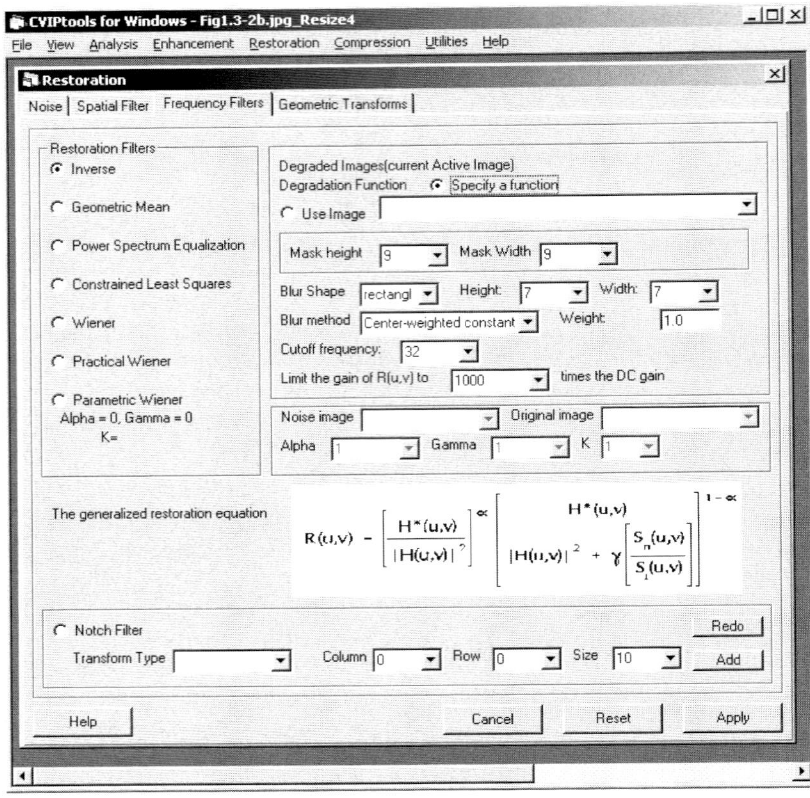

$$R(u,v) = \left[\frac{H^*(u,v)}{|H(u,v)|^2}\right]^{\alpha}\left[\frac{H^*(u,v)}{|H(u,v)|^2 + \gamma\left[\frac{S_n(u,v)}{S_I(u,v)}\right]}\right]^{1-\alpha}$$

FIGURE 2.3.4
CVIPtools Restoration window. The Restoration window with the Frequency Filters tab selected.

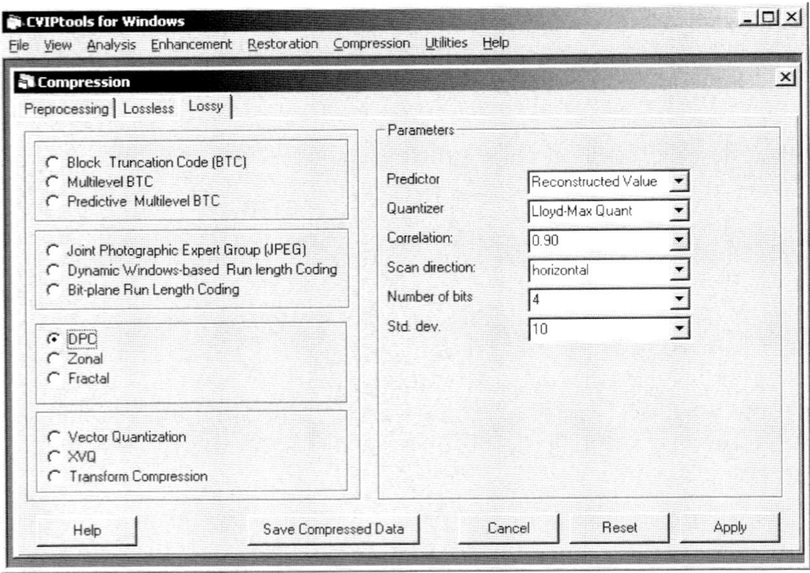

FIGURE 2.3.5
CVIPtools Compression window. The Compression window with the Lossy tab selected.

Lossless and *Lossy*. The preprocessing tab contains functions that can be useful prior to compression. The lossless tab has compression functions that will create images identical to the original, while the lossy tab has the compression functions that will attempt to create the best quality image for a given amount of image data loss.

The image compression functions are used to reduce the file size of images, so experimentation and comparisons can be performed among various compression options. The compression window has an additional button at the bottom—*Save Compressed Data*. This button allows the user to save the image in its compressed format, which is a unique CVIPtools format (except for JPEG, which is a standard file format). If you simply save the displayed image, it will be saved as the decompressed image shown, in any file format that is selected. Compression ratios for the output images are available with the view option *CVIP Function Information*. Details of the compression functions are discussed in Chapter 10.

2.3.7 Utilities Window

The *Utilities* window works differently than the previously discussed windows. This is because it contains functions that are commonly used regardless of the type of processing being performed. It can be accessed with two methods, depending on the user's preferences. The first method is to right click the mouse anywhere in the image viewing area. When this is done a two-level menu will popup, as shown in Figure 2.3.6a. This menu contains various categories of commonly used utilities: *Arithmetic/Logic, Compare, Convert, Create, Enhance, Filter, Size,* and *Stats*. Alternately, the user can select Utilities at the top of the main window, and the previously mentioned menu items will appear across the top of the main window as shown in Figure 2.3.6b.

After using either method to invoke *Utilities* the user selects a menu item, and the necessary information appears in the *Utilities* window for that particular function (see an example in Figure 2.3.6c). By limiting screen usage in this manner, the *Utilities* window is easily accessible when other primary windows are in use. The general philosophy guiding the design of the Utilities GUI is to maximize utility and usage, while minimizing use of screen space. In some cases, for example with *Utilties->Enhancement*, only the most commonly used functions will appear in the *Utilities* window, and the choices for the various parameters may be limited. This allows *Utilities* to be used easily and quickly, and if the user needs more, the main *Enhancement* window can be selected.

2.3.8 Help Window

The CVIPtools *Help* window can be accessed from the top of the main window, or with the button in the lower left of any of the other windows. In Figure 2.3.7 we see the *Help* window which contains information about CVIPtools development, how to use the CVIPtools functions, and documentation for the C and the COM functions. The documentation for the C functions includes a complete description and examples of their use in CVIPlab. Documentation for the COM functions contains the function prototypes, parameter definitions and a description. The *Help* window also contains help for using the CVIPtools functions from the GUI and has links to CVIPtools related websites. The *Help* window has an index of all the documents it contains and allows for keyword searches to assist the user in finding what they need.

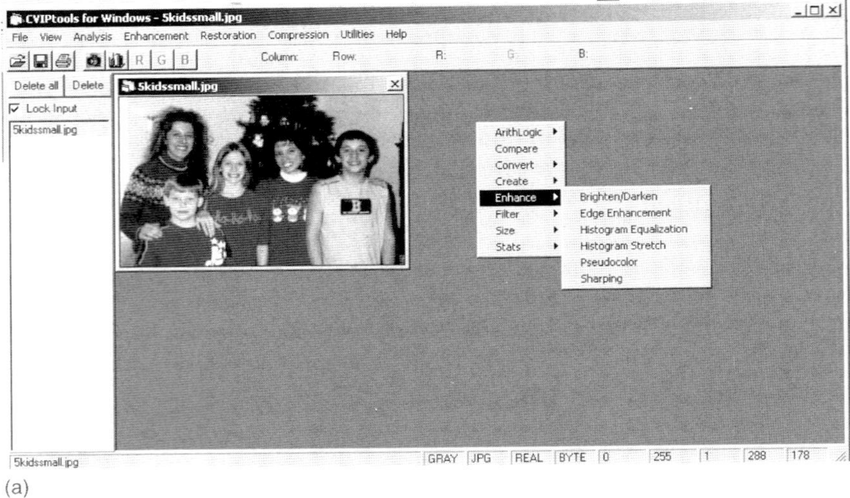

(a)

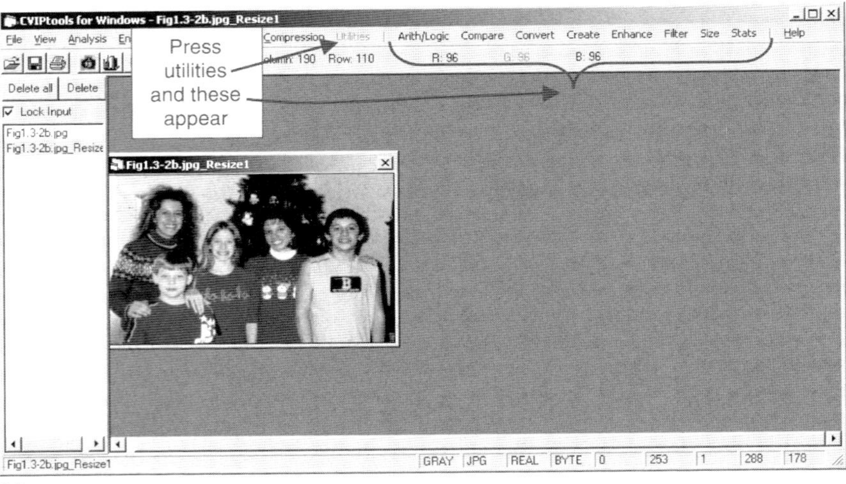

(b)

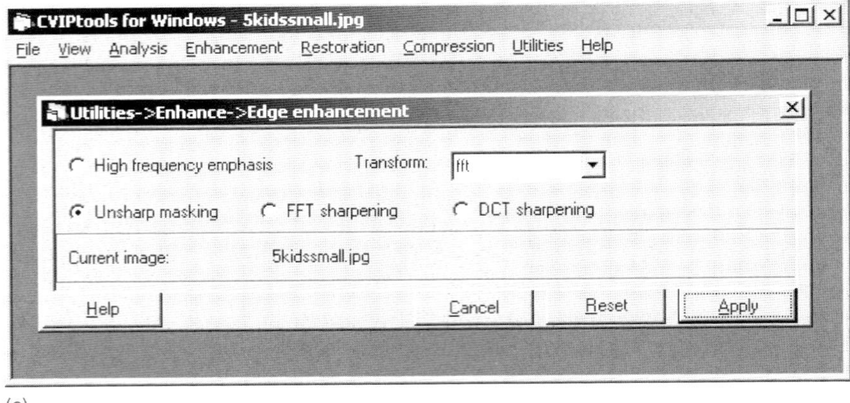

(c)

FIGURE 2.3.6
CVIPtools utilities. The utility functions can be accessed with two methods. (a) the two-level menu for Utilities will popup with a right mouse click in the image viewing area, or (b) click on Utilities at the top of the main window and the primary menu for Utilities will appear across the top. (c) An example Utilities window selection.

2.4 Image Representation

We have seen that an imaging sensor receives an input image as a collection of spatially distributed light energy; this form is called an *optical image*. Optical images are the types we deal with everyday—cameras capture them, monitors display them, and we see them. We have also seen that these optical images are represented as video information in the form of analog electrical signals, and how these are sampled to generate the digital image $I(r, c)$.

The digital image, $I(r, c)$, is represented as a two-dimensional array of data, where each pixel value corresponds to the brightness of the image at the point (r, c). In linear algebra terms, a two-dimensional array like our image model, $I(r, c)$, is referred to as a *matrix*, and one row (or column) is called a *vector*. This image model is for monochrome (one-color, referred to as gray-scale) image data, but we have other types of image data that require extensions or modifications to this model. Typically, these are multi-band images (color, multispectral), and they can be modeled by a

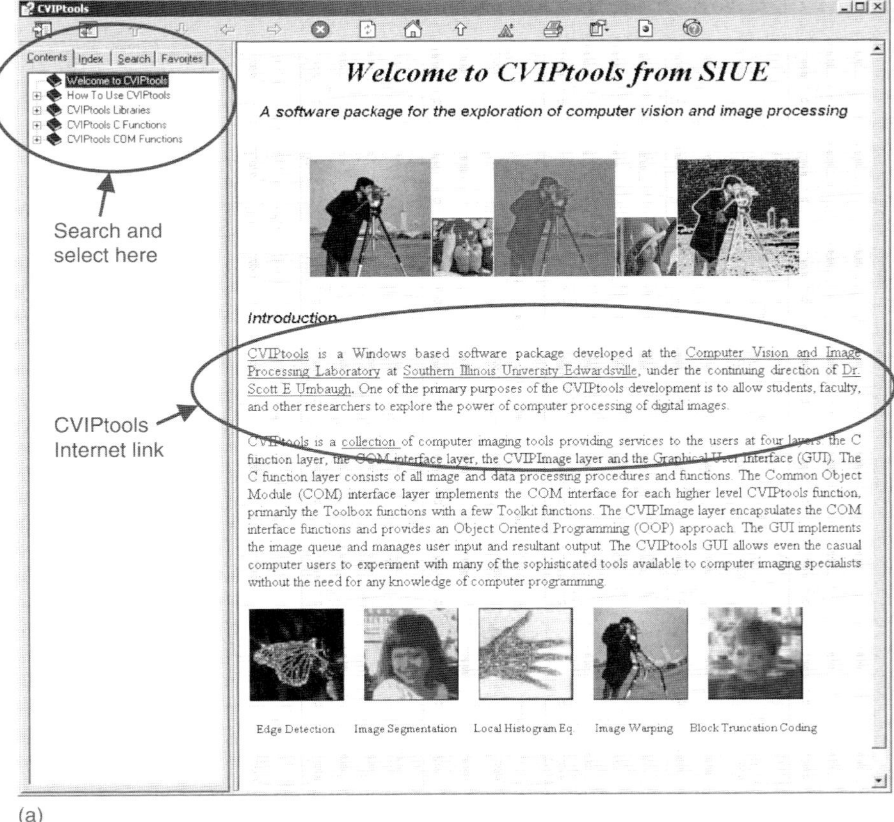

FIGURE 2.3.7
The CVIPtools Help window. The Help window contains information about using CVIPtools and contains documentation for the libraries and C functions, and includes CVIPtools related Internet links. It has an index of all the documents it contains and allows for keyword searches to assist the user in finding what they need. (a) The Help window as it appears when first selected.

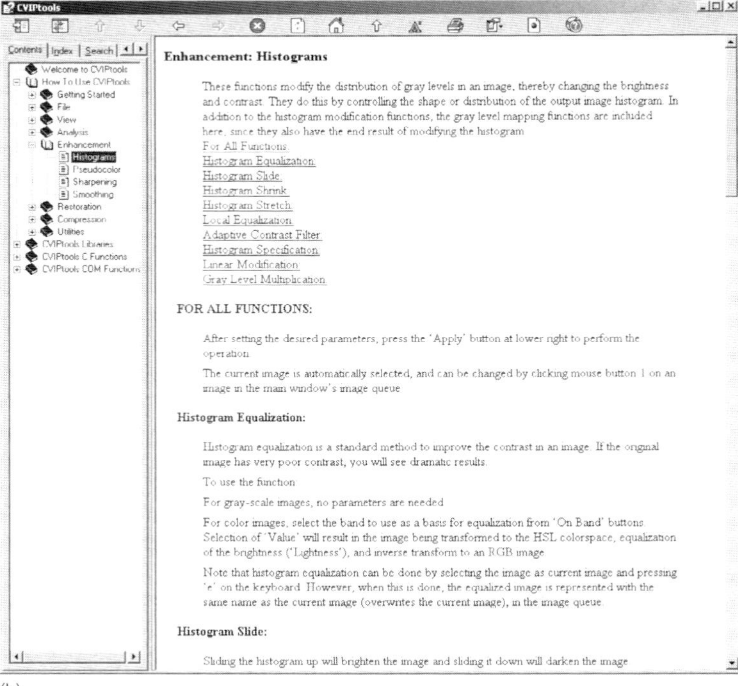

(b)

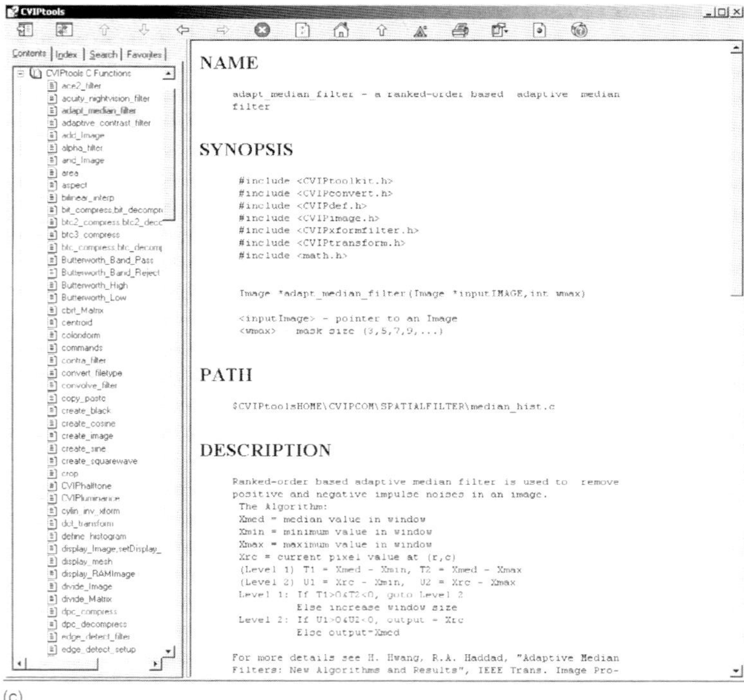

(c)

FIGURE 2.3.7 (Continued)
The CVIPtools Help window. The Help window contains information about using CVIPtools and contains documentation for the libraries and C functions, and includes CVIPtools related Internet links. It has an index of all the documents it contains and allows for keyword searches to assist the user in finding what they need. (b) Help window showing an example of a page under *How to Use CVIPtools*. (c) Help window showing an example of C function documentation.

different $I(r, c)$ function corresponding to each separate band of brightness information. The image types we will consider are: (1) binary, (2) gray-scale, (3) color, and (4) multispectral.

2.4.1 Binary Images

Binary images are the simplest type of images, and can take on two values, typically black and white, or "0" and "1." A binary image is referred to as a 1-bit per pixel image, because it takes only 1 binary digit to represent each pixel. These types of images are most frequently used in computer vision applications where the only information required for the task is general shape, or outline, information. For example, to position a robotic gripper to grasp an object, to check a manufactured object for deformations, for facsimile (FAX) images, or in optical character recognition (OCR).

Binary images are often created from gray-scale images via a threshold operation, where every pixel above the threshold value is turned white ("1"), and those below it are turned black ("0"). Although in this process much information is lost, the resulting image file is much smaller making it easier to store and transmit. In Figure 2.4.1, we see examples of binary images. Figure 2.4.1a is a page of text, such as might be used in an OCR application; Figure 2.4.1b is the outline of an object; and in Figure 2.4.1c we have the results of an edge detection operation (see Section 4.2).

2.4.2 Gray-Scale Images

Gray-scale images are referred to as monochrome, or one color, images. They contain brightness information only, no color information. The number of bits used for each pixel determines the number of different brightness levels available. The typical image contains 8-bit per pixel data, which allows us to have 256 (0–255) different brightness (gray) levels. This representation provides more than adequate brightness resolution, in terms of the human visual system's requirements (see Chapter 7), and provides a "noise margin" by allowing for approximately twice as many gray levels as required. This noise margin is useful in real-world applications due to many different types of noise (false information in the signal) inherent in real systems. Additionally, the 8-bit representation is typical due to the fact that the *byte*, which corresponds to 8-bits of data, is the standard small unit in the world of digital computers.

In certain applications, such as medical imaging or astronomy, 12 or 16-bit per pixel representations are used. These extra brightness levels only become useful when the image is "blown-up," that is, a small section of the image is made much larger. In this case we may be able to discern details that would be missing without this additional brightness resolution. Of course, to be useful, this also requires a higher level of spatial resolution (number of pixels). If we go beyond these levels of brightness resolution, we typically divide the light energy into different bands, where each *band* refers to a specific subsection of the visible image spectrum.

2.4.3 Color Images

Color images can be modeled as three-band monochrome image data, where each band of data corresponds to a different color. The actual information stored in the digital image data is the brightness information in each spectral band. When the image

```
wht2d(3)              C Library Functions              wht2d(3)

NAME
     wht2d - performs Walsh or Hadamard transform

SYNOPSIS
       #include <stdio.h>
       #include <stdlib.h>
       #include <math.h>
       #include "CVIPtools.h"
       #include "CVIPimage.h"
       #include "CVIPdef.h"

     IMAGE *wht2d(IMAGE *in_IMAGE, int ibit, int block_size)

      <in_IMAGE> - pointer to an IMAGE structure
      <ibit> - 0=inverse Walsh transform, 1=Walsh transform
               2=inverse Hadamard transform, 3=Hadamard transform
      <block_size> - block size (4,8,16,...largest_dimension/2)

PATH
     $CVIPHOME/TRANSFORMS/wht2d.c

DESCRIPTION
     This  function   performs  a  fast  Hadamard-ordered  Walsh-
     Hadamard  Transform  on  an image.  The result is then reor-
     dered for display in sequency order.  The routine  works  on
     any  image  with  dimensions that are powers of 2.  Optional
     zero-padding may be performed if input image  has  different
     dimensions.
```

(a)

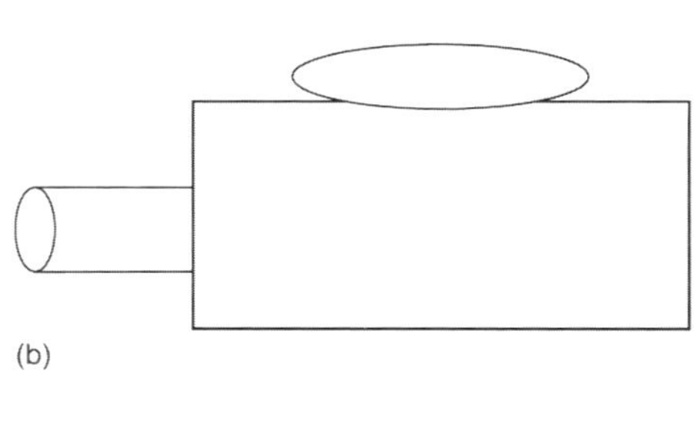

(b)

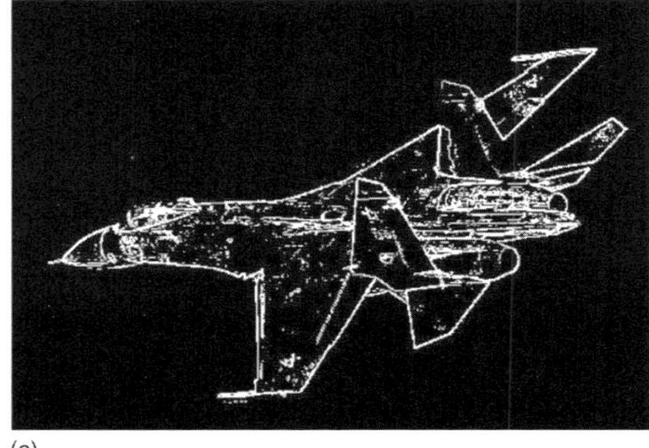

(c)

FIGURE 2.4.1
Binary images. (a) Binary text. (b) Object outline. (c) Edge detection and threshold operation.

is displayed, the corresponding brightness information is displayed on the screen by picture elements that emit light energy corresponding to that particular color. Typical color images are represented as red, green, and blue, or RGB images. Using the 8-bit monochrome standard as a model, the corresponding color image would have 24-bits per pixel—8-bits for each of the three color bands (red, green, and blue). In Figure 2.4.2a we see a representation of a typical RGB color image. Figure 2.4.2b illustrates that, in addition to referring to a row or column as a vector, we can refer to a single pixel's red, green, and blue values as a *color pixel vector*—(R, G, B).

For many applications, RGB color information is transformed into a mathematical space that decouples the brightness information from the color information. This transformation is referred to as a *color model*, a *color transform* or mapping into another *color space*. Once this is done the image information consists of a one-dimensional brightness, or luminance, space and a two-dimensional color space. Now the two-dimensional color space does not contain any brightness information, but typically

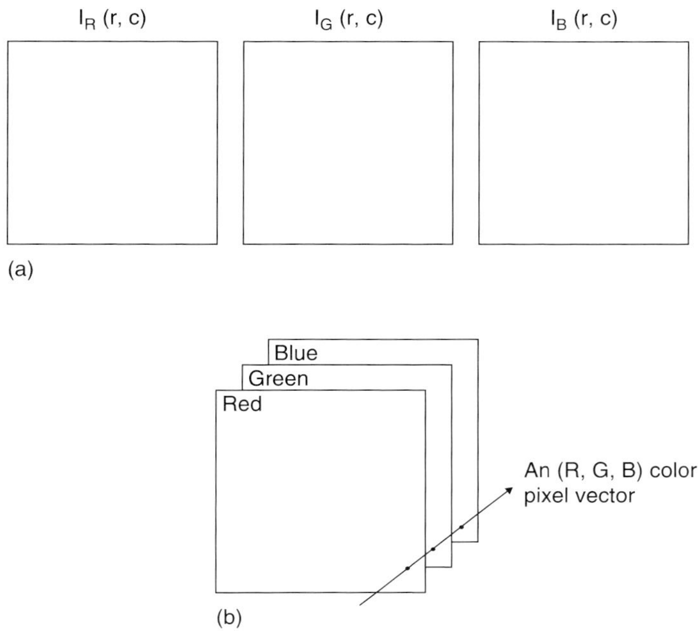

FIGURE 2.4.2
Color image representation. (a) A typical RGB color image can be thought of as three separate images: $I_R(r, c)$, $I_G(r, c)$, and $I_B(r, c)$. (b) A color pixel vector consists of the red, green, and blue pixel values (R, G, B) at one given row/column pixel coordinate (r, c).

contains information regarding the relative amounts of the different colors. An additional benefit of modeling the color information in this manner is that it creates a more people-oriented way of describing the colors.

For example, the Hue/Saturation/Lightness (HSL) color transform allows us to describe colors in terms that we can more readily understand (see Figure 2.4.3). The *lightness* (also referred to as *intensity* or *value*) is the brightness of the color, and the *hue* is what we normally think of as "color"; for example green, blue, or orange. The *saturation* is a measure of how much white is in the color; for example, pink is red with more white, so it is less saturated than a pure red. Most people can relate to this method of describing color, for example "a deep, bright orange" would have a large intensity ("bright"), a hue of "orange," and a high value of saturation ("deep"). We can picture this color in our minds, but if we defined this color in terms of its RGB components, R = 245, G = 110, and B = 20, most people would have no idea how this color appears. Since the HSL color space was developed based on heuristics relating to human perception, various methods are available to transform RGB pixel values into the HSL color space. Most of these are algorithmic in nature and are geometric approximations to mapping the RGB color cube into some HSL-type color space (see Figure 2.4.4). Example equations for mapping RGB to HSL are given below:

$$H = \begin{cases} \theta & \text{if } B \le G \\ 360 - \theta & \text{if } B > G \end{cases}$$

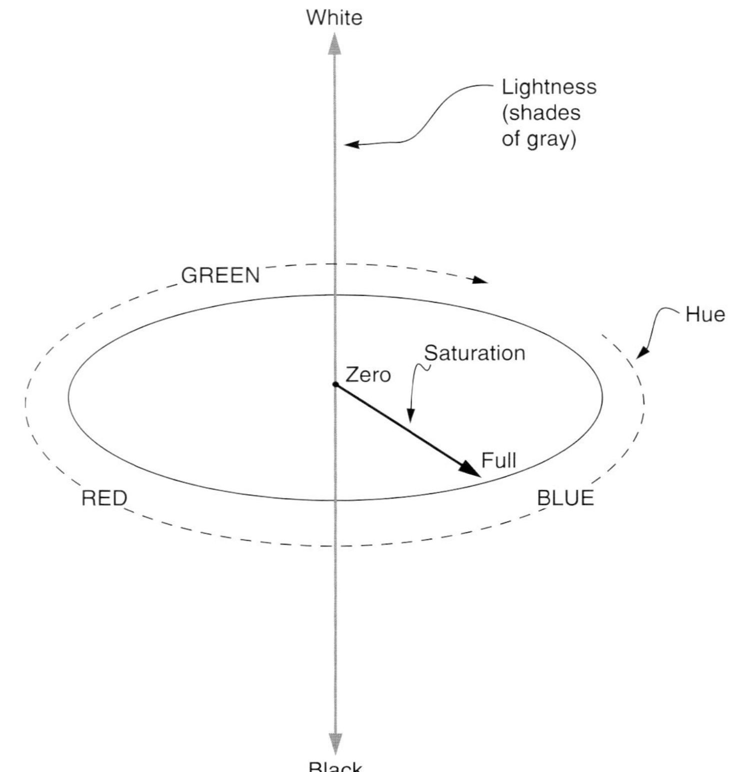

FIGURE 2.4.3
HSL color space.

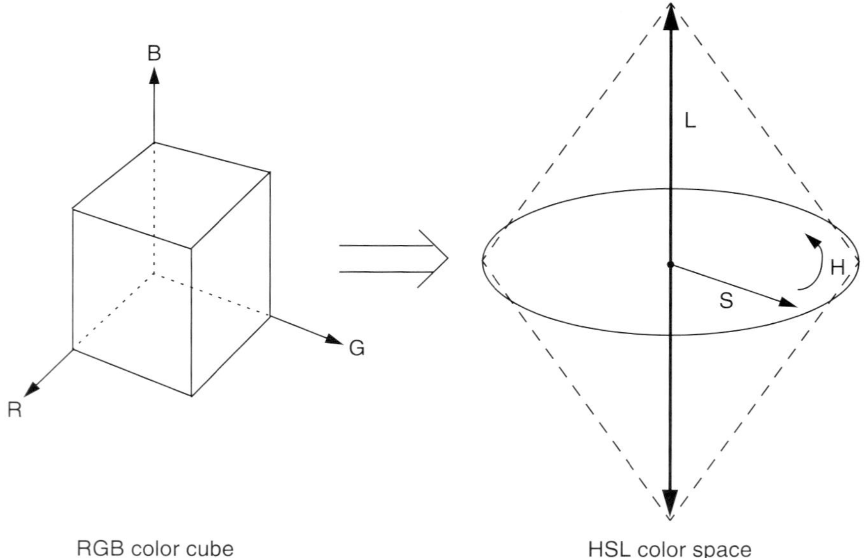

FIGURE 2.4.4
RGB to HSL mapping.

where

$$\theta = \cos^{-1}\left\{\frac{\frac{1}{2}[(R-G)+(R-B)]}{[(R-G)^2+(R-B)(G-B)]^{1/2}}\right\}$$

$$S = 1 - \frac{3}{(R+G+B)}[\min(R, G, B)]$$

$$L = \frac{(R+G+B)}{3}$$

These equations assume that the R, G, and B values are normalized to lie between 0 and 1, and θ is measured in degrees from the red axis. To convert the HSL values back into the RGB coordinates requires consideration of three different sectors in the color space; namely Red–Green (RG), Green–Blue (GB) and the Blue–Red (BR) sector. The following equations apply:

RG Sector $(0^\circ \leq H < 120^\circ)$:

$$R = L\left[1 + \frac{S\cos(H)}{\cos(60^\circ - H)}\right]$$

$$G = 1 - (R+B) \quad \text{(note: find R and B first)}$$

$$B = L(1-S)$$

GB Sector $(120^\circ \leq H < 240^\circ)$:

$$R = L(1-S)$$

$$G = L\left[1 + \frac{S\cos(H - 120^\circ)}{\cos(180^\circ - H)}\right]$$

$$B = 1 - (R+G)$$

BR Sector $(240^\circ \leq H < 360^\circ)$:

$$R = 1 - (G+B) \quad \text{(note: find G and B first)}$$

$$G = L(1-S)$$

$$B = L\left[1 + \frac{S\cos(H - 240^\circ)}{\cos(300^\circ - H)}\right]$$

A color transform can be based on a geometrical coordinate mapping, such as the spherical or cylindrical transforms. With the spherical transform the RGB color space will be mapped to a one-dimensional brightness space and a two-dimensional color space. The spherical coordinate transform (SCT) has been successfully used in a color segmentation algorithm described in Chapter 4. The equations relating the SCT to the RGB components are as follows:

$$L = \sqrt{R^2 + G^2 + B^2}$$

$$\angle A = \cos^{-1}\left[\frac{B}{L}\right]$$

$$\angle B = \cos^{-1}\left[\frac{R}{L\sin(\angle A)}\right]$$

where L is the length of the RGB vector, angle A is the angle from the blue axis to the RG plane, and angle B is the angle between the R and G axes. Here, L contains the brightness information and the two angles contain the color information (see Figure 2.4.5).

The cylindrical coordinate transform (CCT) is different than most color mappings because it does not completely decouple brightness from color information. With this transform we can align the z-axis along the R, G, or B axis of choice; this choice will be application dependent. The cylindrical coordinates are found as follows, assuming the z-axis is aligned along the blue axis:

$$z = B$$

$$d = \sqrt{R^2 + G^2}$$

$$\theta = \tan^{-1}\left(\frac{G}{R}\right)$$

The CCT may be useful in applications where one of the RGB colors is of primary importance, since it can be mapped directly to the z component, and the ratio of the other two is significant. Here, the brightness information is now contained in the d and z coordinates, while the color information is still distributed across all three components, but in a different manner than with the original RGB data. This is illustrated in Figure 2.4.6, where we can see that θ is related to hue in the RG-plane, and d is related to the saturation in the RG plane.

One problem associated with the color spaces previously described is that they are not perceptually uniform. This means that two different colors in one part of the color space will not exhibit the same degree of perceptual difference as two colors in another part of the color space, even though they are the same "distance" apart (see Figure 2.4.7). Therefore, we cannot define a metric to tell us how close, or far apart, two colors are in terms of human perception. In computer imaging applications a perceptually uniform color space could be very useful. For example, if we are trying to identify objects for a computer vision system by color information, we need some method to compare the object's color to a database of the colors of the available objects. Or if we are trying to develop a new image compression algorithm, we need a way to determine if we can map one color to another without losing significant information.

The science of color and how the human visual system perceives color has been studied extensively by an international body, the *Commission Internationale de l'Eclairage*

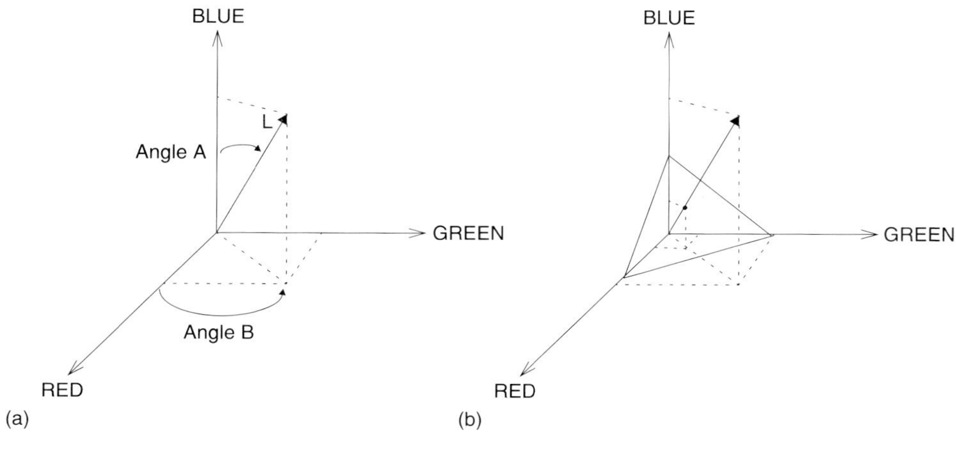

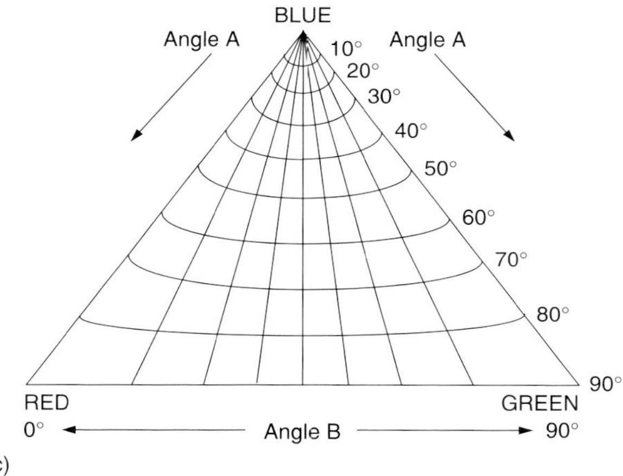

FIGURE 2.4.5
Spherical coordinate transform. (a) The spherical coordinate transform separates the red, green, and blue information into a 2-D color space defined by angles *A* and *B*, and a 1-D brightness space defined by *L*. (b) A color pixel vector (R, G, B). (c) The color triangle showing regions defined by 10 degree increments on Angle *A* and Angle *B*.

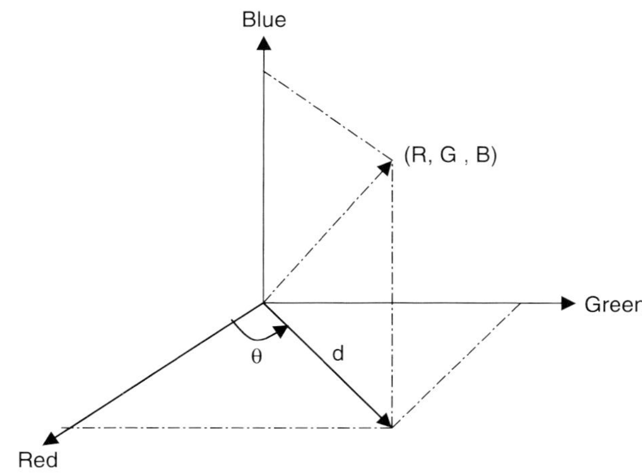

FIGURE 2.4.6
Cylindrical coordinates transform.

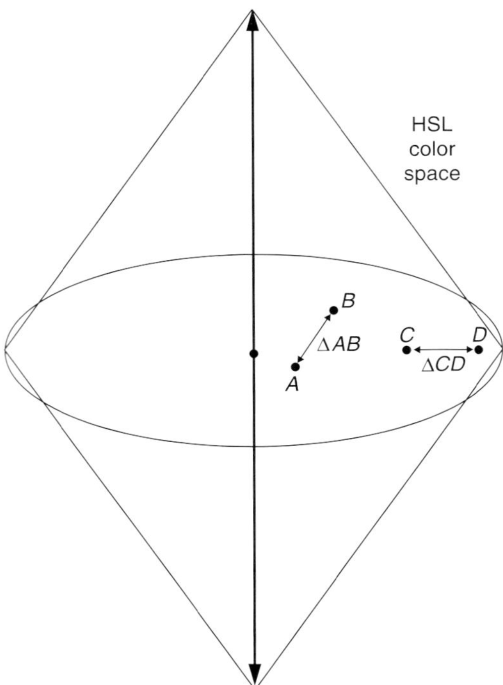

FIGURE 2.4.7
Color perception. Color A may be orange and B may be green; colors C and D may be slightly different shades of the same color. However, $\Delta AB = \Delta CD$.

(CIE). The CIE has defined internationally recognized color standards. One of the basic concepts developed by the CIE involves chromaticity coordinates. For our RGB color space, *chromaticity coordinates* are defined as follows:

$$r = \frac{\mathrm{R}}{\mathrm{R}+\mathrm{G}+\mathrm{B}}$$

$$g = \frac{\mathrm{G}}{\mathrm{R}+\mathrm{G}+\mathrm{B}}$$

$$b = \frac{\mathrm{B}}{\mathrm{R}+\mathrm{G}+\mathrm{B}}$$

These equations basically normalize the individual color components to the sum of the three, which we have seen is one way to represent the brightness information. This decouples the brightness information from the coordinates, and the CIE uses chromaticity coordinates as the basis of the color transforms they define. These include the standard CIE XYZ color space (related to the tristimulus curves discussed in Chapter 7), and the perceptually uniform $L^*u^*v^*$, $L^*a^*b^*$ color spaces. The science of color and human perception is a fascinating topic and can be explored in greater depth with the references.

Another important international committee for developing standards of interest to those involved in computer imaging is the International Telecommunications Union Radio (ITU-R, previously CCIR). This committee has specified the standard for digital video known as ITU-R 601. This standard is based on one luminance signal (Y) and two color difference signals (*Cr* and *Cb*). Given a 24-bit RGB signal, we can find the *Y*, *Cr*, and *Cb* values as follows:

$$Y = 0.299R + 0.587G + 0.114B$$

$$Cb = -0.1687R - 0.3313G + 0.5B + 128$$

$$Cr = 0.5R - 0.4187G - 0.0813B + 128$$

The 128 offset factor is included here to maintain the data range of [0–255] for 8-bit per color band data. This transform is used in many color image compression algorithms, such as MPEG and JPEG, implemented in both hardware and software. This transform is also called YUV encoding and is defined as follows:

$$Y = 0.299R + 0.587G + 0.114B$$

$$U = 0.493(B - Y)$$

$$V = 0.877(R - Y)$$

Note that the 128 offset value for U and V can be added to these equations, if desired.

All the previous color transforms are based on an additive color model such as RGB, where we consider adding red, green or blue light to a black background. For color printing we use a subtractive color model, where we consider subtracting cyan, magenta or yellow (CMY) from white, such as printing on white paper illuminated by white light. The model for white light is that it consists of red, green and blue. The CMY conversion from RGB is defined as follows (these equations assume that the RGB values are normalized to the range of 0 to 1):

$$C = 1 - R$$

$$M = 1 - G$$

$$Y = 1 - B$$

Cyan absorbs red light, magenta absorbs green and yellow absorbs blue. Thus, to print a normalized RGB triple that appears green, $(0, 1, 0)$, we would use CMY $(1, 0, 1)$. For this example the cyan will absorb the red light and the yellow absorbs the blue light, leaving only the green light to be reflected and seen. Also, to print black we print all three (CMY) inks, and all the components of white light, RGB, will be absorbed. In practice, this produces a poor looking black, so black ink is added to the printing process leading to a four-color printing system, called CMYK.

The final color transform we will discuss is called the principal components transform. This mathematical transform allows us to apply statistical methods to put as much of the three-dimensional color information as possible into only one band. This process decorrelates the RGB data components. The *principal components transform* (PCT) works by examining all the RGB vectors within an image and finding the linear transform that aligns the coordinate axes so that most of the information is along one axis, the principal axis. Often, we can get 90% or more of the information into one band. The PCT is used in image segmentation and compression schemes (see Chapters 4 and 10), and the mathematical details of the transform are discussed in more detail in Chapter 5.

2.4.4 Multispectral Images

Multispectral images typically contain information outside the normal human perceptual range, as discussed in Section 2.2.2. They may include infrared, ultraviolet, x-ray, or other bands in the EM spectrum. These are not images in the usual sense, since the information represented is not directly visible by the human visual system. However, the information is often represented in visual form by mapping the different spectral bands to RGB components. If more than three bands of information are in the multispectral image, the dimensionality is reduced for display by applying a principal components transform (see Chapter 5).

Sources for these types of images include: satellite systems, underwater sonar systems, various types of airborne radar, infrared imaging systems, and medical diagnostic imaging systems. The number of bands into which the data is divided is strictly a function of the sensitivity of the imaging sensors used to capture the images. For example, even the visible spectrum can be divided into many more than three bands; three are used because this mimics our visual system. The older satellites currently in orbit collect image information in two to seven spectral bands; typically one to three are in the visible spectrum and one or more in the infrared region, and some have sensors that operate in the microwave range. The newest satellites have sensors that collect image information in 30 or more bands. For example, the NASA/Jet Propulsion Laboratory Airborne Visible/Infrared Imaging Spectrometer (AVRIS) collects information in 224 spectral bands covering the wavelength region from 0.4 to 2.5 μm. As the amount of data that needs to be transmitted, stored and processed increases, the importance of topics such as compression becomes more and more apparent.

2.4.5 Digital Image File Formats

Why do we need so many different types of image file formats? The short answer is that there are many different types of images, and applications with varying requirements. A more complete answer (which we will not go into here) also considers market share, proprietary information, and a lack of coordination within the imaging industry. However, there have been some standard file formats developed, and the ones presented here are widely available. Many other image types can be readily converted to one of the types presented here by easily available image conversion software.

A field related to computer imaging is that of computer graphics. *Computer graphics* is a specialized field within the computer science realm which refers to the reproduction of visual data through the use of the computer. This includes the creation of computer images for display or print, and the process of generating and manipulating any images (real or artificial) for output to a monitor, printer, camera, or any other device that will provide us with an image. Computer graphics can be considered a part of computer imaging, insofar as many of the same tools the graphics artist uses may be used by the computer imaging specialist.

In computer graphics, types of image data are divided into two primary categories: bitmap and vector. *Bitmap images* (also called raster images) can be represented by our image model, $I(r, c)$, where we have pixel data and the corresponding brightness values stored in some file format. *Vector images* refers to methods of representing lines, curves, and shapes by storing only the key points. These *key points* are sufficient to define the shapes, and the process of turning these into an image is called *rendering*. Once the image has been rendered, it can be thought of as being in bitmap format, where each pixel has specific values associated with it.

Most of the types of file formats discussed fall into the category of bitmap images, although some are compressed, so the $I(r,c)$ values are not directly available until the file is decompressed. In general, these types of images contain both header information and the pixel data itself. The *image file header* is a set of parameters normally found at the start of the file and must contain information regarding: (1) the number of rows (height), (2) the number of columns (width), (3) the number of bands, (4) the number of bits per pixel (bpp), and (5) the file type. Additionally, with some of the more complex file formats, the header may contain information about the type of compression used and any other necessary parameters to create the image, $I(r,c)$.

The simplest file formats are the BIN and the PPM file formats. The BIN format is simply the raw image data, $I(r,c)$. This file contains no header information, so the user must know the necessary parameters—size, number of bands, and bits per pixel—to use the file as an image. The PPM formats are widely used and a set of conversion utilities are freely available (pbmplus). They basically contain raw image data with the simplest header possible. The PPM format includes: PBM (binary), PGM (grayscale), PPM (color), and PNM (handles any of the previous types). The headers for these image file formats contains a "magic number" that identifies the file type, the image width and height, the number of bands, and the maximum brightness value (which determines the required number of bits-per-pixel for each band).

The Microsoft Windows bitmap (BMP) format is commonly used today in Windows-based machines. Most imaging and graphics programs in this environment support the BMP format. This file format is fairly simple, with basic headers followed by the raw image data. Another commonly used format is JPEG (Joint Photographic Experts Group). This file format is capable of high degrees of image compression, so is typically used on the Internet to reduce bandwidth requirements—meaning you don't need to wait forever for images to appear. JPEG files come in two main varieties, the original JPEG and the new JPEG2000. The new JPEG2000 provides higher compression ratios, while still maintaining high quality images.

Two image file formats commonly used on many different computer platforms, as well as on the Internet, are the TIFF (Tagged Image File Format) and GIF (Graphics Interchange Format) file formats. GIF files are limited to a maximum of 8 bits per pixel, and allow for a type of compression called LZW (Lempel-Ziv-Welch, see Chapter 10). The 8 bits per pixel limitation does not mean it does not support color images, it simply means that no more than 256 colors (2^8) are allowed in an image. This is typically implemented by means of a look-up-table (LUT), where the 256 colors are stored in a table, and one byte (8 bits) is used as an index (address) into that table for each pixel (see Figure 2.4.8). The concept of LUT-based images is also referred to palette-based. The GIF image header is 13 bytes long, and contains the basic information required.

The TIFF file format is more sophisticated than GIF, and has many more options and capabilities. TIFF files allow a maximum of 24 bits per pixel, and support five types of compression, including RLE (run length encoding), LZW, and JPEG (see Chapter 10). The TIFF header is of variable size and is arranged in a hierarchical manner. The TIFF format is one of the most comprehensive formats available and is designed to allow the user to customize it for specific applications.

Two formats that were initially computer-specific, but have become commonly used throughout the industry, are the Sun Raster and the SGI (Silicon Graphics, Inc) file formats. The Sun Raster file format (RAS) is much more ubiquitous than the SGI (SGI or IRIX), but SGI has become the leader in state-of-the-arts graphics computers. The SGI format handles up to 16 million colors and supports RLE compression. The SGI image header is 512 bytes followed by the image data. The Sun Raster format is defined to allow for any number of bits per pixel and also supports RLE compression.

8-bit index	RED	GREEN	BLUE
0	R_0	G_0	B_0
1	R_1	G_1	B_1
2	R_2	G_2	B_2
⋮	⋮	⋮	⋮
254	R_{254}	G_{254}	B_{254}
255	R_{255}	G_{255}	B_{255}

FIGURE 2.4.8
Look-up table (LUT). One byte is stored for each pixel in $I(r, c)$. When displayed, this 8-bit value is used as an index into the LUT, and the corresponding RGB values are displayed for that pixel.

It has a 32 byte header, followed by the image data, and allows for TIFF format to be embedded in it.

Portable network graphics, PNG, is a file format that supports LUT type images (1, 2, 4, 8-bit) like GIF, as well as full 24-bit color like TIFF. It provides direct support for color correction, which theoretically allows an image to look the same on different computer systems—although in practice this is quite difficult to achieve. The PICT format is unique to the Macintosh computer system, which is widely used in many imaging applications. It allows for both vector and bitmap images. One of the file formats developed specifically for digital cameras is FlashPix format, FPX. This was originally developed by Kodak, but has become widely used due to the creation of a consortium consisting of Kodak, Adobe, Canon, Fuji, Hewlett-Packard, IBM, Intel, Live Picture, and Microsoft.

One file format discussed here, EPS (encapsulated PostScript), is not of the bitmap variety. It is actually a language that supports more than images, and is commonly used in desktop publishing. EPS is directly supported by many printers (in the hardware itself), so is commonly used for data interchange across hardware and software platforms. It is a commonly used standard that allows output devices, monitors, printers, and computer software to communicate regarding both graphics and text. The primary advantage of the EPS format is its wide acceptance. The disadvantage of using EPS is that the files are very big, since it is a general purpose language designed for much more than just images. In computer imaging, EPS is used primarily as a means to generate printed images. The EPS files actually contain text and can be created by any text editor, but are typically generated by applications software. The language itself is very complex and continually evolving.

The final image file type discussed here is the VIP (Visualization in Image Processing) format, developed specifically for the CVIPtools software. When performing computer imaging tasks, temporary images are often created that use floating point representations that are beyond the standard 8-bits-per-pixel capabilities of most display devices. The process of representing this type of data as an image is referred to as *data visualization*,

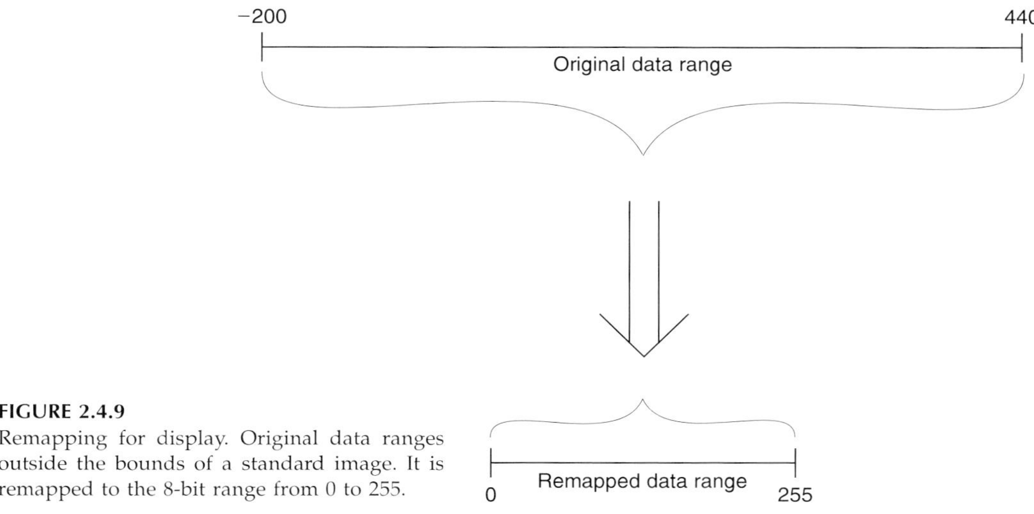

FIGURE 2.4.9
Remapping for display. Original data ranges outside the bounds of a standard image. It is remapped to the 8-bit range from 0 to 255.

and can be achieved by remapping the data to the 8-bit range, 0 to 255. *Remapping* is the process of taking the original data and defining an equation to translate the original data to the output data range, typically 0 to 255 for 8-bit display. The two most commonly used methods in computer imaging are linear and logarithmic mapping. In Figure 2.4.9 we see a graphical representation and example of how this process is performed. In this example the original data ranges from −200 to 440. An equation is found that will map the lowest value (−200) to 0 and the highest value (440) to 255, while all the intermediate values are remapped to values within this range (0 to 255). We can see that this process may result in a loss of information.

The VIP file format was required since we needed to support many nonstandard image formats. This format was defined to allow disk file support for the image data structure used within the CVIPtools software (see Chapter 11). It allows any data type, including floating point and complex numbers, any image size, any number of bands, and has a special history data structure built into it that allows the maintenance of a record of operations that have been performed on the image. More details on the VIP format are included in Part IV of the book, programming with CVIPtools.

2.5 Key Points

IMAGING SYSTEMS

- **Two primary components**—hardware and software

    Hardware—image acquisition subsystem, computer, display devices
    Software—allows for image manipulation, analysis and processing

- **Digital camera interface**—USB, FireWire, Camera Link, or Gigabit Ethernet
- **Frame grabber**—special purpose piece of hardware that converts an analog video signal into a digital image

- **RS-170/NTSC-RS-330**—the video standard used in North America
- **CCIR or PAL**—video standards used in northern Europe
- **SECAM**—video standard used in France and Russia, a CCIR equivalent
- **Frame**—one screen of video information
- **Field**—alternating lines of video information creating one-half of a frame in interlaced video
- **Interlaced video**—two-fields-per-frame video, used in television and video cameras
- **Noninterlaced video**—one field per frame video, used in computer monitors
- **Horizontal synch pulse**—control signal in the video signal that occurs between each line of video information
- **Vertical synch pulse**—control signal in the video signal that occurs between each field or frame of video information
- ***I*(*r*, *c*)**—a two-dimensional array of data, the digital image function, a matrix where the brightness of the image at the point (r, c) is given, with $r = \text{row}$ and $c = \text{column}$
- **Image brightness**—depends on both lighting conditions and the intrinsic object properties
- **Hierarchical image pyramid**—describes the various levels for processing of images (see Figure 2.1.4)

IMAGE FORMATION AND SENSING

- **Sensor**—a device to measure a signal which can be converted into a digital image
- **Electromagnetic spectrum**—electromagnetic signals which, at various wavelengths, consists of gamma rays, x-rays, ultraviolet light, visible light, infrared, microwaves, and radio waves, and can be measured by sensors to produce images
- **Photon**—massless particles which correspond to the minimum amount of energy, the quantum, which can be measured in the EM signal
- **Range image**—created by radar, sonar or lasers to produce an image which depicts distance as brightness
- **Image formation**—two key components: (1) where will the image point appear (the row and column coordinates), and (2) what value will be assigned to that point (the brightness value)
- **Optics**—the physics of light and the study of lenses, required to determine where an image point appears
- **Reflectance function**—the function that determines how an object reflects light
- **Irradiance**—the amount of light energy falling on a surface, measured by a sensor to create an image
- **Radiance**—the amount of light energy emitted, or reflected, from an object into a solid unit angle
- **Lens**—necessary to focus light in an imaging system
- **Lens equation:** $\frac{1}{a} + \frac{1}{b} = \frac{1}{f}$
- **Blur equation:** $c = \frac{d}{b'}|b - b'|$
- **Depth of field**—range of distances over which an object is focused sufficiently well

- **Field of view (FOV)**—angle of the cone of directions from which the device will create the image
- **Lens types**—normal, telephoto, and wide-angle
- **Charge-coupled device (CCD)**—sensor used in digital cameras for imaging
- **Complementary metal-oxide-semiconductor (CMOS) device**—sensor used for imaging; image quality not as good as CCD, but cheaper and requires less power
- **Quantum efficiency** $[q(\lambda)]$—the ratio of the electron flux produced to the incident photon flux
- **Sensor equation:** $N = \delta A \delta t \int b(\lambda) q(\lambda) d\lambda$
- **Imaging outside of visible EM spectrum**—used in medicine, astronomy, microscopy, satellite imaging, military, law enforcement, and industrial applications
- **Acoustic imaging**—measures reflected sound waves, applications in medicine, military, geology, and manufacturing
- **Electron imaging**—using a focused beam of electrons to magnify up to two hundred thousand times
- **Laser imaging**—used to create range images
- **CVIPtools**—a comprehensive computer imaging software package to allow for the exploration of image analysis and processing functions, including algorithm development for applications

IMAGE REPRESENTATION

- **Optical image**—a collection of spatially distributed light energy to be measured by an image sensor to generate $I(r, c)$
- **Binary image**—a simple image type that can take on two values, typically black and white, or "0" and "1"
- **Gray-scale image**—one-color or monochrome image that contains only brightness information, no color information
- **Color image**—modeled as a three-band monochrome image; the three bands are typically red, green, and blue, or RGB
- **Color pixel vector**—a single pixel's values for a color image, (R, G, B)
- **Color transform/color model**—a mathematical method or algorithm to map RGB data into another color space, typically to decouple brightness and color information

    HSL (Hue/Saturation/Lightness)—a color transform that describes colors in terms that we can easily relate to the human visual system's perception, where *hue* is the "color," for example red or yellow, *saturation* is the amount of white in the color, and *lightness* is the brightness

    SCT (spherical coordinate transform)—maps the color information into two angles, and the brightness into the color vector length

    CCT (cylindrical coordinate transform)—does not completely decouple color and brightness, unlike most color transforms, definition tends to be application specific

    Chromaticity coordinates—normalizes RGB values to the sum of all three

CIE $L^*u^*v^*$**/CIE** $L^*a^*b^*$—perceptually uniform color spaces defined by the Commisssion Internationale de l'Eclairage (CIE), the international standards group for color science

YUV/YCbCr—linear transforms of RGB data used in compression algorithms, Y is the luminance and the other two are color difference signals

CMY (Cyan, Magenta, Yellow)/CMYK—color transforms based on a subtractive model, used for color printing; K is added when a separate ink is used for black

PCT (principal components transform)—decorrelates RGB data by finding a linear transform using statistical methods to align the coordinate axes along the path of maximal variance in the data

- **Multispectral image**—images of many bands containing information outside of the visible spectrum

DIGITAL IMAGE FILE FORMATS

- **Bitmap images**—images we can represent by our model, $I(r, c)$, also called raster images
- **Vector images**—artificially generated images by storing only mathematical descriptions of geometric shapes using *key points*
- **Rendering**—the process of changing a vector image into a bitmap image
- **Image file header**—a set of parameters normally found at the start of the image file and must contain information regarding: (1) the number of rows (height), (2) the number of columns (width), (3) the number of bands, (4) the number of bits per pixel (bpp), and (5) the file type; additional information may be included
- **Common image file formats**—BIN, PPM, PBM, PGM, BMP, JPEG, TIFF, GIF, RAS, SGI, PNG, PICT, FPX, EPS, VIP
- **LUT**—look up table, used for storing RGB values for 8-bit color images

2.6 References and Further Reading

More information can be found on imaging in the various EM spectral bands and on image acquisition devices in [Gonzalez/Woods 02], [Shapiro/Stockman 01] and [Sanchez/Canton 99]. Laser-based range images are discussed in more detail in [Forsyth/Ponce 03], [Russ 92] and [Gonzalez/Woods 02] contains information regarding electron imaging. For further study of satellite imaging, see [Sanchez/Canton 99], [Bell 95], and for more information on UV and IR imaging in law enforcement see [Kummer 03]. More on lenses and optics can be found in [Forsyth/Ponce 03], [Horn 86], and [Jain/Kasturi/Schunck 95]. More information on input and output devices for imaging can be found in [Burdick 97].

For further study of digital video processing, see [Orzessek/Sommer 98], [Tekalp 95] and [Sid-Ahmed 95]. [Tekalp 95] has much information on motion estimation methods not available in other texts. For details on video standards and hardware see [Jack 01] and [Poynton 03]. For further study regarding color see [Giogianni/Madden 98], [Durrett 87] and [Wyszecki and Stiles 82]. For more information on JPEG2000, see [Taubman/Marcellin 02]. For further study on computer generated images see [Watt/Policarpo 98], [Foley/van Dam/Feiner/Hughes 95], and [Hill 90].

For other sources of software see [Seul/O'Gorman/Sammon 00], [Parker 97], [Myler/Weeks 93], [Baxes 94], and [Sid-Ahmed 95]. Also, the CVIPtools homepage (*www.ee.siue.edu/CVIPtools*) has useful Internet links. Additionally, the Computer Vision Homepage, sponsored by Carnegie Mellon University, (*www.cs.cmu.edu/~cil/vision.html*) is a great resource for imaging software available on the Internet. Two excellent sources for information on image and graphics file formats, which include code, are [Burdick 97] and [Murray/VanRyper 94].

Baxes, G.A., *Digital Image Processing: Principles and Applications*, NY: Wiley, 1994.

Bell, T.E., "Remote Sensing," *IEEE Spectrum*, March 1995, pp. 25–31.

Burdick, H.E., *Digital Imaging: Theory and Applications*, NY: McGraw-Hill, 1997.

Durrett, H.J., editor, *Color and the Computer*, San Diego, CA: Academic Press, 1987.

Foley, J.D., van Dam, A., Feiner, S.K., Hughes, J.F., *Computer Graphics: Principles and Practice in C*, Reading, MA: Addison Wesley, 1995.

Forsyth, D.A., Ponce, J., *Computer Vision*, Upper Saddle River, NJ: Prentice Hall, 2003.

Giorgianni, E.J., Madden, T.E., *Digital Color Management: Encoding Solutions*, Reading, MA: Addison Wesley, 1998.

Gonzalez, R.C., Woods, R.E., *Digital Image Processing*, Upper Saddle River, NJ: Prentice Hall, 2002.

Hill, F.S., *Computer Graphics*, NY: Macmillan, 1990.

Horn, B.K.P., *Robot Vision*, Cambridge, MA: The MIT Press, 1986.

Jack, K., *Video Demystified: A Handbook for the Digital Engineer*, 3rd Edition, San Diego, CA: HighText Interactive, 1996.

Jain, A.K., *Fundamentals of Digital Image Processing*, Englewood Cliffs, NJ: Prentice Hall, 1989.

Jain, R., Kasturi, R., Schnuck, B.G., *Machine Vision*, NY: McGraw Hill, 1995.

Kummer, S., "The Eye of the Law," *OE Magazine*, October 2003, pp. 22–25.

Myler, H.R., Weeks, A.R., *Computer Imaging Recipes in C*, Englewood Cliffs, NJ: Prentice Hall, 1993.

Murray, J.D., VanRyper, W., *Encyclopedia of Graphics File Formats*, Sebastopol, CA: O'Reilly and Associates, 1994.

Orzessek, M., Sommer, P., *ATM and MPEG-2: Integrating Digital Video into Broadband Networks*, Upper Saddle River, NJ: Prentice Hall PTR, 1998.

Parker, J.R., *Algorithms for Image Processing and Computer Vision*, NY: Wiley, 1997.

Poynton, C., *Digital Video and HDTV Algorithms and Interfaces*, Morgan Kahfmann, 2003.

Russ, J.C., *The Image Processing Handbook*, Boca Raton, FL: CRC Press, 1992.

Sanchez, J., Canton, M.P., *Space Image Processing*, Boca Raton, FL: CRC Press, 1999.

Seul, M., O'Gorman, L., Sammon, M.J., *Practical Algorithms for Image Analysis*, Cambridge, UK: Cambridge University Press, 2000.

Shapiro, L., Stockman, G., *Computer Vision*, Upper Saddle River, NJ: Prentice Hall, 2001.

Sid-Ahmed, M.A. *Image Processing: Theory, Algorithms, and Architectures*, Englewood Cliffs, NJ: Prentice Hall, 1995.

Taubman, D., Marcellin, M., *JPEG2000: Image Compression Fundamentals, Standards and Practice*, Boston, MA: Kluwer Academic Publishers 2002.

Tekalp, A.M., *Digital Video Processing*, Englewood Cliffs, NJ: Prentice Hall, 1995.

Watt, A., Policarpo, F., *The Computer Image*, New York, NY: Addison-Wesley, 1998.

West, M., Barsley, et al., *Journal of Forensic Identification*, 40[5], 249–255, 1990.

Wyszecki, G., and Stiles, W.S., *Color Science: Concepts and Methods, Quantitative Data and Formulae*, New York: Wiley, 1982.

2.7 Exercises

1. What are the two types of components in a computer imaging system?
2. Name four types of video cameras interfaces.
3. Describe how a frame grabber works.
4. What is a sensor? How are they used in imaging systems?
5. What is a range image? How are they created?
6. What is a reflectance function? How does it relate our description of object characteristics?
7. Describe the difference between radiance and irradiance.
8. What is a photon? What does CCD stand for? What is quantum efficiency?
9. Draw a picture and write the equations that shows the blur equation is given by:

$$c = \frac{d}{b'}|b - b'|$$

10. Show that the focal length of a lens can be defined by the distance from the lens at which an object at infinity is focused.
11. Find the number of electrons liberated in a sensor if:

    irradiation $= 600\lambda$ photons/(second)nm^2
    quantum efficiency of device $= 0.95$
    area $= 20\,\text{nm}^2$
    time period $= 10$ milliseconds
    the photon flux is bandlimited to visible wavelengths

    Is this a solid state device? Explain.

12. Show that the image brightness is inversely proportional to the square of the f-number. Consider how light energy is measured related to the surface area of the sensor.
13. A video frame is scanned in 1/30 of a second, using interlaced scanning. If we have 480 lines of interest, and 640 pixels per line, at what rate must we perform the analog to digital conversion? (ignore synch pulse time)
14. Which band in the electromagnetic spectrum has the most energy? Which has the least energy? What significance does this have to human life?
15. Name some applications for UV and IR imaging.
16. How does acoustic imaging work? What is it used for?
17. How does an electron microscope differ from a standard light microscope?
18. What are two methods for lasers to create depth maps?
19. What is an optical image? How are they used to create digital images?
20. What is the difference between a "real" image and a computer generated image?
21. Discuss advantages and disadvantages of binary, gray-scale, color and multi-spectral images.

22. Why would we transform a standard color image consisting of RGB data into another color space? Describe the HSL color space.
23. What does it mean when we say a color space is not perceptually uniform? Name a color space that is perceptually uniform.
24. Find the inverse equations for the SCT and the CCT.
25. Describe the color spaces used in printing. If we had a 24-bit color pixel, (R, G, B) = (100, 50, 200), what amounts of cyan, magenta and yellow would our printer print?
26. Describe the difference between a bitmap and a vector image.
27. Name the elements required in an image file header.
28. Name the image file type used by CVIPtools. Why did we not use a standard file type, such tiff or gif? Why do we sometimes remap image data?
29. Run the CVIPtools software and load a color image. Experiment. Have fun.

2.7.1 Programming Exercise: Introduction to CVIPlab

1. Review the first two sections of Chapter 11, Sections 11.1 and 11.2, to become familiar with the CVIPlab C functions.
2. Review Section 11.3 and compile the CVIPlab.c and threshold_lab.c functions.
3. Run CVIPlab, select menu choice 2, threshold, and experiment with various threshold values.
4. Edit the CVIPlab.c program to include your name in the header. Study the CVIPlab.c file and understand how this program is organized and how the threshold function works. In particular, learn how images are read, written and manipulated. The functions for the following programming exercises are to be added to the CVIPlab program.
5. Run the CVIPtools program. Next open the desired images with the file open icon near the top left of the main window. Select *Utilities*, then select the *Convert->Binary Threshold* option. Use this to compare results to your CVIPlab threshold program to verify correctness. You can use the *Utilities->Compare* option. At the end of this exercise you should understand the CVIPlab environment and be able to write functions by using the threshold function as a prototype.

Section II

Digital Image Analysis

3

Introduction to Digital Image Analysis

3.1 Introduction

Digital image analysis is a key factor in solving any computer imaging problem. Acquisition of a sample image database and examination of these images for the application is the first step in development of an imaging solution. *Image analysis* involves manipulating the image data to determine exactly the information required to develop the computer imaging system. The solution to the problem may require use of existing hardware, software, or may require development of new algorithms and system designs. The image analysis process helps to define the requirements for the system being developed. This analysis is typically part of a larger process, is iterative in nature, and allows us to answer application-specific questions such as: How much spatial and brightness resolution is needed? Will existing methods solve the problem? Is color information needed? Do we need to transform the image data into the frequency domain? Do we need to segment the image to find object information? What are the important features in the images? Is the hardware fast enough for the application?

3.1.1 Overview

Image analysis is primarily a data reduction process. As we have seen, images contain enormous amounts of data, typically on the order of hundreds of kilobytes or even megabytes. Often much of this information is not necessary to solve a specific computer imaging problem, so a primary part of the image analysis task is to determine exactly what information is necessary. With many applications the determining factor in the feasibility of system development are the results of the preliminary image analysis. Image analysis is used in the development of both computer vision and image processing (CVIP) applications.

For computer vision, the end product is typically the extraction of high level information for computer analysis or manipulation. This high level information may include shape parameters to control a robotic manipulator, terrain analysis to enable a vehicle to navigate on Mars, or color and texture features to help in the diagnosis of a skin tumor. Image analysis is central to the computer vision process and is often uniquely associated with computer vision; however, image analysis is an important tool for image processing applications as well.

In image processing applications, image analysis methods may be used to help determine the type of processing required and the specific parameters needed for that processing. For example, developing an enhancement algorithm (Chapter 8), determining the degradation function for an image restoration procedure (Chapter 9), and determining exactly what information is visually important for an image compression method (Chapter 10) are all image analysis tasks. In this chapter we present the system model for the image analysis process, preprocessing methods and a simple example of image analysis using binary images.

3.1.2 System Model

The *image analysis process,* illustrated in Figure 3.1.1, can be broken down into three primary stages: (1) Preprocessing, (2) Data Reduction, and (3) Feature Analysis. Preprocessing is used to remove noise, and eliminate irrelevant, visually unnecessary information. *Noise* is unwanted information that can result from the image acquisition process. Other preprocessing steps might include gray level or spatial quantization (reducing the number of bits per pixel or the image size), or finding regions of interest for further processing. The second stage, data reduction, involves either reducing the data in the spatial domain and/or transforming it into another domain called the frequency domain (Figure 3.1.2) and then extracting features for the analysis process. In the third stage, feature analysis, the features extracted by the data reduction process are examined and evaluated for their use in the application.

A more detailed diagram of this process is shown in Figure 3.1.3. After preprocessing we can perform segmentation on the image in the spatial domain (Chapter 4) or convert it into the frequency domain via a mathematical transform (Chapter 5). Note the dotted line between segmentation and the transform block, this is for extracting spectral features on segmented parts of the image. After either of these processes we may choose to filter the image. This filtering process further reduces the data and allows us to extract the features that we may require for analysis. After the analysis, we have a feedback loop that provides for an application-specific review of the analysis results. This approach often leads to an iterative process that is not complete until satisfactory results are achieved. The application feedback loop is a key aspect of the entire process.

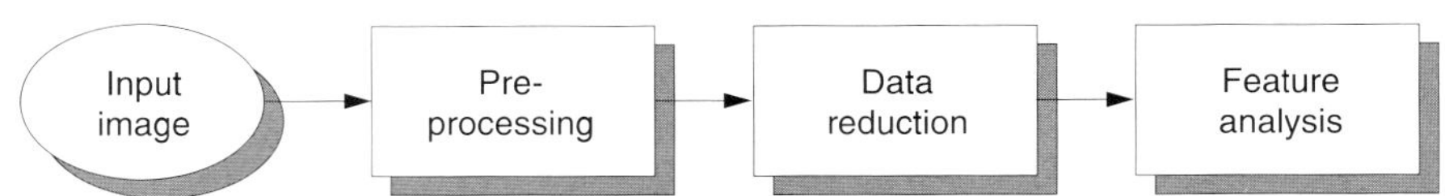

FIGURE 3.1.1
Image analysis.

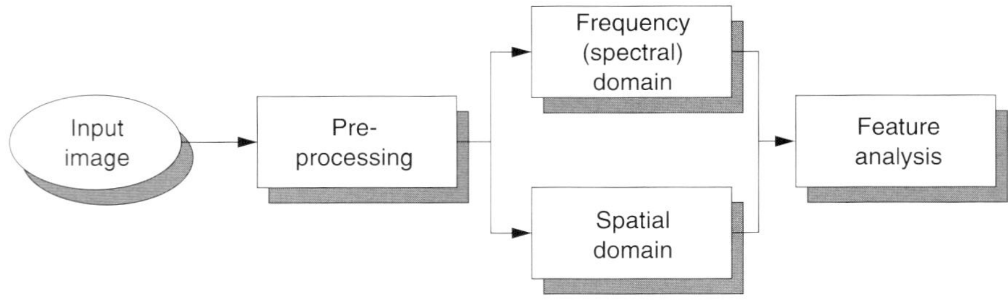

FIGURE 3.1.2
Image analysis domains.

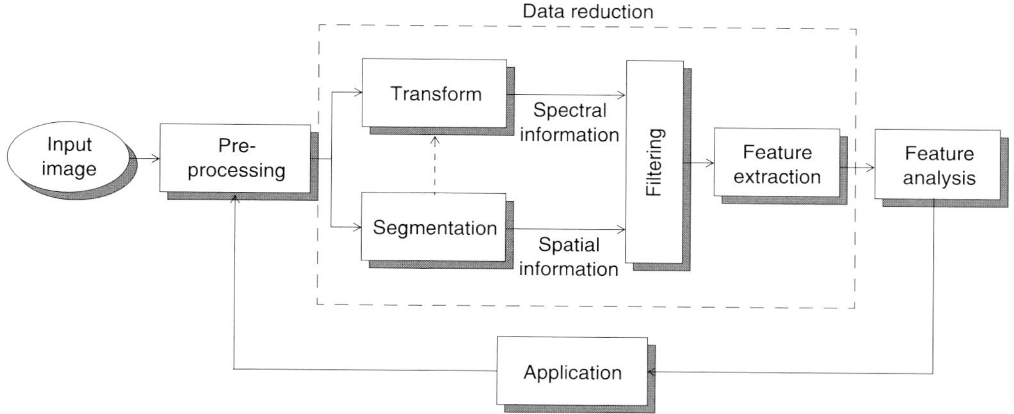

FIGURE 3.1.3
Image analysis.

3.2 Preprocessing

The preprocessing algorithms, techniques and operators are used to perform initial processing that makes the primary data reduction and analysis task easier. They include operations related to extracting regions of interest, performing basic mathematical operations on images, simple enhancement of specific image features (for more on enhancement see Chapter 8), and data reduction in both resolution and brightness. Preprocessing is a stage where the requirements are typically obvious and simple, such as the removal of artifacts from images, or the elimination of image information that is not required for the application. For example, in one application we needed to eliminate borders from the images that resulted from taking the pictures by looking out a window; in another we had to mask out rulers that were present in skin tumor slides. Another example of a preprocessing step involves a robotic gripper that needs to pick and place an object; for this, we reduce a gray level image to a binary (two-valued) image, which contains all the information necessary to discern the object's outline. Two of these examples can be seen in Figure 3.2.1.

3.2.1 Region of Interest Image Geometry

Often, for image analysis, we want to investigate more closely a specific area within the image, called a Region-of-Interest (ROI). To do this we need operations that modify the spatial coordinates of the image, and these are categorized as image geometry operations. The image geometry operations discussed here include crop, zoom, enlarge, shrink, translate, and rotate.

The image *crop* process is the process of selecting a portion of the image, a sub-image, and cutting it away from the rest of the image—that's how the border was removed in Figure 3.2.1(b). Once we have cropped a sub-image from the original image we can *zoom* in on it by enlarging it. Image enlargement is useful in a variety of applications since it can help visual analysis of detailed objects. For example, some imaging applications require that two input images be in tight geometrical alignment prior to their

FIGURE 3.2.1
Preprocessing examples. (a) An image needing border removal. (b) The image after the border is removed. (c) An image where only shape information is necessary in order to control a robotic gripper. (d) The image after unnecessary information removal which leaves only the object shape.

combination; since improper alignment can produce distortion at object boundaries. Enlargement of the images eases the task of manual alignment. Additionally, enlargement may allow visible recognition of image degradation, helping in the selection of a restoration model (see Chapter 9).

This zoom process can be done in numerous ways, but typically a zero-order hold or a first-order hold is used. A zero-order hold is performed by repeating previous pixel values, thus creating a blocky effect. To extend the image size with a first-order hold we do linear interpolation between adjacent pixels. A comparison of the images resulting from these two methods is shown in Figure 3.2.2.

Although the implementation of the zero-order hold is straightforward, the first-order hold is more complicated. The easiest way to do this is to find the average value between two pixels and use that as the pixel value between those two; we can do this for the rows first, as follows:

ORIGINAL IMAGE ARRAY IMAGE WITH ROWS EXPANDED

$$\begin{bmatrix} 8 & 4 & 8 \\ 4 & 8 & 4 \\ 8 & 2 & 8 \end{bmatrix} \qquad \begin{bmatrix} 8 & 6 & 4 & 6 & 8 \\ 4 & 6 & 8 & 6 & 4 \\ 8 & 5 & 2 & 5 & 8 \end{bmatrix}$$

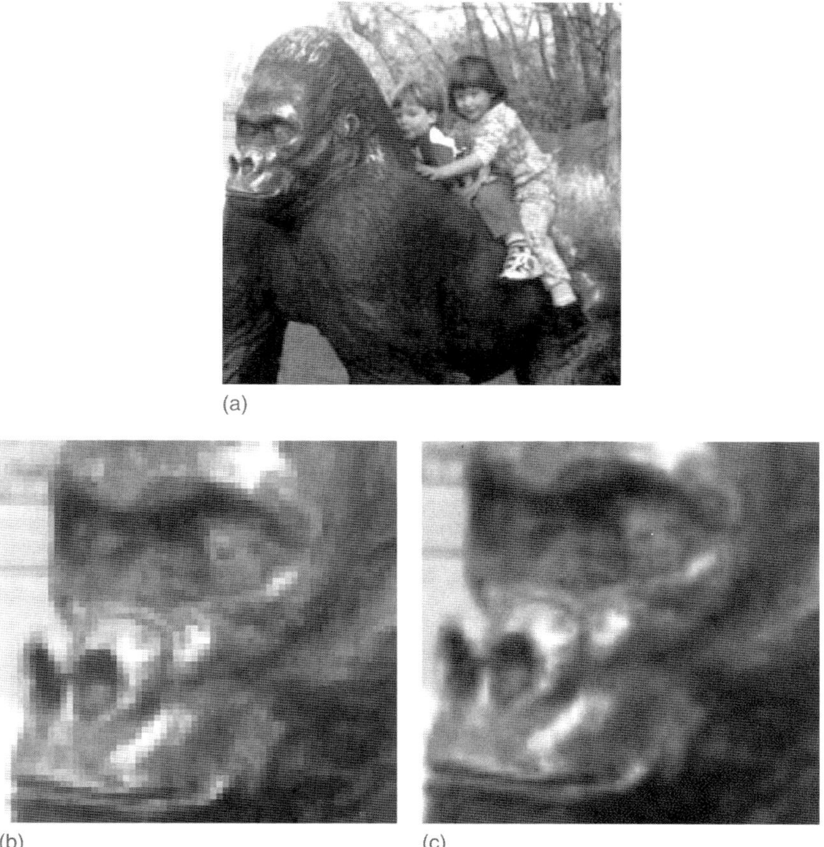

FIGURE 3.2.2
Zooming methods. (a) Original image. The ape's face will be zoomed. (b) Image enlarged by zero-order hold, notice the blocky effect. (c) Image enlarged by first order hold. Note the smoother effect.

The first two pixels in the first row are averaged, $(8+4)/2=6$, and this number is inserted in between those two pixels. This is done for every pixel pair in each row. Next, take that result and expand the columns in the same way, as follows:

IMAGE WITH ROWS AND COLUMNS EXPANDED

$$\begin{bmatrix} 8 & 6 & 4 & 6 & 8 \\ 6 & 6 & 6 & 6 & 6 \\ 4 & 6 & 8 & 6 & 4 \\ 6 & 5.5 & 5 & 5.5 & 6 \\ 8 & 5 & 2 & 5 & 8 \end{bmatrix}$$

This method will allow us to enlarge an $N \times N$ sized image to a size of $(2N-1) \times (2N-1)$, and can be repeated as desired.

Another method that achieves a similar result requires a mathematical process called convolution. With this method of image enlargement a two-step process is required: (1) extend the image by adding rows and columns of zeros between the existing rows and columns, and (2) perform the convolution. The image is extended as follows:

ORIGINAL IMAGE ARRAY

$$\begin{bmatrix} 3 & 5 & 7 \\ 2 & 7 & 6 \\ 3 & 4 & 9 \end{bmatrix}$$

IMAGE EXTENDED WITH ZEROS

$$\begin{bmatrix} 0 & 0 & 0 & 0 & 0 & 0 & 0 \\ 0 & 3 & 0 & 5 & 0 & 7 & 0 \\ 0 & 0 & 0 & 0 & 0 & 0 & 0 \\ 0 & 2 & 0 & 7 & 0 & 6 & 0 \\ 0 & 0 & 0 & 0 & 0 & 0 & 0 \\ 0 & 3 & 0 & 4 & 0 & 9 & 0 \\ 0 & 0 & 0 & 0 & 0 & 0 & 0 \end{bmatrix}$$

Next, we use what is called a convolution mask which is slid across the extended image and a simple arithmetic operation is performed at each pixel location.

CONVOLUTION MASK FOR FIRST-ORDER HOLD

$$\begin{bmatrix} 0 & \frac{1}{2} & 0 \\ \frac{1}{2} & 1 & \frac{1}{2} \\ 0 & \frac{1}{2} & 0 \end{bmatrix}$$

The *convolution process requires* us to overlay the mask on the image, multiply the coincident values, and sum all these results. This is equivalent to finding the *vector inner product* of the mask with the underlying sub-image. For example, if we put the mask over the upper left corner of the image, we obtain (from right to left, and top to bottom):

$$0(0) + 1/2(0) + 0(0) + 1/2(0) + 1(3) + 1/2(0) + 0(0) + 1/2(0) + 0(0) = 3$$

Note that the existing image values do not change. The next step is to slide the mask over by one pixel and repeat the process, as follows:

$$0(0) + 1/2(0) + 0(0) + 1/2(3) + 1(0) + 1/2(5) + 0(0) + 1/2(0) + 0(0) = 4$$

Note this is the average of the two existing neighbors. This process continues until we get to the end of the row, each time placing the result of the operation in the location corresponding to the center of the mask. Once the end of the row is reached, the mask is moved down one row and the process is repeated row by row until this procedure has been performed on the entire image; the process of sliding, multiplying, and summing is called convolution (see Figure 3.2.3). Note that the output image must be put in a separate image array, called a buffer, so that the existing values are not over-written during the convolution process. If we call the convolution mask $M(r, c)$, and the image $I(r, c)$, the convolution equation is given by:

$$\sum_{x=-\infty}^{+\infty} \sum_{y=-\infty}^{+\infty} I(r - x, c - y)M(x, y)$$

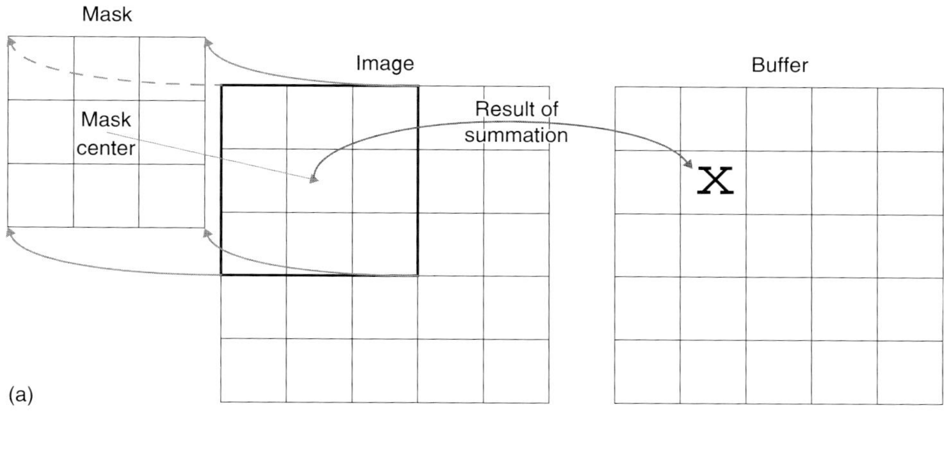

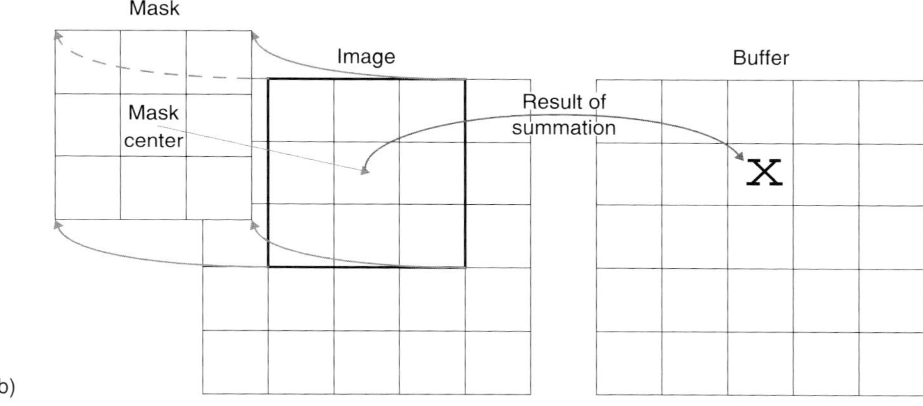

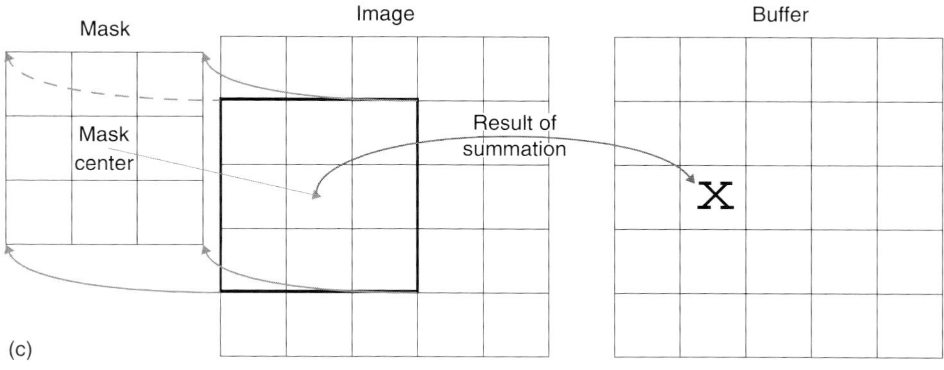

FIGURE 3.2.3
The convolution process. (a) Overlay the convolution mask in upper left corner of the image. Multiply coincident terms, sum, put result into the image buffer at the location that corresponds to the mask's current center, which is $(r, c) = (1, 1)$. (b) Move the mask one pixel to the right, multiply coincident terms, sum, and place the new result into the buffer at the location that corresponds to the new center location of the convolution mask, now at $(r, c) = (1, 2)$. Continue to the end of the row. (c) Move the mask down one row and repeat the process until the mask is convolved with the entire image. Note that we "lose" the outer row(s) and column(s).

For theoretical reasons beyond the scope of this discussion this equation assumes that the image and mask are extended with zeros infinitely in all directions, and that the origin of the mask is at its center. Also, for theoretical reasons, the previous description of convolution assumes that the convolution mask is symmetric; meaning if it is flipped about its center it will remain the same. If it is not symmetric, it must be flipped before the procedure given can be followed. For computer imaging applications these convolution masks are typically symmetric.

At this point a good question would be: Why use this convolution method when it requires so many more calculations than the basic averaging-of-neighbors method? The answer is that many imaging systems can perform convolution in hardware, which is generally very fast, typically much faster than applying a faster algorithm in software. Not only can first-order hold be performed via convolution, but zero-order hold can also be achieved by extending the image with zeros and using the following convolution mask:

ZERO-ORDER HOLD CONVOLUTION MASK

$$\begin{bmatrix} 1 & 1 \\ 1 & 1 \end{bmatrix}$$

Note that for this mask we will need to put the result in the pixel location corresponding to the lower right corner, since there is no center pixel.

The above methods will only allow us to enlarge an image by a factor of $(2N-1)$, but what if we want to enlarge an image by something other than a factor of $(2N-1)$? To do this we need to apply a more general method; we take two adjacent values and linearly interpolate more than one value between them. This *linear interpolation* technique is equivalent to finding the line that connects the two values in the brightness space and sampling it faster to get more samples, thus artificially increasing the resolution. This is done by defining an enlargement number K, then following this process: (1) subtract the two adjacent values, (2) divide the result by K, (3) add that result to the smaller value, and keep adding the result from (2) in a running total until all $(K-1)$ intermediate pixel locations are filled.

Example 3.2.1

We want to enlarge an image to three times its original size, and we have two adjacent pixel values 125 and 140.

1. Find the difference between the two values, $140 - 125 = 15$.
2. Enlargement desired is $K = 3$, so we get $15/3 = 5$.
3. Next determine how many intermediate pixel values we need: $K - 1 = 3 - 1 = 2$. The two pixel values between the 125 and 140 are:

$$125 + 5 = 130 \quad \text{and} \quad 125 + 2 \times 5 = 135$$

We do this for every pair of adjacent pixels, first along the rows and then along the columns. This will allow us to enlarge the image to a size of $K(N-1)+1$, where K is an integer and $N \times N$ is the image size. Typically, N is large and K is small, so this is approximately equal to KN.

Image enlargement methods that use brightness values in both the row and column direction are also available. This technique is called *bilinear interpolation* and is explored in Chapter 9, as applied to geometric restoration. More sophisticated methods that fit curves and surfaces to the existing points, and then sample these surfaces to obtain more points, can be explored in the references.

The process opposite to enlarging an image is shrinking it. This is not typically done to examine an ROI more closely, but to reduce the amount of data that needs to be processed. Shrinking is explored more in Section 3.2.4 Image Quantization.

Two other operations of interest for the ROI image geometry are translation and rotation. These processes may be performed for many application-specific reasons, for example to align an image with a known template in a pattern matching process, or to make certain image details easier to see. The translation process can be done with the following equations:

$$r' = r + r_0$$
$$c' = c + c_0$$

where r' and c' are the new coordinates, r and c are the original coordinates and r_0 and c_0 are the distances to move, or translate, the image.

The rotation process requires the use of these equations:

$$\hat{r} = r(\cos\theta) + c(\sin\theta)$$
$$\hat{c} = -r(\sin\theta) + c(\cos\theta)$$

where $\hat{r}$ and $\hat{c}$ are the new coordinates, r and c are the original coordinates, and θ is the angle to rotate the image. θ is defined in a clockwise direction from the horizontal axis at the image origin in the upper left corner.

The rotation and translation process can be combined into one set of equations:

$$\hat{r}' = (r + r_0)(\cos\theta) + (c + c_0)(\sin\theta)$$
$$\hat{c}' = -(r + r_0)(\sin\theta) + (c + c_0)(\cos\theta)$$

where $\hat{r}'$ and $\hat{c}'$ are the new coordinates and r, c, r_0, c_0 and θ are defined as above.

There are some practical difficulties with the direct application of these equations. When translating, what is done with the "left-over" space? If we move everything one row down, what do we put in the top row? There are two basic options: fill the top row with a constant value, typically black (0) or white (255), or wrap-around by shifting the bottom row to the top, shown in Figure 3.2.4. Rotation also creates some practical difficulties. As Figure 3.2.5a illustrates, the image may be rotated off the "screen" (image plane). Although this can be fixed by a translation back to the center (Figure 3.2.5b, c), we have leftover space in the corners. We can fill this space with a constant or extract the central, rectangular portion of the image and enlarge it to the original image size.

3.2.2 Arithmetic and Logic Operations

Arithmetic and logic operations are often applied as preprocessing steps in image analysis in order to combine images in various ways. Addition, subtraction, division and multiplication comprise the arithmetic operations, while AND, OR, and NOT make up the logic operations. These operations are performed on two images, except the NOT logic operation which requires only one image, and are done on a pixel by pixel basis.

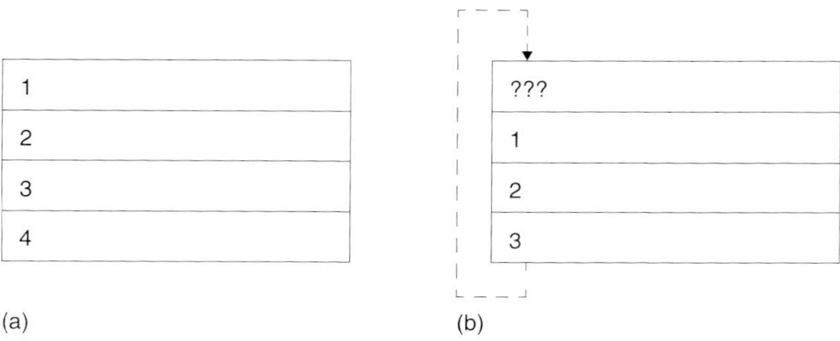

FIGURE 3.2.4
Translation. (a) *Before*: A 4-row image translating down by one row $r_0 = 1$. (b) *After*: If we wrap-around, row 4 goes into ???. Otherwise the top row is filled with a constant, typically zero.

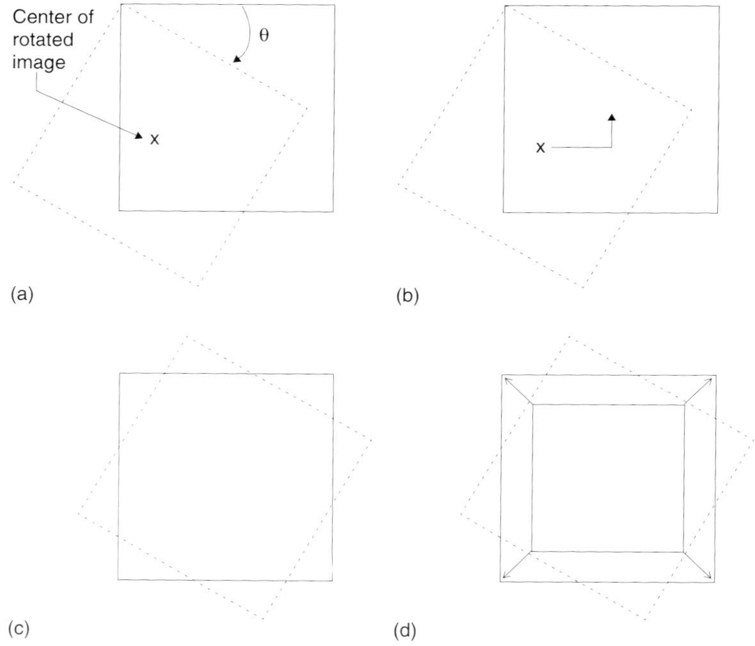

FIGURE 3.2.5
Rotation. (a) Image is rotated off the screen. (b) Fix by translating towards center. (c) Translation complete. (d) Crop and enlarge if desired.

To apply the arithmetic operations to two images, we simply operate on corresponding pixel values. For example, to add images I_1 and I_2 to create I_3:

Example 3.2.2

$I_1(r, c) + I_2(r, c) = I_3(r, c)$

$$I_1 = \begin{bmatrix} 3 & 4 & 7 \\ 3 & 4 & 5 \\ 2 & 4 & 6 \end{bmatrix} \quad I_2 = \begin{bmatrix} 6 & 6 & 6 \\ 4 & 2 & 6 \\ 3 & 5 & 5 \end{bmatrix} \quad I_3 = \begin{bmatrix} 3+6 & 4+6 & 7+6 \\ 3+4 & 4+2 & 5+6 \\ 2+3 & 4+5 & 6+5 \end{bmatrix} = \begin{bmatrix} 9 & 10 & 13 \\ 7 & 6 & 11 \\ 5 & 9 & 11 \end{bmatrix}$$

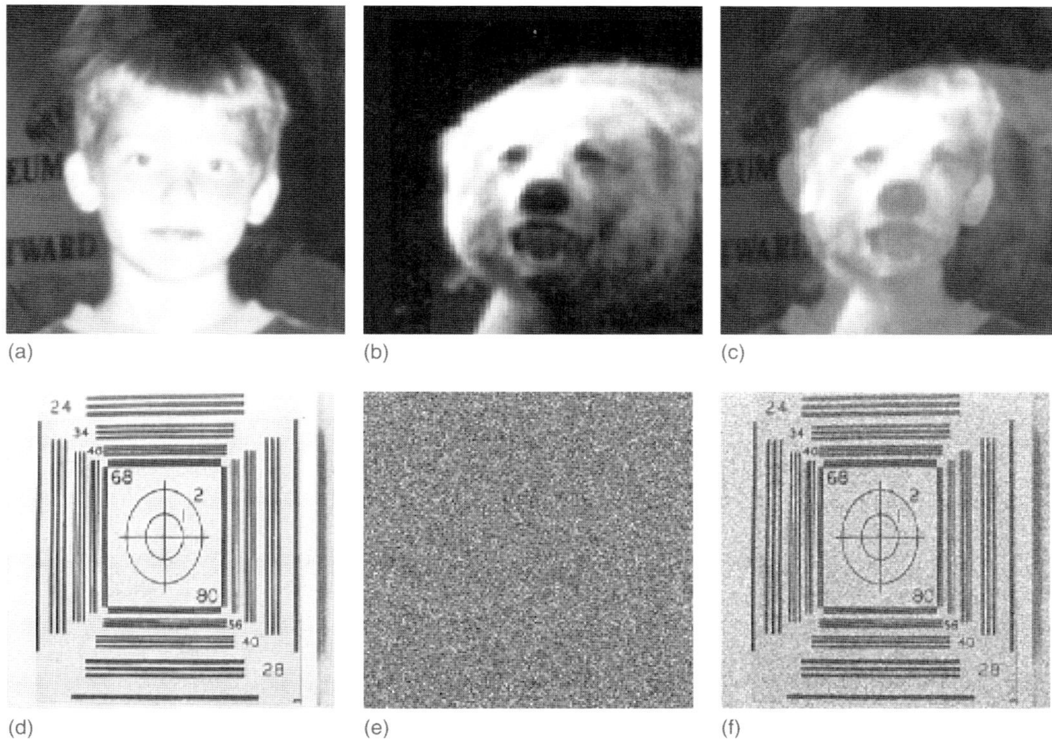

FIGURE 3.2.6
Image addition examples. This example shows one step in the *image morphing* process where an increasing percentage of the second image is slowly added to the first, and a geometric transformation is usually required to align the images. (a) First original. (b) Second original. (c) Addition of images (a) and (b). This example shows adding noise to an image which is often useful for developing image restoration models. (d) Original image. (e) Gaussian noise, variance = 400, mean = 0, (f) addition of images (d) and (e).

Addition is used to combine the information in two images. Applications include development of image restoration algorithms for modeling additive noise, as one step of image sharpening algorithms, and for special effects, such as image morphing, in motion pictures (Figure 3.2.6).

Note that true *image morphing* may also require the use of geometric transforms (see Chapter 9), to align the two images. Image morphing is also usually a time-based operation, so that a proportionally increasing amount of the second image is usually added to the first image over time.

Subtraction of two images is often used to detect motion. Consider the case where nothing has changed in a scene; the image resulting from subtraction of two sequential images is filled with zeros—a black image. If something has moved in the scene, subtraction produces a nonzero result at the location of movement, enabling detection of both the motion and the direction. If the time between image acquisition is known, the moving object's speed can also be calculated. Figure 3.2.7 illustrates the use of subtraction for motion detection. Here we can learn two things: (1) we must threshold the result, and (2) the process is imperfect and will require some further processing.

Another term for image subtraction is *background subtraction*, since we are really simply removing the parts that are unchanged, the background. Although the process is the

FIGURE 3.2.7
Image subtraction. (a) Original scene. (b) Same scene later. (c) Subtraction of *scene a from scene b*. (d) The subtracted image with a threshold of 50.

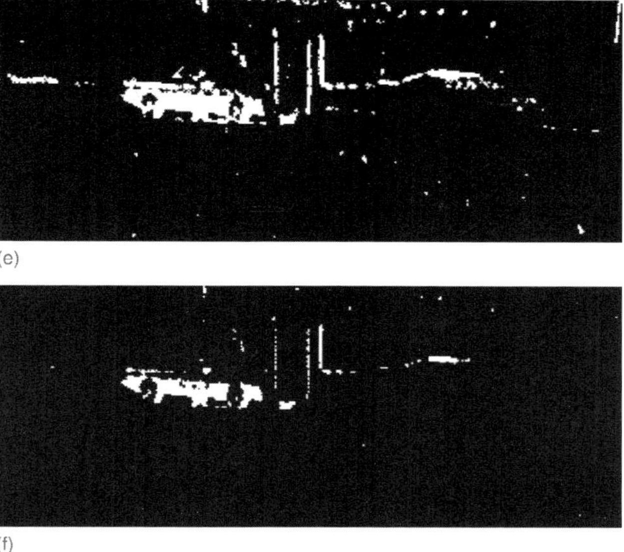

(e)

(f)

FIGURE 3.2.7 (Continued)
Image subtraction. (e) The subtracted image with a threshold of 100. (f) The subtracted image with a threshold of 150. Theoretically, only image elements that have moved should show up in the resultant image. Due to imperfect alignment between the two images, other artifacts appear. Additionally, if an object that has moved is similar in brightness to the background it will cause problems—in this example the brightness of the car is similar to the grass.

same as in motion detection, it is thought of differently. In comparing complex images, it may be difficult to see small changes. By subtracting out common background image information, the differences are more easily detectable. Medical imaging often uses this type of operation to allow the doctor to more readily see changes which are helpful in the diagnosis. The technique is also used in law enforcement and military applications; for example, to find an individual in a crowd or to detect changes in a military installation. The complexity of the image analysis is greatly reduced when working with an image enhanced through this process.

Multiplication and division are used to adjust the brightness of an image. This is done on a pixel by pixel basis and the options are to multiply or divide an image by a constant value, or by another image. Multiplication of the pixel values by a value greater than one will brighten the image (or division by a value less than 1), and division by a factor greater than one will darken the image (or multiplication by a value less than 1). Brightness adjustment by a constant is often used as a preprocessing step in image enhancement and is shown in Figure 3.2.8.

Applying multiplication or division to two images can be done to model a multiplicative noise process (see Chapter 9), or to combine two images in unique ways for special effects. In Figure 3.2.9 we see the results of multiplying two images together. The first set of images superimposes an x-ray of a hand onto another image, and the second set shows how multiplication can be used to add texture to a computer generated image. In both cases the output image has been remapped to byte data range (0–255) for display purposes. Note that multiplication and division of images can also be used for image filtering in the spectral domain (see Chapter 5).

The logic operations AND, OR, and NOT form a complete set, meaning that any other logic operation (XOR, NOR, NAND) can be created by a combination of these basic operations. They operate in a bit-wise fashion on pixel data.

(a)

(b)

(c)

FIGURE 3.2.8
Image division. (a) Original image. (b) Image divided by a value less than 1 to brighten. (c) Image divided by a value greater than 1 to darken.

Example 3.2.3

We are performing a logic AND on two images. Two corresponding pixel values are 111_{10} in one image and 88_{10} in the second image. The corresponding bit strings are:

$$111_{10} = 01101111_2 \qquad 88 = 01011000_2$$

$$\begin{array}{rr} & 01101111_2 \\ \text{AND} & 01011000_2 \\ \hline & 01001000_2 \end{array}$$

The logic operations AND and OR are used to combine the information in two images. This may be done for special effects, but a more useful application for image analysis is to perform a masking operation. AND and OR can be used as a simple method to extract a Region-of-Interest from an image. For example, a white mask ANDed with an image will allow only the portion of the image coincident with the mask to appear in the output image, with the background turned black; and a black mask ORed with an image will allow only the part of the image corresponding to the black mask to appear in the output image, but will turn the rest of the image white. This process is called *image masking* and Figure 3.2.10 illustrates the results of these operations. The NOT operation creates a negative of the original image, by inverting each bit within each pixel value, and is shown in Figure 3.2.11.

3.2.3 Spatial Filters

Spatial filtering is typically applied for noise mitigation or to perform some type of image enhancement. These operators are called spatial filters since they operate on the raw image data in the (r, c) space, the spatial domain. This is in contrast to the frequency or spectral domain filters discussed in Chapter 5. They operate on the image data by considering small neighborhoods in an image, such as 3×3, 5×5, etc., and returning a result based on a linear or nonlinear operation; moving sequentially across and down the entire image.

(a)

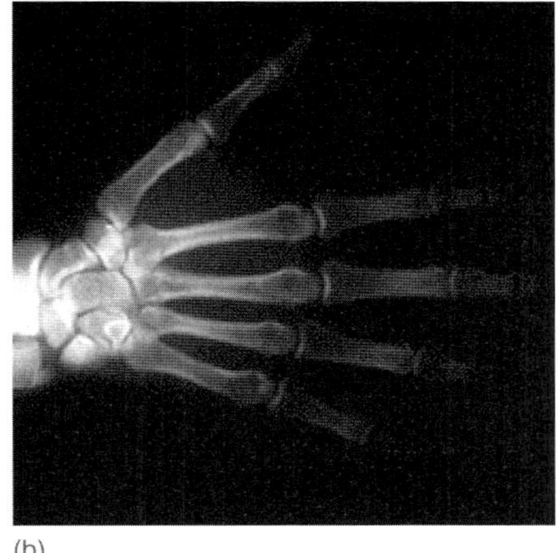

(b)

(c)

FIGURE 3.2.9
Image multiplication. (a) Cameraman image. (b) X-ray image of a hand. (c) Images (a) and (b) multiplied together which superimposes the hand onto the cameraman—the output image from this operation must be remapped for display.

(d)

(e)

(f)

FIGURE 3.2.9 (Continued)
Image multiplication. (d) A computer generated image. (e) Gaussian noise image. (f) The result of multiplying image (d) and image (e), this operation adds texture to a computer generated image—the output is remapped for display.

The three types of filters discussed here include: (1) mean filters, (2) median filters and (3) enhancement filters (for more on these filters, see Chapter 8). The first two are used primarily to deal with noise in images, although they may also be used for special applications. For instance, a mean filter adds a "softer" look to an image, as in Figure 3.2.12. The enhancement filters highlight edges and details within the image.

Many spatial filters are implemented with convolution masks. Since a convolution mask operation provides a result that is a weighted sum of the values of a pixel and

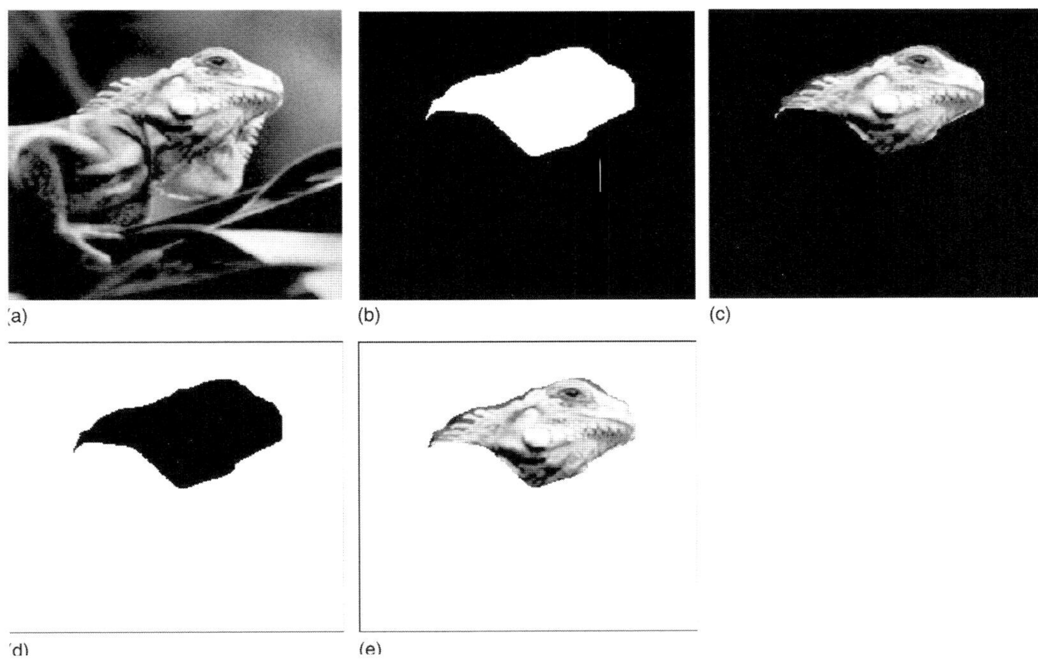

FIGURE 3.2.10
Image masking. (a) Original image. (b) Image mask for AND operation. (c) Resulting image from (a) AND (b), (d). Image mask for OR operation, created by performing a NOT on mask (b), (e) Resulting image from (a) OR (d).

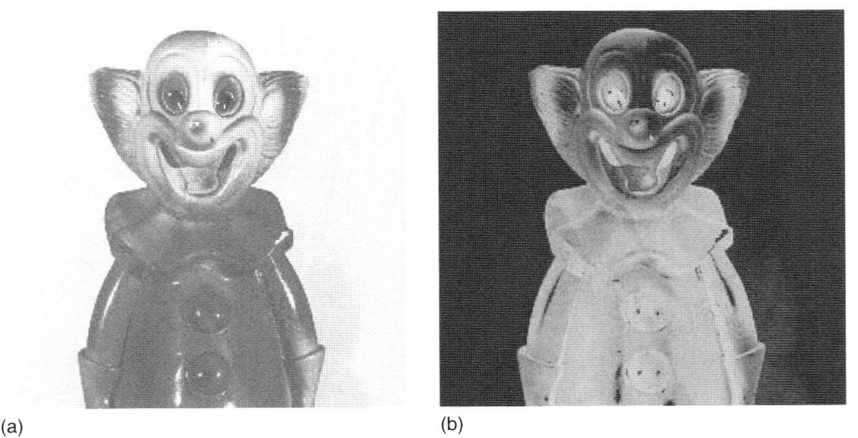

FIGURE 3.2.11
Complement image—NOT operation. (a) Original. (b) NOT operator applied to the image.

its neighbors, it is called a *linear filter*. One interesting aspect of convolution masks is that the overall effect can be predicted based on their general pattern. For example, if the coefficients of the mask sum to one, the average brightness of the image will be retained. If the coefficients sum to zero, the average brightness will be lost and will return a dark image. Furthermore, if the coefficients are alternating positive and negative, the mask is a filter that will sharpen an image; if the coefficients are all positive, it is a filter that will blur the image.

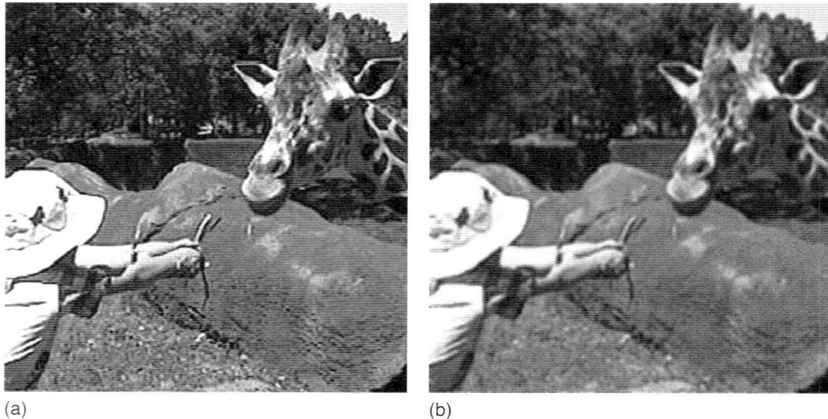

(a) (b)

FIGURE 3.2.12
Mean filter. (a) Original image. (b) Mean filtered image, 3 × 3 kernel. Note the softer appearance.

The mean filters are essentially averaging filters. They operate on local groups of pixels called neighborhoods, and replace the center pixel with an average of the pixels in this neighborhood. This replacement is done with a convolution mask such as the following 3 × 3 mask:

$$\frac{1}{9}\begin{bmatrix} 1 & 1 & 1 \\ 1 & 1 & 1 \\ 1 & 1 & 1 \end{bmatrix}$$

The result is normalized by multiplying by 1/9, so overall mask coefficients sum to one. It is more computationally efficient to perform the integer operations and only multiply by the 1/9 after the image has been processed. Often, with convolution masks, the normalization factor is implied and may not appear in the mask itself. Since the mask coefficients sum to one the average image brightness will be retained, and, since the coefficients are all positive, it will tend to blur the image. Other more complex mean filters are available which are designed to deal with specific types of noise. These are discussed in Chapter 9.

The median filter is a nonlinear filter. A *nonlinear filter* has a result that cannot be found by a weighted sum of the neighborhood pixels, such as is done with a convolution mask. However, the median filter does operate on a local neighborhood. After the size of the local neighborhood is defined, the center pixel is replaced with the median, or middle, value present among its neighbors, rather than by their average.

Example 3.2.4

Given the following 3 × 3 neighborhood:

$$\begin{bmatrix} 5 & 5 & 6 \\ 3 & 4 & 5 \\ 3 & 4 & 7 \end{bmatrix}$$

We first sort the values in order of size, (3, 3, 4, 4, 5, 5, 5, 6, 7), then we select the middle value, in this case it is 5. This 5 is then placed in the center location.

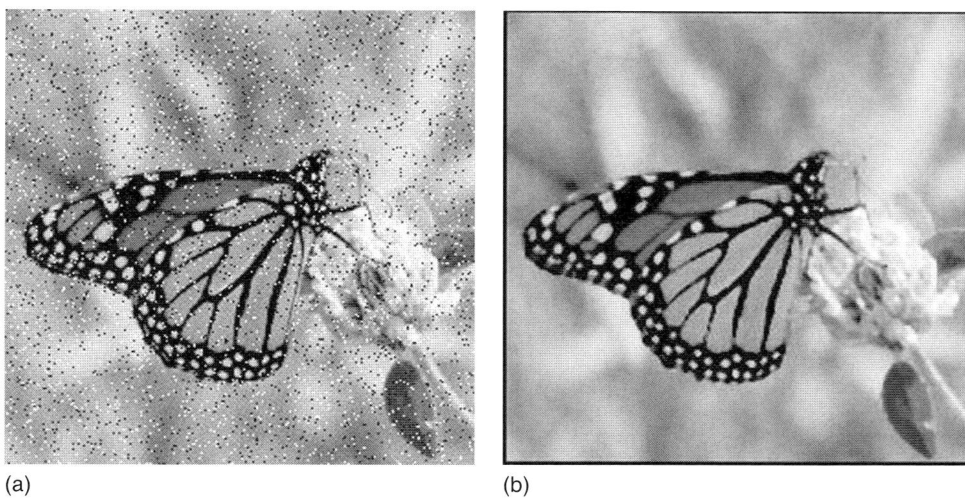

FIGURE 3.2.13
Median filter. (a) Original image with added salt-and-pepper noise. (b) Median filtered image using a 3×3 mask. (Original butterfly photo courtesy of Mark Zuke.)

A median filter can use a neighborhood of any size, but 3×3, 5×5 and 7×7 are typical. Note that the output image must be written to a separate image (a buffer), so that the results are not corrupted as this process is performed. Figure 3.2.13 illustrates the use of a median filter for noise removal.

The enhancement filters are linear filters, implemented with convolution masks having alternating positive and negative coefficients, so they will enhance image details. Many enhancement filters can be defined, here we include Laplacian-type and difference filters. Three 3×3 convolution masks for the Laplacian-type filters are:

$$\begin{array}{ccc} \text{FILTER 1} & \text{FILTER 2} & \text{FILTER 3} \\ \begin{bmatrix} 0 & -1 & 0 \\ -1 & 5 & -1 \\ 0 & -1 & 0 \end{bmatrix} & \begin{bmatrix} -1 & -1 & -1 \\ -1 & 9 & -1 \\ -1 & -1 & -1 \end{bmatrix} & \begin{bmatrix} -2 & 1 & -2 \\ 1 & 5 & 1 \\ -2 & 1 & -2 \end{bmatrix} \end{array}$$

The Laplacian-type filters are called rotationally invariant, which means they tend to enhance details in all directions equally. The difference filters, also called emboss filters, will enhance details in the direction specific to the mask selected. There are four primary difference filter convolution masks, corresponding to lines in the vertical, horizontal, and two diagonal directions:

$$\begin{array}{cccc} \text{VERTICAL} & \text{HORIZONTAL} & \text{DIAGONAL 1} & \text{DIAGONAL 2} \\ \begin{bmatrix} 0 & 1 & 0 \\ 0 & 1 & 0 \\ 0 & -1 & 0 \end{bmatrix} & \begin{bmatrix} 0 & 0 & 0 \\ 1 & 1 & -1 \\ 0 & 0 & 0 \end{bmatrix} & \begin{bmatrix} 1 & 0 & 0 \\ 0 & 1 & 0 \\ 0 & 0 & -1 \end{bmatrix} & \begin{bmatrix} 0 & 0 & 1 \\ 0 & 1 & 0 \\ -1 & 0 & 0 \end{bmatrix} \end{array}$$

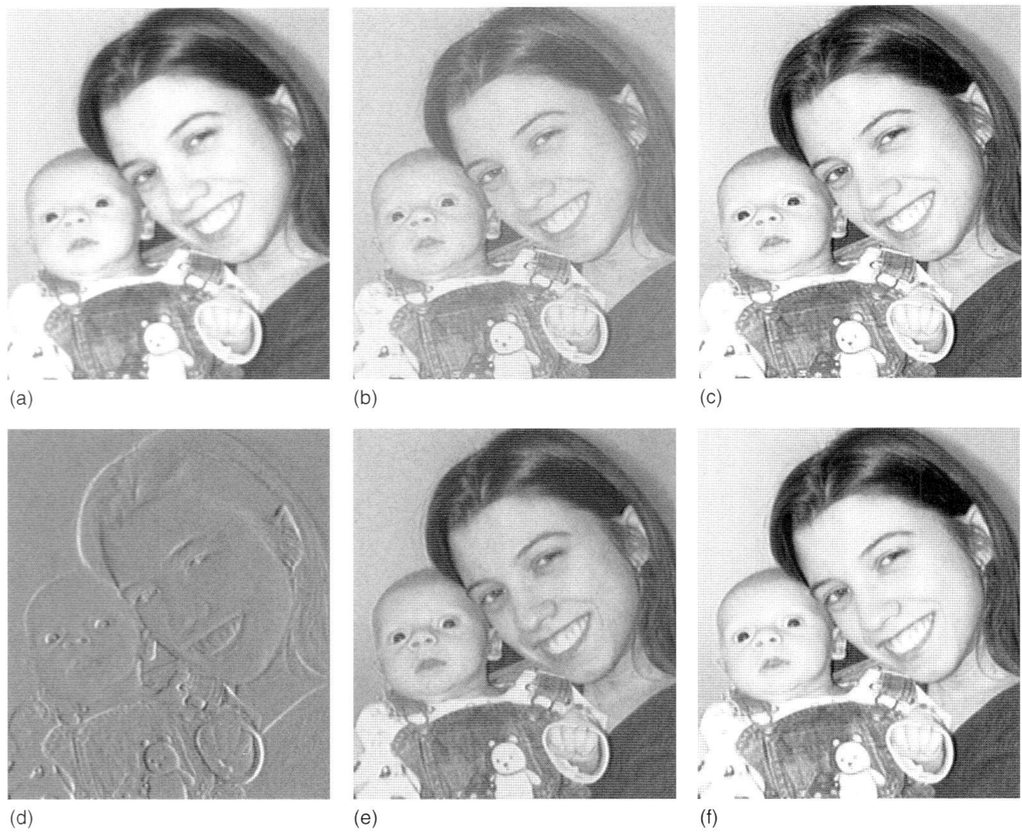

FIGURE 3.2.14
Enhancement filter. (a) Original image. (b) Image after Laplacian filter. (c) Contrast enhanced version of Laplacian filtered image, compare with (a) and note the improvement in fine detail information. (d) Result of a difference (emboss) filter applied to image (a). (e) Difference filtered image added to the original. (f) Contrast enhanced version of image (e).

Note that these are all simply rotated versions of the first mask. By completing the rotation we have four more difference filter masks:

$$\begin{bmatrix} 0 & -1 & 0 \\ 0 & 1 & 0 \\ 0 & 1 & 0 \end{bmatrix} \begin{bmatrix} 0 & 0 & 0 \\ -1 & 1 & 1 \\ 0 & 0 & 0 \end{bmatrix} \begin{bmatrix} -1 & 0 & 0 \\ 0 & 1 & 0 \\ 0 & 0 & 1 \end{bmatrix} \begin{bmatrix} 0 & 0 & -1 \\ 0 & 1 & 0 \\ 1 & 0 & 0 \end{bmatrix}$$

The results of applying the Laplacian-type and difference filters are shown in Figure 3.2.14. A more detailed discussion of these and related filters is given in Chapters 4 (edge detection) and Chapter 8 (sharpening).

3.2.4 Image Quantization

Image quantization is the process of reducing the image data by removing some of the detail information by mapping groups of data points to a single point. This can be done to either the pixel values themselves, $I(r, c)$, or to the spatial coordinates, (r, c). Operation on the pixel values is referred to as *gray level reduction*, while operating on the spatial coordinates is called *spatial reduction*.

The simplest method of gray level reduction is thresholding. We select a threshold gray level and set everything above that value equal to "1" (255 for 8-bit data), and everything at or below the threshold equal to "0." This effectively turns a gray level image into a binary (two-level) image and is often used as a preprocessing step in the extraction of object features such as shape, area, or perimeter.

A more versatile method of gray level reduction is the process of taking the data and reducing the number of bits per pixel, which allows for a variable number of gray levels. This can be done very efficiently by masking the lower bits via an AND operation. With this method, the number of bits that are masked determine the number of gray levels available.

Example 3.2.5

We want to reduce 8-bit information containing 256 possible gray-level values down to 32 possible values. This can be done by ANDing each eight-bit value with the bit-string 11111000_2. This is equivalent to dividing by eight (2^3), corresponding to the lower three bits that we are masking, and then shifting the result left three times. Now, gray level values in the range of 0–7 are mapped to 0, gray levels in the range of 8–15 are mapped to 8, and so on.

We can see that by masking the lower three bits, by setting those bits to 0 in the mask, we reduce 256 gray levels to 32 gray levels: $256 \div 8 = 32$. The general case requires us to mask k bits, where 2^k is divided into the original gray level range to get the quantized range desired. Using this method we can reduce the number of gray levels to any power of 2.

The AND-based method maps the quantized gray level values to the low end of each range; alternately, if we want to map the quantized gray level values to the high end of each range we use an OR operation. The number of "1" bits in the OR mask determine how many quantized gray levels are available.

Example 3.2.6

To reduce 256 gray levels down to 32 we use a mask of 00000111_2. Now, values in the range of 0–7 are mapped to 7, those ranging from 8 to 15 are mapped to 15, and so on.

Example 3.2.7

To reduce 256 gray levels down to 16 we use a mask of 00001111_2. Now, values in the range of 0–15 are mapped to 15, those ranging from 16 to 31 are mapped to 31, and so on.

To determine the number of "1" bits in our OR mask we apply a method similar to the AND mask method. We set the lower k bits equal to "1", where 2^k is divided into the original gray level range to get the quantized range desired. Note that the OR mask can also be found by negating (NOT) the AND mask previously described.

Another potentially useful variation is to map the quantized values to the midpoint of the range. This is done by an AND after the OR operation, or an OR after the AND operation, to either shift the values up or down.

Example 3.2.8

If we performed the quantization down to 16 levels by an OR with a mask of 00001111_2, which maps the values to the high end of the range, we could shift the values down to the middle of the range by ANDing with a mask of 11111100_2.

Using these AND/OR techniques for gray level quantization the number of gray levels can be reduced to any power of 2, such as 2, 4, 8, 16, 32, 64, or 128, as illustrated in Figure 3.2.15. As the number of gray levels decreases we see an increase in a phenomenon

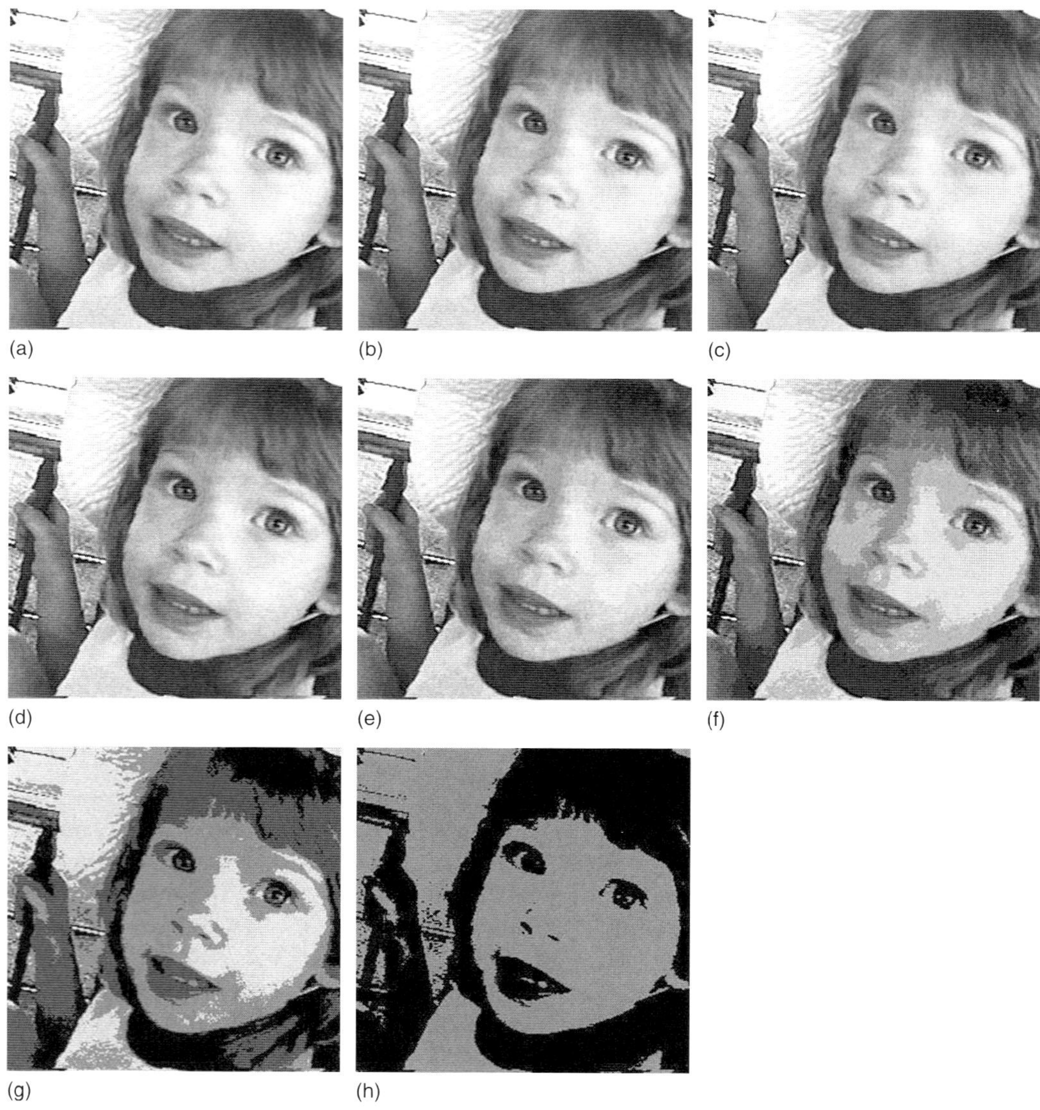

FIGURE 3.2.15
False contouring. (a) Original 8-bit image, 256 gray levels. (b) Quantized to 7 bits, 128 gray levels. (c) Quantized to 6 bits, 64 gray levels. (d) Quantized to 5 bits, 32 gray levels. (e) Quantized to 4 bits, 16 gray levels. (f) Quantized to 3 bits, 8 gray levels. (g) Quantized to 2 bits, 4 gray levels. (h) Quantized to 1 bit, 2 gray levels.

called false contouring. Contours appear in the images as false edges, or lines, as a result of the gray level quantization. We can see in the figure that these contour lines do not become very visible until we get down to about 4 bits per pixel, and then they become very prominent as we use fewer bits.

The false contouring effect can be visually improved upon by using an IGS (improved gray scale) quantization method. The IGS method takes advantage of the human visual system's sensitivity to edges by adding a small random number to each pixel before quantization, which results in a more visually pleasing appearance (see Figure 3.2.16). If we look at Figure 3.2.16c closely we can see that IGS eliminates the appearance of false contours by breaking the sharp edges into smaller random pieces, so the human visual system does a better job of blending the false contours together.

The way IGS works is similar to *dithering*, or *halftoning*, which is typically used in printing or in any application where we desire to reduce the number of gray levels or colors. For example, newspapers are printed in only two levels but we still get the illusion of varying shades of gray in newspaper photographs. Many dithering algorithms have been created and are based on the idea of diffusing the quantization error across edges, where changes occur in the image. In Figure 3.2.17 we see the results

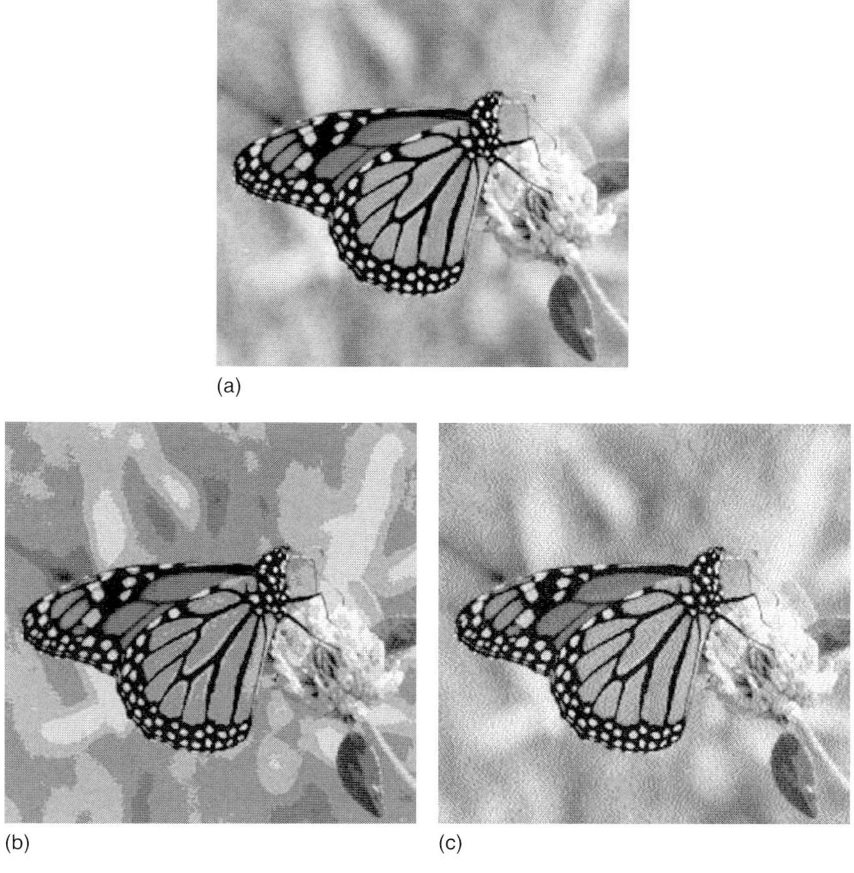

FIGURE 3.2.16
IGS quantization. (a) Original image. (b) Uniform quantization to 8 gray levels (3 bits). (c) IGS quantization to 8 gray levels (3 bits). (Original butterfly photo courtesy of Mark Zuke.)

FIGURE 3.2.17
Halftoning and dithering. (a) Original image, 8-bits per pixel. (b) Floyd-Steinberg error diffusion, 1-bit per pixel. (c) Bayer's ordered dither, 1-bit per pixel. (d) 45-degree clustered-dot dither, 1-bit per pixel.

of applying three algorithms which are representative of the types in use. With these techniques various gray levels are represented by different geometric patterns or various size dots, so the effective spatial resolution is reduced. Looking closely at the examples in Figure 3.2.17 we can see that the closer the black pixels are spaced together, the darker the area appears. As a result of this it requires multiple pixels to represent different gray levels and this is what causes the reduction in effective spatial resolution.

Gray level quantization using the previously discussed AND/OR method is very efficient for quantization, but it is not flexible since the size of the quantization bins is uniform, which is *uniform bin width quantization* (see Figure 3.2.18a). There are other methods of gray-level quantization that allow for variable bin sizes called *variable bin width quantization* (Figure 3.2.18b). These methods are more complicated than, and not as fast as, those used with uniform bins. One such use is in simulating the response of the human visual system by using logarithmically spaced bins. The use of variable bin size is application-dependent, and requires application-specific information. For example, in Figure 3.2.19 we can see the result of an application where four gray levels provided optimal results. Here we are applying varying bin sizes and mapping them to specific gray levels. In Figure 3.2.19, the gray levels in the range 0–101 were mapped to 79, 102–188 mapped to 157, 189–234 mapped to 197, and 235–255 mapped to 255. These numbers were determined as the result of application specific feedback, an important

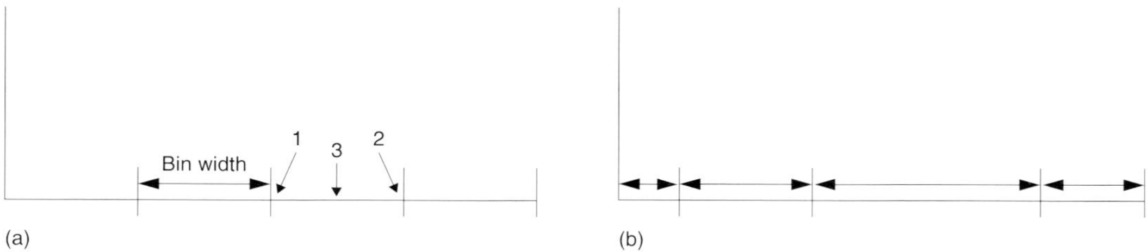

FIGURE 3.2.18
Quantization bins. (a) Uniform quantization bins: all bins are the same width. Values that fall within the same bin can be mapped to the low end (1), high end (2), or the middle (3). (b) Variable quantization bins are of different widths.

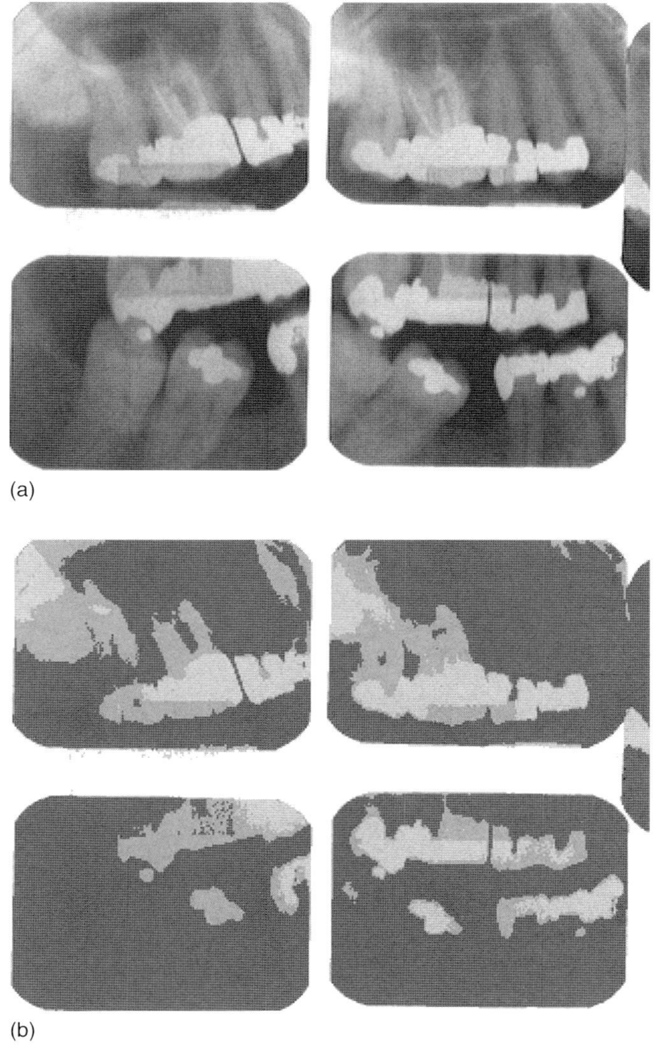

FIGURE 3.2.19
Variable bin-width quantization. (a) Original image. (b) After variable bin-width quantization.

aspect of image analysis as shown in Figure 3.1.3. For this application, the second brightest gray level (197) was used to identify fillings in dental x-rays.

Quantization of the spatial coordinates, *spatial quantization*, results in reducing the actual size of the image. This is accomplished by taking groups of pixels that are spatially adjacent and mapping them to one pixel. This can be done in one of three ways: (1) averaging, (2) median, or (3) decimation. For the first method, averaging, we take all the pixels in each group and find the average gray level by summing the values and dividing by the number of pixels in the group. With the second method, median, we sort all the pixel values from lowest to highest and then select the middle value. The third approach, decimation, also known as sub-sampling, entails simply eliminating some of the data. For example, to reduce the image by a factor of two, we simply take every other row and column and delete them.

To perform spatial quantization we specify the desired size, in pixels, of the resulting image. For example, to reduce a 512×512 image to 1/4 its size, we specify that we want the output image to be 256×256 pixels. We now take every 2×2 pixel block in the original image and apply one of the three methods listed above to create a reduced image. It should be noted that this method of spatial reduction allows for simple forms of geometric distortion, specifically, stretching or shrinking along the horizontal or vertical axis. Geometric distortion is explored more fully in Chapter 9. If we take a 512×512 image and reduce it to a size of 64×128, we will have shrunk the image as well as squeezed it horizontally. This result is shown in Figure 3.2.20, where we can see that the averaging method blurs the image, and the median and decimation method produces some visible artifacts.

To improve the image quality when applying the decimation technique, we may want to preprocess the image with an averaging, or mean, spatial filter—this type of filtering is called *anti-aliasing filtering*. In Figure 3.2.21, we can compare reduction done with or without an anti-aliasing filter. Here, the decimation technique was applied to a text image with a factor of four reduction; note that without the anti-aliasing filter the letter

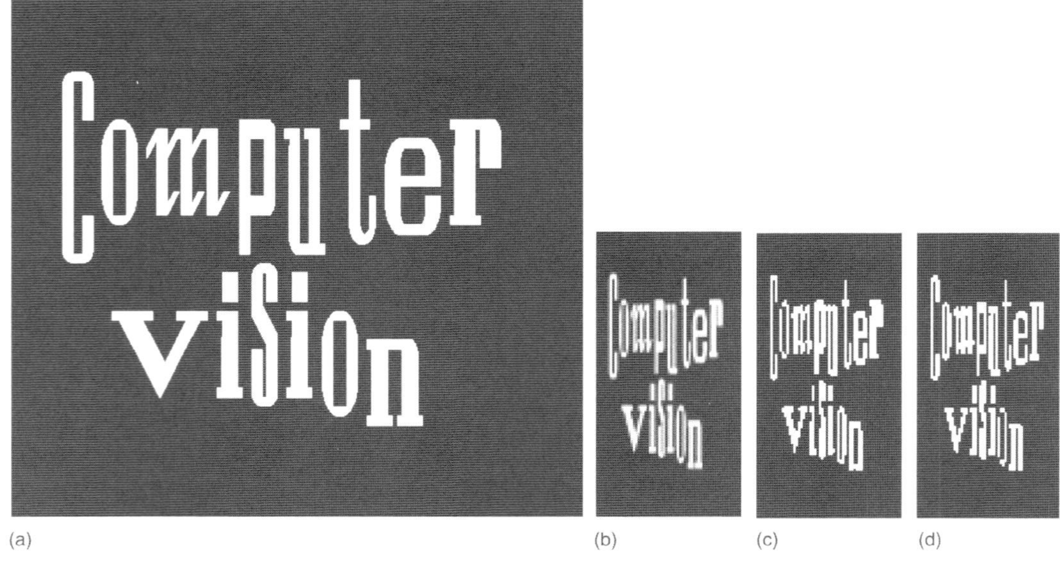

FIGURE 3.2.20
Spatial reduction. (a) Original 512×512 image. (b) Spatial reduction to 64×128 via averaging. (c) Spatial reduction to 64×128 via median method. (d) Spatial reduction to 64×128 via decimation method.

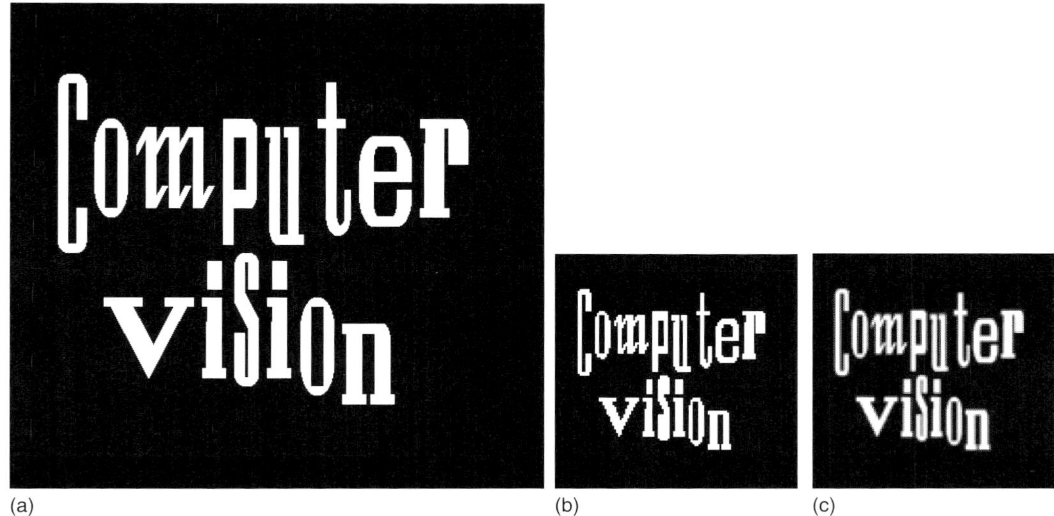

FIGURE 3.2.21
Decimation and anti-aliasing filter. (a) Original 512 × 512 image. (b) Result of spatial reduction to 128 × 128 via decimation. (c) Result of spatial reduction to 128 × 128 via decimation, but the image was first preprocessed by a 5 × 5 averaging filter, an anti-aliasing filter.

"S" becomes enclosed. The cost of retaining this information is that the output image is slightly blurred.

3.3 Binary Image Analysis

To complete this introductory chapter on image analysis we will look at basic binary object features and examine how they can be applied to our image analysis process shown in Figure 3.1.3. Since most cameras will give us color or gray level images, we will first consider how to create binary images from gray level images; followed by extraction of simple binary features; and finally look at some simple methods to classify binary objects. This will clarify the image analysis process and lay the groundwork for Chapters 4, 5, and 6.

3.3.1 Thresholding via Histogram

In order to create a binary image from a gray level image we must perform a threshold operation. This is done by specifying a *threshold value* and will set all values above the specified gray level to "1" and everything below the specified value to "0". Although the actual values for the "0" and "1" can be anything, typically 255 is used for "1" and 0 is used for the "0" value.

In many applications the threshold value will be determined experimentally and is highly dependent on lighting conditions and object to background contrast. It will be much easier to find an acceptable threshold value with proper lighting and good contrast between the object and the background. Figure 3.3.1a,b shows an example of good lighting and high object to background contrast, while Figure 3.3.1c,d illustrates a poor example. Imagine trying to identify the object based on the poor example compared to the good example.

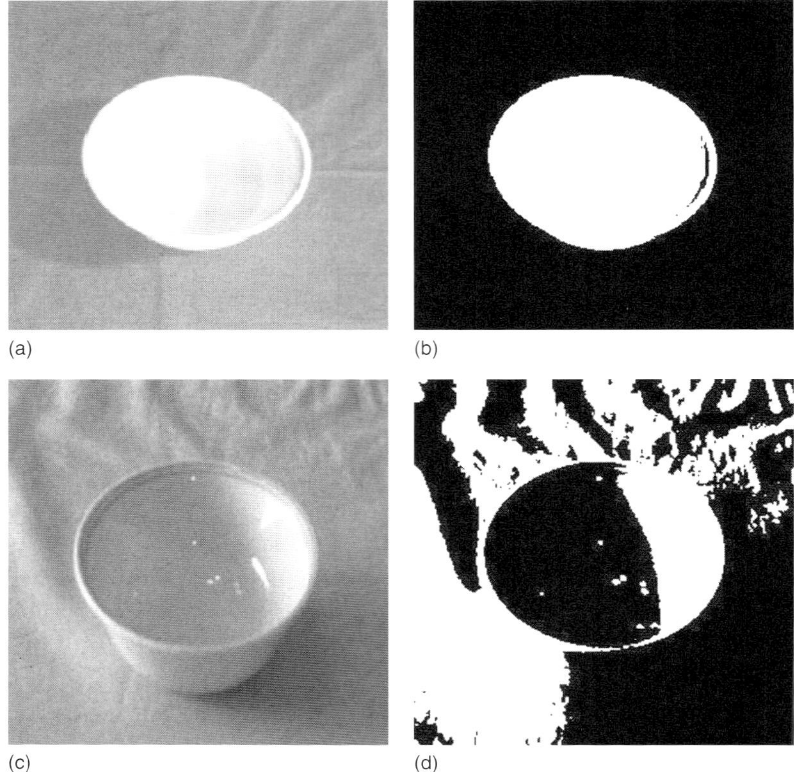

FIGURE 3.3.1
Effects of lighting and object to background contrast on thresholding. (a) An image of a bowl with high object to background contrast and good lighting. (b) Result of thresholding image (a). (c) An image of a bowl with poor object to background contrast and poor lighting. (d) Result of thresholding image (c).

To select the proper threshold value, the histogram is examined. The *histogram* of an image is a plot of gray level versus the number of pixels in the image at each gray level (see Section 8.2 for more details). Figure 3.3.2 shows the two bowl images and their corresponding histograms. The peaks and valleys in the histogram are examined and a threshold is experimentally selected that will best separate the object from the background. Notice the peak in Figure 3.3.2b on the far right; this corresponds to the maximum gray level value and has the highest number of pixels at that value. This peak and the two small peaks to its left represent the bowl. Although many nice valleys can be seen in the histogram for the poor example (Figure 3.3.2d), none will separate the object from the background successfully. Using the histogram to separate objects in an image will be examined more thoroughly in Chapter 4.

3.3.2 Connectivity and Labeling

The images considered in the previous section contained only one object. What will happen if the image contains more than one object? In order to handle images with more than one object we need to consider exactly how pixels are connected to make an object, and then we need a method to label the objects separately. Since we are dealing with digital images, the process of spatial digitization (sampling) can cause problems regarding connectivity of objects. These problems can be resolved with careful connectivity definitions and heuristics applicable to the specific domain. Connectivity

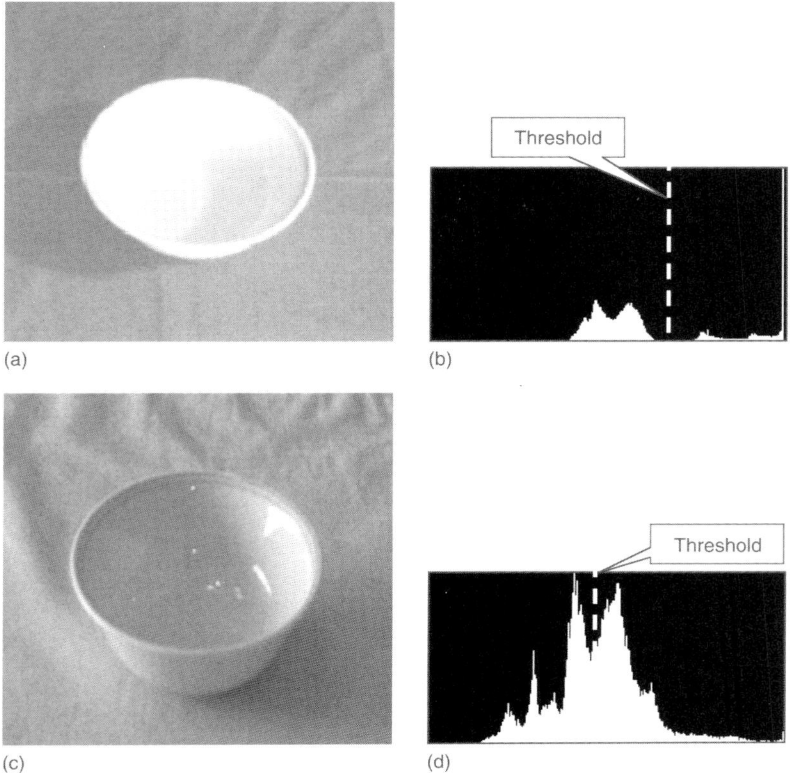

FIGURE 3.3.2
Histograms. (a) An image of a bowl with high object to background contrast and good lighting. (b) The histogram of image (a), showing the threshold that separates object and background (see Figure 3.3-1b). (c) An image of a bowl with poor object to background contrast and poor lighting. (d) The histogram of image (c), showing what appears to be a good threshold, but it does not successfully separate object and background (see Figure 3.3-1d).

refers to the way in which we define an object; once we performed a threshold operation on an image, which pixels should be connected to form an object? Do we simply let all pixels with value of "1" be the object? What if we have two overlapping objects?

First, we must define which of the surrounding pixels are considered to be neighboring pixels. A pixel has eight possible neighbors: two horizontal neighbors, two vertical neighbors, and four diagonal neighbors. We can define connectivity in three different ways: (1) four-connectivity, (2) eight-connectivity, and (3) six-connectivity. Figure 3.3.3 illustrates these three definitions. With four-connectivity the only neighbors considered connected are the horizontal and vertical neighbors; with eight-connectivity all of the eight possible neighboring pixels are considered connected, and with six-connectivity the horizontal, vertical and two of the diagonal neighbors are connected. Which definition is chosen depends on the application, but the key to avoiding problems is to be consistent.

If we select four or eight-connectivity the *connectivity dilemma* arises. Consider the following binary image segment:

$$\begin{matrix} 0 & 1 & 0 \\ 1 & 0 & 1 \\ 0 & 1 & 0 \end{matrix}$$

Assuming four-connectivity there are four separate objects and five separate background objects. The dilemma is that if the objects are separated, shouldn't the background be

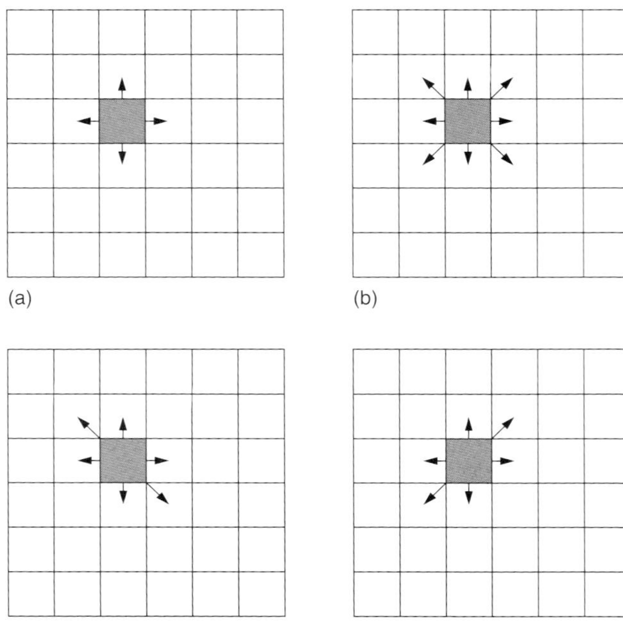

FIGURE 3.3.3
Connectivity. (a) 4-connectivity. (b) 8-connectivity. (c) 6-connectivity NW/SE. (d) 6-connectivity NE/SW.

connected? Alternately, if we assume eight-connectivity we have one connected object, a closed curve, but the background is also connected. This creates another dilemma because a closed curve should separate the background into distinct objects. How do we resolve this issue? These are our choices:

1. Use eight-connectivity for background and four-connectivity for the objects.
2. Use four-connectivity for background and eight-connectivity for the objects.
3. Use six-connectivity.

The first two choices are acceptable for binary images, but get quite complicated when extended to gray level and color images, and we want a standard definition we can use throughout this book. The third choice is a good compromise in most situations, as long as we are aware of the bias created by selection of one diagonal direction. That is, connection by a single diagonal pixel will only be defined in one of two possible directions. For most real images, this is not a problem. We will use the definition of six-connectivity as shown in Figure 3.3.3c, with the northwest (NW) and southeast (SE) diagonal neighbors.

After the definition of connectivity is chosen a labeling algorithm is needed to differentiate between multiple objects within an image. The labeling process requires us to scan the image and label connected objects with the same symbol. With the definition of six-connectivity selected, we can apply the algorithm given in Figure 3.3.4 to label the objects in the image. (Note that this flowchart will label objects in images with more than two gray levels if we assume that any areas not of interest have been masked out by setting the pixels equal to zero.) The UPDATE block in the flowchart refers to a function that will keep track of objects that have been given multiple labels. This can occur with a sequential scanning of the image if the connecting pixels are not encountered until after different parts of the object have already been labeled (see Figure 3.3.5).

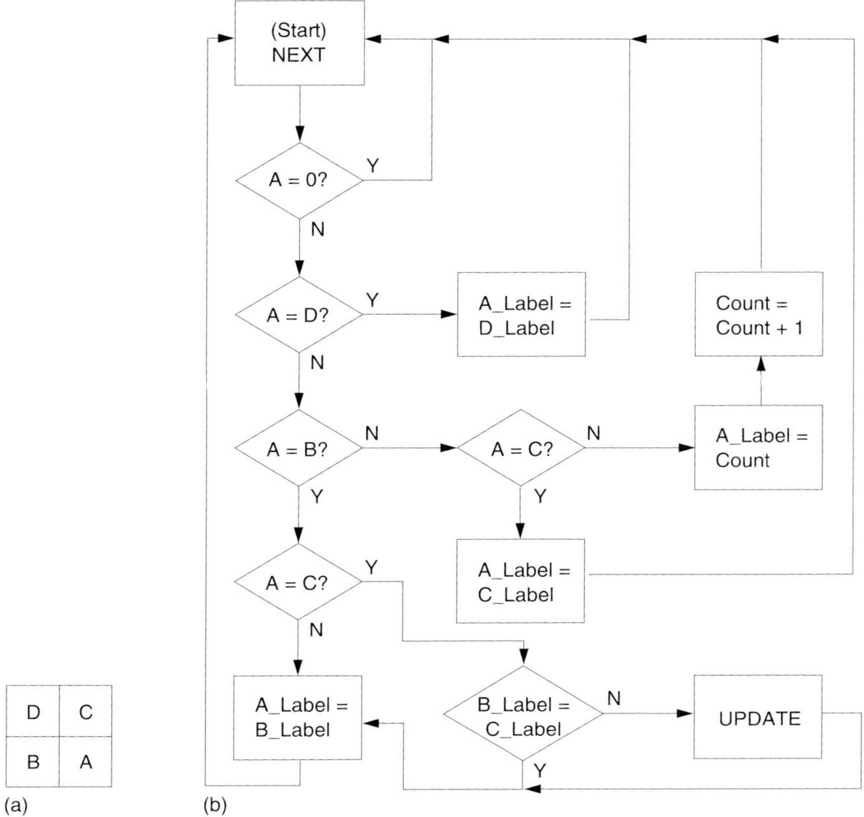

FIGURE 3.3.4
Labeling algorithm.

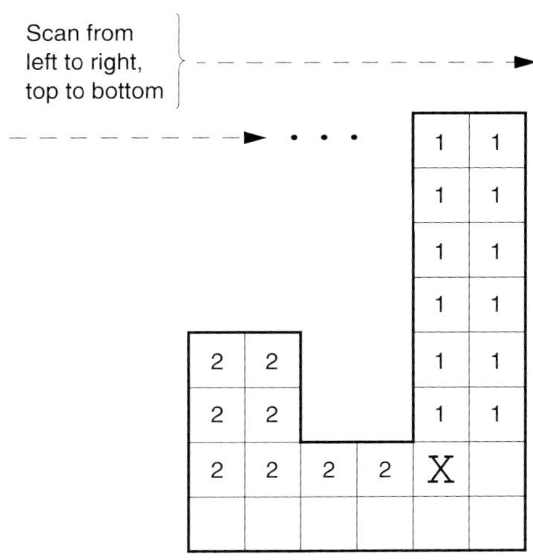

FIGURE 3.3.5
Multiple labels. The labeling algorithm requires an UPDATE function to keep track of objects with more than one label. Multiple labeling can occur during sequential scanning, as shown on the "J" shaped object. We label two different objects until we reach the pixel marked "X", where we discover that objects 1 and 2 are connected.

By labeling the objects, an image filled with object numbers is created. With this labeled image we can extract features specific to each object. These features are used to locate and classify the binary objects. The binary object features defined here include area, center of area, axis of least second moment, projections, and Euler number. The first three tell us something about where the object is, and the latter two tell us something about the shape of the object. More features are provided in Chapter 6.

3.3.3 Basic Binary Object Features

In order to provide general equations for area, center of area, and axis of least second moment, we define a function, $I_i(r, c)$:

$$I_i(r,c) = \begin{cases} 1 & \text{if } I(r,c) = i\text{th object number} \\ 0 & \text{otherwise} \end{cases}$$

Now we can define the *area* of the ith object as:

$$A_i = \sum_{r=0}^{N-1}\sum_{c=0}^{N-1} I_i(r,c)$$

The area, A_i, is measured in pixels, and indicates the relative size of the object. We can then define the *center of area* (centroid in the general case), which finds the midpoint along each row and column axis corresponding to the "middle" based on the spatial distribution of pixels within the object. It can be defined by the pair $(\bar{r}_i, \bar{c}_i)$:

$$\bar{r}_i = \frac{1}{A_i}\sum_{r=0}^{N-1}\sum_{c=0}^{N-1} r\, I_i(r,c); \qquad \bar{c}_i = \frac{1}{A_i}\sum_{r=0}^{N-1}\sum_{c=0}^{N-1} c\, I_i(r,c)$$

These correspond to the row coordinate of the center of area for the ith object, $\bar{r}_i$, and the column coordinate of the center of area for the ith object, $\bar{c}_i$. This feature will help to locate an object in the two-dimensional image plane. The next feature we will consider, the *axis of least second moment*, provides information about the object's orientation. This axis corresponds to the line about which it takes the least amount of energy to spin an object of like shape, or the axis of least inertia. If we move our origin to the center of area, (r, c), the axis of least second moment is defined as follows:

$$\tan(2\,\theta_i) = 2\frac{\sum_{r=0}^{N-1}\sum_{c=0}^{N-1} rcI_i(r,c)}{\sum_{r=0}^{N-1}\sum_{c=0}^{N-1} r^2\, I_i(r,c) - \sum_{r=0}^{N-1}\sum_{c=0}^{N-1} c^2\, I_i(r,c)}$$

This is shown in Figure 3.3.6. The origin is moved to the center of area for the object, and the angle is measured from the r-axis counterclockwise.

The *projections* of a binary object, which also provide shape information, are found by summing all the pixels along rows or columns. If we sum the rows we have the *horizontal projection*, if we sum the columns we have the *vertical projection*. We can define the horizontal projection, $h_i(r)$, as follows:

$$h_i(r) = \sum_{c=0}^{N-1} I_i(r,c)$$

And the vertical projection, $v_i(c)$:

$$v_i(c) = \sum_{r=0}^{N-1} I_i(r,c)$$

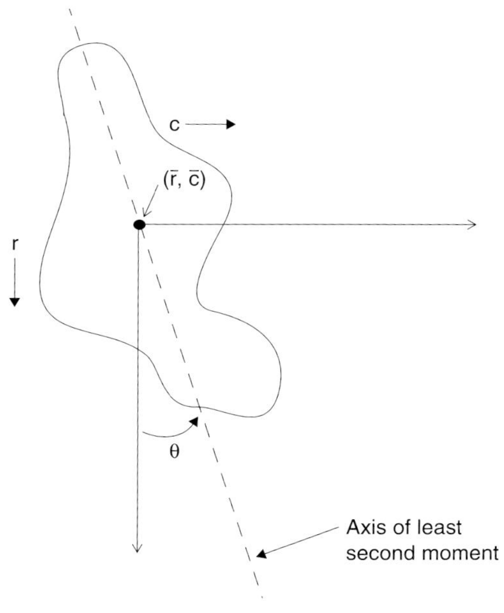

FIGURE 3.3.6
Axis of least second moment.

r ↓ / c →									h (r)
0	0	0	0	1	0	0	0	0	1
0	0	0	0	1	1	0	0	0	2
0	0	1	1	1	1	1	0	0	5
0	1	1	1	1	1	1	0	0	6
0	0	0	0	0	1	1	0	0	2
0	0	0	0	0	0	0	0	0	0
v (c) → 0	1	2	2	4	4	3	0	0	

FIGURE 3.3.7
Projection. To find the projections, we sum the number of 1s in the rows and columns.

An example of the horizontal and vertical projection for a binary image is shown in Figure 3.3.7. Projections are useful in applications like character recognition, where the objects of interest can be normalized with regard to size.

With the projection equations we can define the equations for the center of area as follows:

$$\bar{r}_i = \frac{1}{A_i} \sum_{r=0}^{N-1} \sum_{c=0}^{N-1} r\, I_i(r, c) = \frac{1}{A_i} \sum_{r=0}^{N-1} r h_i(r)$$

$$\bar{c}_i = \frac{1}{A_i} \sum_{r=0}^{N-1} \sum_{c=0}^{N-1} r\, I_i(r, c) = \frac{1}{A_i} \sum_{c=0}^{N-1} c v_i(c)$$

Given these equations we can more easily understand their meaning. Referring to Figure 3.3.7 and the above equations, we can see that a larger projection value along a

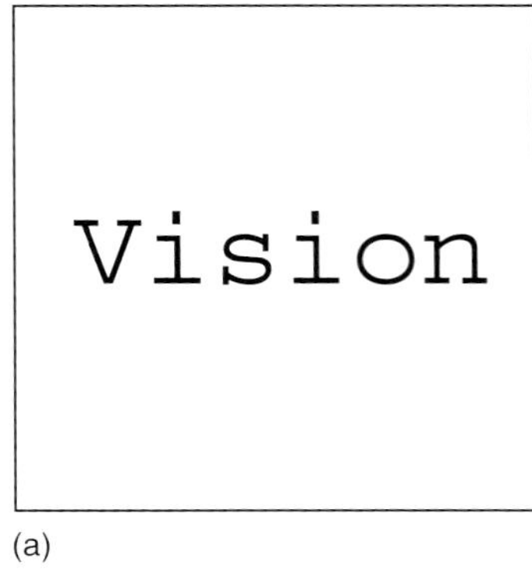

(a)

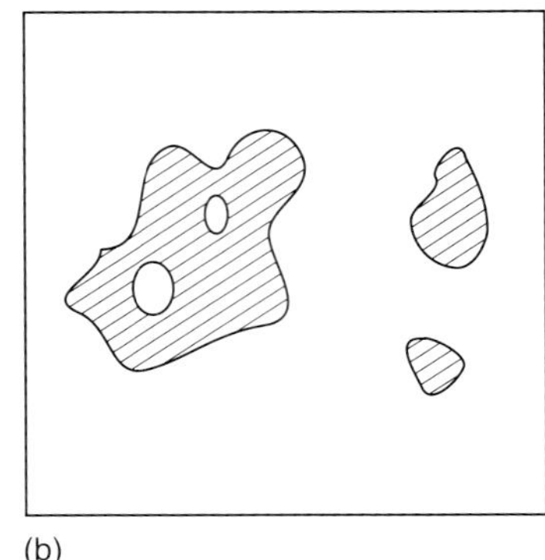

(b)

FIGURE 3.3.8
Euler number. (a) This image has eight objects and one hole, so its Euler number is $8 - 1 = 7$. The letter "V" has Euler number of 1, "1" = 2, "s" = 1, "o" = 0, and "n" = 1. (b) This image has three objects and two holes, so the Euler number is $3 - 2 = 1$.

given row or column will weigh that particular row or column value more heavily in the equation. This will tend to move the center of area coordinate toward that particular row or column; note that all values are normalized by the object area.

The *Euler number* of an image is defined as the number of objects minus the number of holes. For a single object, it relates to the number of closed curves the object contains. It is often useful in tasks such as optical character recognition (OCR), as shown by the example in Figure 3.3.8. In Figure 3.3.8a we have eight objects (don't forget the dots on the i's) and one hole; Figure 3.3.8b has three objects and two holes. Note that we can find the Euler number for the entire image, or for a single object within the image. For example, the letter "i" has an Euler number of 2, and the letter "o" has an Euler number of 0.

Using the connectivity definition we defined when we labeled the image, we can find the Euler number by finding *convexities* and *concavities*. The Euler number will be equal to the number of convexities minus the number of concavities, which are found by scanning the image for the following patterns:

$$\begin{array}{cc} \text{CONVEXITIES} & \text{CONCAVITIES} \\ \begin{bmatrix} 0 & 0 \\ 0 & 1 \end{bmatrix} & \begin{bmatrix} 0 & 1 \\ 1 & 1 \end{bmatrix} \end{array}$$

Each time one of these patterns is found the count is increased for the corresponding pattern.

$$\begin{aligned} \text{Euler number} &= (\text{Count of CONVEXITIES}) - (\text{Count of CONCAVITIES}) \\ &= (\text{Number of objects}) - (\text{Number of holes}) \end{aligned}$$

The number of convexities and concavities can also be useful features for binary objects.

3.3.4 Binary Object Classification

To complete our introduction to image analysis we will apply the process (see Figure 3.1.3) to the development of an algorithm for classifying geometric shapes.

We will use CVIPtools to create the objects and analyze the images. In this process we will explore the *Utilities* functions, which include the preprocessing utilities and other utilities, and the *Features* tab of the *Analysis* window. The binary features discussed previously will be used for the classification. For this experiment we will develop an algorithm to classify the following shapes: (1) circles, (2) ellipses, (3) rectangles, and (4) ellipses with holes.

To create our objects we first invoke CVIPtools, and select the *Utilities* functions (Figure 3.3.9, remember this can also be done with a right click on the image viewing area). Next, we select *Create* and then click on *Circle* (Figure 3.3.10), and select an image size of 512 × 512 by a mouse click on the arrow next to the *Image width* and *Image height* boxes. Note that these text boxes allow for selection via the mouse and the arrow, or allow the user to type in any value. With a mouse click on the *Apply* button in the lower right corner of the window the circle image is created, as shown in Figure 3.3.11. We want to have two of each type of object, so we create another circle but select a different location and size.

Next, we OR these two images together by using *Arith/Logic->OR* (this is not required, we could use separate images, but it will streamline the processing and help to illustrate some important CVIPtools concepts). This is done by selecting one circle as the current image by clicking on the image, or by clicking on the image name in the image queue—the names of the images in the image queue are listed on the left side of the main window. The second image is selected via the mouse and the arrow on the right of the *Second image* box. The result is shown in Figure 3.3.12. Now, we create four ellipses and two rectangles of various sizes and locations, and OR each new object with the current composite image which contains the previous objects. Note that it is easy to select a

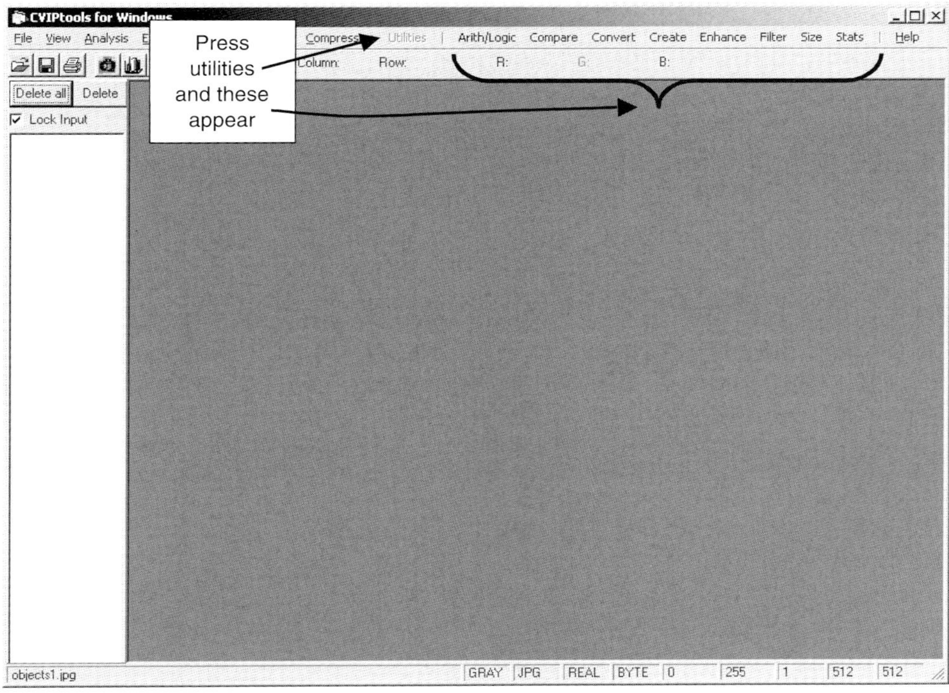

FIGURE 3.3.9
CVIPtools main window and Utilities functions.

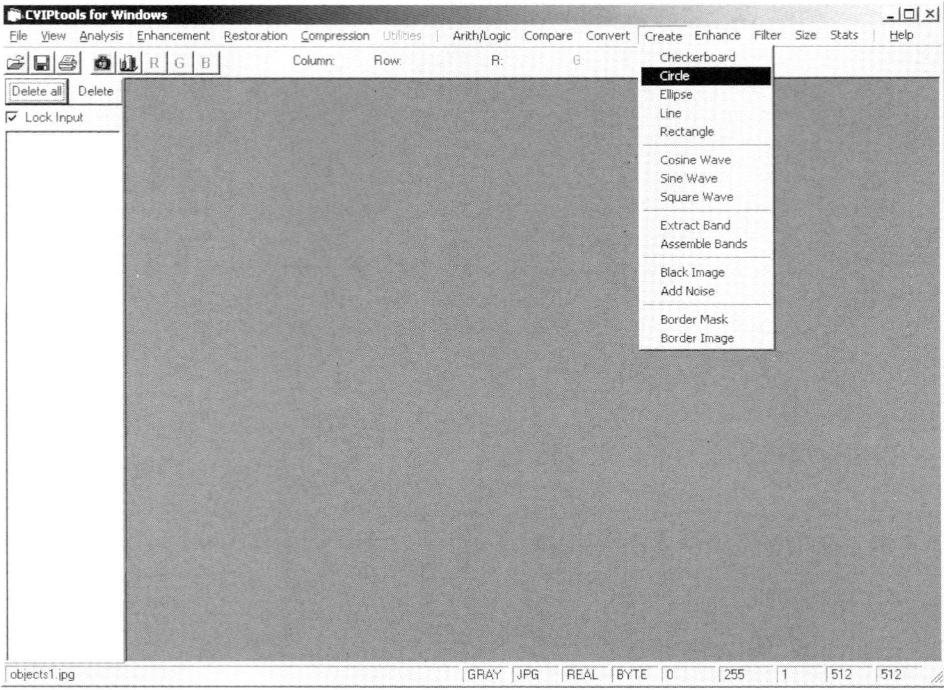

FIGURE 3.3.10
Selection for creating a circle with the Utilities.

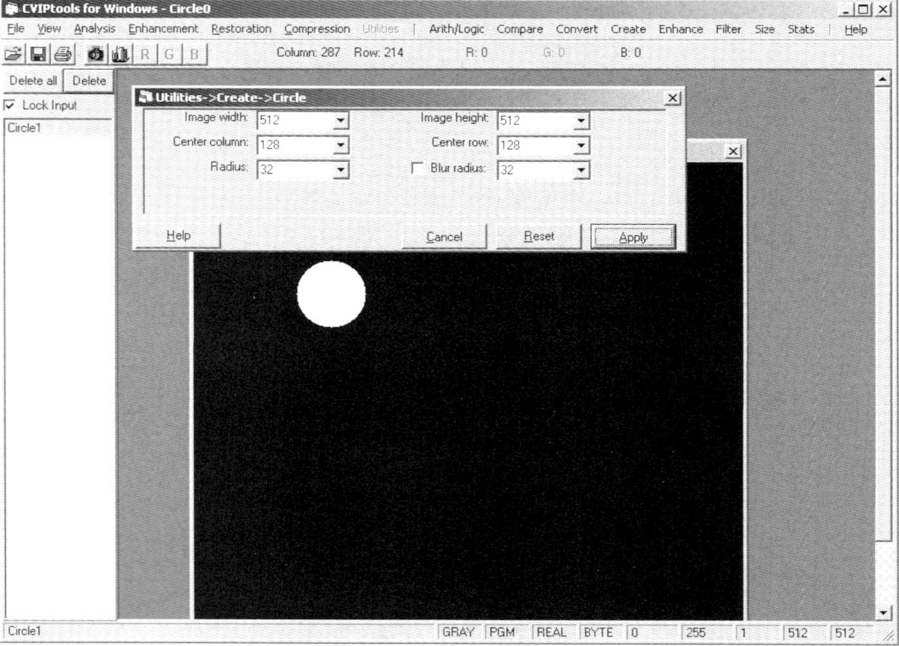

FIGURE 3.3.11
CVIPtools after creating the circle image.

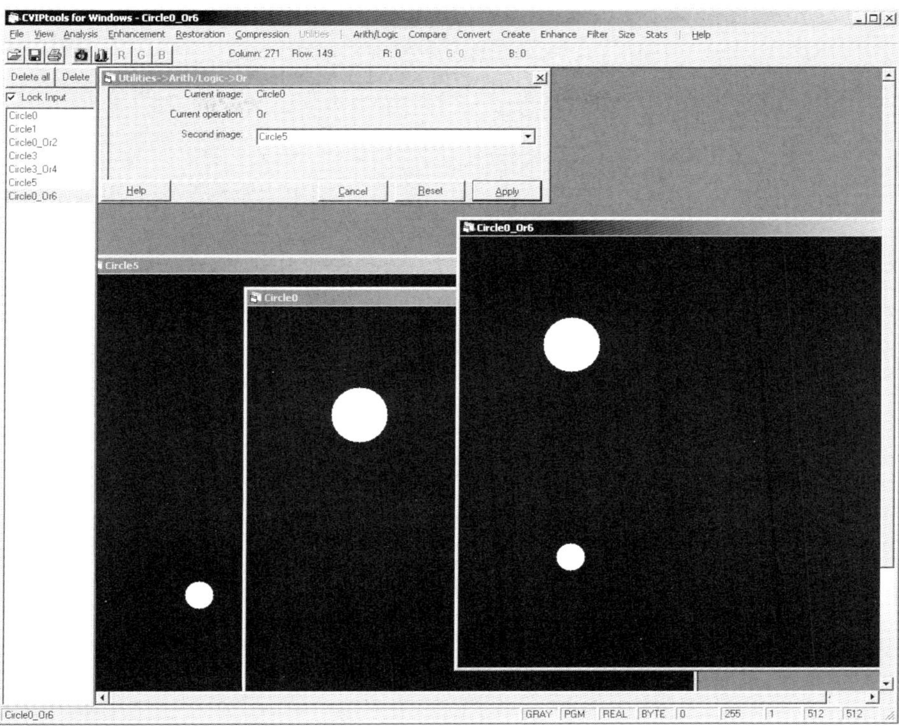

FIGURE 3.3.12
ORing two circles together to create a composite image with both objects.

location for a new object by moving the mouse pointer on the current composite image and observing the row and column coordinates on the top of the main window.

The next task is to create the ellipses with holes. This is done by creating a small circle in and then performing an XOR operation with the circle and the ellipse. Note that the circles need to be in a location within, and smaller than, the ellipses to create these objects. To do this select *Arith/Logic->XOR* on the *Utilities* window (Figure 3.3.13). Perform the XOR to create an ellipse with a hole as shown in Figure 3.3.14, followed by repeating the process to make the second example. Next create the composite image by ORing the ellipses with holes with the previous composite image containing all the other objects (Figure 3.3.15).

In order to better simulate a real application we will blur and add noise to the image containing the objects. To blur the image select *Filter->Specify a Blur*. Use the default parameters and click *Apply*. To add noise, select *Create->Add Noise*. Select *Salt and Pepper* noise, and click *Apply*. The result is shown in Figure 3.3.16. We now have an image with two circles, two ellipses, two rectangles, and two ellipses with holes; and we have blurred and added noise to better simulate a real application.

Now that our example image database has been created, we are ready to analyze the images and develop our classification algorithm. Referring to Figure 3.1.3, we will try the following steps:

1. *Preprocessing*: noise removal with a median filter
2. *Segmentation*: thresholding
3. *Filtering*: none required (we hope!)

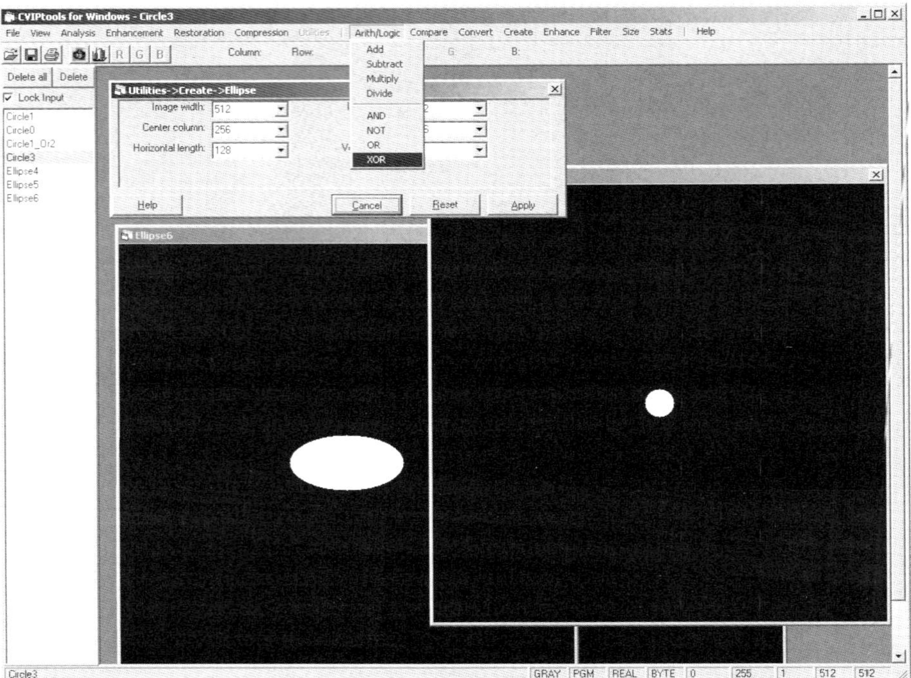

FIGURE 3.3.13
Selection for XOR of two images.

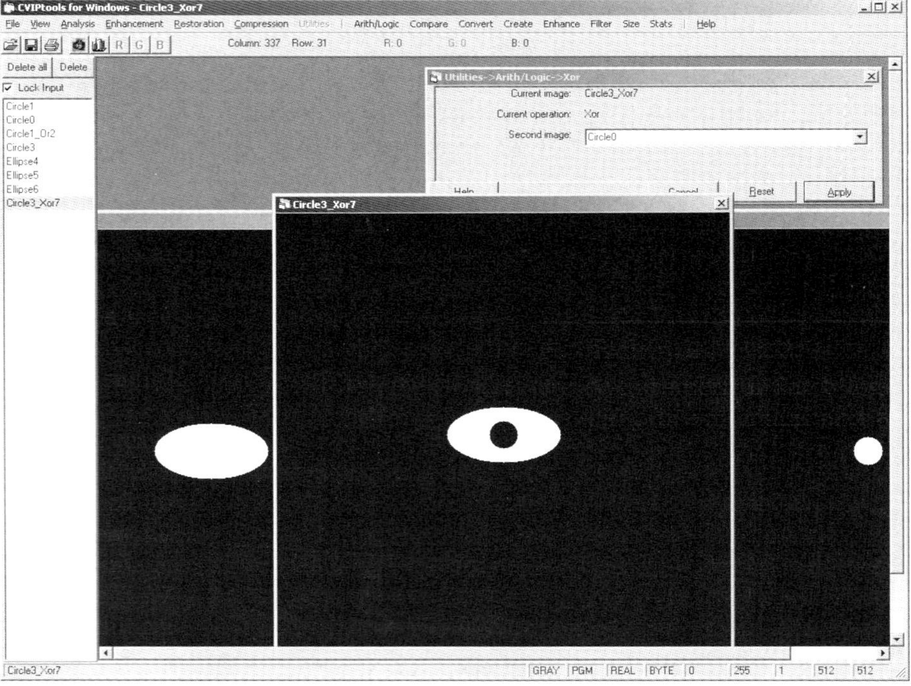

FIGURE 3.3.14
XOR of circle and ellipse to create a new object—ellipses with hole.

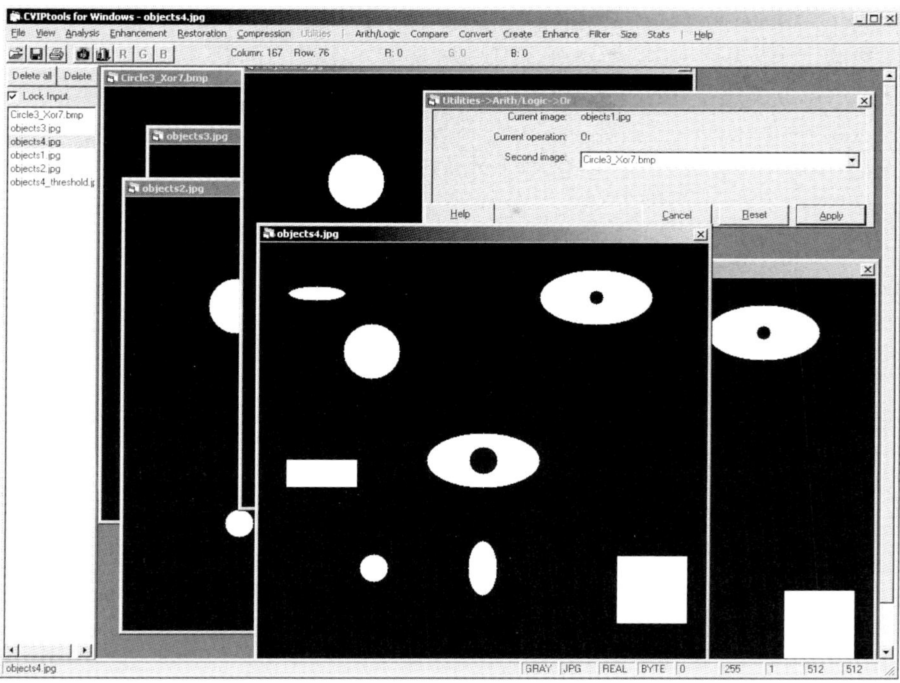

FIGURE 3.3.15
Composite image created by ORing individual object together.

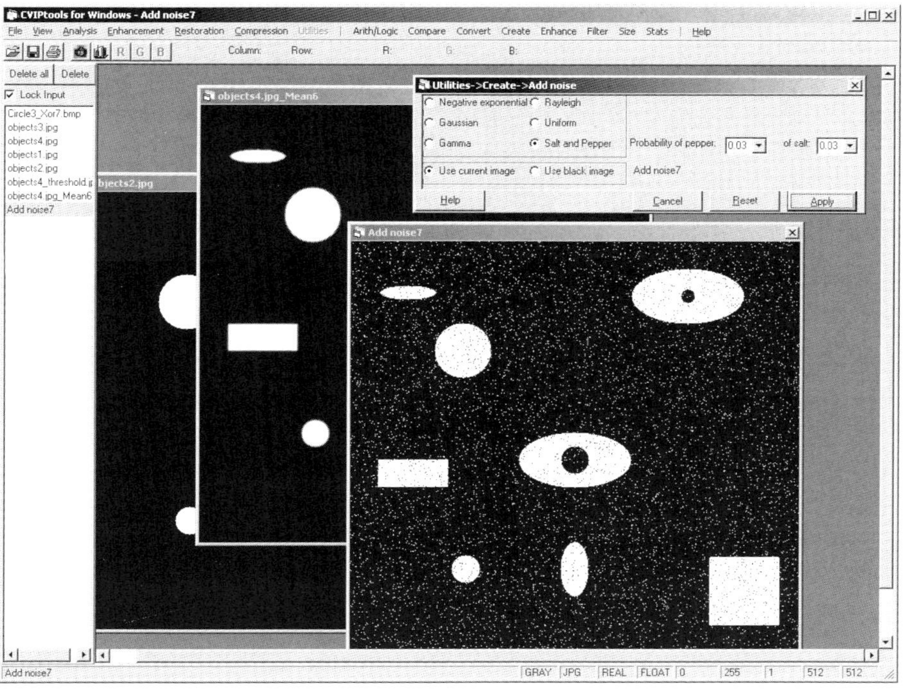

FIGURE 3.3.16
Adding salt and pepper noise to the blurred composite image.

4. *Feature Extraction*: area, center of area, axis of least second moment, projections, and Euler number.
5. *Feature Analysis*: we will do this manually by examining the feature file
6. *Application feedback*: are we successful in developing an algorithm that will identify the objects? If not, go back to Step 1 and modify the algorithm based on our results.

For Step 1, select *Filter->Median*. Apply this to our composite image. This is shown in Figure 3.3.17. Here we see that the noise has been successfully removed. For Step 2 we want to find a proper threshold so that the blurring is mitigated and the objects are clearly defined. This is done with *Convert->Binary Threshold*. After some experimentation we determine a threshold of 155 gives us the desired results, shown in Figure 3.3.18. Now we are ready to extract the features.

From the main CVIPtools window, select the *Analysis* window, and select the *Features* tab. For the original image we want to use the image after noise removal, and for the segmented image we will use the image after thresholding. Next, we enter a feature file name with the 'Save as' button, a class (circle, ellipse, rectangle, or ellipse_hole), and any coordinates within the object of interest. The coordinates can also be selected with a mouse click on the object in the original image. Next, we select the features of interest by clicking on the checkboxes for area, centroid, orientation (axis of least second moment), Euler number, and projections (see Figure 3.3.19). Note that for the projections feature we need to specify the normalizing height and width. The default normalizing size is 10 × 10 and will shrink the object into a 10 × 10 box and then extract the projections. This is done so that the number of projections does not get too large, and so that the values will relate to object shape and not object size.

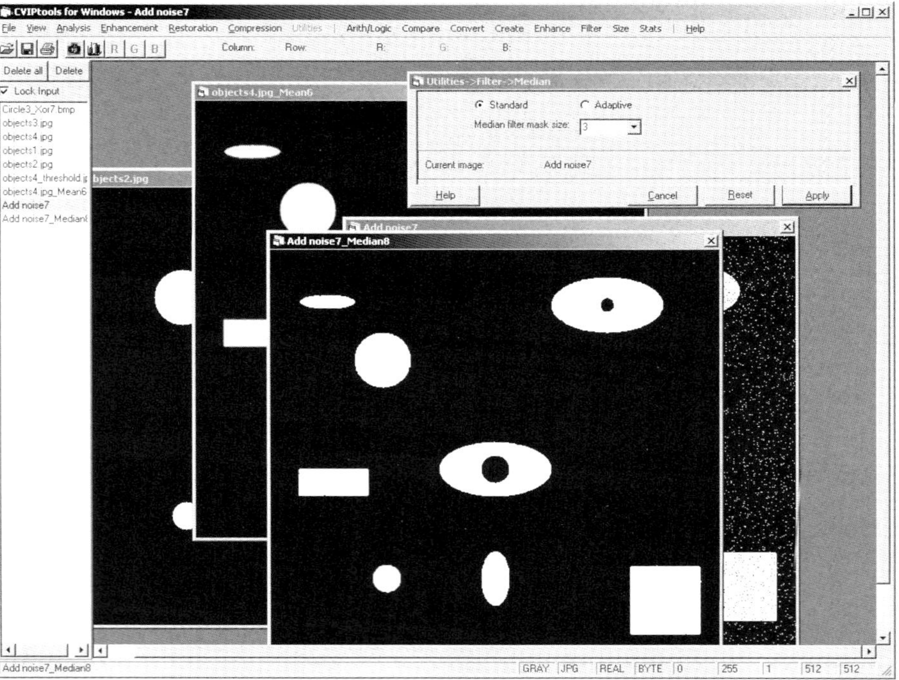

FIGURE 3.3.17
Blurry, noisy composite image after median filtering.

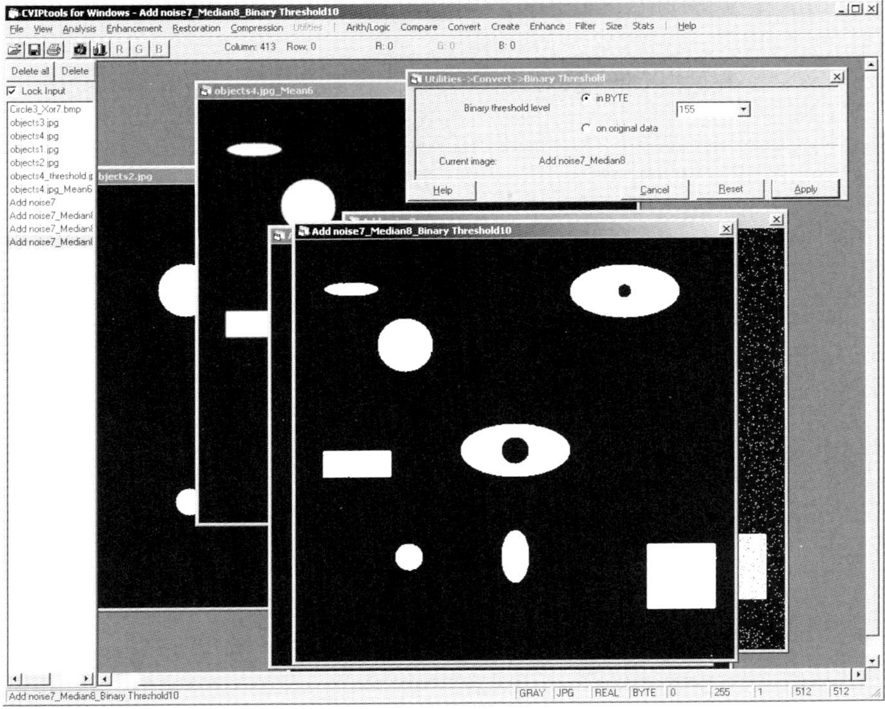

FIGURE 3.3.18
Image after thresholding, note the output shapes still have some minor distortion.

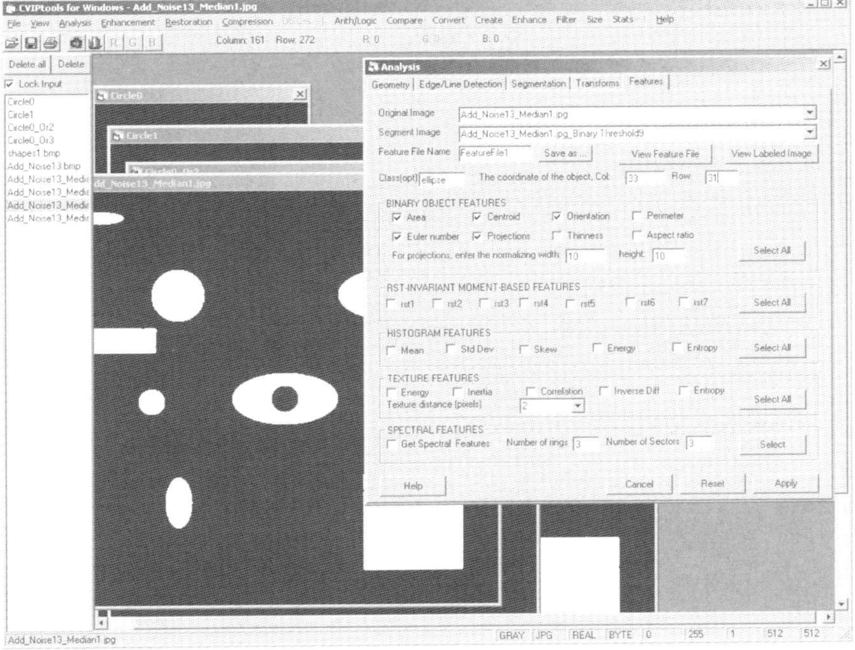

FIGURE 3.3.19
Feature tab with features selected.

Now, we are ready to extract the features by clicking on the *Apply* button on the lower right of the *Analysis* window. We do this for each object in our image by selecting the object coordinates and typing in the desired class name. When we have extracted the features for all the objects we can look at the feature file with the *View Feature File* button. The feature file, shown in Figure 3.3.20, consists of two main parts: (1) the header, which lists the details for each sample, and (2) the data for the extracted feature samples. Every feature file includes the header with the image name, and the object's row and column coordinates, and a list of the selected features. In the data for the extracted samples, each file is separated by a space, and each sample entry is separated by a new line. The task now is to examine this data and look for features that will differentiate the classes.

First, we deduce that area and centroid may be useful for some applications, but will not help us in classification. In Table 3.1 the data from the feature file for the orientation,

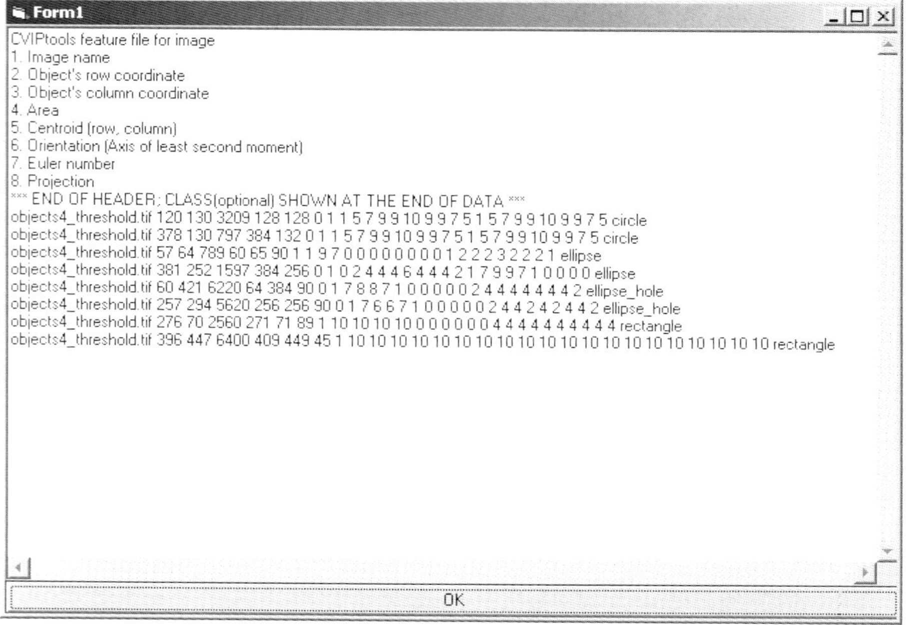
Form1

CVIPtools feature file for image
1. Image name
2. Object's row coordinate
3. Object's column coordinate
4. Area
5. Centroid (row, column)
6. Orientation (Axis of least second moment)
7. Euler number
8. Projection
*** END OF HEADER; CLASS(optional) SHOWN AT THE END OF DATA ***
objects4_threshold.tif 120 130 3209 128 128 0 1 1 5 7 9 9 10 9 9 7 5 1 5 7 9 9 10 9 9 7 5 circle
objects4_threshold.tif 378 130 797 384 132 0 1 1 5 7 9 9 10 9 9 7 5 1 5 7 9 9 10 9 9 7 5 circle
objects4_threshold.tif 57 64 789 60 65 90 1 1 9 7 0 0 0 0 0 0 0 0 1 2 2 2 3 2 2 2 1 ellipse
objects4_threshold.tif 381 252 1597 384 256 0 1 0 2 4 4 4 6 4 4 4 2 1 7 9 9 7 1 0 0 0 0 ellipse
objects4_threshold.tif 60 421 6220 64 384 90 0 1 7 8 8 7 1 0 0 0 0 0 2 4 4 4 4 4 4 4 2 ellipse_hole
objects4_threshold.tif 257 294 5620 256 256 90 0 1 7 6 6 7 1 0 0 0 0 0 2 4 4 2 4 2 4 4 2 ellipse_hole
objects4_threshold.tif 276 70 2560 271 71 89 1 10 10 10 10 0 0 0 0 0 0 0 4 4 4 4 4 4 4 4 4 rectangle
objects4_threshold.tif 396 447 6400 409 449 45 1 10 10 10 10 10 10 10 10 10 10 10 10 10 10 10 10 10 10 10 10 rectangle

OK

FIGURE 3.3.20
CVIPtools features file.

TABLE 3.1
Feature Data

Object type	Orientation	Euler Number	Projections
Circle	0 degrees	1	1 5 7 9 9 10 9 9 7 5 1 5 7 9 9 10 9 9 7 5
Circle	0 degrees	1	1 5 7 9 9 10 9 9 7 5 1 5 7 9 9 10 9 9 7 5
Ellipse	90 degrees	1	1 9 7 0 0 0 0 0 0 0 0 1 2 2 2 3 2 2 2 1
Ellipse	0 degrees	1	0 2 4 4 4 6 4 4 4 2 1 7 9 9 7 1 0 0 0 0
Ellipse_hole	90 degrees	0	1 7 8 8 7 1 0 0 0 0 0 2 4 4 4 4 4 4 4 2
Ellipse_hole	90 degrees	0	1 7 6 6 7 1 0 0 0 0 0 2 4 4 2 4 2 4 4 2
Rectangle	89 degrees	1	10 10 10 10 0 0 0 0 0 0 0 4 4 4 4 4 4 4 4 4
Rectangle	45 degrees	1	10 10 10 10 10 10 10 10 10 10 10 10 10 10 10 10 10 10 10 10

Euler number, and projection data is shown. The next observation is that the orientation will not help in the classification of these objects; although it, along with the area and centroid, would be useful to control a robot in finding and placing the objects.

Next, we observe the Euler number feature will identify the class ellipse_hole, since it is 0 for this class and 1 for all others. Upon close examination of the projections, we can see that they can be used to differentiate the circles, ellipses, and rectangle. In general, the ellipses have some zeros and increasing and decreasing projections, the circles have increasing and decreasing projections, and the rectangles have constant projections, possibly with some zeros. Thus, we have our algorithm:

```
If Euler number = 0
  Then Object = ellipse_hole
Else (Euler number = 1)
  If projections are increasing and decreasing
    If projections has zeros
      Then Object = ellipse
    Else (projections has no zeros)
      Then Object = circle
  Else (projections not increasing and decreasing)
    Then Object = rectangle
```

What we have done is develop a classification algorithm by the use of a training set. A *training set* is a set of sample images used to develop an algorithm. To complete Step 6 in the image analysis process, application feedback, we need to generate some test images. This *test set* of images is then used to see how well the algorithm actually works on a different set of images. The idea is that these results will simulate the real application in practice, and will not be biased by the training process—it is easy to get 100% success on the training set! Success on the test set increases our confidence that the algorithm will work in practice. Test sets can be created with CVIPtools and it is left as an exercise for the reader to validate the algorithm. Pattern classification will be explored further in Chapter 6.

3.4 Key Points

IMAGE ANALYSIS PROCESS MODEL

Image analysis—manipulating the image data to determine exactly the information necessary to help solve a computer imaging problem, primarily a data reduction process

Image analysis process model (see Figure 3.1.3)—consists of three primary stages: (1) preprocessing, (2) data reduction, (3) feature analysis

- **Preprocessing**—used to remove noise and artifacts, visually irrelevant information; preliminary data reduction
- **Noise**—unwanted information from the data acquisition process
- **Data reduction**—reducing data in the spatial domain or transforming into the spectral domain, followed by filtering and feature extraction
- **Feature analysis**—examining the extracted features to see how well they will solve the application problem

- **Application feedback loop**—key aspect of the image analysis process that incorporates application-based information in the development process

PREPROCESSING

Region of interest geometry—to inspect more closely a specific area of an image

- **Crop**—process of selecting a portion of an image, a subimage, and cutting it away from the image
- **Zoom**—enlarging a section of an image. Zero-order hold or first order may be used

  Zero-order hold—repeating pixels
  First-order hold—linear interpolation between adjacent pixels

- **Convolution**—overlay the mask, multiply coincident values, sum results, move to next pixel, across entire image (see Figure 3.2.3), equation:

$$\sum_{x=-\infty}^{\infty} \sum_{y=-\infty}^{\infty} I(r-x, c-y)M(x, y)$$

- **Vector inner product**—multiplying coincident terms of two vectors and summing results
- **Translation**—moving the image data along the row and/or column axes, equations:

$$r' = r + r_0$$
$$c' = c + c_0$$

- **Rotation**—clockwise rotation through a fixed angle θ, given by these equations:

$$\hat{r} = r(\cos\theta) + c(\sin\theta)$$
$$\hat{c} = -r(\sin\theta) + c(\cos\theta)$$

Arithmetic and logic operations—performed on a pixel by pixel basis; arithmetic operations: add, subtract, multiply, divide; logic operations: AND, OR, NOT

- **Addition**—used to combine information in two images, applications include creating models for restoration algorithm development, image sharpening algorithms, and special effects such as image morphing
- **Subtraction**—used for motion detection and background subtraction, applications include object tracking, medical imaging
- **Multiplication**—used to brighten or darken an image, or to combine two images
- **Division**—used to darken or brighten an image
- **AND**—a logical operation performed on a pixel by pixel basis, using two images, by a logical AND of the corresponding bits in each corresponding pixel, defined for BYTE-type images; used to combine two images or for image masking
- **Image masking**—extracting a portion of an image with an AND or OR operation using a binary image mask; masking out image artifacts by setting to 0
- **OR**—a logical operation performed on a pixel by pixel basis, using two images, by a logical OR of the corresponding bits in each corresponding pixel, defined for BYTE-type images; used to combine two images or for image masking
- **NOT**—creates a negative on an image by performing a logical NOT on each bit

Spatial filters—operate on the image data by considering small neighborhoods in an image, such as 3×3, 5×5, etc., and returning a result based on a linear or nonlinear operation; moving sequentially across and down the entire image

- **Linear filters**—can be implemented with a convolution mask, since the output is a linear combination of the (neighborhood) inputs
- **Mask coefficients**—all positive will blur an image, alternating positive and negative will sharpen an image; if they sum to one will tend to retain original image brightness, if they sum to zero will tend to lose original image brightness
- **Mean filters**—averaging filters, will blur an image, all mask coefficients are positive
- **Median filter**—sorts the pixel values in a small neighborhood and replaces the center pixel with the middle value in the sorted list, is a nonlinear filter
- **Nonlinear filter**—cannot be implemented with a convolution mask since the result cannot be represented as a weighted sum of the neighborhood pixel values
- **Enhancement filters**—linear filters, the convolution masks have alternating positive and negative coefficients; will enhance image details via image sharpening, able to enhance details in a specific direction by careful mask selection
- **Laplacian filters**—enhancement filters with convolution masks of alternating positive and negative coefficients, will bring out image details equally in all directions
- **Difference (emboss) filters**—enhancement filters with convolution masks of alternating positive and negative coefficients, will bring out image details in a specific direction based on the mask used

Image quantization—the process of reducing image data by removing some of the detail information by mapping groups of data points to a single point, performed in the spatial or gray level domain

- **Gray level reduction**—reducing the number of gray levels, typically from 256 levels for 8-bit per pixel data to fewer than 8 bits, can be performed with AND or OR masks (see examples)

    Thresholding—the simplest method of gray level reduction performed by setting a threshold value and setting all pixels above it to "1" (typically 255), and those below it to "0", output is a binary image
    False contouring—artificial lines that appear in images with reduced number of gray levels (Figure 3.2.15)
    IGS—improved gray scale, a method to visually improve the results of gray level reduction by adding a random number to each pixel value before the quantization (Figure 3.2.16)
    Halftoning/dithering—methods for reducing the number of gray levels by creating dot patterns or dither patterns to represent various gray levels, reduces effective spatial resolution also (Figure 3.2.17)
    Uniform bin width quantization—the size of the bins for quantization is equal (Figure 3.2.18)
    Variable bin width quantization—the size of the bins for quantization is not equal but may be assigned on an application specific basis (Figure 3.2.18)

- **Spatial quantization**—reducing image size by taking groups of spatially adjacent pixels and mapping them to one pixel, can be done by: (1) averaging, (2) median, or (3) decimation (Figure 3.2.20)

    Averaging—performing size reduction by averaging groups of pixels and replacing the group by the average

Median—sorting the pixel gray values in small neighborhood and replacing the neighborhood with the middle value
Decimation—also known as sub-sampling, reduces image size by eliminating rows and columns
Anti-aliasing filtering—a technique to improve image quality by averaging before decimation (Figure 3.2.21)

BINARY IMAGE ANALYSIS

- **Threshold via histogram**—examining the histogram to find clusters by looking at peaks and valleys and thresholding the image gray values at one of the valleys in the histogram, effects of lighting, and background contrast is important (Figure 3.3.1)
- **Histogram**—a plot of gray values versus numbers of pixels at each gray value (Figure 3.3.2)
- **Connectivity**—defining how pixels are connected by selecting which of the eight neighboring pixels, assuming a square gird, are connected to the center pixel (Figure 3.3.3)

  Four-connectivity—the connected neighbors are the two horizontal neighbors, to the left and right, and the two vertical neighbors, above and below
  Eight-connectivity—horizontal, vertical and all diagonal neighbors are considered connected
  Six-connectivity—horizontal, vertical and two diagonal neighbors are considered connected, this type of connectivity is used in this book
  Connectivity dilemma—the dilemma that arises when we use four- or eight-connectivity for both objects and background where closed curves do not separate the background (eight-connectivity), or we do not have a closed curve and the background is separated (four-connectivity)
- **Labeling**—the process of assigning labels to connected objects in an image

  Labeling algorithm flowchart—see Figure 3.3.4
  UPDATE—a method needed in a sequential labeling algorithm to deal with the situation when two pixels are found connected, but connected neighbors have different labels

- **Binary object features**—features that can be extracted from labeled objects in binary images which can be used to classify the objects

  Area—the size in pixels of a binary object, indicating the relative size of the object, found by summing all the pixels in the object:

$$A_i = \sum_{r=0}^{N-1} \sum_{c=0}^{N-1} I_i(r,c)$$

  Center of area (centroid): the midpoint along each row and column axis corresponding to the "middle" based on the spatial distribution of pixels within the object, used to locate the object spatially, defined by:

$$\bar{r}_i = \frac{1}{A_i} \sum_{r=0}^{N-1} \sum_{c=0}^{N-1} r\, I_i(r,c); \qquad \bar{c}_i = \frac{1}{A_i} \sum_{r=0}^{N-1} \sum_{c=0}^{N-1} c\, I_i(r,c)$$

Axis of least second moment—defines the object's orientation, given by:

$$\tan(2\theta_i) = 2\frac{\sum_{r=0}^{N-1}\sum_{c=0}^{N-1} rcI_i(r,c)}{\sum_{r=0}^{N-1}\sum_{c=0}^{N-1} r^2 I_i(r,c) - \sum_{r=0}^{N-1}\sum_{c=0}^{N-1} c^2 I_i(r,c)}$$

Projections—found by summing pixels along each row or column, provides information about an object's shape, and provide simpler equations for center of area

Horizontal projection—sum of pixels along the rows
Vertical projection—sum of pixels along the columns

Euler number—defined as the number of objects minus the number of holes, or the number of convexities minus the number of concavities

- **Binary object classification**—the process of identifying binary objects through application of the image analysis process given in Figure 3.1.3, consisting of the following steps: (1) Preprocessing, (2) Thresholding, (3) Filtering (optional), (4) Feature extraction, (5) Feature analysis, (6) Application feedback

3.5 References and Further Reading

For more information on image preprocessing, see [Sonka/Hlavac/Boyle 99]. The method of zooming via convolution masks is described in [Sid-Ahmed 95]. For spatial filtering, [Gonzalez/Woods 02], [Sonka/Hlavac/Boyle 99], [Jain/Kasturi/Schnuck 95], [Galbiati 90], [Pratt 91], and [Myler/Weeks 93] contain additional information. More on connectivity can be found in [Gonzalez/Woods 02], [Jain/Kasturi/Schnuck 95], [Horn 86], and [Haralick/Shapiro 92]. For more background on Euler number see [Horn 86]. More information on halftoning and dithering can be found in [Watt/Policarpo 98], [Hill 90], and [Durrett 87]. More labeling algorithms can be found in [Shapiro/Stockman 01], [Sonka/Hlavac/Boyle 99], and [Jain/Kastuiri/Schnuck 95]. More on thresholding techniques can be found in [Shapiro/Stockman 01] and [Davies 97]. Additional information on the processing of binary images can be found in [Shapiro/Stockman 01], [Jain/Kasturi/Schunck 95], [Davies 97] and [Russ 99].

Davies, E.R., *Machine Vision*, San Diego, CA: Academic Press, 1997.
Durrett, H.J., Editor, *Color and the Computer*, Boston, MA, Academic Press, 1987.
Galbiati, L.J., *Machine Vision and Digital Image Processing Fundamentals*, Englewood Cliffs, NJ: Prentice Hall, 1990.
Gonzalez, R.C., Woods, R.E., *Digital Image Processing*, Upper Saddle River, NJ: Prentice Hall, 2002.
Haralick, R.M., Shapiro, L.G., *Computer and Robot Vision*, Reading, MA: Addison-Wesley, 1992.
Hill, F.S., *Computer Graphics*, New York, NY: Macmillan Publishing Company, 1990.
Horn, B.K.P., *Robot Vision*, Cambridge, MA: The MIT Press, 1986.
Jain, R., Kasturi, R., Schnuck, B.G., *Machine Vision*, NY: McGraw Hill, 1995.
Myler, H.R., Weeks, A.R., *Computer Imaging Recipes in C*, Englewood Cliffs, NJ: Prentice Hall, 1993.
Pratt, W.K., *Digital Image Processing*, NY: Wiley, 1991.
Russ, J.C., *The Image Processing Handbook*, Boca Raton, FL: CRC Press, 1999.
Shapiro, L., Stockman, G., *Computer Vision*, Upper Saddle River, NJ: Prentice Hall, 2001.
Sid-Ahmed, M.A. *Image Processing: Theory, Algorithms, and Architectures*, NY: McGraw Hill, 1995.
Sonka, M., Hlavac, V., Boyle, R., *Image Processing, Analysis and Machine Vision*, Pacific Grove, CA: Brooks/Cole Publishing Company, 1999.
Watt, A., Policarpo, F., *The Computer Image*, New York, NY: Addison-Wesley, 1998.

3.6 Exercises

1. What is image analysis? How is it used in computer vision? How is it used in image processing? Give examples of each.

2. What are the three primary stages of image analysis? The second stage can be done in two different domains; what are they?

3. Draw a detailed figure of the image analysis process. Explain each block. Why do we need feedback?

4. List and describe the image geometry operations used in preprocessing for image analysis. Run CVIPtools and experiment with the functions under *Analysis->-Geometry*, and *Utilties->Size*.

5. Use zero-order to increase the size of following image by a factor of 2.

$$\begin{bmatrix} 6 & 7 & 8 \\ 2 & 6 & 4 \\ 6 & 3 & 8 \end{bmatrix}$$

6. Use first-order hold to increase the image by a factor of about 3. Apply the method which will increase the image size to $K(N-1)+1$. What is the resulting image size? Is this "about a factor of 3?" Why or why not?

$$\begin{bmatrix} 2 & 5 & 9 \\ 5 & 6 & 4 \\ 9 & 3 & 8 \end{bmatrix}$$

7. We want to translate image1 by 45 columns to the right and 18 rows up; what are the new coordinates for the point $(r, c) = (120, 22)$? We want to rotate image2 50 degrees in the clockwise direction, what are the new coordinates for the point $(r, c) = (42,100)$? We want to rotate *and* translate image3 the same as we did for image1 and image2; what are the new coordinates for the point $(r, c) = (100, 66)$? Use CVIPtools to verify your answers; note that the origin given by the equations is different than the output equation in the CVIPtools image.

8. Subtract the following two images. What is an example of an application for image subtraction? How do we display negative numbers?

$$\begin{bmatrix} 2 & 5 & 9 \\ 5 & 6 & 4 \\ 9 & 3 & 8 \end{bmatrix} \begin{bmatrix} 1 & 2 & 7 \\ 3 & 4 & 2 \\ 9 & 3 & 9 \end{bmatrix}$$

In CVIPtools subtract a dark image from a brighter image. Observe the data range shown in the lower right of the main window. What does CVIPtools do with the negative numbers for display?

9. Perform a logical OR with the following two images. What can this operation be used for?

$$\begin{bmatrix} 2 & 5 & 9 \\ 5 & 6 & 4 \\ 9 & 3 & 8 \end{bmatrix} \begin{bmatrix} 1 & 2 & 7 \\ 3 & 4 & 2 \\ 9 & 3 & 9 \end{bmatrix}$$

Use CVIPtools to OR images together. During this process consider potential applications.

10. How is image masking performed? What are its uses? Use *Utilities->Create->Border Mask* in CVIPtools to create mask images and then use AND to mask the original image.

11. What does a NOT operation do to the appearance of an image? Perform the NOT operation in CVIPtools on a color image, are the results what you expected? Multiply an image by 1.8 without byte clipping. What is the output image data type? Now perform a NOT on the image multiplied by 1.8. What is the data type of this output image? Explain.

12. What does a convolution filter do that has all positive coefficients? What does a convolution filter do that has alternating positive and negative coefficients? How about one where the coefficients sum to zero? What happens if a filter mask coefficients sum to one? Use CVIPtools *Utilities->Filter->Specify a filter* to verify your answers.

13. Are convolution filters linear? Name a nonlinear filter. Given the following 3×3 neighborhood in an image, what is the result of applying a 3×3 median filter to the center pixel?

$$\begin{bmatrix} 1 & 2 & 7 \\ 3 & 4 & 2 \\ 9 & 3 & 9 \end{bmatrix}$$

Use CVIPtools *Utilities->Create->Add Noise* to add salt and pepper noise to an image. Now, perform a median filter on the noisy image using *Utilities->Filter->Median*. How does it affect the appearance of the image as you increase mask size of the filter?

14. What are the coefficients for a typical 3×3 mean convolution filter? What are the coefficients for a typical 3×3 enhancement filter? Use CVIPtools to verify your results.

15. Why does Figure 3.2.7c look mostly gray? (hint: *remap*) Since the gray level of the car is similar to that of the grass, we lost the lower half of the car in its second location in the results. What could we do to avoid this? Open the image in CVIPtools and experiment with various threshold values.

16. What are example applications for image multiplication and division? Demonstrate your examples with CVIPtools.

17. What is the bit string we would use for an AND mask to reduce 8-bit image information to 64 gray levels? Does this map the data to the low or high end of the range? Use CVIPtools *Utilities->Convert->Gray Level Quantization* to reduce the number of gray levels of an 8-bit image to 32 gray levels. Look at the histogram of the output image by selecting the bar graph icon just to the left of the RGB icons. Does CVIPtools map the output data to the low or high end of the range?

18. What is false contouring? How can we visually improve this effect? Explain. Use CVIPtools to reduce the number of gray levels on an 8-bit image to 8 gray levels, select standard method. Now, select the IGS method and compare the results.

19. What is halftoning and dithering? Why is it used? Use CVIPtools *Utilities->Convert->Halftone* and compare the various methods. Which one do you think works the best? Do you think this is true for all images?

20. Describe variable bin width quantization. Why is it used?

21. Describe the three methods used for spatial reduction. Which method do you think is the fastest? The slowest? When using the decimation technique how can we improve the results? Use CVIPtools *Utilities->Size ->Spatial Quant* to compare the three methods.

22. What is a histogram? How can it be useful?

23. In CVIPtools you can threshold an image with *Utilties->Convert->Binary Threshold*. Using CVIPtools try to find a good threshold to separate the object from background in Figure 3.3.1c. Are you successful? Why or why not? Look at the histogram of various images with CVIPtools (the histogram icon looks like a bar graph).

24. Draw a binary image to illustrate the dilemma that arises when using four- or eight-connectivity. Explain three ways to avoid this dilemma. Label all objects and background objects. Remember a connected line should separate the objects on either side of the line.

25. What is the UPDATE block for in the flowchart in Figure 3.3.4?

26. Given an application where we need to control a robotic gripper to pick and place items on an assembly line, what are the most useful binary features?

27. Find the horizontal and vertical projections for the following binary image:

$$\begin{bmatrix} 0 & 1 & 1 & 1 & 0 \\ 0 & 0 & 1 & 1 & 1 \\ 0 & 0 & 0 & 1 & 1 \end{bmatrix}$$

28. The Euler number for an object is equal to the number of objects minus the number of holes, and equal to the number of convexities minus the number of concavities. Is the number of objects necessarily equal to the number of convexities? Is the number of holes necessarily equal to the number of concavities? Explain.

29. Use CVIPtools to create a test set of images for the algorithm developed in Section 3.3.4.Vary the size of objects, the amount of noise added, and the degree of blurring. Extract the features of interest using CVIPtools. Examine the feature file. Does the classification developed in this section work successfully? How is the success rate affected as the amount of blur and added noise is increased?

3.6.1 Programming Exercise: Image Geometry

1. Write a function to implement an image crop. Incorporate this function call into the case statement at the beginning of the CVIPlab program (refer to Chapter 11, as required), so that it can be accessed via the menu (do this for all the functions written).

2. Write a function to implement an image zoom, have the user specify the starting (r, c) coordinates, the height and width and the zoom factor. Let the user specify zero-order or first-order hold.

3. Incorporate the CVIPtools *zoom* (in the Geometry library) into your CVIPlab program. Experiment with enlarging an image by different factors. The minimum and maximum factors allowed are 1 and 10 respectively. You have the option of

choosing the whole of the image, or any particular quadrant, or you can specify the starting row and column, and the width and height for the enlargement of the particular region of the image.

4. Write a function to rotate an image. Experiment with various degrees of rotation. Incorporate the CVIPtools *rotate* (Geometry library) function into your CVIPlab program. Does this differ from how your rotate function works?

5. Incorporate the CVIPtools *crop* and *bilinear_interp* (Geometry lib) into your CVIPlab program. These two will provide similar functionality to the *zoom* function. Compare the results of using *bilinear_interp* to *zoom*. The *zoom* function performs a zero-order hold, while *bilinear_interp* performs a bilinear interpolation, providing a smoother appearance in the resulting image.

6. Enhance your rotate function to select the center portion of rotated image and enlarge it to the original image size.

7. Put the CVIPtools *spatial_quant* into your CVIPlab program. Compare using the three different reduction methods available: average, median, and decimation.

3.6.2 Programming Exercise: Arithmetic/Logic Operations

1. Write C functions that perform the following logical operations on two images: AND, OR, NOT.

2. Write a C function to subtract two images, put this function in a separate file from the logic functions. Initially, use BYTE data types which will result in clipping at zero for negative results. Next, modify the function to use FLOAT data types (use *cast_Image* in the Image library) and then remap when the process is completed (use *remap_Image* in the Mapping library). Note that these two methods will result in different output images.

3. Extend the logic operations to work with data types other than BYTE.

4. Extend the logic operations to include NAND, NOR, and more complex boolean expressions.

5. Extend the subtraction function to perform addition, multiplication and division.

6. Experiment with different methods of handling overflow and underflow with the arithmetic operations.

3.6.3 Programming Exercise: Spatial Filters

1. Write a program to implement spatial convolution masks. Let the user select from one of the following masks:

Mean filter masks:

$$\frac{1}{9}\begin{bmatrix} 1 & 1 & 1 \\ 1 & 1 & 1 \\ 1 & 1 & 1 \end{bmatrix} \quad \frac{1}{10}\begin{bmatrix} 1 & 1 & 1 \\ 1 & 2 & 1 \\ 1 & 1 & 1 \end{bmatrix} \quad \frac{1}{16}\begin{bmatrix} 1 & 2 & 1 \\ 2 & 4 & 2 \\ 1 & 2 & 1 \end{bmatrix}$$

Enhancement filter masks:

$$\begin{bmatrix} -1 & -1 & -1 \\ -1 & 9 & -1 \\ -1 & -1 & -1 \end{bmatrix} \quad \begin{bmatrix} 1 & -1 & 1 \\ -2 & 5 & -2 \\ 1 & -2 & 1 \end{bmatrix} \quad \begin{bmatrix} 0 & -2 & 0 \\ -1 & 5 & -1 \\ 0 & -1 & 0 \end{bmatrix}$$

2. Modify the program to allow the user to input the coefficients for a 3×3 mask.
3. Experiment with using the masks. Try images with and without added noise.
4. Modify the program to handle larger masks.
5. Write a median filtering function. Compare the median filter to the mean filter masks for image smoothing.
6. Incorporate the CVIPtools function *median_filter* (SpatialFilter library) into your CVIPlab program. Is it faster or slower than your median filtering function?

3.6.4. Programming Exercise: Image Quantization

1. Write a function to reduce the number of gray levels in an image by uniform quantization. Allow the user to specify: (1) how many gray levels in the output image, and (2) to map the gray levels to the beginning, middle, or end of the range.
2. Write a function to reduce the number of gray levels in an image by non-uniform quantization. Allow the user to specify the input ranges and the output value for up to eight output gray levels
3. Write a function to perform spatial quantization. Allow the user to specify the method: decimation, median, averaging. Incorporate an anti-aliasing filter option for the decimation method.
4. Write a function to perform IGS quantization, see the Noise library in Chapter 12 for the noise functions.

3.6.5 Programming Exercise: Connectivity and Labeling

1. Write a function to implement the labeling algorithm described. You may assume that row and column 0 do not contain objects, so start the scan with row 1 and column 1. Follow the flowchart for the labeling algorithm. Define a two-dimensional array for the labels using a fixed size, for example for a 256×256 image:

int label[256][256]; / ∗ declaration ∗ /

NOTE: Be sure to initialize the array elements (for example, via "for" loops), if needed. When the memory is allocated for the array it may contain garbage depending on the compiler and the operating system.

2. Test this function using images you create with CVIPtools. Use the *Utilities* to create test images with the *Create* option (*Utilties->Create*). To create images with multiple objects, use the *AND* and *OR* logic functions available from *Utilties->Arith/Logic*.
3. Once you are certain that your function implements the algorithm correctly, modify the label function using a Matrix structure for the label array. This will allow

the use of any size image, without the need to change the size of the array. This is done as follows:

```
Matrix * label_ptr; /* declaration of pointer to Matrix data structure */
int* * label;   /* declaration of pointer to matrix data */
...
label_ptr = new_Matrix(no_of_rows, no_of_cols,
CVIP_INTEGER, REAL); /* allocating the memory for the matrix structure */
label = (int * *) getData_Matrix(label_ptr); /* getting the matrix}
data into the label array */
...
label[r][c]                  /* accessing the array elements */
...
delete_Matrix(label_ptr);    /* freeing the memory space used by the matrix */
```

4. Modify the function so that it will handle objects on the edges of the image.
5. Modify the Update function (see flowchart, Figure 3.3.4 and 3.3.5) so that it does not require multiple image scans, for example keep a linked list or table of equivalent labels and rescan the image only once (after all the labeling is done).
6. Modify the function to work with gray level images.
7. Modify the function to work with color images.
8. Modify the function to handle any number of objects.
9. Modify the function so that it will output the labeled image (that is, the label array written to disk as an image, with appropriate gray levels to make all the objects visible).

3.6.6 Programming Exercise: Binary Object Features

1. Write a C function to find the area, and coordinates of center of area of a binary image. Assume the image only contains one object. Remember that the value that represents "1" for the binary images are actually 255, and "0" is 0.
2. Test this function using images you create with CVIPtools. Use *Utilities* to create test images with the *Create* option (*Utilties->Create*). To create images with multiple objects, use the *AND* and *OR* logic functions available from *Utilties->Arith/Logic*.
3. Modify the label function so that it will find the area and center of area for each object. Note that you can modify the variables for area and center of area to be arrays and use the label as the index into the array (be sure to initialize the array elements to 0). The information for each object should be printed to the screen, along with the object number.
4. Write a C function to find the number of upstream facing convexities (X), upstream facing concavities (V), and the Euler number for a binary image. Use the method discussed in section 3.3.3, assuming six-connectivity. The function should display the following:

$$\text{The number of upstream facing convexities} =< X >$$

$$\text{The number of upstream facing convexities} =< V >$$

$$\text{The Euler number for the image} =< X - V >$$

5. Test this function using images you create with CVIPtools. Use *Utilities* to create test images with the *Create* option (*Utilities->Create*). To create images with multiple objects, use the *AND* and *OR* logic functions available from *Utilities->Arith/Logic*.

6. Modify the Euler function to find the Euler number for each object in a binary image containing multiple objects.

7. Modify your Euler function to handle other connectivity types (four, eight and four/eight).

8. Modify your functions to handle gray level images.

9. Modify your functions to handle color images.

4

Segmentation and Edge/Line Detection

4.1 Introduction and Overview

The image analysis process requires us to take vast amounts of low level pixel data and extract useful information. In this chapter we will explore methods to divide the image into meaningful regions which represent higher level information. We will discuss edge detection, line detection, and finally image segmentation. We will see that edge and line detection are important steps in one category of image segmentation methods.

The goal of image segmentation is to find regions that represent objects or meaningful parts of objects. Division of the image into regions corresponding to objects of interest is necessary before any processing can be done at a level higher than that of the pixel. Identification of real objects, pseudo-objects, shadows, or actually finding anything of interest within the image, requires some form of segmentation.

Image segmentation methods will look for objects that either have some measure of homogeneity within themselves, or have some measure of contrast with the objects on their border. Most image segmentation algorithms are modifications, extensions, or combinations of these two basic concepts. The homogeneity and contrast measures can include features such as gray level, color, and texture. Once we have performed some preliminary segmentation we may incorporate higher-level object properties, such as shape or color features, into the segmentation process.

We can divide image segmentation techniques into three main categories (see Figure 4.1.1): (1) region growing and shrinking, (2) clustering methods, and (3) boundary detection. The region growing and shrinking methods use the row and column, (r, c), based image domain; while the clustering techniques can be applied to any domain, such as any N-dimensional color or feature space, whose components may even include the spatial domain's (r, c) coordinates. From this perspective, the region growing and shrinking category can be considered a subset of the clustering methods, but is limited to the spatial domain. We separate them here since the spatial domain is of primary significance in images. The boundary detection methods are extensions of the edge detection techniques.

Edge detection techniques are discussed in Section 4.2, as well as metrics to measure edge detector performance. Section 4.2 concludes with a discussion of the Hough transform for line finding, with a brief consideration of its extension to curves. Section 4.3 will explore various representative examples of the many image segmentation algorithms and this chapter concludes with a discussion of morphological filtering in Section 4.3.5.

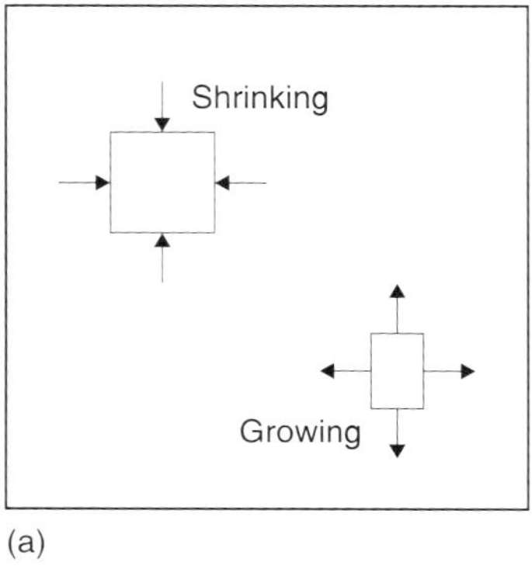

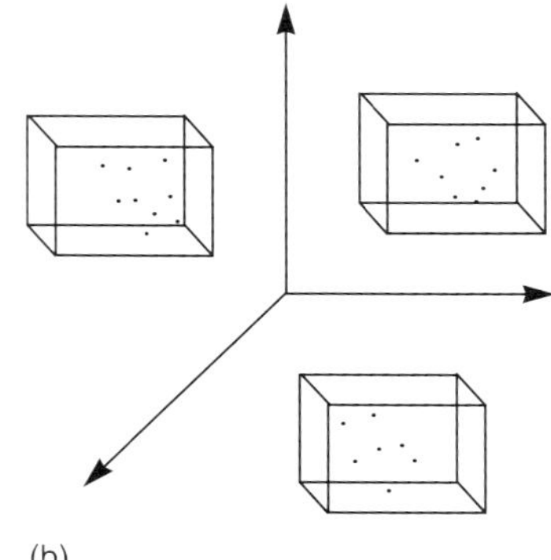

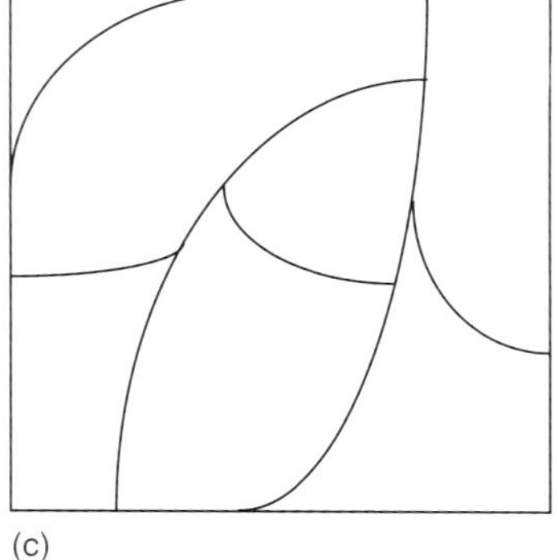

FIGURE 4.1.1
Image segmentation categories. (a) Region growing/shrinking is performed by finding homogeneous regions and changing them until they no longer meet the homogeneity criteria. (b) Clustering looks for data that can be grouped in domains other than the spatial domain. (c) Boundary detection is often achieved using a differentiation operator to find lines or edges, followed by postprocessing to connect the points into borders.

4.2 Edge/Line Detection

The edge and line detection operators presented here represent the various types of operators in use today. Many are implemented with convolution masks, and most are based on discrete approximations to differential operators. Differential operations measure the rate of change in a function, in this case, the image brightness function. A large change in image brightness over a short spatial distance indicates the presence of an edge. Some edge detection operators return orientation information (information about the direction of the edge), while others only return information about the existence of an edge at each point. Also included in this section is a special transform, the Hough transform, which is specifically defined to find lines.

Edge detection methods are used as a first step in the line detection process. Edge detection is also used to find complex object boundaries by marking potential edge points corresponding to places in an image where rapid changes in brightness occur. After these edge points have been marked, they can be merged to form lines and object outlines. Often people are confused about the difference between an edge and a line. This is illustrated in Figure 4.2.1 where we see that an edge occurs at a point and is perpendicular to the line. The edge is where the sudden change occurs, and a line or curve is a continuous collection of edge points along a certain direction.

With many of the edge detection operators, noise in the image can create problems. That is why it is best to preprocess the image to eliminate, or at least minimize, noise

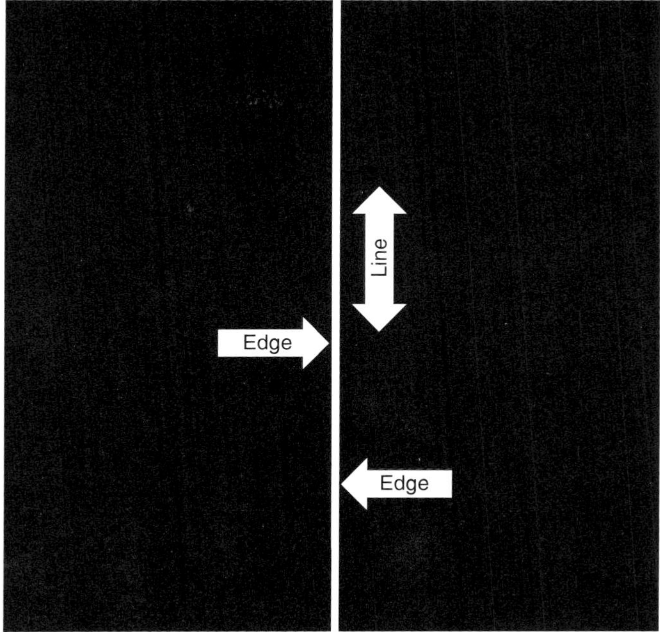

FIGURE 4.2.1
Edges and lines are perpendicular. The line shown here is vertical and the edge direction is horizontal. In this case the transition from black to white occurs along a row, this is the edge direction, but the line is vertical along a column.

effects. To deal with noise effects we must make tradeoffs between the sensitivity and the accuracy of an edge detector. For example, if the parameters are set so that the edge detector is very sensitive, it will tend to find many potential edge points that are attributable to noise. If we make it less sensitive, it may miss valid edges. The parameters that we can set include the size of the edge detection mask and the value of the gray level threshold. A larger mask or a higher gray level threshold will tend to reduce noise effects, but may result in a loss of valid edge points. The tradeoff between sensitivity and accuracy is illustrated in Figure 4.2.2.

Edge detection operators are based on the idea that edge information in an image is found by looking at the relationship a pixel has with its neighbors. If a pixel's gray level value is similar to those around it, there is probably not an edge at that point. However, if a pixel has neighbors with widely varying gray levels, it may represent an edge point. In other words, an edge is defined by a discontinuity in gray level values. Ideally, an edge separates two distinct objects. In practice, apparent edges are caused by changes in color, texture, or by the specific lighting conditions present during the image acquisition process. This means that what we refer to as image objects may actually be only parts of the objects in the real world, see Figure 4.2.3.

Figure 4.2.4 illustrates the differences between an ideal edge and a real edge. Figure 4.2.4a shows a representation of one row in an image of an ideal edge. The vertical axis represents brightness, and the horizontal axis shows the spatial coordinate. The abrupt change in brightness characterizes an ideal edge. In the corresponding image, the edge appears very distinct. In Figure 4.2.4b we see the representation of a real edge, which changes gradually. This gradual change is a minor form of blurring caused by the imaging device, the lenses, or the lighting, and is typical for real-world (as opposed to computer generated) images. In the figure, where the edge has been exaggerated for illustration purposes, note that from a visual perspective this image contains the same information as does the ideal image: black on one side, white on the other, with a line down the center.

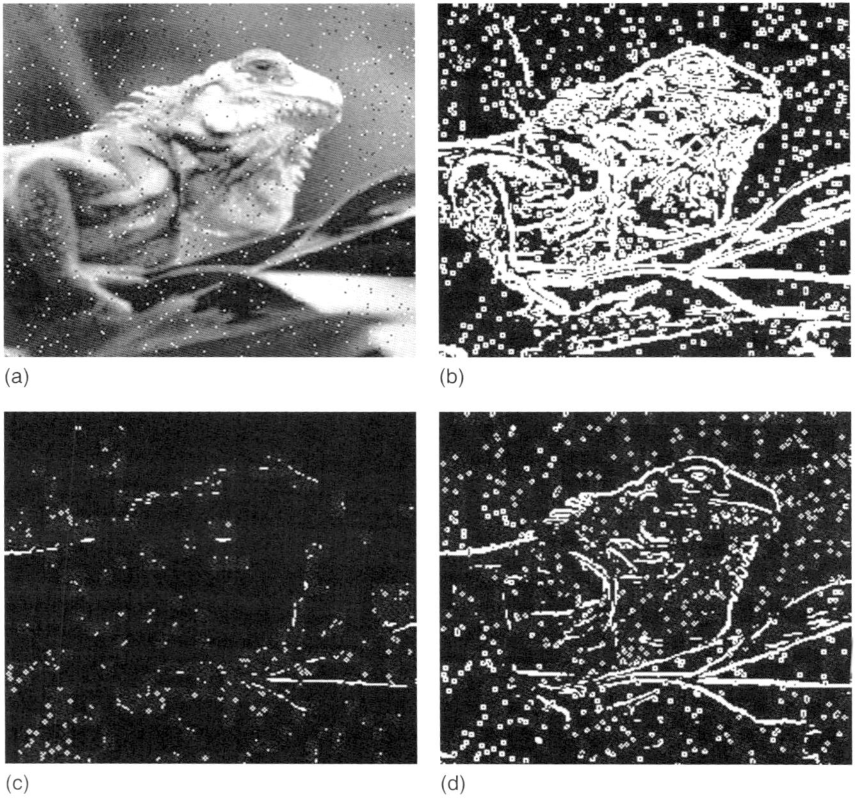

FIGURE 4.2.2
Noise in images requires tradeoffs between sensitivity and accuracy for edge detectors. (a) Noisy image. (b) Edge detector too sensitive, many edge points found that are attributable to noise. (c) Edge detector not sensitive enough, loss of valid edge points. (d) Reasonable result obtained by compromise between sensitivity and accuracy, may mitigate noise via postprocessing. (Original lizard photo courtesy of Mark Zuke.)

4.2.1 Gradient Operators

Gradient operators are based on the idea of using the first or second derivative of the gray level function as an edge detector. Remember from calculus that the derivative measures the rate of change of a line, or the slope of the line. If we model the gray level transition of an edge by a ramp function (which is a good approximation to a real edge), we can see what the first and second derivatives look like in Figure 4.2.5. When the gray level is constant the first derivative is zero, and when it is linear it is equal to the slope of the line. With the following operators we will see that this is approximated with a difference operator, similar to the methods used to derive the definition of the derivative. The second derivative is positive at the change on the dark side of the edge, negative at the change on the light side, and zero elsewhere.

In Figure 4.2.5c we can see that the magnitude of the first derivative will mark edge points, with steeper gray level changes corresponding to stronger edges and larger magnitudes from the derivative operators. In Figure 4.2.5d we can see that applying a second derivative operator to an edge returns two impulses, one on either side of the edge. An advantage of this is that if a line is drawn between the two impulses the position where this line crosses the zero axis is the center of the edge, which theoretically allows us to measure edge location to sub-pixel accuracy. Sub-pixel accuracy refers to

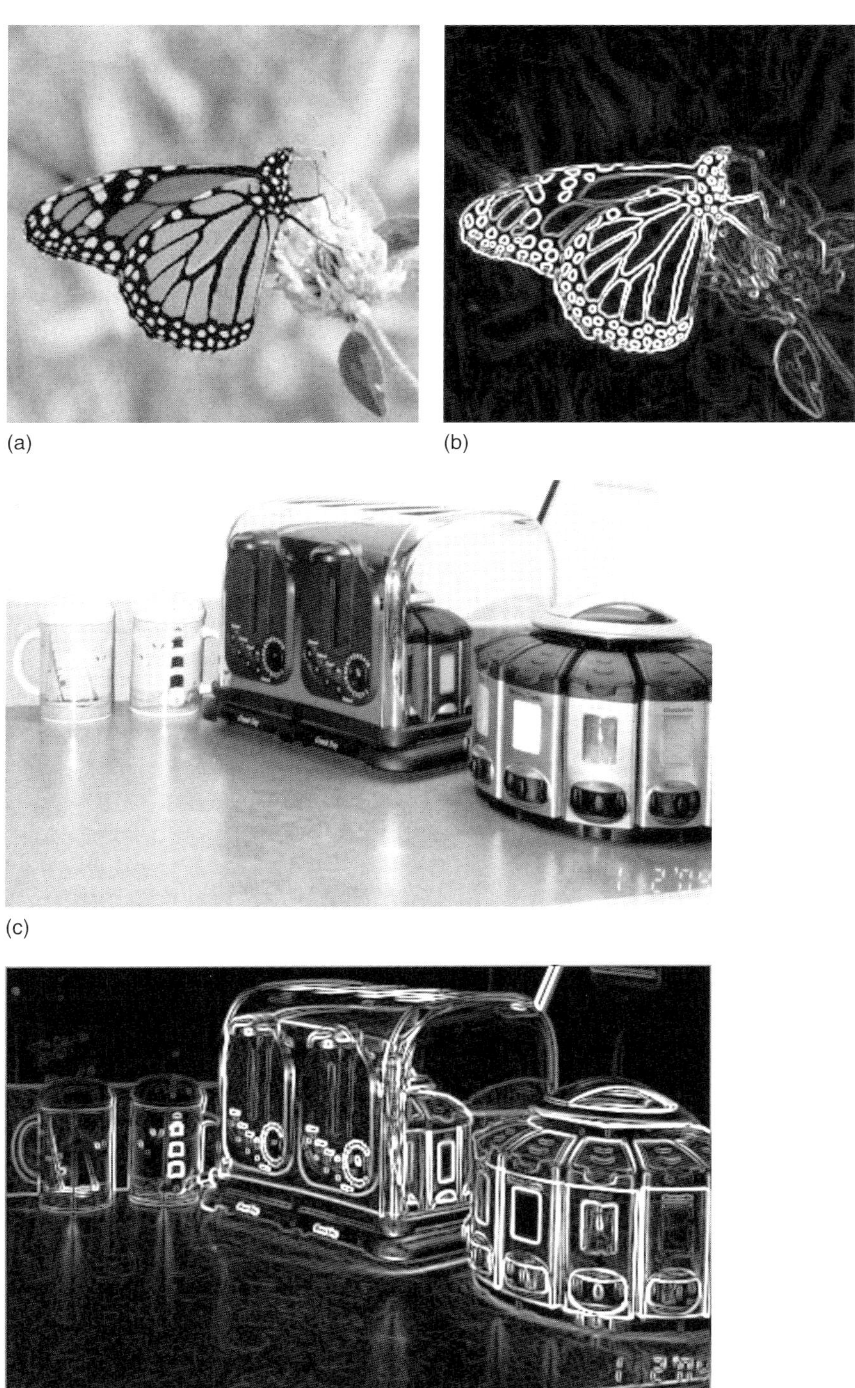

FIGURE 4.2.3
Image objects may be parts of real objects. (a) Butterfly image (original photo courtesy of Mark Zuke). (b) Butterfly after edge detection. Note that image objects are separated by color, or gray level, changes. (c) Image of objects in kitchen corner. (d) Image after edge detection. Note that some image objects are created by reflections in the image due to lighting conditions and object properties.

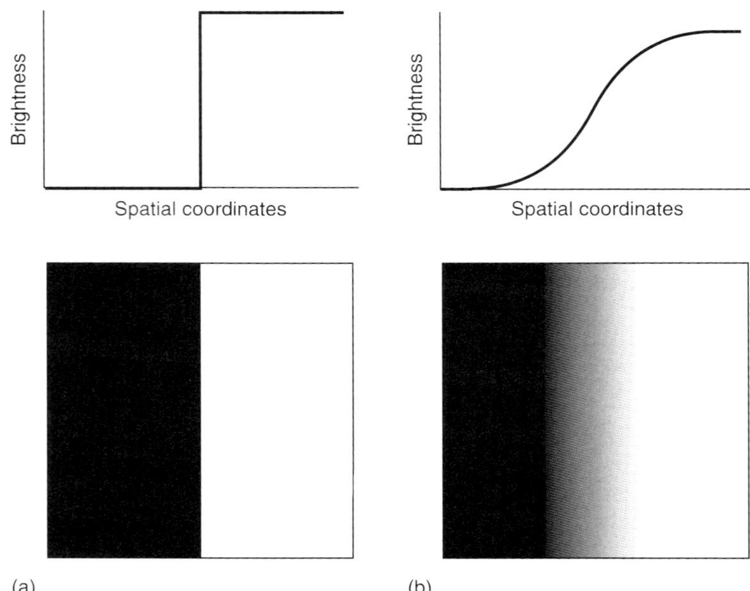

FIGURE 4.2.4
Ideal versus real edge. (a) Ideal edge. (b) Real edge.

the fact that the zero-crossing may be at a fractional pixel distance, for example halfway between two pixels, so we could say the edge is at, for instance, $c = 75.5$.

The ***Roberts operator*** is a simple approximation to the first derivative. It marks edge points only; it does not return any information about the edge orientation. It is the simplest of the edge detection operators and will work best with binary images (gray-level images can be made binary by a threshold operation). There are two forms of the Roberts operator. The first consists of the square root of the sum of the differences of the diagonal neighbors squared, as follows:

$$\sqrt{[I(r,c) - I(r-1,c-1)]^2 + [I(r,c-1) - I(r-1,c)]^2}$$

The second form of the Roberts operator is the sum of the magnitude of the differences of the diagonal neighbors, as follows:

$$|I(r,c) - I(r-1,c-1)| + |I(r,c-1) - I(r-1,c)|$$

The second form of the equation is often used in practice due to its computational efficiency—it is typically faster for a computer to find an absolute value than to find square roots.

The ***Sobel operator*** approximates the gradient by using a row and a column mask, which will approximate the first derivative in each direction. The Sobel edge detection masks look for edges in both the horizontal and vertical directions, and then combine this information into a single metric. The masks are as follows:

VERTICAL EDGE HORIZONTAL EDGE

$$\begin{bmatrix} -1 & -2 & -1 \\ 0 & 0 & 0 \\ 1 & 2 & 1 \end{bmatrix} \qquad \begin{bmatrix} -1 & 0 & 1 \\ -2 & 0 & 2 \\ -1 & 0 & 1 \end{bmatrix}$$

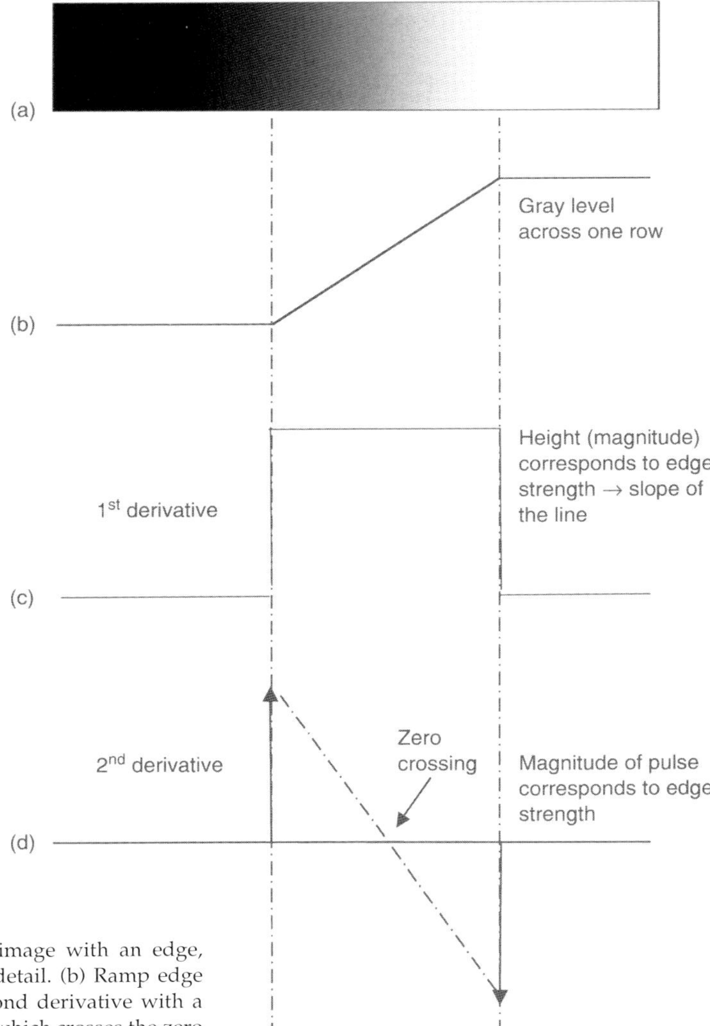

FIGURE 4.2.5
Edge model. (a) A portion of an image with an edge, which has been enlarged to show detail. (b) Ramp edge model. (c) First derivative. (d) Second derivative with a line drawn between the two pulses which crosses the zero axis at the edge center.

These masks are each convolved with the image. At each pixel location we now have two numbers: s_1, corresponding to the result from the vertical edge mask, and s_2, from the horizontal edge mask. We use these numbers to compute two metrics, the edge magnitude and the edge direction, defined as follows:

$$\text{EDGE MAGNITUDE} \quad \sqrt{s_1^2 + s_2^2}$$

$$\text{EDGE DIRECTION} \quad \tan^{-1}\left[\frac{s_1}{s_2}\right]$$

As seen in Figure 4.2.1, the edge direction is perpendicular to the line (or curve), because the direction specified is the direction of the gradient, along which the gray levels are changing.

The ***Prewitt*** is similar to the Sobel, but with different mask coefficients. The masks are defined as:

$$\text{VERTICAL EDGE} \quad \begin{bmatrix} -1 & -1 & -1 \\ 0 & 0 & 0 \\ 1 & 1 & 1 \end{bmatrix} \qquad \text{HORIZONTAL EDGE} \quad \begin{bmatrix} -1 & 0 & 1 \\ -1 & 0 & 1 \\ -1 & 0 & 1 \end{bmatrix}$$

These masks are each convolved with the image. At each pixel location we find two numbers: p_1, corresponding to the result from the vertical edge mask, and p_2, from the horizontal edge mask. We use these results to determine two metrics, the edge magnitude and the edge direction, which are defined as follows:

$$\text{EDGE MAGNITUDE} \quad \sqrt{p_1^2 + p_2^2}$$

$$\text{EDGE DIRECTION} \quad \tan^{-1}\left[\frac{p_1}{p_2}\right]$$

As with the Sobel edge detector, the direction lies 90 degrees from the apparent direction of the line or curve. The Prewitt is easier to calculate than the Sobel, since the only coefficients are 1's, which makes it easier to implement in hardware. However, the Sobel is defined to place emphasis on the pixels closer to the mask center, which may be desirable for some applications.

The ***Laplacian operators*** described here are similar to the ones used for preprocessing as described in Section 3.2.3 Spatial Filters. The three Laplacian masks presented below represent different approximations of the Laplacian, which is the two-dimensional version of the second derivative. Unlike the Sobel and Prewitt edge detection masks, the Laplacian masks are rotationally symmetric, which means edges at all orientations contribute to the result. They are applied by selecting *one* mask and convolving it with the image. The sign of the result (positive or negative) from two adjacent pixel locations provides directional information, and tells us which side of the edge is brighter.

LAPLACIAN MASKS

$$\begin{bmatrix} 0 & -1 & 0 \\ -1 & 4 & -1 \\ 0 & -1 & 0 \end{bmatrix} \begin{bmatrix} -1 & -1 & -1 \\ -1 & 8 & -1 \\ -1 & -1 & -1 \end{bmatrix} \begin{bmatrix} -2 & 1 & -2 \\ 1 & 4 & 1 \\ -2 & 1 & -2 \end{bmatrix}$$

These masks differ from the Laplacian-type previously described in that the center coefficients have been decreased by one. With these masks, we are trying to find edges, and are not interested in the image itself—if we increase the center coefficient by one it is equivalent to adding the original image to the edge detected image.

An easy way to picture the difference is to consider the effect each mask would have when applied to an area of constant value. The above convolution masks would return a value of zero. If we increase the center coefficients by one, each would return the original gray level. Therefore, if we are only interested in edge information, the sum of the coefficients should be zero. If we want to retain most of the information that is in the original image, the coefficients should sum to a number greater than zero. The larger the center value, the less the processed image is changed from the original image. Consider an extreme example in which the center coefficient is very large compared with the other coefficients in the mask. The resulting pixel value will depend most heavily upon the current value, with only minimal contribution from the surrounding pixel values.

4.2.2 Compass Masks

The Kirsch and Robinson edge detection masks are called compass masks since they are defined by taking a single mask and rotating it to the eight major compass orientations: North, Northwest, West, Southwest, South, Southeast, East, and Northeast. The ***Kirsch compass masks*** are defined as follows:

$$k_0 \begin{bmatrix} -3 & -3 & 5 \\ -3 & 0 & 5 \\ -3 & -3 & 5 \end{bmatrix} \quad k_1 \begin{bmatrix} -3 & 5 & 5 \\ -3 & 0 & 5 \\ -3 & -3 & -3 \end{bmatrix} \quad k_2 \begin{bmatrix} 5 & 5 & 5 \\ -3 & 0 & -3 \\ -3 & -3 & -3 \end{bmatrix} \quad k_3 \begin{bmatrix} 5 & 5 & -3 \\ 5 & 0 & -3 \\ -3 & -3 & -3 \end{bmatrix}$$

$$k_4 \begin{bmatrix} 5 & -3 & -3 \\ 5 & 0 & -3 \\ 5 & -3 & -3 \end{bmatrix} \quad k_5 \begin{bmatrix} -3 & -3 & -3 \\ 5 & 0 & -3 \\ 5 & 5 & -3 \end{bmatrix} \quad k_6 \begin{bmatrix} -3 & -3 & -3 \\ -3 & 0 & -3 \\ 5 & 5 & 5 \end{bmatrix} \quad k_7 \begin{bmatrix} -3 & -3 & -3 \\ -3 & 0 & 5 \\ -3 & 5 & 5 \end{bmatrix}$$

The edge magnitude is defined as the maximum value found at each point by the convolution of each of the masks with the image. The edge direction is defined by the mask that produces the maximum magnitude; for instance, k_0 corresponds to a horizontal edge, whereas k_5 corresponds to a diagonal edge in the Northeast/Southwest direction (remember edges are perpendicular to the lines). We also see that the last four masks are actually the same as the first four, but flipped about a central axis.

The ***Robinson compass masks*** are used in a manner similar to the Kirsch masks, but are easier to implement, as they rely only on coefficients of 0, 1, and 2, and are symmetrical about their directional axis—the axis with the zeros which corresponds to the line direction. We only need to compute the results on four of the masks; the results from the other four can be obtained by negating the results from the first four. The masks are as follows:

$$r_0 \begin{bmatrix} -1 & 0 & 1 \\ -2 & 0 & 2 \\ -1 & 0 & 1 \end{bmatrix} \quad r_1 \begin{bmatrix} 0 & 1 & 2 \\ -1 & 0 & 1 \\ -2 & -1 & 0 \end{bmatrix} \quad r_2 \begin{bmatrix} 1 & 2 & 1 \\ 0 & 0 & 0 \\ -1 & -2 & -1 \end{bmatrix} \quad r_3 \begin{bmatrix} 2 & 1 & 0 \\ 1 & 0 & -1 \\ 0 & -1 & -2 \end{bmatrix}$$

$$r_4 \begin{bmatrix} 1 & 0 & -1 \\ 2 & 0 & -2 \\ 1 & 0 & -1 \end{bmatrix} \quad r_5 \begin{bmatrix} 0 & -1 & -2 \\ 1 & 0 & -1 \\ 2 & 1 & 0 \end{bmatrix} \quad r_6 \begin{bmatrix} -1 & -2 & -1 \\ 0 & 0 & 0 \\ 1 & 2 & 1 \end{bmatrix} \quad r_7 \begin{bmatrix} -2 & -1 & 0 \\ -1 & 0 & 1 \\ 0 & 1 & 2 \end{bmatrix}$$

The edge magnitude is defined as the maximum value found at each point by the convolution of each of the masks with the image. The edge direction is defined by the mask that produces the maximum magnitude. It is interesting to note that masks r_0 and r_6 are the same as the Sobel masks. We can see that any of the edge detection masks can be extended by rotating them in a manner like these compass masks, which will allow us to extract explicit information about edges in any direction.

4.2.3 Advanced Edge Detectors

The edge detectors considered here include the Laplacian of a Gaussian (LoG), the Canny algorithm, the Boie–Cox algorithm, the Shen–Castan algorithm and the Frei–Chen masks. They are considered to be advanced because they are algorithmic in nature, which basically means they require multiple steps. Except for the Frei–Chen masks, these algorithms begin with the idea that, in general, most edge detectors are too sensitive to noise and by blurring the image prior to edge detection we can mitigate these noise effects. The noise considered here includes irrelevant image detail, as well as a combination of blurring from camera optics and signal corruption from camera electronics.

The simplest of these is the ***Laplacian of a Gaussian*** (LoG) which requires two steps:

1. Convolve the image with a Gaussian smoothing filter
2. Convolve the image with a Laplacian mask

By pre-processing with a smoothing filter we can mitigate noise effects (see Section 4.2.5), and then use the Laplacian to enhance the edges. This operator is also called the *Mexican hat* operator, since the function resembles a sombrero (see Figure 4.2.6). Since the process requires the successive convolution of two masks, they can be combined into one *LoG* mask. A common 5×5 mask that approximates the combination of the Gaussian and Laplacian into one convolution mask is as follows:

$$\begin{bmatrix} 0 & 0 & -1 & 0 & 0 \\ 0 & -1 & -2 & -1 & 0 \\ -1 & -2 & 16 & -2 & -1 \\ 0 & -1 & -2 & -1 & 0 \\ 0 & 0 & -1 & 0 & 0 \end{bmatrix}$$

Another interesting aspect of the *LoG* operator is that it is believed to closely model biological vision systems.

The ***Canny algorithm,*** developed by John Canny in 1986, is an optimal edge detection method based on a specific mathematical model for edges. The edge model is a step edge corrupted by Gaussian noise. The algorithm consists of four primary steps:

1. *Apply a Gaussian filter mask to smooth the image* to mitigate noise effects. This can be performed at different scales, by varying the size of the filter mask which corresponds to the variance of the Gaussian function. A larger mask will blur the image more and will find fewer, but more prominent, edges.

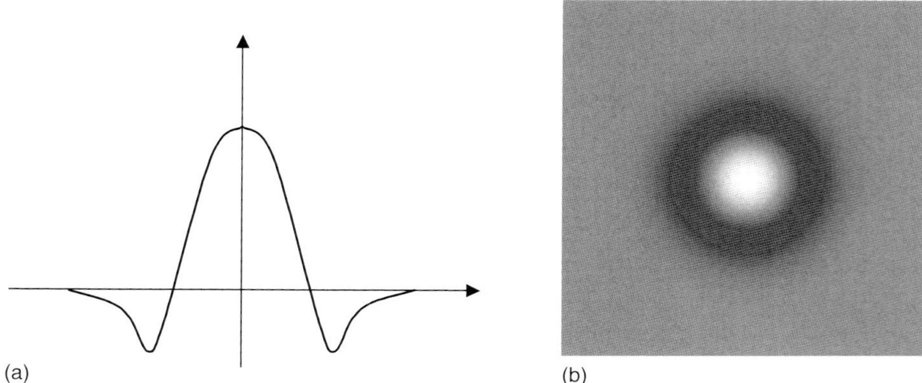

FIGURE 4.2.6
Laplacian of a Gaussian. (a) One-dimensional plot of the LoG function. (b) The LoG as an image with white representing positive numbers, black negative numbers, and gray representing zero.

2. *Find the magnitude and direction of the gradient* using equations similar to the Sobel or Prewitt edge detectors, for example:

$$\text{VERTICAL} \quad \frac{1}{2}\begin{bmatrix} -1 & -1 \\ 1 & 1 \end{bmatrix} \qquad \text{HORIZONTAL} \quad \frac{1}{2}\begin{bmatrix} -1 & 1 \\ -1 & 1 \end{bmatrix}$$

   These masks are each convolved with the image. At each pixel location we find two numbers: c_1, corresponding to the result from the vertical edge mask, and c_2, from the horizontal edge mask. We use these results to determine two metrics, the edge magnitude and the edge direction, which are defined as follows:

$$\text{EDGE MAGNITUDE} \quad \sqrt{c_1^2 + c_2^2}$$

$$\text{EDGE DIRECTION} \quad \tan^{-1}\left[\frac{c_1}{c_2}\right]$$

3. *Apply nonmaxima suppression* which results in thinned edges. This is done by considering small neighborhoods in the magnitude image, for example 3×3, and comparing the center value to its neighbors in the direction of the gradient. If the center value is not larger than the neighboring pixels along the gradient direction, then set it to zero. Otherwise, it is a local maximum, so we keep it. In Figure 4.2.7 we see an example of a 3×3 neighborhood showing the magnitude at each location, and using an arrow to show the gradient direction. The center pixel has a value of 100 and the gradient direction is horizontal (corresponding to a vertical line), so it is compared to the pixels to the right and left; which are 40 and 91. Since it is greater than both, it is retained as an edge pixel; if it was less than either one it would be removed as an edge point. Note that this will have the effect of making thick edges thinner, by selecting the "best" point along a gradient direction.

$$\begin{bmatrix} \leftarrow 50 & 112 \rightarrow & 20 \rightarrow \\ \leftarrow 40 & 100 \rightarrow & 91 \rightarrow \\ \leftarrow 88 & 95 \rightarrow & 92 \rightarrow \end{bmatrix}$$

FIGURE 4.2.7
Nonmaxima suppression. A 3×3 subimage of the magnitude image, which consists of the magnitude results in an image grid. The arrows show the gradient directions. This particular subimage has a vertical line (a horizontal edge). To apply nonmaxima suppression we compare the center pixel magnitude along the gradient direction. Here the 100 is compared with the 40 and the 91. Since it is a local maximum, it is retained as an edge pixel.

4. *Apply two thresholds* to obtain the final result. This technique, known as *hysteresis thresholding* helps to avoid false edges caused by too low a threshold value or missing edges caused by too high a value. It is a two step thresholding method, which first marks edge pixels above a high threshold; and then applies a low threshold to pixels connected to the pixels found with the high threshold. This can be performed multiple times, as either a recursive or iterative process.

The ***Boie–Cox algorithm***, developed in 1986 and 1987, is a generalization of the Canny algorithm. It consists of similar steps, but uses matched filters and Wiener filters (see Chapter 9) to allow for a more generalized edge model. The ***Shen–Castan algorithm***, developed in 1992, was developed as an optimal solution to a specific mathematical model. Shen and Casten claim that their filter does better than the Canny at finding the precise location of the edge pixels. Like the Canny, it uses a smoothing filter followed by a search for the edge pixels. The search includes steps similar to the Canny, but with modifications and extensions (for more details see the references).

The ***Frei–Chen masks*** are unique in that they form a complete set of *basis vectors*. This means we can represent any 3×3 subimage as a weighted sum of the nine Frei–Chen masks (Figure 4.2.8). These weights are found by projecting a 3×3 subimage onto each of these masks. This projection process is similar to the convolution process in that both overlay the mask on the image, multiply coincident terms, and sum the results (also called a *vector inner product*). This is best illustrated by example.

Example 4.2.1

Suppose we have the following subimage, I_s:

$$I_s = \begin{bmatrix} 1 & 0 & 1 \\ 1 & 0 & 1 \\ 1 & 0 & 1 \end{bmatrix}$$

To project this subimage onto the Frei–Chen masks, start by finding the projection onto f_1. Overlay the subimage on the mask and consider the first row. The 1 in the upper left corner of the subimage coincides with the 1 in the upper left corner of the mask, the 0 is over the $\sqrt{2}$, and the 1 on the upper right corner of the subimage coincides with the 1 in the mask. Note that all these must be summed and then multiplied by the $1/(2\sqrt{2})$ factor to normalize the masks. The projection of I_s onto f_1 is equal to:

$$\frac{1}{2\sqrt{2}}\left[1(1) + 0\left(\sqrt{2}\right) + 1(1) + 1(0) + 0(0) + 1(0) + 1(-1) + 0\left(-\sqrt{2}\right) + 1(-1)\right] = 0$$

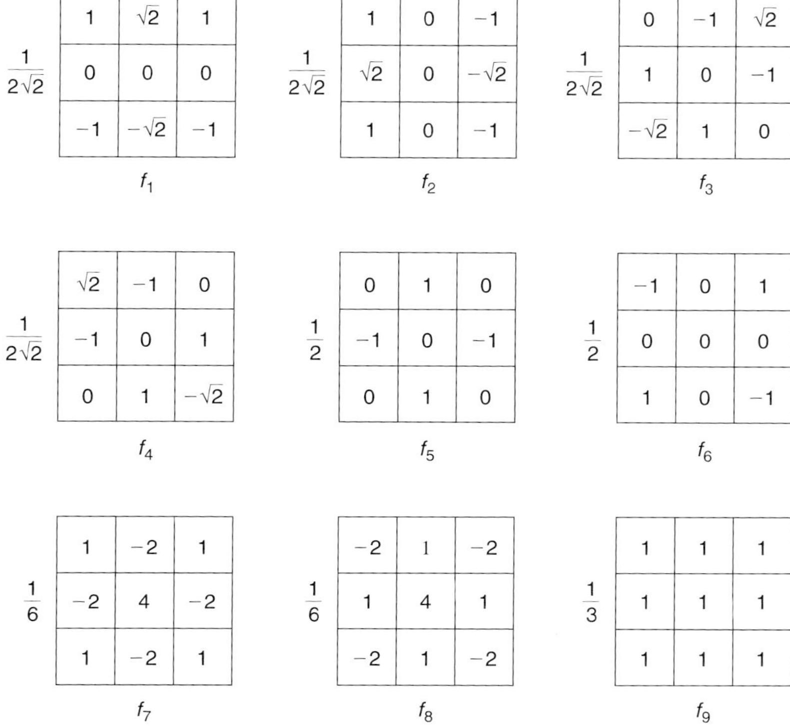

FIGURE 4.2.8
Frei–Chen masks. The first four masks comprise the edge subspace. The next four masks comprise the line subspace. The final mask is the average subspace.

If we follow this process and project the subimage, I_s, onto each of the Frei–Chen masks, we get the following:

$$f_1 \to 0, \quad f_2 \to 0, \quad f_3 \to 0, \quad f_4 \to 0, \quad f_5 \to -1, \quad f_6 \to 0, \quad f_7 \to 0, \quad f_8 \to -1, \quad f_9 \to 2$$

We can now see what is meant by a complete set of basis vectors allowing us to represent a subimage by a weighted sum. The basis vectors in this case are the Frei–Chen masks, and the weights are the projection values. Take the weights and multiply them by each mask, then sum the corresponding values. For this example the only nonzero terms correspond to masks f_5, f_8 and f_9, and we find the following:

$$(-1)\left(\frac{1}{2}\right)\begin{bmatrix} 0 & 1 & 0 \\ -1 & 0 & -1 \\ 0 & 1 & 0 \end{bmatrix} + (-1)\left(\frac{1}{6}\right)\begin{bmatrix} -2 & 1 & -2 \\ 1 & 4 & 1 \\ -2 & 1 & -2 \end{bmatrix} + (2)\left(\frac{1}{3}\right)\begin{bmatrix} 1 & 1 & 1 \\ 1 & 1 & 1 \\ 1 & 1 & 1 \end{bmatrix} = \begin{bmatrix} 1 & 0 & 1 \\ 1 & 0 & 1 \\ 1 & 0 & 1 \end{bmatrix}$$
$$= I_s$$

We have seen how the Frei–Chen masks can be used to represent a subimage as a weighted sum, but how are they used for edge detection? The Frei–Chen masks can be grouped into a set of four masks for an edge subspace, four masks for a line subspace, and one mask for an average subspace. To use them for edge detection, select a particular subspace of interest and find the relative projection of the image onto the particular subspace. This is given by the following equation:

$$\cos(\Theta) = \sqrt{\frac{M}{S}}$$

where:

$$M = \sum_{k \in \{e\}} (I_s \cdot f_k)^2$$

$$S = \sum_{k=1}^{9} (I_s \cdot f_k)^2$$

The set $\{e\}$ consists of the masks of interest. The (I_s, f_k) notation refers to the process of overlaying the mask on the subimage, multiplying coincident terms, and summing the results (a vector inner product). An illustration of this is shown in Figure 4.2.9. The advantage of this method is that we can select particular edge or line masks of interest, and consider the projection of those masks only.

4.2.4 Edges in Color Images

We saw in Chapter 2 that color images are described as three bands of monochrome image data, and typical images use red, green, and blue (RGB) bands. We also saw that various color transforms exist to map these RGB images into different color spaces. Given these choices, we have more than one possible definition of what constitutes a color edge. The simplest method is to extract the luminance, or brightness, information and use the previously defined methods. Or we may wish to map the RGB data into another color space and find edges in one of those bands. For example, for a particular application we may not be interested in changes in brightness, but changes in what we classically think of as "color"; so we could map the RGB data into the HSL color spaces and search for edges in hue or saturation.

Alternately, we can use all three bands and require that in order for an edge to exist, it must be present in all three bands at the same location. With this scheme we can use any of the color spaces, depending on the application, and we may want to define a quantum for "location error", and not require the edge to be at *exactly* the same

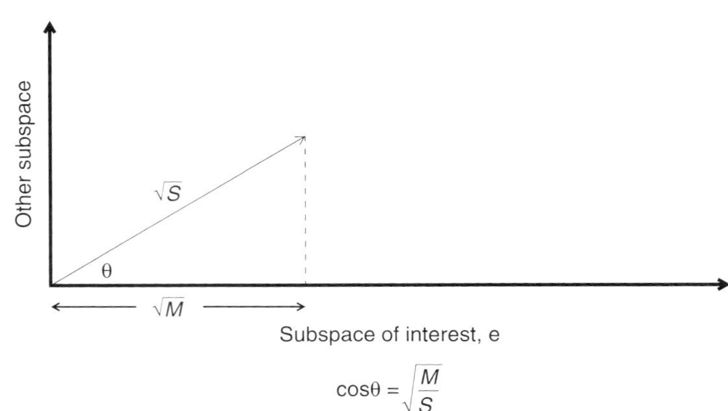

FIGURE 4.2.9
Frei–Chen projection. A 2-D representation of the Frei–Chen projection concept. The actual Frei–Chen space is nine-dimensional, where each dimension is given by one of the masks, f_k.

pixel location in all three bands. Also, with this scheme, we could use any of the previously defined edge detection methods on each of the three bands individually. We can then combine the results from all three bands into a three-band image, or simply retain the maximum value at each pixel location from all three bands and output a monochrome image, or use a linear combination of all three results to create a monochrome image.

Another method that uses all three bands simultaneously is to consider the color pixel vectors and search through the image marking edge points only if two neighboring color pixel vectors differ by some minimum distance measure. Here we can use any vector distance measure, such as Euclidean distance (see Chapter 6 for definitions of other distance measures).

One specific method for finding edges in multispectral images uses pixel values in all the image bands. This can be applied to three-band color images, as well as multispectral satellite images. It uses equations similar to the Roberts gradient, but is applied to all the image bands with a simple set of equations. The result of this edge detector is the minimum given by the following two equations:

$$\frac{\sum_{b=1}^{n}\left[I_b(r,c)-\bar{I}(r,c)\right]\left[I_b(r+1,c+1)-\bar{I}(r+1,c+1)\right]}{\sqrt{\sum_{b=1}^{n}\left[I_b(r,c)-\bar{I}(r,c)\right]^2\sum_{b=1}^{n}\left[I_b(r+1,c+1)-\bar{I}(r+1,c+1)\right]^2}}$$

$$\frac{\sum_{b=1}^{n}\left[I_b(r+1,c)-\bar{I}(r+1,c)\right]\left[I_b(r,c+1)-\bar{I}(r,c+1)\right]}{\sqrt{\sum_{b=1}^{n}\left[I_b(r+1,c)-\bar{I}(r+1,c)\right]^2\sum_{b=1}^{n}\left[I_b(r,c+1)-\bar{I}(r,c+1)\right]^2}}$$

where:

$\bar{I}(r,c)$ is the arithmetic average of all the pixels in all bands at pixel location (r,c)

$I_b(r,c)$ is the value at location (r,c) in the bth band, with a total of n bands

4.2.5 Edge Detector Performance

In evaluating the performance of many processes, we can consider both objective and subjective evaluation. The objective metric allows us to compare different techniques with fixed analytical methods, whereas the subjective methods often have unpredictable results. However, for many computer imaging applications, the subjective measures tend to be the most useful. We will examine the types of errors encountered with edge detection, look at an objective measure based on these criteria, and review results of the various edge detectors for our own subjective evaluation.

To develop a performance metric for edge detection operators, we need to define what constitutes success. For example, the Canny algorithm was developed considering three important edge detection success criteria:

Detection—the edge detector should find all real edges and not find any false edges.

Localization—the edges should be found in the correct place.

Single Response—there should not be multiple edges found for a single edge.

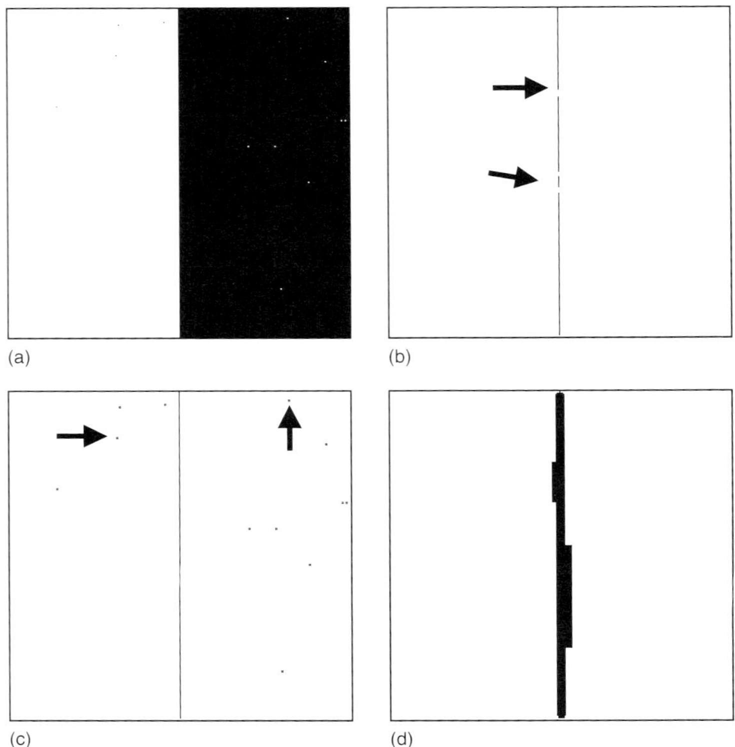

FIGURE 4.2.10
Errors in edge detection. (a) Original image. (b) Missed edge points, examples marked with arrows. (c) Noise misclassified as edge points, examples marked with arrows. (d) Smeared edge.

These correlate nicely with *Pratt's Figure of Merit (FOM)* defined in 1978. Pratt first considered the types of errors that can occur with edge detection methods. The types of errors are: (1) missing valid edge points, (2) classifying noise pulses as valid edge points, and (3) smearing of edges (see Figure 4.2.10). If these errors do not occur, we can say that we have achieved success.

The ***Pratt FOM***, is defined as follows:

$$FOM = \frac{1}{I_N} \sum_{i=1}^{I_F} \frac{1}{1 + \alpha\, d_i^2}$$

where

I_N = the maximum of I_I and I_F

I_I = the number of ideal edge points in the image

I_F = the number of edge points found by the edge detector

α = a scaling constant that can be adjusted to adjust the penalty for offset edges

d_i = the distance of a found edge point to an ideal edge point

For this metric, *FOM* will be 1 for a perfect edge. Normalizing to the maximum of the ideal and found edge points guarantees a penalty for smeared edges or missing edge points. In general, this metric assigns a better rating to smeared edges than to offset or missing edges. This is done because techniques exist to thin smeared edges, but it is difficult to determine when an edge is missed. The distance, d, can be defined in more than one way and typically depends on the connectivity definition used. The possible definitions for d are as follows:

Let the (r, c) *values for two pixels be* (r_1, c_1) *and* (r_2, c_2).

1. ***City block distance***, based on four connectivity:

$$d = |r_1 - r_2| + |c_1 - c_2|$$

With this distance measure we can only move horizontally and vertically.

2. ***Chessboard distance***, based on eight-connectivity:

$$d = \max(|r_1 - r_2|, |c_1 - c_2|)$$

With this distance measure we can move diagonally, as well as horizontally or vertically.

3. ***Euclidean distance***, based on actual physical distance:

$$d = \left[(r_1 - r_2)^2 + (c_1 - c_2)^2\right]^{1/2}$$

Example 4.2.2

Given the following image array, find the Figure of Merit for the following found edge points, designated by 1's, in (a), (b), and (c). Let $\alpha = 0.5$, and use the city block distance measure. We assume that actual edge in the locations where the line appears, that is, at the 100's.

$$\text{Image Array}\begin{bmatrix} 0 & 0 & 0 & 0 & 0 \\ 0 & 0 & 0 & 0 & 0 \\ 0 & 100 & 100 & 100 & 0 \\ 0 & 0 & 0 & 0 & 0 \\ 0 & 0 & 0 & 0 & 0 \end{bmatrix}$$

$$\text{(a)}\begin{bmatrix} 0 & 0 & 0 & 0 & 0 \\ 0 & 0 & 0 & 0 & 0 \\ 0 & 1 & 1 & 1 & 0 \\ 0 & 0 & 0 & 0 & 0 \\ 0 & 0 & 0 & 0 & 0 \end{bmatrix} \qquad \text{(b)}\begin{bmatrix} 0 & 0 & 0 & 0 & 0 \\ 0 & 1 & 1 & 1 & 0 \\ 0 & 1 & 1 & 1 & 0 \\ 0 & 0 & 0 & 0 & 0 \\ 0 & 0 & 0 & 0 & 0 \end{bmatrix} \qquad \text{(c)}\begin{bmatrix} 0 & 0 & 0 & 0 & 0 \\ 0 & 0 & 0 & 0 & 0 \\ 0 & 0 & 0 & 0 & 0 \\ 0 & 1 & 1 & 1 & 1 \\ 0 & 0 & 0 & 0 & 0 \end{bmatrix}$$

(a) $$FOM = \frac{1}{I_N}\sum_{i=1}^{I_F}\frac{1}{1+\alpha d_i^2} = \frac{1}{3}\left[\frac{1}{1+0.5(0)^2} + \frac{1}{1+0.5(0)^2} + \frac{1}{1+0.5(0)^2}\right] = 1$$

(b) $$FOM = \frac{1}{I_N}\sum_{i=1}^{I_F}\frac{1}{1+\alpha d_i^2} = \frac{1}{6}\left[\frac{1}{1+0.5(0)^2} + \frac{1}{1+0.5(0)^2} + \frac{1}{1+0.5(0)^2} + \frac{1}{1+0.5(1)^2} + \frac{1}{1+0.5(1)^2} + \frac{1}{1+0.5(1)^2}\right]$$

$$\approx 0.8333$$

$$\text{(c)}\quad FOM = \frac{1}{I_N}\sum_{i=1}^{I_F}\frac{1}{1+\alpha d_i^2} = \frac{1}{4}\left[\frac{1}{1+0.5(1)^2}+\frac{1}{1+0.5(1)^2}+\frac{1}{1+0.5(1)^2}+\frac{1}{1+0.5(2)^2}\right]$$
$$\approx 0.5833$$

With result (a), we find a perfect edge. In result (b), we see that a smeared edge provides us with about 83%, and an offset edge in (c) gives us about 58%. Note that the α parameter can be adjusted to determine the penalty for offset edges.

Applying the Pratt FOM to selected edge detectors from each category—gradient operators, compass masks and the advanced edge detectors—results are shown in Figure 4.2.11 and Figure 4.2.12. Figure 4.2.11 shows example test images, and the Pratt FOM results are plotted as the noise variance increases. The original test image has a gray level of 127 on the left and 102 on the right side, and then Gaussian noise was added. Figure 4.2.12 shows resulting images with noise variances of 50 and 100 added to the test image. As expected, the advanced algorithms will have the best result as shown here with the Canny.

As previously mentioned, the objective metrics are often of limited use in practical applications, so we will take a subjective look at the results of the edge detectors. The human visual system is still superior, by far, to any computer vision system that

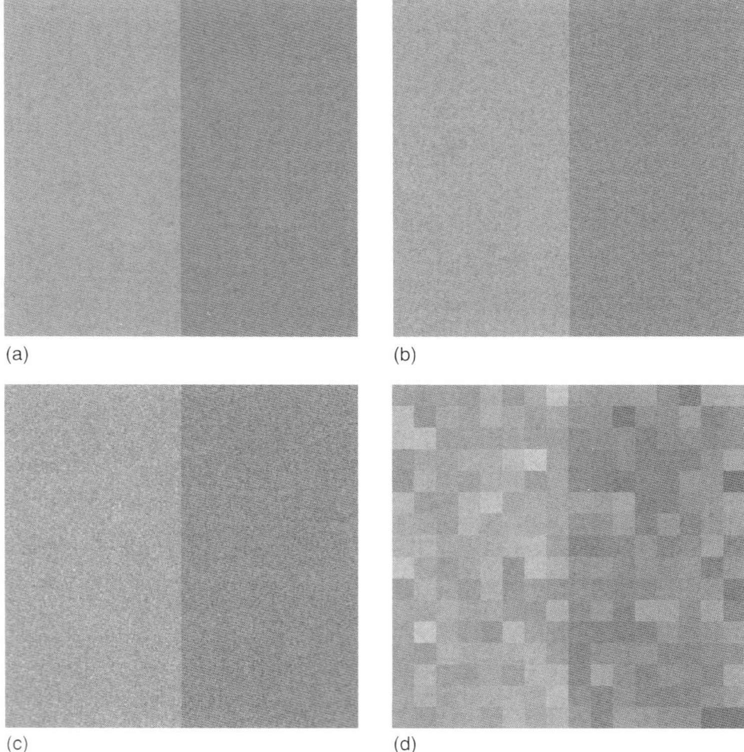

FIGURE 4.2.11
Pratt Figure of Merit. (a) the original test image, 256 × 256 pixels. (b) Test image with added Gaussian noise with a variance of 25. (c) Test image with added Gaussian noise with a variance of 100. (d) A 16 × 16 subimage cropped from image (c), enlarged to show that the edge is not as easy to find at the pixel level.

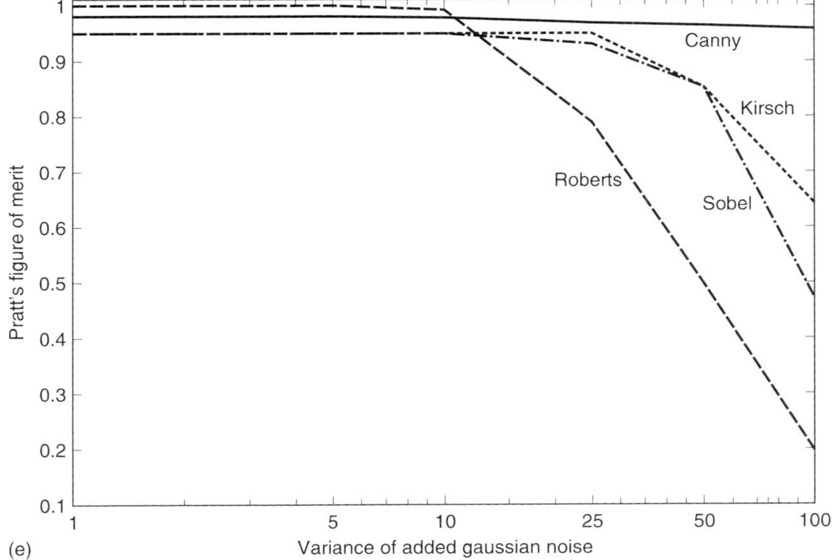

FIGURE 4.2.11 (Continued)
Pratt Figure of Merit. (e) This graph shows that as the noise variance increases the Canny has the best performance. We also see that the Roberts has the worst performance at high noise levels. The Roberts does poorly due to being based on a 2×2 mask, as opposed to the Sobel and Kirsch which are based on 3×3 masks. As we have seen with noisy images, a larger mask will perform better because it tends to spread the noise out—it is effectively a lowpass filter. The disadvantage of this is that fine details will be missed. This is the tradeoff that occurs with all edge detection—sensitivity versus accuracy. The test image was a step edge with Gaussian noise, so it is expected that the Canny performs the best because its development was based on this model.

has yet been devised, and is often used as the final judge in application development. Figure 4.2.13 shows the results of the basic gradient and compass mask edge detection operators. Here we see similar results from all the operators, but the Laplacian. This results from the Laplacian returning negative and positive numbers which get linearly remapped to 0 to 255 (for 8-bit display), which means that the background value of 0 is mapped to some intermediate gray level. For the other edge detection operators, only the magnitude is used for displaying the results. Similar results can be achieved with the Laplacian if we apply a threshold operation to the results (thresholding for boundary detection is explored in Section 4.3).

If we add noise to the image, the edge detection results are not as good. As mentioned before, we can preprocess the image with mean, or averaging, spatial filters to mitigate the effects from noise (this is explored more in Chapter 9), or we can expand the edge detection operators themselves to mitigate noise effects. One way to do this is to extend the size of the edge detection masks. An example of this method is to extend the Prewitt edge mask as follows:

EXTENDED PREWITT EDGE DETECTION MASK

$$\begin{bmatrix} 1 & 1 & 1 & 0 & -1 & -1 & -1 \\ 1 & 1 & 1 & 0 & -1 & -1 & -1 \\ 1 & 1 & 1 & 0 & -1 & -1 & -1 \\ 1 & 1 & 1 & 0 & -1 & -1 & -1 \\ 1 & 1 & 1 & 0 & -1 & -1 & -1 \\ 1 & 1 & 1 & 0 & -1 & -1 & -1 \\ 1 & 1 & 1 & 0 & -1 & -1 & -1 \end{bmatrix}$$

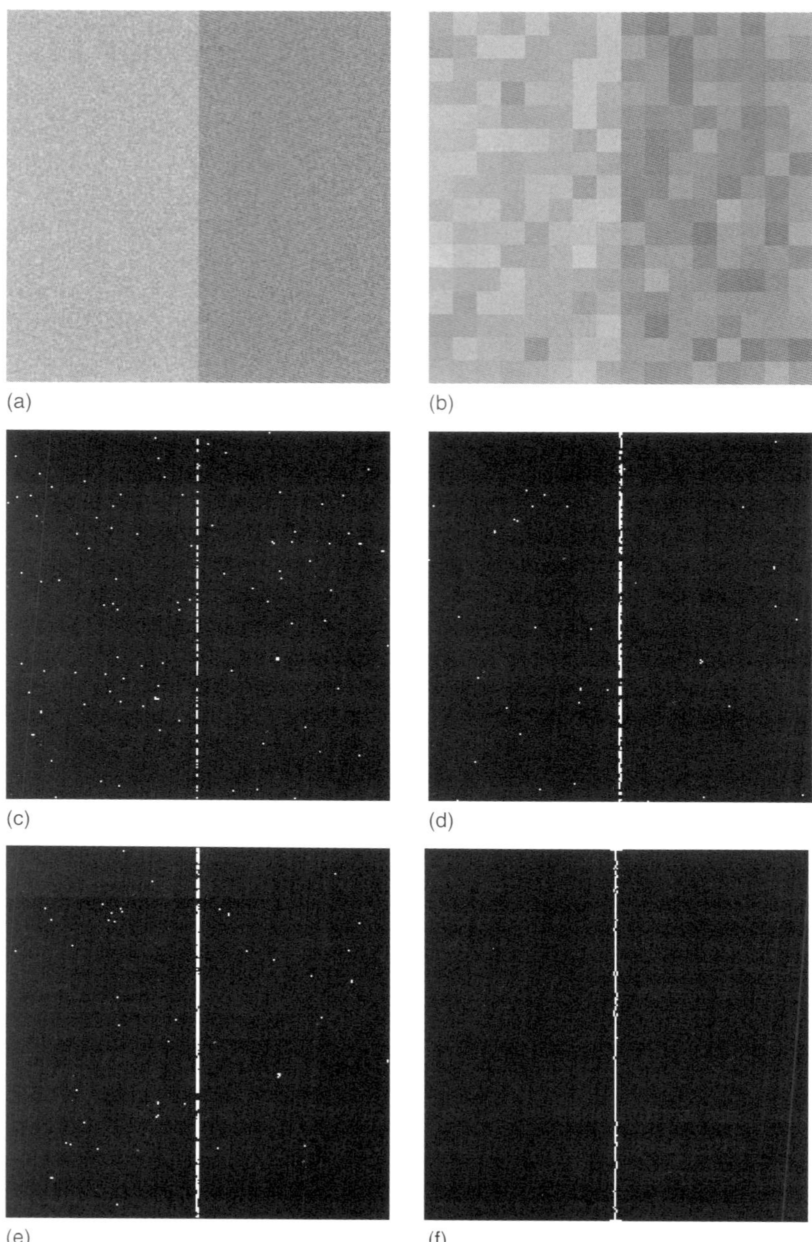

FIGURE 4.2.12
Pratt Figure of Merit images. (a) Test image with added Gaussian noise with a variance of 50. (b) A 16×16 subimage cropped from image (a), enlarged to show that the edge is not as easy to find at the pixel level. (c) Roberts result, $FOM = 0.853$. (d) Sobel result, $FOM = 0.853$. (e) Kirsch result, $FOM = 0.851$. (f) Canny result, $FOM = 0.963$.

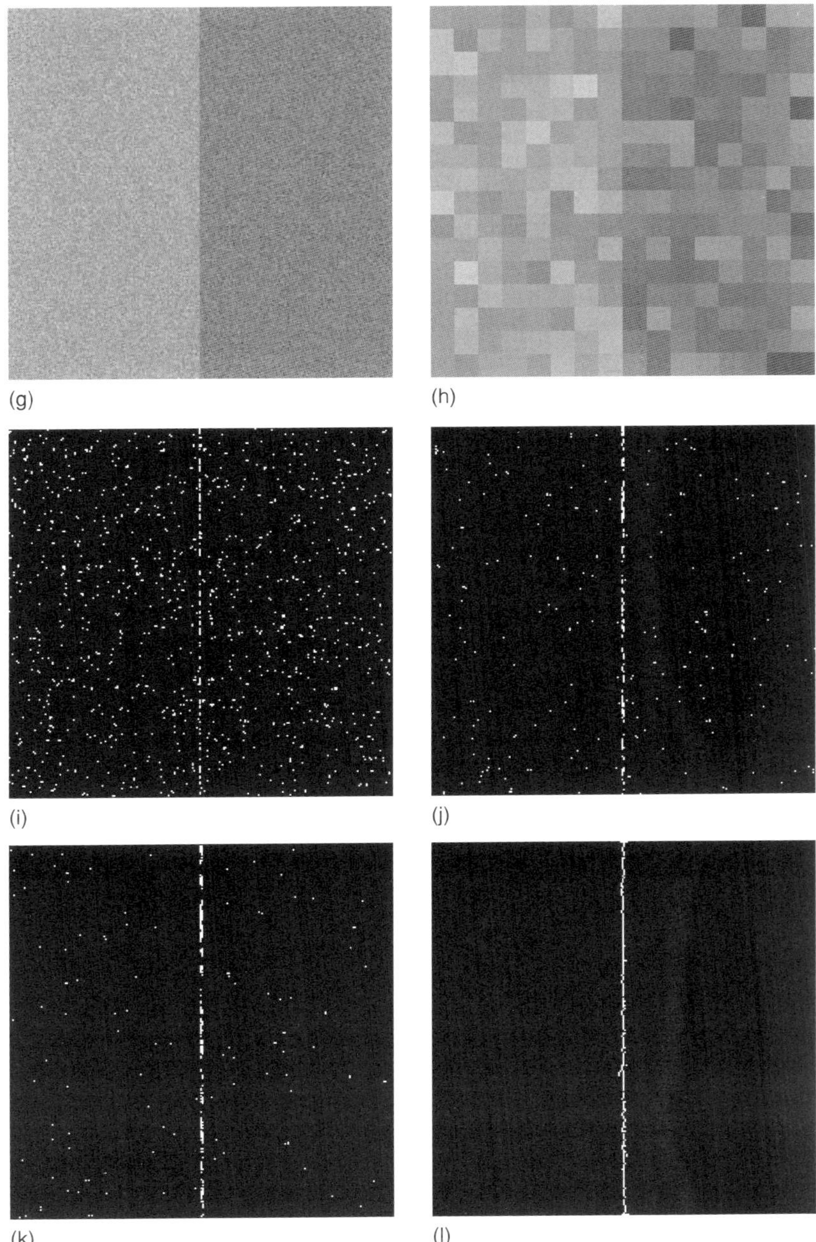

FIGURE 4.2.12 (Continued)
Pratt Figure of Merit images. (g) Test image with added Gaussian noise with a variance of 100. (h) A 16 × 16 subimage cropped from image (g), enlarged to show that the edge is not as easy to find at the pixel level. (i) Roberts, *FOM* = 0.1936. (j) Sobel, *FOM* = 0.470. (k) Kirsch, *FOM* = 0.640. (l) Canny, *FOM* = 0.956.

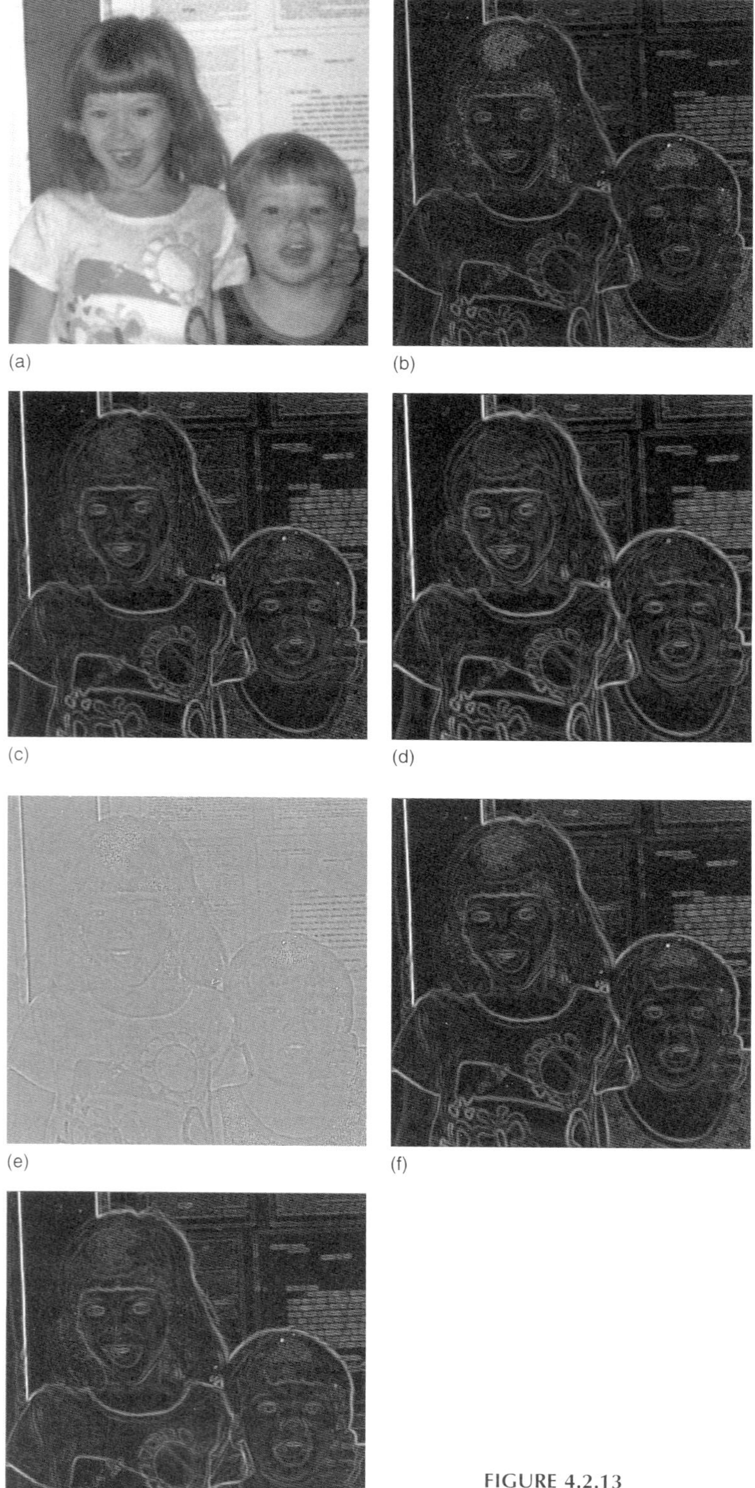

FIGURE 4.2.13
Edge detection examples. (a) Original image. (b) Roberts operator. (c) Sobel operator. (d) Prewitt operator. (e) Laplacian operator. (f) Kirsch operator. (g) Robinson operator.

We then can rotate this by 90 degrees and have both row and column masks which can be used like the Prewitt operators to return the edge magnitude and gradient. These types of operators are called boxcar operators and can be extended to any size, although 7×7, 9×9 and 11×11 are typical. The Sobel operator can be extended in a similar manner:

EXTENDED SOBEL EDGE DETECTION MASK

$$\begin{bmatrix} -1 & -1 & -1 & -2 & -1 & -1 & -1 \\ -1 & -1 & -1 & -2 & -1 & -1 & -1 \\ -1 & -1 & -1 & -2 & -1 & -1 & -1 \\ 0 & 0 & 0 & 0 & 0 & 0 & 0 \\ 1 & 1 & 1 & 2 & 1 & 1 & 1 \\ 1 & 1 & 1 & 2 & 1 & 1 & 1 \\ 1 & 1 & 1 & 2 & 1 & 1 & 1 \end{bmatrix}$$

If we approximate a linear distribution we obtain the *truncated pyramid* operator, as follows:

$$\begin{bmatrix} 1 & 1 & 1 & 0 & -1 & -1 & -1 \\ 1 & 2 & 2 & 0 & -2 & -2 & -1 \\ 1 & 2 & 3 & 0 & -3 & -2 & -1 \\ 1 & 2 & 3 & 0 & -3 & -2 & -1 \\ 1 & 2 & 3 & 0 & -3 & -2 & -1 \\ 1 & 2 & 2 & 0 & -2 & -2 & -1 \\ 1 & 1 & 1 & 0 & -1 & -1 & -1 \end{bmatrix}$$

This operator provides weights that decrease as we get away from the center pixel, which will smooth the result in a more natural manner. These operators are used in the same manner as the Prewitt and Sobel—we define a row and column mask, then find a magnitude and direction at each point. A comparison of applying the extended operators and the standard operators to a noisy image is shown in Figure 4.2.14. Comparing Figure 4.2.14c,d and Figure 4.2.14e,f we see that with noisy images the extended operators exhibit better performance than the 3×3 masks. However, they require more computations and will blur the edges. In Figure 4.2.14g,h we can compare the results of thresholding the Prewitt edge detector; we see that the extended mask, 7×7, does a much better job of ignoring the noise than the 3×3 mask. Figure 4.2.14i,j show the results of a 7×7 truncated pyramid operator followed by a threshold operation.

The advanced edge detectors can also be used effectively in noisy images. Results from the Canny, Frei–Chen and Shen–Castan algorithms are shown in Figure 4.2.15. The first group of images shows results from an image with added salt-and-pepper noise, and the second group with added Gaussian noise. Here we see that these algorithms perform well in the presence of these types of noise, except for the Frei–Chen with salt-and-pepper noise and the Shen–Castan with Gaussian noise.

4.2.6 Hough Transform

The Hough transform is designed specifically to find lines. A line is a collection of edge points that are adjacent and have the same direction. The Hough transform is an

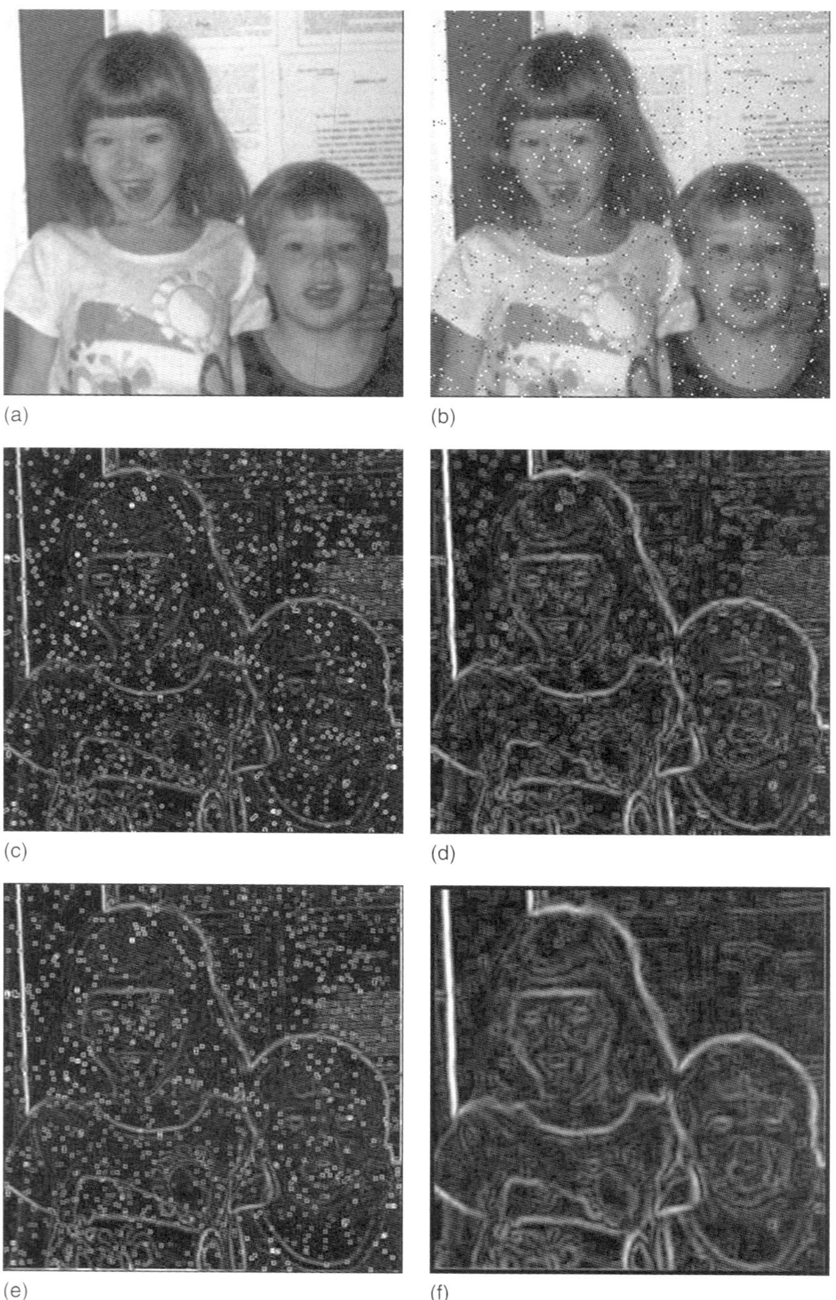

FIGURE 4.2.14
Edge detection examples—noise. (a) Original image. (b) Image with added noise. (c) Sobel with a 3×3 mask. (d) Sobel with a 7×7 mask. (e) Prewitt with a 3×3 mask. (f) Prewitt with a 7×7 mask.

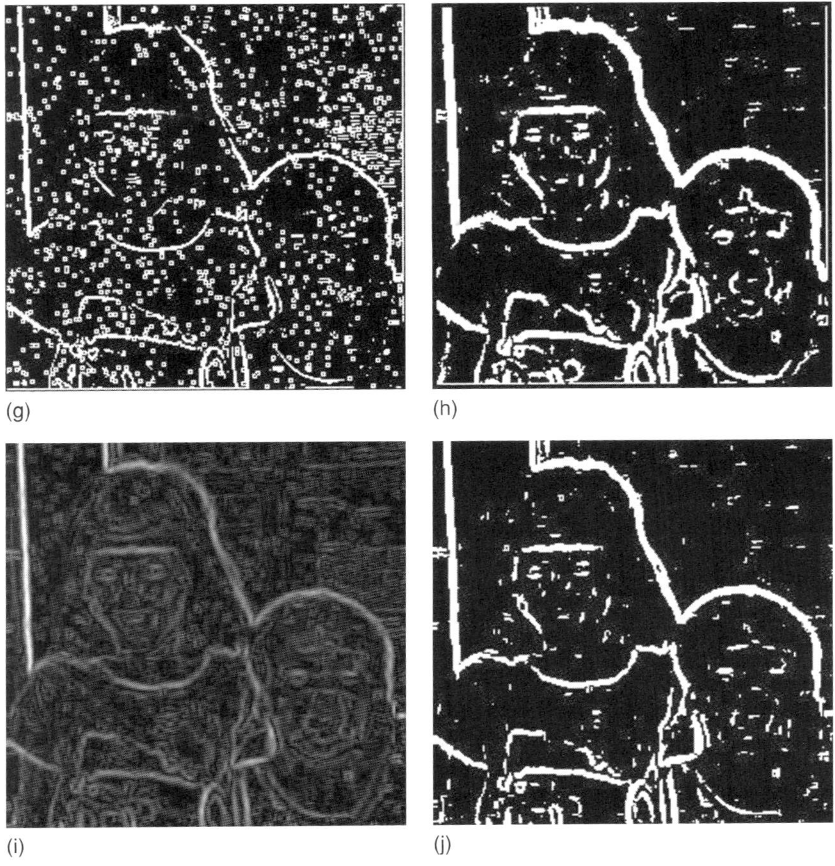

FIGURE 4.2.14 (Continued)
Edge detection examples—noise. (g) Result from applying a threshold to the 3×3 Prewitt. (h) Result from applying a threshold to the 7×7 Prewitt. (i) Truncated pyramid with a 7×7 mask. (j) Results from applying a threshold to the 7×7 truncated pyramid.

algorithm that will take a collection of n edge points, as found by an edge detector, and efficiently find all the lines on which these edge points lie. Although a brute force search method can be used that will find all the lines associated with each pair of points, then check every point with every possible line, it involves finding $n(n-1)/2$ (on the order of n^2) lines, and comparing every point to all the lines, which is $(n)(n(n-1))/2$ or about n^3 comparisons. This heavy computational burden is certainly not practical for real time applications, and provides much more information than is necessary for most applications. The advantage of the Hough transform is that it provides parameters to reduce the search time for finding lines based on a set of edge points, and that these parameters can be adjusted based on application requirements.

In order to understand the Hough transform we will first consider the normal (perpendicular) representation of a line:

$$\rho = r\cos(\theta) + c\sin(\theta)$$

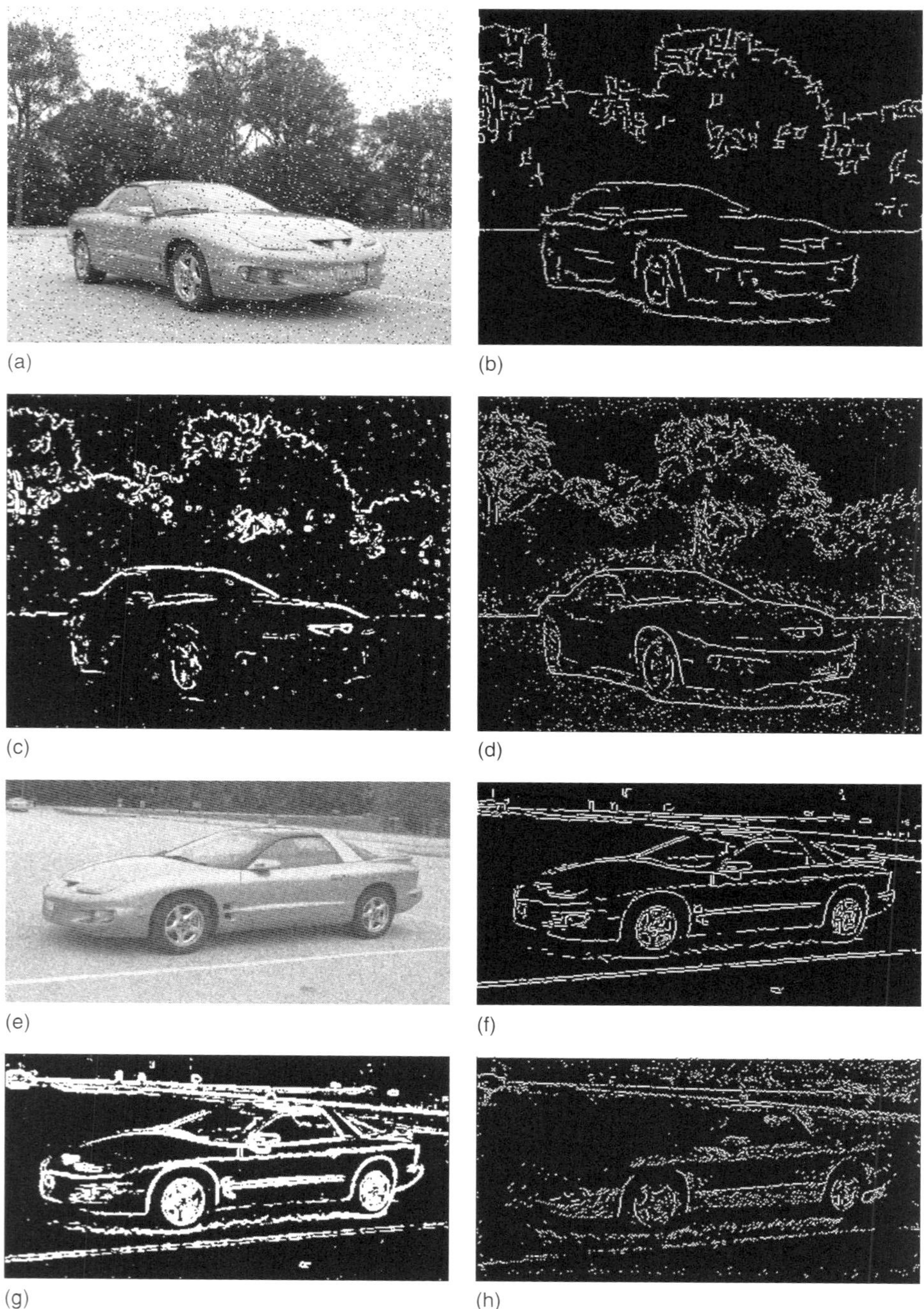

FIGURE 4.2.15

Advanced edge detectors with noisy images. (a) Original image with salt-and-pepper noise added with a probability of 3% each. (b) Canny resuts, parameters: % Low Threshold = 1, % High Threshold = 2, Variance = 2. (c) Frei–Chen results, parameters: Gaussian2 prefilter, max(edge,line), post-threshold = 190. (d) Shen–Castan results, parameters: % Low Threshold = 1, % High Threshold = 2, Smooth factor = 0.9, Window size = 7, Thin Factor = 1. (e) Original image with zero-mean Gaussian noise with a variance of 200 added. (f) Canny results, parameters: % Low Threshold = 1, % Hight Threshold = 1, Variance = 0.8. (g) Frei–Chen results, parameters: Gaussian2 prefilter, max (edge,line), post-threshold = 70. (h) Shen–Castan results, parameters: % Low Threshold = 1, % High Threshold = 2, Smooth factor = 0.9, Window size = 7, Thin Factor = 1.

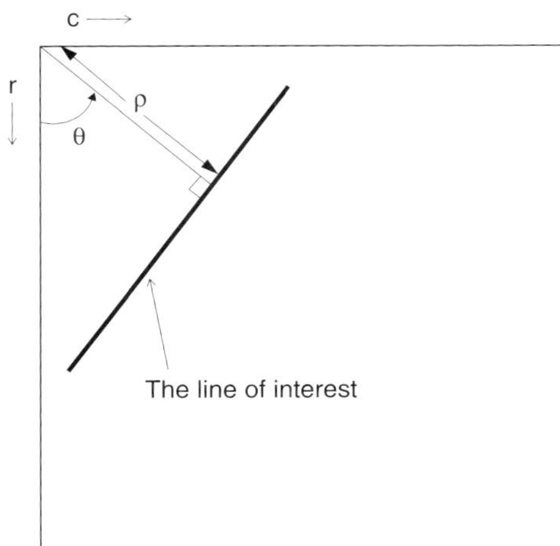

FIGURE 4.2.16
Hough transform.

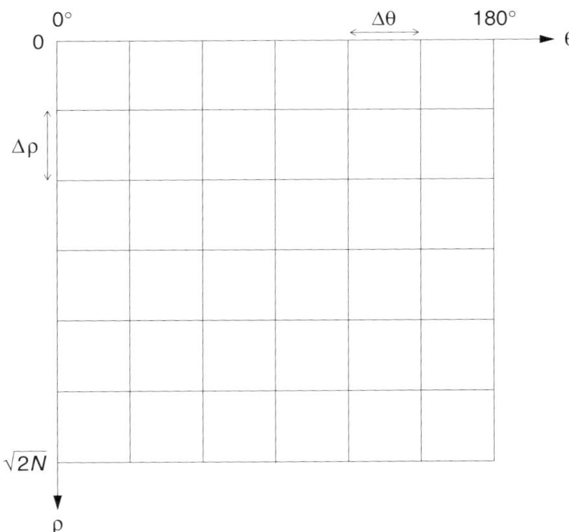

FIGURE 4.2.17
Hough space.

If we have a line in our row and column, (r, c) based image space, we can define that line by ρ, the distance from the origin to the line along a perpendicular to the line, and θ, the angle between the r-axis and the ρ-line (see Figure 4.2.16). Now, for each pair of values of ρ and θ we have defined a particular line. Consequently, just as one point in the $\rho\theta$ parameter space corresponds to a line in rc-based image space, one (r, c) point corresponds to a sinusoid in the $\rho\theta$ parameter space. A set of (r, c) points on a line will thus create a series of sinusoids in $\rho\theta$ space that intersect at the point corresponding to ρ and θ for that particular line. The range on θ is 0 to 180°, and ρ ranges from 0 to $\sqrt{2}N$, where N is the image size. Next, we can take this $\rho\theta$ parameter-space and quantize it, to reduce our search time. We quantize the $\rho\theta$ parameter-space, as shown in Figure 4.2.17, by dividing the space into a specific number of blocks. Each block corresponds to a line, or group of possible lines, with ρ and θ varying across the increment as defined by the size of the block. The size of these blocks corresponds to the coarseness of the quantization; bigger blocks provide less line resolution.

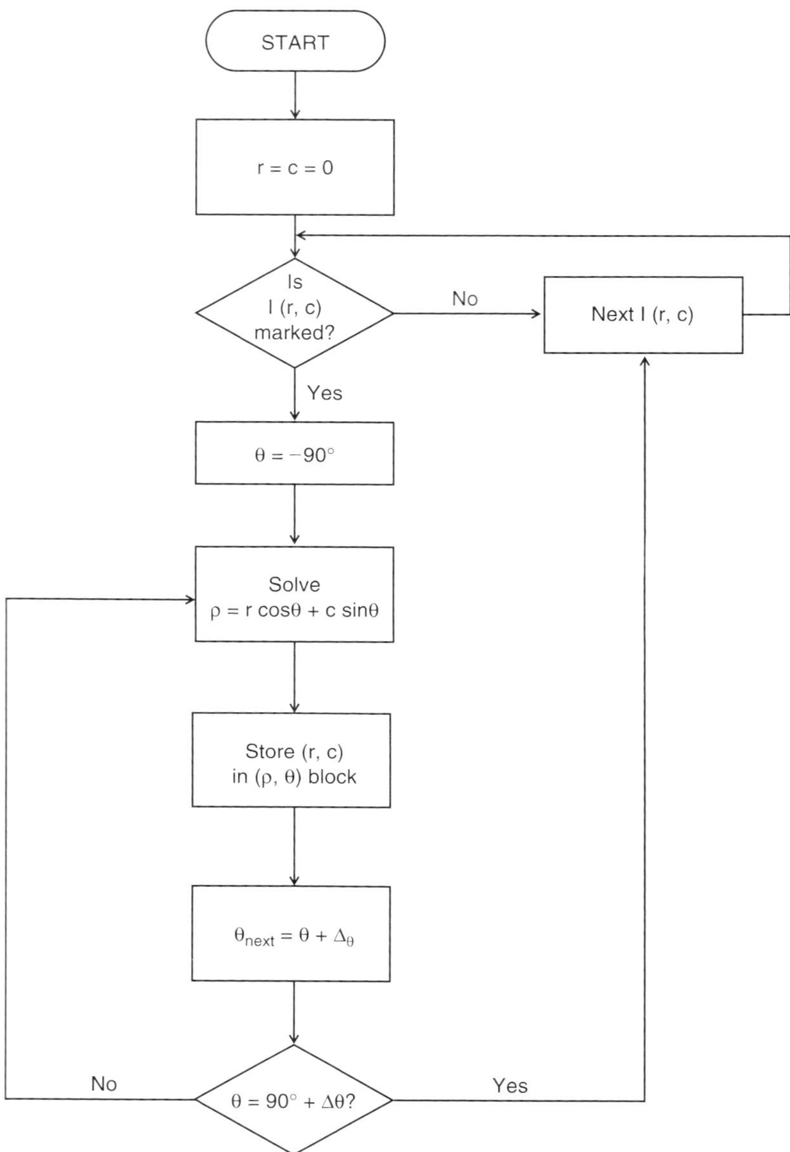

FIGURE 4.2.18
Hough transform flowchart. The flowchart is followed until *I(r, c)* have been examined.

The algorithm used for the Hough transform (see Figure 4.2.18 for a flowchart of the process) will help understand what this means. The algorithm consists of three primary steps:

1. Define the desired increments on ρ and θ, Δ_ρ *and* Δ_θ, and quantize the space accordingly.
2. For every point of interest (typically points found by edge detectors that exceed some threshold value), plug the values for r and c into the line equation:

$$\rho = r\cos(\theta) + c\sin(\theta)$$

Then, for each value of θ in the quantized space, solve for ρ.

3. For each $\rho\theta$ pair from Step 2, record the r and c pair in the corresponding block in the quantized space. This constitutes a hit for that particular block.

When this process is completed, the number of hits in each block corresponds to the number of pixels on the line as defined by the values of ρ and θ in that block. The advantage of large quantization blocks is that the search time is reduced, but the price paid is less line resolution in the image space. Examining Figure 4.2.19, we can see that this means the line of interest in the image space can vary more. One block in the Hough Space corresponds to all the solid lines in this figure—this is what we mean by reduced line resolution.

Next, select a threshold and examine the quantization blocks that contain more points than the threshold. Here, we look for continuity by searching for gaps in the line by looking at the distance between points on the line (remember the points on a line correspond to points recorded in the block). When this process is completed, the lines are marked in the output image. Note that the Hough transform will allow us to look for lines of specific orientation, if desired.

A more advanced post-processing algorithm is implemented in CVIPtools with the Hough transform. Images resulting from this algorithm searching for lines at 45 degrees are shown in Figure 4.2.20, and any these intermediate images is available as output in CVIPtools with the *Output Image* select box for the Hough transform. The algorithm works as follows:

1. Perform the Hough transform on the input image containing marked edge points, which we will call image1. The result, image2, is an image in Hough space quantized by the parameter *delta length* (ρ) and delta angle (fixed at one degree in CVIPtools).

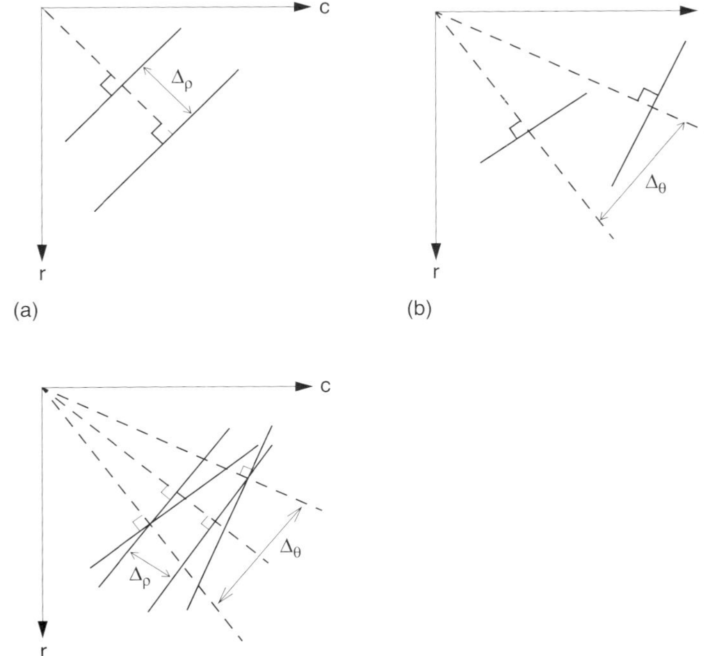

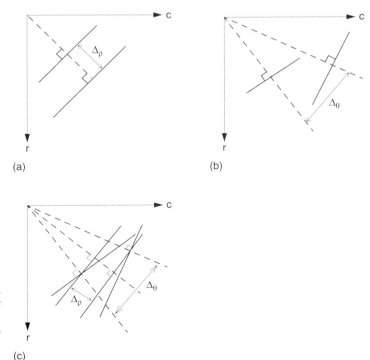

FIGURE 4.2.19
Effects of quantization block size for Hough transform. (a) Range of lines included by choice of Δ_ρ. (b) Range of lines included by choice of Δ_θ. (c) Range of lines included by choice of block size.

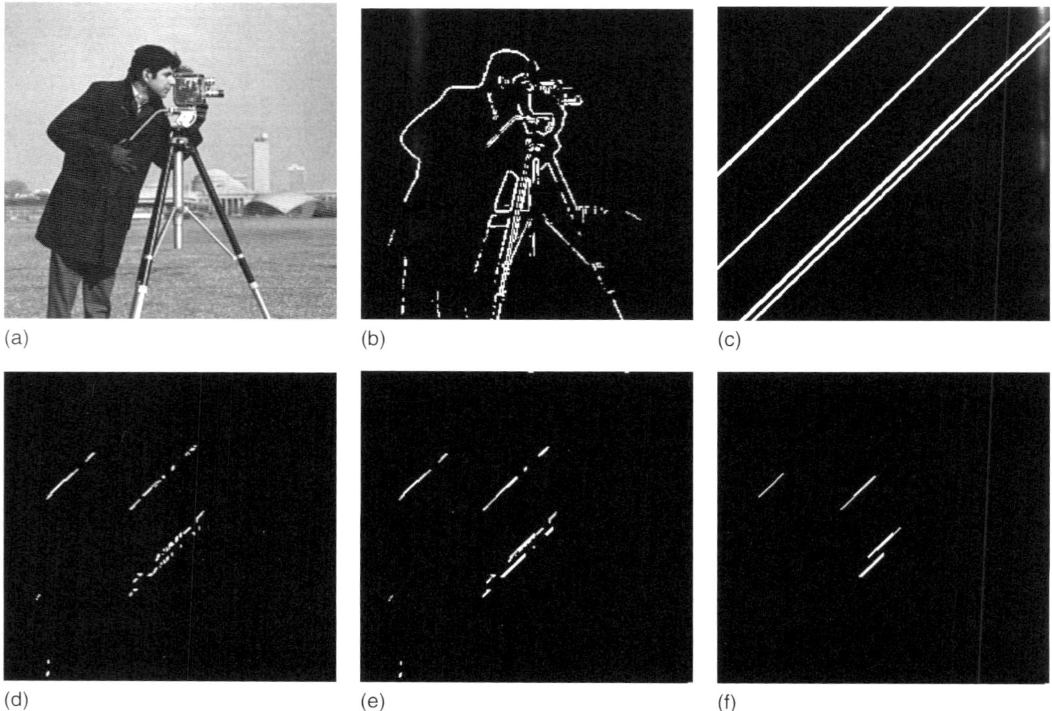

FIGURE 4.2.20
Hough transform post processing algorithm details. The Hough parameters used are as follows: Line Angles: 45 degrees, Line Pixels (min): 25, Connect distance (max): 5, Delta length: 1, Segment Length (min): 15. (a) Original image. (b) Image after applying the Kirsch edge operator and a threshold operation, (c) The mask image created from the Hough result for lines at 45 degrees. (d) Result of Logical AND of the images in (b) and (c). (e) Image (d) after snake eating, see that the camera's handle has been connected. (f) The final result after snake extinction, small dashed lines are removed. Note that we have four lines, starting from the upper left: one line corresponding to the lower part of the arm above the elbow, note that the upper part of the arm is missing as it is not quite at 45 degrees; one line for the camera handle; the next line corresponds to the part of his other arm from elbow to wrist, and the last line (the lower one) that is not a true line in the image but is created by a combination of the edge detail in that area and using a connect distance of 5.

2. Threshold image2 by using the parameter *line pixels*, which is the minimum number of pixels in a line (or in one quantization box in Hough space), and do the inverse Hough transform. This result, image3, is a mask image with lines found in the input image at the specified angle(s), illustrated in Figure 4.2.20c. Note that these lines span the entire image.
3. Perform a logical operation, image1 AND image3. The result is image4, see Figure 4.2.20d.
4. Apply an *edge linking* process to image4 to connect line segments; specifically we implemented a *snake eating algorithm*. This works as follows:

   a. A line segment is considered to be a snake. It can eat another snake within *connect distance* along *line angles*, and becomes longer (see Figure 4.2.20e). This will connect disjoint line segments.
   b. If a snake is too small, less than *segment length*, it will be extinct. This will remove small segments. The output from the snake eating algorithm is the final result, illustrated in Figure 4.2.20f.

4.2.6.1 CVIPtools Parameters for the Hough Transform

Line Angles: The range of angles for which the Hough transform will search. In CVIPtools Δ_θ is fixed at one degree.

Line Pixels (min): The minimum number of pixels a line must possess to be retained, also referred to as the threshold value in the Hough image.

Connect distance (max): Controls how far apart two line segments can be and still be connected.

Delta Length: Quantizes the Hough space ρ parameter. Controls how "thick" a line can be; note that a "thick" line might consist of multiple separate lines if they are in close proximity.

Segment Pixels (min): The minimum number of pixels in a line segment for it to be retained.

Segment Pixels controls how many pixels a solid line must have while *Line Pixels* controls how many pixels a dashed line must have.

The result of applying the Hough transform to an aircraft image is shown in Figure 4.2.21. The Sobel edge detection operator was used on the original image to provide input to the Hough transform. The Sobel edge detection results were thresholded at a gray level of about 200. The Hough transform parameter *delta length* (*rho*) was set at 1, *line pixels* (the number-of-points threshold) was set at a minimum of 20 pixels per line, and *segment length* set to 10. The figure illustrates the effects of changing the range of line angles and the *connect distance* between line segments. Although the Hough transform is an efficient line finding algorithm, when a post-processing algorithm is applied as defined above we have a boundary detection segmentation method, and these are discussed more in Section 4.3.3.

4.3 Segmentation

As was discussed at the beginning of this chapter, the goal of image segmentation is to find regions that represent objects or meaningful parts of objects. Image segmentation methods will look for objects that either have some measure of homogeneity within themselves, or have some measure of contrast with the objects on their border. The homogeneity and contrast measures can include features such as gray level, color, and texture. In Figure 4.1.1 we saw the three categories of image segmentation methods: (1) region growing/shrinking, (2) clustering, and (3) boundary detection. In this section we will examine algorithms that are representative of each of these categories.

4.3.1 Region Growing and Shrinking

Region growing and shrinking methods segment the image into regions by operating principally in the row and column, (r, c), based image space. Some of the techniques used are local, in which small areas of the image are processed at a time; others are global, with the entire image considered during processing. Methods that can combine local and global techniques, such as ***split and merge***, are referred to as state space techniques and use graph structures to represent the regions and their boundaries.

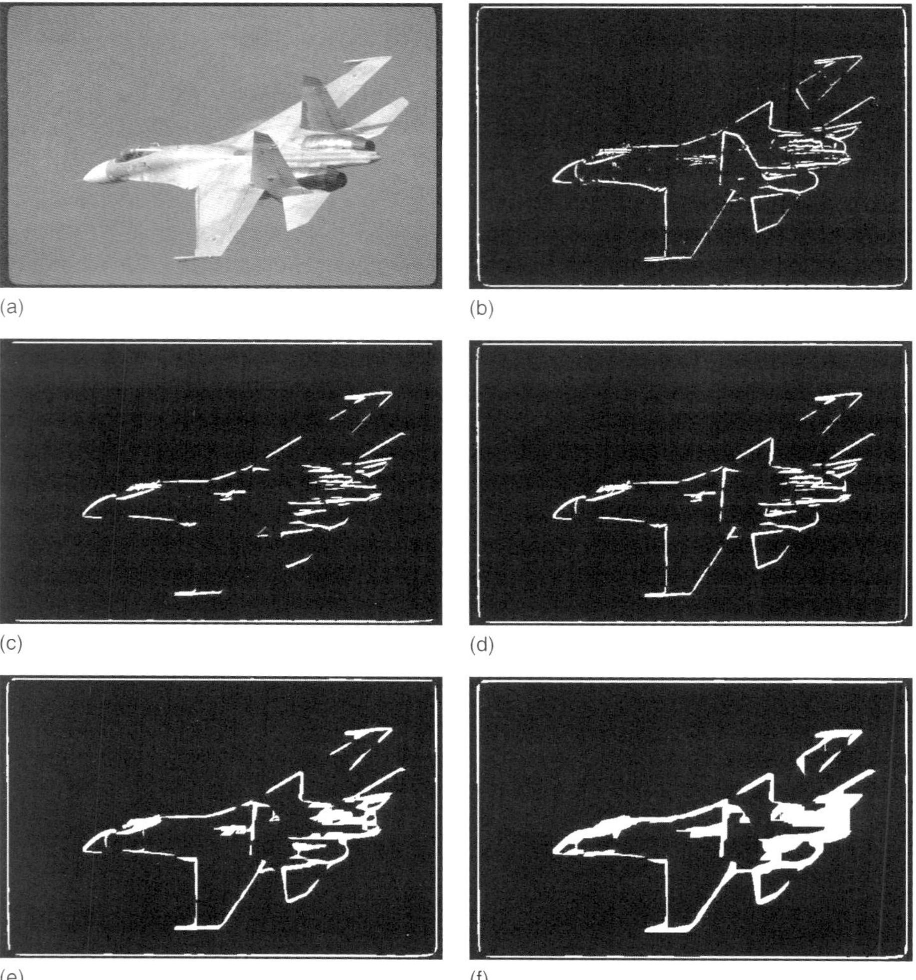

FIGURE 4.2.21
Hough transform. (a) Original image. (b) Sobel edge operator followed by a thresholding. (c) Hough output with the range of line angles = 0 to 45 degrees, delta length (ρ) = 1, minimum number of pixels per line = 20, maximum connect distance = 2, minimum segment size = 10. (d) Hough output with same parameters as (c) except range of angles from 0 to 90 degrees. (e) Hough output with same parameters as (d) except connect distance = 5. (f) Hough output with same parameters as (e) except connect distance = 10.

The data structure most commonly used for this is the quadtree. A tree is a data structure which has nodes that point to (connect) the elements. The top element is called the parent, and the connected elements are called children. In a *quadtree* each node can have four children; this is illustrated in Figure 4.3.1. This data structure facilitates the splitting and merging of regions.

Various split and merge algorithms have been described, but they all are most effective when heuristics applicable to the domain under consideration can be applied. This gives a starting point for the initial split. In general, the ***split and merge*** technique proceeds as follows:

1. Define a homogeneity test. This involves defining a homogeneity measure, which may incorporate brightness, color, texture, or other application-specific

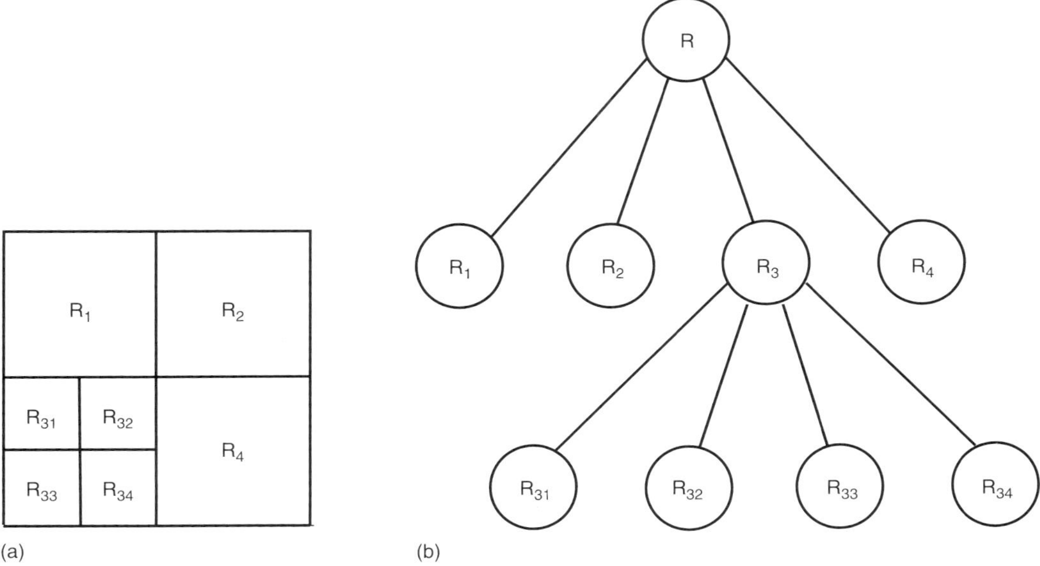

FIGURE 4.3.1
Quadtree data structure. (a) A partitioned image where R_i represents different regions. (b) The corresponding quadtree data structure.

information, and determining a criterion the region must meet to pass the homogeneity test.

2. Split the image into equal sized regions.
3. Calculate the homogeneity measure for each region.
4. If the homogeneity test is passed for a region, then a merge is attempted with its neighbor(s). If the criterion is not met, the region is split.
5. Continue this process until all regions pass the homogeneity test.

There are many variations of this algorithm. For example, we can start out at the global level, where we consider the entire image as our initial region, and then follow an algorithm similar to the above, but without any region merging. Algorithms based on splitting only are called ***multiresolution*** algorithms. Alternately, we can start at the smallest level and only merge, with no region splitting. This merge-only approach is one example of region growing methods. Often the results from all of these approaches will be quite similar, with the differences apparent only in computation time. Parameter choice, such as the minimum block size allowed for splitting, will heavily influence the computational burden as well as the resolution available in the results.

The user-defined homogeneity test is largely application-dependent, but the general idea is to look for features that will be similar within an object and different from the surrounding objects. In the simplest case we might use gray level as our feature of interest. Here we could use the gray level variance as our homogeneity measure and define a homogeneity test that required the gray level variance within a region to be less than some threshold. We can define *gray level variance* as:

$$\text{Gray level variance} = \frac{1}{N-1} \sum_{(r,c) \in \text{REGION}} [I(r,c) - \bar{I}]^2$$

where

$$\bar{I} = \frac{1}{N} \sum_{(r,c) \in \text{REGION}} [I(r,c)]$$

Note that the sum is taken over the region of interest and N is the number of pixels in the region. The variance is basically a measure of how widely the gray levels within a region vary. Higher order statistics can be used for features such as texture, and are explored in Chapter 6.

A similar approach involves searching the image for a homogeneous region and enlarging it until it no longer meets the homogeneity criteria. At this point, a new region is found that exhibits homogeneity and is grown. This process continues until the entire image is divided into regions. With this technique the initial regions are called *seed regions*, and their selection can heavily influence the resulting segmented image.

Many different homogeneity criteria have been defined, and the choice is often application dependent. In CVIPtools we have the following homogeneity criteria available:

1. *Pure uniformity*: A region is considered homogeneous if the gray levels are constant.
2. *Local mean versus global mean*: A region is considered homogeneous if the local mean is greater than the global mean.
3. *Local standard deviation versus global mean*: A region is considered homogeneous if the local standard deviation, which is the square root of the variance, is less than 10% of the global mean.
4. *Variance*: A region is considered homogeneous if a minimum percentage of the pixels, specified by the CVIPtools parameter *Percentage*, are within two standard deviations of the local mean, unless the standard deviation exceeds a maximum *Threshold* value.
5. *Weighted gray level distance*: A region is considered homogeneous if the weighted gray level value, which is based on the mode and the gray level distance from the mode weighted by the distribution, is less than a specified *Threshold* value.
6. *Texture*: A region is considered homogeneous if the four quadrants of the region have similar texture, based on five of the textural features defined in Chapter 6; specifically *energy*, *inertia*, *correlation*, *inverse difference* and *entropy*. The parameters specified are *pixel distance* and *similarity*.

Figure 4.3.2 shows results of applying the split and merge algorithm to an image with the various homogeneity criteria. The original image is 256×256 pixels and the *Entry level* parameter determines the size of the initial regions. For example, if the entry level is 1, the image is divided once (see Figure 4.3.1), so the initial region size is 128×128 for a 256×256 image. If the entry level is 2, the initial region is 64×64 for a 256×256 image, and so on. In this figure the entry level was set to 6, which provides an initial region size of $256/2^6 \times 256/2^6$, or 4×4 pixels. Note that this particular image is probably not a good candidate for texture based segmentation.

Another segmentation method we include in the region growing and shrinking category is the ***watershed segmentation algorithm.*** This method is often classified as a morphological technique because it is implemented with morphological methods (see Section 4.3.5). We include it here since it operates in the row and column based image space. The watershed algorithm is a morphological technique based on the idea of modeling a gray level image as a topographic surface, with higher gray levels

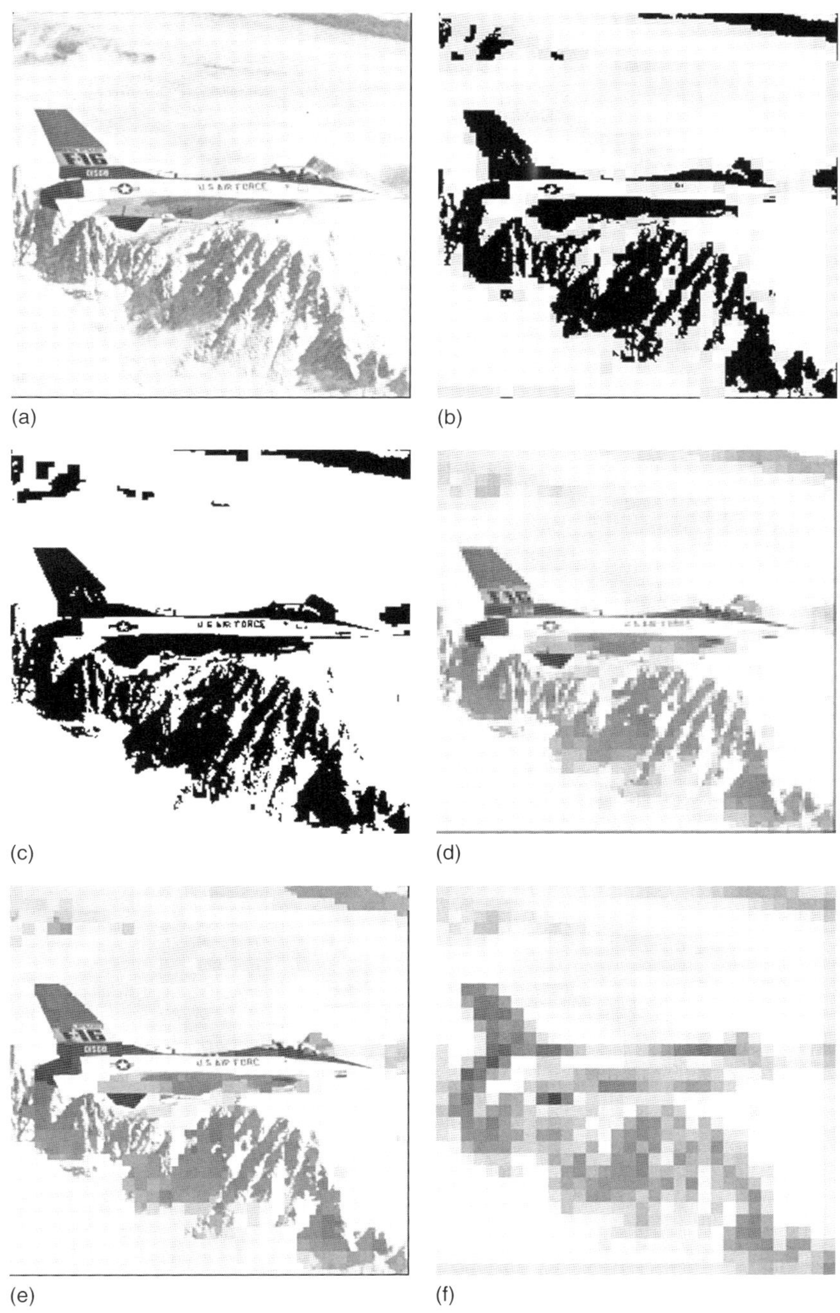

(a) (b) (c) (d) (e) (f)

FIGURE 4.3.2
Segmentation with split and merge algorithm, various homogeneity criteria. The original image is 256 × 256 pixels, and the *Entry level* parameter was set to 6. (a) Original image. (b) Local mean versus global mean. (c) Local standard deviation versus global mean. (d) Variance with Threshold = 25, Percentage = 0.7 (70%). (e) Weighted gray level distance with Threshold = 25. (f) Texture homogeneity with Similarity = 50, and Pixel distance = 2.

corresponding to higher elevations. The image is then flooded with a rainfall simulation, and pools of water are created corresponding to segments within the image. When rising water reaches a point where two pools will merge, a dam is built to prevent the merging.

Many different variations of the watershed algorithm can be implemented. The watershed segmentation algorithm as implemented in CVIPtools was initially designed to separate a single object from the background in color images. It provides the user with two parameters—merge and threshold. The *merge* parameter has a checkbox, to merge or not to merge. If merge is selected, the *threshold* parameter determines the amount of merging that will occur. The threshold parameter works by creating a histogram using the average gray value within each watershed segment. Next, it finds the maximum value in the histogram and merges this group with adjacent lower and higher gray levels until the threshold is reached. The threshold represents the percent of total area in the image.

In Figure 4.3.3 we show a skin lesion image where the goal is to separate the lesion (tumor) from normal skin. The results of the watershed segmentation are shown along

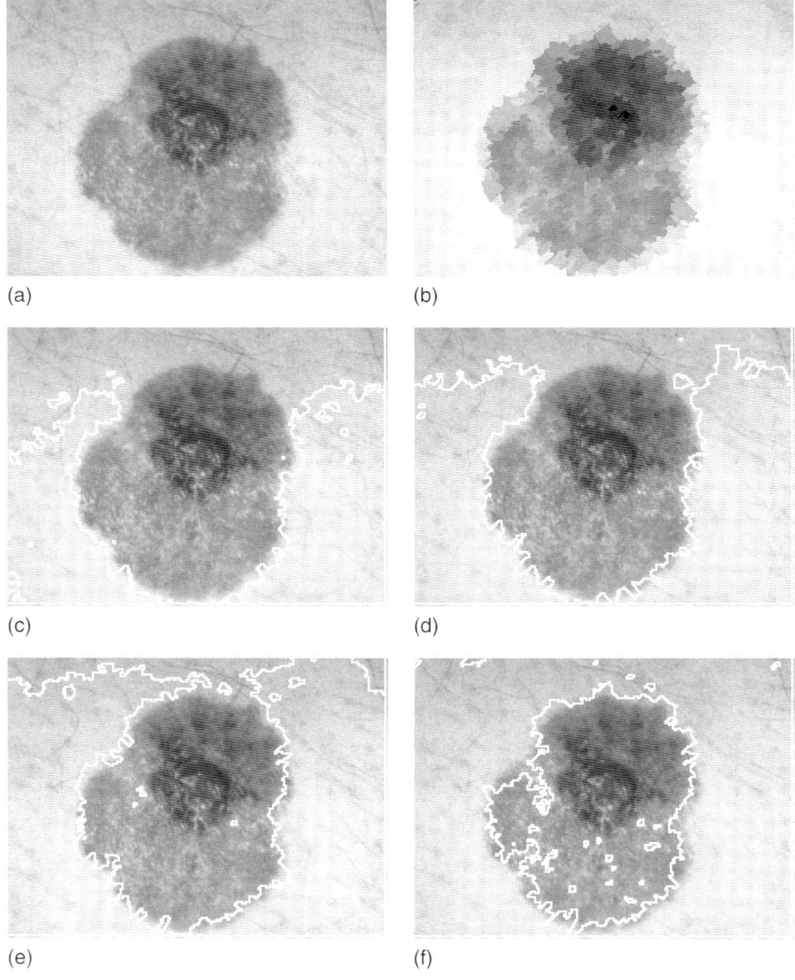

FIGURE 4.3.3
Watershed segmentation. (a) Original image of a skin lesion. (b) Result of watershed segmentation without merging. (c) Borders shown after merging with a threshold of 0.4. (d) Merge with a threshold of 0.5, (e) Merge with a threshold of 0.6. (f) Merge with a threshold of 0.7.

with various values for the threshold. In Figure 4.3.3c we can tell that the maximum histogram value corresponds to the bright area to the left and right of the lesion. As the threshold is increased in the following three images, we can see this area expand as it is merged with neighboring gray level values. Figure 4.3.4 shows the watershed segmentation algorithm applied a natural scene; and also shows borders that have been extracted after the merge process. In CVIPtools the borders are extracted with a simple threshold at 254 of the output merged image.

4.3.2 Clustering Techniques

Clustering techniques are image segmentation methods by which individual elements are placed into groups; these groups are based on some measure of similarity within the group. The major difference between these techniques and the region growing techniques is that domains other than the row and column, (r, c), based image space (the spatial domain) may be considered as the primary domain for clustering. Some of these other domains include color spaces, histogram spaces or complex feature spaces. (Note that the terms *domain* and *space* are used interchangeably here, these terms both refer to some abstract N-dimensional mathematical space, not to be confused with the spatial domain, which refers to the row and column, (r, c), image space).

FIGURE 4.3.4
Watershed segmentation. (a) Original image. (b) Result of watershed segmentation without merging. (c) Image with borders after merging with a threshold of 0.3. (d) Borders only with threshold of 0.3.

FIGURE 4.3.4 (Continued)
Watershed segmentation. (e) Image with borders after merging with a threshold of 0.6. (f) Borders only with threshold of 0.6. (g) Image with borders after merging with a threshold of 0.8. (h) Borders only with threshold of 0.8.

What is done is to look for clusters in the domain, or mathematical space, of interest. The simplest method is to divide the space of interest into regions by selecting the center or median along each dimension and splitting it there; this can be done iteratively until the space is divided into the specific number of regions needed. This method is used in the SCT/Center and PCT/Median segmentation algorithms. This method will only be effective if the space we are using and the entire algorithm is designed intelligently because the center or median split alone may not find good clusters.

The next level of complexity uses an adaptive and intelligent method to decide where to divide the space. These methods include histogram thresholding and other, more complex feature-space-based statistical methods. Representative algorithms will be discussed conceptually here, and a detailed look will be taken at two application-specific algorithms.

Recursive region splitting is a clustering method that has become a standard technique. This method uses a thresholding of histograms technique to segment the image. A set of histograms is calculated for a specific set of features, and then each of these histograms is searched for distinct peaks (see Figure 4.3.5). The best peak is selected and the image

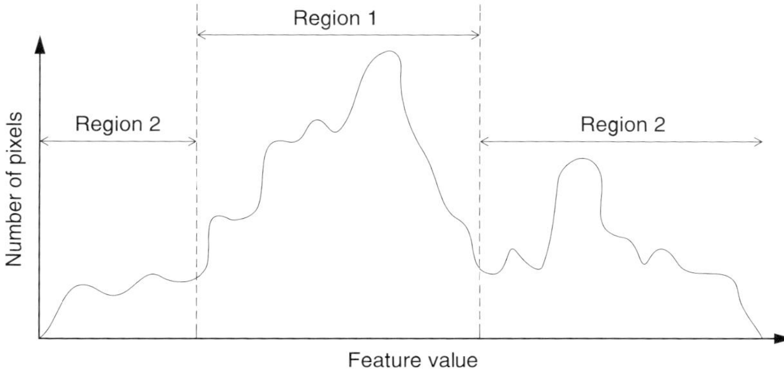

FIGURE 4.3.5
Histogram peak finding. Two thresholds are selected, one on each side of the best peak. The image is then split into two regions. Region 1 corresponds to those pixels with feature values between the selected thresholds, known as those in the peak. Region 2 consists of those pixels with feature values outside the threshold.

is split into regions based on this thresholding of the histogram. One of the first algorithms based on these concepts proceeds as follows:

1. Consider the entire image as one region and compute histograms for each component of interest (for example red, green, and blue for a color image).
2. Apply a peak finding test to each histogram. Select the best peak and put thresholds on either side of the peak. Segment the image into two regions based on this peak.
3. Smooth the binary thresholded image so only a single connected subregion is left.
4. Repeat Steps 1–3 for each region until no new subregions can be created, that is, no histograms have significant peaks.

Many of the parameters of this algorithm are application-specific. For example, what peak finding test do we use and what is a "significant" peak? An example of histogram-thresholding-based image segmentation is shown in Figure 4.3.6. In CVIPtools we have two histogram thresholding based segmentation methods, called ***histogram thresholding*** and ***fuzzy c-means***. These are explored in the exercises at the end of this chapter and details of these particular algorithms can be found in the references.

The ***SCT/Center*** color segmentation algorithm was initially developed for the identification of variegated coloring in skin tumor images. Variegated coloring is a feature believed to be highly predictive in the diagnosis of melanoma, the deadliest form of skin cancer. The spherical coordinate transform (SCT) was chosen for this segmentation method, as it decouples the color information from the brightness information. The brightness levels may vary with changing lighting conditions, so by using the two-dimensional color subspace defined by two angles (Figure 4.3.7a) we have a more robust algorithm.

If we slice a plane through the RGB color space, we can model a color triangle (Figure 4.3.7b). The vertices of the color triangle were chosen to bear some correlation to the human visual system. The placement of blue at the top of the triangle, and the way in which the spherical transform was defined, relates to the physiological fact that the cones in the human visual system that see blue are more discriminatory than the red or green sensitive cones.

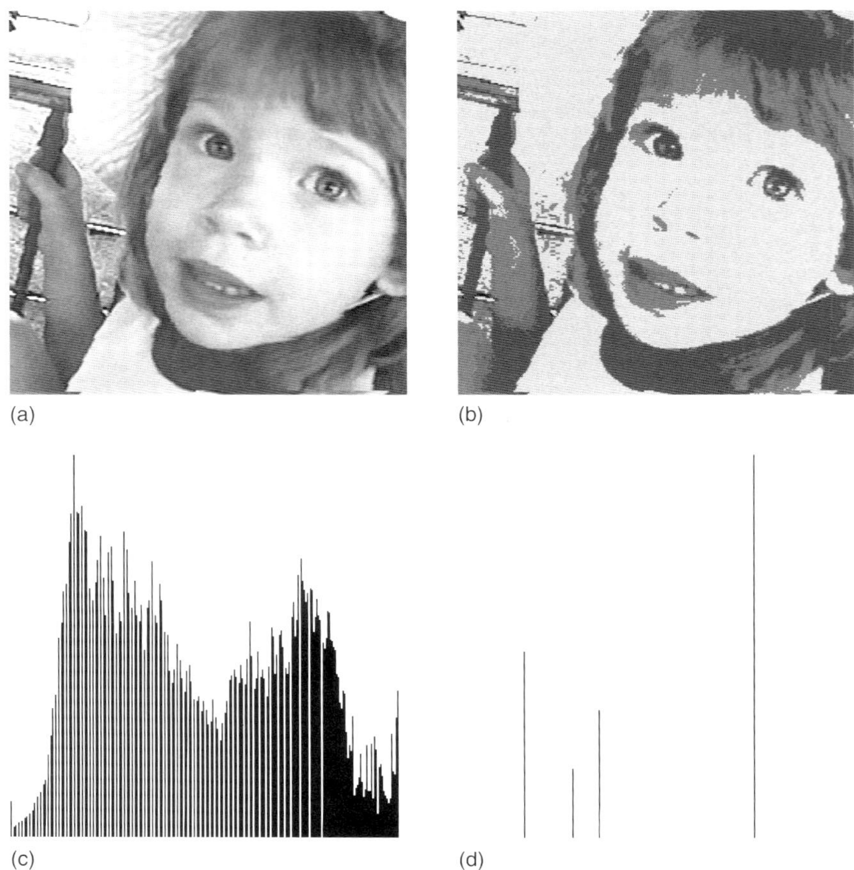

FIGURE 4.3.6
Histogram thresholding segmentation. (a) Original image. (b) Image after histogram thresholding segmentation using four gray levels. (c) Histogram of image (a). (d) Histogram of image (b).

We can segment the image by taking the color triangle and dividing it into blocks based on limits on the two angles. Figure 4.3.7c shows the shape of the resulting blocks. We can see that for a region defined by a range of minima and maxima on the two angles, the side of the region that is closest to the blue vertex is shorter than the side that is closest to the line that joins the red and green vertices.

Also, the distortion caused by the transform facilitates the perception-based aspect of the image segmentation; the closer to the perimeter of the triangle, the larger the region that is defined by a fixed angle range. This is analogous to the observation that as the white point is approached in the color space, a greater number of hues will be observable in a fixed area by the human visual system than on the perimeter of the color triangle. This observation is application-specific, since it only applies to colors from white (in the center of the triangle) to the green and red vertices. Skin tumor colors typically range from white out to the red vertex.

The SCT/Center segmentation algorithm is outlined as follows:

1. Convert the (R, G, B) triple into spherical coordinates—(L, angle A, angle B).
2. Find the minima and maxima of angle A and angle B.

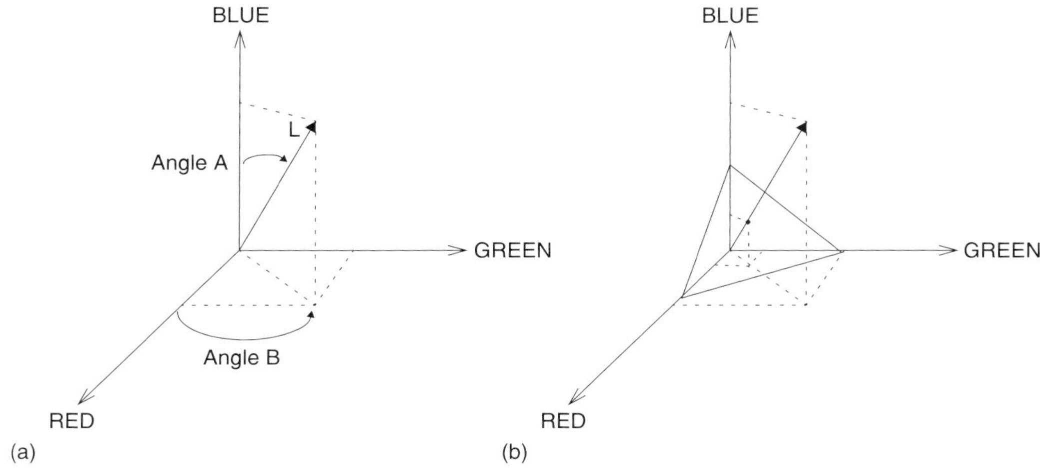

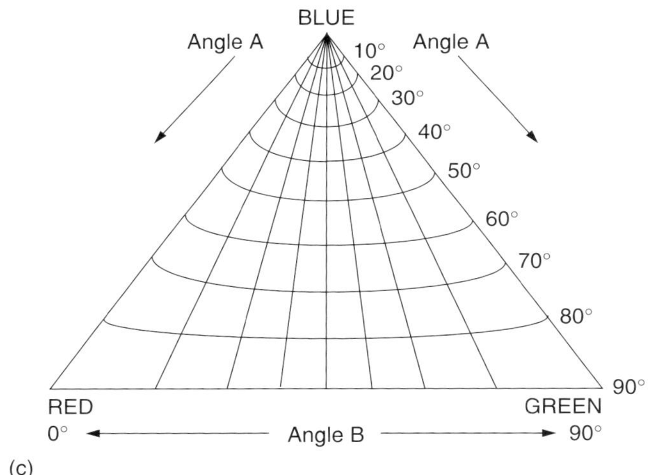

FIGURE 4.3.7
SCT/CENTER and color triangles. (a) The spherical coordinate transform separates the red, green, and blue information into a 2-D color space defined by angles A and B, and a 1-D brightness space defined by *L*. (b) The color triangle. (c) The color triangle showing regions defined by 10 degree increments on Angle A and Angle B.

3. Divide the subspace, defined by the maxima and minima, into equal-sized blocks.
4. Calculate the RGB means for the pixel values in each block.
5. Replace the original pixel values with the corresponding RGB means.

For the identification of variegated coloring in the skin tumor application it was determined that segmenting the image into four colors was optimal. An example of this segmentation method is shown in Figure 4.3.8.

The ***PCT/Median*** color segmentation algorithm was developed because, for certain features other than variegated coloring, the results provided by the previously described algorithm were not totally satisfactory. This algorithm is based around the principal components transform (PCT). The median split part of the algorithm is based on an

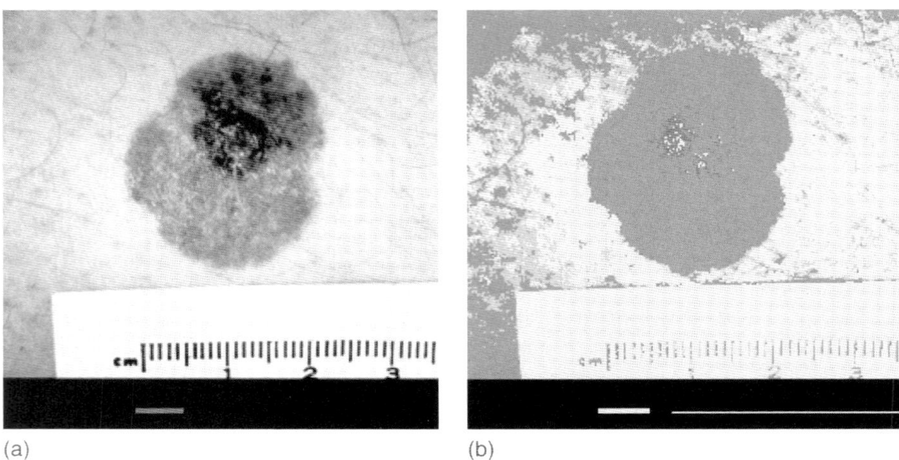

FIGURE 4.3.8
(See color insert following page 362.) SCT/Center segmentation algorithm. (a) Original image. (b) SCT/Center segmentation of skin tumor using four colors.

algorithm developed for color compression to map 24-bits per pixel color images into images requiring an average of 2-bits per pixel.

The principal components transform (defined in Chapter 5) is based on statistical properties of the image, and can be applied to any K-dimensional mathematical space. In this case, the PCT is applied to the three-dimensional color space. It was believed that the PCT used in conjunction with the median split algorithm would provide a satisfactory color image segmentation, since the PCT aligns the main axis along the maximum variance path in the data set (see Figure 4.3.9). In pattern recognition theory a feature with large variance is said to have large discriminatory power. Once we have transformed the color data so that most of the information (variance) lies along a principal axis, we proceed to divide the image into different colors by using a median split on the transformed data.

The PCT/Median segmentation algorithm proceeds as follows:

1. Find the PCT for the RGB image. Transform the RGB data using the PCT.
2. Perform the median split algorithm: find the axis that has the maximal range (initially it will be the PCT axis). Divide the data along this axis so that there are equal numbers of points on either side of the split—the median point. Continue splitting at the median along the maximum range segment until the desired number of colors is reached.
3. Calculate averages for all the pixels falling within a single parallelepiped (box).
4. Map each pixel to the closest average color values, based on a Euclidean distance measure.

For the skin tumor application it was determined that the optimum number of colors was dependent upon the feature of interest. Results of this segmentation algorithm are shown in Figure 4.3.10. Here we observe that if the image is segmented with more colors, then more of the details in the image are retained (as expected), while a smaller number of colors will segment the image on a coarser scale, leaving only relatively large features. Selection of the number of colors for segmentation has

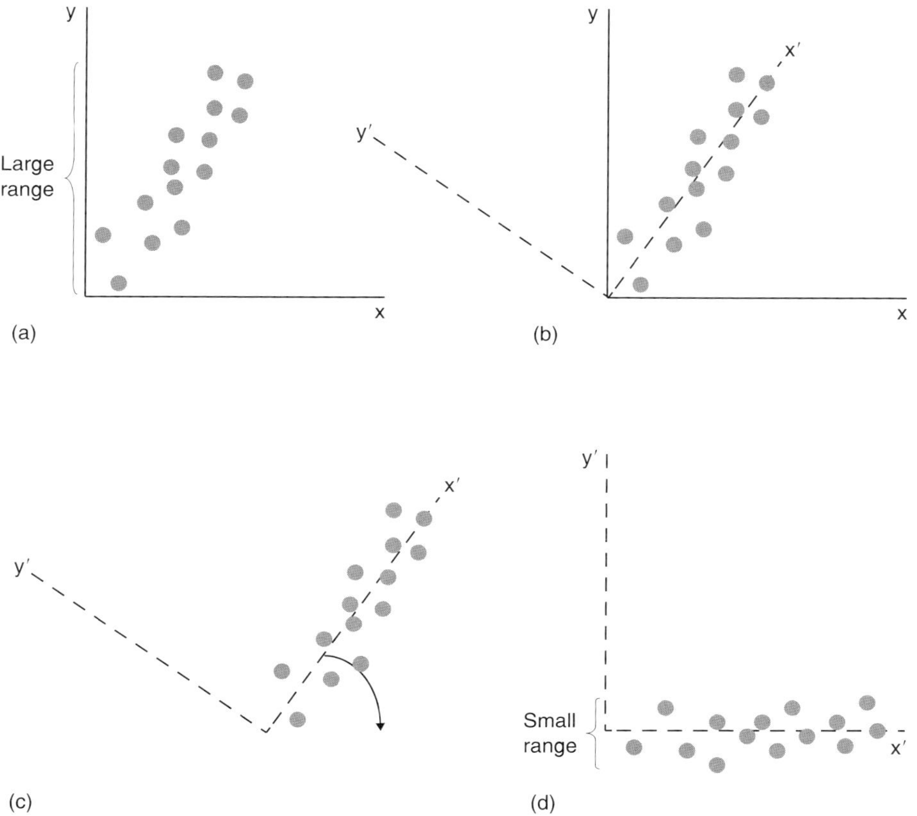

FIGURE 4.3.9
Principal components transform. (a) Original data exhibits a large range. (b) PCT aligns the main axis (x') along the maximum variance path. (c) The new axes are rotated. (d) Transformed data now has a small range. Most of the variance, or information is along the x' axis, in one dimension rather than two, as in (a).

a significant impact on the difficulty of the feature identification task—if the proper number of colors is selected for a specific feature it can make the feature identification process relatively easy.

4.3.3 Boundary Detection

Boundary detection, as a method of image segmentation, is performed by finding the boundaries between objects, thus indirectly defining the objects. This method is usually begun by marking points that may be a part of an edge. These points are then merged into line segments, and the line segments are then merged into object boundaries. The edge detectors previously described are used to mark points of rapid change, thus indicating the possibility of an edge. These edge points represent local discontinuities in specific features, such as brightness, color or texture.

After the edge detection operation has been performed, the next step is to threshold the results. One method to do this is to consider the histogram of the edge detection results, looking for the best valley manually (Figure 4.3.11). With a bimodal histogram, a histogram with two major peaks, an analytical solution is available to find a good threshold value. A bimodal histogram is typical for computer applications where we

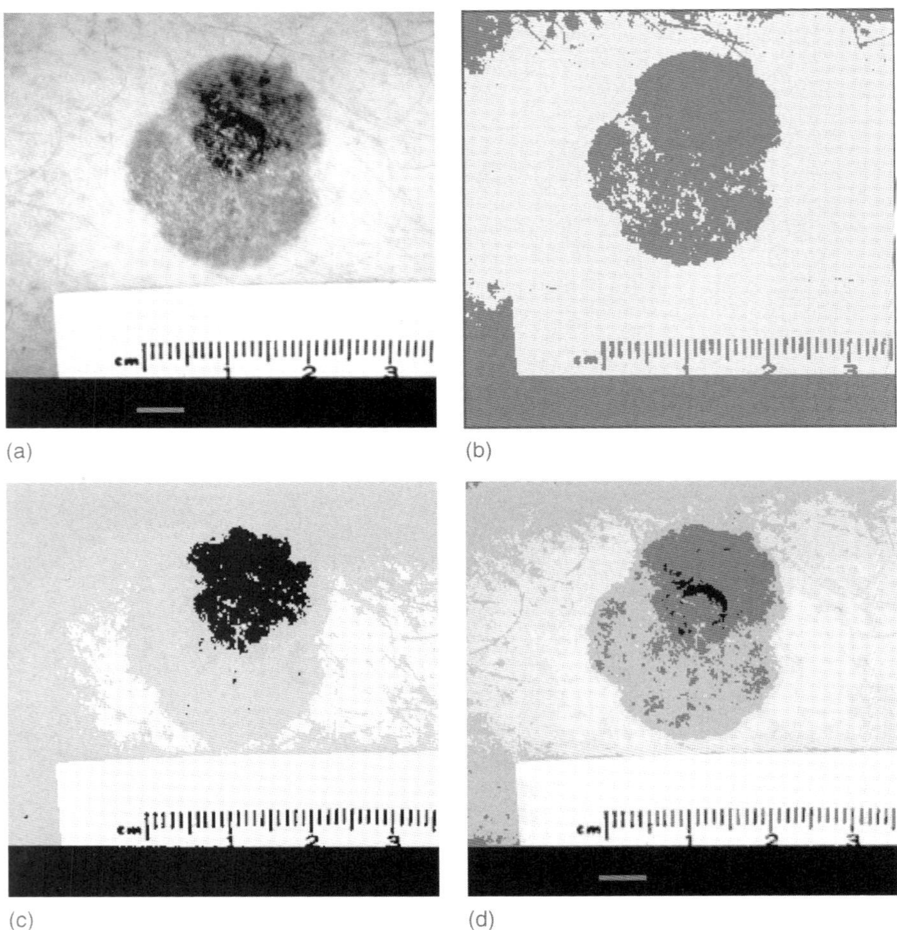

FIGURE 4.3.10
(See color insert following page 362.) PCT/Median segmentation algorithm. (a) Original image. (b) PCT/Median segmented image with two colors. (c) PCT/Median segmented image with four colors. (d) PCT/Median segmented image with eight colors.

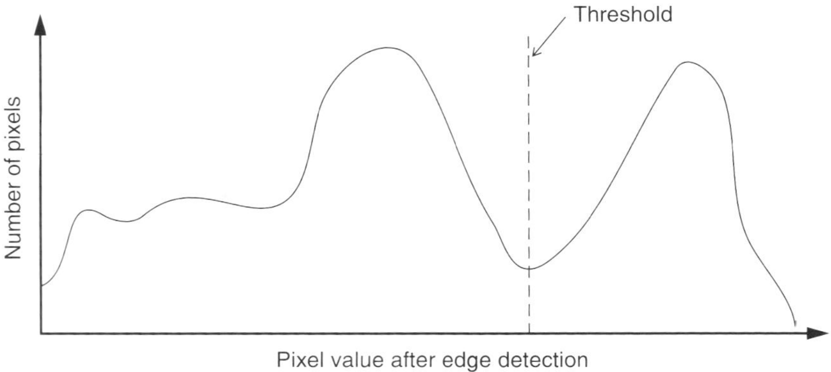

FIGURE 4.3.11
Edge detection threshold. This method works best with a bimodal (two peaks) histogram.

have one object against a background of high contrast. This method provides a theoretically good solution based on the assumption that each peak has a Gaussian shape and the peaks are fairly well separated. This method is called *minimizing within group variance*, or the *Otsu method*, and works as follows:

Let $P(i)$ be the histogram probability for gray level i, which is simply the count of the number of pixels at gray level g normalized by the total number of pixels in the image, and is given by:

$$P(g) = \frac{1}{(\#\text{ Rows})(\#\text{ Columns})} \sum_{I(r,c)=g} \frac{I(r,c)}{g};$$

where (# Rows)(# Columns) is the total number of pixels

Let $\sigma_w^2(t)$ be the within group variance, which is a weighted sum of the variance of the two groups, as a function of the threshold t, defined as follows:

$$\sigma_w^2(t) = P_1(t)\sigma_1^2(t) + P_2(t)\sigma_2^2(t)$$

where

$$P_1(t) = \sum_{g=1}^{t} P(g)$$

$$P_2(t) = \sum_{g=t+1}^{Maxgray} P(g)$$

$$\mu_1(t) = \sum_{g=1}^{t} g \times P(g)/P_1(t)$$

$$\mu_2(t) = \sum_{g=t+1}^{Maxgray} g \times P(g)/P_2(t)$$

$$\sigma_1^2(t) = \sum_{g=1}^{t} [g - \mu_1(t)]^2 P(g)/P_1(t)$$

$$\sigma_2^2(t) = \sum_{g=t+1}^{Maxgray} [g - \mu_2(t)]^2 P(g)/P_2(t)$$

where *Maxgray* is the maximum gray level value.

Now we simply find the value of the threshold t that will minimize the within group variance, $\sigma_w^2(t)$. This can done calculating the values for $\sigma_w^2(t)$ for each possible gray level value and selecting the one that provides the smallest $\sigma_w^2(t)$. We can usually streamline this search by limiting the possible threshold values to those between the modes, the two peaks, in the histogram.

Often, the histogram of an image that has been operated on by an edge detector is unimodal (one peak), so it may be difficult to find a good valley. A method that provides reasonable results for unimodal histograms is to use the *average value* for the threshold,

FIGURE 4.3.12
Average value thresholding. (a) Original image, (b) Image after Sobel edge detector. (c) Unimodal histogram of image after Sobel. (d) Sobel image after thresholding with average value.

as in Figure 4.3.12. With very noisy images and a unimodal histogram, a good rule of thumb is to use 10% to 20% of the maximum value as a threshold. An example of this is shown in Figure 4.3.13.

After we have determined a threshold for the edge detection, we need to merge the existing edge segments into boundaries. This is done by edge linking. The simplest approach to edge linking involves looking at each point that has passed the threshold test, and connecting it to all other such points that are within a maximum distance. This method tends to connect many points and is not useful for images where too many points have been marked; it is most applicable to simple images.

Instead of thresholding and then edge linking, we can perform edge linking on the edge detected image before we threshold it. If this approach is used, we look at small neighborhoods (3×3 or 5×5) and link similar points. Similar points are defined as having close values for both magnitude and direction. The entire image undergoes this process, while keeping a list of the linked points. When the process is complete, the boundaries are determined by the linked points.

The *Hough transform* combined with the *snake eating edge linking algorithm* described in the previous section is one method to use the Hough transform for segmentation

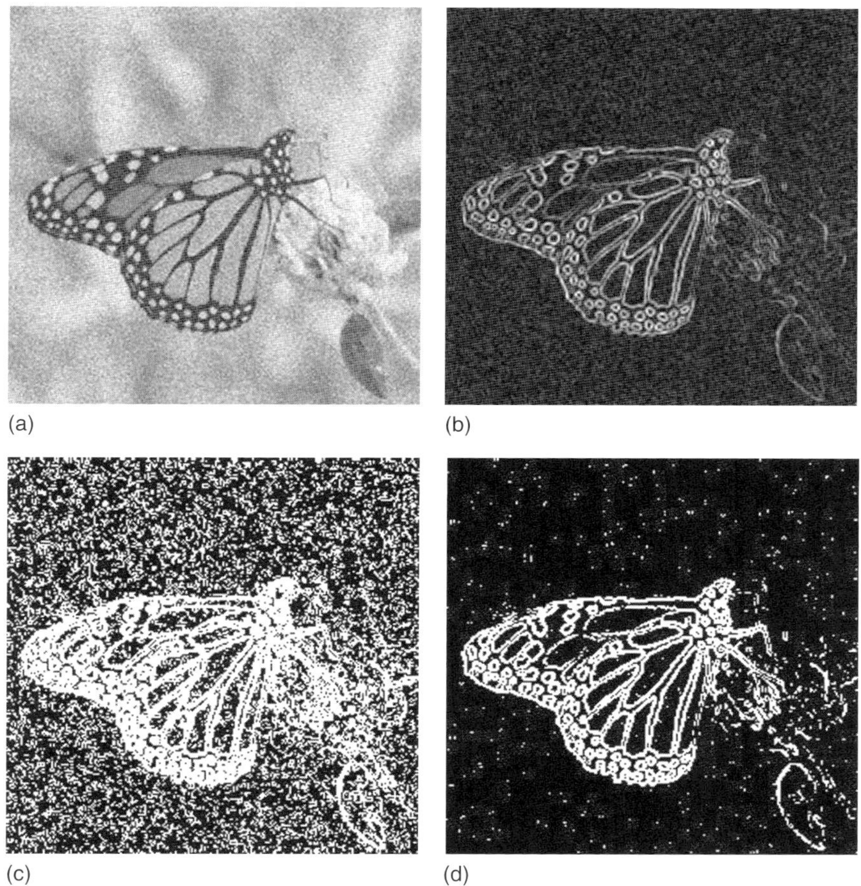

FIGURE 4.3.13
Thresholding noisy images. (a) Original image with Gaussian noise added (zero mean, variance = 400). (b) Sobel edge detector results. (c) Threshold on Sobel at 10% of maximum value. (d) Threshold on Sobel at 20% of maximum.

via boundary detection. However, if we are searching for specific geometric shapes we can extend the Hough transform to search for any geometric shape that can be described by mathematical equations, such as circles, ellipses or parabolas. The line finding Hough transform we discussed previously was defined by quantizing the parameter space that defined the lines, specifically the mathematical space defined by the parameters ρ and θ. To extend this concept we simply define a parameter vector and apply the Hough algorithm to this new parameter space. The extended Hough transform can be applied to any geometric shape that can be described by an equation of the following form:

$$f(r, c; \bar{p}) = 0$$

where $f(\cdot)$ is any function of the row and column coordinates, (r, c), and a parameter vector $\bar{p}$.

In the case of the line finding Hough transform, the function is:

$$\rho = r\ cos(\theta) + c\ sin(\theta)$$

and the parameter vector is:

$$\bar{p} = \begin{bmatrix} \rho \\ \theta \end{bmatrix}$$

In the case of a circle, with the equation of a circle as follows, where a and b are the center coordinates of the circle and d is the diameter:

$$(r - a)^2 + (c - b)^2 = \left(\frac{d}{2}\right)^2$$

The parameter vector is:

$$\bar{p} = \begin{bmatrix} a \\ b \\ d \end{bmatrix}$$

If we applied the Hough transform to find circles we would follow the same procedure as we did for line finding, but with an increased dimensionality to the search space—it is now a three-dimensional parameter space.

If we desire to search for general geometric shapes that are not readily described by parametric equations, such as the circle in the previous example, a generalized Hough transform can be used. The generalized Hough works by creating a description of the shape defined by a reference point and a table of lines, called an R-table. The reference point is chosen inside the sample shape, and a random line is found from the reference point to a point on the border. This intersection information is recorded in the table. The shape is then described by a multitude of line intersection information in the R-table. The generalized Hough algorithm is then used to search for shapes described by this table; details of this algorithm can be found in the references.

4.3.4 Combined Segmentation Approaches

Image segmentation methods may actually be a combination of region growing methods, clustering methods, and boundary detection. As previously mentioned, we could consider the region growing methods, to be a subset of the clustering methods, by allowing the space of interest to include the row and column parameters. Quite often, in boundary detection, heuristics applicable to the specific domain must be employed in order to find the true object boundaries. What is considered noise in one application may be the feature of interest in another application. Finding boundaries of different features, such as texture, brightness, or color, and applying artificial intelligence techniques at a higher level to correlate the feature boundaries found to the specific domain may give the best results. Optimal image segmentation is likely to be achieved by focusing on the application, and on how the different methods can be used, singly or in combination, to achieve the desired results.

4.3.5 Morphological Filtering

Morphology relates to the structure or form of objects. Morphological filtering simplifies a segmented image to facilitate the search for objects of interest. This is done by smoothing out object outlines, filling small holes, eliminating small projections, and with other

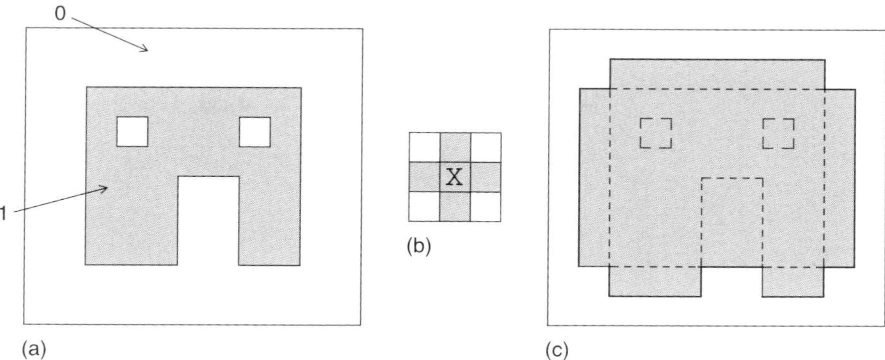

FIGURE 4.3.14
Dilation. (a) Original image. (b) Structuring element; x = origin. (c) Image after dilation; original in dashes.

similar techniques. While this section will focus on applications to binary images, the extension of the concepts to gray level images will also be discussed. We will look at the different types of operations available and at some examples of their use.

The two principal morphological operations are dilation and erosion. *Dilation* allows objects to expand, thus potentially filling in small holes and connecting disjoint objects. *Erosion* shrinks objects by etching away (eroding) their boundaries. These operations can be customized for an application by the proper selection of the *structuring element*, which determines exactly how the objects will be dilated or eroded. Basically, the structuring element is used to probe the image to find how it will fit, or not fit, into the image object(s).

The dilation process is performed by laying the structuring element on the image and sliding it across the image in a manner similar to convolution. The difference is in the operation performed. It is best described in a sequence of steps:

1. If the origin of the structuring element coincides with a "0" in the image, there is no change; move to the next pixel.
2. If the origin of the structuring element coincides with a "1" in the image, perform the OR logic operation on all pixels within the structuring element.

An example is shown in Figure 4.3.14. Note that with a dilation operation, all the "1" pixels in the original image will be retained, any boundaries will be expanded, and small holes will be filled.

Example 4.3.1

Given the following image and structuring element, perform a dilation operation. We assume the origin of the structuring element is in the center and ignore cases where the structuring element extends beyond the image. Note that since the holes are all smaller than the structuring element, they are all filled.

STRUCTURING ELEMENT

$$\begin{bmatrix} 1 & 0 & 0 \\ 1 & 1 & 1 \\ 1 & 0 & 0 \end{bmatrix}$$

IMAGE

$$\begin{bmatrix} 1 & 1 & 1 & 1 & 1 & 1 & 1 \\ 1 & 0 & 0 & 1 & 0 & 0 & 1 \\ 1 & 1 & 1 & 1 & 1 & 1 & 1 \\ 1 & 0 & 0 & 0 & 0 & 0 & 1 \\ 1 & 1 & 1 & 1 & 1 & 1 & 1 \\ 1 & 1 & 1 & 1 & 1 & 1 & 1 \\ 1 & 1 & 0 & 0 & 1 & 1 & 1 \end{bmatrix}$$

RESULT

$$\begin{bmatrix} 1 & 1 & 1 & 1 & 1 & 1 & 1 \\ 1 & 1 & 1 & 1 & 1 & 1 & 1 \\ 1 & 1 & 1 & 1 & 1 & 1 & 1 \\ 1 & 1 & 1 & 1 & 1 & 1 & 1 \\ 1 & 1 & 1 & 1 & 1 & 1 & 1 \\ 1 & 1 & 1 & 1 & 1 & 1 & 1 \\ 1 & 1 & 1 & 1 & 1 & 1 & 1 \end{bmatrix}$$

The erosion process is similar to dilation, but we turn pixels to "0", not "1". As before, slide the structuring element across the image, and:

1. If the origin of the structuring element coincides with a "0" in the image, there is no change; move to the next pixel.
2. If the origin of the structuring element coincides with a "1" in the image, and any of the "1" pixels in the structuring element extend beyond the object ("1" pixels) in the image, then change the "1" pixel in the image, whose location corresponds to the origin of the structuring element, to a "0".

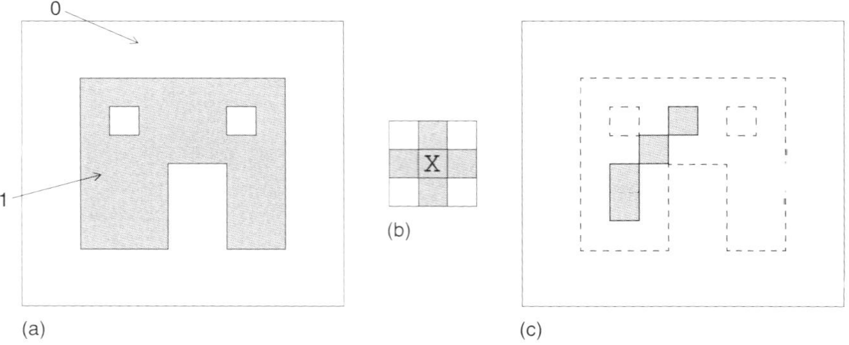

FIGURE 4.3.15
Erosion. (a) Original image. (b) Structuring element; x = origin. (c) Image after erosion; original in dashes.

In Figure 4.3.15, the only remaining pixels are those that coincide to the origin of the structuring element where the entire structuring element was contained in the existing object. Since the structuring element is three pixels wide, the two-pixel-wide right "leg" of the image object was eroded away, but the three-pixel-wide left "leg" retained some of its center pixels.

Example 4.3.2

Given the following image and structuring element, perform an erosion operation. We assume the origin of the structuring element is in the center and ignore cases where the structuring element extends beyond the image. Note the only 1's left inside the image mark places where the shape of the structuring element exists in the image.

STRUCTURING ELEMENT

$$\begin{bmatrix} 1 & 0 & 0 \\ 1 & 1 & 1 \\ 1 & 0 & 0 \end{bmatrix}$$

IMAGE

$$\begin{bmatrix} 1 & 1 & 1 & 1 & 1 & 1 & 1 \\ 1 & 0 & 0 & 1 & 0 & 0 & 1 \\ 1 & 1 & 1 & 1 & 1 & 1 & 1 \\ 1 & 0 & 0 & 0 & 0 & 0 & 1 \\ 1 & 1 & 1 & 1 & 1 & 1 & 1 \\ 1 & 1 & 1 & 1 & 1 & 1 & 1 \\ 1 & 1 & 0 & 0 & 1 & 1 & 1 \end{bmatrix}$$

RESULT

$$\begin{bmatrix} 1 & 1 & 1 & 1 & 1 & 1 & 1 \\ 1 & 0 & 0 & 0 & 0 & 0 & 1 \\ 1 & 1 & 0 & 0 & 1 & 0 & 1 \\ 1 & 0 & 0 & 0 & 0 & 0 & 1 \\ 1 & 0 & 0 & 0 & 1 & 0 & 1 \\ 1 & 1 & 1 & 0 & 0 & 1 & 1 \\ 1 & 1 & 0 & 0 & 1 & 1 & 1 \end{bmatrix}$$

These two basic operations, dilation and erosion, can be combined into more complex sequences. The most useful of these for morphological filtering are called opening and closing. *Opening* consists of an erosion followed by a dilation, and can be used to eliminate all pixels in regions that are too small to contain the structuring element. In this case the structuring element is often called a *probe*, as it is probing the image looking for small objects to filter out of the image. In Figure 4.3.16 we can see that opening expands holes ("opens" them up) and may erode edges, in a way that depends

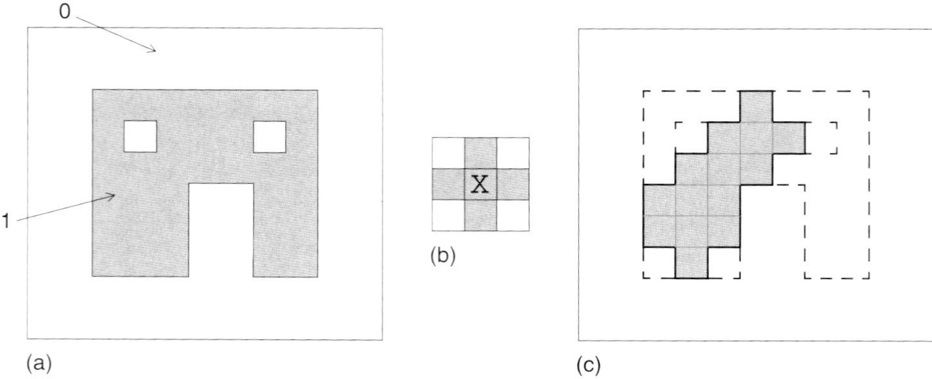

FIGURE 4.3.16
Opening. (a) Original image. (b) Structuring element; x = origin. (c) Image after opening = erosion followed by dilation.

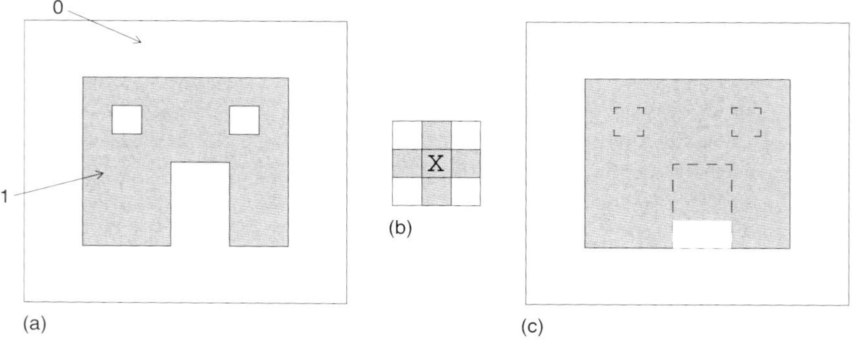

FIGURE 4.3.17
Closing. (a) Original image. (b) Structuring element x = origin. (c) Image after closing = dilation followed by erosion; original in dashes.

on the structuring element shape. The output image tends to take a shape similar to the structuring element itself.

Closing consists of a dilation followed by erosion, and can be used to fill in holes and small gaps as shown in Figure 4.3.17. Here we see that the two small holes have been closed and the gap has been partially filled; if a different structuring element is used the results will be similar but different (see some of the following figures). Comparing Figure 4.3.16 to Figure 4.3.17, we see that the order of operation is important. Closing and opening will have different results even though both consist of an erosion and a dilation.

The following two figures show results of dilation from varying the shape and size of the structuring element. The original image is a microscopic image of a cell that has undergone a threshold operation to create a binary image. Figure 4.3.18 illustrates dilation using different shape structuring elements. Here we see that the small objects, as well as edges on larger objects, will take on the shape of the structuring element itself. In Figure 4.3.19 we see that effect of using the same shape structuring element, but increasing the size of the structuring element—as it gets larger the size of the holes that get filled increases. Here we also see that small objects are merged together by dilation, and that the degree of the merging depends on the size of the structuring element. Also

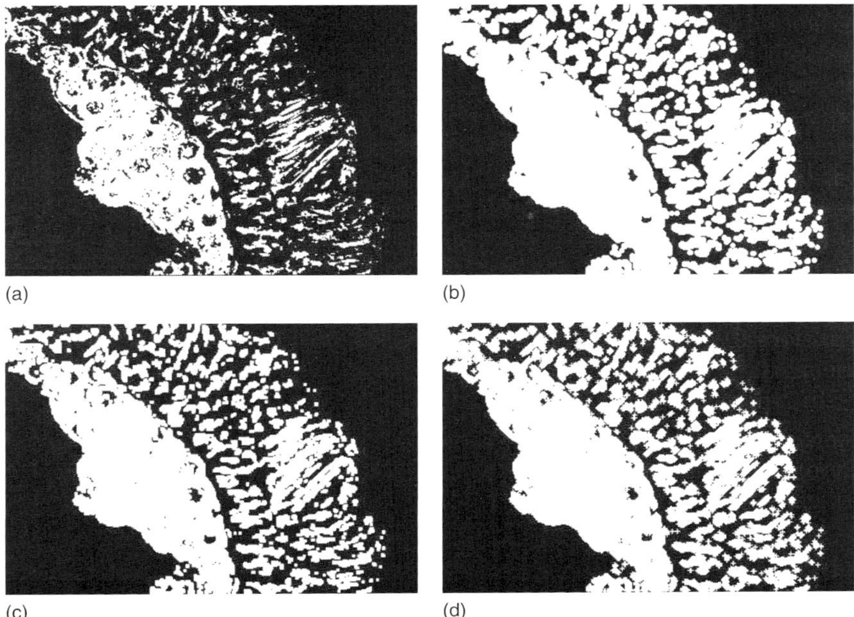

FIGURE 4.3.18
Binary dilation with various shape structuring elements. (a) Original image, a microscopic cell image that has undergone a threshold operation (Original image courtesy of Sara Sawyer, SIUE). (b) Dilation with a circular structuring element. (c) Dilation with a square structuring element. (d) Dilation with a cross shape structuring element.

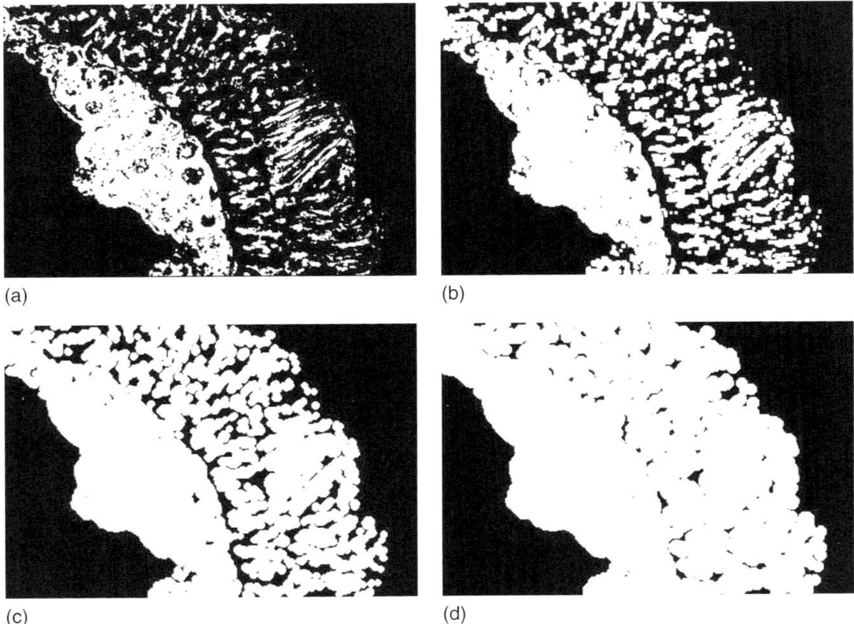

FIGURE 4.3.19
Dilation with different size structuring elements. (a) Original image. (b) Dilation with a circular structuring element of size 3. (c) Dilation with a circular structuring element of size 7. (d) Dilation with a circular structuring element of size 11.

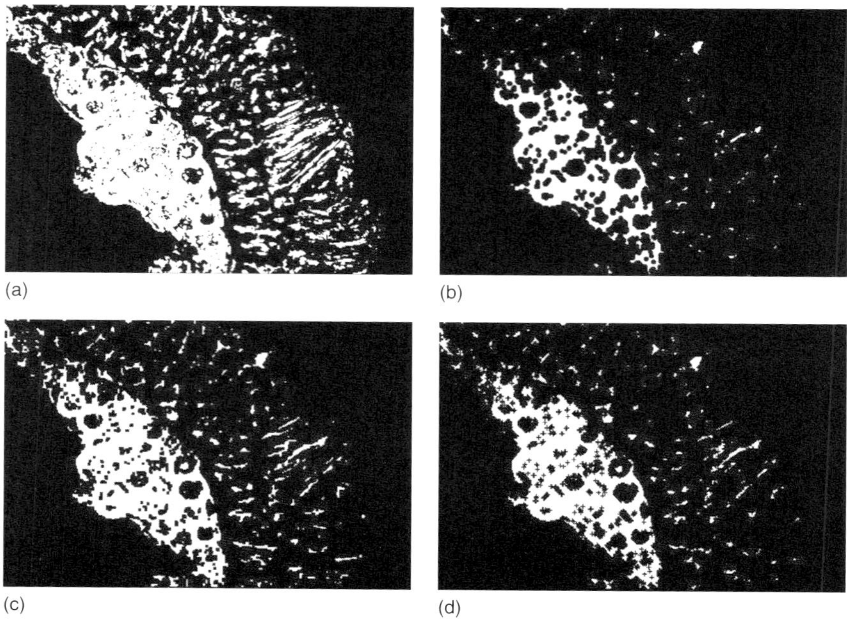

FIGURE 4.3.20
Binary erosion with various shape structuring elements. (a) Original image, a microscope cell image that has undergone a threshold operation. (b) Erosion with a circular structuring element. (c) Erosion with a square structuring element. (d) Erosion with a cross shape structuring element.

note that in Figure 4.3.19b, even though a circular structuring element was used, the small objects appear to be rectangular—why is this? (Hint: consider the shape of a binary circle on a 4×4 rectangular grid.)

Figure 4.3.20 illustrates erosion using different shape structuring elements. Here we see that the holes, as well as edges on larger objects, will take on the shape of the structuring element itself. Figures 4.3.21 and 4.3.22 show the results of opening and closing using various shape structuring elements. Here we can see how the shape of the structuring elements affects the results of these operations. In Figure 4.3.23 we see a comparison of opening and closing with different size circular structuring elements.

Another approach to binary morphological filtering is based on an iterative approach. The usefulness of this approach lies in its flexibility. It is based on a definition of six-connectivity, in which each pixel is considered connected to its horizontal and vertical neighbors, but to only two diagonal neighbors (the two on the same diagonal). This connectivity definition is equivalent to assuming that the pixels are laid out on a hexagonal grid, which can be simulated on a rectangular grid by assuming that each row is shifted by one-half a pixel (see Figure 4.3.24). With this definition a pixel can be surrounded by 14 possible combinations of 1's and 0's, as seen in Figure 4.3.25; we call these different combinations surrounds. For this approach to morphological filtering, we define:

1. The set of surrounds S, where $a = 1$.
2. A logic function, $L(a, b)$, where b is the current pixel value, and the function specifies the output of the morphological operation.
3. The number of iterations, n.

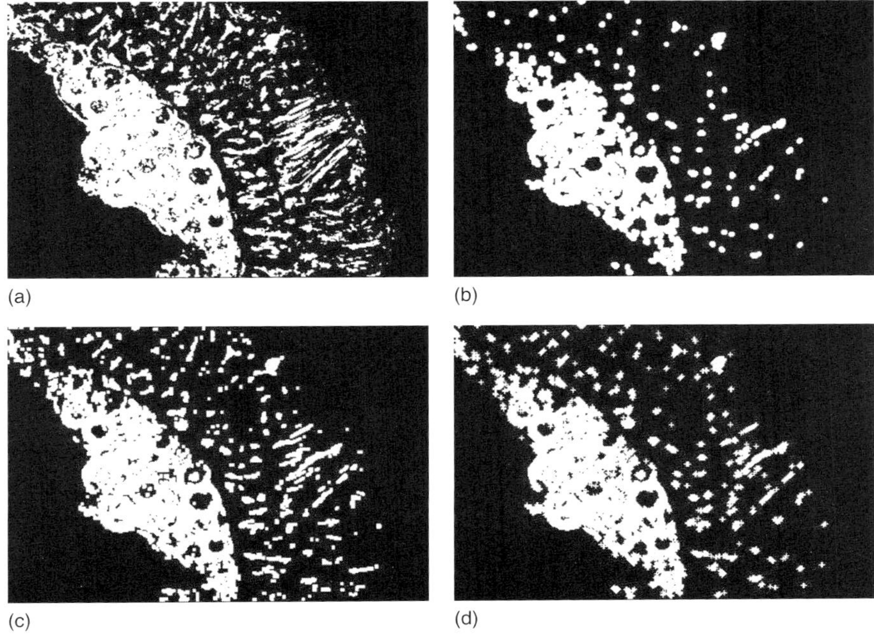

(a) (b) (c) (d)

FIGURE 4.3.21
Binary opening with various shape structuring elements. (a) Original image, a microscope cell image that has undergone a threshold operation. (b) Opening with a circular structuring element. (c) Opening with a square structuring element. (d) Opening with a cross shape structuring element.

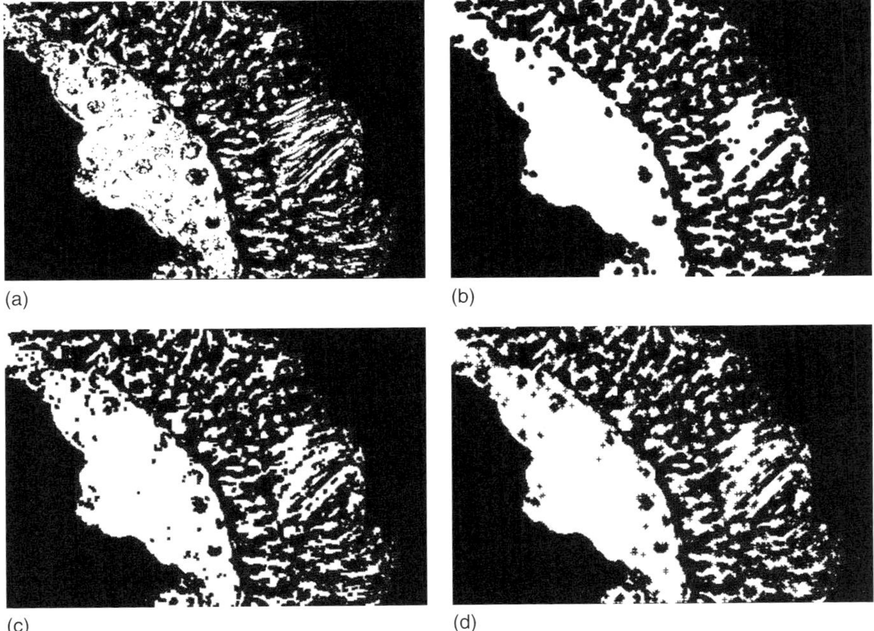

(a) (b) (c) (d)

FIGURE 4.3.22
Binary closing with various shape structuring elements. (a) Original image, a microscope cell image that has undergone a threshold operation. (b) Closing with a circular structuring element. (c) Closing with a square structuring element. (d) Closing with a cross shape structuring element.

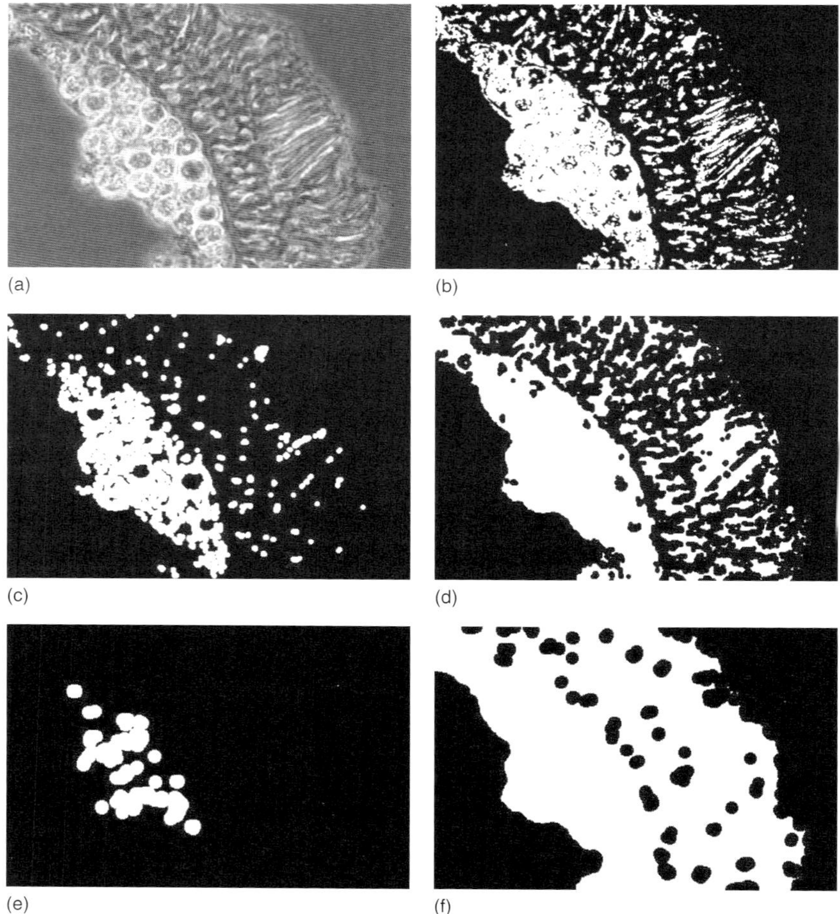

FIGURE 4.3.23
Opening and closing with different size structuring elements. (a) Original microscopic cell image. (b) Image after undergoing a threshold operation. (c) Opening with a circular structuring element of size 5. (d) Closing with a circular structuring element of size 5. (e) Opening with a circular structuring element of size 13. (f) Closing with a circular structuring element of size 13.

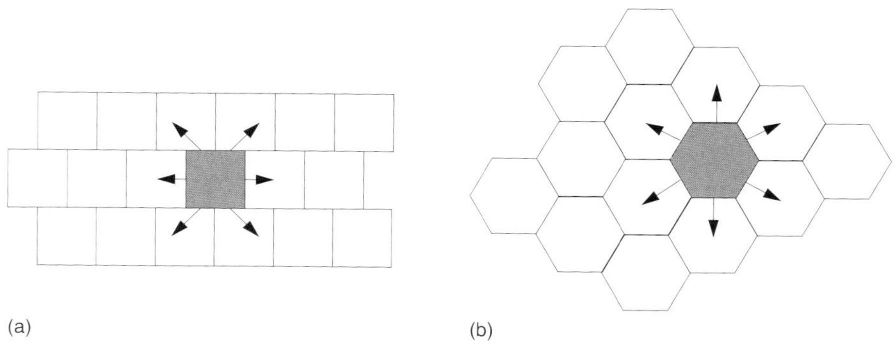

FIGURE 4.3.24
Hexagonal grid. (a) Rectangular image grid with every other row shifted by one-half pixel. (b) Hexagonal grid.

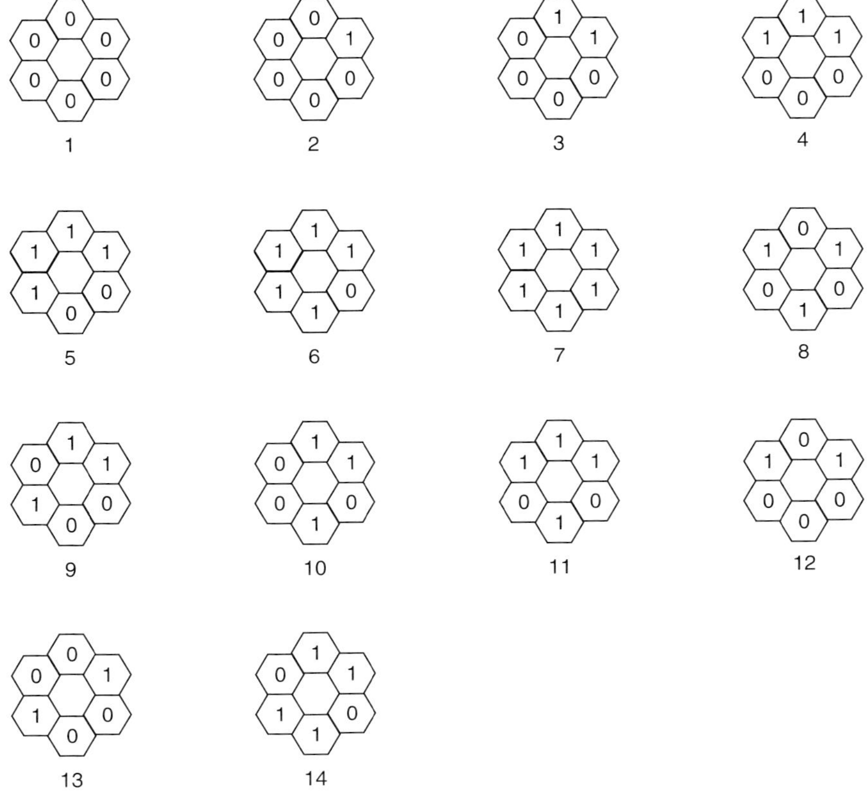

FIGURE 4.3.25
Surrounds for iterative morphological filtering.

The function $L(a, b)$, and the values of a and b are all functions of the row and column, (r, c), but for concise notation this is implied. Set S can contain any or all of the 14 surrounds defined in Figure 4.3.25. $L(a, b)$ can be any logic function, but it turns out that the most useful are the AND and OR functions. The AND function tends to etch away at object boundaries (erosion) and the OR function tends to grow objects (dilation). The following examples illustrate iterative morphological filtering. In these examples we will not change the outer rows and columns, since the image is undefined beyond the borders and the 3×3 surrounds will not fit within the image in these cases.

Example 4.3.3

let $L(a, b) = ab$ (logical AND operation).

$$\text{IMAGE} \quad \begin{bmatrix} 0 & 0 & 0 & 0 & 0 & 0 & 0 & 0 \\ 0 & 1 & 1 & 1 & 1 & 1 & 1 & 0 \\ 0 & 1 & 0 & 1 & 1 & 1 & 1 & 0 \\ 0 & 0 & 0 & 0 & 1 & 1 & 1 & 0 \\ 0 & 1 & 1 & 1 & 1 & 1 & 1 & 0 \\ 0 & 0 & 0 & 0 & 0 & 0 & 0 & 0 \end{bmatrix}$$

$$S = \{5\} = \begin{bmatrix} 1 & 1 & x \\ 1 & x & 1 \\ x & 0 & 0 \end{bmatrix}; \quad \text{assume the origin is in the center}$$

Notes: this means the set S contains surround number 5 from Figure 4.3.25 and the x's are not neighbors, since we are using six-connectivity

The window S (a 3×3 window) is scanned across the image. If a match is found, then $a = 1$ and the output is computed by performing the specified $L(a, b)$ function, in this case by ANDing a with b (b is the center pixel of the subimage under the window). This gives the value of our new image, which will equal $ab = (1)b = b$. If the window S does not match the underlying subimage, then $a = 0$ (false) and $L(a, b) = ab = (0)b = 0$. In either case, the resulting value is written to the new image at the location corresponding to the center of the window.

The window S is scanned across the entire image in this manner and the resultant image is as follows:

$$\begin{bmatrix} 0 & 0 & 0 & 0 & 0 & 0 & 0 & 0 \\ 0 & 0 & 0 & 0 & 0 & 0 & 0 & 0 \\ 0 & 0 & 0 & 0 & 1 & 1 & 0 & 0 \\ 0 & 0 & 0 & 0 & 0 & 1 & 0 & 0 \\ 0 & 0 & 0 & 0 & 0 & 1 & 0 & 0 \\ 0 & 0 & 0 & 0 & 0 & 0 & 0 & 0 \end{bmatrix}$$

Here we see that the AND operation erodes the object. Also note that the set S can contain more than one surround; if it does, then $a = 1$ when the underlying neighborhood matches *any* of the surrounds in the set S. Another parameter which can be considered is the rotation of the surrounds in S. For example, rotating surround $S = \{5\}$ counterclockwise we have the following five possibilities:

$$\begin{bmatrix} 1 & 1 & x \\ 1 & x & 1 \\ x & 0 & 0 \end{bmatrix}, \begin{bmatrix} 1 & 1 & x \\ 1 & x & 0 \\ x & 1 & 0 \end{bmatrix}, \begin{bmatrix} 1 & 0 & x \\ 1 & x & 0 \\ x & 1 & 1 \end{bmatrix}, \begin{bmatrix} 0 & 0 & x \\ 1 & x & 1 \\ x & 1 & 1 \end{bmatrix}, \begin{bmatrix} 0 & 1 & x \\ 0 & x & 1 \\ x & 1 & 1 \end{bmatrix}, \begin{bmatrix} 1 & 1 & x \\ 0 & x & 1 \\ x & 0 & 1 \end{bmatrix}$$

With iterative morphological filtering, normally it is implied that the surrounds in S can be rotated when looking for a match. Additionally, since this is an iterative approach, n is used to define the number of iterations. Following are more examples of this technique.

Example 4.3.4

$$S = \{\ \}, \quad L(a, b) = 0, \quad n = 1$$

The set of surrounds (neighbors) is a null set. This implies $a = 0$; since a surround is not specified. The Boolean function $L(a, b) = 0$. For this combination, all the cells of the image are set to zero, i.e., we have a black image as output.

Example 4.3.5

$$S = \{\ \}, \quad L(a,b) = (!b), \quad n = 1$$

In this case $a = 0$, but this is irrelevant since $L(a,b) = !b$, which implies that the center pixel is negated (complimented).

If $b = 1$, $L(a,b)(!1) = 0$;
Elseif $b = 0$, $L(a,b) = (!0) = 1$.

Example 4.3.6

$$S = \{7\}, \quad L(a,b) = ab, \quad n = 1$$

Consider the following image with the surround S as follows:

$$\begin{bmatrix} 0 & 0 & 0 & 0 & 0 & 0 & 0 & 0 \\ 0 & 1 & 1 & 1 & 1 & 1 & 1 & 0 \\ 0 & 1 & 1 & 1 & 1 & 1 & 1 & 0 \\ 0 & 1 & 1 & 1 & 1 & 1 & 1 & 0 \\ 0 & 1 & 1 & 1 & 1 & 1 & 1 & 0 \\ 0 & 1 & 1 & 0 & 0 & 0 & 0 & 0 \\ 0 & 1 & 1 & 0 & 0 & 0 & 0 & 0 \end{bmatrix}$$

Let:

$$S = \{7\} = \begin{bmatrix} 1 & 1 & x \\ 1 & x & 1 \\ x & 1 & 1 \end{bmatrix}$$

In this case, $a = 1$ for the surround shown above. If the surround does not match then $L(a,b) = 0(b) = 0$. If there is a match then $L(a,b) = 1(b) = b$. The resultant image is as follows:

$$\begin{bmatrix} 0 & 0 & 0 & 0 & 0 & 0 & 0 & 0 \\ 0 & 0 & 0 & 0 & 0 & 0 & 0 & 0 \\ 0 & 0 & 1 & 1 & 1 & 1 & 0 & 0 \\ 0 & 0 & 1 & 1 & 1 & 1 & 0 & 0 \\ 0 & 0 & 0 & 0 & 0 & 0 & 0 & 0 \\ 0 & 0 & 0 & 0 & 0 & 0 & 0 & 0 \\ 0 & 0 & 0 & 0 & 0 & 0 & 0 & 0 \end{bmatrix}$$

Since the logic function is a logical AND operation, if the edge pixels are not "1's", the edges are removed. This operation retains a cluster of "1's" with the edge pixels removed. So, the appendages (thin lines) are removed from the original image—this is an erosion operation.

Example 4.3.7

$$S = \{1, 7\}, \quad L(a, b) = (!a)b, \quad n = 1$$

Consider the following image with the surrounds $\{S\}$ as follows:

$$\begin{bmatrix} 0 & 0 & 0 & 0 & 0 & 0 & 0 & 0 \\ 0 & 1 & 1 & 1 & 1 & 1 & 1 & 0 \\ 0 & 1 & 1 & 1 & 1 & 1 & 1 & 0 \\ 0 & 1 & 1 & 1 & 1 & 1 & 1 & 0 \\ 0 & 1 & 1 & 1 & 0 & 0 & 0 & 0 \\ 0 & 1 & 1 & 1 & 0 & 0 & 0 & 0 \\ 0 & 0 & 0 & 0 & 0 & 0 & 0 & 0 \end{bmatrix}$$

Let $S = \{1, 7\}$, that is

$$S = \left\{ \begin{bmatrix} 0 & 0 & x \\ 0 & x & 0 \\ x & 0 & 0 \end{bmatrix}, \begin{bmatrix} 1 & 1 & x \\ 1 & x & 1 \\ x & 1 & 1 \end{bmatrix} \right\}$$

If $b = 1$, $L(a, b) = (!a)1 = !a$;
Elseif $b = 0$, $L(a, b) = (!a)0 = 0$.

The new image after the above operation is:

$$\begin{bmatrix} 0 & 0 & 0 & 0 & 0 & 0 & 0 & 0 \\ 0 & 1 & 1 & 1 & 1 & 1 & 1 & 0 \\ 0 & 1 & 0 & 0 & 0 & 0 & 1 & 0 \\ 0 & 1 & 0 & 1 & 1 & 1 & 1 & 0 \\ 0 & 1 & 0 & 1 & 0 & 0 & 0 & 0 \\ 0 & 1 & 1 & 1 & 0 & 0 & 0 & 0 \\ 0 & 0 & 0 & 0 & 0 & 0 & 0 & 0 \end{bmatrix}$$

We can see that this operation removes interior pixels and keeps the edges only. Hence, this is an edge detection operation.

Example 4.3.8

Let $S = \{2, 3, 4, 5, 6, 7\}$ and $L = a + b$. ($+ =$ OR)

$$\text{IMAGE} \quad \begin{bmatrix} 0 & 0 & 0 & 0 & 0 & 0 & 0 & 0 \\ 0 & 1 & 1 & 0 & 0 & 0 & 0 & 0 \\ 0 & 1 & 0 & 0 & 0 & 0 & 0 & 0 \\ 0 & 0 & 0 & 0 & 0 & 0 & 0 & 0 \\ 0 & 0 & 0 & 1 & 1 & 1 & 0 & 0 \\ 0 & 0 & 0 & 0 & 0 & 0 & 0 & 0 \end{bmatrix}$$

$$S = \left\{ \begin{bmatrix} 0 & 1 & x \\ 0 & x & 0 \\ x & 0 & 0 \end{bmatrix}, \begin{bmatrix} 1 & 1 & x \\ 0 & x & 0 \\ x & 0 & 0 \end{bmatrix}, \begin{bmatrix} 1 & 1 & x \\ 1 & x & 0 \\ x & 0 & 0 \end{bmatrix}, \begin{bmatrix} 1 & 1 & x \\ 1 & x & 0 \\ x & 1 & 0 \end{bmatrix}, \begin{bmatrix} 1 & 1 & x \\ 1 & x & 1 \\ x & 1 & 0 \end{bmatrix}, \begin{bmatrix} 1 & 1 & x \\ 1 & x & 1 \\ x & 1 & 1 \end{bmatrix} \right\}$$

Because $L(a, b)$ is an OR operation, all pixels that are 1 in the original will remain 1. That is:

$$L = a + b = a + 1 = 1$$

The only pixels that will change are those that are 0 in the original image and have a surround that is S (this means that $a = 1$). That is:

$$L = a + b = a + 0 = a$$

If we examine the set S we see that this set contains all pixels that are surrounded by a connected set of 1's. This operation will expand the object, and illustrates that the OR operation results in a dilation. The resultant image is:

$$\begin{bmatrix} 0 & 0 & 0 & 0 & 0 & 0 & 0 & 0 \\ 0 & 1 & 1 & 1 & 0 & 0 & 0 & 0 \\ 0 & 1 & 1 & 1 & 0 & 0 & 0 & 0 \\ 0 & 0 & 1 & 1 & 1 & 1 & 0 & 0 \\ 0 & 0 & 0 & 1 & 1 & 1 & 0 & 0 \\ 0 & 0 & 0 & 0 & 0 & 0 & 0 & 0 \end{bmatrix}$$

We can see from these examples that this iterative morphological approach is quite versatile. The process can be iterated, or repeated, to any degree desired. We can use this technique to define methods for dilation, erosion, opening, closing, marking corners, finding edges, and other binary morphological operations. For this technique the selection of the set S is comparable to defining the structuring element in the previously described approaches, and the operation $L(a, b)$ defines the type of filtering that occurs. In general, if $L(a, b)$ is an OR operation it will tend to grow, or dilate, objects. When $L(a, b)$ is an AND operation, it will tend to shrink, or erode, objects.

One important operation, which is a controlled erosion process, is called skeletonization. It is often used in optical character recognition and in many other applications. A *skeleton* is what is left of an object when it has been eroded to the point of being only one pixel wide. As illustrated in Figure 4.3.26, we can use the iterative modification approach to skeleton a binary image by using these parameters: $L(a, b) = (a!)b$, and $S = (3, 4)$.

In Figure 4.3.27 we use the same operation, but change the set S and see that it now works as an edge detector. In this case these are the parameters: $L(a, b) = (a!)b$, and $S = (1, 7)$.

The morphological operations described (dilation, erosion, opening, and closing) can be extended to gray level images in different ways. The easiest method is to simply threshold the gray level image to create a binary image, and then apply the existing operators. For many applications this is not desired, as too much information is lost during the thresholding process. Another method that allows us to retain more information is to treat the image as a sequence of binary images by operating on each gray

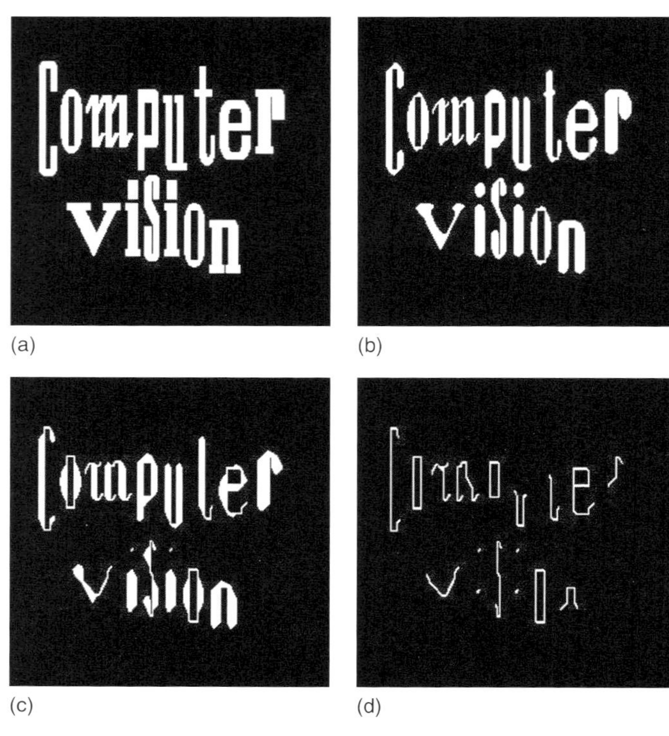

FIGURE 4.3.26
Skeletonization via iterative modification morphological filtering. In this example: $L(a, b) = (!a)b$, and $S = (3, 4)$. (a) Original image. (b) After 10 iterations. (c) After 20 iterations. (d) Iterating until no more changes occur.

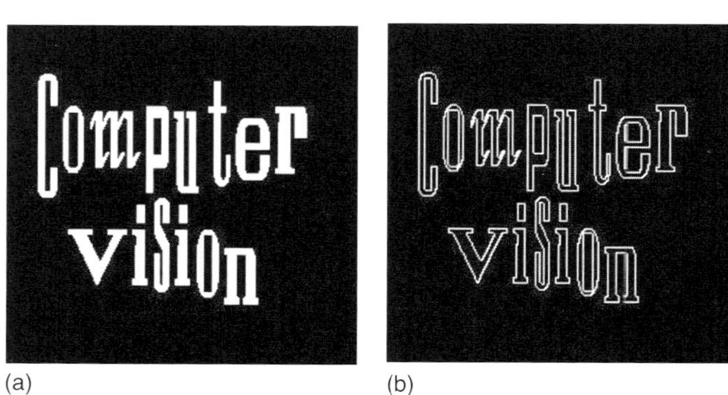

FIGURE 4.3.27
Edge detection via iterative modification morphological filtering. In this example: $L(a, b) = (!a)b$, and $S = (1, 7)$. (a) Original image. (b) Resultant image after one iteration.

level as if it were the "1" value and assuming everything else to be "0". The resulting images can then be combined by laying them on top of each other and "promoting" each pixel to the highest gray level value coincident with that location.

An example of results from gray level morphological filtering is shown in Figure 4.3.28. For this application an opening operation followed by a closing operation was performed. A circular structuring element was used, as the object of interest was the skin tumor border. The opening procedure served to smooth the contours of the object, break narrow isthmuses, and eliminate thin protrusions and small objects. Next, the closing was performed to fill in gaps and eliminate small holes. To fully understand gray level morphology, we must remember that with two adjacent gray levels, the brightest one is considered to be the object (the equivalent of "1" in a binary image) and the darker is the background (the "0" equivalent in binary morphology). In this figure we see the tremendous data reduction achieved, thus simplifying the process of identifying the tumor features of interest.

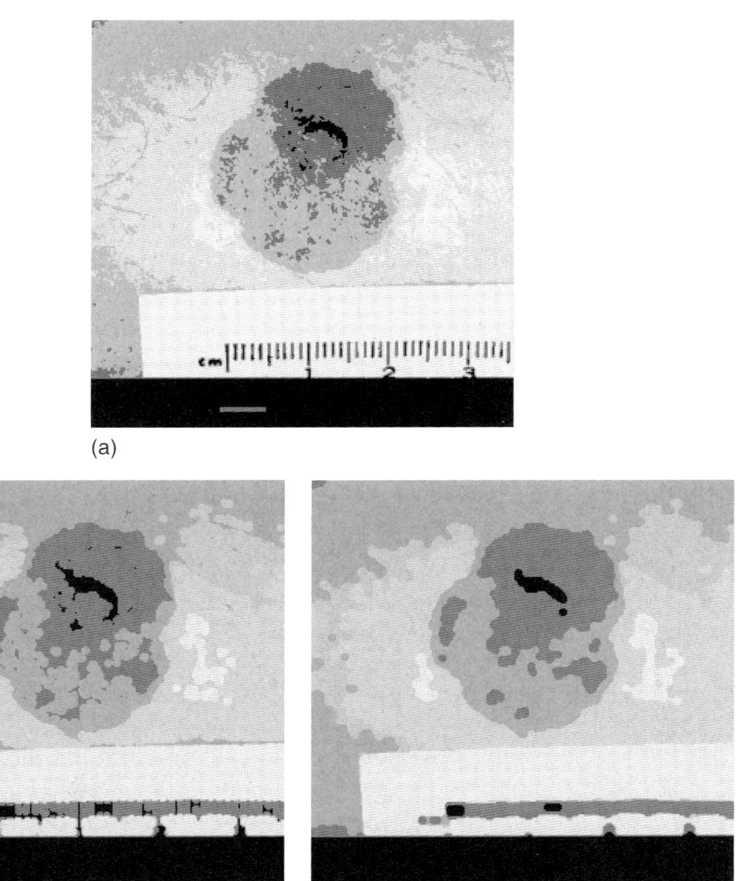

FIGURE 4.3.28
Gray level morphological filtering. (a) Original segmented skin tumor image, contains 1708 objects. (b) Image (a) after morphological opening using a 5×5 circular structuring element, contains 443 objects. (c) Image (b) after morphological closing using a 5×5 circular element, contains 136 objects.

4.4 Key Points

OVERVIEW: IMAGE ANALYSIS AND SEGMENTATION

- Image analysis requires data reduction and segmentation is the primary method in the spatial domain to accomplish it.
- The goal of segmentation is to find regions that represent objects or meaningful parts of objects.
- Image segmentation methods look for regions that have some measure of homogeneity within themselves, or some measure of contrast with objects on their border.
- Three categories for image segmentation methods: (1) region growing and shrinking, (2) clustering methods, (c) boundary detection.

EDGE/LINE DETECTION

- Edge detection operators are often implemented with convolution masks.
- Edge detection operators are often discrete approximations to differential operators.
- Edge detection operators may return magnitude and direction information, some return magnitude only.
- The Hough transform is used for line finding, but can be extended to find arbitrary shapes.
- Edge direction and lines are perpendicular to each other, because the edge direction is the direction of change in gray level.
- There is tradeoff between sensitivity and accuracy in edge detection (see Figure 4.2.2).
- Potential edge points are found by examining the relationship a pixel has with its neighbors; an edge implies a change in gray level.
- Edges may exist anywhere and be defined by color, texture, shadow, etc., and may not necessarily separate real world objects.
- A real edge in an image tends to change slowly, compared to the ideal edge model which is abrupt (see Figure 4.2.4).

Gradient Operators

- Gradient operators are based on the idea of using the first or second derivative of the gray level
- The first derivative will mark edge points, with steeper gray level changes providing stronger edge points (larger magnitudes)
- The second derivative returns two impulses, one on either side of the edge

Roberts operator—a simple approximation to the first derivative, two forms of the equations:

$$\sqrt{[I(r,c) - I(r-1,c-1)]^2 + [I(r,c-1) - I(r-1,c)]^2}$$

$$|I(r,c) - I(r-1,c-1)| + |I(r,c-1) - I(r-1,c)|$$

Sobel operator—approximates the gradient with a row and column mask, and returns both magnitude and direction:

$$\begin{bmatrix} -1 & -2 & -1 \\ 0 & 0 & 0 \\ 1 & 2 & 1 \end{bmatrix} \quad \begin{bmatrix} -1 & 0 & 1 \\ -2 & 0 & 2 \\ -1 & 0 & 1 \end{bmatrix}$$

EDGE MAGNITUDE $\sqrt{s_1^2 + s_2^2}$

EDGE DIRECTION $\tan^{-1}\left[\frac{s_1}{s_2}\right]$

Prewitt operator—approximates the gradient with a row and column mask, and returns both magnitude and direction. It is easier to calculate or implement in hardware than the Sobel, as it uses only 1's in the masks:

$$\begin{bmatrix} -1 & -1 & -1 \\ 0 & 0 & 0 \\ 1 & 1 & 1 \end{bmatrix} \quad \begin{bmatrix} -1 & 0 & 1 \\ -1 & 0 & 1 \\ -1 & 0 & 1 \end{bmatrix}$$

$$\text{EDGE MAGNITUDE} \quad \sqrt{p_1^2 + p_2^2}$$

$$\text{EDGE DIRECTION} \quad \tan^{-1}\left[\frac{p_1}{p_2}\right]$$

Laplacian operators—these are two-dimensional discrete approximations to the second derivative, it is implemented by applying *one* of the following convolution masks:

$$\begin{bmatrix} 0 & -1 & 0 \\ -1 & 4 & -1 \\ 0 & -1 & 0 \end{bmatrix} \quad \begin{bmatrix} -1 & -1 & -1 \\ -1 & 8 & -1 \\ -1 & -1 & -1 \end{bmatrix} \quad \begin{bmatrix} -2 & 1 & -2 \\ 1 & 4 & 1 \\ -2 & 1 & -2 \end{bmatrix}$$

Compass Masks

- The compass mask edge detectors are created by taking a single mask and rotating it to the eight major compass orientations.
- The edge magnitude is found by convolving each mask with the image and selecting the largest value at each pixel location.
- The edge direction at each point is defined by the direction of the edge mask that provides the maximum magnitude.

Kirsch compass masks:

$$k_0\begin{bmatrix} -3 & -3 & 5 \\ -3 & 0 & 5 \\ -3 & -3 & 5 \end{bmatrix} \quad k_1\begin{bmatrix} -3 & 5 & 5 \\ -3 & 0 & 5 \\ -3 & -3 & -3 \end{bmatrix} \quad k_2\begin{bmatrix} 5 & 5 & 5 \\ -3 & 0 & -3 \\ -3 & -3 & -3 \end{bmatrix} \quad k_3\begin{bmatrix} 5 & 5 & -3 \\ 5 & 0 & -3 \\ -3 & -3 & -3 \end{bmatrix}$$

$$k_4\begin{bmatrix} 5 & -3 & -3 \\ 5 & 0 & -3 \\ 5 & -3 & -3 \end{bmatrix} \quad k_5\begin{bmatrix} -3 & -3 & -3 \\ 5 & 0 & -3 \\ 5 & 5 & -3 \end{bmatrix} \quad k_6\begin{bmatrix} -3 & -3 & -3 \\ -3 & 0 & -3 \\ 5 & 5 & 5 \end{bmatrix} \quad k_7\begin{bmatrix} -3 & -3 & -3 \\ -3 & 0 & 5 \\ -3 & 5 & 5 \end{bmatrix}$$

Robinson compass masks:

$$r_0\begin{bmatrix} -1 & 0 & 1 \\ -2 & 0 & 2 \\ -1 & 0 & 1 \end{bmatrix} \quad r_1\begin{bmatrix} 0 & 1 & 2 \\ -1 & 0 & 1 \\ -2 & -1 & 0 \end{bmatrix} \quad r_2\begin{bmatrix} 1 & 2 & 1 \\ 0 & 0 & 0 \\ -1 & -2 & -1 \end{bmatrix} \quad r_3\begin{bmatrix} 2 & 1 & 0 \\ 1 & 0 & -1 \\ 0 & -1 & -2 \end{bmatrix}$$

$$r_4\begin{bmatrix}1 & 0 & -1\\ 2 & 0 & -2\\ 1 & 0 & -1\end{bmatrix}\quad r_5\begin{bmatrix}0 & -1 & -2\\ 1 & 0 & -1\\ 2 & 1 & 0\end{bmatrix}\quad r_6\begin{bmatrix}-1 & -2 & -1\\ 0 & 0 & 0\\ 1 & 2 & 1\end{bmatrix}\quad r_7\begin{bmatrix}-2 & -1 & 0\\ -1 & 0 & 1\\ 0 & 1 & 2\end{bmatrix}$$

Advanced Edge Detectors

Laplacian of a Gaussian (LoG)—consists of two steps: (1) convolve the image with a Gaussian smoothing filter, (2) convolve the image with a Laplacian mask. These two steps can be combined into one convolution filter, such as:

$$\begin{bmatrix}0 & 0 & -1 & 0 & 0\\ 0 & -1 & -2 & -1 & 0\\ -1 & -2 & 16 & -2 & -1\\ 0 & -1 & -2 & -1 & 0\\ 0 & 0 & -1 & 0 & 0\end{bmatrix}$$

Canny algorithm—an optimal edge detector based on a specific mathematical model, it is a four step process: (1) apply a Gaussian filter mask to smooth the image to mitigate noise effects, (2) find the magnitude and direction of the gradient, (3) apply nonmaxima suppression which results in thinned edges, (4) apply two thresholds known as hysteresis thresholding

Hysteresis thresholding—mark pixels above a high threshold, and then apply a low threshold to connected pixels

Boie–Cox algorithm—a generalization of the Canny algorithm using matched filters and Wiener filters

Shen–Castan algorithm—developed as an optimal solution to a specific mathematical model, similar to Canny, but with modifications and extensions

Frei–Chen masks—they form a complete set of basis vectors, which means any 3×3 subimage can be represented as a weighted sum of the basis vectors. The weights are found by projecting the subimage onto each basis vector; that is, perform a vector inner product. Can be used to find edges or lines of specific orientation.

Vector inner product—found by multiplying coincident terms of two vectors and summing the results

Edges in Color Images

Edge detection in color images can be performed on the original RGB data, or on the data after mapping into another color space. The edges are found using different methods:

1. Extract the luminance or brightness information and apply a monochrome edge detection method. The brightness information can be found by averaging the RGB components $L = (R + G + B)/3$, or by the luminance equation $Y = 0.299R + 0.587G + 0.114B$, or by the vector length $L = \sqrt{R^2 + G^2 + B^2}$.
2. Apply a monochrome edge detection method to each of the RGB bands separately and then combine the results into a composite image.
3. Apply a monochrome edge detection method to each of the RGB bands separately and then retain the maximum value at each location.

4. Apply a monochrome edge detection method to each of the RGB bands separately and then use a linear combination of the three results at each location.
5. Apply a monochrome edge detection method to each of the RGB bands separately and then select specific criteria at each pixel location in order to find an edge point. For example, define an edge only if the magnitude is above a specified threshold in each band, and the direction is within a specified range in each band. this is done at each pixel location, or can be extended to allow a range for "location error."
6. Equations for multispectral edges, find the minimum of the following two equations:

$$\frac{\sum_{b=1}^{n}\left[I_b(r,c)-\bar{I}(r,c)\right]\left[I_b(r+1,c+1)-\bar{I}(r+1,c+1)\right]}{\sqrt{\sum_{b=1}^{n}\left[I_b(r,c)-\bar{I}(r,c)\right]^2\sum_{b=1}^{n}\left[I_b(r+1,c+1)-\bar{I}(r+1,c+1)\right]^2}}$$

$$\frac{\sum_{b=1}^{n}\left[I_b(r+1,c)-\bar{I}(r+1,c)\right]\left[I_b(r,c+1)-\bar{I}(r,c+1)\right]}{\sqrt{\sum_{b=1}^{n}\left[I_b(r+1,c)-\bar{I}(r+1,c)\right]^2\sum_{b=1}^{n}\left[I_b(r,c+1)-\bar{I}(r,c+1)\right]^2}}$$

where

$\bar{I}(r,c)$ is the arithmetic average of all the pixels in all bands at pixel location (r,c)

$I_b(r,c)$ is the value at location (r,c) in the bth band, with a total of n bands.

Edge Detector Performance

- Objective and subjective evaluations can be useful.
- Success criteria must be defined, such as was used to develop the Canny algorithm: (1) Detection—the edge detector should find all real edges and not find any false edges. (2) Localization—the edges should be found in the correct place. (3) Single response—there should not be multiple edges found for a single edge.
- Extended edge detection masks can be defined to improve performance in the presence of noise (see Figure 4.2.14)

Pratt Figure of Merit (FOM)—an objective measure developed by Pratt in 1978, which ranges from 0 (0%) for a missing edge to 1 (100%) for a perfectly found edge. It is defined as follows:

$$FOM = \frac{1}{I_N}\sum_{i=1}^{I_F}\frac{1}{1+\alpha d_i^2}$$

I_N = the maximum of I_I and I_F

I_I = the number of ideal edge points in the image

I_F = the number of edge points found by the edge detector

α = a scaling constant that can be adjusted to adjust the penalty for offset edges

d_i^2 = the distance of a found edge point to an ideal edge point

The distance measure can be defined in one of three ways:

1. ***City block distance***, based on four connectivity:

$$d = |r_1 - r_2| + |c_1 - c_2|$$

With this distance measure we can only move horizontally and vertically.

2. ***Chessboard distance, based on eight-connectivity:***

$$d = \max(|r_1 - r_2|, |c_1 - c_2|)$$

With this distance measure we can move diagonally, as well as horizontally or vertically.

3. ***Euclidean distance***, based on actual physical distance:

$$d = \left[(r_1 - r_2)^2 + (c_1 - c_2)^2\right]^{1/2}$$

Hough Transform

Designed as an efficient method to find lines from marked edge points, consisting of three primary steps based on using the normal representation of a line:

$$\rho = r\cos(\theta) + c\sin(\theta)$$

1. Define the desired increments on ρ and θ, Δ_ρ, and Δ_θ and quantize the space accordingly.
2. For every point of interest (typically points found by edge detectors that exceed some threshold value), plug the values for r and c into the line equation:

$$\rho = r\cos(\theta) + c\sin(\theta)$$

Then, for each value of θ in the quantized space, solve for ρ.

3. For each $\rho\theta$ pair from Step 2, record the r and c pair in the corresponding block in the quantized space. This constitutes a hit for that particular block.

After performing the Hough transform, post processing must be done to extract the line information.

SEGMENTATION

- The goal of segmentation is to find regions that represent objects or meaningful parts of objects.
- Image segmentation methods look for regions that have some measure of homogeneity within themselves, or some measure of contrast with objects on their border.
- Three categories for image segmentation methods: (1) region growing and shrinking, (2) clustering methods, (c) boundary detection

Region Growing and Shrinking

- Operate principally on the row and column, (r, c), based image space.
- Methods can be local, operating on small neighborhoods, global, operating on the entire image, or a combination of both.

Split and merge—a segmentation method that divides regions that do not pass a homogeneity test, and combines regions that pass the homogeneity test. This technique proceeds as follows:

1. Define a homogeneity test. This involves defining a homogeneity measure, which may incorporate brightness, color, texture, or other application-specific information, and determining a criterion the region must meet to pass the homogeneity test.
2. Split the image into equal sized regions.
3. Calculate the homogeneity measure for each region.
4. If the homogeneity test is passed for a region, then a merge is attempted with its neighbor(s). If the criterion is not met, the region is split.
5. Continue this process until all regions pass the homogeneity test.

Quadtree—a data structure used in split and merge in which each node can have four children.
Homogeneity criteria—a measure of similarity within a region in an image. In CVIPtools these are available: (1) pure uniformity, (2) local mean versus global mean, (3) local standard deviation versus global, (4) variance, (5) weighted gray level distance, (6) texture.
Watershed algorithm—a morphological technique based on the ideas of modeling a gray level image as a topographic surface, with higher gray levels corresponding to higher elevations. The image is then flooded with a rainfall simulation, and pools of water are created corresponding to segments within the image.

Clustering Techniques

- Segments the image by placing similar elements into groups, or clusters, based on some similarity measure.
- Differs from region growing and shrinking methods in that the mathematical space includes dimensions beyond the row and column image space.
- The mathematical space used for clustering may include, as examples, color spaces, histogram spaces or complex feature spaces.

Recursive region splitting—uses a thresholding of histograms to segment the image. An example of this type of algorithm:

1. Consider the entire image as one region and compute histograms for each component of interest (for example red, green and blue for a color image)
2. Apply a peak finding test to each histogram. Select the best peak and put thresholds on either side of the peak. Segment the image into two regions based on this peak.
3. Smooth the binary thresholded image so only a single connected subregion is left.
4. Repeat Steps 1–3 for each region until no new subregions can be created, that is, no histograms have significant peaks.

SCT/Center algorithm—a color segmentation algorithm initially developed for use in skin tumor identification, defined based on the human visual system response. The algorithm proceeds as follows:

1. Convert the (R, G, B) triple into spherical coordinates—(L, angle A, angle B).
2. Find the minima and maxima of angle A and angle B.

3. Divide the subspace, defined by the maxima and minima, into equal-sized blocks.
4. Calculate the RGB means for the pixel values in each block.
5. Replace the original pixel values with the corresponding RGB means.

PCT/Median algorithm—a color segmentation method initially developed for use in skin tumor identification, based on the principal components transform (PCT). The PCT provides a linear transform that will align the primary axis along the path of maximum variance. The algorithm proceeds as follows:

1. Find the PCT for the RGB image. Transform the RGB data using the PCT.
2. Perform the median split algorithm: find the axis that has the maximal range (initially it will be the PCT axis). Divide the data along this axis so that there are equal numbers of points on either side of the split—the median point. Continue splitting at the median along the maximum range segment until the desired number of colors is reached.
3. Calculate averages for all the pixels falling within a single parallelepiped (box).
4. Map each pixel to the closest average color values, based on a Euclidean distance measure.

Boundary Detection

- Boundary detection for image segmentation is performed by finding boundaries between objects, thus indirectly defining the objects.
- The general steps are: (1) mark potential edge points by finding discontinuities in features such as brightness, color or texture, (2) threshold the results, (3) merge edge segments into boundaries via edge linking.

Thresholding—a technique where pixels are marked above a specified value. Various algorithms are available, including: (1) by manually examining the histogram and looking for the best valley, (2) *minimizing within group variance, Otsu method*—an analytical algorithm that works well for bimodal histograms based on the assumption that each peak is Gaussian shaped and the peaks are well separated, (3) use the average value, (4) use 10% to 20% of the maximum value for noisy images.

Edge linking—methods to link the edge points into segments and boundaries, including: (1) consider points that have passed the threshold test and connect them to other marked points within some maximum distance, (2) consider small neighborhoods and link points with similar magnitude and direction, then link points together to form boundaries, (3) the snake eating algorithm described in Section 4.2.6.

Extended Hough transform—is used to find shapes and mark boundaries that can be defined by analytical equations, such as circles or ellipses. The search space is a parameter space, where the parameters are found in the equation describing the shape of interest.

Generalized Hough transform—is used find any arbitrary shape. It works by creating a description of the shape defined by a reference point and a table of lines, the R-table.

Combined Segmentation Approaches

- Image segmentation methods may actually be a combination of region growing methods, clustering methods and boundary detection.
- Optimal image segmentation is likely to be achieved by focusing on the application.
- Finding boundaries of different features, such as texture, brightness, or color, and applying artificial intelligence techniques at a higher level to correlate the feature boundaries found to the specific domain may give the best results.

Morphological Filtering

- Morphology relates to structure or form of objects.
- Morphological filtering simplifies segmented images by smoothing out object outlines, filling small holes, eliminating small projections, or *skeletonizing* a binary object down to lines that are a single pixel wide.
- Primary operations are *dilation* and *erosion*.
- These operations use a structuring element which determines exactly how the object will be dilated or eroded.
- *Opening* and *closing* are useful combination of dilation and erosion.

Dilation—the process of expanding image objects by changing pixels with value of "0" to "1". It can be done in two steps:

1. If the origin of the structuring element coincides with a "0" in the image, there is no change; move to the next pixel.
2. If the origin of the structuring element coincides with a "1" in the image, perform the OR logic operation on all pixels within the structuring element.

Erosion—the process of shrinking binary objects by changing pixels with a value of "1" to "0". It can be done in two steps:

1. If the origin of the structuring element coincides with a "0" in the image, there is no change; move to the next pixel.
2. If the origin of the structuring element coincides with a "1" in the image, and any of the '1' pixels in the structuring element extend beyond the object ("1" pixels) in the image, then change the "1" pixel in the image, whose location corresponds to the origin of the structuring element, to a "0".

Opening—an erosion followed by a dilation. This will eliminate all pixels in regions too small to contain the structuring element. It will expand holes, erode edges, and eliminate small objects. It may split objects that are connected by narrow strips, and eliminate peninsulas.

Closing—a dilation followed by an erosion. It can be used to fill holes and small gaps. It will also connect separate objects, if the gap is smaller than the structuring element.

Iterative morphological filtering—as defined here, it is based on a definition of six-connectivity, so a pixel can be surrounded by 14 possible combinations (allowing for rotation). This approach can be used to dilate, erode, open, close, skeletonize, mark corners, find edges, and perform other binary morphological operations. To do this, we define: (1) the set of surrounds S, where $a = 1$, (2) a logic function, $L(a, b)$, where b is the current pixel value, and the function specifies the output of the morphological operation, (3) the number of iterations, n.

Gray level morphological filtering—the previously defined binary operations can be extended to gray level images in various ways: (1) threshold the image to create a binary image and apply binary operators, (2) treat the image as sequence of binary images by operating on each gray level as if it were the "1" value and assuming everything else to be "0". The resulting images can then be combined by laying them on top of each other and "promoting" each pixel to the highest gray level value coincident with that location.

4.5 References and Further Reading

More information regarding edge detection can be found in [Gonzalez/Woods 02], [Forsyth/Ponce 03], [Shapiro/Stockman 01], [Sonka/Hlavac/Boyle 99], and [Pratt 91]. For more information on the LoG and its relationship to biological vision systems see [Marr 82] and [Shapiro/Stockman 01]. For more detail on the Canny algorithm see [Jain/Kasturi/Schunck 95] and [Sonka/Hlavac/Boyle 99], for more on the Shen–Castan algorithms see [Parker 97], and for more on the Boie–Cox algorithm see [Seul/O'Gorman/Sammon 00]. [Shapiro/Stockman 01] have a different, and potentially useful, approach based on energy for using the Frei–Chen masks. The multispectral edge detection equations were found in [Sonka/Hlavac/Boyle 99]. The Hough transform as described here can be found in [Gonzalez/Woods 02]. More details on the generalized Hough transform can be found in [Sonka/Hlavac/Boyle 99]. Details on improved gray scale (IGS) quantization can be found in [Gonzalez/Woods 02].

The definitions for connectivity are described in [Horn 86], and further information can be found in [Haralick/Shapiro 92]. Detailed information on tree data structures can be found in [Weiss 97] and [Shapiro/Stockman 01]. The PCT/Median and SCT/Center image segmentation methods presented are described in [Umbaugh 90], and applied in [Umbaugh/Moss/Stoecker 89], [Umbaugh/Moss/Stoecker 93], and [Umbaugh/Moss/Stoecker 92]. More on thresholding algorithms and the watershed algorithm can be found in [Gonzalez/Woods 02], [Baxes 94], [Dougherty/Lotufo 03], and [Sonka/Hlavac/Boyle 99]. The histogram thresholding segmentation algorithms in CVIPtools are based on the [Carlotto 87] and [Lim/Lee 90] papers. Additional information about image segmentation methods can be found in [Gonzalez/Woods 02], [Shapiro/Stockman 01], [Sonka/Hlavac/Boyle 99], [Haralick/Shapiro 92], [Schalkoff 89], [Castleman 96], and [Jain/Kasturi/Schnuck 95]. More information on image morphology is found in [Gonzalez/Woods 02], [Shapiro/Stockman 01], [Sonka/Hlavac/Boyle 99], and [Jain/Kasturi/Schnuck 95]. For a practical approach with numerous examples of morphological processing see [Dougherty/Lotufo 03]. The iterative method to morphological filtering is described in [Horn 86].

Baxes, G.A., *Digital Image Processing: Principles and Applications*, NY: Wiley, 1994.

Castleman, K.R., *Digital Image Processing*, Upper Saddle River, NJ: Prentice Hall, 1996.

Carlotto, M., Histogram Analysis Using a Scale-State Approach, *IEEE Transactions on Pattern Analysis and Machine Intelligence*, Vol. 9, No. 1, pp. 121–129, 1987.

Dougherty, E., Lotufo, *Hands-on Morphological Image Processing*, Bellingham, WA: SPIE Press, 2003.

Gonzalez, R.C., Woods, R.E., *Digital Image Processing*, Upper Saddle River, NJ: Prentice Hall, 2002.

Haralick, R.M., Shapiro, L.G., *Computer and Robot Vision*, Reading, MA: Addison-Wesley, 1992.

Horn, B.K.P., *Robot Vision*, Cambridge, MA: The MIT Press, 1986.

Jain, R., Kasturi, R., Schnuck, B.G., *Machine Vision*, NY: McGraw Hill, 1995.

Lim, Y., Lee, S., On Color Segmentation Algorithm Based on the Thresholding and Fuzzy c-Means Techniques, *Pattern Recognition*, Vol. 23, No. 9, pp. 935–952, 1990.

Marr, D., *Vision*, NY: Freeman and Company, 1982.

Parker, J.R., *Algorithms for Image Processing and Computer Vision*, NY: Wiley, 1997.

Pratt, W.K., *Digital Image Processing*, NY: Wiley, 1991.

Schalkoff, R.J., *Digital Image Processing and Computer Vision*, NY: Wiley, 1989.

Seul, M., O'Gorman, L., Sammon, M.J., *Practical Algorithms for Image Analysis*, Cambridge, UK: Cambridge University Press, 2000.

Shapiro, L., Stockman, G., *Computer Vision*, Upper Saddle River, NJ: Prentice Hall, 2001.

Sonka, M., Hlavac, V., Boyle, R., *Image Processing, Analysis and Machine Vision*, Pacific Grove, CA: Brooks/Cole Publishing Company, 1999.

Umbaugh S.E., Moss, R.H., Stoecker, W.V., Automatic Color Segmentation of Images with Application to Detection of Variegated Coloring in Skin Tumors, *IEEE Engineering in Medicine and Biology*, Vol. 8, No. 4, Dec. 1989.

Umbaugh, S.E., *Computer Vision in Medicine: Color Metrics and Image Segmentation Methods for Skin Cancer Diagnosis*, PhD dissertation, UMI Dissertation Service, 1990.

Umbaugh S.E., Moss, R.H., Stoecker, W.V., An Automatic Color Segmentation Algorithm with Application to Identification of Skin Tumor Borders, *Computerized Medical Imaging and Graphics*, Vol. 16, No. 3, 1992.

Umbaugh S.E., Moss, R.H., Stoecker, W.V., Automatic Color Segmentation Algorithms with Application to Skin Tumors Feature Identification, *IEEE Engineering in Medicine and Biology*, Vol. 12, No. 3, Sept. 1993.

Weiss, M., *Data Structures and Algorithm Analysis in C*, Reading, MA: Addison-Wesley, 1997.

4.6 Exercises

1. (a) What is the goal of image segmentation? (b) What type of objects do segmentation methods look for? (c) List the three categories of segmentation methods.
2. What does a differential operator measure, and how does this relate to edge detectors?
3. In dealing with noise in edge detection there is a tradeoff between sensitivity and accuracy. Explain what this means.
4. Compare and contrast an ideal edge and a real edge in an image. Draw a picture of both.
5. (a) Explain the idea on which gradient edge detection operators are based. (b) How do the results differ if we use a second-order compared to a first-order derivative operator? (c) Explain what is meant by sub-pixel accuracy and how it relates to gradient-based edge detectors.
6. Find the results of applying the Robert's edge detector to the following image. Use the absolute value form of the operator. For the result, don't worry about top row and left column.

$$\begin{bmatrix} 5 & 7 & 4 & 3 \\ 4 & 0 & 0 & 0 \\ 6 & 1 & 2 & 1 \end{bmatrix}$$

7. Find the results, magnitude and direction, of applying the Prewitt edge detector to the following image. For the result, don't worry about the outer rows and columns.

$$\begin{bmatrix} 0 & 8 & 0 & 0 \\ 0 & 8 & 0 & 0 \\ 0 & 8 & 0 & 0 \\ 0 & 8 & 0 & 0 \end{bmatrix}$$

8. Two of the three Laplacian masks given are based on eight-connectivity, the other one is based on four-connectivity—which one? Devise a Laplacian type edge detection mask based on six-connectivity.
9. Find the results of applying the Robinson compass masks to the following image. For the result, don't worry about the outer rows and columns. Keep track of the maximum magnitude and which mask corresponds to it.

$$\begin{bmatrix} 0 & 0 & 0 & 0 \\ 0 & 0 & 0 & 0 \\ 5 & 5 & 5 & 5 \\ 0 & 0 & 0 & 0 \end{bmatrix}$$

10. Use CVIPtools to explore the basic edge detection operators. (a) Run CVIPtools, and load test images of your choice. As one of the test images, create a binary image of a circle with *Utilties->Create->Circle*. (b) Select *Utilties->Create->Add noise*. Add noise to your test images. Use both Gaussian and salt-and-pepper noise. (c) Select *Analysis->Edge/Line Detection*. Compare thresholding the output at different levels with the Kirsch, Pyramid, and Robinson edge detection operators. In CVIPtools: Select the desired edge detector, select none for pre-filtering, select the desired threshold with post-threshold option. Alternately, select none for post-threshold and use the Threshold button at the bottom of the window—this allows for the testing of different threshold levels without the need to rerun the edge detection operation. (d) For the images containing noise, compare the resultant images with and without applying a low-pass filter as a preprocessing step (use the pre-filter option in CVIPtools).
11. Use CVIPtools to explore the basic edge detection operators. (a) Run CVIPtools, and load test images of your choice. As one of the test images, create a binary image of a circle with *Utilties->Create->Circle*. (b) Select *Utilties->Create->Add noise*. Add noise to your test images. Use both Gaussian and salt-and-pepper noise. (c) Select *Analysis->Edge/Line Detection*. (d) Compare using different size kernels with the Sobel and Prewitt operators. In CVIPtools: select the desired edge detection operator, select the kernel size. (e) For the images containing noise, compare the resultant images with and without applying a low-pass filter as a preprocessing step (use the pre-filter option in CVIPtools). Also, compare the results with different size kernels (masks). Does the 3×3 or 7×7 provide better results in the presence of noise?
12. Use CVIPtools to explore the basic edge detection operators. (a) Run CVIPtools, and load test images of your choice. As one of the test images, create a binary image of a circle with *Utilties->Create->Circle*. (b) Select *Utilties->Create->Add noise*. Add noise to your test images. Use both Gaussian and salt-and-pepper noise. (c) Select *Analysis->Edge/Line Detection*. (d) Compare keeping the DC bias

versus not keeping it, using the Roberts and Laplacian. (e) For the images containing noise, compare the resultant images with and without applying a low-pass filter as a preprocessing step (use the pre-filter option in CVIPtools).

13. The Canny algorithm for edge detection was developed based on a specific edge model; describe the model. What are the three criteria used to develop the algorithm? How do the four algorithmic steps relate to these three criteria?
14. Use CVIPtools to explore the Laplacian and Frei–Chen edge detection operators. Use real images, and create some simple geometric images with *Utilities->Create*. Additionally, add noise to the images. (a) In *Analysis->Edge/Line Detection*, select the Frei–Chen, compare the line subspace versus the edge subspace, with the projection option in CVIPtools. Experiment with various threshold angles, as well as post threshold values. Examine the histogram to select good threshold values. Compare the threshold values that work the best to the images with and without noise, are they the same? Why or why not? (b) Using the Laplacian in *Analysis->Edge/Line Detection*, select prefiltering with a Gaussian—this will perform a LoG filter. Examine the histogram to select good threshold values. (c) Use *Utilities->Filter->Specify a Filter* to input the values for the 5×5 LoG filter given in the text. Do this by selecting the 5×5, and then entering the values in the box (note: the <tab> key, or the mouse, can be used to move around the box), then select OK. Compare these results to the results from (b). Are they similar? Why or why not?
15. Use CVIPtools to explore the Canny, Boie–Cox, and Shen–Casten algorithms. Use *Utilities->Create* to create a circle with a radius of 32 and a blur radius of 128. (a) Select the Canny algorithm. Use the Low Threshold Factor = 1, High Threshold Factor = 0.5 and compare results with a variance of 0.5, 1.5, and 3. Next, use Low Threshold Factor = 1, High Threshold Factor = 1 and compare results with a variance of 0.5, 1.5, and 3. Now apply the Canny to a complex image of your choice. Experiment with the parameters until you achieve a good result. (b) Select the Boie–Cox algorithm and apply it to your blurred circle image. Experiment with the various parameters. Next, apply it to the complex image you used with the Canny. Which settings work best for your images? (c) Select the Shen–Castan algorithm and apply it to your blurred circle image. Use the Low Threshold Factor = 1, High Threshold Factor = 1, Smoothing Factor = 0.9, Window Size = 5 and compare results with a Thin Factor of 1, 3 and 6. Next, apply it to the complex image you used with the Canny. Which settings work best for your images? (d) Repeat (a)–(c), but add noise to the images.
16. Select a color image of your choice. Use *Utilties->Convert->Color Space* to explore edge detection using the various color spaces, and *Utilities->Create->Extract Band* to operate on individual bands, and *Utilties->Create->Assemble Bands* to combine resulting bands into composite images. Devise an algorithm that works best for your particular image.
17. In the following image find the distance between the points labeled a and b in the following image:

$$\begin{bmatrix} 5 & 7 & 8 & b & 12 \\ 3 & 6 & 7 & 6 & 6 \\ 6 & a & 10 & 10 & 11 \\ 7 & 7 & 0 & 0 & 0 \\ 0 & 7 & 9 & 1 & 0 \end{bmatrix}$$

(a) Use city block distance, (b) use chessboard distance, (c) use Euclidean distance.

18. Given the following image, apply a Robert's edge detector, absolute value format, to the image (do not worry about the top row and leftmost column). Next, threshold the image with the following values and find Pratt's Figure of Merit for the found edge points. Let $\alpha = 0.5$, and use the chessboard distance measure. Threshold values: (a) 5, (b) 12, (c) 22.

$$\begin{bmatrix} 0 & 0 & 0 & 0 & 0 \\ 0 & 0 & 0 & 0 & 0 \\ 0 & 10 & 10 & 10 & 0 \\ 0 & 0 & 0 & 0 & 0 \\ 0 & 0 & 0 & 0 & 0 \end{bmatrix}$$

19. (a) List all the steps in the Hough transform for finding lines. (b) Use CVIPtools to visualize the Hough space by finding the Hough transform of a checkerboard. First, create a checkerboard with *Utilities->Create-Checkerboard*, using the default values (32×32 squares on a 256×256 image). (c) Next, select an edge detector of your choice and an appropriate threshold for your output image. (d) Select the Hough transform, and set the *Output Image* to *Hough image*, and the *Line Angles* to 0–180. Now, apply this to your thresholded image from (c). What size is the output image? Can you explain why? Describe what you see in the output image and explain any bright points in this image. (e) Change the *Line Angles* to 0–10. Apply the Hough and describe the Hough output now. (f) Change *Output Image* to *Final result (after extinction)*. Is the result what you expected? Why or why not?
20. Use CVIPtools to explore the Hough transform and the post-processing edge linking algorithm on an artificial image. (a) Use *Utilities->Create* to create a circle and perform an edge detector of your choice, (b) threshold the resultant image, (c) perform the Hough transform on the thresholded image, input 5 for both *line pixels* and *segment pixels*, then select line angles 0, (d) 90, (e) 0–20, (f) 0–45, (g) 0–90, (h) 0–150, and (i) 0–180. Do the results make sense? How are the results affected by increasing the *connect distance*? How are the results affected by increasing the *segment pixels*?
21. Use CVIPtools to explore the Hough transform and the post-processing edge linking algorithm on a real image. (a) Select an image of your choice and perform an edge detection, (b) threshold the resultant image, (c) perform the Hough transform on the thresholded image, input 5 for both *line pixels* and *segment pixels*, then select line angles 0, (d) 90, (e) 0–20, (f) 0–45, (g) 0–90, (h) 0–150, and (i) 0–180. Do the results make sense? How are the results affected by increasing the *connect distance*? Find a *connect distance* that gives you the desired results. How are the results affected by increasing the *segment pixels*? Find a *segment pixels* that gives you the desired results.
22. What is a quadtree and for which segmentation algorithms is it used? Why is it useful for these algorithms?
23. Compare and contrast region growing and shrinking segmentation methods from clustering methods.
24. Regarding split and merge segmentation algorithms, what is a homogeneity test? Describe three different homogeneity criteria.
25. Briefly describe the watershed algorithm. To which category of segmentation methods does it belong? Explain why or why not it belongs in this category.

26. Use CVIPtools to explore histogram thresholding segmentation methods; including the *Histogram Thresholding* and *Fuzzy C-Means* algorithms. Select an image of your choice and, (a) examine the histogram, (b) perform *histogram thresholding* segmentation, (c) compare the histogram of the image from *histogram thresholding* segmentation to the histogram of the original image. If you had manually selected the peaks, are these the ones you would have selected? Looking at the segmented image, do you think the segmentation was effective? (d) Do parts (a)–(c) using the *fuzzy c-means* algorithm. Note that with this algorithm we can control the degree of segmentation with the *Gaussian kernel variance*. What happens as we increase this parameter?
27. Use CVIPtools to explore the various segmentation methods and their associated parameters. Select an image that allows you to judge when the segmentation has been successful.
28. (a) In image analysis, what do we call the type of spatial filtering typically performed after segmentation? (b) What are the two principal operations called? Briefly describe each.
29. Given the following image and structuring element, perform an opening operation. Assume the origin of the structuring element is in the center. Ignore cases where the structuring element extends beyond the image.

STRUCTURING ELEMENT

$$\begin{bmatrix} 1 & 0 & 0 \\ 1 & 1 & 1 \\ 1 & 0 & 0 \end{bmatrix}$$

IMAGE

$$\begin{bmatrix} 1 & 1 & 1 & 1 & 1 & 1 & 1 \\ 1 & 0 & 0 & 1 & 0 & 0 & 1 \\ 1 & 1 & 1 & 1 & 1 & 1 & 1 \\ 1 & 0 & 0 & 0 & 0 & 0 & 1 \\ 1 & 1 & 1 & 1 & 1 & 1 & 1 \\ 1 & 1 & 1 & 1 & 1 & 1 & 1 \\ 1 & 1 & 0 & 0 & 1 & 1 & 1 \end{bmatrix}$$

30. Given the following, what will be the resultant pixel values after operating on the following image? Assume all rotations of the surrounds are included in *S*.

IMAGE

$$\begin{array}{c} \\ 0 \\ 1 \\ 2 \\ 3 \\ 4 \\ 5 \\ 6 \end{array} \begin{array}{c} \begin{array}{ccccccc} 0 & 1 & 2 & 3 & 4 & 5 & 6 \end{array} \\ \begin{bmatrix} 1 & 1 & 1 & 1 & 1 & 1 & 1 \\ 1 & 0 & 0 & 1 & 0 & 0 & 1 \\ 1 & 1 & 1 & 1 & 1 & 1 & 1 \\ 1 & 0 & 0 & 0 & 0 & 0 & 1 \\ 1 & 1 & 1 & 1 & 1 & 1 & 1 \\ 1 & 1 & 1 & 1 & 1 & 1 & 1 \\ 1 & 1 & 0 & 0 & 1 & 1 & 1 \end{bmatrix} \end{array}$$

(a) $S=\{2, 3, 4, 5, 6\}$, $L(a,b)=a\bar{b}$, $n=1$, $f=1$.

Resultant pixel value at $(r,c)=(3, 2)$—

Resultant pixel value at $(r,c)=(3, 3)$—

Resultant pixel value at $(r,c)=(4, 5)$—

Resultant pixel value at $(r,c)=(3, 5)$—

(b) $S=\{7\}$, $L(a,b)=a+b$, $n=1$, $f=1$.

Resultant pixel value at $(r,c)=(4, 5)$—

Resultant pixel value at $(r,c)=(2, 2)$—

Resultant pixel value at $(r,c)=(4, 2)$—

Resultant pixel value at $(r,c)=(4, 4)$—

4.6.1 Programming Exercise: Edge Detection—Roberts and Sobel

1. Write a function to implement the Roberts edge detector. Let the user select either the square root, or absolute value form. The C functions for absolute value and square root are abs( ) and sqrt( ). Note that you will need to deal with potential overflow problems, as the results may be greater than 255. This may be dealt with by using a floating point image structure as an intermediate image, and then remapping the image when completed. This is done as follows:

```
Image *outputimage; /*declaration of image structure pointer*/
Float **image_data; /* declaration of image data pointer*/
.....
outputimage=new_Image(PGM,GRAY_SCALE,no_of_bands,no_of_rows,
no_of_cols,CVIP_FLOAT,REAL);/*creatinganewimagestructure*/
image_data=getData_Image(outputimage, bands); /*getting the
data into an array that can be accessed as: image_data[r][c] */
.....
outputimage=remap_Image(outputimage,CVIP_BYTE,0,255);/
*remapping a float image to byte size, this is done before writing
the image to disk with the write_Image function*/
```

   Test the function on gray level images. Compare the results from the two methods by using the *Utilities->Compare* selection in CVIPtools.
2. Write a function to implement the Sobel edge detector. The function should output an image that contains the Sobel magnitude, and is remapped as with the Roberts. Test the function on gray level images of your choice.
3. Modify the functions to handle multiband (color) images.
4. Use the *Analysis->Edge/Line detection->Edge link* selection in CVIPtools to connect the lines in the output images. Note that this requires a binary image, so be sure to apply a threshold operation to the images first. Thresholding can be performed directly in this window by typing the threshold value in the entry box and clicking on the *Threshold current image* button at the bottom of the window.

4.6.2 Programming Exercise: Hough Transform

1. Write a C function to implement a Hough transform. Let the user enter the line angles of interest and the minimum number of pixels per line.
2. Compare your results to those obtained in CVIPtools.

4.6.3 Programming Exercise: SCT/Center Segmentation

1. Write a C function to implement a SCT/Center segmentation algorithm. Let the user enter the number of colors along the *angle A* and *angle B* axes.
2. Compare your results to those obtained in CVIPtools.

4.6.4 Programming Exercise: Histogram Thresholding Segmentation

1. Use the CVIPtools libraries to put the *fuzzyc_segment* and *hist_thresh_segment* functions into your CVIPlab.
2. Compare your results to those obtained in CVIPtools.

4.6.5 Programming Exercise: Morphological Filters

1. Write C functions to implement dilation, erosion, opening, and closing. Let the user enter the nine 0's and 1's for a 3×3 structuring element.
2. Compare your results to those in obtained CVIPtools.

4.6.6 Programming Exercise: Iterative Morphological Filters

1. Use the CVIPtools libraries to put the *morpho* function into your CVIPlab. This function implements the iterative morphological operations as described in this chapter.
2. Compare your results to those in obtained CVIPtools.

5

Discrete Transforms

5.1 Introduction and Overview

A transform is simply another term for a mathematical mapping process. Most of the transforms discussed in this chapter are used in image analysis and processing to provide information regarding the rate at which the gray levels change within an image—the spatial frequency or sequency. However, the principal components transform (PCT) is included and its primary purpose is to decorrelate the data between image bands. Additionally, the wavelet and the Haar transforms are different in that they retain both spatial and frequency information.

In general, a transform maps image data into a different mathematical space via a transformation equation. In Chapter 2, we discussed transforming image data into alternate color spaces. However, those color transforms mapped data from one color space to another color space with a one-to-one correspondence between a pixel in the input and the output. Basically, most of these transforms map the image data from the spatial domain to the frequency domain (also called the spectral domain), where *all* the pixels in the input (spatial domain) contribute to *each* value in the output (frequency domain). This is illustrated in Figure 5.1.1.

These transforms are used as tools in many areas of engineering and science, including computer imaging. Originally defined in their continuous forms, they are commonly used today in their discrete (sampled) forms. The large number of arithmetic operations required for the discrete transforms, combined with the massive amounts of data in an image, require a great deal of computer power. The ever-increasing compute power, memory capacity, and disk storage available today make the use of these transforms much more feasible than in the past.

The discrete form of these transforms is created by sampling the continuous form of the functions on which these transforms are based, that is, the *basis functions*. The functions used for these transforms are typically sinusoidal or rectangular, and the sampling process, for the one-dimensional (1-D) case, provides us with *basis vectors*. When we extend these into two-dimensions, as we do for images, they are *basis matrices* or *basis images* (see Figure 5.1.2). The process of transforming the image data into another domain, or mathematical space, amounts to projecting the image onto the basis images. The mathematical term for this projection process is an *inner product*, and is identical to what was done with Frei–Chen edge and line masks in Chapter 4. The Frei–Chen projections are performed to uncover edge and line information in the image, and use 3×3 image blocks. The frequency transforms considered here use the entire image, or blocks that are typically at least 8×8, and are used to discover spatial frequency information.

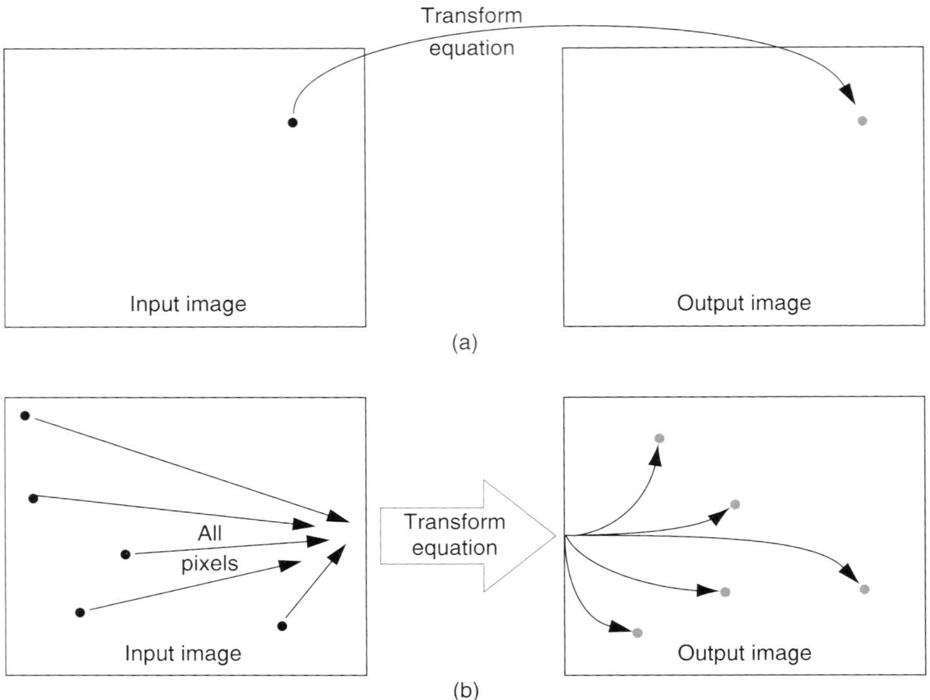

FIGURE 5.1.1
Discrete Transforms. (a) Color transforms use a single-pixel to single-pixel mapping. (b) All pixels in the input image contribute to each value in the output image for frequency transforms.

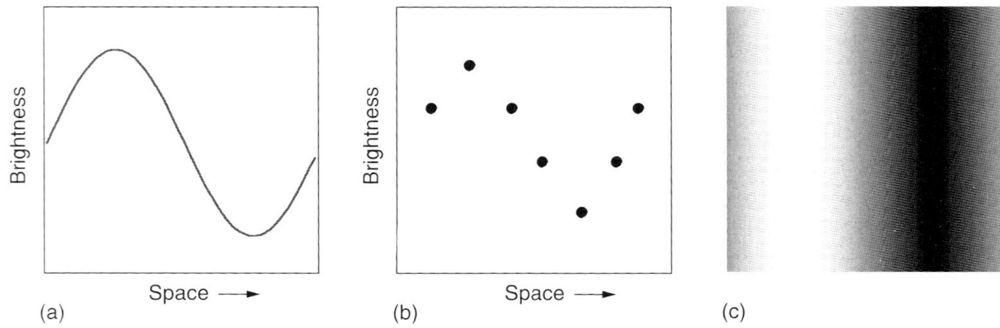

FIGURE 5.1.2
Basis vectors and images. (a) A basis function: a 1-D sinusoid. (b) A basis vector: a sampled 1-D sinusoid. (c) A basis image: a sampled sinusoid shown in 2-D as an image. The pixel brightness in each row corresponds to the sampled values of the 1-D sinusoids, which are repeated along each column.

The ways in which the image brightness levels change in space define the spatial frequency. For example, rapidly changing brightness corresponds to high spatial frequency, whereas slowly changing brightness levels relate to low frequency information. The lowest spatial frequency, called the zero frequency term, corresponds to an image with a constant value. These concepts are illustrated in Figure 5.1.3, using square waves and sinusoids as basis vectors.

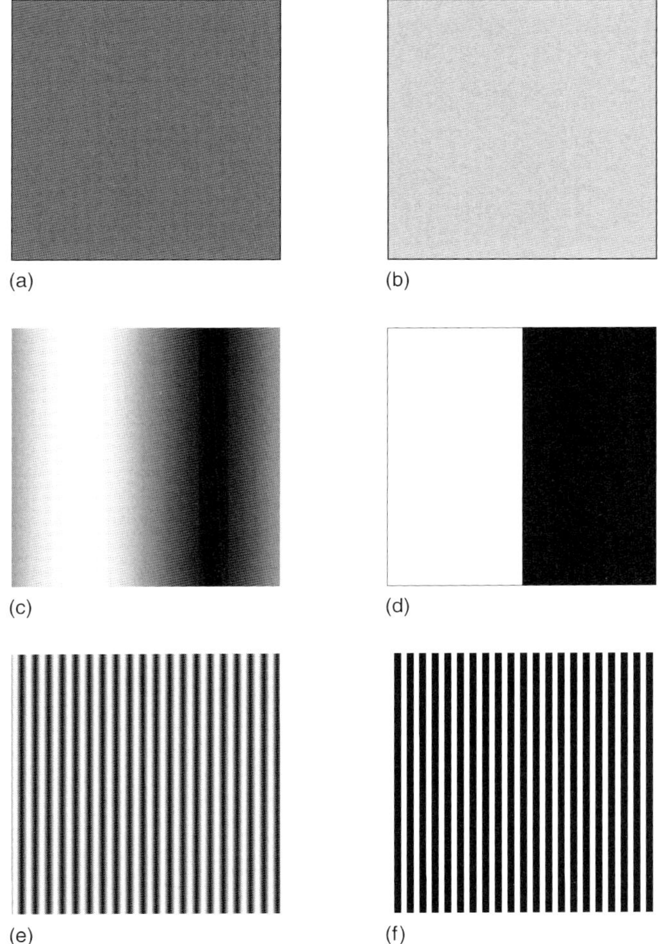

FIGURE 5.1.3
Spatial frequency. (a) Frequency = 0, gray level = 54. (b) Frequency = 0, gray level = 202. (c) Frequency = 1, horizontal sine wave. (d) Frequency = 1, horizontal square wave. (e) Frequency = 20, horizontal sine wave. (f) Frequency = 20, horizontal square wave.

The general form of the transformation equation, assuming an $N \times N$ image, is given by:

$$T(u, v) = k \sum_{r=0}^{N-1} \sum_{c=0}^{N-1} I(r, c)B(r, c; u, v)$$

Here, u and v are the frequency domain variables, k is a constant that is transform dependent, $T(u, v)$ are the transform coefficients, and $B(r, c; u, v)$ correspond to the basis images. The notation $B(r, c; u, v)$ defines a set of basis images, corresponding to each different value for u and v, and the size of each is r by c (Figure 5.1.4). The transform coefficients, $T(u, v)$, are the projections of $I(r, c)$ onto each $B(u, v)$. This is illustrated in Figure 5.1.5. These coefficients tell us how similar the image is to the basis image; the more alike they are, the bigger the coefficient. This transformation process amounts to decomposing the image into a weighted sum of the basis images, where the coefficients $T(u, v)$ are the weights.

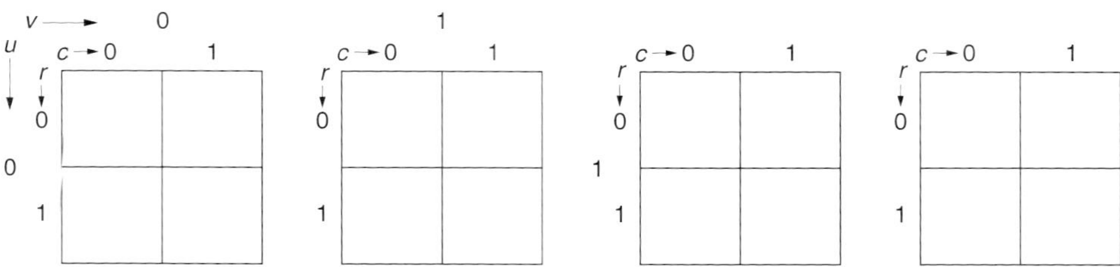

FIGURE 5.1.4
A set of basis vectors $B(r, c; u, v)$. Size of generic basis vectors for a 2×2 transform.

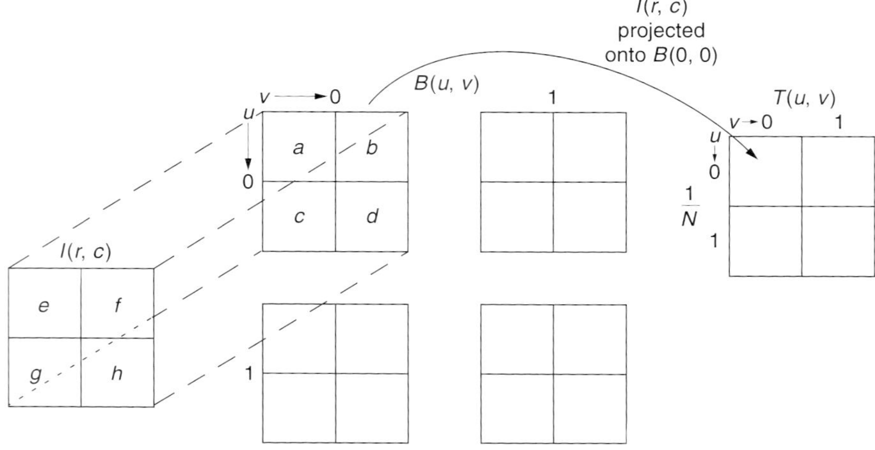

FIGURE 5.1.5
Transform coefficients. To find $T(u, v)$, we project $I(r, c)$ onto the basis vectors of $B(u, v)$. For example, $T(0, 0)$ is the projection of $I(r, c)$ onto $B(0, 0)$, which equals $(ea + fb + gc + hd)$.

Example 5.1.1

Let

$$I(r, c) = \begin{bmatrix} 5 & 3 \\ 1 & 2 \end{bmatrix}$$

and let

$$B(u, v, r, c) = \left\{ \begin{matrix} \begin{bmatrix} +1 & +1 \\ +1 & +1 \end{bmatrix} \begin{bmatrix} +1 & -1 \\ +1 & -1 \end{bmatrix} \\ \begin{bmatrix} +1 & +1 \\ -1 & -1 \end{bmatrix} \begin{bmatrix} +1 & -1 \\ -1 & +1 \end{bmatrix} \end{matrix} \right.$$

Then

$$T(u, v) = \left\{ \begin{matrix} T(0, 0) = 5(1) + 3(1) + 1(1) + 2(1) = 11 \\ T(0, 1) = 5(1) + 3(-1) + 1(1) + 2(-1) = 1 \\ T(1, 0) = 5(1) + 3(1) + 1(-1) + 2(-1) = 5 \\ T(1, 1) = 5(1) + 3(-1) + 1(-1) + 2(1) = 3 \end{matrix} \right\} = \begin{bmatrix} 11 & 1 \\ 5 & 3 \end{bmatrix}$$

To obtain the image from the transform coefficients we apply the inverse transform equation:

$$I(r, c) = T^{-1}[T(u, v)] = k' \sum_{u=0}^{N-1} \sum_{v=0}^{N-1} T(u, v) B^{-1}(r, c; u, v)$$

Here the $T^{-1}[T(u, v)]$ represents the inverse transform, and the $B^{-1}(r, c; u, v)$ represents the inverse basis images, and the k' is a constant that is transform dependent. In many cases the inverse basis images are the same as the forward ones, and in cases where they are not, they are very similar.

Example 5.1.2

From the previous example, we have

$$T(u, v) = \begin{bmatrix} 11 & 1 \\ 5 & 3 \end{bmatrix}$$

and let

$$B^{-1}(u, v, r, c) = \left\{ \begin{matrix} \begin{bmatrix} +1 & +1 \\ +1 & +1 \end{bmatrix} \begin{bmatrix} +1 & -1 \\ +1 & -1 \end{bmatrix} \\ \begin{bmatrix} +1 & +1 \\ -1 & -1 \end{bmatrix} \begin{bmatrix} +1 & -1 \\ -1 & +1 \end{bmatrix} \end{matrix} \right.$$

Then

$$I(r, c) = \left\{ \begin{matrix} I(0, 0) = 11(1) + 1(1) + 5(1) + 3(1) = 20 \\ I(0, 1) = 11(1) + 1(-1) + 5(1) + 3(-1) = 12 \\ I(1, 0) = 11(1) + 1(1) + 5(-1) + 3(-1) = 4 \\ I(1, 1) = 11(1) + 1(-1) + 5(-1) + 3(1) = 8 \end{matrix} \right\} = \begin{bmatrix} 20 & 12 \\ 4 & 8 \end{bmatrix} ?$$

Is this correct? No, since

$$I(r, c) = \begin{bmatrix} 5 & 3 \\ 1 & 2 \end{bmatrix}$$

Comparing our results we see that we must multiply our answer by $\frac{1}{4}$. What does this tell us?—It tells us that the transform pair, $B(u,v; r,c)$ and $B^{-1}(u,v; r,c)$ are not properly defined, we need to be able to recover our original image to have a proper transform pair. We can solve this by letting $k' = \frac{1}{4}$, or by letting $k = k' = \frac{1}{2}$. Note that $\frac{1}{2}$ will normalize the magnitude of the basis images to 1. Remember that the magnitude of a vector can be found by the square root of the sum of the squares of the vector components; in this case:

$$\text{Magnitude of the basis images} = \sqrt{1^2 + (\pm 1)^2 + (\pm 1)^2 + (\pm 1)^2} = \sqrt{4} = 2$$

Therefore, to normalize the magnitude to 1, we need to divide by 2, or multiply by $\frac{1}{2}$.

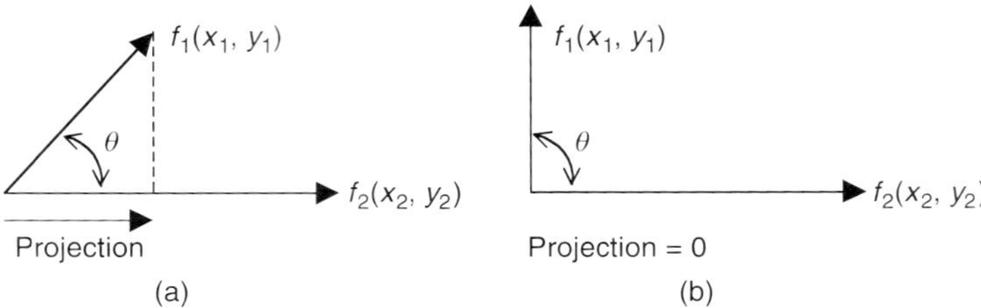

FIGURE 5.1.6
Vector inner product projection. (a) Given two vectors. $f_1(x_1, y_1)$ and $f_2(x_2, y_2)$, we can find the vector inner product by the equation $|f_1||f_2|\cos\theta = x_1x_2 + y_1y_2$. Here we see the projection of f_1 onto f_2, with θ less than 90 degrees. (b) If the two vectors are perpendicular, then the inner product is zero because the $\cos(90°) = 0$, and the two vectors have nothing in common.

Two important attributes for basis images is that they be orthogonal and orthonormal. If basis images are *orthogonal*, it means the vector inner product of each one with every other one is equal to zero. Basis images that are *orthonormal* are orthogonal and have magnitudes equal to one. In the above example we saw why we want the basis images to have a magnitude of one, but what does orthogonality really mean and why is it important for basis images? Orthogonality means that the projection of one basis image onto another has a zero result—the two have nothing in common, they are uncorrelated. In Figure 5.1.6 we see an illustration of the vector inner product in a two-dimensional mathematical (x, y) space. Given two vectors, $f_1(x_1, y_1)$ and $f_2(x_2, y_2)$, we can find the vector inner product by the following equation:

$$\text{Inner product or projection} = |f_1||f_2|\cos\theta = x_1x_2 + y_1y_2$$

In the figure we see that the projection consists of what is common between the two vectors, and that if they are perpendicular, then the inner product is zero and they have nothing in common. This is important for basis images because we are decomposing a complex function into a weighted sum of these basis images, and if the basis images are not orthogonal then these weights, $T(u, v)$, will contain redundant information. This will become clearer as we look at the specific transforms.

5.2 Fourier Transform

The Fourier transform is the best known, and the most widely used, transform. It was developed by Baptiste Joseph Fourier (1768–1830) to explain the distribution of temperature and heat conduction. Since that time the Fourier transform has found numerous uses, including vibration analysis in mechanical engineering, circuit analysis in electrical engineering, and here in computer imaging. The Fourier transform decomposes a complex signal into a weighted sum of a zero frequency term (the DC term which is related to the average value), and sinusoidal terms, the basis functions, where each sinusoid is a harmonic of the fundamental. The *fundamental* is the basic or lowest frequency, and the *harmonics* are frequency multiples of the fundamental (the fundamental is also called the first harmonic). We can recreate the original signal by adding the fundamental and all the harmonics, with each term weighted by its corresponding transform coefficient. This is shown in Figure 5.2.1.

Fourier transform theory begins with the one-dimensional continuous transform, defined as follows:

$$F(v) = \int_{-\infty}^{\infty} I(c)\, e^{-j2\pi vc}\, \mathrm{d}c$$

The basis functions, $e^{-j2\pi vc}$, are complex exponentials and will be defined in the next section, but for now suffice it to say that they are sinusoidal in nature. Also note that continuous Fourier transform theory assumes that the functions start at $-\infty$ and go to $+\infty$, so they are continuous and everywhere. This aspect of the underlying theory is important for the periodic property of the Fourier transform discussed later.

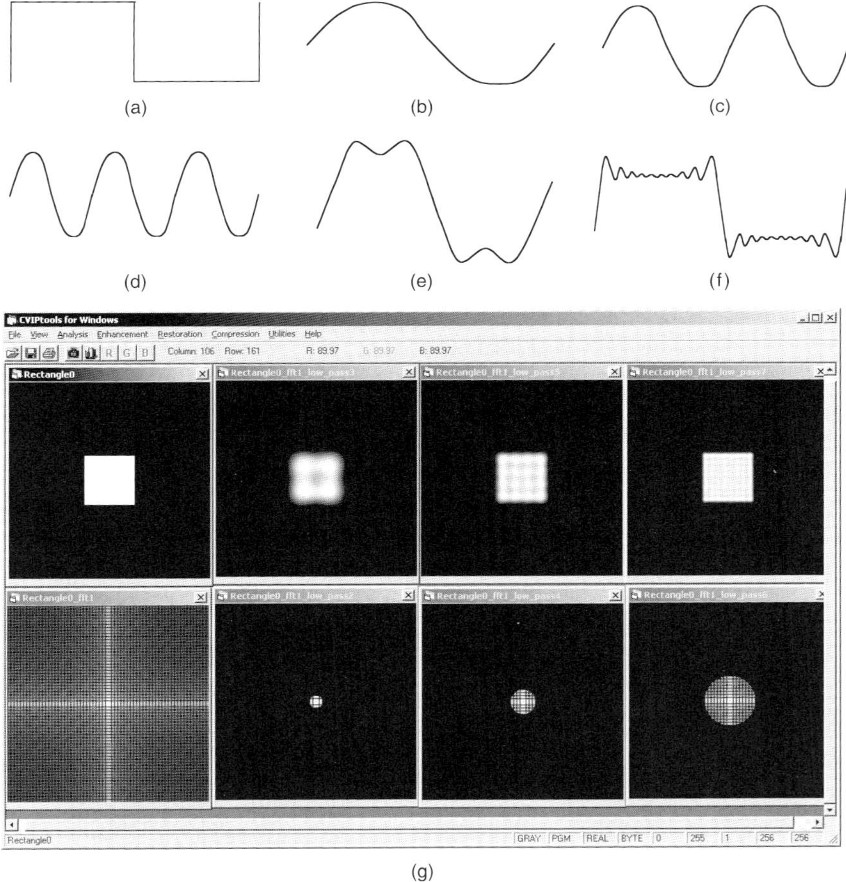

FIGURE 5.2.1
Decomposing a square wave with a Fourier transform. (a) The square wave. (b) The fundamental. (c) The first harmonic. (d) The second harmonic. (e) Approximation to the sum of the fundamental and the first three harmonics. (f) Approximation to the sum of the first 20 harmonics. (g) CVIPtools screen capture of a square and successively adding more harmonics. Across the top are the reconstructed squares with 8, 16 and then 32 harmonics. Across the bottom are the corresponding Fourier transform magnitude images.

Example 5.2.1

Given the simple rectangle function shown in Figure 5.2.2a, we can find the Fourier transform by applying the equation defined above:

$$\begin{aligned}
F(v) &= \int_{-\infty}^{\infty} I(c)e^{-j2\pi vc}\,\mathrm{d}c \\
&= \int_{0}^{C} Ae^{-j2\pi vc}\,\mathrm{d}c \\
&= -\frac{-A}{j2\pi v}\left[e^{-j2\pi vc}\right]_0^C = \frac{-A}{j2\pi v}\left[e^{-j2\pi C} - 1\right] \\
&= \frac{A}{j2\pi v}\left[e^{j\pi vC} - e^{-j\pi vC}\right]e^{-j\pi vC} \\
&= \frac{A}{\pi v}\sin(\pi vC)\,e^{-j\pi vC}
\end{aligned}$$

This result is a complex function and here we are interested in the magnitude (defined in the next section), which is:

$$\begin{aligned}
|F(v)| &= \left|\frac{A}{\pi v}\right|\left|\sin(\pi vC)\right|\left|e^{-j\pi vC}\right| \\
&= AC\left|\frac{\sin(\pi vC)}{(\pi vC)}\right|
\end{aligned}$$

Figure 5.2.2b shows this result.

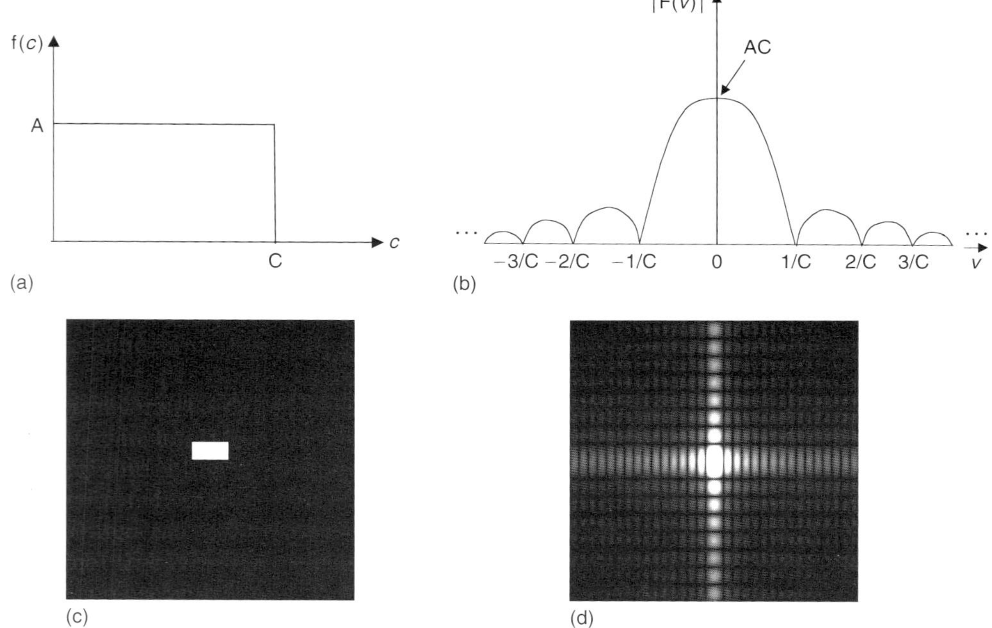

FIGURE 5.2.2
Fourier transform example. (a) The one-dimensional rectangle function. (b) The Fourier transform of the 1-D rectangle function. (c) Two-dimensional rectangle function as an image. (d) The Fourier spectrum of the 2-D rectangle.

Figure 5.2.2c shows the two-dimensional rectangle function, with the brightness of the image representing the magnitude of the function. In Figure 5.2.2d we see the magnitude of the Fourier spectrum in image form. The reasons for introducing this example here are as follows: (1) to illustrate the continuous and infinite nature of the basis functions in the underlying theory, (2) to illustrate that when we have a function that ends abruptly in one domain, such as the function $F(c)$, it leads to a continuous series of decaying ripples in the other domain as shown in Figure 5.2.2b,d, and (3) to show that the width of the rectangle in one domain is inversely proportional to the spacing of the ripples in the other domain. As you will see, this will be useful in understanding the nature of phenomena that occurs in images at object boundaries, especially when we apply filters; but first we will explore the details of the discrete Fourier transform.

5.2.1 The One-Dimensional Discrete Fourier Transform

The equation for the one-dimensional discrete Fourier transform (DFT) is:

$$F(v) = \frac{1}{N} \sum_{c=0}^{N-1} I(c)\, e^{-j2\pi(vc/N)}$$

The inverse DFT is given by:

$$F^{-1}[F(v)] = I(r, c) = \frac{1}{N} \sum_{c=0}^{N-1} F(v)\, e^{j2\pi(vc/N)}$$

where the F^{-1} notation represents the inverse transform. These equations correspond to one row of an image; note that as we move across a row, the column coordinate is the one that changes. The base of the natural logarithmic function, e, is about 2.71828; j, the imaginary coordinate for a complex number, equals $\sqrt{-1}$. The basis functions are sinusoidal in nature, as can be seen by Euler's identity:

$$e^{j\theta} = \cos(\theta) + j\sin(\theta)$$

Putting this equation into the DFT equation by substituting $\theta = -2\pi vc/N$, and remembering that $\cos(\theta) = \cos(-\theta)$ and $\sin(-\theta) = -\sin(\theta)$, the one-dimensional DFT equation can be written as:

$$F(v) = \frac{1}{N} \sum_{c=0}^{N-1} I(c)[\cos(2\pi vc/N) - j\sin(2\pi vc/N)] = R(v) + jI(v)$$

In this case, $F(v)$ is also complex, with the real part corresponding to the cosine terms, and the imaginary part corresponding to the sine terms. If we represent a complex spectral component by $F(v) = R(v) + jI(v)$, where $R(v)$ is the real part and $I(v)$ is the imaginary part, then we can define the magnitude and phase of a complex spectral component as:

$$\text{MAGNITUDE} = |F(v)| = \sqrt{[R(v)]^2 + [I(v)]^2}$$

and

$$\text{PHASE} = \phi(v) = \tan^{-1}\left[\frac{I(v)}{R(v)}\right]$$

The magnitude of a sinusoid is simply its peak value, and the phase determines where the origin is, or where the sinusoid starts (see Figure 5.2.3). Keep in mind that the basis functions are simply sinusoids at varying frequencies, the complex exponential notation, e^{jx}, is simply a mathematical notational tool to make it easier to write and manipulate the equations. In Figure 5.2.4 we see that a complex number can be expressed in rectangular form, described by the real and imaginary part; or in exponential form, by the magnitude and phase. A memory aid for evaluating $e^{j\theta}$ is given in Figure 5.2.5.

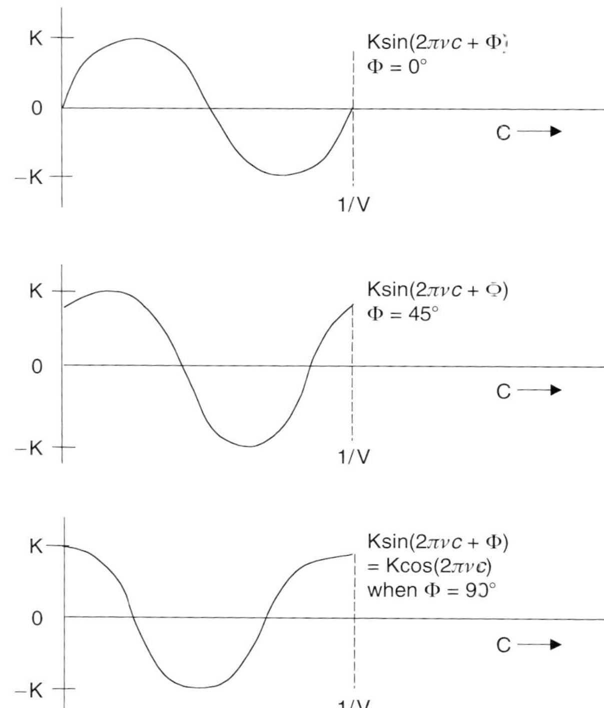

FIGURE 5.2.3
Magnitude and phase of sinusoidal waves.

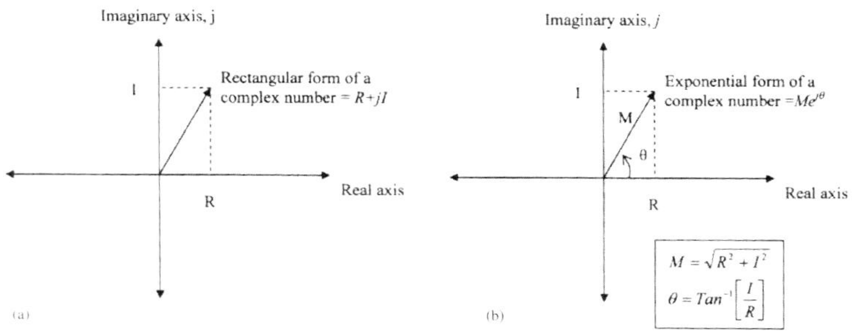

FIGURE 5.2.4
Complex numbers. (a) A complex number shown as a vector and expressed in rectangular form, in terms of the real, R, and imaginary components, I. (b) A complex number expressed in exponential form in terms of magnitude, M, and angle, θ. Note that θ is measured from the real axis counterclockwise.

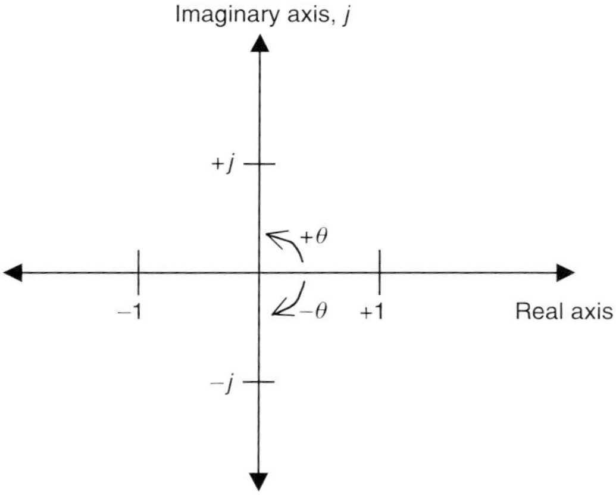

FIGURE 5.2.5
A memory aid for evaluating $e^{j\theta}$. The angle is measured from the real axis counterclockwise, so:

$$\theta = 0 \Rightarrow e^{j\theta} = +1$$
$$\theta = \pi/2 \Rightarrow e^{j\theta} = +j$$
$$\theta = \pi \Rightarrow e^{j\theta} = -1$$
$$\theta = 3\pi/2 \Rightarrow e^{j\theta} = -j$$
$$\theta = 2\pi \Rightarrow e^{j\theta} = +1$$
$$\theta = 5\pi/2 \Rightarrow e^{j\theta} = +j$$

and so on . . .

Example 5.2.2

Given $I(c) = [3, 2, 2, 1]$, corresponding to the brightness values of one row of a digital image. Find $F(v)$ in both rectangular form, and in exponential form.

$$F(v) = \frac{1}{N}\sum_{c=0}^{N-1} I(c)\, e^{-j2\pi(vc/N)}$$

$$F(0) = \tfrac{1}{4}\sum_{c=0}^{3} I(c)e^{-j2\pi vc/4} = \tfrac{1}{4}\sum_{c=0}^{3} I(c)e^{0} = \tfrac{1}{4}[I(0) + I(1) + I(2) + I(3)]1$$
$$= \tfrac{1}{4}[3 + 2 + 2 + 2 + 1] = 2$$

$$F(1) = \tfrac{1}{4}\sum_{c=0}^{3} I(c)e^{-j2\pi(1)c/4} = \tfrac{1}{4}\left[3e^{0} + 2e^{-j\pi/2} + 2e^{-j\pi} + 1e^{-j\pi 3/2}\right]$$
$$= \tfrac{1}{4}[3 + 2(-j) + 2(-1) + 1(j)] = \tfrac{1}{4}[1 - j]$$

$$F(2) = \tfrac{1}{4}\sum_{c=0}^{3} I(c)e^{-j2\pi(2)c/4} = \tfrac{1}{4}\left[3e^{0} + 2e^{-j\pi} + 2e^{-j2\pi} + 1e^{-j3\pi}\right] = \tfrac{1}{4}[3 + (-2) + 2 + (-1)] = \tfrac{1}{2}$$

$$F(3) = \tfrac{1}{4}\sum_{c=0}^{3} I(c)e^{-j2\pi(3)c/4} = \tfrac{1}{4}\left[3e^{0} + 2e^{-j\pi 3/2} + 2e^{-j3\pi} + 1e^{-j\pi 9/2}\right]$$
$$= \tfrac{1}{4}[3 + 2j + 2(-1) + 1(-j)] = \tfrac{1}{4}[1 + j]$$

Therefore we have:

$$F(v) = \left[2, \tfrac{1}{4}[1+j], \tfrac{1}{2}, \tfrac{1}{4}[1+j]\right]$$

Next, put these into exponential form:

$$F(0) = 2 = 2 + 0j \Rightarrow M = \sqrt{2^2 + 0^2} = 2; \quad \theta = \tan^{-1}\left[\frac{0}{2}\right] = 0$$

$$F(1) = \tfrac{1}{4}[1-j] = \tfrac{1}{4} - \tfrac{1}{4}j \Rightarrow M = \sqrt{(1/4)^2 + (-1/4)^2} \cong 0.35; \quad \theta = \tan^{-1}\left[\frac{-1/4}{1/4}\right] = -\pi/4$$

$$F(2) = 1/2 = 1/2 + 0j \Rightarrow M = \sqrt{(1/2)^2 + 0^2} = 0.5; \quad \theta = \tan^{-1}\left[\frac{0}{1/2}\right] = 0$$

$$F(3) = \tfrac{1}{4}[1+j] = \tfrac{1}{4} + \tfrac{1}{4}j \Rightarrow M = \sqrt{(1/4)^2 + (1/4)^2} \cong 0.35; \quad \theta = \tan^{-1}\left[\frac{1/4}{1/4}\right] = \pi/4$$

Therefore, we have:

$$F(v) = \left[2, 0.35e^{-j\pi/4}, 0.5, 0.35\,e^{j\pi/4}\right]$$

5.2.2 The Two-Dimensional Discrete Fourier Transform

Extending the discrete Fourier transform to the two-dimensional (2-D) case for images, we can decompose an image into a weighted sum of 2-D sinusoidal terms. The physical interpretation of a 2-D sinusoid is shown in Figure 5.2.6. Here we see that a sinusoid that is not directly on the u or the v axis can be broken done into separate frequency terms by finding the period along each axis. Assuming a square $N \times N$ image, the equation for the 2-D discrete Fourier transform is:

$$F(u, v) = \frac{1}{N} \sum_{r=0}^{N-1} \sum_{c=0}^{N-1} I(r, c) \exp\left[-j2\pi \frac{(ur + vc)}{N}\right]$$

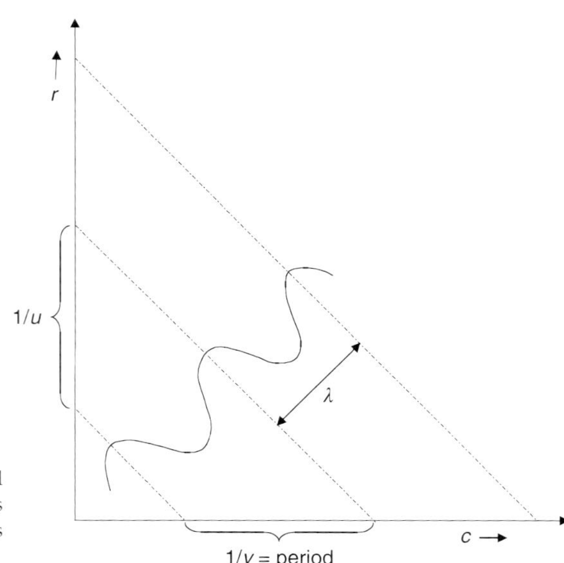

FIGURE 5.2.6
Physical interpretation of a two-dimensional sinusoid. The wavelength of the sinusoid is $\lambda = 1/\sqrt{u^2 + v^2}$, where (u, v) are the frequencies along (r, c) and the periods $1/u$ and $1/v$.

As before, we can also write the Fourier transform equation as:

$$F(u, v) = \frac{1}{N} \sum_{r=0}^{N-1} \sum_{c=0}^{N-1} I(r, c) \left[\cos\left(\frac{2\pi}{N}(ur + vc) \right) - j \sin\left(\frac{2\pi}{N}(ur + vc) \right) \right]$$

Now, $F(u, v)$ is also complex, with the real part corresponding to the cosine terms, and the imaginary part corresponding to the sine terms. If we represent a complex spectral component by $F(u,v) = R(u,v) + jI(u,v)$, where $R(u,v)$ is the real part and $I(u,v)$ is the imaginary part, then we can define the magnitude and phase of a complex spectral component as:

$$\text{MAGNITUDE} = |F(u, v| = \sqrt{[R(u, v)]^2 + [I(u, v)]^2}$$

and

$$\text{PHASE} = \phi(u, v) = \tan^{-1}\left[\frac{I(u, v)}{R(u, v)} \right]$$

Figure 5.2.7 shows an image recovered with the phase information only, which illustrates its importance in images. Although we lose the relative magnitudes, which results in a loss of contrast (Figure 5.2.7b), we retain the relative placement of objects—in other words, the phase data contains information about *where objects are* in an image.

Once we perform the transform, if we want to get our original image back, we need to apply the *inverse transform*. The inverse 2-D DFT is given by:

$$F^{-1}[F(u, v)] = I(r, c) = \frac{1}{N} \sum_{u=0}^{N-1} \sum_{v=0}^{N-1} F(u, v) \, e^{j2\pi(ur+vc)/N}$$

The F^{-1} notation represents the inverse transform. This equation illustrates that the function, $I(r,c)$, is represented by a weighted sum of the basis functions, and that the transform coefficients, $F(u,v)$, are the weights. With the inverse Fourier transform,

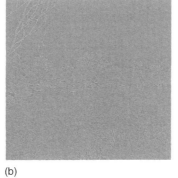

(a)

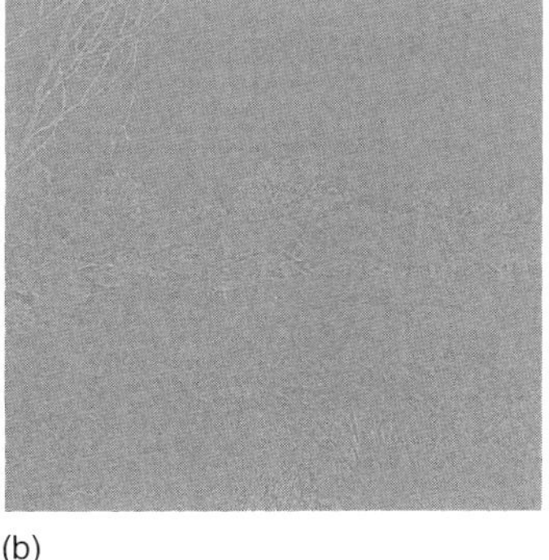
(b)

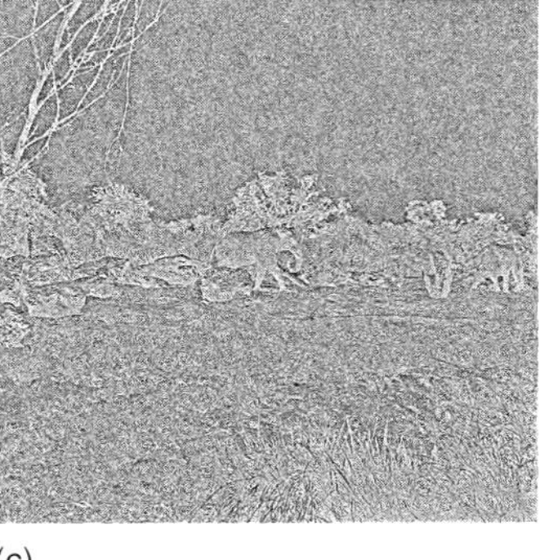
(c)

FIGURE 5.2.7
Fourier transform phase information. (a) Original image. (b) Phase only image. (c) Contrast enhanced version of image (b) to show detail.

the sign on the basis functions' exponent is changed from -1 to $+1$. However, this only corresponds to the phase and not the frequency and magnitude of the basis functions (see Figure 5.2.3 and the magnitude and phase equations above).

One important property of the Fourier transform is called *separability*; which means that the two-dimensional basis image can be decomposed into two product terms where each term depends only on the rows or columns. Also, if the basis images are *separable*, then the result can be found by successive application of two, one-dimensional transforms. This is illustrated by first separating the basis image term (also called the transform kernel) into a product, as follows:

$$e^{-j2\pi(ur+vc)/N} = e^{-j2\pi(ur/N)}\, e^{-j2\pi(vc/N)}$$

Next, we write the Fourier transform equation in the following form:

$$F(u,v) = \frac{1}{N}\sum_{r=0}^{N-1}\left(e^{-j2\pi(vc/N)}\right)\sum_{c=0}^{N-1} I(r,c)\, e^{-j2\pi(vc/N)}$$

The advantage of the separability property is that $F(u,v)$ or $I(r,c)$ can be obtained in two steps by successive applications of the one-dimensional Fourier transform or its inverse.

Expressing the equation as:

$$F(u,v) = \frac{1}{N}\sum_{r=0}^{N-1} F(r,v)\, e^{-j2\pi(ur/N)}$$

where:

$$F(r,v) = (N)\left(\frac{1}{N}\right)\sum_{c=0}^{N-1} I(r,c)\, e^{-j2\pi(vc/N)}$$

For each value of r, the expression inside the brackets is a one-dimensional transform with frequency values $v = 0, 1, 2, 3, \ldots N-1$. Hence the two-dimensional function $F(r,v)$ is obtained by taking a transform along each row of $I(r,c)$ and multiplying the result by N. The desired result, $F(u,v)$ is obtained by taking a transform along each column of $F(r,v)$.

Often, the discrete Fourier transform is implemented as a Fast Fourier Transform (FFT). There are fast algorithms for most of the transforms described here, and many are based on the input data having a number of elements that are a power of 2, which is common for images. In general, these algorithms take advantage of the many redundant calculations involved and operate to eliminate this redundancy. The transforms in CVIPtools are implemented with fast algorithms based on powers of 2, which means that any image that is not a power of 2 will be zero-padded. Details of these algorithms can be found in the references.

5.2.3 Fourier Transform Properties

A Fourier transform pair refers to an equation in a one domain, either spatial or spectral, and its corresponding equation in the other domain. This implies that if we know what is done in one domain, we know what will occur in the other domain.

5.2.3.1 Linearity

The Fourier transform is a linear operator and is shown by the following equations:

$$F[aI_1(r, c) + bI_2(r, c)] = aF_1(u, v) + bF_2(u, v)$$

$$aI_1(r, c) + bI_2(r, c) = F^{-1}[aF_1(u, v) + bF_2(u, v)]$$

where a and b are constants.

5.2.3.2 Convolution

Convolution in one domain is the equivalent of multiplication in the other domain. This is what allows us to perform filtering in the spatial domain with convolution masks (see Section 5.7). Using $*$ to denote the convolution operation, and $F[\ ]$ for the forward Fourier transform and $F^{-1}[\ ]$ for the inverse Fourier transform, these equations define this property:

$$F[I_1(r, c) * I_2(r, c)] = F_1(u, v)F_2(u, v)$$

$$I_1(r, c) * I_2(r, c) = F^{-1}[F_1(u, v)F_2(u, v)]$$

$$F[I_1(r, c)I_2(r, c)] = F_1(u, v) * F_2(u, v)$$

$$I_1(r, c)I_2(r, c) = F^{-1}[F_1(u, v) * F_2(u, v)]$$

Note that it may be computationally less intensive to apply filters in the spatial domain of the image rather than the frequency domain of the image, especially if parallel hardware is available.

5.2.3.3 Translation

The translation property of the Fourier transform is given by the following equations:

$$F[I(r - r_0, c - c_0)] = F(u, v)\, e^{-j2\pi(ur_0 + vc_0)/N}$$

$$I(r - r_0, c - c_0) = F^{-1}\left(F(u, v)\, e^{-j2\pi(ur_0 + vc_0)/N}\right)$$

These equations tell us that if the image is moved, the resulting Fourier spectrum undergoes a phase shift, but the magnitude of the spectrum remains the same. This is shown in Figure 5.2.8.

5.2.3.4 Modulation

The modulation property, also called the frequency translation property, is given by:

$$F\left[I(r, c)\, e^{j2\pi(u_0 r + v_0 c)}\right] = F(u - u_o, v - v_0)$$

$$I(r, c)\, e^{j2\pi(u_0 r + v_0 c)} = F^{-1}[F(u - u_o, v - v_0)]$$

These equations tell us that if the image is multiplied by a complex exponential (remember this is really a form of a sinusoid), its corresponding spectrum is shifted. This property is illustrated in Figure 5.2.9.

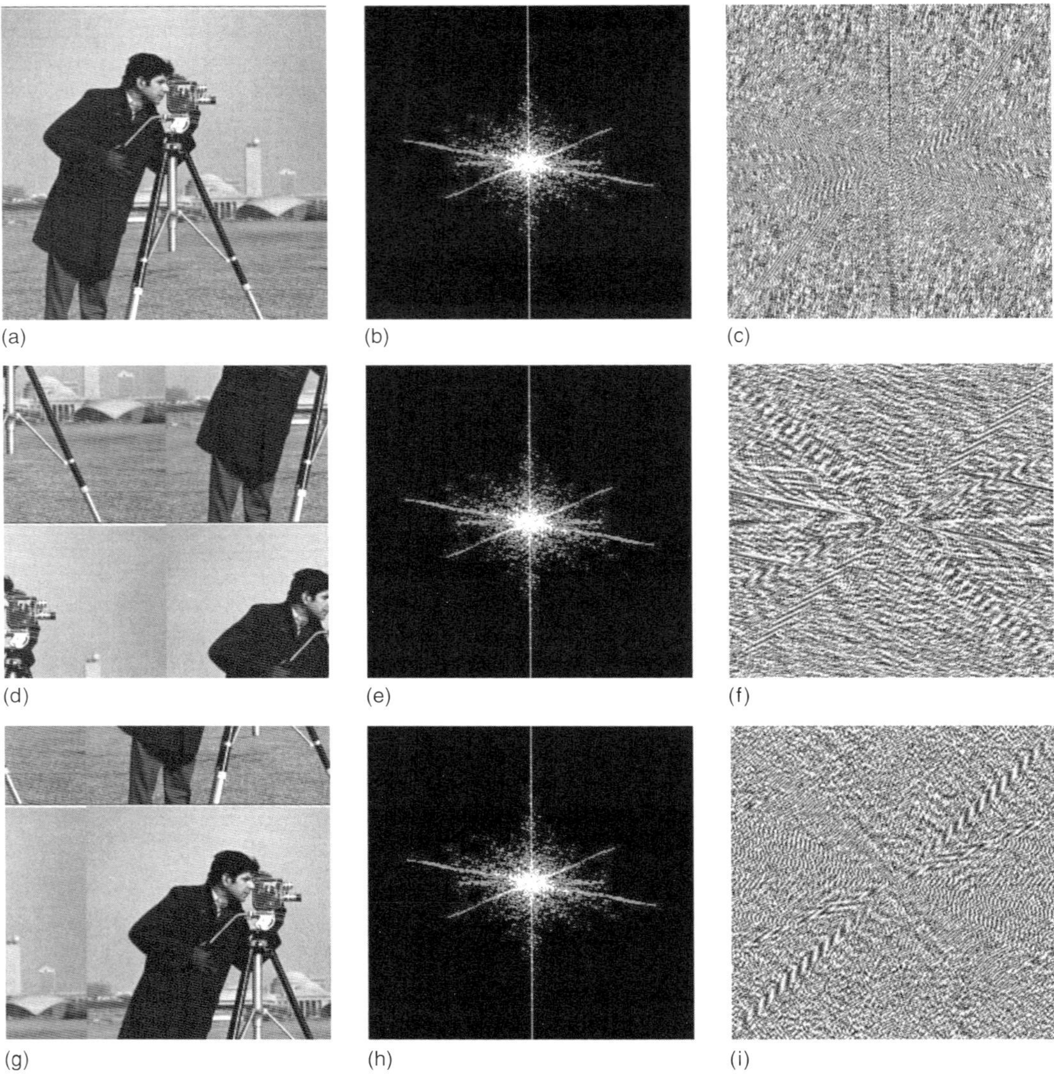
(a) (b) (c)
(d) (e) (f)
(g) (h) (i)

FIGURE 5.2.8
Translation property results in a phase shift of the spectrum. (a) Original image. (b) The magnitude of the Fourier spectrum from (a) represented as an image. (c) The phase of the Fourier spectrum from (a) represented by an image. (d) Original image shifted by 128 rows and 128 columns. (e) The magnitude of the Fourier spectrum from (d) represented as an image. (f) The phase of the Fourier spectrum from (d) represented by an image. (g) The original image shifted by 64 columns and 64 rows. (h) The magnitude of the Fourier spectrum from (g) represented as an image. (i) The phase of the Fourier spectrum from (g) represented by an image. These images illustrate that when an image is translated, the phase changes, even though magnitude remains the same.

5.2.3.5 Rotation

The rotation property can be easily illustrated by using polar coordinates:

$$r = x\cos(\theta), \quad c = x\sin(\theta)$$
$$u = w\cos(\phi), \quad v = w\sin(\phi)$$

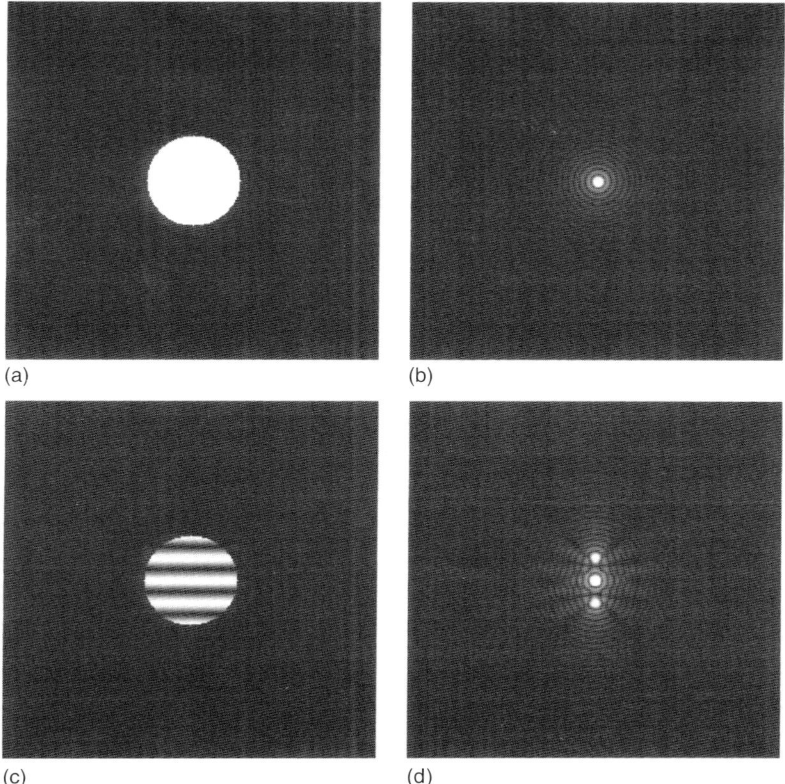

FIGURE 5.2.9
Modulation property results in frequency shift. (a) Original image. (b) Magnitude of Fourier spectrum of (a) represented as an image. (c) Original image multiplied by a vertical cosine wave at a relative frequency of 16 (16 cycles per image). (d) Magnitude of Fourier spectrum of (c) represented as an image. Note that the spectrum has been shifted by 16 above and below the origin (in these spectral images the origin is in the center of the image).

The Fourier transform pair $I(r, c)$ and $F(u, v)$ become $I(x, \theta)$ and $F(w, \phi)$, respectively, and we can write a Fourier transform pair to illustrate the rotation property as follows:

$$I(x, \theta + \theta_0) = F^{-1}[F(w, \phi + \theta_0)]$$

$$F[I(x, \theta + \theta_0)] = F(w, \phi + \theta_0)$$

This property tell us that if an image is rotated by an angle θ_0, then $F(u, v)$ is rotated by the same angle, and vice versa. This is shown in Figure 5.2.10.

5.2.3.6 Periodicity

The DFT is periodic with period N, for an $N \times N$ image. This means:

$$F(u, v) = F(u + N, v) = F(u, v + N) = F(u + N, v = N) \ldots$$

This is shown in Figure 5.2.11a. This figure shows nine periods, but the theoretical implication is that it continues in all directions to infinity. This property defines the implied symmetry in the Fourier spectrum that results from certain theoretical

FIGURE 5.2.10
Rotation property results in corresponding rotations with image and spectrum. (a) Original image. (b) Fourier spectrum image of original image. (c) Original image rotated by 90 degrees. (d) Fourier spectrum image of rotated image.

considerations, which have not been rigorously developed here. We will, however, examine the practical implications of these theoretical aspects.

5.2.4 Displaying the Fourier Spectrum

The Fourier spectrum consists of complex floating point numbers, which are stored in CVIPtools as a two band image—one band for the real plane data, and one band for the imaginary plane data. What we usually see in a spectral image is actually the magnitude data that has been remapped in a way that makes visualization easier. For displaying the magnitude of the Fourier spectrum, we usually shift the origin to the center. Applying the periodicity property and the modulation property with $u_0 = v_0 = N/2$, we obtain:

$$I(r, c)e^{j2\pi(Nr/2+Nc/2)/N} = I(r, c)\,e^{j\pi(r+c)} = I(r, c)(-1)^{(r+c)}$$

In other words, we can shift the spectrum by $N/2$ by multiplying the original image by $(-1)^{(r+c)}$, which will shift the origin to the center of the image (shown in Figure 5.2.11). This is how it is done in CVIPtools for various reasons: (1) it is easier to understand the spectral information with the origin in the center and frequency increasing from the

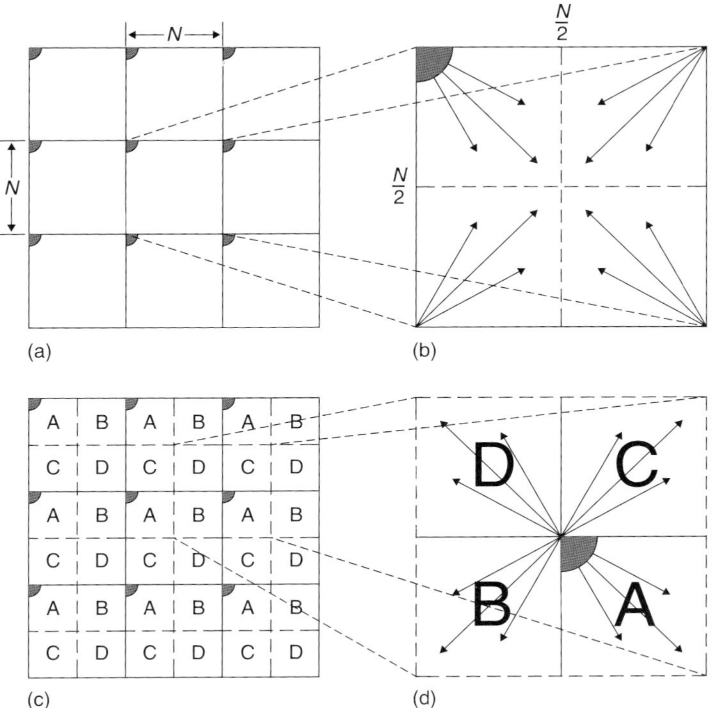

FIGURE 5.2.11
Periodicity and implied symmetry for the Fourier transform. (a) Implied symmetry with origin in upper-left corner. Each $N \times N$ block represented all the transform coefficients, and is repeated infinitely in all directions. (b) Increasing frequency in direction of arrows. (c) Periodic spectrum, with quadrants labeled A, B, C, and D. (d) Spectrum shifted to center. Frequency increases in all directions as we move away from the origin.

center out towards the edges, (2) it makes it easier to visualize the filters (Section 5.7), and (3) it looks better.

The actual dynamic range of the Fourier spectrum is much greater than the 256 gray levels (8-bits) available with most image display devices. Thus, when we remap it to 256 levels, we can only see the largest values, which are typically the low frequency terms around the origin and/or terms along the u and v axis. Figure 5.2.12a shows a Fourier magnitude image that has been directly remapped to 0–255 where all we see is the zero frequency term. We can apply contrast enhancement techniques to show more information, as in Figure 5.2.12c–f, but we are still missing much of the visual information due to the limited dynamic range and the human visual system's response.

To take advantage of the human visual system's response to brightness we can greatly enhance the visual information available by displaying the following log transform of the spectrum:

$$\log(u, v) = k \log[1 + |F(u, v)|]$$

The log function compresses the data, and the scaling factor k remaps the data to the 0–255 range. In Figure 5.2.13 we compare displaying the magnitude of the spectrum by direct remapping and contrast enhancement versus the log remap method. Here we see that the log remap method shows much more information visually. This effect is most prominent with the spectra from natural images (corresponding to a–c and g–i), as

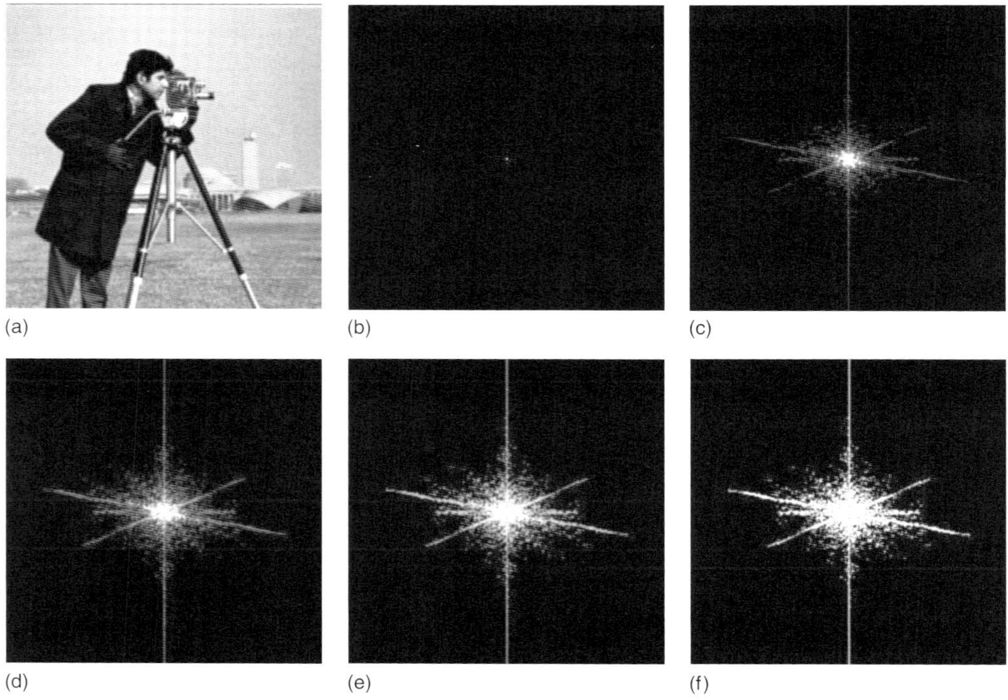

FIGURE 5.2.12
Direct mapping of Fourier magnitude data. (a) Original image. (b) The Fourier magnitude directly remapped to 0–255 without any enhancement. (c)–(f) Contrast enhanced versions of (b). Note that in (f), where we can see the most, the image is visually reduced to being either black or white, most of the dynamic range is lost.

compared with artificial images (corresponding to d–f and j–l). Can you guess the shapes of the artificial images that created the spectra in j–l and d–f? Remember the first example we saw with the continuous Fourier transform, where we learned that a function that ends abruptly in one domain results in rippling in the other domain; and that the spacing in one domain is inversely proportional to the spacing in the other domain (see Figure 5.2.2). For images of simple geometric objects this means that the ripples occur as a result of abrupt changes in gray level at object boundaries, and that the width of the ripples is inversely proportional to the object size. In Figure 5.2.14 are images of simple geometric shapes and their corresponding spectral images. Examine them carefully and apply what you learned thus far to understand them.

In addition to the magnitude information, the phase information is available in the Fourier spectrum. Typically, this information is not displayed as an image, but we have found it useful to illustrate phase changes, as was shown in Figure 5.2.8. This information has a range of 0 to 360 degrees, or 0 to 2π radians. It is floating point data, so it has a larger dynamic range than the 256 levels typically available for display.

5.3 Cosine Transform

The cosine transform, like the Fourier transform, uses sinusoidal basis functions. The difference is that the cosine transform basis functions are not complex; they use only

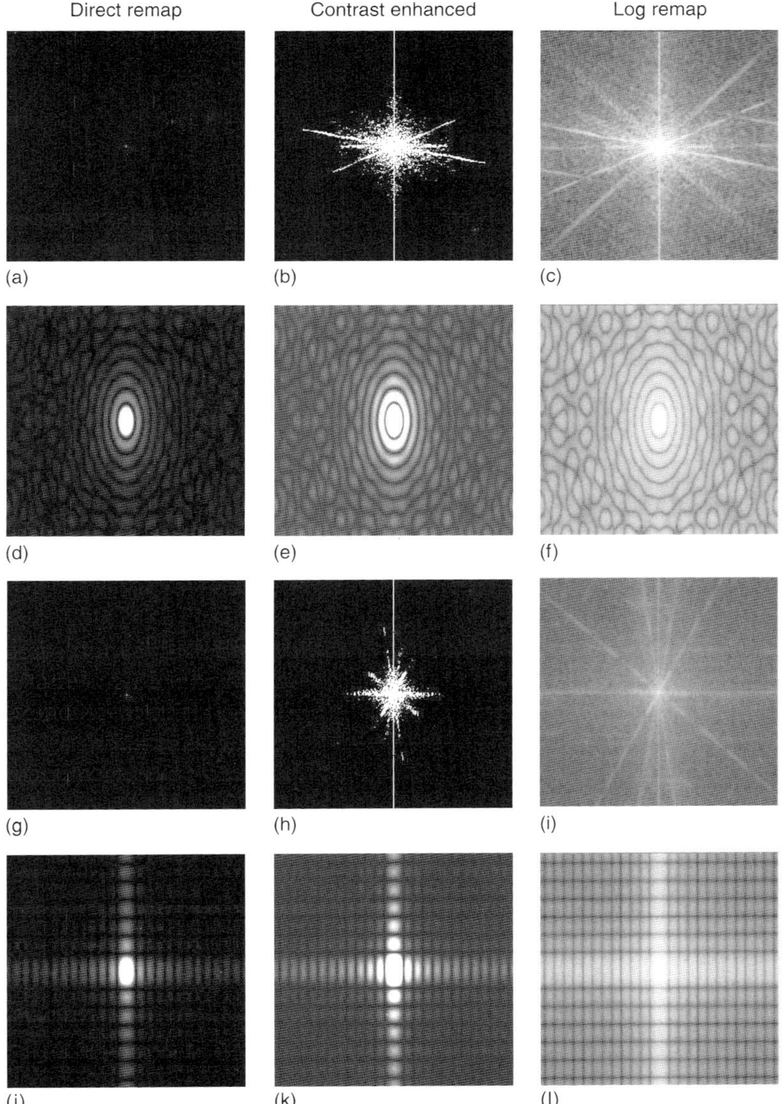

FIGURE 5.2.13
Displaying DFT spectrum with various remap methods. (a) Fourier magnitude spectrum of cam.pgm, direct remap to byte. (b) Contrast enhanced version of (a). (c) Log remapped version of cam.pgm DFT spectrum. (d) Fourier magnitude spectrum of an ellipse, direct remap to byte. (e) Contrast enhanced version of (d). (f) Log remapped version, of an ellipse's DFT spectrum. (g) Fourier magnitude spectrum of house.pgm, direct remap to byte. (h) Contrast enhanced version of (g). (i) Log remapped version of house.pgm DFT spectrum. (j) Fourier magnitude spectrum of a rectangle, direct remap to byte. (k) Contrast version of (j). (l) Log remapped version of a rectangle's DFT spectrum.

cosine functions, and not sine functions. The two-dimensional discrete cosine transform (DCT) equation for an $N \times N$ image is given by:

$$C(u, v) = \alpha(u)\alpha(v) \sum_{r=0}^{N-1} \sum_{c=0}^{N-1} I(r, c) \cos\left[\frac{(2r+1)u\pi}{2N}\right] \cos\left[\frac{(2c+1)v\pi}{2N}\right]$$

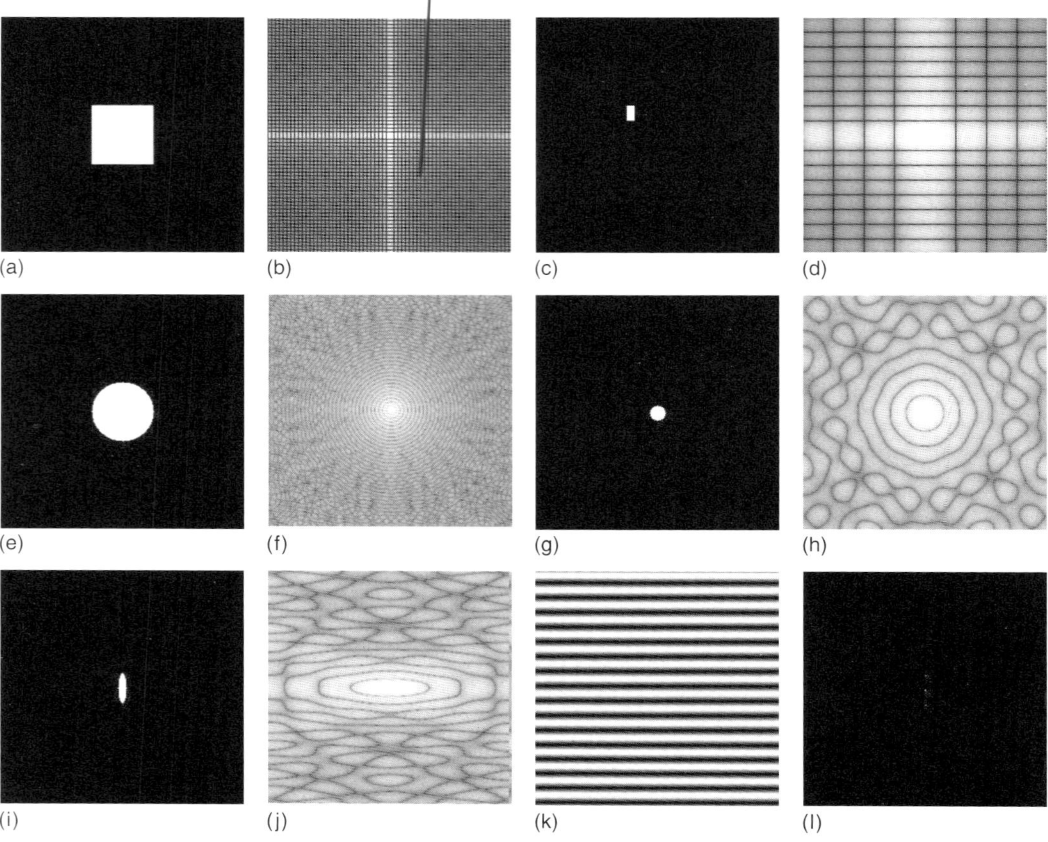

FIGURE 5.2.14
Images of simple geometric shapes and their Fourier spectral images. (a) An image of a square. (b) The log remapped spectrum of the square. (c) A small rectangle. (d) The log remapped spectrum of the small rectangle. (e) An image of a circle. (f) The log remapped spectrum of the circle image. (g) A small circle. (h) The log remapped spectrum of the small circle. (i) A small ellipse. (j) The log remapped spectrum of the small ellipse. (k) An image of a vertical sine wave. (l) The magnitude of the spectrum of the sine wave.

where

$$\alpha(u),\ \alpha(v) = \begin{cases} \sqrt{\frac{1}{N}} & \text{for } u,\ v = 0 \\ \sqrt{\frac{2}{N}} & \text{for } u,\ v = 1,\ 2,\ \ldots,\ N-1 \end{cases}$$

Since this transform uses only the cosine function it can be calculated using only real arithmetic, instead of complex arithmetic as the DFT requires. The cosine transform can be derived from the Fourier transform by assuming that the function (the image) is mirrored about the origin, thus making it an even function, which means it is symmetric about the origin. This has the effect of canceling the odd terms, which correspond to the sine terms (imaginary terms) in the Fourier transform. This also affects the implied symmetry of the transform, where we now have a function that is implied to be $2N \times 2N$. In Figure 5.3.1 we see the meaning of mirroring, or folding, a function about the origin creating a $2N \times 2N$ function from an original $N \times N$ one. Now, we are only interested in an $N \times N$ portion of this spectrum, which corresponds to our image, since the other

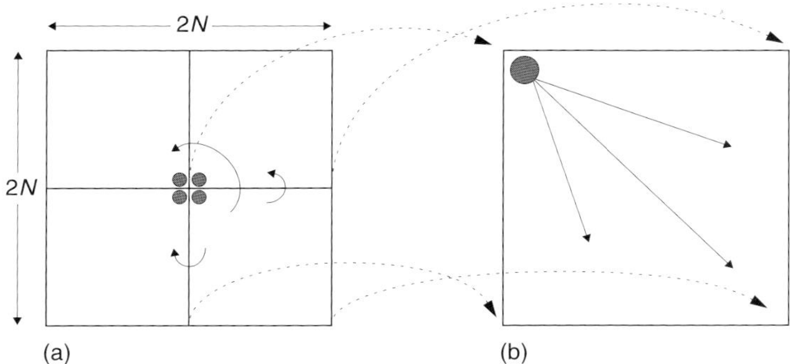

FIGURE 5.3.1
Cosine symmetry. (a) Spectrum folded about origin, represented by the ●. The $2N \times 2N$ block is repeated infinitely in all directions. (b) Arrows indicate direction of increasing frequency for cosine spectrum.

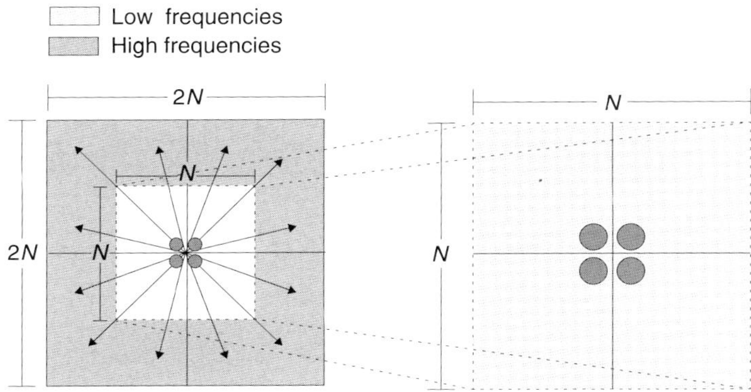

FIGURE 5.3.2
Cosine spectrum should not be shifted to center. (a) Cosine spectrum with arrows in direction of increasing frequency. (b) Extracting the central $N \times N$ portion, we lose the high frequency information.

quadrants are redundant. Understand that we do not want to shift the origin to the center for the cosine transform, or we lose information (see Figure 5.3.2).

The cosine transform is often used in image compression, in particular in the first version of the Joint Photographers Expert Group (JPEG) image compression method, which has been established as an international standard (the newer JPEG2000 method uses the wavelet transform). In computer imaging we often represent the basis matrices as images, called basis images, where we use various gray values to represent the different values in the basis matrix. The 2-D basis images for the cosine transform are shown in Figure 5.3.3 for a 4×4 image, where the actual values have been remapped for illustration purposes by the legend at the bottom of the figure. Remember that the transform actually projects the image onto each of these basis images (see Figure 5.1.5), so the transform coefficients, $C(u, v)$, tell us the amount of that particular basis image that the original image, $I(r, c)$, contains.

The inverse cosine transform is given by:

$$C^{-1}[C(u, v)] = I(r, c) = \sum_{u=0}^{N-1} \sum_{v=0}^{N-1} \alpha(u)\alpha(v)C(u, v) \cos\left[\frac{(2r+1)u\pi}{2N}\right] \cos\left[\frac{(2c+1)v\pi}{2N}\right]$$

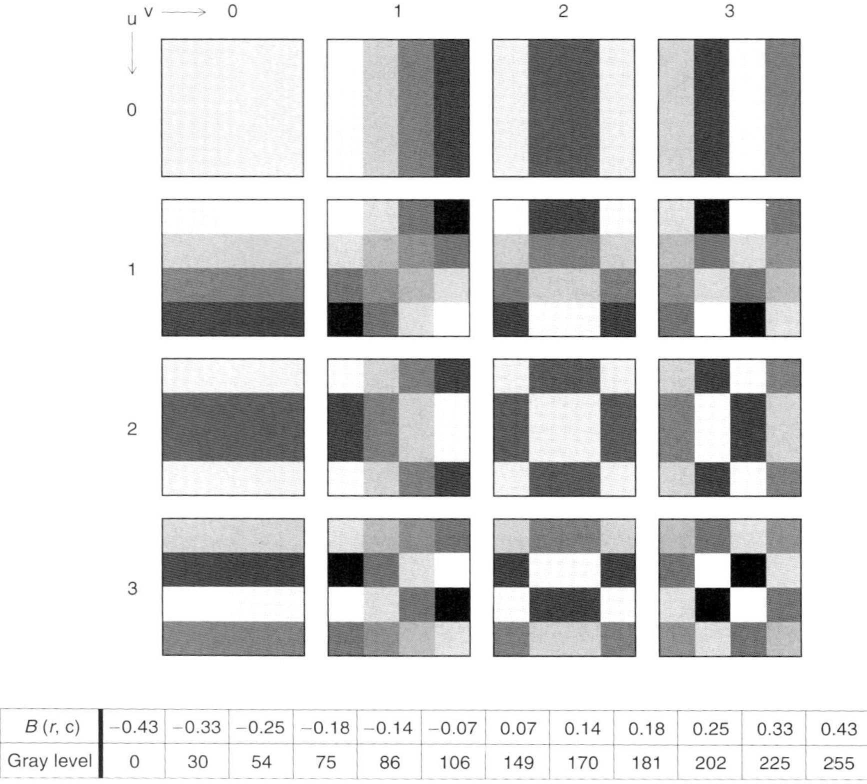

B (r, c)	−0.43	−0.33	−0.25	−0.18	−0.14	−0.07	0.07	0.14	0.18	0.25	0.33	0.43
Gray level	0	30	54	75	86	106	149	170	181	202	225	255

FIGURE 5.3.3
Discrete cosine transform basis images.

5.4 Walsh–Hadamard Transform

The Walsh–Hadamard transform (WHT) differs from the Fourier and cosine transforms in that the basis functions are not sinusoids. The basis functions are based on square or rectangular waves with peaks of ±1 (see Figure 5.4.1). Here the term rectangular wave refers to any function of this form, where the width of the pulse

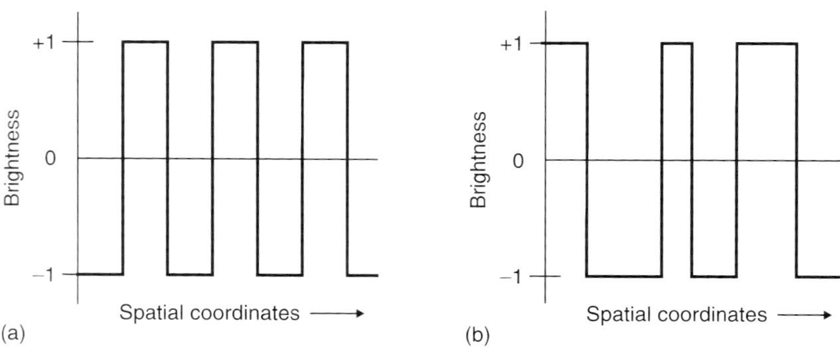

FIGURE 5.4.1
Form of the Walsh–Hadamard basis functions. (a) A square wave. (b) Representation of a rectangular wave. The width of each pulse may vary.

may vary. One primary advantage of a transform with these type of basis functions is that the computations are very simple. When we project the image onto the basis functions, all we need to do is to multiply each pixel by ± 1, as in seen in the WHT equation:

$$WH(u, v) = \frac{1}{N} \sum_{r=0}^{N-1} \sum_{c=0}^{N-1} I(r, c)(-1)^{\sum_{i=0}^{n-1} [b_i(r)p_i(u)+b_i(c)p_i(v)]}$$

where $N = 2^n$, the exponent on the (-1) is performed in modulo 2 arithmetic, and $b_i(r)$ is found by considering r as a binary number, and finding the ith bit.

Example 5.4.1

$n = 3$ (3 bits, so $N = 8$), and $r = 4$
r in binary is 100, so $b_2(r) = 1$, $b_1(r) = 0$, and $b_0(r) = 0$.

Example 5.4.2

$n = 4$, (4 bits, so $N = 16$), and $r = 2$
r in binary is 0010, so $b_3(r) = 0$, $b_2(r) = 0$, $b_1(r) = 1$, and $b_0(r) = 0$.

$p_i(u)$ is found as follows:

$$p_0(u) = b_{n-1}(u)$$

$$p_1(u) = b_{n-1}(u) + b_{n-2}(u)$$

$$p_2(u) = b_{n-2}(u) + b_{n-3}(u)$$

$$\vdots$$

$$p_{n-1}(u) = b_1(u) + b_0(u)$$

The sums are performed in modulo 2 arithmetic, and the values for $b_i(c)$ and $p_i(v)$ are found in a similar manner. Strictly speaking we cannot call the Walsh–Hadamard transform a frequency transform, as the basis functions do not exhibit the frequency concept in the manner of sinusoidal functions. However, we define an analogous term for use with these types of functions. If we consider the number of zero crossings (or sign changes) we have a measure that is comparable to frequency, and we call this *sequency*. In Figure 5.4.2 we see the 1-D Walsh–Hadamard basis functions for $N = 4$, and the corresponding sequency. We can see that the basis functions are in the order of increasing sequency, much like the sinusoidal functions are in order of increasing frequency. In Figure 5.4.3, we have the basis images for the WHT for a 4×4 image; we use white for the $+1$ and black for the -1.

It may be difficult to see how the 2-D basis images are generated from the 1-D basis vectors. For the terms that are along the u or v axis, we simply repeat the 1-D function along all the rows or columns. For the basis images that are not along the u or v axis we perform a *vector outer product* on the corresponding 1-D vectors. We have seen that a *vector inner product* is what we call a projection, and is performed by overlaying, multiplying coincident terms, and summing the results—this gives us a

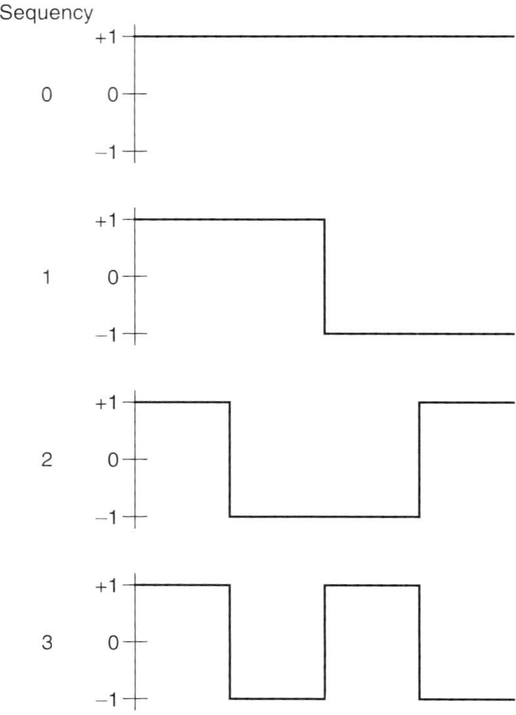

FIGURE 5.4.2
1-D Walsh–Hadamard basis functions.

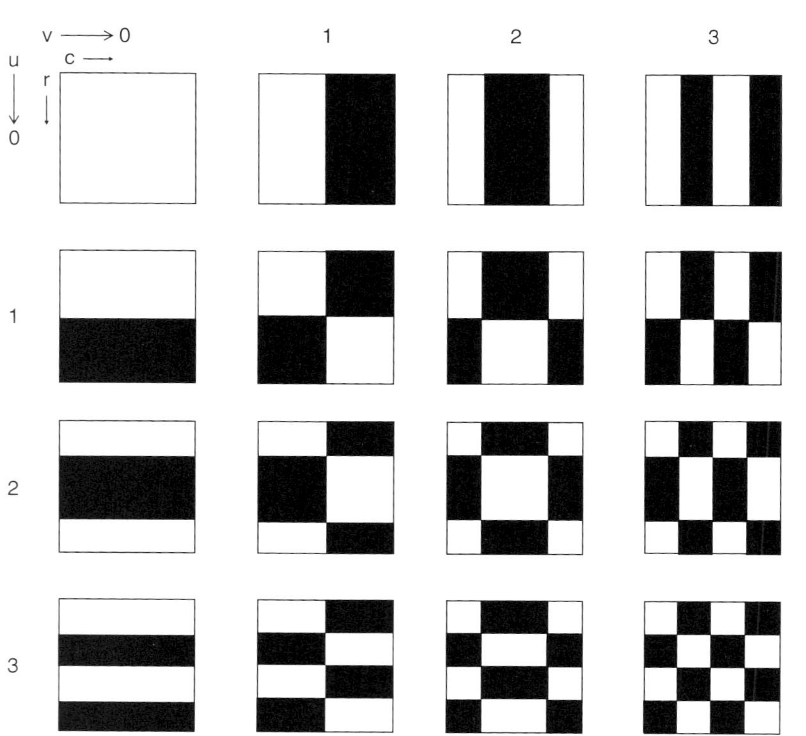

FIGURE 5.4.3
Walsh–Hadamard basis images.

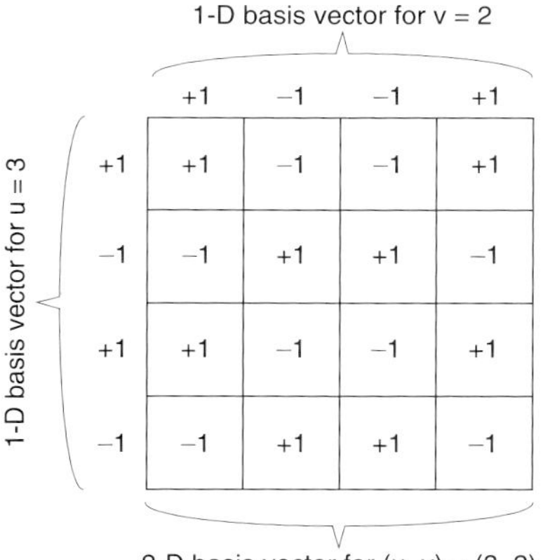

FIGURE 5.4.4
Vector outer product.

scalar, or a single number, for a result. The vector outer product gives us a matrix, which is obtained as follows:

Example 5.4.3

For $(u, v) = (3, 2)$, see Figure 5.4.4. If we look along one row of the $v = 2$ $(u = 0)$ basis image in Figure 2.5.10 we find the following numbers: $+1\ -1\ -1\ +1$. Then if we look along one column in the u direction for $u = 3$ $(v = 0)$, we see $+1\ -1\ +1\ -1$. These are the corresponding 1-D basis vectors. We then put the row vector across the top and the column vector down the left side and fill in the matrix by multiplying the column by the corresponding row element, as in Figure 5.4.4. The resulting matrix is the vector outer product. Compare this to the corresponding basis image in Figure 5.4.3.

This process can be used to generate the 2-D basis images for any function that has a separable basis. Remember that *separable* means that the basis function can be expressed as a product of terms that depend only on one of the variable pairs, r, u or c, v, and that this separability also allows us to perform the 2-D transformation by two 1-D transforms. This is accomplished by first doing a 1-D transform on the rows, and then performing the 1-D transform on the resulting columns as was shown in the DFT section.

It is interesting to note that with the Walsh-Hadamard transform there is another visual method to find the off axis basis images, by assuming that the black in Figure 5.4.3 corresponds to 0 and the white corresponds to 1. The basis images not along the u or v axis can be obtained by taking the corresponding basis images on these axes, overlaying them, and performing an XOR followed by a NOT. For example, to find the Walsh–Hadamard basis image corresponding to $(u, v) = (3, 2)$, we take the basis image along the u axis for $u = 3$, and the basis image along the v axis for $v = 2$, overlay them, XOR the images and then perform a NOT. This is illustrated in Figure 5.4.5.

The inverse Walsh–Hadamard transform equation is:

$$WH^{-1}[WH(u, v)] = I(r, c) = \frac{1}{N} \sum_{u=0}^{N-1} \sum_{v=0}^{N-1} WH(u, v)(-1)^{\sum_{i=0}^{n-1} [b_i(r)p_i(u) + b_i(c)p_i(v)]}$$

In CVIPtools there is a separate Walsh transform and a separate Hadamard transform. Even though (if N is power of 2) they both have the same basis functions, as initially defined the basis functions were ordered differently. The Hadamard ordering was not sequency based, so this ordering is not really that useful for image processing. It was

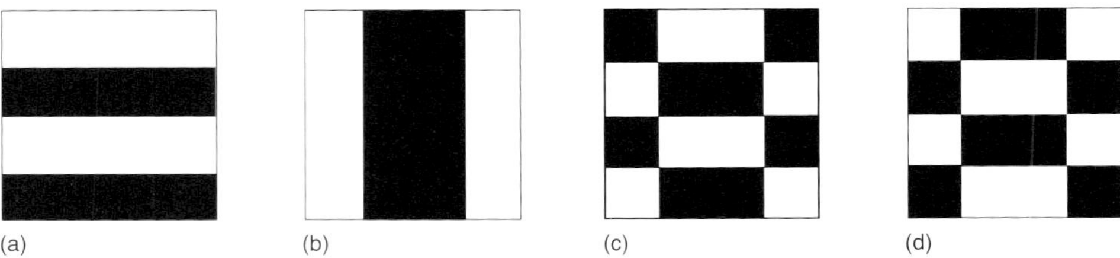

FIGURE 5.4.5
Finding an off-axis Walsh–Hadamard basis image. (a) Basis $u=3$ $(v=0)$. (b) Basis $v=2$ $(u=0)$. (c) XOR of images a, b. (d) NOT of image (c). This is the Walsh–Hadamard basis image for $(u,v)=(3,2)$.

originally defined for the ease of generating the basis vectors. The standard now is to use the sequency ordered basis functions as defined here and call it the Walsh–Hadamard transform (WHT). In the CVIPtools the transform called the Walsh is sequency ordered, and the one called the Hadamard is in standard "Hadamard ordering"—not sequency based.

5.5 Haar Transform

The Haar transform has rectangular waves as basis functions, similar to the Walsh–Hadamard transform. The primary differences are that the basis vectors contain not just +1 and −1, but also contain zeros. The Haar transform is derived from the Haar matrices; in these matrices each row represents a one-dimensional basis vector. The following shows the basis vectors for a Haar transform of two basis vectors ($N=2$), four basis vectors ($N=4$) and eight basis vectors ($N=8$).

$$Haar2 \quad \Rightarrow \quad \frac{1}{\sqrt{2}}\begin{Bmatrix} +1 & +1 \\ +1 & -1 \end{Bmatrix}$$

$$Haar4 \quad \Rightarrow \quad \frac{1}{\sqrt{4}}\begin{Bmatrix} +1 & +1 & +1 & +1 \\ +1 & +1 & -1 & -1 \\ \sqrt{2} & -\sqrt{2} & 0 & 0 \\ 0 & 0 & \sqrt{2} & -\sqrt{2} \end{Bmatrix}$$

$$Haar8 \quad \Rightarrow \quad \frac{1}{\sqrt{8}}\begin{Bmatrix} +1 & +1 & +1 & +1 & +1 & +1 & +1 & +1 \\ +1 & +1 & +1 & +1 & -1 & -1 & -1 & -1 \\ \sqrt{2} & \sqrt{2} & -\sqrt{2} & -\sqrt{2} & 0 & 0 & 0 & 0 \\ 0 & 0 & 0 & 0 & \sqrt{2} & \sqrt{2} & -\sqrt{2} & -\sqrt{2} \\ +2 & -2 & 0 & 0 & 0 & 0 & 0 & 0 \\ 0 & 0 & +2 & -2 & 0 & 0 & 0 & 0 \\ 0 & 0 & 0 & 0 & +2 & -2 & 0 & 0 \\ 0 & 0 & 0 & 0 & 0 & 0 & +2 & -2 \end{Bmatrix}$$

(a)

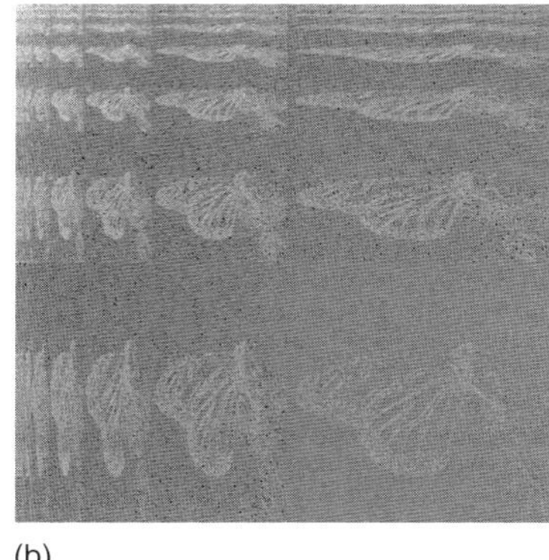

(b)

FIGURE 5.5.1
5.1 Haar transform. (a) Original image. (b) Haar transform image.

The Haar basis vectors can be extended to higher orders by following the same patterns shown in the above. Note that as the order increases the number of zeros in the basis vectors increase. This has the unique effect of allowing a multiresolution decomposition of an image (explored more in Section 5.8), and is best illustrated by example. In Figure 5.5.1 we see the log remapped Haar spectrum. Here we see that the Haar provides edge information at increasing levels of resolution.

5.6 Principal Components Transform

The principal components transform (PCT) is also referred to as the Hotelling, Karhunen–Loeve or eigenvector transform. This is because it was first derived by Karhunen and Loeve for continuous signals, and later developed for discrete signals by Hotelling. Mathmatically it involves finding eigenvectors of covariance matrices, hence eigenvector transform, and it results in decomposing the image into its principal components, hence PCT. It differs from the transforms that we have considered thus far, as it is not related to extracting frequency or sequency information from images, but is a mathematical transform that decorrelates multiband image data. It can, however, be used to find optimal basis images for a specific image, but this use of it is not very practical due to the extensive processing required.

Applying the PCT to multiband images, color images or multispectral images, provides a linear transform matrix that will decorrelate the input data. In most color images there is a high level of correlation between the red, green, and blue bands. This can be seen in Figure 5.6.1 where we show the brightness values in the red, green, and blue bands of a color image with each band presented as a monochrome image, and the three bands after the PCT. Here we see that the red, green, and blue bands are highly correlated—they look similar; whereas with the PCT bands most of the visual information is in band 1, some information is in band 2 and practically no visual information in band 3. This is what it means when we say that the PCT decorrelates the data and puts most of the information into the principal component band.

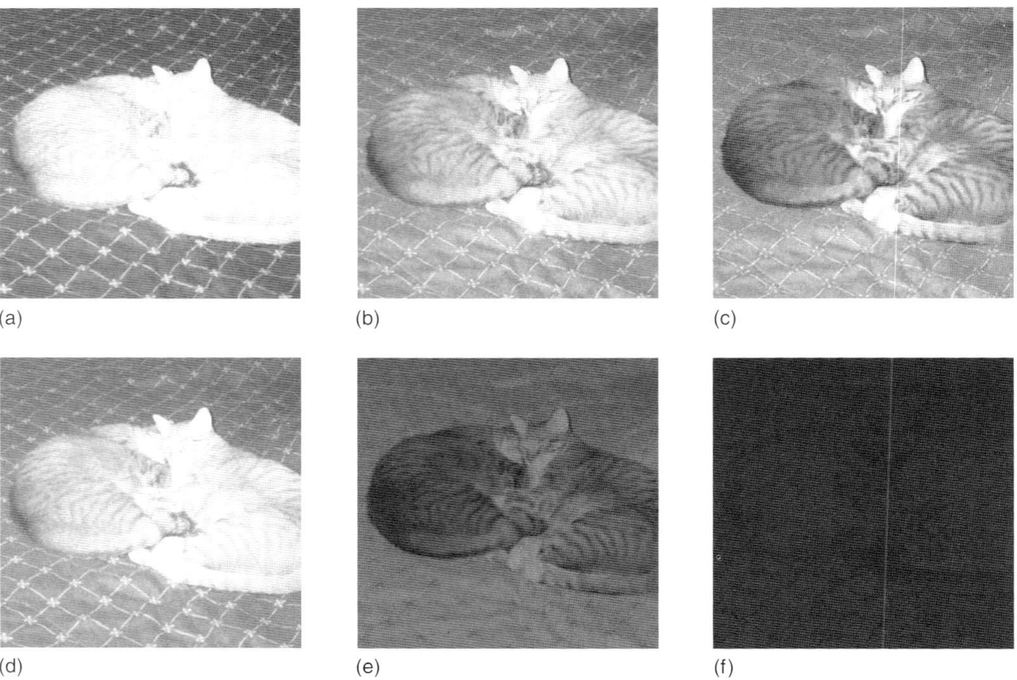

FIGURE 5.6.1
Principal components transform—PCT. (a) Red band of a color image. (2) Green band. (c) Blue band. (d) Principal component band 1. (e) Principal component band 2. (f) Principal component band 3. Note that the red, green, and blue bands are highly correlated—they look similar; whereas with the PCT bands most of the visual information is in band 1, some in band 2 and none in band 3. This is what it means when we say that the PCT decorrelates the data and puts most of the information into the principal component band.

The three step procedure for finding the PCT for a color, RGB, image is as follows:

1. Find the covariance matrix in RGB space, given by:

$$[COV]_{\text{RGB}} = \begin{bmatrix} C_{\text{RR}} & C_{\text{GR}} & C_{\text{BR}} \\ C_{\text{RG}} & C_{\text{GG}} & C_{\text{BG}} \\ C_{\text{RB}} & C_{\text{GB}} & C_{\text{BB}} \end{bmatrix}$$

where:

$$C_{\text{RR}} = \frac{1}{P} \sum_{i=1}^{P} (R_i - m_{\text{R}})^2$$

P = the number of pixels in the image
R_i = the red value for the ith pixel

$$m_{\text{R}} = \text{red mean (average)} = \frac{1}{P} \sum_{i=1}^{P} R_i$$

Similar equations are used for C_{GG} and C_{BB} (the autocovariance terms). The elements of the covariance matrix that involve more than one of the RGB variables, C_{GR}, C_{BR}, C_{RG}, C_{BG}, C_{RB}, and C_{GB}, are called cross-covariance terms and are found as follows:

$$C_{XY} = \frac{1}{P}\left[\sum_{i=1}^{P} X_i Y_i\right] - m_x m_y$$

with the means defined as above.

2. Find the eigenvalues of the covariance matrix, e_1, e_2, and e_3, and their corresponding eigenvectors:

$$e_1 \Rightarrow [E_{11}, E_{12}, E_{13}]$$

$$e_2 \Rightarrow [E_{21}, E_{22}, E_{23}]$$

$$e_3 \Rightarrow [E_{31}, E_{32}, E_{33}]$$

Order them such that e_1 is the largest eigenvalue, and e_3 is the smallest.

3. Perform the linear transform on the RGB data by using the eigenvectors as follows:

$$\begin{bmatrix} P_1 \\ P_2 \\ P_3 \end{bmatrix} = \begin{bmatrix} E_{11} & E_{12} & E_{13} \\ E_{21} & E_{22} & E_{23} \\ E_{31} & E_{32} & E_{33} \end{bmatrix} \begin{bmatrix} i \\ i \\ i \end{bmatrix} = \begin{bmatrix} E_{11}i + E_{12}i + E_{13}i \\ E_{21}i + E_{22}i + E_{23}i \\ E_{31}i + E_{32}i + E_{33}i \end{bmatrix}$$

Now the PCT data is P_1, P_2 and P_3 where the P_1 data is the principal component and contains the most variance (as illustrated in Figure 4.3.9). In pattern recognition theory the measure of variance is considered to be a measure of information, so we can say that the principal component data contains the most information—as shown in Figure 5.6.1. It is interesting to note that the PCT is also used in image compression (see Chapter 10), since this transform is optimal in the least-square-error sense. This means that most of the information, assumed to be directly correlated with variance, is in a reduced dimensionality. For example, in one application involving a database of medical images, it was experimentally determined that the dimension with the largest variance after the PCT was performed contained approximately 91% of the variance. This would allow at least a 3:1 compression and still retain 91% of the information.

The PCT is easily extended to data of any dimensionality (the above example is three-dimensional, RGB), so it can be applied to multispectral images in a similar manner. In some textbooks it is defined to be extended to $N \times N$ data and is performed on the image itself (see the references) to find optimal basis images.

5.7 Filtering

After the image has been transformed into the frequency or sequency domain, we may want to modify the resulting spectrum. Filtering modifies the frequency or sequency spectrum by selectively retaining, removing or scaling the various components of the spectrum. High frequency information can be removed with a lowpass filter, which will have the effect of blurring an image, or low frequency information can be removed with a highpass filter, which will tend to sharpen the image. We may want to extract the frequency information in specific parts of the spectrum by bandpass filtering. Alternately, band-reject filtering can be employed to eliminate specific parts of the spectrum, for example, to remove unwanted noise. All of these types of filters will be explored here.

Before we consider the filters we need to be aware of the implied symmetry for each of the transforms of interest. We will assume that the Fourier spectrum has been shifted to the center and exhibits the symmetry shown in Figure 5.2.11b. Both the cosine and the Walsh–Hadamard are assumed to have the symmetry shown in Figure 5.3.1. The Haar transform is unique, but also has the origin in the upper left corner (see Figure 5.5.1) as the cosine and Walsh–Hadamard. The PCT as defined does not lend itself to the type of filtering under discussion here.

5.7.1 Lowpass Filters

Lowpass filters tend to blur images. They pass low frequencies, and attenuate or eliminate the high frequency information. They are used for image compression, or for dealing with noise in images. Visually they blur the image, although this blur is sometimes considered an enhancement as it imparts a softer effect to the image (see Figure 5.7.1). Lowpass filtering is performed by multiplying the spectrum by a filter, and then applying the inverse transform to obtain the filtered image. The ideal filter function is shown in Figure 5.7.2; note the two types of symmetry in the filter match the type of symmetry in the spectrum. The frequency at which we start to eliminate information is called the *cutoff frequency*, f_0. The frequencies in the spectrum that are not filtered out are in the *passband*, while the spectral components that do get filtered out are in the *stopband*. We can represent the filtering process by the following equation:

$$I_{\text{fil}}(r, c) = T^{-1}[T(u, v)H(u, v)]$$

where $I_{\text{fil}}(r,c)$ is our filtered image, $H(u, v)$ is the filter function, $T(u,v)$ is the transform, and T^{-1} represents the inverse transform. The multiplication, $T(u,v)H(u,v)$, is performed with a point-by-point method. That is, $T(0,0)$ is multiplied by $H(0,0)$, $T(0,1)$ is multiplied by $H(0,1)$, and so on. The resulting products are placed into an array at the same (r,c) location.

(a)

(b)

(c)

FIGURE 5.7.1
Lowpass filtering. (a) Original image. (b) Filtered image, using a non-ideal lowpass filter. Note the blurring that softens the image. (c) Ideal lowpass-filtered image shows the ripple artifacts at boundaries. Frequency cutoff = 32.

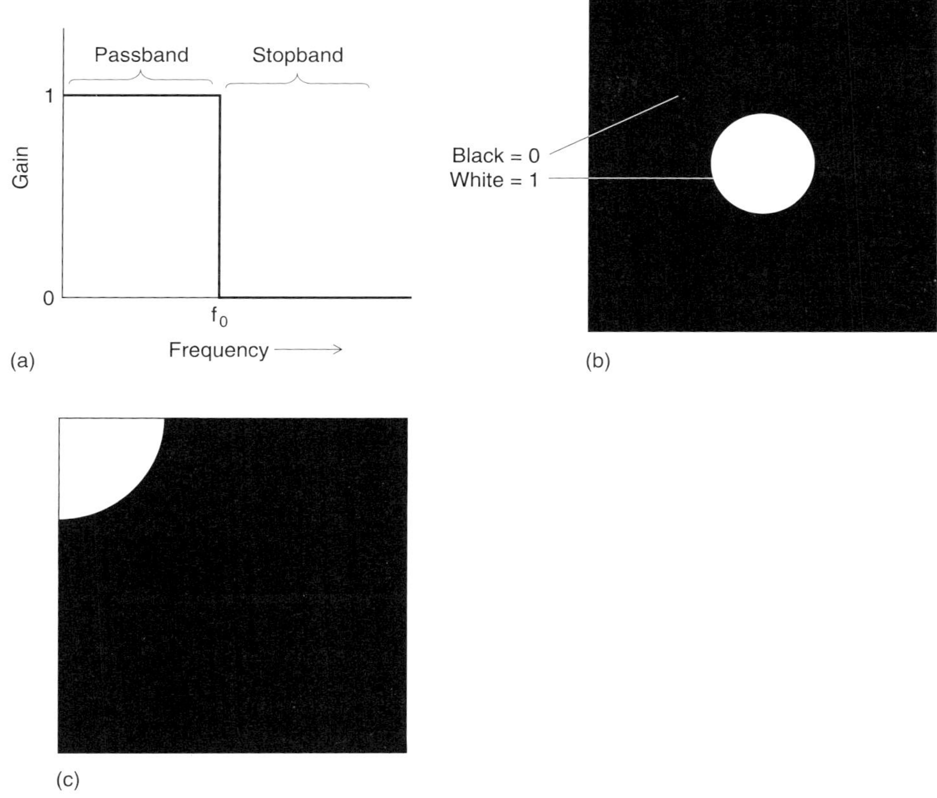

FIGURE 5.7.2
Ideal Lowpass filters. (a) 1-D lowpass ideal filter. (b) 2-D lowpass ideal filter shown as an image for Fourier transform. (c) 2-D lowpass ideal filter for Walsh–Hadamard and cosine transforms.

Example 5.7.1

Let $H(u, v)$ and $T(u, v)$ be the following 2×2 images.

$$H(u, v) = \begin{bmatrix} 2 & -3 \\ 4 & 1 \end{bmatrix} \qquad T(u, v) = \begin{bmatrix} 4 & 6 \\ -5 & 8 \end{bmatrix}$$

Then $T(u,v)H(u,v)$ is equal to:

$$\begin{bmatrix} 8 & -18 \\ -20 & 8 \end{bmatrix}$$

Note that for the ideal filters in Figure 5.7.2 the $H(u,v)$ matrix will contain only 1's and 0's, but, as in the above example, the matrix can contain any numbers.

The ideal filter is called ideal because the transition from the passband to the stopband in the filter is perfect—it goes from 0 to 1 instantly. Although this type of filter is not realizable in physical systems, such as with electronic filters, it is a reality for digital image processing applications, where we simply multiply numbers in software. However, the ideal filter leaves undesirable artifacts in images. This artifact appears in

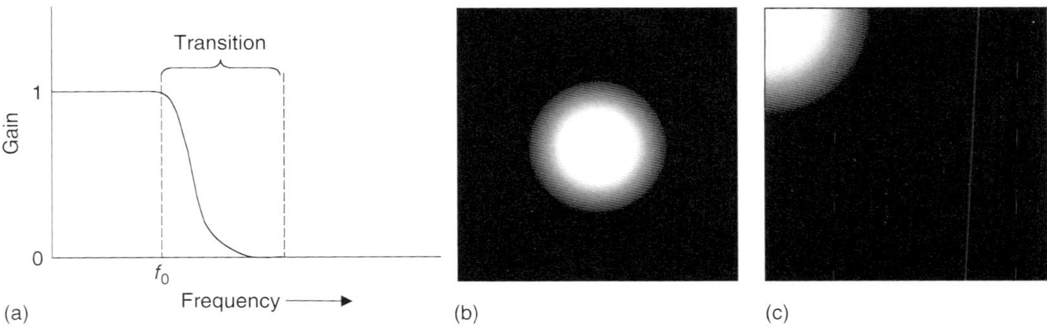

FIGURE 5.7.3
Nonideal lowpass filters. (a) 1-D nonideal filter. (b) 2-D lowpass nonideal filter shown as an image for Fourier symmetry, in the image shown back = 0, white = 1, and the gray value in between represent the transition band. (c) 2-D lowpass nonideal filter shown as an image for cosine and Walsh–Hadamard symmetry, in the image shown black = 0, white = 1, and the gray values in between represent the transition band.

the lowpass filtered image in Figure 5.7.1c as ripples, or waves, wherever there is a boundary in the image. This problem can be avoided by using a "non-ideal" filter that does not have perfect transition, as is shown in Figure 5.7.3. The image created in Figure 5.7.1b was generated using a nonideal filter of a type called a Butterworth filter.

With the Butterworth filter we can specify the *order* of the filter, which determines how steep the slope is in the transition of the filter function. A higher order to the filter creates a steeper slope, and the closer we get to an ideal filter. The filter function of a Butterworth lowpass filter of order n is given by the following equation:

$$H(u, v) = \frac{1}{1 + \left[\sqrt{u^2 + v^2}/f_0\right]^{2n}}$$

Note that $\sqrt{u^2 + v^2}$ is the distance from the origin, so the gain falls off as we get farther away from the zero frequency term, which is what we expect for a lowpass filter—to cut the high frequencies. Also note that for the Fourier spectrum shifted to the center, but still indexing our matrix with $(0, 0)$ in the upper left corner (as we may do in a computer program), we need to replace u with $(u - N/2)$ and v with $(v - N/2)$ in the above equation.

In Figure 5.7.4 we compare the results of different orders of Butterworth filters. We see that as we get closer to an ideal filter, the blurring effect becomes more prominent due to the elimination of even partial high frequency information. Another effect, that is most noticeable in the 8th-order filter, is the appearance of waves, or ripples, wherever boundaries occur in the image. This artifact is called ringing and increases as the Butterworth filter's order increases.

In Figure 5.7.5 we see the result of using a third-order Butterworth filter, but decreasing the cutoff frequency. As the cutoff frequency is lowered the image becomes more and more blurry because we are keeping less and less of the high frequency information.

5.7.2 Highpass Filters

Highpass filters will keep high frequency information, which corresponds to areas of rapid change in brightness, such as edges or fine textures. The highpass filter functions

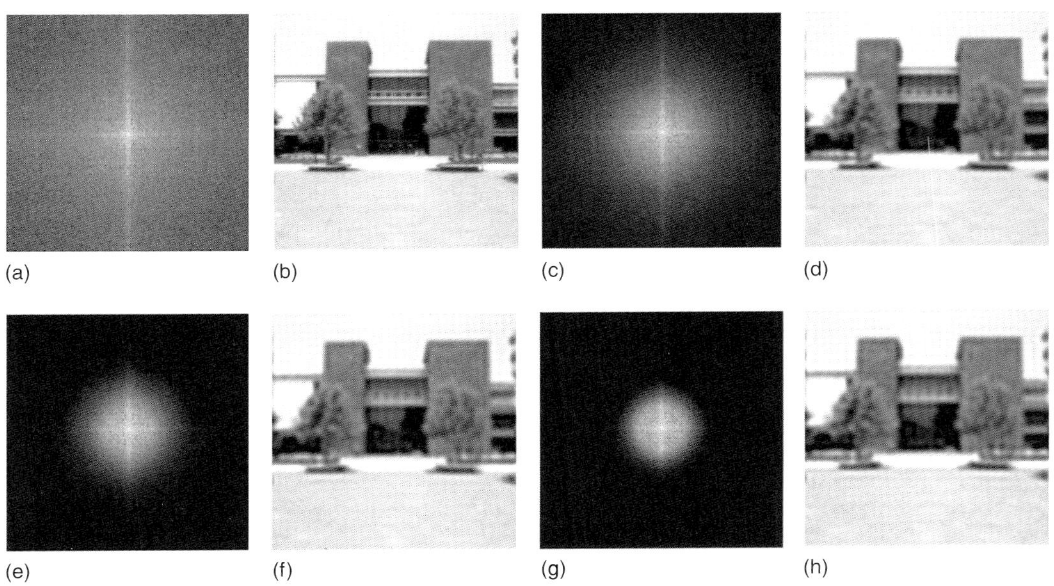

FIGURE 5.7.4
Lowpass Butterworth filters. (a) Fourier spectrum, filter order = 1. (b) Resultant image with order = 1. (c) Fourier spectrum, filter order = 3. (d) Resultant image with order = 3. (e) Fourier spectrum, filter order = 5. (f) Resultant image with order = 5. (g) Fourier spectrum, filter order = 8. (h) resultant image with order = 8.

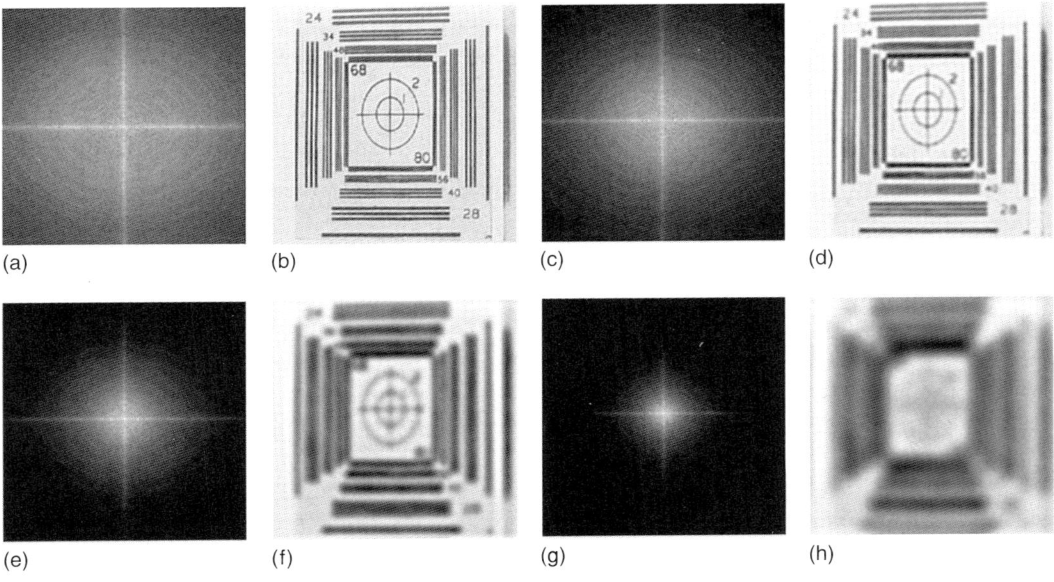

FIGURE 5.7.5
Butterworth lowpass filtering, filter order = 3,various cutoff frequencies. (a) Fourier spectrum, cutoff frequency = 64. (b) Resultant image with cutoff frequency = 64. (c) Fourier spectrum, cutoff frequency = 32. (d) Resultant image with cutoff frequency = 32. (e) Fourier spectrum, cutoff frequency = 16. (f) Resultant image with cutoff frequency = 16. (g) Fourier spectrum, cutoff frequency = 8. (h) Resultant image with cutoff frequency = 8.

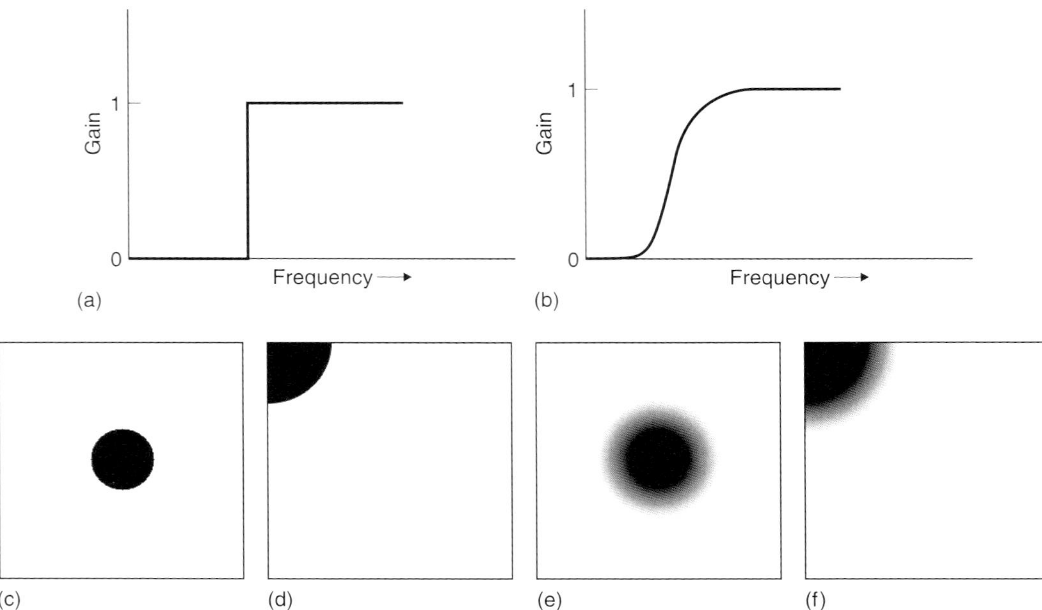

FIGURE 5.7.6
Highpass filter function. (a) 1-D ideal highpass filter. (b) 1-D nonideal highpass filter. (c) 2-D ideal highpass filter shown as an image. (d) 2-D nonideal highpass filter shown as an image. (e) 2-D ideal highpass filter for cosine and Walsh–Hadamard symmetry. (f) 2-D nonideal highpass filter for cosine and Walsh–Hadamard symmetry. *Note*: for the nonideal filters, white = 1, black = 0, and gray values in between represent the transition band.

are shown in Figure 5.7.6, where we see both ideal and Butterworth filter functions. A highpass filter can be used for edge enhancement, since it passes only high frequency information, corresponding to places where gray levels are changing rapidly (edges in images are characterized by rapidly changing gray levels). The Butterworth filter of order n for the highpass filter is:

$$H(u, v) = \frac{1}{1 + \left[f_0/\sqrt{u^2 + v^2}\right]^{2n}}$$

Note that this filter gain is very small for frequencies much smaller than f_0, and approaches a gain of one as the frequencies get much larger than f_0. Also note that for the Fourier spectrum shifted to the center, but still indexing our matrix with $(0, 0)$ in the upper left corner (as we may do in a computer program), we need to replace u with $(u - N/2)$ and v with $(v - N/2)$ in the above equation.

The function for a special type of highpass filter, called a high-frequency emphasis filter, is shown in Figure 5.7.7. This filter function boosts the high frequencies and retains some of the low frequency information and by adding an offset value to the function, so we do not lose the overall image information. The results from applying these types of filters are shown in Figure 5.7.8. The original is shown in Figure 5.7.8a, and Figure 5.7.8b, c show the results from a Butterworth and an ideal filter function. Here we can see the edges enhanced, and the ripples that occur from using an ideal filter (Figure 5.7.8c), but note a loss in the overall contrast of the image. In Figure 5.7.8d,e, we see the contrast added back to the image by using the high frequency emphasis filter function.

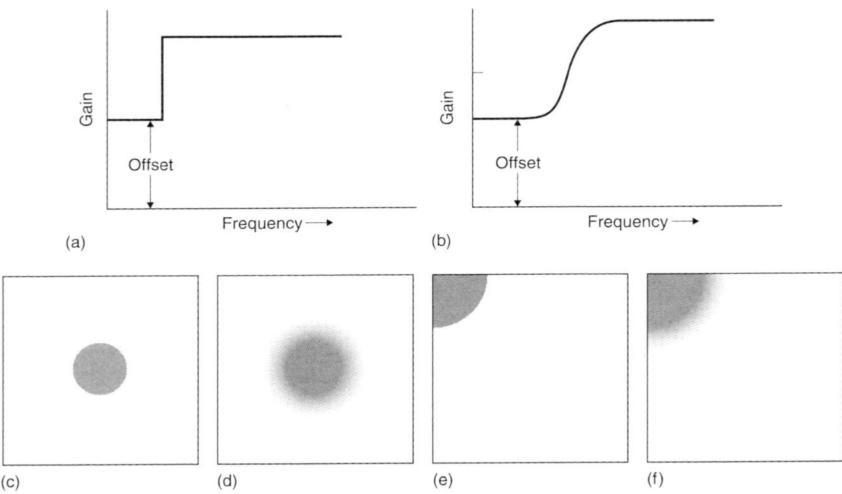

FIGURE 5.7.7
High-frequency emphasis filter. (a) 1-D ideal high frequency emphasis filter. (b) Nonideal high-frequency emphasis filter. (c) 2-D ideal high frequency emphasis filter shown as an image for Fourier symmetry. (d) 2-D nonideal higth frequency emphasis filter shown as an image for cosine and Walsh–Hadamard symmetry. (f) 2-D nonideal high frequency emphasis filter shown as an image for cosine and Walsh–Hadamard symmetry.

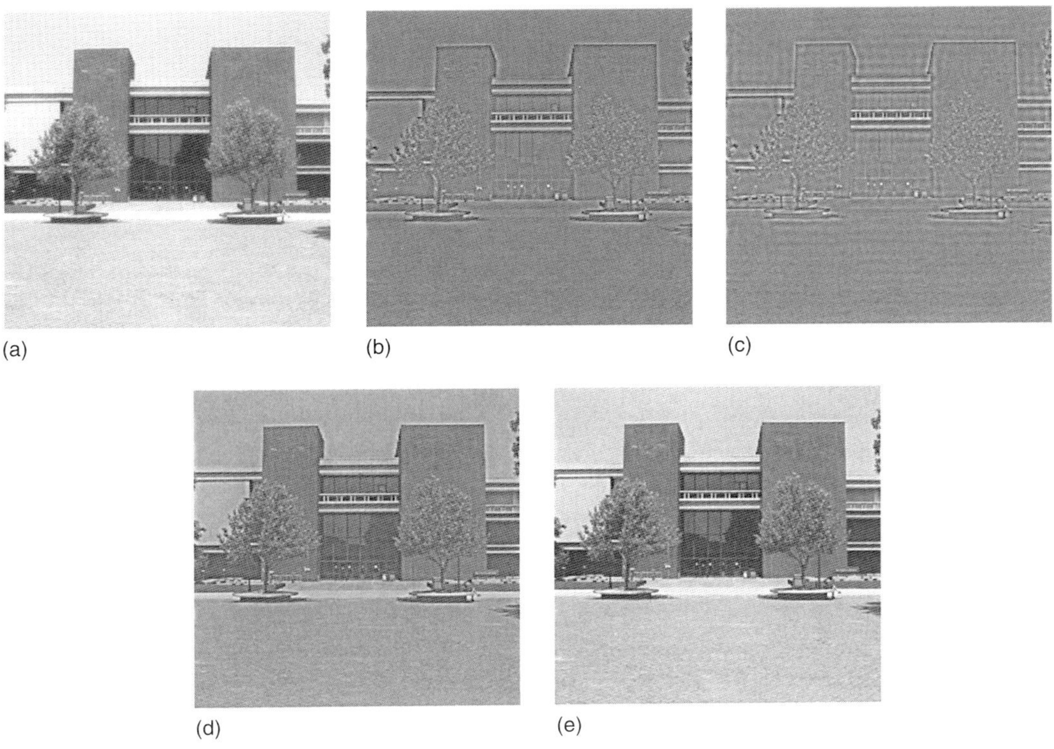

FIGURE 5.7.8
Highpass filtering. (a) Original image. (b) Butterworth filter; order = 2; cutoff = 32. (c) Ideal filter; cutoff = 32. (d) High-frequency emphasis; offset = 0.5, order = 2, cutoff = 32. (e) High frequency emphasis filter; offset = 1.5, order = 2, cutoff = 32.

5.7.3 Bandpass and Bandreject Filters

The bandpass and bandreject filters are specified by two cutoff frequencies, a low cutoff and a high cutoff, shown in Figure 5.7.9. These filters can be modified into nonideal filters by making the transitions gradual at the cutoff frequencies, as was done for the lowpass filter in Figure 5.7.3 and the highpass filter in Figure 5.7.6. A special form of these filters is called a notch filter, because it only notches out, or passes, specific frequencies (see Figure 5.7.9g,h). These filters are useful for retaining, bandpass, or removing, bandreject, specific frequencies of interest which are typically application

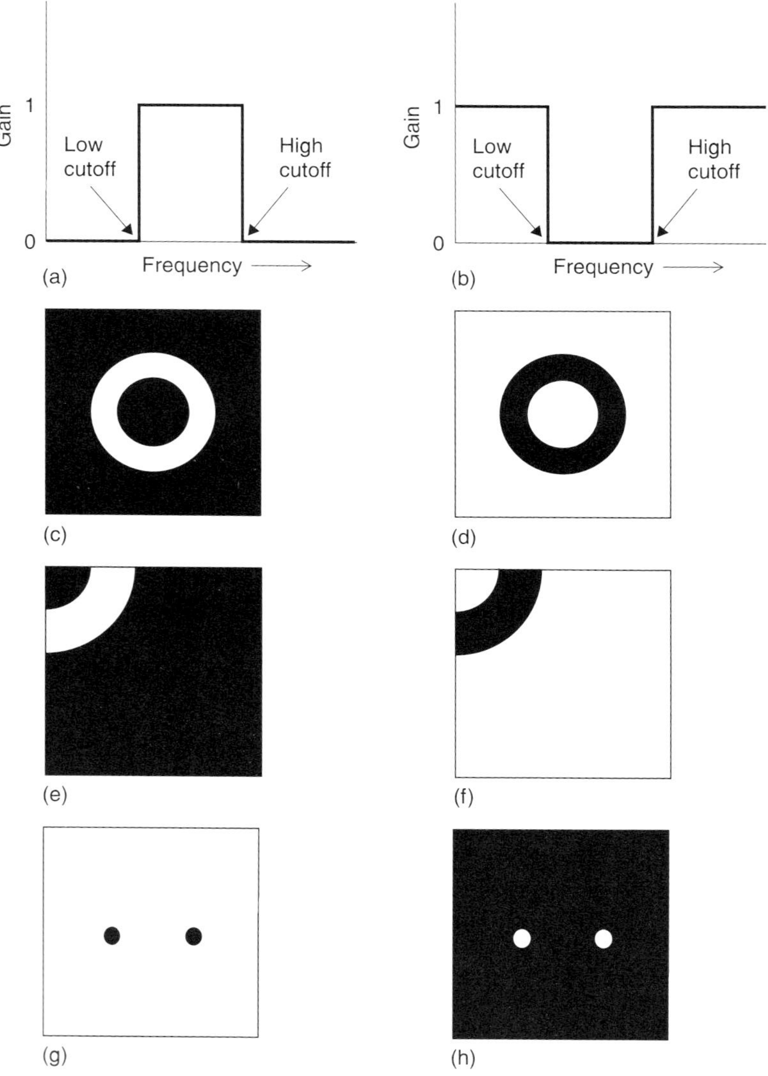

FIGURE 5.7.9
Bandpass, bandreject, and notch filters. (a) 1-D ideal bandpass filter. (b) 1-D ideal bandreject filter. (c) 2-D ideal bandpass filter shown as an image (Fourier). (d) 2-D ideal bandreject filter shown as an image (Fourier). (e) 2-D ideal bandpass filter for Walsh–Hadamard and cosine functions. (f) 2-D ideal banreject filter for Walsh–Hadamard and cosine functions. (g) 2-D ideal notch filter for rejecting specific frequencies. (h) 2-D ideal notch filter for passing specific frequencies.

dependent—one common application is for noise removal. These three types of filters are typically used in image enhancement restoration, and compression, and examples can be seen in Chapters 8, 9, and 10.

5.8 Wavelet Transform

The wavelet transform is really a family of transforms that satisfy specific conditions. From our perspective we can describe the *wavelet transform* as a transform that has basis functions that are shifted and expanded versions of themselves. Because of this, the wavelet transform contains not just frequency information, but spatial information as well. One of the most common models for a wavelet transform uses the Fourier transform and highpass and lowpass filters. To satisfy the conditions for a wavelet transform, the filters must be *perfect reconstruction filters*, which means that any distortion introduced by the forward transform will be canceled in the inverse transform (an example of these types of filters are *quadrature mirror filters*).

The wavelet transform breaks an image down into four subsampled, or decimated, images. They are subsampled by keeping every other pixel. The results consist of one image that has been highpass filtered in both the horizontal and vertical directions, one that has been highpass filtered in the vertical and lowpassed in the horizontal, one that has been highpassed in the vertical and lowpassed in the horizontal, and one that has been lowpass filtered in both directions.

This transform is typically implemented in the spatial domain by using 1-D convolution filters. In the section on edge detection we looked at 2-D convolution masks that mark places in the image where the gray levels are changing rapidly. These rapid changes correspond to high frequency information, so edge detectors are basically highpass filters. To implement the wavelet transform we apply the convolution theorem, which is an important Fourier transform property. As we have seen, the *convolution theorem* states that convolution in the spatial domain is the equivalent of multiplication in the frequency domain. We have seen that multiplication in the frequency domain is used to perform filtering; the convolution theorem tells us that we can also perform filtering in the spatial domain via convolution, such as we have already seen with spatial convolution masks. Therefore, if we can define convolution masks that satisfy the wavelet transform conditions, we can implement the wavelet transform in the spatial domain. We have also seen that if the transform basis functions are separable, we can perform the 2-D transform by using two 1-D transforms. An additional benefit of convolution versus frequency domain filtering is that, if the convolution mask is short, it is much faster.

To perform the wavelet transform with convolution filters, a special type of convolution called circular convolution must be used. *Circular convolution* is performed by taking the underlying image array and extending it in a periodic manner to match the symmetry implied by the discrete Fourier transform (see Figure 5.8.1a,b). The convolution process starts with the origin of the image and the convolution mask aligned, so that the first value contains contributions from the "previous" copy of the periodic image (see Figure 5.8.1c). In Figure 5.8.1c we see that the last value(s) contain contributions from the "next" copy of the extended, periodic image. Performing circular convolution allows us to retain the outer rows and columns, unlike the previously used method where the outer rows and columns were ignored. This is important since we may want to

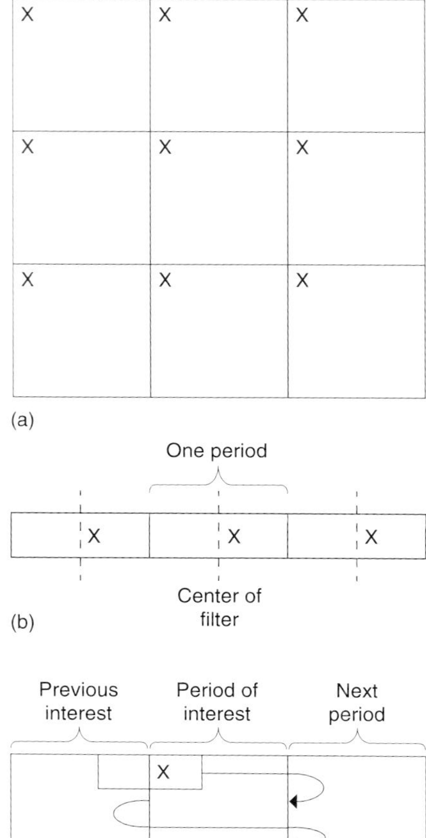

FIGURE 5.8.1
Circular convolution. (a) Extended, periodic image (X = origin). (b) Extended, periodic 1-D convolution filter (X = origin). (c) With circular convolution, the outer rows and columns include products of both the previous and next periods.

perform the wavelet transform on small blocks, and eliminating the outer row(s) and column(s) is not practical.

Many different convolution filters are available for use with the wavelet transform. Here we will consider two examples based on the Daubechies and the Haar functions. These are separable, so they can be used to implement a wavelet transform by first convolving them with the rows and then the columns. The Haar basis vectors are simple:

$$\text{LOWPASS: } \frac{1}{\sqrt{2}}[1 \quad 1]$$

$$\text{HIGHPASS: } \frac{1}{\sqrt{2}}[1 \quad -1]$$

An example of Daubechies basis vectors are:

$$\text{LOWPASS: } \frac{1}{4\sqrt{2}}[1+\sqrt{3},\ 3+\sqrt{3},\ 3-\sqrt{3},\ 1-\sqrt{3}]$$

$$\text{HIGHPASS: } \frac{1}{4\sqrt{2}}[1-\sqrt{3},\ \sqrt{3}-3,\ 3+\sqrt{3},\ -1-\sqrt{3}]$$

To use the basis vectors to implement the wavelet transform they must be zero-padded to be the same size as the image (or subimage). Also note that the origin of the basis vectors is in the center, corresponding to the value to the right of the middle of the vector.

Example 5.8.1

We want to use the Haar basis vectors to perform a wavelet transform on an image by dividing it into 4×4 blocks. The basis vectors need to be zero-padded so that they have a length of 4, as follows:

$$\text{LOWPASS: } \frac{1}{\sqrt{2}}[1 \quad 1 \quad 0 \quad 0]$$

$$\text{HIGHPASS: } \frac{1}{\sqrt{2}}[1 \quad -1 \quad 0 \quad 0]$$

$$\uparrow$$

origin

These are aligned with the image so that the origins coincide, and the result from the first vector inner product is placed into the location corresponding to the origin. Note that when the vector is zero-padded on the right, the origin is no longer to the right of the center of the resulting vector. The origin is determined by selecting the coefficient corresponding to the right of center *before* zero-padding.

Example 5.8.2

To use the Daubechies basis vectors to do a wavelet transform on an image by dividing it into 8×8 blocks, we need to zero-pad them to a length of 8, as follows:

$$\text{LOWPASS: } \frac{1}{4\sqrt{2}}[1+\sqrt{3},\ 3+\sqrt{3},\ 3-\sqrt{3},\ 1-\sqrt{3},\ 0,\ 0,\ 0,\ 0]$$

$$\text{HIGHPASS: } \frac{1}{4\sqrt{2}}[1-\sqrt{3},\ \sqrt{3}-3,\ 3+\sqrt{3},\ -1-\sqrt{3},\ 0,\ 0,\ 0,\ 0]$$

$$\uparrow$$

origin

Note that the origin is the value to the right of the center of the original vector before zero-padding. Since these are assumed periodic for circular convolution, we could zero-pad equally on both ends, then the origin *is* to the right of the center of the zero-padded vector, as follows:

$$\text{LOWPASS: } \frac{1}{4\sqrt{2}}[0,\ 0,\ 1+\sqrt{3},\ 3+\sqrt{3},\ 3-\sqrt{3},\ 1-\sqrt{3},\ 0,\ 0]$$

$$\text{HIGHPASS: } \frac{1}{4\sqrt{2}}[0,\ 0,\ 1-\sqrt{3},\ \sqrt{3}-3,\ 3+\sqrt{3},\ -1-\sqrt{3},\ 0,\ 0]$$

$$\uparrow$$

origin

After the basis vectors have been zero-padded (if necessary), the wavelet transform is performed by doing the following:

1. Convolve the lowpass filter with the rows (remember that this is done by sliding, multiplying coincident terms, and summing the results) and save the results. (*Note*: For the basis vectors as given, they do *not* need to be reversed for convolution).

2. Convolve the lowpass filter with the columns (of the results from Step 1), and subsample this result by taking every other value; this gives us the lowpass–lowpass version of the image.
3. Convolve the result from Step 1, the lowpass filtered rows, with the highpass filter on the columns. Subsample by taking every other value to produce the lowpass–highpass image.
4. Convolve the original image with the highpass filter on the rows, and save the result.
5. Convolve the result from Step 4 with the lowpass filter on the columns; subsample to yield the highpass–lowpass version of the image.
6. To obtain the highpass–highpass version, convolve the columns of the result from Step 4 with the highpass filter.

In practice the convolution sum of every other pixel is not performed, since the resulting values are not used. This is typically done by shifting the basis vector by 2, instead of by 1 at each convolution step. Note that with circular convolution the basis vector will overlap the extended periodic copies of the image when both the first and last convolution sums are calculated.

The convention for displaying the wavelet transform results, as an image, is shown in Figure 5.8.2. In Figure 5.8.3, we see the results of applying the wavelet transform to an image. In Figure 5.8.3b we can see the lowpass–lowpass image in the upper left corner, the lowpass–highpass images on the diagonals, and the highpass–highpass in the lower right corner. We can continue to run the same wavelet transform on the lowpass–lowpass version of the image to get seven subimages, as in Figure 5.8.3c, or perform it another time to get ten subimages, as in Figure 5.8.3d. This process is called *multiresolution decomposition*, and can continue to achieve 13, 16, or as many subimages as are practical. The *decomposition level* refers to how many times we have performed the wavelet transform, where each successive decomposition level means that the wavelet is performed on the lowpass–lowpass version of the image as shown in Figure 5.8.3.

We can see in the resulting images that the transform contains spatial information, as the image itself is still visible in the transform domain. This is similar to what we saw with the Haar transform (Figure 5.5.1), but the Fourier, Walsh–Hadamard, and cosine spectrum does not necessarily have any visible correlation to the image itself when performed on the entire image (Figure 5.8.4). However, if we perform these transforms using small blocks, the resulting spectrum will resemble the image primarily due to the zero frequency term's magnitude (Figure 5.8.5).

The inverse wavelet transform is performed by enlarging the wavelet transform data to its original size. Insert zeros between each value, convolve the corresponding

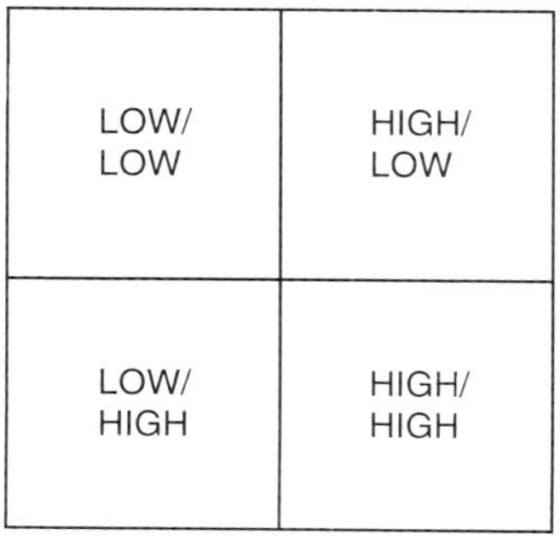

LOW/ LOW	HIGH/ LOW
LOW/ HIGH	HIGH/ HIGH

FIGURE 5.8.2
Wavelet transform display. Location of frequency bands in a four-band wavelet transformed image. Designation is row/column.

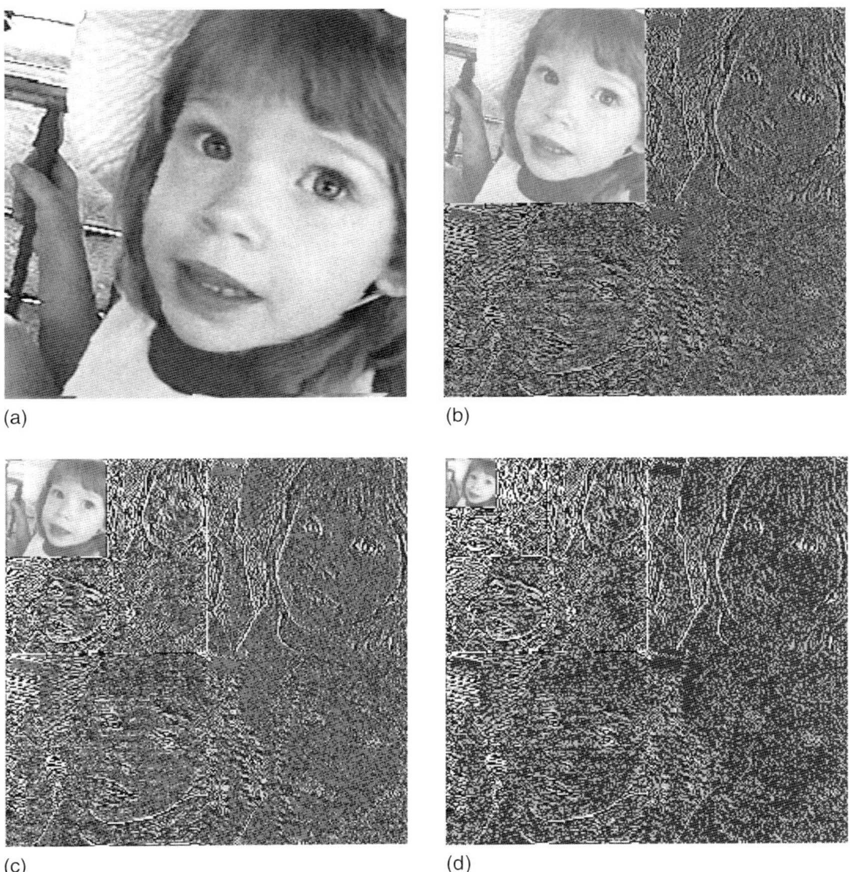

FIGURE 5.8.3
Wavelet transform. (a) Original image. (b) Wavelet transform using Daubechies basic vectors, 1 level decomposition, 4 bands. (c) Wavelet transform using Daubechies basic vectors, 2 level decomposition, 7 bands. (d) Wavelet transform using Daubechies basis vectors, 3 level decomposition, 10 bands.

(lowpass and highpass) inverse filters to each of the four subimages, and sum the results to obtain the original image. For the Haar filter, the inverse wavelet filters are identical to the forward filters; for the Daubechies example given, the inverse wavelet filters are:

$$\text{LOWPASS}_{\text{inv}}: \frac{1}{4\sqrt{2}}[3-\sqrt{3},\ 3+\sqrt{3},\ 1+\sqrt{3},\ 1-\sqrt{3}]$$

$$\text{HIGHPASS}_{\text{inv}}: \frac{1}{4\sqrt{2}}[1-\sqrt{3},\ -1-\sqrt{3},\ 3+\sqrt{3},\ -3+\sqrt{3}]$$

The use of the wavelet transform is increasingly popular for image compression, a very active research area today. The computer revolution, along with the increasing ubiquity of the Internet, multi-media applications, and high definition television, all contribute to the high level of interest in image compression. The multiresolution decomposition property of the wavelet transform, which separates low-resolution information from more detailed information, makes it useful in applications where it is desirable to have coarse information available fast such as perusing an image database or progressively transmitting images on the Internet. The wavelet transform is one of the relatively new

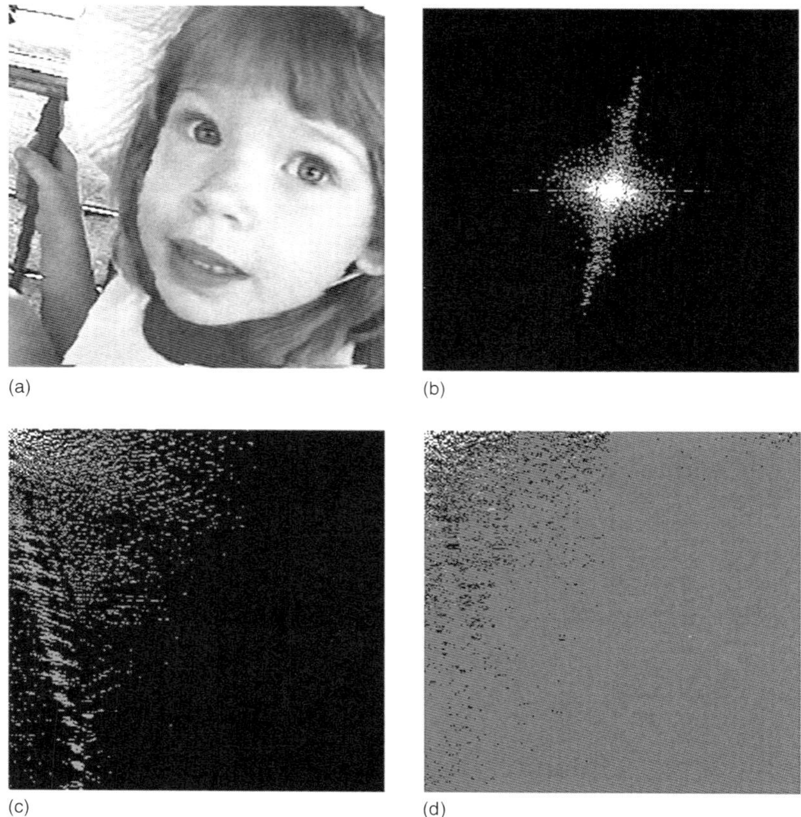

FIGURE 5.8.4
Fourier, Walsh–Hadamard, cosine spectra contain no obvious spatial information. (a) Original image. (b) Fourier spectrum. (c) Walsh–Hadamard spectrum. (d) Cosine spectrum.

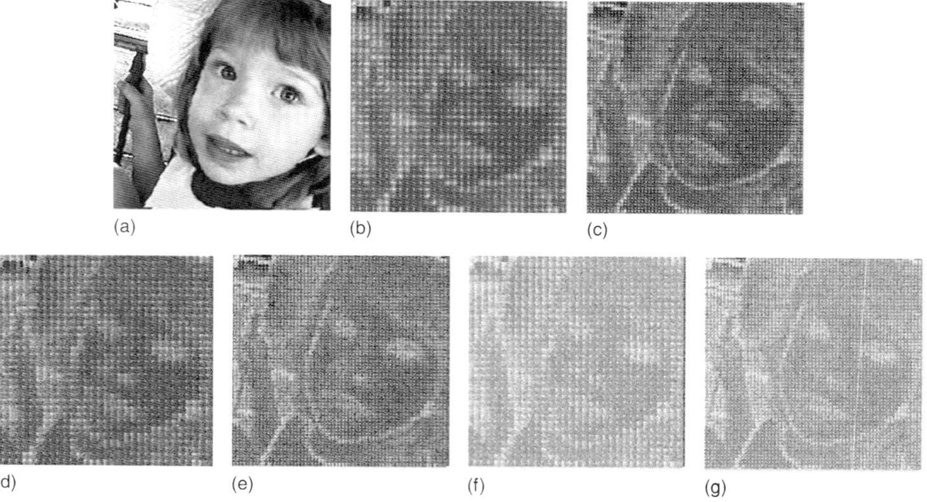

FIGURE 5.8.5
Fourier Walsh–Hadamard, cosine spectra performed on small blocks resembles image. (a) Original image. (b) Fourier spectrum 8×8 blocks. (c) Fourier spectrum 4×4 blocks. (d) Walsh–Hadamard spectrum 8×8 blocks. (e) Walsh–Hadamard spectrum 4×4 blocks. (f) Cosine spectrum, 8×8 blocks. (g) Cosine spectrum 4×4 blocks. *Note:* all spectra are log remapped.

transforms being explored for image compression applications; as mentioned before it is used in the new JPEG compression standard JPEG2000.

5.9 Key Points

OVERVIEW: DISCRETE TRANSFORMS

- Most of the discrete transforms provide information regarding spatial frequency content of an image
- The principal component transform decorrelates multiband image data
- The wavelet and the haar transforms retain both spatial and frequency information
- Most of the discrete transforms map image data into a frequency or sequency mathematical space where all the pixels contribute to each value in the transform domain
- Spatial frequency and sequency relates to how brightness levels change relative to spatial coordinates
- Frequency is the term for sinsuiodal transforms, sequency for rectangular wave transforms
- Rapidly changing brightness values correspond to high frequency (or sequency) terms, slowly changing brightness values correspond to low frequency (or seqency) terms
- A constant brightness value is called the zero frequency (sequency) term, or the DC term
- Most of the discrete transforms decompose an image into a weighted sum of basis images
- Basis images are two-dimensional (2-D) versions of basis vectors
- Basis vectors are sampled versions of basis functions
- The weights for the basis images are found by projecting the basis image onto the image being transformed
- Mathematically, the projection process is performed by calculating the vector inner product of the basis image and the image being transformed
- Basis images should be orthogonal and orthonormal
- Orthogonal basis images have vector inner products equal to zero–they have nothing in common, they are uncorrelated
- Orthonormal basis images are orthogonal and have magnitudes of one

FOURIER TRANSFORM

- The Fourier transform decomposes an image into complex sinusoidal terms
- These terms include a zero frequency term, also called the DC term, related to the average value

- The higher order terms include the fundamental or lowest frequency term, and harmonics which are multiples of the fundamental

One-dimensional DFT

- The one-dimensional (1-D) DFT corresponds to one row (or column) of an image
- Basis vectors are complex sinusoids, defined by Euler's Identity:
  $e^{jx} = \cos(\theta) + j\ \sin(\theta)$
- FORWARD:

$$F(v) = \frac{1}{N}\sum_{c=0}^{N-1} I(c)\ e^{-j2\pi(vc/N)}$$
$$= \frac{1}{N}\sum_{c=0}^{N-1} I(c)\,[\cos(2\pi vc/N) - j\sin(2\pi vc/N)] = R(v) + jI(v)$$

- INVERSE:

$$F^{-1}[F(v)] = I(r,\, c) = \frac{1}{N}\sum_{c=0}^{N-1} F(v)\,e^{j2\pi(vc/N)}$$

- The $F(v)$ terms can be broken down into a magnitude and phase component:
- MAGNITUDE $= |Fv| = \sqrt{[R(v)]^2 + [I(v)]^2}$
- PHASE $= \phi(v) = \tan^{-1}\left[\frac{I(v)}{R(v)}\right]$

Two-dimensional DFT

- Basis images are complex sinusoids:

$$e^{-j2\pi(ur+vc)/N} = \cos\left(\frac{2\pi}{N}(ur+vc)\right) - j\ \sin\left(\frac{2\pi}{N}(ur+vc)\right)$$

- FORWARD:

$$F(u,\, v) = \frac{1}{N}\sum_{r=0}^{N-1}\sum_{c=0}^{N-1} I(r,\, c)\,e^{-j2\pi(ur+vc)/N}$$
$$= \frac{1}{N}\sum_{r=0}^{N-1}\sum_{c=0}^{N-1} I(r,\, c)\left[\cos\left(\frac{2\pi}{N}(ur+vc)\right) - j\sin\left(\frac{2\pi}{N}(ur+vc)\right)\right]$$

- INVERSE: $F^{-1}[F(u,\, v)] = I(r,\, c) = \frac{1}{N}\sum_{u=0}^{N-1}\sum_{v=0}^{N-1} F(u,\, v)\,e^{2\pi(ur+vc/N)}$
- The 2-D DFT is separable, which means the basis image can be broken down into product terms where each term depends only on the rows or columns:

$$e^{-j2\pi(ur+vc/N)} = e^{-j2\pi(ur/N)} = e^{-j2\pi(vc/N)}$$

- Separabliity also implies that the 2-D DFT can be found by successive application of two 1-D DFTs

Fourier Transform Properties

- LINEARITY:

$$F[aI_1(r, c) + bI_2(r, c)] = aF_1(u, v) + bF_2(u, v)$$
$$aI_1(r, c) + bI_2(r, c) = F^{-1}[aF_1(u, v) + bF_2(u, v)]$$

where a and b are constants

- CONVOLUTION:

$$F[I_1(r, c) * I_2(r, c)] = F_1(u, v)F_2(u, v)$$
$$I_1(r, c) * I_2(r, c) = F^{-1}[F_1(u, v)F_2(u, v)]$$
$$F[I_1(r, c)I_2(r, c)] = F_1(u, v) * F_2(u, v)$$
$$I_1(r, c)I_2(r, c) = F^{-1}[F_1(u, v) * F_2(u, v)]$$

- TRANSLATION:

$$F[I(r - r_0, c - c_0)] = F(u, v)\,^{-j2\pi(ur_0+vc_0)/N}$$
$$I(r - r_0, c - c_0) = F^{-1}\left[F(u, v)\,^{-j2\pi(ur_0+vc_0)/N}\right]$$

- MODULATION:

$$F\left[I(r, c)\,^{j2\pi(u_0r+v_0c)}\right] = F(u - u_o, v - v_0)$$
$$I(r, c)\,^{j2\pi(u_0r+v_0c)} = F^{-1}[F(u - u_o, v - v_0)]$$

- ROTATION: Let

$$r = x\cos(\theta), \quad c = x\sin(\theta)$$
$$u = w\cos(\phi), \quad v = w\sin(\phi)$$

$$I(x, \theta + \theta_0) = F^{-1}[F(w, \phi + \theta_0)]$$
$$F[I(x, \theta + \theta_0)] = F(w, \phi + \theta_0)$$

- PERIODICITY:

$$F(u, v) = F(u + N, v) = F(u, v + N) = F(u + N, v = N)\ldots$$

Displaying the Fourier Spectrum

- The Fourier spectrum consists of complex floating point numbers, stored in CVIPtools as a two band image—one for the real part and one for the imaginary part
- In CVIPtools we shift the origin to the center of the image by applying the properties of periodicity and modulation with $u_0 = v_0 = N/2$:

$$I(r, c)\,^{j2\pi(Nr/2+Nc/2)/N} = I(r, c)\,^{j\pi(r+c)} = I(r, c)(-1)^{(r+c)}$$

- To take advantage of the human visual system's response to brightness we can greatly enhance the visual information available by performing a log remap by displaying the following log transform of the spectrum:

$$\log(u, v) = k \log[1 + |F(u, v|]$$

- The phase can be displayed primarily to illustrate phase changes

COSINE TRANSFORM

- The cosine transform uses cosine functions as basis functions and can be derived by using a Fourier transform and extending the original $N \times N$ image to an image that is $2N \times 2N$ by folding it about the origin
- Extending the image to $2N \times 2N$ has the effect of creating an even function, one that is symmetric about the origin, so the imaginary terms in the DFT cancel out
- The cosine transform requires only real arithmetic
- The basis images are separable
- The DCT has been used historically in image compression, such as JPEG
- FORWARD:

$$C(u, v) = \alpha(u)\alpha(v) \sum_{r=0}^{N-1} \sum_{c=0}^{N-1} I(r, c) \cos\left[\frac{(2r+1)u\pi}{2N}\right] \cos\left[\frac{(2c+1)v\pi}{2N}\right]$$

where:

$$\alpha(u), \alpha(v) = \begin{cases} \sqrt{\frac{1}{N}} & \text{for } u, v = 0 \\ \sqrt{\frac{2}{N}} & \text{for } u, v = 1, 2, \ldots, N-1 \end{cases}$$

- INVERSE:

$$C^{-1}[C(u, v)] = I(r, c)$$

$$= \sum_{u=0}^{N-1} \sum_{v=0}^{N-1} \alpha(u)\alpha(v) C(u, v) \cos\left[\frac{(2r+1)u\pi}{2N}\right] \cos\left[\frac{(2c+1)v\pi}{2N}\right]$$

WALSH–HADAMARD TRANSFORM

- The Walsh–Hadamard transform (WHT) uses rectangular functions for basis functions
- Instead of frequency terms, we have sequency terms
- Sequency is the number of zero crossings
- A 2-D basis image is found from two 1-D basis vectors by performing a vector outer product of the two
- The WHT is separable

- FORWARD:

$$WH(u, v) = \frac{1}{N}\sum_{r=0}^{N-1}\sum_{c=0}^{N-1} I(r, c)(-1)^{\sum_{i=0}^{n-1}[b_i(r)p_i(u)+b_i(c)p_i(v)]}$$

- INVERSE:

$$WH^{-1}[WH(u, v)] = I(r, c) = \frac{1}{N}\sum_{u=0}^{N-1}\sum_{v=0}^{N-1} WH(u, v)(-1)^{\sum_{i=0}^{n-1}[b_i(r)p_i(u)+b_i(c)p_i(v)]}$$

HAAR TRANSFORM

- The Haar transform has rectangular waves as basis functions
- The Haar transform is derived from the Haar matrices
- The Haar transform retains both spatial and sequency information
- In the Haar matrices each row is a 1-D basis vector, for example:

$$Haar8 \quad \Rightarrow \quad \frac{1}{\sqrt{8}}\begin{bmatrix} +1 & +1 & +1 & +1 & +1 & +1 & +1 & +1 \\ +1 & +1 & +1 & +1 & -1 & -1 & -1 & -1 \\ \sqrt{2} & \sqrt{2} & -\sqrt{2} & -\sqrt{2} & 0 & 0 & 0 & 0 \\ 0 & 0 & 0 & 0 & \sqrt{2} & \sqrt{2} & -\sqrt{2} & -\sqrt{2} \\ +2 & -2 & 0 & 0 & 0 & 0 & 0 & 0 \\ 0 & 0 & +2 & -2 & 0 & 0 & 0 & 0 \\ 0 & 0 & 0 & 0 & +2 & -2 & 0 & 0 \\ 0 & 0 & 0 & 0 & 0 & 0 & +2 & -2 \end{bmatrix}$$

- The Haar transform allows for multiresolution decomposition of an input image (see Figure 5.5.1)

PRINCIPAL COMPONENTS TRANSFORM

- The principal components transform (PCT) differs from the previous transforms, as it is not related to extracting frequency or sequency information from images, but is a mathematical transform that decorrelates multiband image data
- The PCT provides a linear transform matrix that will decorrelate the bands in the input image
- The linear transform matrix is found by a three step procedure; for example in a three band color, RGB, image: (1) find the covariance matrix in RGB space, (2) find the eigenvalues of the covariance matrix, and their corresponding eigenvectors, (3) use the eigenvectors as a linear transform on the RGB data
- The PCT data will have the most variance in the principal band

- In pattern recognition theory variance is a measure of information, in this sense most information is in the principal band
- Another use of the PCT is to find optimal basis images for a specific image, but this use is not very practical due to the extensive processing required

FILTERING

- Filtering modifies the frequency or sequency spectrum by selectively retaining, removing or scaling the various components of the spectrum
- The shape of the filter depends on the implied symmetry in the transform used
- Ideal filters have abrupt transitions in the filter function
- Ideal filters cause artifacts that appear as ripples or waves at edges in the image
- Nonideal filters have gradual changes in the filter function
- A commonly used nonideal filter is called a Butterworth filter
- For a Butterworth filter the order determines the slope of the transition, a higher order is a steeper slope

Lowpass Filters

- A lowpass filter keeps low frequencies and attenuates high frequencies
- Lowpass filters will blur the image by removing fast brightness changes which correspond to image detail
- Butterworth lowpass filter function of order n, and cutoff frequency f_0:

$$H(u, v) = \frac{1}{1 + \left[\sqrt{u^2 + v^2}/f_0\right]^{2n}}$$

Highpass Filters

- A highpass filter keeps the high frequencies and attenuates low frequencies
- Highpass filters will tend to sharpen the image by retaining areas of rapid change in brightness which correspond to edges
- Butterworth highpass filter function of order n, and cutoff frequency f_0:

$$H(u, v) = \frac{1}{1 + \left[f_0/\sqrt{u^2 + v^2}\right]^{2n}}$$

- A high frequency emphasis filter is a highpass filter which retains some of the low frequency information and boosts the gain of the high frequencies by including an offset value in the filter function (Figure 5.7.7)

Bandpass and Bandreject Filters

- Bandpass filtering will retain specific parts of the spectrum
- Bandreject filters will remove specific parts of the spectrum
- Bandpass and bandreject filters require high and low frequency cutoff values
- Bandreject filters are often used for noise removal
- A special type of bandreject filter is a notch filter which only removes specific frequencies

WAVELET TRANSFORM

- The discrete wavelet transform (DWT) is a family of transforms that satisfy specific conditions
- The *wavelet transform* has basis functions that are shifted and expanded versions of themselves
- The wavelet transform contains not just frequency information, but also spatial information
- The wavelet transform breaks an image down into four subsampled, or decimated, images by keeping every other pixel
- The wavelet results consist of one subsampled image that has been highpass filtered in both the horizontal and vertical directions, one that has been highpass filtered in the vertical and lowpassed in the horizontal, one that has been highpassed in the vertical and lowpassed in the horizontal, and one that has been lowpass filtered in both directions
- One of the most common models for a wavelet transform uses the Fourier transform and highpass and lowpass filters
- This model uses the convolution property of the Fourier transform to perform the wavelet transform in the spatial domain
- This model uses the separable property to perform a 2-D wavelet with two 1-D filters
- Circular convolution must be used which requires zero-padding
- The filters discussed include the Haar and Daubechies

$$\text{LOWPASS: } \frac{1}{\sqrt{2}}[1 \quad 1]$$

HARR (Inverse same as forward)

$$\text{HIGHPASS: } \frac{1}{\sqrt{2}}[1 - 1]$$

$$\text{LOWPASS: } \frac{1}{4\sqrt{2}}\left[1+\sqrt{3},\ 3+\sqrt{3},\ 3-\sqrt{3},\ 1-\sqrt{3}\right]$$

DAUBECHIES:

$$\text{HIGHPASS: } \frac{1}{4\sqrt{2}}\left[1-\sqrt{3},\ \sqrt{3}-3,\ 3+\sqrt{3},\ -1-\sqrt{3}\right]$$

$$\text{LOWPASS}_{\text{inv}}: \frac{1}{4\sqrt{2}}\left[3-\sqrt{3},\ 3+\sqrt{3},\ 1+\sqrt{3},\ 1-\sqrt{3}\right]$$

$$\text{HIGHPASS}_{\text{inv}}: \frac{1}{4\sqrt{2}}\left[1-\sqrt{3},\ -1-\sqrt{3},\ 3+\sqrt{3},\ -3+\sqrt{3}\right]$$

- The algorithm described for the wavelet transform can be performed in six steps:

  1. Convolve the lowpass filter with the rows (remember that this is done by sliding, multiplying coincident terms and summing the results) and save the results. (*Note*: For the basis vectors as given, they *do not* need to be reversed for convolution.)
  2. Convolve the lowpass filter with the columns (of the results from Step 1), and subsample this result by taking every other value; this gives us the lowpass–lowpass version of the image.
  3. Convolve the result from Step 1, the lowpass filtered rows, with the highpass filter on the columns. Subsample by taking every other value to produce the lowpass–highpass image.
  4. Convolve the original image with the highpass filter on the rows, and save the result.
  5. Convolve the result from Step 4 with the lowpass filter on the columns; subsample to yield the highpass–lowpass version of the image.
  6. To obtain the highpass–highpass version, convolve the columns of the result from Step 4 with the highpass filter.
- The wavelet transform is used in image compression, for example in JPEG2000

5.10 References and Further Reading

For discrete transforms, many excellent texts are available, including [Gonzalez/Woods 02], [Sonka/Hlavac/Boyle 99], [Castleman 96], [Bracewell 95], [Pratt 91], [Jain 89], and [Rosenfeld/Kak 82]. For details regarding the fast implementation of the transforms see [Gonzalez/Woods 02], [Petrou/Bosdogianni 99], and [Press/Teukolsky/Vetterling/Flannery 92]. See [Gonzalez/Woods 92] and [Pratt 91] for details on the separate Walsh and Hadamard transforms. More details on the Haar transform can be found in [Gonzalez/Woods 02], [Castleman 96], [Pratt 91] and [Jain 89]. Additional detail on the PCT can be found in [Gonzalez/Woods 02], and (applied to feature analysis) [Sonka/Hlavac/Boyle 99].

For more information on filters see [Gonzalez/Woods 02], [Pratt 91], [Lim 90], and [Banks 90]. For an excellent and detailed discussion on the need for zero-padding with convolution filters see [Gonzalez/Woods 02]. For mathematical details on implementing filters in the spatial domain from the frequency domain specifications see [Petrou/Bosdogianni99] and [Gonzalez/Woods 92]. The wavelet transform as implemented in CVIPtools is described in [Kjoelen 95]. More information on wavelet transforms is found in [Gonzalez/Woods 02], [Masters 94], and [Castleman 96]. Implementation details for the wavelet as applied in JPEG200 are found in [Taubman/Marcellin 02].

Banks, S., *Signal Processing, Image Processing and Pattern Recognition*, Upper Saddle River, NJ: Prentice Hall, 1990.

Bracewell, R.N., *Two-Dimensional Imaging*, Upper Saddle River, NJ: Prentice Hall, 1995.

Castleman, K.R., *Digital Image Processing*, Upper Saddle River, NJ: Prentice Hall, 1996.

Gonzalez, R.C., Woods, R.E., *Digital Image Processing*, Upper Saddle River, NJ: Prentice Hall 2002.

Gonzalez, R.C., Woods, R.E., *Digital Image Processing*, Reading, MA: Addison Wesley, 1992.

Jain, A.K., *Fundamentals of Digital Image Processing*, Upper Saddle River, NJ: Prentice Hall, 1989.
Kjoelen, A., *Wavelet Based Compression of Skin Tumor Images*, Master's Thesis in Electrical Engineering, Southern Illinois University at Edwardsville, 1995.
Lim, J.S., *Two-Dimensional Signal and Image Processing*, Upper Saddle River, NJ: PTR Prentice Hall, 1990.
Masters, T., *Signal and Image Processing with Neural Networks*, NY: Wiley, 1994.
Petrou, M., Bosdogianni, P., *Image Processing: The Fundamentals*, NY: Wiley, 1999.
Pratt, W.K., *Digital Image Processing*, NY: Wiley, 1991.
Press, W.H., Teukolsky, S.A., Vetterling, W.T., Flannery, B.P, *Numerical Recipes in C*, NY: Cambridge University Press, 1992.
Rosenfeld, A., Kak, A.C., *Digital Picture Processing*, San Diego, CA: Academic Press, 1982.
Sonka, M., Hlavac, V., Boyle, R., *Image Processing, Analysis and Machine Vision*, Pacific Grove, CA: Brooks/Cole Publishing Company, 1999.
Taubman, D.S., Marcellin, M.W., *JPEG2000: Image Compression Fundamentals, Standards and Practice*, Norwell, MA: Kluwer Academic Publishers.

5.11 Exercises

1. When transforming an image into the frequency domain, how does this differ from a color transform?
2. Define basis function, basis vector, basis image and vector inner product. Explain how these relate to discrete image transforms.
3. (a) What is spatial frequency? (b) What frequency is an area of constant brightness in an image? (c) Are edges in an image primarily high or low frequency?
4. (a) Find the projection, $T(u, v)$, of the image, $I(r, c)$, onto the basis images:

Let $I(r, c) = \begin{bmatrix} 7 & 3 \\ 2 & 8 \end{bmatrix}$

And let $B(u, v, r, c) = \left\{ \begin{matrix} \begin{bmatrix} +1 & +1 \\ +1 & +1 \end{bmatrix} \begin{bmatrix} +1 & -1 \\ +1 & -1 \end{bmatrix} \\ \begin{bmatrix} +1 & +1 \\ -1 & -1 \end{bmatrix} \begin{bmatrix} +1 & -1 \\ -1 & +1 \end{bmatrix} \end{matrix} \right.$

(b) Using the following inverse basis images, project $T(u, v)$ from above onto them to recover the image.

$$B^{-1}(u, v, r, c) = \left\{ \begin{matrix} \begin{bmatrix} +1 & +1 \\ +1 & +1 \end{bmatrix} \begin{bmatrix} +1 & -1 \\ +1 & -1 \end{bmatrix} \\ \begin{bmatrix} +1 & +1 \\ -1 & -1 \end{bmatrix} \begin{bmatrix} +1 & -1 \\ -1 & +1 \end{bmatrix} \end{matrix} \right.$$

Did you get the original image back? Why or why not? Are these basis images orthogonal? Are they orthonormal?

5. (a) Find the DFT of the following row of an image: [2 2 2 2] (b) Do the inverse DFT on the result. Did you get your image row back? Why or why not? (c) Find the DFT of the following row of an image: [2 4 4 2], d) Do the inverse DFT on the result. Did you get your image row back? Why or why not?,
6. For the 4×4 image shown below, do the following:

$$\begin{bmatrix} 2 & 2 & 2 & 2 \\ 2 & 4 & 4 & 2 \\ 2 & 4 & 4 & 2 \\ 2 & 2 & 2 & 2 \end{bmatrix}$$

   (a) Perform the 2-D DFT. Show the results after transforming each row, and after each column. Leave answers in $R+jI$ form.
   (b) Perform the inverse DFT on result of (a). Did you get the same data back? Why or why not?
   (c) Multiply each element in the original image by $(-1)^{(r+c)}$ and repeat (a). Calculate $F(u, v)$. Is it shifted to the center? Where is the "center" on a 4×4 grid?
7. Use CVIPtools to explore the Fourier spectra of simple objects.
   (a) Create an image of a vertical line with *Utilities->Create*. Perform the FFT on this image (use *Analysis->Transforms*). In which direction do you see the frequency components?
   (b) Create an image of a horizontal line, and perform the FFT. Now which direction do you see the frequency components?
   (c) Create a 256×256 image of a rectangle, using the default values. Perform an FFT on this image. In which direction are the frequency components? Use *File->Show spectrum* to look at just the magnitude image. Next, select the magnitude image with the mouse, and press 'e' on the keyboard (this performs a contrast enhancement). Does this help to see the primary frequency components?
   (d) Create an image of a vertical cosine wave of frequency 64, perform the FFT. Now do the same with a horizontal cosine. Do the resulting spectra look correct? Explain.
   (e) Create circles, ellipses, and checkerboard images. Perform the FFT and view the spectra and phase images. Are the results what you expected? Explain.
8. Use CVIPtools to illustrate the linearity property of the Fourier transform.
   (a) Create a horizontal line and a rectangle image using *Utilities->Create*. Add these two images together with *Utilities->Arith/Logic*. Perform the FFT on the resultant image.
   (b) Perform the FFT on original line and original rectangle image. Add the two resulting spectra. Does the result look like the spectrum from (a)? Why or why not? (*Hint*: log remap)
   (c) Perform the inverse FFT on the spectra from (a) and (b). Did you get the image back?
9. Use CVIPtools to compare filters in the frequency and spatial domain, which illustrates the convolution property. Use a square image whose size is a power of 2;

for example 256 × 256. Note that an image can be resized with *Utilities->Size*. Use the default blocksize, which is equal to the image size.

(a) Apply a lowpass spatial filter mask using the mean filter under *Utilities->Filters*, to an image. Apply lowpass filtering in the frequency domain with *Analysis->Transforms*. This is done by first performing the Fourier transform (FFT), followed by the filter on the output Fourier spectrum. Note that the filter automatically performs the inverse transform. Compare the resultant images. Experiment with the mask size of the spatial filter, and the cutoff frequency, and type, of the frequency domain lowpass filter. Adjust these parameters until the resultant images look similar. Perform a Fourier transform on the resultant images and compare the spectra.

(b) Apply a highpass spatial filter mask, using *Utilities->Filters->Specify a filter*, to an image. This is done by holding the mouse button on the drop-down arrow to select a filter type, then select the mask size, followed by a click on the *OK* button. Apply highpass filtering in the frequency domain. Follow a process similar to what was done in (a) above, and then compare the resultant images and spectra.

10. Use CVIPtools to create frequency domain filters.
    (a) Open an image of your choice and resize it to 256 × 256 with *Utilities->Resize*. Perform the FFT on the 256 × 256 image.
    (b) Use *Utilities->Create* to create a circle. Use the default parameter values to create circles—if they have been modified, click the RESET button on the Utilities window, this will reset all the parameters to the default values—if in doubt, killing the window with a click on the **X** in the upper right corner will do a hard reset on the window. Create a circle in the center of the image with a radius of 32. Next, check the *Blur radius* checkbox, set blur radius value to 64, click *Apply*.
    (c) Use *Utilties->Arith/Logic* to multiply the FFT spectrum with the circle images. Select the FFT spectrum as the *current image* (click on the image itself or the name in the image queue), and select the circle as the *second image*. Perform the multiplication with the spectrum and both circles. Note that these multiplied images are both filtered spectra—use the "e" option on the keyboard to enhance these images.
    (d) Perform the inverse FFT transform on the two multiplied images. Look at the output images and compare. How do they differ? What type of filters are these? Why do the two output images differ?
    (e) Repeat steps (c) and (d) but first perform a logical NOT on the two circle images, using *Utilities->Arith/Logic*.
11. Use CVIPtools to illustrate the translation property of the Fourier transform.
    (a) Translate an image in the spatial domain, using *Analysis->Geometry->Translate an image* with the default Wrap-around option. Now perform an FFT on the original image and the translated image. Compare the spectra of these images. Are they the same? In addition to the log remapped magnitude (the default spectral display), use the *File->Show spectrum* in the main window to compare the phase images. Are the phase images the same?
    (b) Change the translation values, the amount you move the image right and down, and repeat part (a).

12. Use CVIPtools to illustrate the modulation property of the Fourier transform.
    (a) Use *Utilities->Create* to create an image of a circle and an image of a horizontal cosine wave of frequency 32. Multiply these two images together.
    (b) Find the FFT spectra of the circle and of the two images multiplied together. Examine the modulation property of the Fourier transform. Do these images look correct? Why or why not?
    (c) Use *File->Show spectrum* to look at the magnitude images of the spectra. Do these images look correct? Why or why not?
13. Use CVIPtools to illustrate the rotation property of the Fourier transform.
    (a) Create a 256×256 image of a rectangle, using the default values. Perform an FFT on this image.
    (b) Rotate the rectangle image, using *Analysis->Geometry*, by 45 degrees. Next, crop a 256×256 image from the center of this rotated image with *Utilities->Size* (e.g., use $(r, c) = (64, 64)$ for the upper left corner and a size of 256×256), and perform the FFT.
    (c) Compare the resulting spectra. Did the rotation cause any artifacts that may be affecting the spectrum? Look at the phase image by using *File->Show Spectrum* for both the original and rotated rectangle spectra, what do you see? Does this seem reasonable?
14. (a) Is the cosine an even or an odd function? (b) Do you think that the Fourier or the cosine transform is faster to compute? Explain.
15. For an $N \times N$ image, we assume a Fourier symmetry that repeats the $N \times N$ pattern. For an $N \times N$ image, what size is the pattern that repeats for the cosine transform?
16. (a) What is the general form of the Walsh–Hadamard basis functions? (b) Do you think that the Walsh–Hadamard or the cosine transform is faster to compute? Explain.
17. Let the rows of the following matrix be basis vectors:

$$\begin{bmatrix} +1 & +1 & +1 & +1 \\ +1 & -1 & -1 & +1 \\ +1 & -1 & +1 & -1 \\ +1 & +1 & -1 & -1 \end{bmatrix}$$

    (a) What is the sequency of each row? Going from top to bottom are they in sequency order?
    (b) Are they orthogonal? Why or why not?
    (c) Are the orthonormal? Why or why not? If not, how can we make them orthonormal?
    (d) To which transform do these basis vectors belong?
    (e) Find the vector outer product of the first and last row.
18. Of the transforms described in this chapter, which ones are separable? Explain what this means.
19. Name the two transforms discussed that provide a multiresolution decomposition of the input image. Explain what this means.

20. Sketch the (a) Fourier, (b) cosine, (c) Walsh–Hadamard, transform spectral image of the following image:

$$\begin{bmatrix} 255 & 255 & 255 & 255 \\ 255 & 255 & 255 & 255 \\ 255 & 255 & 255 & 255 \\ 255 & 255 & 255 & 255 \end{bmatrix}$$

21. (a) Sketch the Fourier spectrum for a horizontal sine wave of frequency 32 on a 256×256 image, (b) The spectrum for a vertical sine wave of frequency 32 on a 256×256 image, (c) Apply the convolution theorem to find the image that results from the convolution of the horizontal and vertical sine wave.
22. (a) Where does the PCT get its name? Explain. (b) What are the three steps to performing the PCT?
23. Use CVIPtools to explore the PCT transform with color images.
    (a) Open a color image of your choice. Use *Utilities->Convert->Color space* to apply the PCT to the image. Next, do the inverse PCT on the output image, by unchecking the *Forward* checkbox. Is the PCT an invertible transform?—in other words, did you get your image back OK?
    (b) Use *Utilities->Create->Extract band* to get all three bands from the original image as well as the PCT image. Place the RGB bands above and the three PCT bands below and compare. Are the RGB bands correlated, that is, do they look similar? Are the PCT bands correlated, that is, do they look similar? Examine the data range on each image, which band has the largest data range? (*Note*: the data range is seen at the bottom of the main window next to the data type) Why do you think this band has the largest data range?
24. (a) What is the difference between an ideal filter and a nonideal filter?, (b) Are there any advantages to one over the other? What are they?
25. (a) What is the *order* of a Butterworth filter? (b) How does increasing the order of the filter affect the image?
26. Name an application of a band reject, or a notch, filter. Explain.
27. (a) What is the difference between a highpass filter and a high frequency emphasis filter? (b) How does this difference affect the resultant image?
28. Use CVIPtools to explore a bandreject filter.
    (a) Create a 256×256 image of a horizontal sinusoidal wave using *Utilities->Create* in CVIPtools. Select *Analysis->Transforms* and perform the FFT on the image with a blocksize of 256 (the entire image). View the magnitude of the FFT output using *File->Show spectrum->Magnitude* on the main window. Notice the frequency component corresponding to the horizontal frequency of the sinusoid. What is the approximate location of this frequency component? Notice also the location of the DC component (center). As is the case in most images, the DC component is the largest component in the transformed image.
    (b) Now filter the transformed image with a Band-reject filter. Check the *Keep DC* checkbox, so the average value is retained. Choose an ideal band reject filter. Choose cutoff frequencies to remove the horizontal sinusoidal

frequency. Perform the inverse FFT on the image. Compare the resulting image with the original image. Did you succeed in removing most of the sinusoid? (Perfect results are not attainable for this experiment, because of quantization noise.)

29. Use CVIPtools to see how the DC terms affects filtering. Use the *Analysis->Transforms* window.
    (a) Load a 256×256 image of your choice and perform an FFT using a block size of 64. Filter the spectrum with an ideal lowpass filter with a cutoff of 16, and keep the DC by checking the *Keep DC* checkbox. Next, use the same filter, but without keeping the DC term. Compare the data range of the two resulting images. How does the DC term affect the data range? Compare how the images look—do they look different? Why or why not?
    (b) Perform a DCT using a block size of 64. Filter the spectrum with an ideal lowpass filter with a cutoff of 32, and keep the DC by checking the *Keep DC* checkbox. Next, use the same filter, but without keeping the DC term. Compare the data range of the two resulting images. How does the DC term affect the data range? Compare how the images look—do they look different? Why or why not? Compare the DCT and FFT results, what do you see? Explain the results.
    (b) Perform a WHT using a block size of 64. Filter the spectrum with an ideal lowpass filter with a cutoff of 32, and keep the DC by checking the *Keep DC* checkbox. Next, use the same filter, but without keeping the DC term. Compare the data range of the two resulting images. How does the DC term affect the data range? Compare how the images look—do they look different? Why or why not? Compare these results with the DCT and FFT results, what do you see? Explain the results.
30. Use CVIPtools to illustrate the effects of low-pass filtering on a real image and compare the effects of ideal filters to those of Butterworth filters.
    (a) Load a 256×256 complex gray-scale image and perform the FFT with the *Analysis->Transforms* window, using a blocksize of 16. Filter the transformed image using an ideal low-pass filter with a cutoff of 4 and keep the DC value. Notice the absence of high-frequency information and the ringing effect, which appears as waves emanating from edges in the image, caused by the sharp frequency cutoff.
    (b) Now apply a Butterworth low-pass filter of order 1 to the Fourier-transformed image. Use the same cutoff as in part (a). Keep the DC during filtering. Compare the result with that of part (a). Is the ringing effect as noticeable now? Why/why not?
    (c) Repeat part (b) using a sixth order Butterworth filter instead of a first order filter. Compare the result to the result from parts (a) and (b). Because the frequency response of a sixth-order Butterworth filter is close to that of an ideal filter, the image should be similar to that of part (a). (d) Repeat (a), (b), and (c) using a blocksize of 256×256 and cutoff of 64.
31. Use CVIPtools to see the relationship between transform-domain filtering in the DCT, FFT and WHT domain. Use the *Analysis->Transforms* window. Note that the origin for the FFT, corresponding to the zero-frequency term (DC component), is shifted to the center, while for all other transforms the origin is in the upper left corner. Since all the transforms implemented are fast transforms based on powers of 2, the blocksize must be a power of 2. If the selected blocksize will not evenly cover the image, then the image is zero-padded as required.

For DC components located in the upper-left-hand corner, CVIPtools lets you specify cutoff frequencies ranging from 1 to *blocksize*. For the FFT, with the DC term in the center, the range is from 1 to *blocksize/2*.

(a) Load any image of your choice. If the size is not 256 × 256, use *Utilities->Size* to resize it to 256 × 256.

(b) Choose the ideal lowpass filter type, any blocksize, and a cutoff frequency (CF) divisible by 2. Apply this filter to the image in the Walsh domain. Repeat this procedure using the DCT and the FFT, but use a cutoff frequency equal to CF/2 for the FFT.

(c) Compare the images resulting from filtering with different transforms.

(d) Which transform resulted in the best quality of the filtered image? Which transform resulted in the poorest quality filtered image? Compare your answers with what you know about the properties of the DCT, FFT, and Walsh transform. Do your answers agree with what you would expect?

32. Use CVIPtools to compare highpass and high frequency emphasis filters.
   (a) Perform a DCT on a 256 × 256 image using 256 as the block size.
   (b) Apply a butterworth highpass filter with a cutoff of 64.
   (c) Apply a high frequency emphasis filter with a cutoff of 64, use the same order used in (b), and compare results to (b). What do you see? Is this what you expected? Explain.
   (d) Add the original image to the result from (b), and compare the added image to the one from (c). Are they similar? Why or why not?

33. Use CVIPtools to see how the DC term affects highpass filters, review data remapping. Perform a DCT on a 256 × 256 image, and apply a highpass filter with a cutoff of 64. Do it with and without keeping the DC term. Do the two filtered images look similar? Explain. Compare the data range on the two images by clicking on each image (the image information appears at the bottom of the main window).

34. (a) Does the wavelet transform have a unique set of basis functions? (b) Name two functions commonly used for wavelet filters. (c) Of these two, with which one is the wavelet transform faster to calculate? (d) Does the wavelet transform provide spatial or frequency information?

35. (a) Describe the process to implement a wavelet transform using one-dimensional filters. (b) What is the criterion that the filters must meet for a wavelet transform, and why? (c) How is circular convolution performed? (d) What does the term *decomposition level* mean as related to the wavelet transform?

36. Use CVIPtools to explore the wavelet transform.
   (a) Open an image of your choice and resize it to 256 × 256. Perform the wavelet transform using *Analysis->Transforms*. Use decomposition levels of 1, 2, 3, and 4. Compare the wavelet images. What exactly is the *decomposition level*?
   (b) Perform the wavelet transform twice with a decomposition level of 9 on your 256 × 256 image, once with the Haar basis and once with the Daubechies. Note that the Haar uses filters with two coefficients, and the Daubechies uses four. Perform an ideal lowpass filter with a cutoff of 32 on the wavelet images and compare the filtered images. Are they different? How? Why?
   (c) Repeat (b), but use a first order Butterworth filter.

5.11.1 Programming Exercise: Filtering

1. Write a function to create frequency domain ideal filter masks that will work with the Fourier transform in CVIPtools. Let the user specify the size, the filter type: highpass, lowpass, or bandreject, and the cutoff frequencies. Be sure the output images are floating point.
2. Test these filter mask images using CVIPtools. Use the FFT, and multiply to test the filters.
3. Repeat (1) and (2) for DCT symmetry, and test with the DCT.
4. Repeat 1–3, but create Butterworth filter masks. Let the user specify the filter order.

5.11.2 Programming Exercise: Fourier Transform

1. Use the CVIPtools libraries functions to put the *fft_transform* and *ifft_transform* functions into your CVIPlab program.
2. Compare your results to those obtained with CVIPtools.

5.11.3 Programming Exercise: Discrete Cosine Transform

1. Use the CVIPtools libraries functions to put the *dct_transform* and *idct_transform* functions into your CVIPlab program.
2. Compare your results to those obtained with CVIPtools.

5.11.4 Programming: Exercises: Walsh–Hadamard Transform

1. Use the CVIPtools libraries functions to put the *walhad_transform* function into your CVIPlab program.
2. Compare your results to those obtained with CVIPtools.

5.11.5 Programming Exercise: Haar Transform

1. Use the CVIPtools libraries functions to put the *haar_transform* function into your CVIPlab program.
2. Compare your results to those obtained with CVIPtools.

5.11.6 Programming Exercise: Wavelet Transform

1. Use the CVIPtools libraries functions to put the *wavhaar_transform* and *wavdaub4_transform* functions into your CVIPlab program.
2. Compare your results to those obtained with CVIPtools.

5.11.7 Programming Exercise: CVIPtools Library Filter Functions

1. Use the CVIPtools functions in the *TransformFilter* library to perform filtering in your CVIPlab program.
2. Compare your results to those obtained with CVIPtools, and with the filter functions you wrote.

6

Feature Analysis and Pattern Classification

6.1 Introduction and Overview

Feature analysis and pattern classification are often the final steps in the image analysis process. *Feature analysis* involves examining the features extracted from the images and determining if and how they can be used to solve the imaging problem under consideration. In some cases the extracted features may not solve the problem and the information gained by analyzing the features can be used to determine further analysis methods that may prove helpful, including additional features that may be needed. *Pattern classification*, often called pattern recognition, involves the classification of objects into categories. For many imaging applications this classification needs to be done automatically, via computer. The patterns to be classified consist of the extracted feature information, which are associated with image objects and the classes or categories will be application dependent.

As discussed in Chapter 3, the goal in image analysis is to extract information useful for solving application-based problems. This is done by intelligently reducing the amount of image data with the tools we have explored, including image segmentation (Chapter 4) and transforms (Chapter 5). Once we have performed these operations, we have modified the image from the lowest level of pixel data into higher-level representations. Now, we can consider extraction of features that can be useful for solving computer imaging problems. Image segmentation allows us to look at object features, and the image transforms provide us with features based on spatial frequency information—spectral features. The object features of interest include the geometric properties of binary objects, histogram features, spectral features, texture features, and color features. Once we have extracted the features of interest, we can analyze the image.

Exactly what we do with the features will be application-dependent. If we are working on a computer vision problem, the end goal may be the generation of a classification rule in order to identify objects. If we are working to develop a new image compression algorithm, we may want to determine the specific image data that is important; the insignificant information can be compressed or eliminated completely. For image restoration we may want to determine the type of noise that exists in the image, or how the image has been degraded. Image analysis may help us to solve an image enhancement problem by allowing us to determine exactly what it is that makes images visually pleasing.

As was shown in Figure 3.1.3, feature extraction is part of the data reduction process and is followed by feature analysis. One of the important aspects of feature analysis is to determine exactly which features are important, so the analysis is not complete until we incorporate application-specific feedback into the system (see Figure 6.1.1). In this chapter we will discuss feature extraction and analysis, as well as provide an introduction to pattern classification. Although pattern classification is used primarily

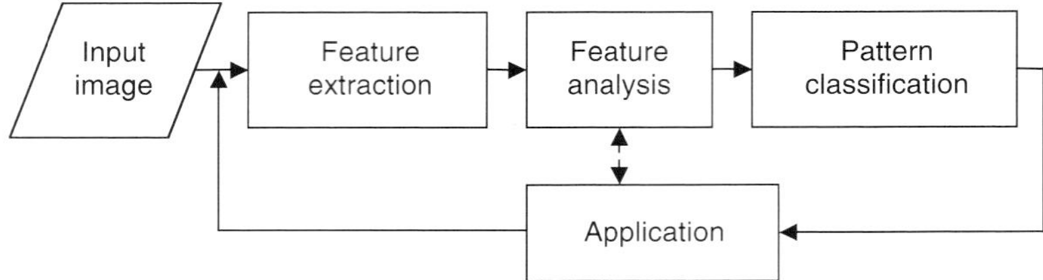

FIGURE 6.1.1
Feature extraction, feature analysis and pattern classification. To be effective the application-specific feedback loop is of paramount importance.

in computer vision applications, it can be helpful in solving any type of computer imaging problem.

6.2 Feature Extraction

Feature extraction is a process that begins with feature selection. The selected features will be the major factor that determines the complexity and success of the analysis and pattern classification process. Initially, the features are selected based on the application requirements and the developer's experience. After the features have been analyzed, with attention to the application, the developer may gain insight into the application's needs which will lead to another iteration of feature selection, extraction, and analysis. The overall process shown in Figure 6.1.1 will continue until an acceptable success rate is achieved for the application.

When selecting features for use in a computer imaging application, an important factor is the robustness of a feature. A feature is robust if it will provide consistent results across the entire application domain. For example, if we are developing a system to work under any lighting conditions, we do not want to use features that are lighting-dependent—they will not provide consistent results in the application domain. Another type of robustness, especially applicable to object features, is called RST-invariance, where the RST means rotation, size, and translation. A very robust feature will be RST-invariant, meaning that if the image object is rotated, shrunk or enlarged, or translated (shifted left/right or up/down), the value for the feature will not change. As we explore the binary object features, consider the invariance of each feature to these simple geometric operations.

6.2.1 Shape Features

Shape features depend on a silhouette of the image object under consideration, so all that is needed is a binary image. We can think of this binary image as a mask of the image object, as shown in Figure 6.2.1. The basic binary object features are in Section 3.3.3; including area, center of area, axis of least second moment, projections, and euler number. Here we will add perimeter, thinness ratio, irregularity, aspect ratio, moments, and a moment related set of RST invariant features.

The *perimeter* of the object can help provide us with information about the shape of the object. The perimeter can be found in the original binary image by counting the number

FIGURE 6.2.1
Shape features need a simple binary image. (a) The original image. (2) The image divided into image objects via segmentation. (c) The segmented image with an outline drawn on one of the drumhead image objects. (d) The binary mask image for the marked image object which is used for extraction of features related to object shape.

of "1" pixels that have "0" pixels as neighbors. Perimeter can also be found by application of an edge detector to the object, followed by counting the "1" pixels. Note that counting the "1" pixels is the same as finding the area, but in this case we are finding the "area" of the border. Since the digital images are typically mapped onto a square grid, curved outlines tend to be jagged, so these methods only give an estimate to the actual perimeter for objects with curved edges. An improved estimate to the perimeter can be found by multiplying the results from either of the above methods by $\pi/4$. If better accuracy is required, more complex methods which use chain codes for finding perimeter can be used (see references). An illustration of perimeter is shown in Figure 6.2.2.

In Chapter 3 we found the area of a binary object by counting the number of "1" pixels in the object. Given the area, A, and perimeter, P, we can calculate the *thinness ratio* T:

$$T = 4\pi\left(\frac{A}{P^2}\right)$$

This measure has a maximum value of 1, which corresponds to a circle, so this also is used as a measure of roundness. The closer to 1, the more like a circle the object is. As the perimeter becomes larger relative to the area, this ratio decreases, and the object is getting thinner. This metric is also used to determine the regularity of an object: regular objects

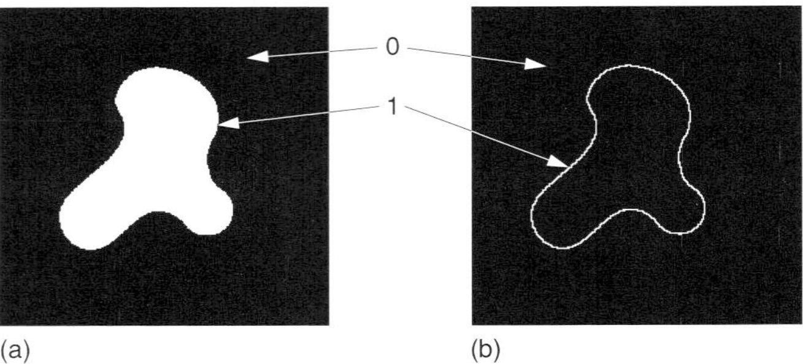

FIGURE 6.2.2
Perimeter. (a) Image with binary object. We can find the perimeter by counting the "1" pixels that have a "0" neighbor. (b) Image after Roberts edge detection. We find the perimeter by counting the "1" pixels.

have higher thinness ratios than similar but irregular objects. The inverse of this metric, $1/T$, is sometimes called the *irregularity* or *compactness* ratio. The area to perimeter ratio, A/P, has properties similar to the thinness ratio, but is easier to calculate.

A related feature is the *aspect ratio* (also called *elongation* or *eccentricity*), defined by the ratio of the bounding box of an object. This can be found by scanning the image and finding the minimum and maximum values on the row and columns where the object lies. This ratio is then defined by:

$$\frac{c_{\max} - c_{\min} + 1}{r_{\max} - r_{\min} + 1}$$

Note that this definition is not rotationally invariant, so to be useful as a comparative measure the objects should be rotated to some standard orientation; such as orientating the axis of least second moment in the horizontal direction.

Moments can be used to generate a set of RST-invariant features. Given a binary image, where $I(r,c)$ can only be "0" or "1," the *moment of order* $(p+q)$ is:

$$m_{pq} = \sum_r \sum_c r^p c^q I(r, c)$$

In order to be translationally invariant we use the *central moments* defined by:

$$\mu_{pq} = \sum_r \sum_c (r - \bar{r})^p (c - \bar{c})^q I(r, c)$$

where

$$\bar{r} = \frac{m_{10}}{m_{00}} \quad \text{and} \quad \bar{c} = \frac{m_{01}}{m_{00}}$$

Note that these central moments are simply the standard moments shifted to the center of area of the object—compare the equations for $\bar{r}$ and $\bar{c}$ to the center of area as defined in Chapter 3. To create the RST-invariant moment-based features we need the *normalized central moments*:

$$\eta_{pq} = \frac{\mu_{pq}}{\mu_{00}^{\gamma}}$$

TABLE 6.1

$$\phi_1 = \eta_{20} + \eta_{02}$$
$$\phi_2 = (\eta_{20} - \eta_{02})^2 + 4\eta_{11}^2$$
$$\phi_3 = (\eta_{30} - 3\eta_{12})^2 + (3\eta_{21} - \eta_{03})^2$$
$$\phi_4 = (\eta_{30} + \eta_{12})^2 + (\eta_{21} + \eta_{03})^2$$
$$\phi_5 = (\eta_{30} - 3\eta_{12})(\eta_{30} + \eta_{12})\left[(\eta_{30} + \eta_{12})^2 - 3(\eta_{21} + \eta_{03})^2\right] + (3\eta_{21} - \eta_{03})(\eta_{21} + \eta_{03})\left[3(\eta_{30} + \eta_{12})^2 - (\eta_{21} + \eta_{03})^2\right]$$
$$\phi_6 = (\eta_{20} - \eta_{02})\left[(\eta_{30} + \eta_{12})^2 - (\eta_{21} + \eta_{03})^2\right] + 4\eta_{11}(\eta_{30} + \eta_{12})(\eta_{21} + \eta_{03})$$
$$\phi_7 = (3\eta_{21} - \eta_{03})(\eta_{30} + \eta_{12})\left[(\eta_{30} + \eta_{12})^2 - 3(\eta_{21} + \eta_{03})^2\right] - (\eta_{30} - 3\eta_{12})(\eta_{21} + \eta_{03})\left[3(\eta_{30} + \eta_{12})^2 - (\eta_{21} + \eta_{03})^2\right]$$

where

$$\gamma = \frac{p+q}{2} + 1, \quad \text{for } (p+q) = 2, 3, 4, \ldots$$

Given these normalized central moments a set of RST-invariant features, ϕ_1–ϕ_7, can be derived using the second and third moments. These *invariant moment features* are shown in Table 6.1.

An example image with binary objects showing the results of extracting these features is shown in Figure 6.2.3. Here we see two squares of different sizes (scales), two rotated rectangles, and two objects that are scaled and rotated. All the objects are translated since they are all in different locations. We see that these features for the same objects are identical. In Figure 6.2.4 we have added noise to the image, performed a simple threshold at 128 to get the segmented image, and extracted the RST-invariant features. Here we see that we can still classify the objects with the first one or two features.

Fourier descriptors (FDs) represent a group of methods often used in shape analysis which require representing the shape as a one or two-dimensional signal, and then taking the Fourier transform of the signal. For imaging applications the simplest method is to use the binary image of the object, and use the spectral features defined in Section 6.2.4. Other FD methods include representing the outline of the object in various mathematical forms and finding the one or two-dimensional Fourier transform of the signal; details of these methods can be explored with the references.

6.2.2 Histogram Features

The *histogram* of an image is a plot of the gray level values versus the number of pixels at that value. The shape of the histogram provides us with information about the nature of the image, or subimage if we are considering an object within the image. For example, a very narrow histogram implies a low contrast image, a histogram skewed toward the high end implies a bright image, and a histogram with two major peaks, called bimodal, implies an object that is in contrast with the background. Examples of the different types of histograms are shown in Figure 6.2.5.

The histogram features that we will consider are statistical-based features, where the histogram is used as a model of the probability distribution of the gray levels. These statistical features provide us with information about the characteristics of the gray level distribution for the image or subimage. We define the first-order histogram probability, $P(g)$, as:

$$P(g) = \frac{N(g)}{M}$$

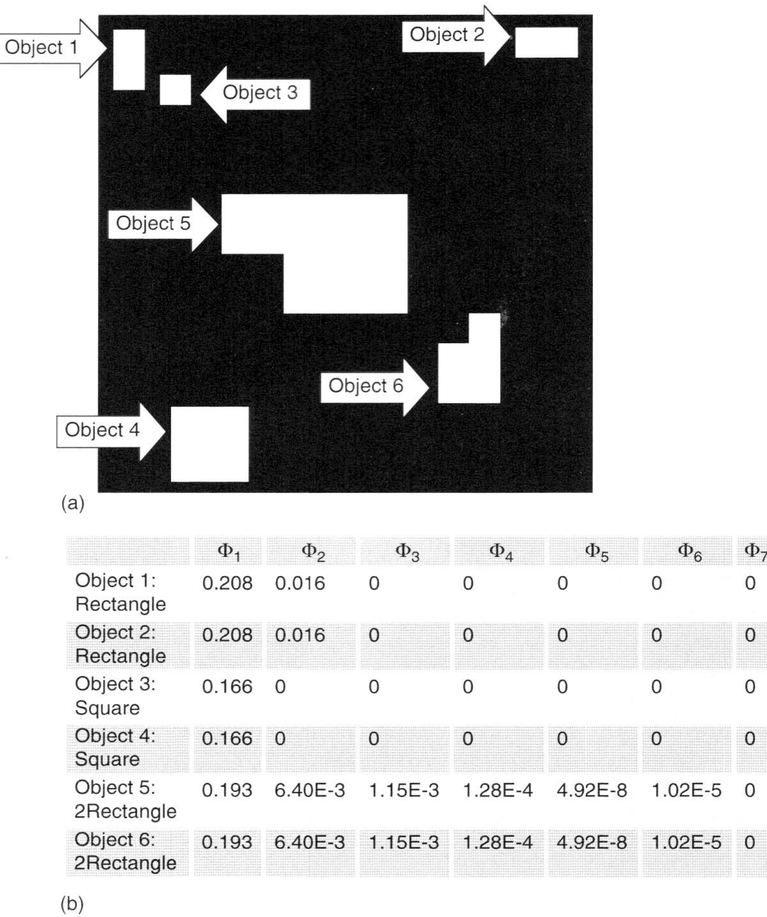

	Φ_1	Φ_2	Φ_3	Φ_4	Φ_5	Φ_6	Φ_7
Object 1: Rectangle	0.208	0.016	0	0	0	0	0
Object 2: Rectangle	0.208	0.016	0	0	0	0	0
Object 3: Square	0.166	0	0	0	0	0	0
Object 4: Square	0.166	0	0	0	0	0	0
Object 5: 2Rectangle	0.193	6.40E-3	1.15E-3	1.28E-4	4.92E-8	1.02E-5	0
Object 6: 2Rectangle	0.193	6.40E-3	1.15E-3	1.28E-4	4.92E-8	1.02E-5	0

FIGURE 6.2.3
RST invariant features. (a) The image with the six objects. (b) The extracted feature data.

M is the number of pixels in the image or subimage (if the entire image is under consideration then $M = N^2$ for an $N \times N$ image), and $N(g)$ is the number of pixels at gray level g. As with any probability distribution all the values for $P(g)$ are less than or equal to 1, and the sum of all the $P(g)$ values is equal to 1. The features based on the first order histogram probability are the mean, standard deviation, skew, energy, and entropy.

The *mean* is the average value, so it tells us something about the general brightness of the image. A bright image will have a high mean, and a dark image will have a low mean. We will use L as the total number of gray levels available, so the gray levels range from 0 to $L-1$. For example, for typical 8-bit image data, L is 256 and ranges from 0 to 255. We can define the mean as follows:

$$\bar{g} = \sum_{g=0}^{L-1} gP(g) = \sum_r \sum_c \frac{I(r, c)}{M}$$

If we use the second form of the equation we sum over the rows and columns corresponding to the pixels in the image or subimage under consideration.

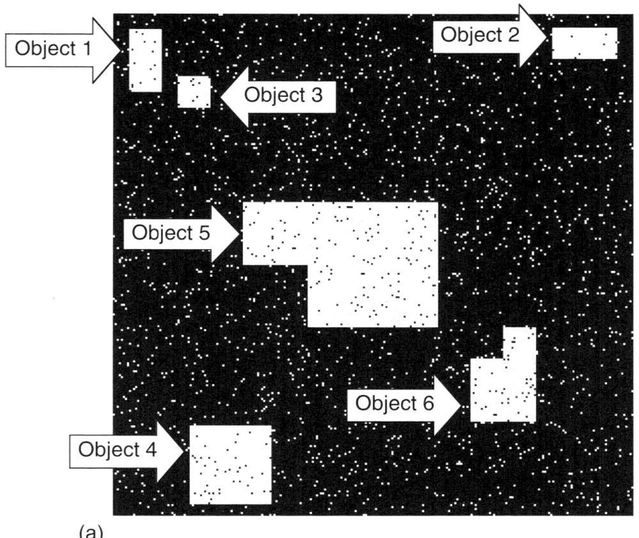

(a)

	Φ_1	Φ_2	Φ_3	Φ_4	Φ_5	Φ_6	Φ_7
Object 1: Rectangle	0.212 (0.208)	0.016 (0.016)	8.79E-7 (0)	8.75E-8 (0)	0 (0)	-1.19E-9 (0)	0 (0)
Object 2: Rectangle	0.213 (0.208)	0.017 (0.016)	9.01E-7 (0)	2.07E-7 (0)	0 (0)	2.83E-9 (0)	0 (0)
Object 3: Square	0.172 (0.166)	2.17E-6 (0)	4.84E-6 (0)	8.69E-7 (0)	0 (0)	0 (0)	0 (0)
Object 4: Square	0.172 (0.166)	3.19E-6 (0)	8.29E-8 (0)	1.17E-7 (0)	0 (0)	0 (0)	0 (0)
Object 5: 2Rectangle	0.199 (0.193)	6.88E-3 (6.40E-3)	1.25E-3 (1.15E-3)	1.39E-4 (1.28E-4)	5.81E-8 (4.92E-8)	1.15E-5 (1.02E-5)	-1.60E-9 (0)
Object 6: 2Rectangle	0.198 (0.193)	6.60E-3 (6.40E-3)	1.22E-3 (1.15E-3)	1.34E-4 (1.28E-4)	5.42E-8 (4.92E-8)	1.08E-5 (1.02E-5)	-3.95E-9 (0)

(b)

FIGURE 6.2.4
RST invariant features with noise. (a) The image with the six objects and noise added. (b) The extracted feature data, with the data from the images without noise in parenthesis.

The *standard deviation*, which is also known as the square root of the variance, tells us something about the contrast. It describes the spread in the data, so a high contrast image will have a high variance, and a low contrast image will have a low variance. It is defined as follows:

$$\sigma_g = \sqrt{\sum_{g=0}^{L-1} (g - \bar{g})^2 P(g)}$$

The *skew* measures the asymmetry about the mean in the gray level distribution. It is defined as:

$$\text{SKEW} = \frac{1}{\sigma_g^3} \sum_{g=0}^{L-1} (g - \bar{g})^3 P(g)$$

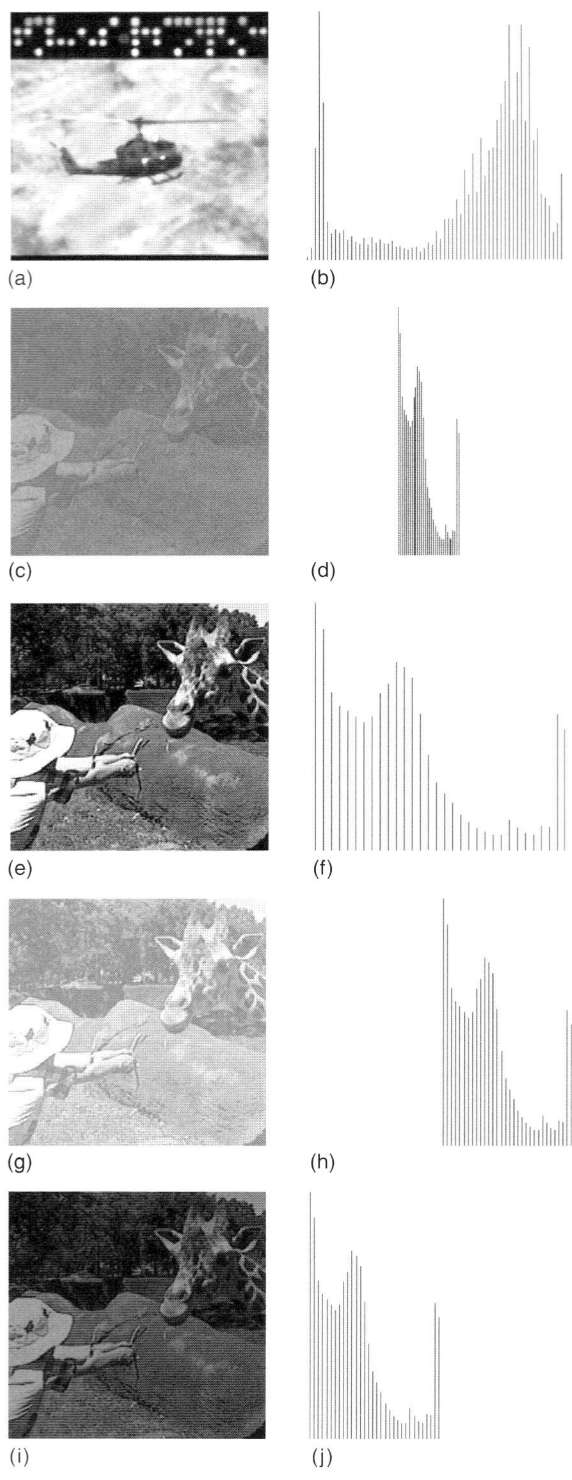

FIGURE 6.2.5
Histograms. (a) Object in contrast with background. (b) Histogram of (a) shows bimodal shape. (c) Low contrast image. (d) Histogram of (c) appears clustered. (e) High contrast image. (f) Histogram of (e) appears spread out. (g) Bright image. (h) Histogram of (g) appears shifted to the right. (i) Dark image. (j) Histogram of (i) appears shifted to the left.

The skew will be positive if the tail of the histogram spreads to the right (positive), and negative if the tail of the histogram spreads to the left (negative). Another method to measure the skew uses the mean, mode, and standard deviation, where the *mode* is defined as the peak, or highest, value:

$$\text{SKEW}' = \frac{\bar{g} - \text{mode}}{\sigma_g}$$

This method of measuring skew is more computationally efficient, especially considering that, typically, the mean and standard deviation have already been calculated.

The *energy* measure tells us something about how the gray levels are distributed:

$$\text{ENERGY} = \sum_{g=0}^{L-1} [P(g)]^2$$

The energy measure has a maximum value of 1 for an image with a constant value, and gets increasingly smaller as the pixel values are distributed across more gray level values (remember all the $P(g)$ values are less than or equal to 1). The larger this value is, the easier it is to compress the image data. If the energy is high it tells us that the number of gray levels in the image is few, that is, the distribution is concentrated in only a small number of different gray levels.

The *entropy* is a measure that tells us how many bits we need to code the image data, and is given by:

$$\text{ENTROPY} = -\sum_{g=0}^{L-1} P(g)\log_2[P(g)]$$

As the pixel values in the image are distributed among more gray levels, the entropy increases. A complex image has higher entropy than a simple image. This measure tends to vary inversely with the energy.

Figure 6.2.6 shows images and the corresponding histogram features. In Figure 6.2.6a–d we see what occurs when an image is segmented. In the segmented image, the mean, standard deviation and skew remain about the same, but the energy goes up and the entropy goes down. The energy goes up as the image is simplified and the individual probabilities increase, which also causes the entropy to decrease. In Figure 6.2.6e–h we see what occurs when an image is enhanced with a histogram stretch. In the enhanced image the energy and entropy remain about the same, but the mean, standard deviation and skew are changed. The mean increases due to an increase in average brightness, the standard deviation increases from the spread in the histogram increasing which also causes the skew to decrease. Note that, in general, the histogram energy is the opposite of what might be expected—a simpler image has more histogram energy than a complex image.

Second-order histogram features, which contain information about the relationship *between pixels*, are used to obtain texture information, so these are in *Section 6.2.5: Texture Features*.

6.2.3 Color Features

Color is useful in many applications. Typical color images consisting of three color planes, red, green, and blue, can be treated as three separate gray scale images. This

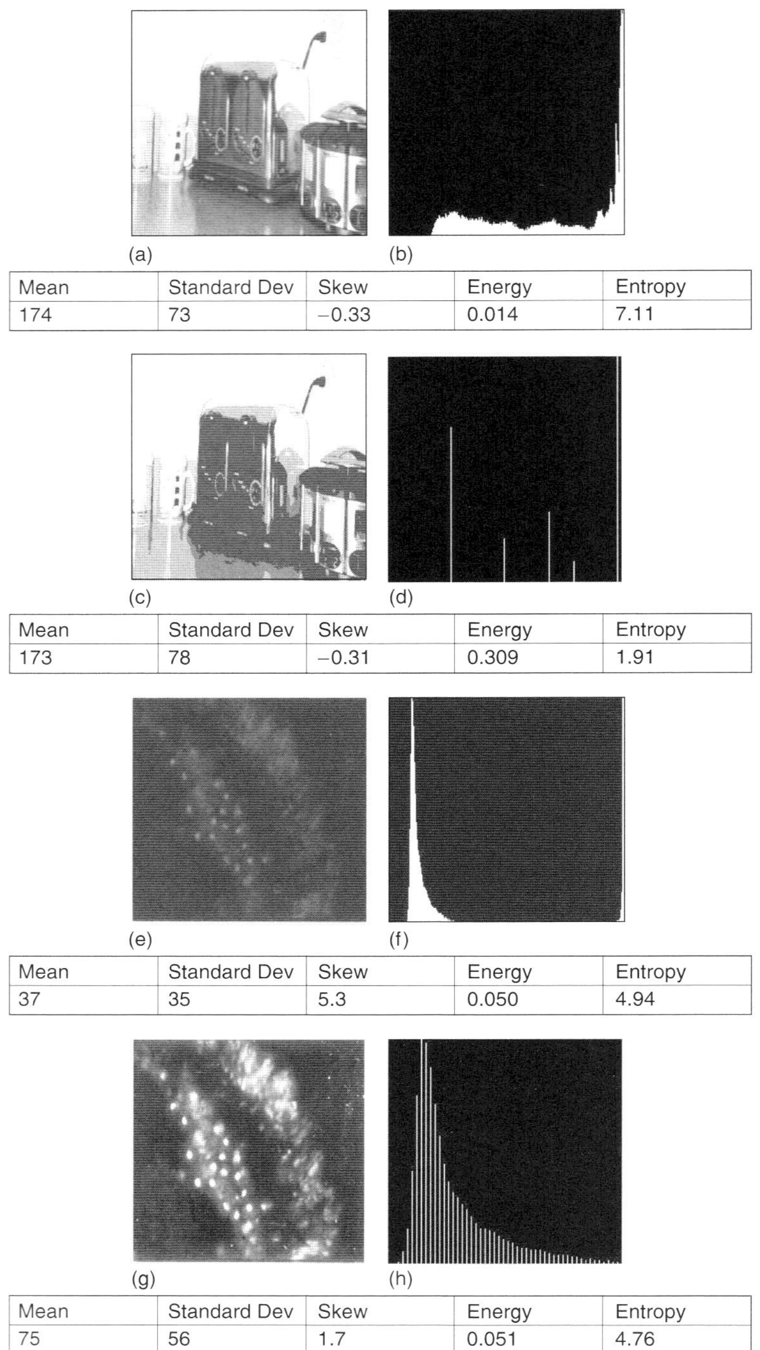

(a) (b)

Mean	Standard Dev	Skew	Energy	Entropy
174	73	−0.33	0.014	7.11

(c) (d)

Mean	Standard Dev	Skew	Energy	Entropy
173	78	−0.31	0.309	1.91

(e) (f)

Mean	Standard Dev	Skew	Energy	Entropy
37	35	5.3	0.050	4.94

(g) (h)

Mean	Standard Dev	Skew	Energy	Entropy
75	56	1.7	0.051	4.76

FIGURE 6.2.6
Histogram features. (a) Original bright image. (b) Histogram of image (a). (c) Image (a) after segmentation. (d) Histogram of image (c). (e) Original dark image. (f) Histogram of image (e). (g) Image (e) after histogram stretch. (h) Histogram of image (g). Comparing images (a) and (c), we observe that as the image is simplified the energy goes up and the entropy goes down; also note that these images have negatively skewed histograms. Comparing images (e) and (g), we observe that as we stretch the histogram, the energy and entropy do not change much, but the standard deviation increases; also note these images have positively skewed histograms.

approach allows us to use any of the gray level features, but with three times as many, one for each of the three color bands. By using this approach we may be able to determine that information useful for the application is contained in one, two or all three of the color bands.

Often, when interested in color features, we want to incorporate information into the feature vector pertaining to the relationship *between* the color bands. These relationships are found by considering normalized color, or color differences. This is done by using the color transforms defined in Chapter 2, and then applying to this new representation the features previously defined. For example, the chromaticity transform provides a normalized color representation, which will decouple the image brightness from the color itself. Many color transforms, including HSI, spherical, cylindrical, $L^*u^*v^*$, and $L^*a^*b^*$, will provide us with two color components and a brightness component. The YIQ and YCrCb provide us with color difference components that signify the relative color. After performing a color transform, depending on the application, we may be interested in a specific aspect of the color information, such as hue or saturation. If this is the case, we can extract features from the band of interest.

The color features chosen will be primarily application-specific, but caution must be taken in selecting color features. Typically, some form of relative color is best, because most absolute color measures are not very robust. In many applications the environment is not carefully controlled, so a system developed under specific color conditions using absolute color may not function properly in a different environment. Remember all the factors that contribute to the color—the lighting, the sensors, any optical filtering, any print or photographic process in the system model. If any of these factors change then any absolute color measures, such as red, green, or blue, will change. An application specific relative color measure can be defined, or a known color standard can be used for comparison. When using a known color standard, the system can be calibrated if the conditions change.

An example of the problem caused by using absolute color arose during development of a system to automatically diagnosis skin tumors. An algorithm was found that seemed to always correctly identify melanoma (a deadly form of skin cancer). At one point in the research, the algorithm ceased to work. What had happened? A big mistake had been made in developing the algorithm—it had relied on some absolute color measures. The initial set of melanoma images had been digitized from Ektachrome slides, and the non-melanoma tumor images had been digitized from Kodachrome slides. Due to the types of film involved, all the melanomas had a blue tint (Ektachrome), while all the other tumor images had a red tint (Kodachrome). Thus, with the first set of tumor images, the use of average color alone provided an easy way to differentiate between the melanoma and nonmelanoma tumors. As more tumor images became available, both melanoma and nonmelanoma tumors were digitized from Kodachrome (red tint), so the identification algorithm ceased to work. A senior member of the research team had a similar experience while developing a tank recognition algorithm based on Ektachrome images of Soviet tanks, and Kodachrome images of U.S. tanks. Avoid absolute color measures for features, except under very carefully controlled conditions.

6.2.4 Spectral Features

With regard to spectral features, or frequency/sequency-domain based features, the primary metric is *power*. How much spectral power do we find in various parts of the spectrum? Texture is often measured by looking for peaks in the power spectrum,

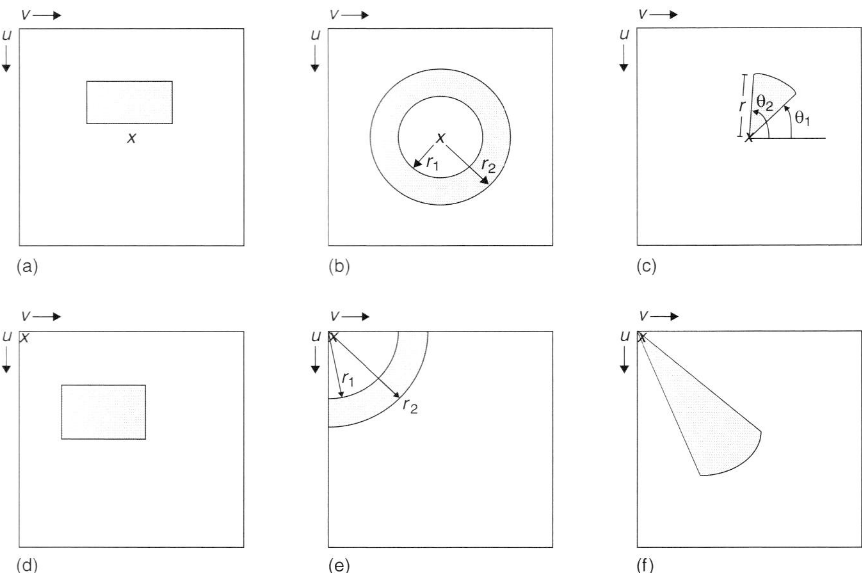

FIGURE 6.2.7
Spectral regions. The first three, a–c, illustrate the regions for Fourier symmetry, and the next three, d–f, show cosine symmetry, x designates the origin, or DC term. (a) Box is defined by limits on u and v. (b) Ring is defined by limits on the radii from origin x. (c) Sector is defined by radius "r" and angles θ_1 and θ_2. (d) Box symmetry. (e) Ring symmetry. (f) Sector symmetry.

especially if the texture is periodic or directional. The power spectrum is defined by the magnitude of the spectral components squared:

$$\text{POWER} = \left|T(u,v)\right|^2$$

Although it is typical to use the Fourier transform for these features, we have used the generic $T(u,v)$ as any of the transforms can be used. The standard approach for spectral features is to find power in various spectral regions, and these regions can be defined as rings, sectors, or boxes. In Figure 6.2.7 we see examples of these types of spectral regions, for both types of symmetry that we have considered. We then measure the power in a region of interest by summing the power over the range of frequencies of interest:

$$\text{SPECTRAL REGION POWER} = \sum_{u \in \text{REGION}} \sum_{v \in \text{REGION}} |T(u,v)|^2$$

The *box* is the easiest to define, by setting limits on u and v.

Example 6.2.1

we may be interested in all spatial frequencies at a specific horizontal frequency, $v = 20$. So we define a spectral region as:

$$\text{Region of interest} = \begin{cases} -\dfrac{N}{2} < u < \dfrac{N}{2} \\ 19 < v < 21 \end{cases}$$

Then we calculate the power in this region by summing over this range of u and v. Note that u should vary from 0 to $N-1$ for non-Fourier symmetry.

The *ring* is defined by two radii, r_1 and r_2. These are measured from the origin, and the summation limits on u and v, for Fourier symmetry, are:

$$u \Rightarrow -r_2 \leq u < r_2$$

$$v \Rightarrow \pm\sqrt{r_1^2 - u^2} \leq v < \pm\sqrt{r_2^2 - u^2}$$

(*Note*: For non-Fourier symmetry u will range from 0 *to* r_2, and v ranges over the positive square roots only.) The *sector* is defined by a radius, r, and two angles, θ_1 and θ_2. The limits on the summation are defined by:

$$\theta_1 < \tan^{-1}\left(\frac{v}{u}\right) < \theta_2$$

$$u^2 + v^2 \leq r^2$$

The sector measurement will find spatial frequency power of a specific orientation whatever the frequency (limited only by the radius), while the ring measure will find spatial frequency power at specific frequencies regardless of orientation. In terms of image objects, the sector measure will tend to be size invariant, and the ring measure will tend to be rotation invariant.

Due to the redundancy in the Fourier spectral symmetry we often measure the sector power over one-half of the spectrum, and the ring power over the other half of the spectrum (see Figure 6.2.8). In practice we may want to normalize these numbers, as they get very large, by dividing by the DC (average) value—this is done in CVIPtools spectral feature extraction. (*Note*: In CVIPtools if the DC value in the magnitude image of a Fourier transform is examined, it needs to be divided by $N \times N$ to get the true average value, due to the implementation of the Fourier transform.)

6.2.5 Texture Features

Texture is related to properties such as smoothness, coarseness, roughness, and regular patterns. Spectral features can be used as texture features; for example, the ring power

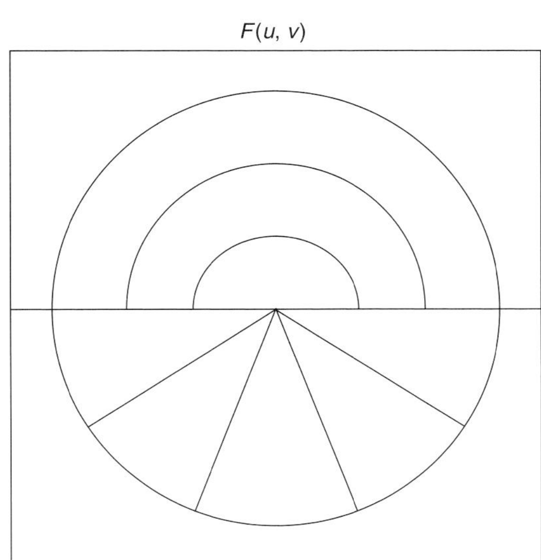

FIGURE 6.2.8
Fourier spectrum power. With Fourier spectrum symmetry, which contains redundant information, we often measure ring power over half the spectrum and sector power over the other half.

can be used to find texture. High power in small radii (ring 1) corresponds to low frequency and thus coarse textures—those with large element sizes; as the ring number increases the frequencies are higher and correspond to finer textures. As the frequency gets very high, for example in the outer ring, the textures will appear very fine and may actually appear smooth. This is really a function of the human visual system's perception—we see texture as rapidly changing variation in the brightness due to the object scattering the light. At some point the variation in brightness becomes too fast for us to perceive (see Chapter 7), so the texture appears smooth.

Texture is also a function of image size relative to the object, as well as magnification of the original image. Remember that the frequency as we have defined it is relative to the image size. Figure 6.2.9 shows a corduroy material at different magnifications, along with the corresponding spectra. Here we see that a higher magnification corresponds to larger element size and lower frequency energy, and as we "zoom out" the element size decreases and the energy spreads out to higher frequencies.

If the magnification is unknown or variable, but we have a known orientation, the spectral sector measures may be useful for providing us with textural information. The power in a sector includes all frequencies, which corresponds to all sizes of elements or magnifications, but has a fixed orientation. In practice, the spectral features can be calculated for 10 or 20 (or more) rings and sectors and the magnitudes plotted to look for signature shapes which will correspond to specific textures.

Another approach to measuring texture is to use the second-order histogram of the gray levels based on a joint probability distribution model. The *second-order histogram* provides statistics based on pairs of pixels and their corresponding gray levels.

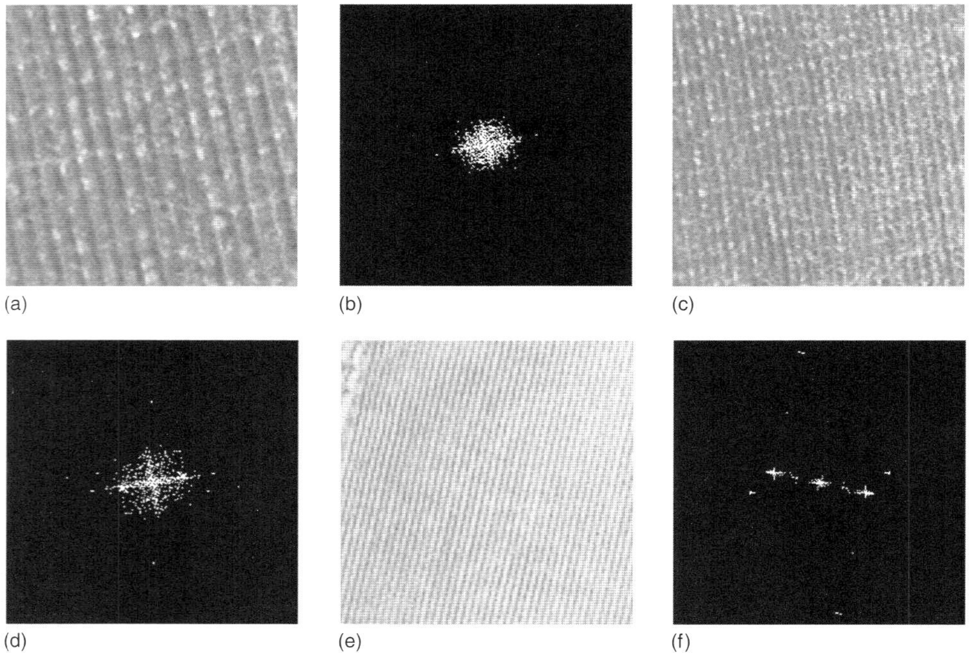

FIGURE 6.2.9
Texture at varying magnification and their spectra. (a) Image 1 at a high magnification corresponding to lower frequency. (b) The Fourier magnitude spectrum of image 1. (c) Image 2 at a medium magnification corresponding to medium frequency. (d) The Fourier magnitude spectrum of image 2. (e) Image 3 at a low magnification corresponding to higher frequency. (f) The Fourier magnitude spectrum of image 3 (*Note*: the Fourier spectra were remapped to BYTE, and then histogram equalized.)

The second-order histogram methods are also referred to as *gray level co-occurrence matrix* or *gray level dependency matrix* methods. These features are based on two parameters: distance and angle. The distance is the pixel distance between the pairs of pixels that are used for the second-order statistics, and the angle refers to the angle between the pixel pairs. Typically, four angles are used corresponding to vertical, horizontal, and two diagonal directions. The pixel distance chosen depends on the resolution of the image and the coarseness of the texture of interest, although it is typical to use 1 or 2. To make the features rotationally invariant they can be calculated for all angles and then averaged (in CVIPtools the average and the range of these features are returned for the four angles).

Numerous features have been derived via these methods, but these five have been found to be the most useful: energy (homogeneity), inertia (contrast), correlation (linearity), inverse difference (local homogeneity) and entropy. If we let c_{ij} be the elements in the co-occurrence matrix normalized by dividing by the number of pixel pairs in the matrix, and assume a given distance and angle (direction), the equations are as follows:

$$Energy = \sum_i \sum_j c_{ij}^2$$

$$Inertia = \sum_i \sum_j (i-j)^2 c_{ij}$$

$$Correlation = \frac{1}{\sigma_x \sigma_y} \sum_i \sum_j (i-\mu_x)(j-\mu_y) c_{ij}$$

where:

$$\mu_x = \sum_i i \sum_j c_{ij}$$

$$\mu_y = \sum_j j \sum_i c_{ij}$$

$$\sigma_x^2 = \sum_i (1-\mu_x)^2 \sum_j c_{ij}$$

$$\sigma_y^2 = \sum_j (1-\mu_y)^2 \sum_i c_{ij}$$

and

$$InverseDifference = \sum_i \sum_j \frac{c_{ij}}{|i-j|}; \quad \text{for } i \neq j$$

$$Entropy = -\sum_i \sum_j c_{ij} \log_2 c_{ij}$$

An example of the gray level co-occurence matrices is shown in Figure 6.2.10. Note in the calculation of these matrices that each pixel pair, with coordinates $[(r_1, c_1), (r_2, c_2)]$, actually represents two pixel pairs where the second one is represented by $[(r_2, c_2), (r_1, c_1)]$. In other words, for example, when counting horizontal pixel pairs first look left to right (0 degrees), and then right to left (180 degrees) across the image. Also remember, before calculating the texture features, to normalize by dividing by the number of pixel pairs in the matrix. The figure illustrates the complexity involved with

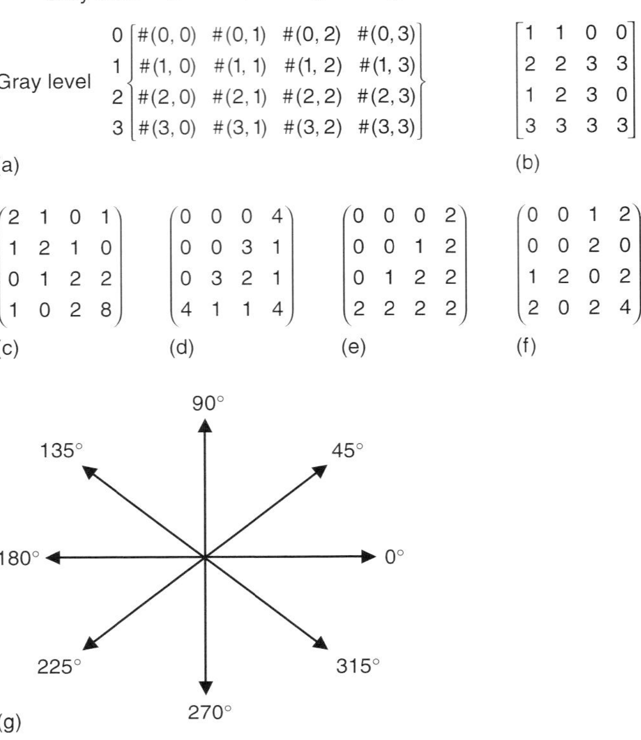

FIGURE 6.2.10
Example of gray level co-occurrence matrices. Given a 4×4 image with four possible gray levels, 2-bits per pixel, and using a distance $d = 1$, we have: (a) general form of the matrix, where each entry is the number (#) of occurrences of the pair listed, (b) an example 4×4 image, (c) the matrix corresponding to the horizontal direction (0° and 180°), (d) the matrix corresponding to the vertical direction (90° and 270° degrees), (e) the matrix corresponding to the left diagonal direction (135° and 315°), (f) the matrix corresponding to the right diagonal direction (45° and 225°), (g) angle definitions. Remember it is *important to normalize* the values in the co-occurrence matrix by dividing by the number of pixel pairs in the matrix before calculating the texture features.

a small image and a small number of gray levels, so in practice the number of gray levels may be quantized to reduce the number of calculations involved, and to reduce effects caused by noise in the images.

Laws texture energy masks are another method for measuring texture. They work by finding the average gray *L*evel, *E*dges, *S*pots, *R*ipples, and *W*aves in the image. They are based on the following five vectors:

$$L_5 = (1, 4, 6, 4, 1)$$

$$E_5 = (-1, -2, 0, 2, 1)$$

$$S_5 = (-1, 0, 2, 0, -1)$$

$$R_5 = (1, -4, 6, -4, 1)$$

$$W_5 = (-1, 2, 0, -2, -1)$$

These are used to generate the Laws 5×5 filter masks by finding the vector outer product of each pair of vectors. For example, using L_5 and S_5:

$$\begin{bmatrix} -1 & 0 & 2 & 0 & -1 \\ -4 & 0 & 8 & 0 & -4 \\ -6 & 0 & 12 & 0 & -6 \\ -4 & 0 & 8 & 0 & -4 \\ -1 & 0 & 2 & 0 & -1 \end{bmatrix}$$

The first step to applying these masks is to preprocess the image to remove artifacts caused by uneven lighting (actually this technique is useful as a preprocessing step for all texture measures). The easiest method for this is to subtract the local average from every pixel, using, for example, a 15×15 pixel size window. To do this move the window across the image, such as is done with convolution, find the average gray level value in the window and then subtract this average from the current pixel in the center of the window. Be sure to put the output into another image buffer (structure), so the current image is not over-written. This will create an image with average local gray levels close to zero.

The next step is to convolve the masks with the image to produce the texture filtered images, $F_k(r, c)$ for the kth filter mask. These texture filtered images are used to produce a *texture energy map*, E_k for the kth filter:

$$E_k(r, c) = \sum_{j=c-7}^{c+7} \sum_{i=r-7}^{r+7} |F_k(i, j)|$$

For these energy maps, the range on the summations depend on the window size, here we specified a window size of 15×15. These energy maps are then used to generate a texture feature vector for each pixel, which can be used for texture classification.

6.2.6 Feature Extraction with CVIPtools

CVIPtools allows the user to extract features from objects within the image. This is done by using the original image and a segmented or mask image to define the location of the object. Figure 6.2.11 shows the CVIPtools main window and the *Analysis->Features* window. To extract features we need to enter the original image, the segmented image, a feature file name, select the desired features, and select the image object by clicking on it with the mouse. Note that a name for the object *class* can be entered, but this is optional.

The CVIPtools software can be used for feature extraction in three primary ways: (1) extract features for the entire image, (2) extract features for an image object using a segmented image, or (3) extract features for an image object using a mask image. To extract features from the entire image, create an all white image with *Utilities->Create->Rectangle* that is the same size as the original image, and use that as the segmented image. To use a segmented image, select the *Segmentation* tab on the *Analysis* window, perform the segmentation method along with any post-segmentation morphological filtering to get the desired objects, and use the output image as the segmented image in the *Features* window. If segmentation does not provide the desired results, use *Utilties->Create->Border Mask* to create an image with an outline of the desired object (see Figure 6.2.12).

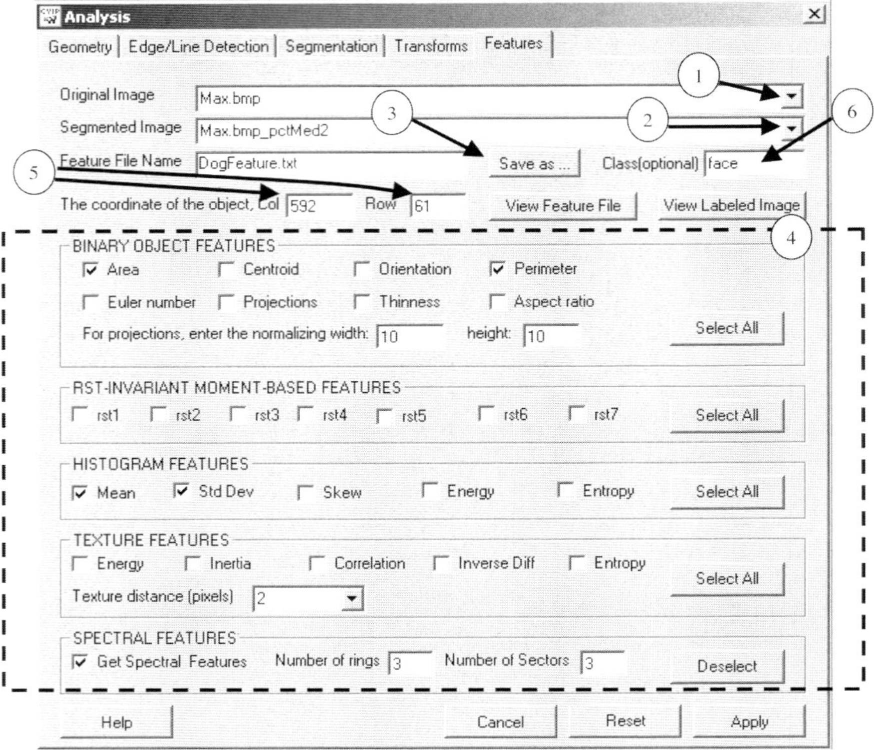

FIGURE 6.2.11
Using CVIPtools *Analysis->Features* window for feature extration. (1) Select an original image from which you want to extract features. (2) Select a segmented image, this can be a rectangle that has been created, an image that has been segmented, or a border mask that has been created. (3) Enter a name for the feature file. (4) Select the desired features, all features of a specific type can be selected with the *Select All* button, or individual features can be selected with the checkboxes. (5) Enter any coordinates within the object by clicking on the object in the image with the mouse. (6) A *Class* for the object is optional.

After the segmented or mask image is created, the selected features are extracted with the *Apply* button. CVIPtools does this by labeling the segmented image, selecting the object corresponding to the row and column coordinates, and then using the labeled image as a mask on the original image to extract the features for the selected object (see Table 6.2 for details). The features will be written to a feature file, as described in Chapter 3 (see Figure 3.3.20). The default directory for feature files is in the CVIP tools main directory under *bin feature*, but the user can specify the location with the *Save as* button. During processing the feature file can be viewed with the *View Feature File* button, and the labeled image can be viewed with the *View Labeled Image* button.

6.3 Feature Analysis

After the features have been extracted, feature analysis is important to aid in the feature selection process. Initially, the features were selected based on the understanding of the problem, and the developer's experience. Now that the features have been extracted, they

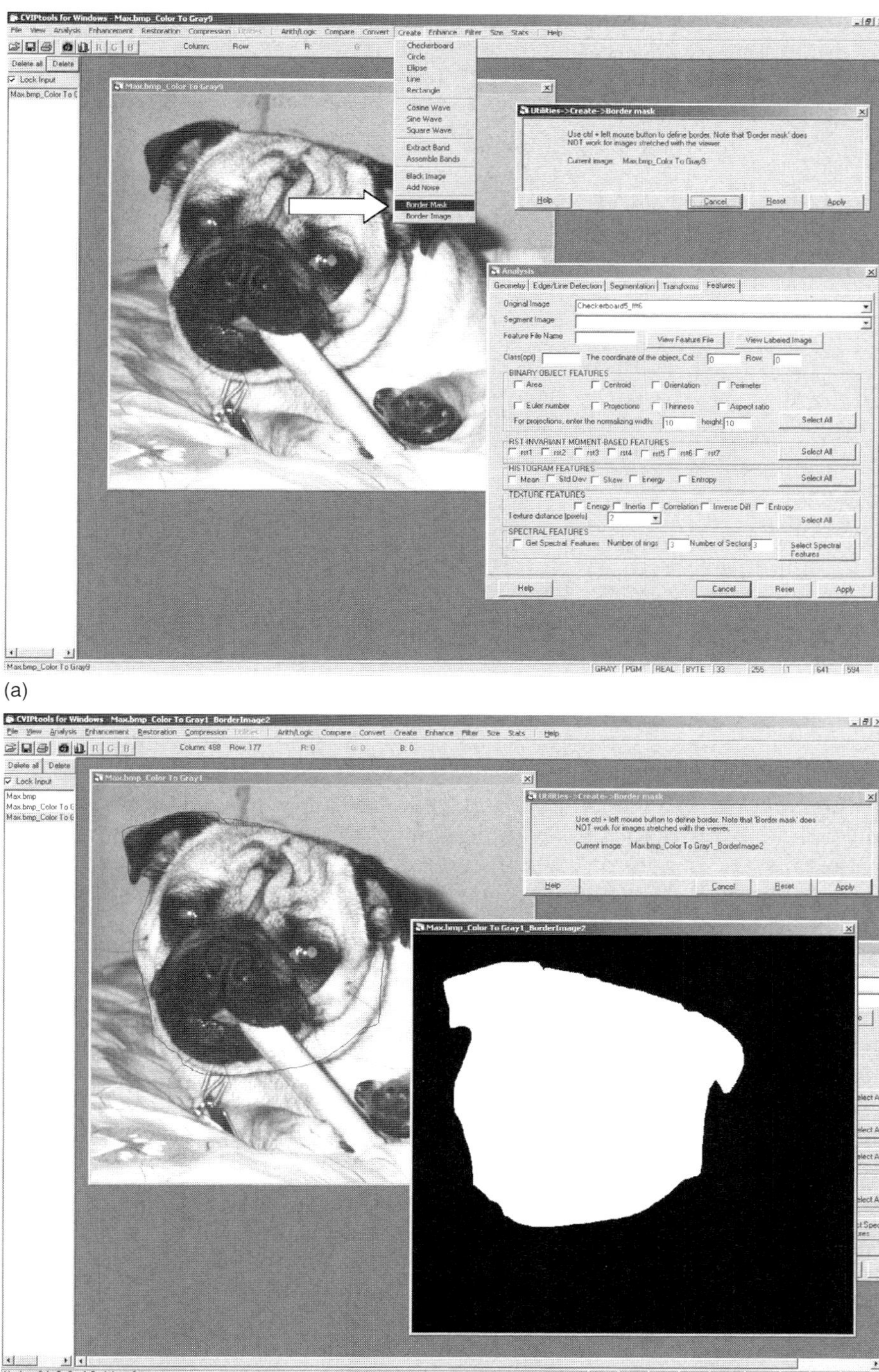

FIGURE 6.2.12
Creating a border mask image to extract features. (a) Selecting *Utilities->Create->Border Mask* function. (b) After using the mouse to draw a border, by holding the *Control* key on the keyboard and using the left mouse button, and then clicking on *Apply*. The border mask image can now be used as the *Segmented image* in the Features window to extract features relating to the outlined object.

TABLE 6.2
Feature Extraction with CVIPtools

Feature category	How the features are extracted
Binary object	The labeled image is used by selecting the object corresponding to the row and column coordinates and treating the object as a binary image with the object = "1" and the background = "0."
RST-invariant moment-based	The labeled image is used by selecting the object corresponding to the row and column coordinates and treating the object as a binary image with the object = "1" and the background = "0."
Histogram	The labeled image is used by selecting the object corresponding to the row and column coordinates, and then this binary object is used as a mask on the original image to extract features. This is done by only including in the calculations pixels that are part of the object.
Texture	The labeled image is used by selecting the object corresponding to the row and column coordinates, and then this binary object is used as a mask on the original image to extract features. This is done by only including in the calculations pixels that are part of the object.
Spectral	The labeled image is used by selecting the object corresponding to the row and column coordinates, and then this binary object is used as a mask on the original image to extract features. This is done by creating a black image (all zeros) with dimensions a power of 2, imbedding the object from the original image within the black image, and then calculating the Fourier transform on this image.

can be carefully examined to see which ones are the most useful and put back through the application feedback loop (see Figure 6.1.1) in the development process. To understand the feature analysis process, we need to define the mathematical tools to use; including feature vectors, feature spaces, distance and similarity measures to compare feature vectors, and various methods needed to preprocess the data for development of pattern classification algorithms. After these are understood, the feature analysis process begins with selection of the tools and methods that will be used for our specific imaging problem.

6.3.1 Feature Vectors and Feature Spaces

A feature vector is one method to represent an image, or part of an image (an object), by finding measurements on a set of features. The *feature vector* is an n-dimensional vector that contains these measurements, where n is the number of features. The measurements may be symbolic, numerical, or both. An example of a symbolic feature is color such a "blue" or "red"; an example of a numerical feature is the area of an object. If we take a symbolic feature and assign a number to it, it becomes a numerical feature. Care must be taken in assigning numbers to symbolic features, so that the numbers are assigned in a meaningful way. For example, with color we normally think of the hue by its name such as "orange" or "magenta." In this case, we could perform an HSL transform on the RGB data, and use the H (hue) value as a numerical color feature. But with the HSL transform the hue value ranges from 0 to 360 degrees, and 0 is "next to" 360, so it would be invalid to compare two colors by simply subtracting the two hue values.

The feature vector can be used to classify an object, or provide us with condensed higher-level image information. Associated with the feature vector is a mathematical abstraction called a *feature space*, which is also n-dimensional and is created to allow visualization of feature vectors, and relationships between them. With two- and

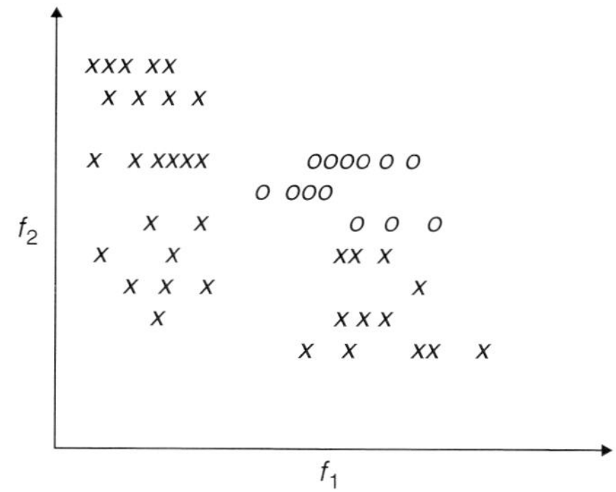

FIGURE 6.3.1
A two-dimensional feature space. This shows a two-dimensional feature space defined by feature vectors, $\mathbf{F} = \begin{bmatrix} f_1 \\ f_2 \end{bmatrix}$, and two classes represented by x and o. Each x and o represents one sample in the feature space defined by its values for f_1 and f_2. One of the goals of feature analysis and pattern classification is to find clusters in the feature space which correspond to different classes.

three-dimensional feature vectors it is modeled as a geometric construct with perpendicular axes and created by plotting each feature measurement along one axis (see Figure 6.3.1). For n-dimensional feature vectors it is an abstract mathematical construction called a *hyperspace*. As we shall see the creation of the feature space allows us to define distance and similarity measures which are used to compare feature vectors and aid in the classification of unknown samples.

Example 6.3.1

We are working on a computer vision problem for robotic control. We need to control a robotic gripper which picks parts from an assembly line and puts them into boxes. In order to do this, we need to determine: (1) where the object is in the two-dimensional plane in which the objects lie; (2) what type of object it is; one type goes into Box A, another type goes into Box B. First, we define the feature vector that will solve this problem. We determine that knowing the area, and center of area of the object, defined by an (r, c) pair, will locate it in space. We determine that if we also know the perimeter we can identify the object. So our feature vector contains four feature measures, and the feature space is four-dimensional. We can define the feature vector as: [area, r, c, perimeter].

6.3.2 Distance and Similarity Measures

The feature vector is meant to represent the object and will be used to classify it. To perform the classification we need methods to compare two feature vectors. The primary methods are to either measure the *difference* between the two, or to measure the *similarity*. Two vectors that are closely related will have a small difference and a large similarity.

The difference can be measured by a *distance measure* in the n-dimensional feature space; the bigger the distance between two vectors, the greater the difference. *Euclidean distance* is the most common metric for measuring the distance between two vectors, and is given by the square root of the sum of the squares of the differences between vector components. Given two vectors **A** and **B**, where:

$$\mathbf{A} = \begin{bmatrix} a_1 \\ a_2 \\ \vdots \\ a_n \end{bmatrix} \quad \text{and} \quad \mathbf{B} = \begin{bmatrix} b_1 \\ b_2 \\ \vdots \\ b_n \end{bmatrix}$$

then the Euclidean distance is given by:

$$\sqrt{\sum_{i=1}^{n}(a_i - b_i)^2} = \sqrt{(a_1 - b_1)^2 + (a_2 - b_2)^2 + (a_3 - b_3)^2 + \cdots + (a_n - b_n)^2}$$

Another distance measure, called the *city block* or *absolute value metric*, is defined as follows (using **A** and **B** as above):

$$\sum_{i=1}^{n} |a_i - b_i|$$

This metric is computationally faster than the Euclidean distance, but gives similar results. A distance metric that considers only largest difference is the *maximum value* metric defined by:

$$\max\{|a_1 - b_1|, |a_2 - b_2|, \ldots, |a_n - b_n|\}$$

We can see that this will measure the vector component with the maximum distance, which is useful for some applications. A generalized distance metric is the *Minkowski distance* defined as:

$$\left[\sum_{i=1}^{n} |a_i - b_i|^r\right]^{1/r} \quad \text{where } r \text{ is a positive integer}$$

The Minkowski distance is referred to as generalized because, for instance, if $r=2$ it is the same as Euclidean distance and when $r=1$ it is the city block metric.

The second type of metric used for comparing two feature vectors is the *similarity measure*. Two vectors that are close in the feature space will have a large similarity measure. The most common form of the similarity measure is one that we have already seen, the *vector inner product*. Using our definitions for the two vectors **A** and **B**, we can define the vector inner product by the following equation:

$$\sum_{i=1}^{n} a_i b_i = (a_1 b_1 + a_2 b_2 + a_n b_n)$$

Another commonly used similarity measure is the *Tanimoto metric*, defined as:

$$\frac{\sum_{i=1}^{n} a_i\, b_i}{\sum_{i=1}^{n} a_i^2 + \sum_{i=1}^{n} b_i^2 - \sum_{i=1}^{n} a_i\, b_i}$$

This metric takes on values between 0 and 1, which can be thought of as a "percent of similarity" since the value is 1 (100%) for identical vectors and gets smaller as the vectors get farther apart.

6.3.3 Data Preprocessing

Now that we have seen methods to compare two vectors, we need to analyze the set of feature vectors and prepare them for use in developing the classification algorithm. The data preprocessing consists of primarily three steps: (1) noise removal, (2) data normalization and/or decorrelation, and (3) insertion of missing data. Many of the classification algorithm development methods are based on mathematical theory that assumes specific distributions in the feature data and require data normalization. Typically the assumption is zero mean, Gaussian distributed data. For some methods it is desirable to have the features be uncorrelated. In either case the first step is noise removal, also called *outlier removal*.

An outlier is a data point that is alone very far from the average value. The assumption in removing it is that it is "bad," or noisy data. Possibly a mistake was made during its measurement, or it does not really represent the underlying structure. In the development of a classification algorithm a major part of what we are trying to do is to find a model for the underlying structure. As you will see the sample feature vectors are used to develop this model, so any "bad" samples will hinder the process. Before any samples are discarded care must be taken that they do not represent a subgroup for which we simply do not have many samples.

The distance and similarity measures defined before may be biased due to the varying range on different components of the vector. For example, one component may only range from 1 to 5 and another may range from 1 to 5000, so a difference of 5 for the first component will be maximum, but a difference of 5 for the second feature may be insignificant. It may help to *range-normalize* the vector components by dividing by the range on each vector component, where the range is simply the maximum value for that component minus the minimum.

Another option is to perform *unit vector normalization* which will modify the feature vectors so that they all have a magnitude of 1. If this is done we will retain only directional information about the vector, which preserves relationships between the features, but loses magnitudes. For example, these two vectors:

$$\mathbf{A} = \begin{bmatrix} 3 \\ 5 \\ 1 \end{bmatrix} \quad \text{and} \quad \mathbf{B} = \begin{bmatrix} 6 \\ 10 \\ 2 \end{bmatrix}$$

have the same direction, but different magnitudes, so their unit vectors are the same. To do this we normalize the vector components to their vector lengths (magnitude) by dividing by the Euclidean distance of the vector itself, which is found by calculating the Euclidean distance of the vector from the origin.

Example 6.3.2

Given the two vectors

$$\mathbf{A} = \begin{bmatrix} 3 \\ 5 \\ 1 \end{bmatrix} \quad \text{and} \quad \mathbf{B} = \begin{bmatrix} 6 \\ 10 \\ 2 \end{bmatrix}$$

The Euclidean distance of vector **A** from the origin, also called the magnitude is:

$$\sqrt{(3-0)^2+(5-0)^2+(1-0)^2} = \sqrt{9+25+1} = \sqrt{35}$$

The distance of vector **B** from the origin, also called the magnitude is:

$$\sqrt{6^2+10^2+2^2} = \sqrt{36+100+4} = \sqrt{140}$$

To normalize these two vectors to unit vectors, we divide each component by the vector magnitude:

$$\mathbf{A}' = \begin{bmatrix} \frac{3}{\sqrt{35}} \\ \frac{5}{\sqrt{35}} \\ \frac{1}{\sqrt{35}} \end{bmatrix} \approx \begin{bmatrix} 0.507 \\ 0.845 \\ 0.169 \end{bmatrix} \quad \text{and} \quad \mathbf{B}' = \begin{bmatrix} \frac{6}{\sqrt{140}} \\ \frac{10}{\sqrt{140}} \\ \frac{2}{\sqrt{140}} \end{bmatrix} \approx \begin{bmatrix} 0.507 \\ 0.845 \\ 0.169 \end{bmatrix}$$

So we can see that $\mathbf{A}' = \mathbf{B}'$. In this case, the second vector is in the same direction as the first, but is twice as long, that is, each component is multiplied by 2.

A commonly used statistical-based method to normalize these measures is to take each vector component and subtract the mean and divide by the standard deviation. This method can be applied to any of the measures, both distance and similarity, but requires knowledge of the probability distribution of the feature measurements. In practice the probability distributions are often estimated by using the existing data. This is done as follows, given a set of k feature vectors, $\mathbf{F}_j = \{\mathbf{F}_1, \mathbf{F}_2, \ldots, \mathbf{F}_k\}$, with n features in each vector:

$$\mathbf{F}_j = \begin{bmatrix} f_{1j} \\ f_{2j} \\ \vdots \\ f_{nj} \end{bmatrix} \quad \text{for } j = 1, 2, \ldots, k$$

$$\text{means} \Rightarrow m_i = \frac{1}{k}\sum_{j=1}^{k} f_{ij} \quad \text{for } i = 1, 2, \ldots, n$$

$$\text{standard deviation} \Rightarrow \sigma_i = \sqrt{\frac{1}{k}\sum_{j=1}^{k}\left(f_{ij} - m_i\right)^2} = \sqrt{\frac{1}{k}\sum_{ji=1}^{k}\left(f_{ij}\right)^2 - m_i^2} \qquad \text{for } i = 1, 2, \ldots, n$$

Now, for each feature component, we subtract the mean and divide by the standard deviation:

$$f_{ij\text{SND}} = \frac{f_{ij} - m_i}{\sigma_i} \quad \text{for all } i, j$$

This will give us new feature vectors where the distribution has been normalized so that the means are 0 and the standard deviations are 1; the resulting distribution on the each vector component is called the *standard normal density* (SND).

Other linear techniques can be used to limit the feature values to specific ranges, such as between 0 and 1, by scaling or shifting. Note that in above equation, we have simply shifted the data by the mean and scaled it by the standard deviation. To map the data to a specified range, S_{MIN} to S_{MAX}, but still retain the relationship between the values, we use *min–max normalization*:

$$f_{ij\text{MINMAX}} = \left(\frac{f_{ij} - f_{\text{MIN}}}{f_{\text{MAX}} - f_{\text{MIN}}}\right)(S_{\text{MAX}} - S_{\text{MIN}}) + S_{\text{MIN}}$$

where S_{MIN} and S_{MAX} are minimum and maximum values for the specified range and f_{MIN} and f_{MAX} are minimum and maximum values on the original feature data.

Nonlinear methods may be desired if the data distribution is skewed; that is, not evenly distributed about the mean. One common method, called *softmax scaling*, requires two steps:

$$\text{STEP1} \Rightarrow y = \frac{f_{ij} - m_i}{r\sigma_i}$$

$$\text{STEP2} \Rightarrow f_{ij\text{SMC}} = \frac{1}{1 + e^{-y}} \quad \text{for all } i, j$$

This is essentially a method that compresses the data into the range of 0 to 1. The first step is similar to mapping the data to the SND, but with a user defined factor, r. The process is approximately linear for small values of y with respect to f_{ij}, and then compresses the data exponentially as it gets farther away from the mean. The factor r determines the range of values for the feature, f_{ij}, that will fall into the linear range. In addition to moving the mean and normalizing the spread of the data, this transform will change the shape of the distribution.

If the data normalization techniques are applied take care that it will serve the application; since they will move the mean, and change the spread and/or shape of the resulting data distribution. In some cases this may not be desired. If useful information is contained in the mean, spread, or shape of the data distribution, be careful not to lose that information since the choice of the wrong normalization method will effectively filter it out. Also, be aware that the results will only be useful if the set of sample vectors represent the entire population, including all the classes. In practice this means that the sample set is large, the more the better. How many? It depends on the application. In general, as many as are practical for the application and as many as the development schedule allows.

Performing a principal components transform (PCT) in the n-dimensional feature space provides new features that are linear transforms of the original features, and are uncorrelated. This is desirable for some classification algorithm development methods, such as neural networks. Use of the PCT, also referred to as principal components analysis (PCA), is also useful for data visualization in feature space. With an n-dimensional feature space it is difficult (that is, impossible) to represent the space visually. A useful tool for feature analysis is to perform the PCT and view the first two or three components graphically.

The final step in data preprocessing is to insert missing data. This means that we analyze the distribution of the sample feature vectors and, based on a desired data distribution, create feature vectors that we think belong and include them in our feature vector set. Once again, care must be taken in this process so that we do not bias the results with our artificially generated sample feature vectors. To fully apply the methods discussed here requires a complete understanding of the problem, including how the features relate to the desired output and the underlying structure of each feature's distribution. Typically this information is unknown, so a trial and error approach is used during development.

6.4 Pattern Classification

Pattern classification, as related to image analysis, involves taking the features extracted from the image and using them to automatically classify image objects. This is done by developing classification algorithms that use the feature information. The distance or similarity measures are used for comparing different objects and their feature vectors. In this section we will define pertinent terms, conceptually discuss the most widely used current methods, and look in detail at some of the basic algorithms.

The primary uses of pattern classification in image analysis are for computer vision and image compression applications development. It can be considered a part of feature analysis, or as a postprocessing step to feature extraction and analysis. Pattern classification is typically the final step in the development of a computer vision algorithm, since in these types of applications the goal is to identify objects (or parts of objects) in order for the computer to perform some vision-related task. These tasks range from computer diagnosis of medical images to object classification for robotic control. In the case of image compression, we want to remove redundant information from the image, and compress important information as much as possible. One way to compress information is to find a higher-level representation of it, which is exactly what feature analysis and pattern classification is all about—finding a single class that will represent many pixel values.

6.4.1 Algorithm Development: Training and Testing Methods

To develop a classification algorithm, we need to divide our data into a training set and a test set. The *training set* is used for algorithm development, and the *test set* is used to test the algorithm for success. If this is not done, and we test the algorithm with the same set with which it was developed, the success we measure may be an invalid indicator of success on any other set of images. In addition, to work properly, both the training and test sets should completely represent all types of images that will be seen in the application domain. If this is not the case, the success we measure with the test set may not be a good predictor of success with the application. The use of two distinct sets of images provides us with results that are unbiased, and allows us to have confidence that the success measured during development is a good predictor of the success we can expect to achieve in the actual application.

The selection of the two sets should be done before development starts, to avoid biasing the test results. The size of the sets depends on many factors, but in practice we typically split the available images into two equal-sized groups. Theoretically, we want to maximize the size of the training set to develop the best algorithm, but the larger the test set is, the more confidence we have that the results are indicative of application success. If time allows, it is often instructive to use increasingly larger training sets (randomly selected), and analyze the results. What we expect to achieve is an increasing success rate as the training set size increases. If this does not happen we need to verify that our training set(s) actually represents the domain of interest. It may be that there are not enough samples in the training set, or it may be that the set of features being used is incomplete. Figure 6.4.1 shows the results from an experiment to classify skin tumors which illustrates this.

Figure 6.4.1a shows the results from using 13 features. The training set size was increased (horizontal axis), with the success plotted on the vertical axis. Here we see that, at least for the YES class, the success rate was not necessarily increasing, which led us to believe that the features in the training set were incomplete. In Figure 6.4.1b,c are the results of increasing the number of features in the feature vectors. We see that, with 18 features (Figure 6.4.1c), we achieved an essentially increasing success rate as we increased the test set size, which is a good indicator that we have a complete set of vectors in our training data.

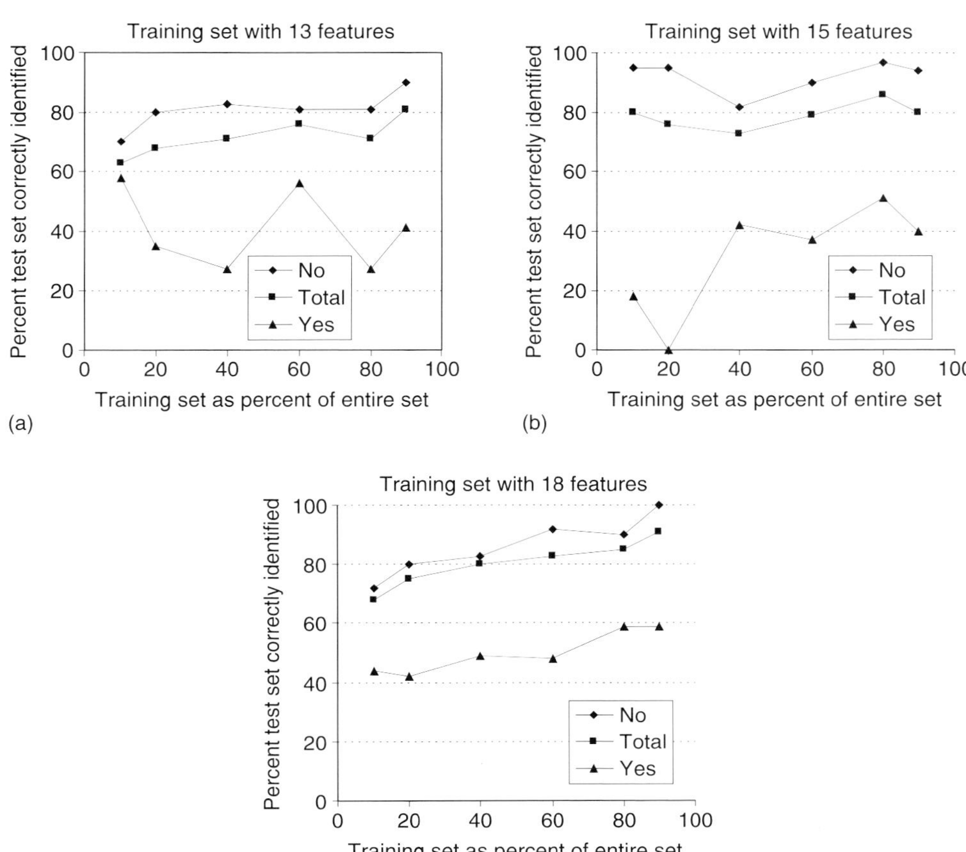

FIGURE 6.4.1
Example of increasing the training set size. These are results from an experiment to classify skin tumor images. (a) Results with 13 features, with the success rate jumping around we are not confident in the feature set being complete. (b) Results with 15 features, better results but still not increasing consistently. (c) Results with 18 features, the consistently increasing success rate as we increase the training set size gives us confidence in the completeness of the feature set being used.

One commonly used method to avoid needing to select the training and test sets is the *leave-one-out-method*. With the technique, all but one of the samples is used for training, and then it is tested on the one that was left out. This is done as many times as there are samples, and the number misclassified represents the error rate. However, this method is only practical by using pattern classification software during development, which will automate the process. Also, it is only useful with certain types of classifiers and applications. It is also valid to devise a *leave-K-out-method*, where K is a developer defined constant. K is defined based on the size of the sample set available, the available resources for testing and development, and the time allowed. In general, the smaller the value for K, the greater the confidence we have in the results. To use this approach we leave K vectors out and train (develop) the classification algorithm, and then test on the K vectors left out. We do this for all sets of K, or as many as is practical, and average the results to predict application success.

Another important concept in the development of a pattern classification scheme is the idea of a *cost function*. Often we may not want to rely on correct classification as the sole criteria in evaluating success of a classification system, because some types of misclassification may be more "costly" than others. For example, if we are developing a medical system to diagnoses cancer, the cost of mistakenly identifying a cancerous tumor as harmless is much higher than the cost of identifying a harmless tumor as cancerous. In the first case, the patient dies, while in the second case the patient is subjected to some temporary stress, but survives. Or consider a system to identify land mines. What are the relevant cost functions?

Once the data has been divided into the training and test sets, work can begin on the development of the classification algorithm. There are many methods available for this; we will consider some basic representative methods. The general approach is to use the information in the training set to classify the "unknown" example in the test set. It is assumed that all the samples available have a known classification. The success rate is measured by correct classification, but how various misclassifications contribute to the failure rate can be weighted by a cost function.

6.4.2 Classification Algorithms and Methods

The simplest algorithm for identifying a sample from the test set is called the *Nearest Neighbor* method. The object of interest is compared to every sample in the training set, using a distance measure, a similarity measure, or a combination of measures. The "unknown" object is then identified as belonging to the same class as the closest sample in the training set. This is indicated by the smallest number if using a distance measure, or the largest number if using a similarity measure. This process is computationally intensive and not very robust.

We can make the Nearest Neighbor method more robust by selecting not just the closest sample in the training set, but by consideration of a group of close feature vectors. This is called the *K-Nearest Neighbor* method, where, for example, $K=5$. Then we assign the unknown feature vector to the class that occurs most often in the set of K-Neighbors. This is still very computationally intensive, since we have to compare each unknown sample to every sample in the training set, and we want the training set as large as possible to maximize success.

We can reduce this computational burden by using a method called *Nearest Centroid*. Here, we find the centroids for each class from the samples in the training set, and then we compare the unknown samples to the representative centroids only. The centroids are calculated by finding the average value for each vector component in the training set.

Example 6.4.1

Suppose we have a training set of four feature vectors, and we have two classes.

$$\text{Class } \mathbf{A} : \mathbf{A}_1 = \begin{bmatrix} 3 \\ 4 \\ 7 \end{bmatrix} \quad \text{and} \quad \mathbf{A}_2 = \begin{bmatrix} 1 \\ 7 \\ 6 \end{bmatrix}$$

$$\text{Class } \mathbf{B} : \mathbf{B}_1 = \begin{bmatrix} 4 \\ 2 \\ 9 \end{bmatrix} \quad \text{and} \quad \mathbf{B}_2 = \begin{bmatrix} 2 \\ 3 \\ 3 \end{bmatrix}$$

The representative vector, centroid, for class **A** is:

$$\begin{bmatrix} (3+1)/2 \\ (4+7)/2 \\ (7+6)/2 \end{bmatrix} = \begin{bmatrix} 2 \\ 5.5 \\ 6.5 \end{bmatrix}$$

The representative vector, centroid, for class **B** is:

$$\begin{bmatrix} (4+2)/2 \\ (2+3)/2 \\ (9+3)/2 \end{bmatrix} = \begin{bmatrix} 3 \\ 2.5 \\ 6 \end{bmatrix}$$

To identify an unknown sample, we need only compare it to these two representative vectors, not the entire training set. The comparison is done using any of the previously defined distance or similarity measures. With a distance measure, the distance between the unknown sample vector and each class centroid is calculated and it is classified as the one it is closest to—the one with the *smallest* distance. With a similarity measure, the similarity between the unknown sample vector and each class centroid is calculated and it is classified as the one it is closest to—the one with the *largest* similarity.

The technique of comparing unknown vectors to classified vectors, and finding the closest one, is also called *template matching*. In addition to applying this technique to feature vectors, template matching can be applied the raw image data. A template is devised, possibly via a training set, which is then compared to subimages by using a distance or similarity measure. Typically, a threshold is set on this measure to determine when we have found a match, that is, a successful classification. This may be useful for applications where the size and orientation of the objects is known, and the objects shapes are regular. For example, for the recognition of computer generated text.

Bayesian theory provides a statistical approach to the development of a classification algorithm. To apply Bayesian analysis we need a complete statistical model of the features, and usually normalize the features so that their distribution is standard normal density (see Section 6.3). The Bayesian approach provides a classifier that has an optimal classification rate. This means that the boundaries that separate the classes provide a minimum average error for the samples in the training set. These boundaries are called *discriminant functions*, and an example is shown in Figure 6.4.2. Here we have two classes in two-dimensional feature space and show a *linear discriminant function* to separate the two classes. This type of plot is called a *scatterplot*, and is a useful visualization technique to find clusters corresponding to classes with the sample feature vectors. With the scatterplot we can find the desired line to separate the two classes. As previously mentioned, the PCT can be used to reduce n-dimensional data into two or three uncorrelated components that contain much of the original information, and these can then be used for visualization.

In practice, the feature space is typically larger than two-dimensional, and since the three-dimensional form of a linear function is a plane, the n-dimensional form of a linear function is called a *hyperplane*. In general, discriminant functions can also be quadratic (curved), or take on arbitrary shapes. Generally, the forms are limited to those that can be defined analytically via equations, such as circles, ellipses, parabolas, or hyperbolas. In n-dimensional vector spaces these decision boundaries are

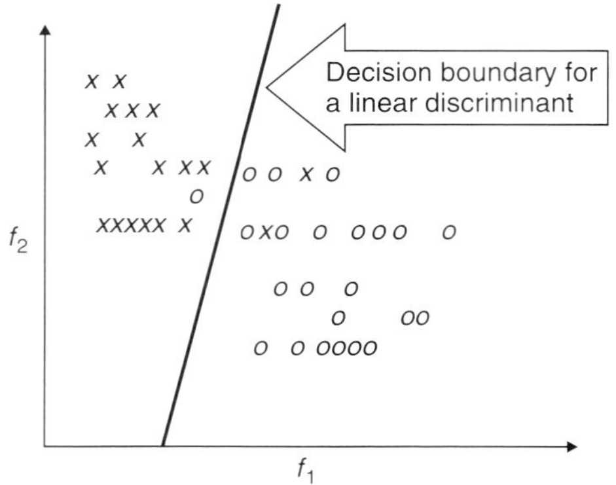

FIGURE 6.4.2
A linear discriminant separating two classes. This shows a two-dimensional feature space defined by feature vectors, $\mathbf{F} = \begin{bmatrix} f_1 \\ f_2 \end{bmatrix}$ and two classes represented by x and o. Note that this linear discriminant misclassifies two x's and one o.

called *hyperquadrics;* specifically hyperspheres, hyperellipsoids, hyperparaboloids, and hyperhyperboloids.

Neural networks represent another category of techniques used for pattern classification. Neural networks are meant to model the nervous system in biological organisms, and are thus referred to as *artificial neural networks*. Mathematical models have been developed for these biolological systems, based on a simple processing element called a neuron. These neurons function by outputting a weighted sum of the inputs, and these weights are generated during the learning or training phase. The element's output function is called the *activation function*, and the basic types are: (1) the identify function, the input equals the output, (2) a threshold function where every input greater than the threshold value outputs a 1, and less than the threshold outputs 0, (3) a sigmoid function, an S-shaped curve, which is nonlinear and is used most often. Single layer networks are limited in their capabilities and nonlinear functions are required to take advantage of the power available in multilayer networks.

The neural network consists of the input layer, the output layer, and possibly hidden layers, see Figure 6.4.3. The main distinguishing characteristics of the neural

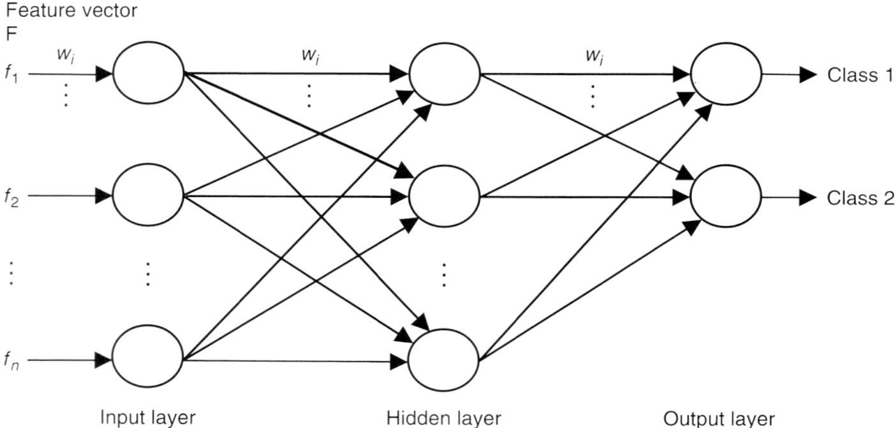

FIGURE 6.4.3
Neural network. Here we show a neural network architecture with an input layer, where the feature values are input, one hidden layer, and the output layer. In this case the output layer corresponds to two classes. The circles are the processing elements, neurons, and the arrows represent the connections. Associated with each connection is a weight (w_i, shown only across the top for clarity) by which the signal is multiplied. These weights are adjusted during the training (learning) phase.

network are: (1) the *architecture*, which includes the number of inputs, outputs, and hidden layers, and how they are all connected, (2) the *activation function*, which is typically used for all processing elements in a layer, but this is not required, and (3) the *learning algorithm*. Many learning algorithms have been developed, but they all work by inputting the training vectors, calculating an error measure and then adjusting the weights to improve the error. The training continues for a specific number of iterations, specified by the user, or until a specified error criterion is reached. To achieve optimal results with neural networks it is important that the feature vectors are preprocessed with a PCT to decorrelate the input data. Neural networks have been used successfully in many applications, for example in the recognition of hand written characters.

Numerous other methods for pattern classification are available, including artificial intelligence approaches, structural approaches, fuzzy logic approaches, and genetic algorithms. More information on these methods, and those briefly discussed here, can be found in the references.

6.5 Key Points

OVERVIEW: FEATURE ANALYSIS AND PATTERN CLASSIFICATION

- Feature analysis involves examining the extracted features and determining if and how they can be used to solve the imaging problem
- Pattern classification uses image object features to identify or categorize the image object

FEATURE EXTRACTION

- Feature extraction starts with feature selection
- Feature selection is important for successful pattern classification
- A good feature is robust, which means it provides consistent results, even with varying conditions
- RST-invariant features do not change under rotation, scale (size), or translation (movement) of the image object

Shape Features

- Shape features depend on a silhouette of the image object, so require only a binary image
- Shape features include area, center of area, axis of least second moment, projections, and Euler number from Section 3.3.3, and perimeter, thinness ratio, irregularity, aspect ratio, moments, set of seven based RST-invariant features, and Fourier descriptors

Perimeter—find by (1) count the number of "1" pixels next to "0" in the binary shape image, or (2) perform an edge detector and count "1" pixels, or (3) use chain code methods for better accuracy. These approximations can be improved by multiplying the result by $\pi/4$ for arbitrary, curved, shapes.

Thinness ratio—has a maximum value of 1, corresponding to a circle, and decreases as object gets thinner or perimeter gets more convoluted:

$$T = 4\pi\left(\frac{A}{P^2}\right)$$

Irregularity or compactness ratio—$1/T$, reciprocal of the thinness ratio

Aspect ratio—also called elongation or eccentricity, ratio of bounding box:

$$\frac{c_{\max} - c_{\min} + 1}{r_{\max} - r_{\min} + 1}$$

To be useful as a comparative measure the objects should be rotated to some standard orientation; such as orientating the axis of least second moment in the horizontal direction

Moments—for binary images, they are used to generate moment based RST-invariant features given in Table 6.1, defined by the *moment order* $(p+q)$:

$$m_{pq} = \sum_r \sum_c r^p c^q I(r, c)$$

In order to be translationally invariant we use the *central moments* defined by:

$$\mu_{pq} = \sum_r \sum_c (r - \bar{r})^p (c - \bar{c})^q I(r, c)$$

where

$$\bar{r} = \frac{m_{10}}{m_{00}} \quad \text{and} \quad \bar{c} = \frac{m_{01}}{m_{00}}$$

Fourier descriptors—use binary image of the object and find spectral features defined in Section 6.2.4

Histogram Features

- The histogram is a plot of gray level values versus the number of pixels at that value
- The histogram tells us something about the brightness and contrast
- A narrow histogram has low contrast, a histogram with a wide spread has high contrast
- A bimodal histogram has two peaks, usually object and background
- A histogram skewed toward the high end is bright, skewed toward the low end is dark
- First-order histogram probability:

$$P(g) = \frac{N(g)}{M}$$

M is the number of pixels in the image or subimage, and $N(g)$ is the number of pixels at gray level g.

Mean—average value, which tells us something about the general brightness of the image:

$$\bar{g} = \sum_{g=0}^{L-1} gP(g) = \sum_r \sum_c \frac{I(r, c)}{M}$$

Standard deviation (SD)—tells us about the contrast, high SD = high contrast, low SD = low contrast:

$$\sigma_g = \sqrt{\sum_{g=0}^{L-1} (g - \bar{g})^2 P(g)}$$

Skew—measures asymmetry about the mean:

$$\text{SKEW} = \frac{1}{\sigma_g^3} \sum_{g=0}^{L-1} (g - \bar{g})^3 P(g)$$

Energy—relates to gray level distribution, with a maximum value of 1 for an image of constant value and decreases as the gray levels are more widely distributed:

$$\text{ENERGY} = \sum_{g=0}^{L-1} [P(g)]^2$$

Entropy—varies inversely with energy, as defined measures how many bits are needed to code the data:

$$\text{ENTROPY} = -\sum_{g=0}^{L-1} P(g) \log_2[P(g)]$$

Color Features

- Color images consist of three bands, one each for red, green, and blue or RGB
- All of the features can be calculated separately in each color band
- Alternately, we desire information about the relationship *between* color bands
- To include between band information preprocess with a color transform (defined in Chapter 2)
- Most color transforms will decouple color and brightness information
- Avoid absolute color measures as they are not robust
- A relative color measure can be used which is typically application specific, or a known color standard can be used for comparison
- When using a known color standard the system can be calibrated if conditions change

Spectral Features

- Primary metric is power
- $\text{POWER} = |T(u, v)|^2$
- Measure power in specific regions in the spectrum
- The regions are box, ring or sector (wedge) shaped
- $\text{SPECTRAL REGION POWER} = \sum_{u \in \text{REGION}} \sum_{v \in \text{REGION}} |T(u, v)|^2$
- Due to the redundancy in the Fourier spectral symmetry we often measure the sector power over one-half the spectrum, and the ring power over the other half of the spectrum (see Figure 6.2.8)

Texture Features

- Spectral features can be used as texture features
- Texture is a function of image size relative to the object, as well as magnification
- In practice, the spectral features can be calculated for 10 or 20 (or more) rings and sectors and the magnitudes plotted to look for signature shapes which will correspond to specific textures
- Second-order histogram methods, also called gray level co-occurrence matrix methods, measure texture by considering relationship pixel pairs and require the parameters *distance* and *angle* between pixel pairs.

  Let c_{ij} be the elements in the co-occurrence matrix normalized by the number of pixel pairs in the matrix, and assume a given distance and angle (direction), the equations are as follows:

$$Energy = \sum_i \sum_j c_{ij}^2$$

$$Inertia = \sum_i \sum_j (i - j)^2 c_{ij}$$

$$Correlation = \frac{1}{\sigma_x \sigma_y} \sum_i \sum_j (i - \mu_x)(j - \mu_y) c_{ij}$$

where

$$\mu_x = \sum_i i \sum_j c_{ij}$$

$$\mu_y = \sum_j j \sum_i c_{ij}$$

$$\sigma_x^2 = \sum_i (1 - \mu_x)^2 \sum_j c_{ij}$$

and $$\sigma_y^2 = \sum_j (1-\mu_y)^2 \sum_i c_{ij}$$

$$InverseDifference = \sum_i \sum_j \frac{c_{ij}}{|i-j|}; \quad \text{for } i \neq j$$

$$Entropy = -\sum_i \sum_j c_{ij} \log_2 c_{ij}$$

- Laws energy masks can be used for measuring texture and are generated as the vector outer product of pairs of the following vectors, which correspond to gray level, edges, spots, ripples, and waves:

$$L_5 = (1, 4, 6, 4, 1)$$
$$E_5 = (-1, -2, 0, 2, 1)$$
$$S_5 = (-1, 0, 2, 0, -1)$$
$$R_5 = (1, -4, 6, -4, 1)$$
$$W_5 = (-1, 2, 0, -2, -1)$$

Laws filters are used by first removing the local average and then convolving the masks with the image to produce the texture filtered images, $F_k(r, c)$ for the kth filter mask. These texture filtered images are used to produce a *texture energy map*, E_k for the kth filter, using, for example, a 15×15 window:

$$E_k(r, c) = \sum_{j=c-7}^{c+7} \sum_{i=r-7}^{r+7} |F_k(i, j)|$$

The texture energy maps are used to generate a texture feature vector for each pixel

Feature Extraction with CVIPtools

- Extraction of features with CVIPtools requires the original image and a segmented or mask image
- A segmented image can be created with *Analysis->Segmentation* followed by any desired morphological filtering
- A mask image can be created with *Utilities->Create->Border Mask*
- Methods used for feature extraction by category are in Table 6.2

FEATURE ANALYSIS

- Feature analysis is important to aid in feature selection
- After feature extraction the feature analysis process includes consideration of the application
- The feature analysis process begins by selection of tools and methods that will be used for the imaging problem

Feature Vectors and Feature Spaces

Feature vector—an n-dimensional vector containing measurements for an image object, where n is the number of features. Feature vectors are symbolic, numeric or both.

Feature space—a mathematical abstraction created to allow visualization of feature vectors, and relationships between them. With two- and three-dimensional feature vectors it is modeled as a geometric construct with perpendicular axes and created by plotting each feature measurement along one axis. For n-dimensional feature vectors it is a mathematical construction called a *hyperspace*.

Distance and Similarity Measures

Distance measures—used to compare two vectors in feature space by finding the distance or error between the two, the smaller the metric the more alike the two are

Euclidean distance—geometric distance in feature space:

$$\sqrt{\sum_{i=1}^{n}(a_i - b_i)^2} = \sqrt{(a_1 - b_1)^2 + (a_2 - b_2)^2 + (a_3 - b_3)^2 + \cdots + (a_n - b_n)^2}$$

City block or absolute value metric—results similar to Euclidean, but faster to calculate:

$$\sum_{i=1}^{n}|a_i - b_i|$$

Maximum value metric—only counts largest vector component distance:

$$\max\{|a_1 - b_1|, |a_2 - b_2|, \ldots, |a_n - b_n|\}$$

Minkowski distance—generalized distance metric:

$$\left[\sum_{i=1}^{n}|a_i - b_i|^r\right]^{1/r} \quad \text{where } r \text{ is a positive integer}$$

Similarity measure—used to compare two vectors in feature space by finding the similarity between the two, the larger the metric the more alike the two are

Vector inner product—

$$\sum_{i=1}^{n} a_i b_i = (a_1 b_1 + a_2 b_2 + \cdots + a_n b_n)$$

Tanimoto metric—takes on values between 0 and 1; 1 for identical vectors:

$$\frac{\sum_{i=1}^{n} a_i b_i}{\sum_{i=1}^{n} a_i^2 + \sum_{i=1}^{n} b_i^2 - \sum_{i=1}^{n} a_i b_i}$$

Data Preprocessing

Data preprocessing—to prepare the feature vectors for use in pattern classification algorithm development; consists of three steps: (1) noise (outlier) removal, (2) data normalization and/or decorrelation, and (3) insertion of missing data

Noise (outlier) removal—removal of feature vectors that are so far from the average as to be considered noise

Data normalization and/or decorrelation—performed to avoid biasing the distance or similarity measures, and to prepare the data for pattern classification methods

Range-normalize—divide each vector component by the data range for that component

Unit vector normalization—divide each vector component by the magnitude of the feature vector, where the magnitude is the vector length or the Euclidean distance from the origin

Standard normal density normalization—creating a distribution with 0 mean and standard deviation of 1:

Given a set of k feature vectors, $\mathbf{F}_j = \{\mathbf{F}_1, \mathbf{F}_2, \ldots, \mathbf{F}_k\}$, with n features in each vector:

$$\mathbf{F}_j = \begin{bmatrix} f_{1j} \\ f_{2j} \\ \vdots \\ f_{nj} \end{bmatrix} \quad \text{for } j = 1, 2, \ldots, k$$

$$\text{means} \Rightarrow m_i = \frac{1}{k}\sum_{j=1}^{k} f_{ij} \quad \text{for } i = 1, 2, \ldots, n$$

$$\text{standard deviation} \Rightarrow \sigma_i$$

$$= \sqrt{\frac{1}{k}\sum_{j=1}^{k}\left(f_{ij} - m_i\right)^2} = \sqrt{\frac{1}{k}\sum_{ji=1}^{k}\left(f_{ij}\right)^2 - m_i^2} \quad \text{for } i = 1, 2, \ldots, n$$

Now, for each feature component, we subtract the mean and divide by the standard deviation:

$$f_{ij\text{SND}} = \frac{f_{ij} - m_i}{\sigma_i} \quad \text{for all } i, j$$

Min–max normalization—to map the data to a specified range:

$$f_{ij\text{MINMAX}} = \left(\frac{f_{ij} - f_{\text{MIN}}}{f_{\text{MAX}} - f_{\text{MIN}}}\right)(S_{\text{MAX}} - S_{\text{MIN}}) + S_{\text{MIN}}$$

Where S_{MIN} and S_{MAX} are minimum and maximum value for the specified range and f_{MIN} and f_{MAX} are minimum and maximum value on the original feature data.

Softmax scaling—a nonlinear normalization method for use with skewed data distributions:

$$\text{STEP1} \Rightarrow y = \frac{f_{ij} - m_i}{r\sigma_i}$$

$$\text{STEP2} \Rightarrow f_{ij\text{SMC}} = \frac{1}{1 + e^{-y}} \quad \text{for all } i, j$$

Principal components transform—performed in the n-dimensional feature space to decorrelate the data; useful preprocessing for neural networks

Insertion of missing data—analyze the distribution of the sample feature vectors and, based on a desired data distribution, create feature vectors that we think belong and include them in our feature vector set

PATTERN CLASSIFICATION

- Pattern classification involves taking the features extracted from the image and using them to automatically classify image objects
- Used primarily in computer vision and image compression applications

Algorithm Development: Training and Testing Methods

- Available feature vector samples are divided into a *training set* and a *test set*
- Each set should represent all types of images in the application domain
- The *training set* is used for algorithm development
- The *test set* is used to test the algorithm that was developed with the training set
- The use of two distinct sets allows us to have confidence that the success measured during development is a good predictor of the success we can expect to achieve in the actual application
- Testing methods: (1) gradually increase training set size and plot (Figure 6.4.1), (2) leave-one-out method, (3) leave-*K*-out-method
- A *cost function* can be used if different misclassifications have different levels of importance

Classification Algorithm and Methods

Nearest neighbor—compare an unknown sample to each vector in the training set using a distance or similarity (or both) metric, and classify it the same as the one it is closest to

***K*-nearest neighbor**—comparing the unknown feature vector to entire training set and finding the *K* nearest, where *K* is an integer such as 5, and classifying it as the class that appears most often in the set of *K* samples

Nearest centroid—finding the centroid for each class and comparing the unknown to the representative centroids, and classifying it the same as the class of the closest centroid

Template matching—general term for comparing vectors, can be used to compare raw image data to sample image objects

Bayesian analysis—provides a statistical approach to the development of a classification algorithm which requires a complete statistical model of the features. Preprocess by normalizing the features with standard normal density. The analysis finds boundaries in feature space to separate the classes called *discriminant functions*, and provides a theoretically optimal classification rate.

Neural networks—modeled after the nervous system in biological systems, based on the processing element the *neuron* (see Figure 6.4.3). The main distinguishing characteristics are: (1) the *architecture*, which includes the number of inputs, outputs and hidden layers, and how they are all connected, (2) the *activation function*, typically identity, threshold or sigmoid, and (3) the *learning algorithm*. Learning algorithms work by inputting the training vectors, calculating an error measure and then adjusting the weights to improve the error. To achieve optimal results with neural networks use a PCT for preprocessing.

6.6 References and Further Reading

More on feature extraction and selection can be found in [Theodoridis/Koutroumbas 03], [Forsyth/Ponce 03], [Ripley 96], and [Duda/Hart 73]. Information regarding chain codes can be found in [Costa/Cesar 01], [Nadler/Smith 93], [Gonzalez/Woods 92], [Jain/Kasturi/Schnuck 95], and [Ballard/Brown 82]. For an excellent book on shape analysis and classification see [Costa/Cesar 01]. More information on shape features can be found in [Forsyth/Ponce 03], [Shapiro/Stockman 01], [Castleman 96], [Schalkoff 89], [Horn 86], and [Levine 85]. More details on color features can be found in [Forsyth/Ponce 03] and [Shapiro/Stockman 01]. Details regarding Fourier descriptors can be found in [Gonzalez/Woods 02], [Sonka/Hlavac/Boyle 99], and [Nadler/Smith 93]. More information for texture-based features can be found in [Gonzalez/Woods 02], [Shapiro/Stockman 01], [Sonka,/Hlavac/Boyle 99], [Castleman 96], [Granlund/Knutsson 95], [Haralick/Shapiro 92], [Pratt 91], and [Rosenfeld/Kak 82]. Details on the co-occurrence matrix for texture can be found in [Gonzalez/Woods 02], [Shapiro/Stockman 01], [Sonka,/Hlavac/Boyle 99], [Haralick/Shapiro 92], and [Nadler/Smith 93]. An example of using spectral feature plots for texture identification is found in [Nadler/Smith 93]. The information regarding the Laws energy mask is found in [Shapiro/Stockman 01] and [Sonkac/Hlavac/Boyle 99]. More on the RST invariant moment features can be found in [Gonzalez/Woods 02], [Costa/Cesar 01], [Nadler/Smith 93], and [Schalkoff 92].

For more on pattern classification see [Theodoridis/Koutroumbas 03], [Shapiro/Stockman 01], [Sonka,/Hlavac/Boyle 99], [Gose/Johnsonbaugh/Jost 96], [Nadler/Smith 93], [Schalkoff 92], and [Duda/Hart 73]. For a practical book on statistics see [Kennedy/Neville 86]. For details on the experiments used as an example for increasing training set size, see [Umbaugh/Moss/Stoecker 91]. More information on template matching is found in [Gose/Johnsonbaugh/Jost 96], [Schalkoff 92], and [Duda/Hart 73]. Neural networks are discussed in more depth in [Forsyth/Ponce 03], [Kulkarni 01], [Sonka,/Hlavac/Boyle 99], and [Gose/Johnsonbaugh/Jost 96]. More detail on the biological models for neural nets are in [Harvey 94]. One book which walks through the various parameters and models for developing pattern classification systems, and includes software, is [Kennedy/Lee/VanRoy/Reed/Lippman 97]. Books that relate classical pattern recognition methods and neural nets include [Kulkarni 01], [Ripley 96], and [Schalkoff 92]. For a practical approach to the use of neural networks in image processing see [Masters 94]. For an excellent handbook on image processing, quite useful for feature extraction, see [Russ 99]. For information on fuzzy logic methods for pattern recognition see [Kulkarni 01], [Sonka/Hlavac/Boyle 99], and [Nadler/Smith 93]. For information on genetic algorithms see [Sonka/Hlavac/Boyle 99].

Ballard, D.H., Brown, C.M., *Computer Vision*, Upper Saddle River, NJ: Prentice Hall, 1982

Banks, S., *Signal Processing, Image Processing and Pattern Recognition*, Upper Saddle River, NJ: Prentice Hall, 1990.

Castleman, K.R., *Digital Image Processing*, Upper Saddle River, NJ: Prentice Hall, 1996.

Costa, L., Cesar, R.M., *Shape Analysis and Classification: Theory and Practice*, Boca Raton, FL: CRC Press, 2001.

Duda, R.O., Hart, P.E., *Pattern Classification and Scene Analysis*, NY: Wiley, 1973.
Forsyth, D.A., Ponce, J., *Computer Vision*, Upper Saddle River, NJ: Prentice Hall, 2003.
Gonzalez, R.C., Woods, R.E., *Digital Image Processing*, Reading, MA: Addison Wesley, 1992.
Gonzalez, R.C., Woods, R.E., *Digital Image Processing*, Upper Saddle River, NJ: Prentice Hall, 2002.
Gose, E., Johnsonbaugh, R., Jost, S., *Pattern Recognition and Image Analysis*, Upper Saddle River, NJ: Prentice Hall PTR, 1996.
Granlund, G., Knutsson, H., *Signal Processing for Computer Vision*, Boston: Kluwer Academic Publishers, 1995.
Haralick, R.M., Shapiro, L.G., *Computer and Robot Vision*, Reading, MA: Addison-Wesley, 1992.
Harvey, R.L., *Neural Network Principles*, Upper Saddle River, NJ: Prentice Hall, 1996.
Horn, B.K.P., *Robot Vision*, Cambridge, MA: The MIT Press, 1986.
Jain, R., Kasturi, R., Schnuck, B.G., *Machine Vision*, NY: McGraw Hill, 1995.
Kennedy, J.B., Neville, A.M., *Basic Statistical Methods for Engineers and Scientists*, NY: Harper and Row, 1986.
Kennedy, R.L, Lee, Y., Van Roy, B., Reed, C.D., Lippman, R.P., *Solving Data Mining Problems Through Pattern Recognition*, Upper Saddle River, NJ: Prentice Hall, 1997.
Kulkarni, A., *Computer Vision and Fuzzy-Neural Systems*, Upper Saddle River, NJ: Prentice Hall, 2001.
Levine, M.D., *Vision in Man and Machine*, NY: McGraw Hill, 1985.
Masters, T., *Signal and Image Processing with Neural Networks*, NY: Wiley, 1994.
Nadler, M., Smith, E.P., *Pattern Recognition Engineering*, NY: Wiley, 1993.
Pratt, W.K., *Digital Image Processing*, NY: Wiley, 1991.
Ripley, B.D., *Pattern Recognition and Neural Networks*, NY: Cambridge University Press, 1996.
Rosenfeld, A., Kak, A.C., *Digital Picture Processing*, San Diego, CA: Academic Press, 1982.
Russ, J.C., *The Image Processing Handbook*, Boca Raton, FL: CRC Press, 1999.
Schalkoff, R.J., *Digital Image Processing and Computer Vision*, NY: Wiley, 1989.
Schalkoff, R.J., *Pattern Recognition: Statistical, Structural and Neural Approaches*, NY: Wiley, 1992.
Shapiro, L., Stockman, G., *Computer Vision*, Upper Saddle River, NJ: Prentice Hall, 2001.
Sonka, M., Hlavac, V., Boyle, R., *Image Processing, Analysis and Machine Vision*, Pacific Grove, CA: Brooks/Cole Publishing Company, 1999.
Theodoridis, S., Koutroumbas, K., *Pattern Recognition*, NY: Academic Press, 2003.
Tou, J.T., Gonzalez, R.C., *Pattern Recognition Principles*, Reading, MA: Addison Wesley, 1974.
Umbaugh, S.E, Moss, R.H., Stoecker, W.V., Applying Artificial Intelligence to the Identification of Variegated Coloring in Skin Tumors, *IEEE Engineering in Medicine and Biology*, December 1991, pp: 57–62.

6.7 Exercises

1. (a) In Figure 6.1.1 there is a dotted line between *Feature Analysis* and *Application*. Explain. (b) In the same figure, there is a feedback loop from *Pattern Classification* and *Application*. Explain.
2. Why might image segmentation be performed before feature extraction and analysis?
3. (a) Name the first step in feature extraction. Why is this important? (b) Why is it important for a feature to be robust?
4. (a) Describe a method to find perimeter of a binary object. (b) How can this estimate be improved for objects with curved edges?
5. (a) What is the thinness ratio of a circle? (b) What is the thinness ratio of a rectangle that is 20 pixels wide by 80 pixels high?

6. (a) What is the aspect ratio of circle with a radius of 25? (b) Why rotate an object before finding the aspect ratio?
7. (a) For the moment based-features defined, why do we need the normalized central moments, instead of the regular moments? (b) Of what use are the RST-invariant moment-based features?
8. Use CVIPtools to explore the RST-invariant features. (a) Create binary objects using *Utilities->Create* and *Utilities->Arith/Logic* to OR objects together. (b) Use *Analysis->Features* to extract the RST-invariant features from the objects. (c) Add noise to your objects with *Utilities->Noise* and extract the features from the noisy objects. Compare the results for the objects with and without noise, can you still classify the objects? Why or why not?
9. (a) What can we say about an image with a narrow histogram? (b) What can we say about a histogram skewed toward the left? (c) What is a histogram with two major peaks called? What do the peaks typically correspond to?
10. (a) What does the standard deviation of the histogram tell us about the image? (b) What is maximum value for histogram energy? What image type does this correspond to? (c) What does histogram entropy tell us? (d) What is the relationship between histogram energy and entropy?
11. (a) Given the following binary checkerboard image, where the image is 256×256 pixels and the squares are 32×32, calculate the histogram features, mean, standard deviation, skew, energy, and entropy. Verify your results with CVIPtools using *Utilities->Stats->Image Statistics*. Are they the same? Why or why not?

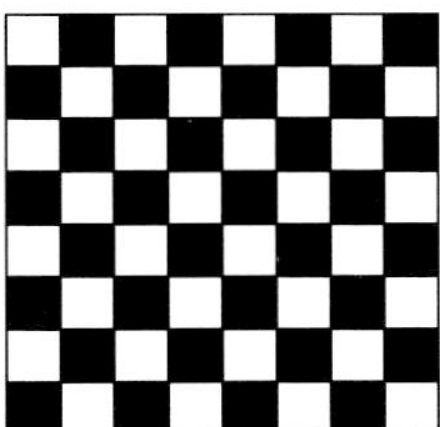

(b) Given the following binary circle image, where the image is 256×256 pixels and the radius of the circle is 32, calculate the histogram features, mean, standard deviation, skew, energy, and entropy by using the equation of the area of a circle (πr^2). Verify your results with CVIPtools using *Utilities->Stats->Image Statistics*. Are they the same? Why or why not?

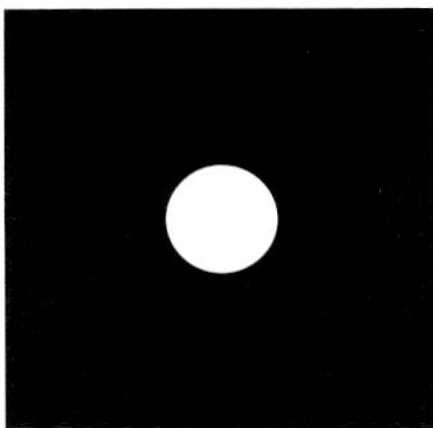

12. (a) Describe the easiest method to obtain color features. (b) Why might this method not be what we want? How can we get this information?

13. (a) What is the primary metric for spectral features? (b) Regarding spectral features explain the statement: "The sector measure will tend to be size invariant, and the ring measure will tend to be rotation invariant." Sketch images to illustrate this. (c) Are the sector or ring spectral features translationally invariant? That is, if an object moves in the image, will the values change?
14. (a) As we zoom in on a textured object, how does this affect the spectral features? (b) As we zoom out on a textured object, how does this affect the spectral features?
15. Using a pixel distance, $d = 1$, find the gray level co-occurrence matrices for the horizontal, vertical, right diagonal and left diagonal directions, for the following image:

$$\begin{bmatrix} 0 & 0 & 1 & 1 \\ 0 & 0 & 1 & 1 \\ 0 & 2 & 2 & 2 \\ 2 & 2 & 3 & 3 \end{bmatrix}$$

16. (a) Find the 5×5 Laws texture energy mask for spots and edges. (b) Find the 5×5 Laws texture energy mask for gray level and ripples. (c) Find the 5×5 Laws texture energy mask for ripples and waves. (d) What, if any, preprocessing is necessary to use the Laws energy masks?
17. Use CVIPtools to explore feature extraction. (a) Select an image(s) of your choice with objects of interest. (b) Use *Utilties->Create->Border Mask* to create mask images for your objects of interest. (c) Use *Analysis->Features* to extract features that you think will be of interest for these objects. Examine the feature file. Are they results what you expected? Why or why not?
18. (a) Define a feature vector that is useful to classify engineers and non-engineers. (b) Define a classification rule for these two classes based on your feature vectors
19. Given the following two features vectors, find the following distance and similarity metrics:

$$F_1 = \begin{bmatrix} 5 \\ 8 \\ 2 \end{bmatrix} \qquad F_2 = \begin{bmatrix} 6 \\ 10 \\ 1 \end{bmatrix}$$

(a) Euclidean distance. (b) City block distance. (c) Maximum value. (d) Minkowski distance, with $r = 2$. (e) Vector inner product. (f) Tanimoto metric.
20. Calculate the same metrics as in Exercise 19, but first normalize the vectors to their length, which is the Euclidean distance from the origin.
21. Calculate the same metrics as in Exercise 19, but first range normalize the vectors, using the following ranges:

$$f_1 \rightarrow \text{range} = 10, \qquad f_2 \rightarrow \text{range} = 20, \qquad f_3 \rightarrow \text{range} = 5$$

22. In the following scatter plot we have a two-dimensional feature space with all our sample vectors shown for two classes. Discuss any reasons to remove or

add any feature vectors to our data set before we begin developing the pattern classification algorithm.

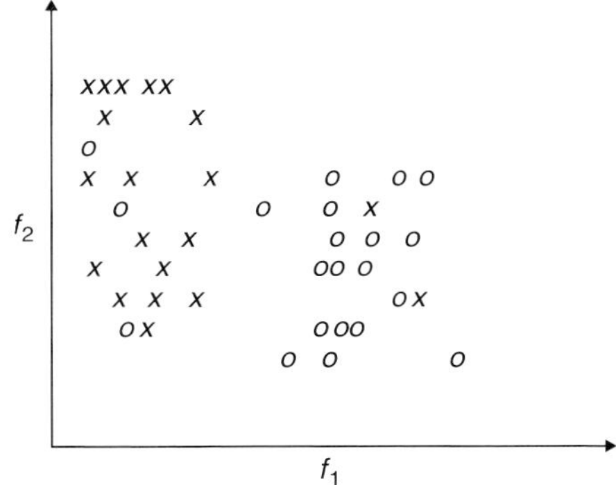

23. (a) When developing a classification algorithm, why do we divide our data into training and test sets? (b) Describe two methods for dividing the data into these two sets. Discuss important factors to consider when selecting the training and test sets.
24. Describe the *leave-one-out* and the *leave-K-out* method of developing and testing pattern classification algorithms.
25. Describe an example, other than the ones in the book, which shows why a cost function is important when developing a pattern classification algorithm.
26. Given the following feature vectors, with two classes:

$$\text{Class 1:} \quad \left\{ F_1 = \begin{bmatrix} 5 \\ 8 \\ 6 \end{bmatrix} F_2 = \begin{bmatrix} 7 \\ 6 \\ 1 \end{bmatrix} F_3 = \begin{bmatrix} 6 \\ 7 \\ 2 \end{bmatrix} \right\}$$

$$\text{Class 2:} \quad \left\{ F_1 = \begin{bmatrix} 1 \\ 8 \\ 7 \end{bmatrix} F_2 = \begin{bmatrix} 3 \\ 6 \\ 8 \end{bmatrix} F_3 = \begin{bmatrix} 2 \\ 7 \\ 6 \end{bmatrix} \right\}$$

(a) Using the Nearest Neighbor classification method, and the absolute value distance metric, classify the following unknown sample vector as Class 1 or Class 2:

$$F = \begin{bmatrix} 4 \\ 6 \\ 9 \end{bmatrix}$$

(b) Use K Nearest Neighbor, with $K = 3$

27. Given the following feature vectors, with two classes:

$$\text{Class 1:} \quad \left\{ F_1 = \begin{bmatrix} 5 \\ 8 \\ 6 \end{bmatrix} F_2 = \begin{bmatrix} 7 \\ 6 \\ 1 \end{bmatrix} F_3 = \begin{bmatrix} 6 \\ 7 \\ 2 \end{bmatrix} \right\}$$

$$\text{Class 2:} \quad \left\{ F_1 = \begin{bmatrix} 1 \\ 8 \\ 7 \end{bmatrix} F_2 = \begin{bmatrix} 3 \\ 6 \\ 8 \end{bmatrix} F_3 = \begin{bmatrix} 2 \\ 7 \\ 6 \end{bmatrix} \right\}$$

Using the Nearest Centroid classification method, and the absolute value distance metric, classify the following unknown sample vector as Class 1 or Class 2:

$$F = \begin{bmatrix} 3 \\ 6 \\ 10 \end{bmatrix}$$

28. Given the following feature vectors, with two classes:

$$\text{Class 1:} \quad \left\{ F_1 = \begin{bmatrix} 5 \\ 8 \\ 6 \end{bmatrix} F_2 = \begin{bmatrix} 7 \\ 6 \\ 1 \end{bmatrix} F_3 = \begin{bmatrix} 6 \\ 7 \\ 2 \end{bmatrix} \right\}$$

$$\text{Class 2:} \quad \left\{ F_1 = \begin{bmatrix} 1 \\ 8 \\ 7 \end{bmatrix} F_2 = \begin{bmatrix} 3 \\ 6 \\ 8 \end{bmatrix} F_3 = \begin{bmatrix} 2 \\ 7 \\ 6 \end{bmatrix} \right\}$$

Using the Nearest Centroid classification method, and the vector inner product similarity measure, classify the following unknown sample vectors as Class 1 or Class 2:

$$\text{(a)} \quad F = \begin{bmatrix} 4 \\ 6 \\ 9 \end{bmatrix} \qquad \text{(b)} \quad F = \begin{bmatrix} 8 \\ 6 \\ 4 \end{bmatrix} \qquad \text{(c)} \quad F = \begin{bmatrix} 3 \\ 6 \\ 10 \end{bmatrix}$$

29. (a) What type of preprocessing normalization should we do to apply Bayesian analysis? (b) What do we call the n-dimensional form of the linear discriminant

function? (c) Given the following scatter plot, draw a linear discriminant function to separate the two classes.

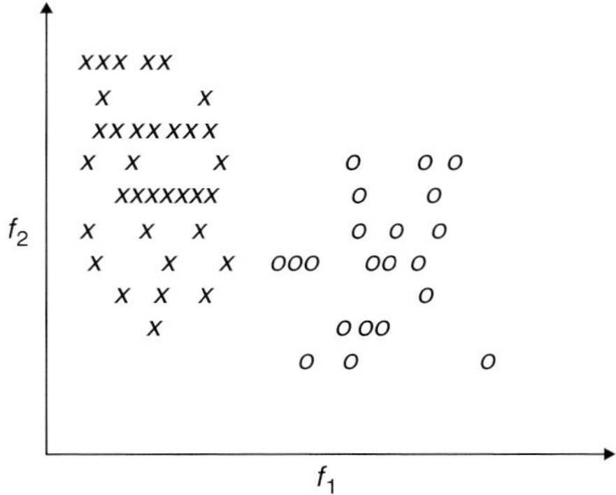

30. (a) What are the main distinguishing characteristics of a neural network? (b) What type of preprocessing should be done to apply a neural network? (c) Describe three types of activation functions. Which one is used most often? (d) In general, how does a learning algorithm work?

6.7.1 Programming Exercise: Perimeter

1. Write a function to find the perimeter of a solid (no holes) binary object. The input parameters to the function are the binary image with the object and a row and column coordinate within the object. The function will label the image and then estimate perimeter by counting the number of "1" pixels next to "0" pixels.
2. Test this function using images you create with CVIPtools. Use *Utilties->Create* to create test images. To create images with multiple objects, use the *AND* and *OR* logic functions available from *Utilties->Arith/Logic*.
3. Modify the function to estimate perimeter by performing a Roberts edge detection followed by counting the number of "1" pixels for the object of interest. Test the function with the images you created.
4. Modify the function to find the perimeter of all objects in the image, if passed $(-1, -1)$ as row and column coordinates. Output the object number along with its perimeter to the user.

6.7.2 Programming Exercise: Thinness Ratio

1. Write a function to find the thinness ratio of a solid (no holes) binary object. The input parameters to the function are the binary image with the object and a row and column coordinate within the object.
2. Test this function using images you create with CVIPtools. Use *Utilties->Create* to create test images. To create images with multiple objects, use the *AND* and *OR* logic functions available from *Utilties->Arith/Logic*.
3. Modify the function to find the thinness ratio of all objects in the image, if passed $(-1, -1)$ as row and column coordinates. Output the object number along with its thinness ratio to the user.

6.7.3 Programming Exercise: Aspect Ratio

1. Write a function to find the aspect ratio of a binary object. The input parameters to the function are the binary image with the object and a row and column coordinate within the object.
2. Test this function using images you create with CVIPtools. Use *Utilties->Create* to create test images. To create images with multiple objects, use the *AND* and *OR* logic functions available from *Utilties->Arith/Logic*.
3. Modify the function to find the aspect ratio after rotating the object so that the axis of least second moment is horizontal. Use the CVIPtools library function *orientation* to find the axis of least second moment.

6.7.4 Programming Exercise: Moment-Based RST-Invariant Features

1. Incorporate the CVIPtools library function *rst_invariant* into your CVIPlab. Note that the function returns a pointer to an array of the seven features, the data type is double.
2. Verify that the function is working properly by comparison to results you obtain with CVIPtools.

6.7.5 Programming Exercise: Histogram Features

1. Write a function to find the histogram features, mean, standard deviation, skew, energy, and entropy for a gray level image.
2. Extend the function to work with color images.
3. Modify the function to find these features for individual image objects by passing the input image, a segmented image, and row and column coordinates within the object. Note that a labeled image must be generated from the segmented image and used in conjunction with the original image.

6.7.6 Programming Exercise: Color Features

1. Incorporate the CVIPtools library function *colorxform* into your CVIPlab.
2. Experiment by performing the HSL and CIE $L^*u^*v^*$ transform and then extracting histogram features on the resulting 3 bands.
3. Verify that the function is working properly by comparison to results you obtain with CVIPtools using *Utilities->Convert->Color Space* and *Analysis->Features*.

6.7.7 Programming Exercise: Spectral Features

1. Incorporate the CVIPtools library function *spectral_features* into your CVIPlab. Note that this function returns a pointer to a POWER data structure, defined as follows:

```
typedef struct
{
int    no_of_sectors;
```

```
int     no_of_bands;
int     imagebands;
double    *dc;
double    *sector;
double    *band;
}
POWER
```

2. Verify that the function is working properly by comparison to results you obtain with CVIPtools.

6.7.8 Programming Exercise: Texture Features

1. Incorporate the CVIPtools library function *texture* into your CVIPlab. Note that this function returns a pointer to a TEXTURE data structure, defined as follows:

```
typedef struct {
/* [0] -> 0 degree, [1] -> 45 degree, [2] -> 90 degree, [3] -> 135 degree,
[4] -> average, [5] -> range (max-min) */
float ASM[6];               /* (1) Angular Second Moment */
float contrast[6];          /* (2) Contrast */
float correlation[6];       /* (3) Correlation */
float variance[6];          /* (4) Variance */
float IDM[6];               /* (5) Inverse Difference Moment */
float sum_avg[6];           /* (6) Sum Average */
float sum_var[6];           /* (7) Sum Variance */
float sum_entropy[6];       /* (8) Sum Entropy */
float entropy[6];           /* (9) Entropy */
float diff_var[6];          /* (10) Difference Variance */
float diff_entropy[6];      /* (11) Difference Entropy */
float meas_corr1[6];        /* (12) Measure of Correlation 1 */
float meas_corr2[6];        /* (13) Measure of Correlation 2 */
float max_corr_coef[6];     /* (14) Maximal Correlation Coefficient */
} TEXTURE;
```

This data structure returns an array of six float numbers for each of the texture features listed. The first four correspond to the four directions, the fifth is the average of the first four, and the sixth one is the range on the feature. For more details on the features themselves, see the online documentation.

2. Verify that the function is working properly by comparison to results you obtain with CVIPtools. Note that CVIPtools provides the average and range of five of the texture features.

6.7.9 Programming Exercise: Distance and Similarity Measures

1. Write a function to calculate the Minkowski distance between two vectors. The input parameters include: the r value, the two vectors as arrays (vectors).

2. Write a function to find the similarity measure, vector inner product, of two vectors.
3. Write a function to normalize the vector parameters to standard normal density by passing the function the mean and standard deviation for each vector component, along with a vector.
4. Incorporate the normalization function as an input parameter option in the distance and similarity functions.

6.7.10 Programming Exercise: Template Matching

1. Write a C function to perform template matching. The function should take two input images: the image of interest, $I(r, c)$, and the template image, $T(r, c)$. The function will move the template across the image of interest, searching for pattern matches by calculating the error at each point. The distance measure to be used for this exercise is the Euclidean distance measure defined by:

$$D(\bar{r}, \bar{c}) = \sqrt{\sum_r \sum_c [I(r', c') - T(r, c)]^2}$$

   If we overlay the template on the image, then $\bar{r}$, $\bar{c}$ are the row and column coordinates of $I(r, c)$ corresponding to the center of the template where a match occurs. The r', c' designation is used to illustrate that as we slide the template across the image, the limits on the row and column coordinates of $I(r, c)$ will vary depending on: (a) where we are in the image, and (b) the size of the template. You need only consider parts of the image that fully contain the template image. Your function should handle any size image and template, but you may assume that the template is smaller than the image. A match will occur when the error measure is less than a specified threshold. In your function, the threshold should be specified by user input. Where a match occurs, the program should display the error and the (r, c) coordinates.
2. Test this function with images you create using CVIPtools. For example, create a small image for the template with a single object, and then create a larger image with multiple objects for the test image.
3. Expand the function by allowing for the rotation of the template. Consider the error to be the minimum error from all rotations.
4. Modify the function for efficiency by comparing the template only to image objects, not every subimage.
5. Make your function more useful by adding size invariance to the template matching. This is done by growing, or shrinking, the object to the size of the template before calculating the error.
6. Experiment with using different error and similarity measures described in Section 6.3.2.

6.7.11 Programming Exercise: Pattern Classification I

1. Write a function that will read a CVIPtools feature file and classify any unknown vectors, those without classes listed, by comparison to all other feature vectors

in the file by using K-Nearest Neighbors method and the absolute value distance metric, where K is an input parameter.

2. Write a function that will read a CVIPtools feature file and calculate the centroid vector for each class contained in the feature file, and write an output file with the class names and the corresponding centroid vectors.
3. Write a function that will read a CVIPtools feature file and a "centroid file" from the previous function, and classify the feature vectors using Nearest Centroid method.

6.7.12 Programming Exercise: Pattern Classification II

1. Modify the functions from the Pattern Classification I exercises to allow the user to specify the type of distance or similarity metric desired. Include Minkowski distance, vector inner product and Tanimoto metric.
2. Modify the functions from the Pattern Classification I exercises to allow the user to specify the type of normalization desired, including range normalization, standard normal density or min-max normalization.

Section III

Digital Image Processing

7

Digital Image Processing and Visual Perception

7.1 Introduction and Overview

In Chapter 1 we discussed the distinction between image processing and image analysis/computer vision, by saying that image analysis/computer vision applications produced images that are for computer use or analysis, and image processing creates images that are for human consumption. In our exploration of image analysis, we discovered that the output is some form of higher level image representation that can be used for analysis or in some applications pattern classification. The area of image processing refers to the process of taking an image and getting a better image out. Better is what sense?—enhanced, restored or compressed.

In this part of the book we will look in detail at each of these three areas, image enhancement, image restoration, and image compression with a chapter devoted to each area. We will see that image enhancement and image restoration both involve techniques to create a better image. Image compression involves the development of techniques to make smaller files while still retaining high quality images. Metrics necessary to measure image quality are explored in this chapter, as well as human visual perception. We must first learn how the human visual system perceives images to determine exactly what it is that makes one image better than another.

7.2 Human Visual Perception

Human visual perception is something most of us take for granted. We do not think about how the makeup of the physiological systems affects what we see and how we see it. Although human visual perception encompasses both physiological and psychological components, we are going to focus primarily on the physiological aspects, which are more easily quantifiable, using the current models available for understanding the systems.

The first question is—why study visual perception? We need to understand how we perceive visual information in order to design compression algorithms that compact the data as much as possible, but still retain all the necessary visual information. This is desired for both transmission and storage economy. Images are often transmitted over the airwaves, and are transmitted more frequently via the Internet, and people do not want to wait minutes or hours for the images. Additionally, the storage requirements can become overwhelming without compression. For example, an 8-bit monochrome image, with a resolution of 512 pixels wide by 512 pixels high, requires 1/4 of a megabyte of data. If we make this a color image, it requires 3/4 of a megabyte of data (1/4 for each of three color planes—red, green, and blue). With many applications requiring the capability to efficiently manage thousands of images, the need to reduce data is apparent.

For the development of image enhancement and restoration algorithms, we also have the need to understand how the human visual system works. For enhancement methods, we need to know the types of operations that are likely to improve an image visually, and this can only be achieved by understanding how the information is perceived. For restoration methods, we must determine aspects of the restoration process that are likely to achieve optimal results based on our perception of the restored image.

7.2.1 The Human Visual System

Vision is our most powerful sense. It allows us to gather information about our environment, and provides us with the opportunity to learn through observation. Vision enables us to interact with our environment safely, by allowing control of our physical movements without direct physical contact (ouch!). It is also our most complicated sense, and although current knowledge of biological vision systems is incomplete we do have a basic understanding of the different parts of the system, and how they interact.

The human visual system has two primary components—the eye and the brain, which are connected by the optic nerve (see Figure 7.2.1). The structure that we know the most about is the image receiving sensor—the human eye. The brain can be thought of as being an information processing unit, analogous to the computer in our computer imaging system. These two are connected by the optic nerve, which is really a bundle of nerves that contains the pathways for the visual information to travel from the receiving sensor, the eye, to the processor, the brain. The way the human visual system works is as follows: (1) light energy is focused by the lens of the eye onto the sensors on the retina at the back of the eye; (2) these sensors respond to this light energy by an electro-chemical reaction that sends an electrical signal down the optic nerve to the brain; and (3) the brain uses these nerve signals to create neurological patterns that we perceive as images.

The visible light energy corresponds to an electromagnetic wave that falls into the wavelength range from about 380 nanometers (nm) for ultraviolet to about 780 nm for infrared, although above 700 nm the response is minimal. In young adults wavelengths

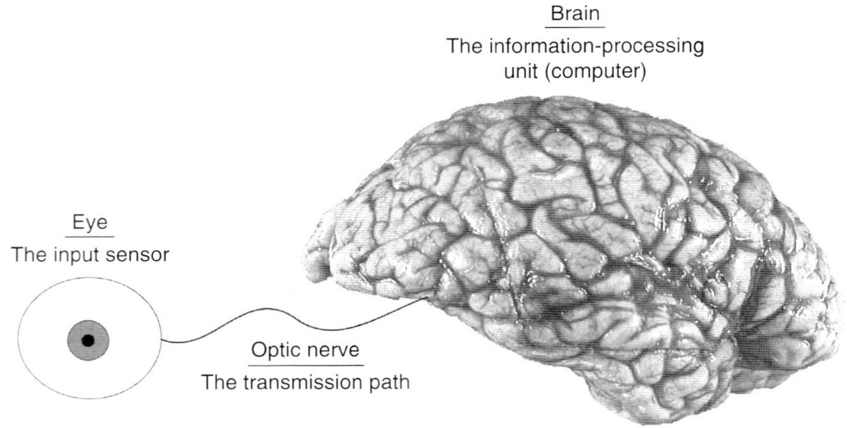

FIGURE 7.2.1
The human visual system.

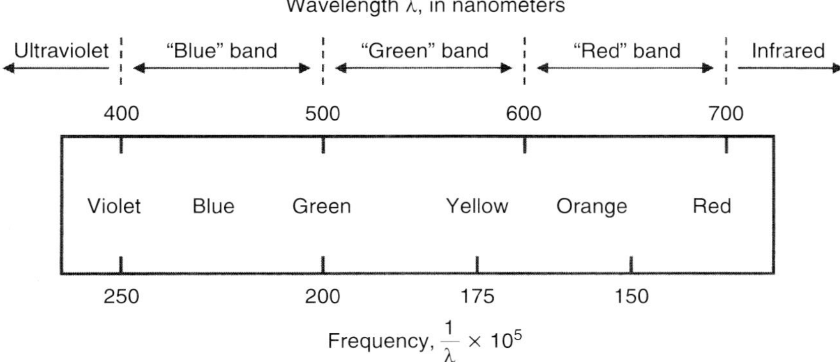

FIGURE 7.2.2
Visible light spectrum. For many imaging applications the visible spectrum is divided into three bands: red, green, and blue (RGB). Red is the longest wavelength and the lowest frequency, and blue (or violet) is the shortest wavelength and highest frequency. Beyond the red end of the visible spectrum is infrared, and below the violet in ultraviolet.

as high as 1000 nm or down to 300 nm may be seen, but the standard range for human vision is typically given as 400 to 700 nm. How this fits in with other parts of the electromagnetic spectrum was shown in Figure 2.2.1. In imaging systems the spectrum is often divided into various *spectral bands*, where each band is defined by a range on the wavelengths (or frequency). For example, it is typical to divide the visible spectrum into roughly three bands corresponding to "blue" (400 to 500 nm), "green" (500 to 600 nm), and "red" (600 to 700 nm) In Figure 7.2.2 we see the visible wavelengths of light and their corresponding colors, and how these relate to the standard separation into red, green, and blue (RGB) color bands.

The eye has two primary types of light energy receptors, or photoreceptors, which respond to the incoming light energy and convert it into electrical energy, or neural signals, via a complex electro-chemical process. These two types of sensors are called rods and cones. The sensors are distributed across the *retina*, the inner backside of the eye where the light energy falls after being focused by the lens (Figure 7.2.3). The *cones* are primarily used for *photopic* (daylight) vision, are sensitive to color, are concentrated in the central region of the eye, and have a high resolution capability. The *rods* are used in *scotopic* (night) vision, see only brightness (not color), are distributed across the retina, and have medium to low level resolution. There are many more rods in the human eye than cones; with an order of magnitude difference—on the order of 10 million cones to 100 million rods.

In Figure 7.2.3 we can see that there is one place on the retina where no light sensors exist; this is necessary to make a place for the optic nerve, and is referred to as the *blind spot*. One of the amazing aspects of the human brain is that we do not perceive this as a blind spot—the brain fills in the missing visual information. By examining Figure 7.2.3b, we can see why an object must be in our central field of vision, which is only a few degrees wide, in order to really perceive it in fine detail. This is where the high-resolution-capability cones are concentrated. They have a higher resolution than the rods because they have individual nerves tied to each sensor (cone), whereas the rods have multiple sensors (rods) connected to each nerve. The distribution of the rods across the retina shows us that they are more numerous than cones, and that they are used for our peripheral vision—there are very few cones away from the central visual axis. The response of the rods to various wavelengths of light is shown in Figure 7.2.4a.

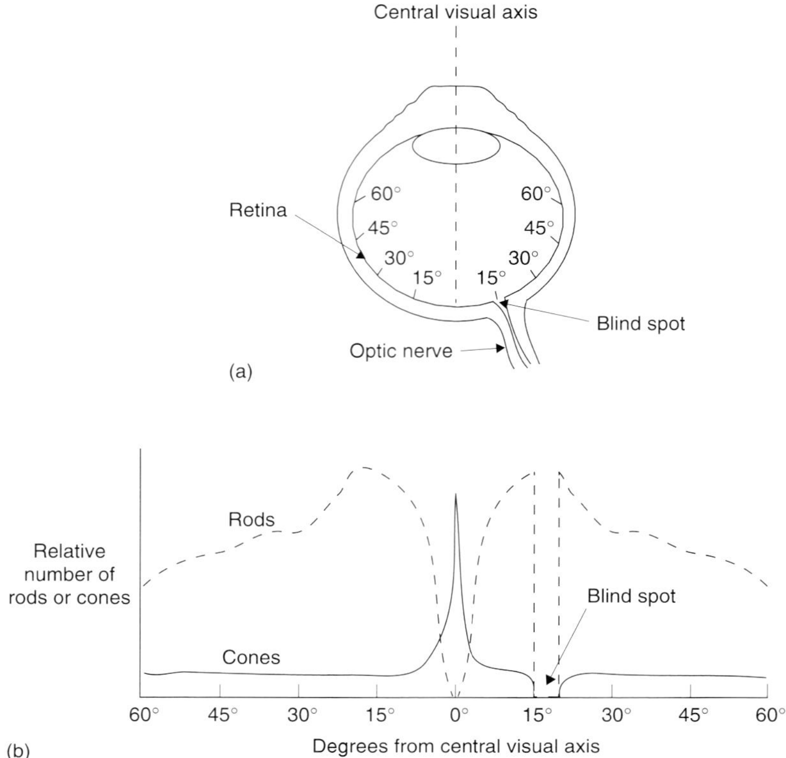

FIGURE 7.2.3
The human eye. (a) Basic eye structure. (b) Concentration of rods and cones across retina.

There are three types of cones, each responding to different wavelengths of light energy. The response to these can be seen in Figure 7.2.4b, the *tristimulus curves*. These are called the tristimulus (three stimuli) curves because all the colors that we perceive are the combined result of the response to these three sensors. These curves plot the wavelength versus the relative intensity (gain) of the sensor's response, or in engineering terms, the transfer function of the sensor. Although there are many more of the "red" and "green" cones than the "blue" cones, we are still able to see blue quite nicely. Apparently, the part of the brain that perceives color can compensate for this and is just one of the many phenomena in the human brain that we do not yet fully understand.

The cones in the eye respond in such a way as to generate three brightness values for each of the red, green, and blue bands. This is why we model color images in this manner—it is a model for human visual perception. We can approximate these RGB values as follows:

$$\mathrm{R} = K \int_{400}^{700} \mathrm{R}(\lambda) b(\lambda)\, \mathrm{d}\lambda$$

$$\mathrm{G} = K \int_{400}^{700} \mathrm{G}(\lambda) b(\lambda)\, \mathrm{d}\lambda$$

$$\mathrm{B} = K \int_{400}^{700} \mathrm{B}(\lambda) b(\lambda)\, \mathrm{d}\lambda$$

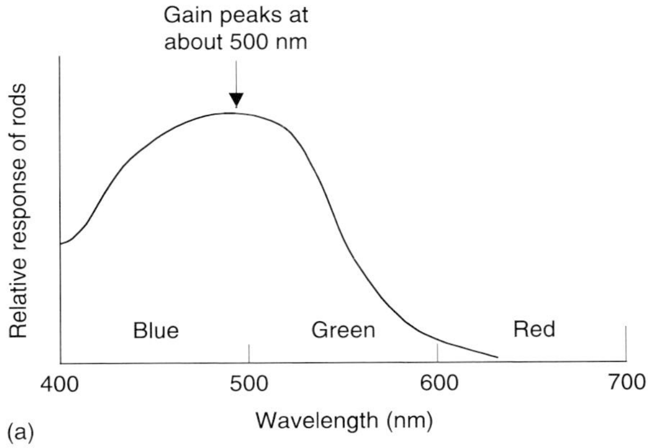

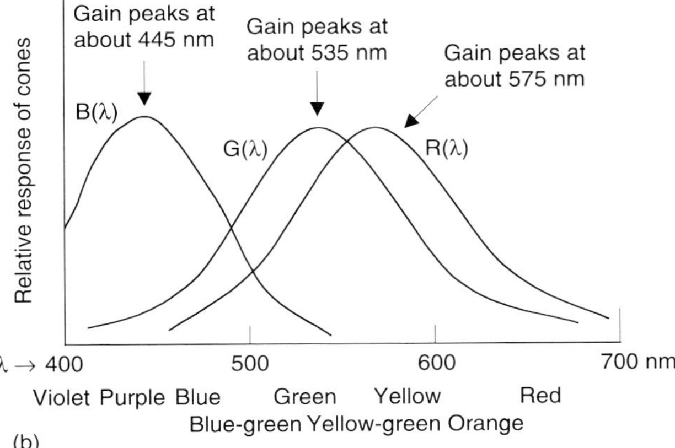

FIGURE 7.2.4
Relative response of rods and cones. (a) Rods react in low light levels, scotopic vision, but respond to only a single spectral band, so cannot distinguish colors. (b) Cones react only to high light intensities, photopic vision, and, since there are three different types which respond to different spectral bands, they enable us to see color.

Where $b(\lambda)$ is the incident photon flux (as in Chapter 2), $R(\lambda)$, $G(\lambda)$, and $B(\lambda)$ are the tristimulus curves, and K is a constant that is dependent on the sensor's area, the length of the time the signal is measured and the units used.

Example 7.2.1

Assume that the tristimulus curves are approximated by bandpass filters with a flat response with a gain of one, using the red, green, and blue bands as shown in Figure 7.2.2. Find the RGB values for the following incident photon flux: (a) $b(\lambda) = 5$, (b) $b(\lambda) = 10\lambda$ (for these calculations, don't worry about units as they depend on the time and area as was shown in the sensor equation in Section 2.2.1, and for color perception the important aspect is the relative amount of R, G, and B)

$$\text{(a)}\quad R = K\int_{400}^{700} R(\lambda)b(\lambda)\,d\lambda = K\int_{600}^{700}(1)(5)\,d\lambda = K5\lambda\Big|_{600}^{700} = 500K$$

$$G = K\int_{400}^{700} G(\lambda)b(\lambda)\,d\lambda = K\int_{500}^{600}(1)(5)\,d\lambda = K5\lambda\Big|_{500}^{600} = 500K$$

$$B = K\int_{400}^{700} B(\lambda)b(\lambda)\,d\lambda = K\int_{400}^{500}(1)(5)\,d\lambda = K5\lambda\Big|_{400}^{500} = 500K$$

$$\text{(b)}\quad R = K\int_{400}^{700} R(\lambda)b(\lambda)\,d\lambda = K\int_{600}^{700}(1)(10\lambda)\,d\lambda = K10\frac{\lambda^2}{2}\bigg|_{600}^{700} = 6.5\times 10^5 K$$

$$G = K\int_{400}^{700} G(\lambda)b(\lambda)\,d\lambda = K\int_{500}^{600}(1)(10\lambda)\,d\lambda = K10\frac{\lambda^2}{2}\bigg|_{500}^{600} = 5.5\times 10^5 K$$

$$B = K\int_{400}^{700} B(\lambda)b(\lambda)\,d\lambda = K\int_{400}^{500}(1)(10\lambda)\,d\lambda = K10\frac{\lambda^2}{2}\bigg|_{400}^{500} = 4.5\times 10^5 K$$

Two colors that have similar R, G, and B values will appear similar, and two colors with identical RGB values will look identical. However, it is possible for colors to have different spectral distributions, $b(\lambda)$, and still appear the same. *Metameric* is the term for two colors with different spectral distributions that have the same RGB values. Two colors that are *metamers* will look identical to the human visual system. This is possible since the RGB values are calculated by multiplying the spectral distribution of the incoming light energy by the sensor response, in this case the tristimulus curves, and then integrating the result—remember that the integral operator is simply the area under the curve.

Example 7.2.2

As before, assume that the tristimulus curves are approximated by bandpass filters with a flat response with a gain of one, using the red, green, and blue bands as shown in Figure 7.2.2. In the previous example, part (a) we had a spectral distribution of $b(\lambda) = 5$, and calculated the three RGB values to be 500K. Given the following spectral distribution, find the RGB values:

$$b(\lambda) = \begin{cases} 10 & \text{for } 425 \le \lambda \le 475 \\ 10 & \text{for } 525 \le \lambda \le 575 \\ 10 & \text{for } 625 \le \lambda \le 675 \\ 0 & \text{elsewhere} \end{cases}$$

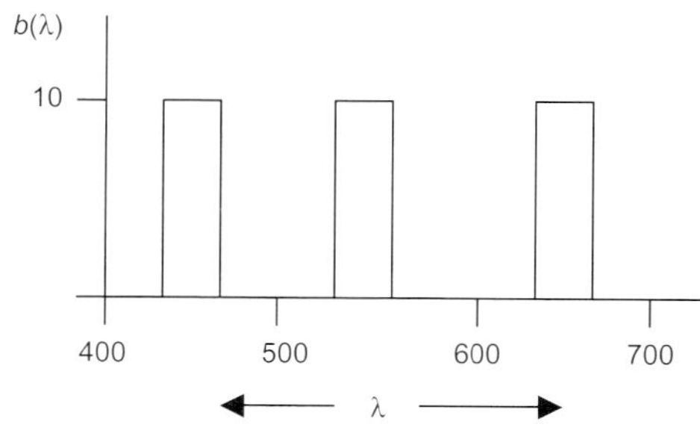

$$R = K\int_{400}^{700} R(\lambda)b(\lambda)\,d\lambda = K\int_{625}^{675}(1)(10)\,d\lambda = K10\lambda\bigg|_{625}^{675} = 500K$$

$$G = K\int_{400}^{700} G(\lambda)b(\lambda)\,d\lambda = K\int_{525}^{575}(1)(10)\,d\lambda = K10\lambda\bigg|_{525}^{575} = 500K$$

$$B = K\int_{400}^{700} B(\lambda)b(\lambda)\,d\lambda = K\int_{425}^{475}(1)(10)\,d\lambda = K10\lambda\bigg|_{425}^{475} = 500K$$

Note that this is the same color as when the spectral distribution is a constant value of 5 across the entire spectrum! Therefore, these two colors are metamers and appear the same.

Since the human visual system (HVS) sees in a manner that is dependent on the sensors, most cameras and display devices are designed with sensors that mimic the HVS response. With this type of design, metamers, which look the same to us, will also appear the same to the camera. However, we can design a camera that will distinguish between metamers, something that we cannot do, by proper specification of the response function of the sensors. This is one example that illustrates that a computer vision system can be designed that has capabilities beyond the HVS—we can even design a machine vision system to "see" x-rays or gamma rays. The strength of the HVS system lies not with the sensors, but with the intricate complexity that we call the human brain.

Another point of comparison between electronic imaging equipment and the human visual system is resolution. In the human eye the maximum resolution available is in the area of highest concentration of the color sensors (see Figure 7.2.3). The area of highest concentration of cones, called the *fovea* and located at the 0° point on the retina, is a circular area with a diameter of about 1.5 millimeters (mm). The density of cones in this area is about 150,000 per square millimeter, so we can estimate the fovea has about 300,000 elements. A CCD imaging chip of medium resolution can have about this many elements in an array of 5 mm by 5 mm, so current electronic sensors have a resolution capability similar to the human eye.

The eye is the input sensor for image information, but the optic nerve and the brain processes the signals. The *neural system model* is shown in Figure 7.2.5a, where a logarithmic response models the known response of the eye which is then multiplied

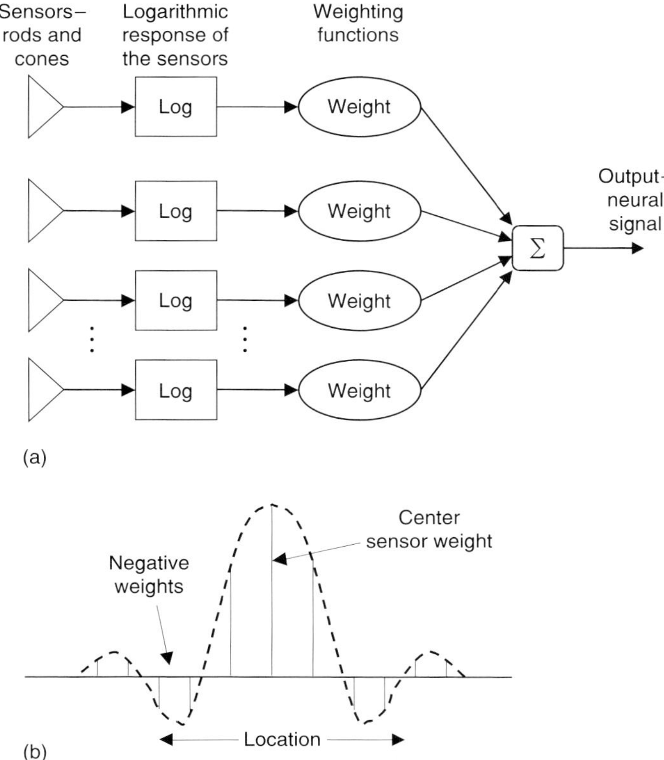

FIGURE 7.2.5
A model for the neural processing system. (a) System model. (b) Lateral inhibition weights.

by weighting factors and summed. With this model the weighting factors apply only to small neighborhoods, so they act like a convolution mask. The weighting factors can control effects such as *lateral inhibition*, which uses positive and negative factors for adjacent sensor weights (see Figure 7.2.5b) thus creating a high-pass filter effect which emphasizes edges. This phenomenon is essential for our visual perception since edges often contain valuable information; such as where one object ends and another begins.

7.2.2 Spatial Frequency Resolution

One of the most important aspects of the human visual system is spatial frequency resolution. How much fine detail can we see? How many pixels do we need on a video screen or in a camera's sensor? In order to understand the concept of spatial frequency resolution we first need to define exactly what we mean by resolution. Resolution has to do with the ability to separate two adjacent pixels—if we can see two adjacent pixels as being separate, then we can resolve the two. If the two appear as one and cannot be seen as separate, then we cannot resolve the two. The concept of resolution is closely tied to the concept of spatial frequency, as discussed in Chapter 5 and illustrated in Figure 7.2.6.

In Figure 7.2.6a we use a square wave to illustrate the concept of spatial frequency resolution, where *spatial frequency* refers to how rapidly the brightness signal is changing in space, and the signal has two values for the brightness—0 and maximum. If we use this signal for one line (row) of an image, and then repeat the line down the entire image, we get an image of vertical stripes, as in Figure 7.2.6b. If we increase this frequency,

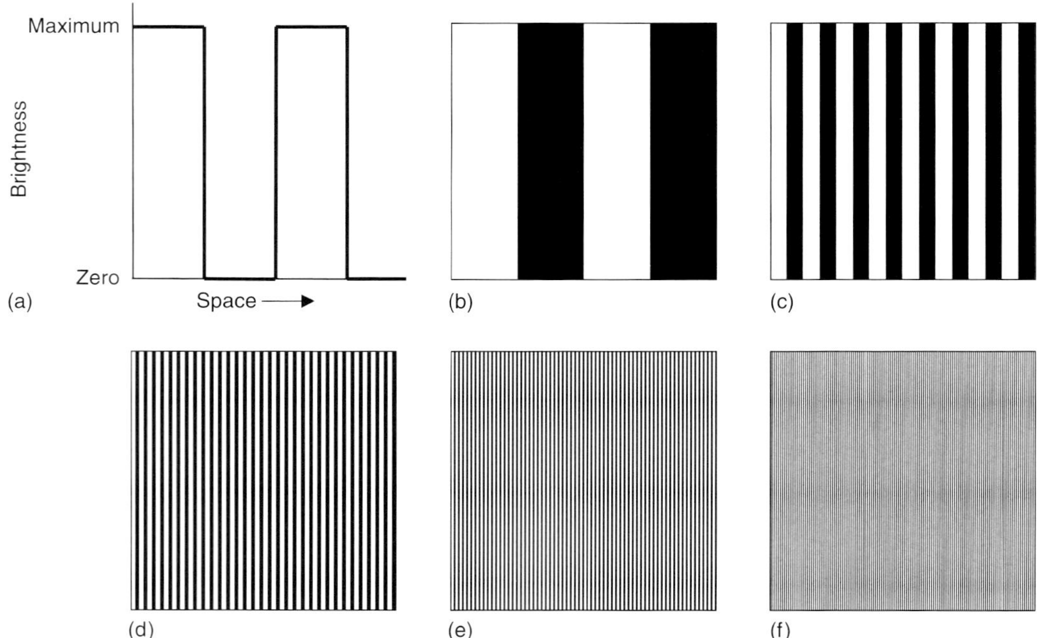

FIGURE 7.2.6
Resolution and spatial frequency. The following images are originally 256 × 256 pixels. (a) One-dimensional square wave, the brightness values correspond to one row of the following images. (b) Low frequency, 2 cycles per image. (c) Higher frequency, 8 cycles per image, (d) 32 cycles per image. (e) At 64 cycles we begin to see the difficult in resolving the lines. (f) Much higher frequencies are difficult to resolve, 128 cycles.

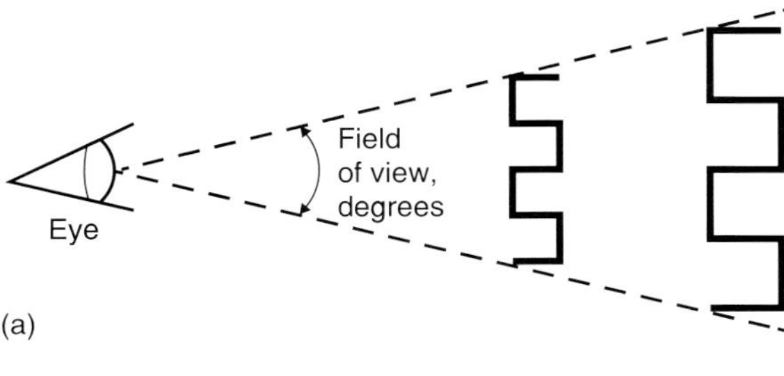

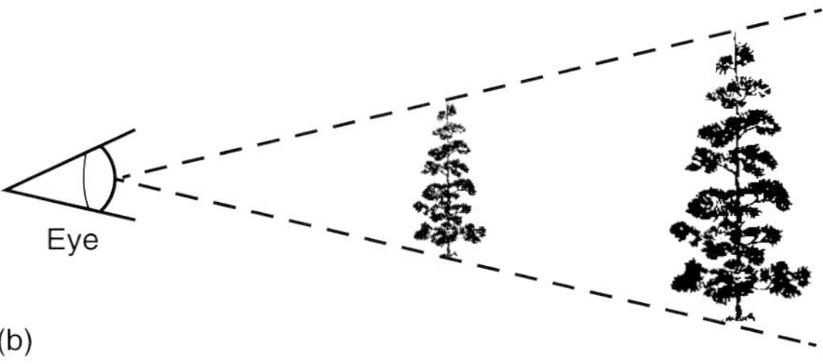

FIGURE 7.2.7
Cycles per degree. (a) With a fixed field of view of a given number of cycles, the farther from the eye, the larger each cycle must be. (b) A larger, more distant object can appear to be the same size as a smaller, closer object.

the stripes get closer and closer together (Figure 7.2.6c), until they start to blend together as in Figure 7.2.6e,f. (Remember we are discussing the resolution of the visual system, not the display device. Here we are assuming that the display device has enough resolution to separate the lines; the resolution of the display device must also be considered in many applications.)

By looking at Figure 7.2.6e,f and moving it away from our eyes, we can see that the spatial frequency concept must include the distance from the viewer to the object as part of the definition. With a typical television image, we cannot resolve the individual pixels unless we get very close, so the distance from the object is important when defining resolution and spatial frequency. We can eliminate the necessity to include distance by defining spatial frequency in terms of *cycles-per-degree*, which provides us with a relative measure. A cycle is one complete change in the signal; for example, in the square wave it corresponds to one high point and one low point, thus we need at least two pixels for a cycle. When we use cycles-per-degree, the "per degree" refers to the field of view—the width of your thumb held at arms length is about one degree, and television sets are typically designed for fields of view of about 5 to 15 degrees.

The cycles-per-degree concept is illustrated in Figure 7.2.7a, where as we get farther away from the eye, the same spatial frequency (in cycles per degree) must have larger cycles. In other words, as in Figure 7.2.7b, in order for a larger object to appear the same size, it must be farther away. This definition decouples the distance of the observer from consideration, and provides a metric for measuring the spatial resolution of the human visual system.

The physical mechanisms that affect the spatial frequency response of the visual system are both optical and neural. We are limited in spatial resolution by the physical size of the image sensors, the rods and cones; we cannot resolve things smaller than the individual sensor. The primary optical limitations are caused by the lens itself—it is of finite size, which limits the amount of light it can gather, and typically contains imperfections that cause distortion in our visual perception. Although gross imperfections can be corrected with lenses (glasses or contacts), subtle flaws cannot be corrected. Additionally, factors such the lens being slightly yellow (which progresses with age) limit the eye's response to various wavelengths of light.

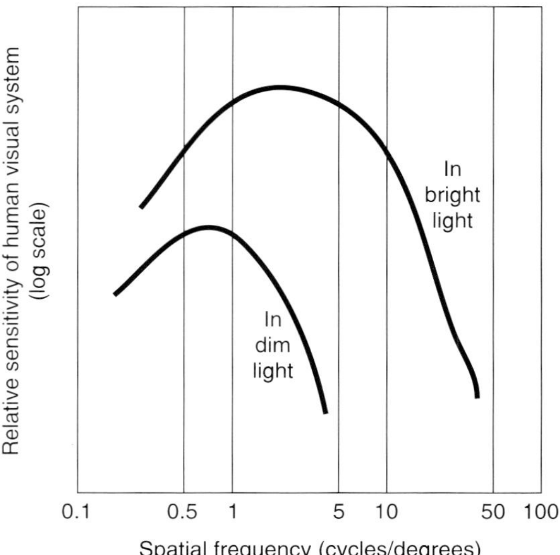

FIGURE 7.2.8
Spatial resolution.

The spatial resolution of the visual system has been empirically determined, and is plotted in Figure 7.2.8. The vertical axis, "relative sensitivity," is measured by adjusting the contrast required to see specific frequencies. Here we see that both low and high frequencies require more contrast (the eye is less sensitive) than the middle frequencies. Note that the spatial resolution is affected by the average (background) brightness of the display; the two plots correspond to different average brightness levels. In general, we have the ability to perceive higher spatial frequencies at brighter levels, but overall the *cutoff frequency* is about 50 cycles per degree, peaking at around 4 cycles per degree.

Example 7.2.3

The conventional format for a television display is a 4 : 3 aspect ratio, where 4 is the horizontal width compared to 3 for the vertical height. The standard is to assume the viewer sits six times the picture height from the screen. Use 53 μs (53×10^{-6} s) as the time it takes to scan one line (ignore blanking intervals), and a cutoff frequency of 50 cycles per degree. (a) What is the maximum frequency a video amplifier needs to pass to use the full resolution of the eye? (b) How many pixels per line are needed? (c) How many lines are needed in the display?

(a) First, find the angle the viewer will see in the horizontal direction:

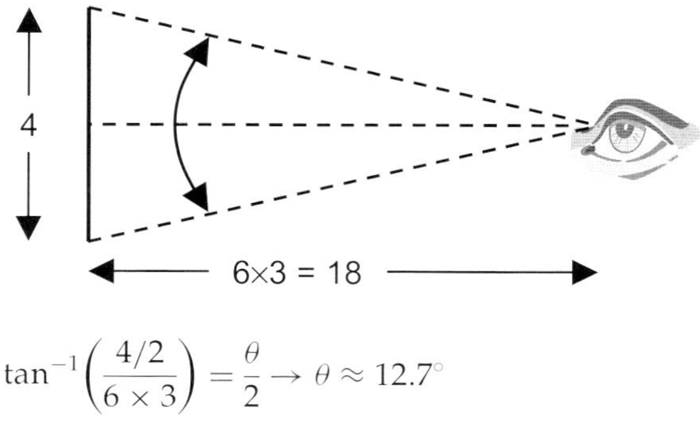

$$\tan^{-1}\left(\frac{4/2}{6 \times 3}\right) = \frac{\theta}{2} \rightarrow \theta \approx 12.7^{\circ}$$

Now, at 50 cycles per degree, the total number of cycles per line is:

$$12.7^\circ \times 50 \text{ cycles/degree} = 635 \text{ cycles}$$

To find maximum frequency, which are in units of cycles per second:

$$\frac{635 \text{ cycles}}{53 \times 10^{-6} \text{ seconds}} = 11{,}981{,}132 \approx 12 \text{ MHz bandwidth}$$

(b) Since we require at least two pixels per cycle:

$$635 \text{ cycles} \times 2 \text{ pixels/cycle} = 1270 \text{ pixels per line}$$

(c) To find the number of lines we need to find the angle in the vertical direction:

$$\tan^{-1}\left(\frac{3/2}{6 \times 3}\right) = \frac{\theta}{2} \to \theta \approx 9.5^\circ$$

Now, at 50 cycles per degree, the total number of cycles in the vertical direction:

$$9.5^\circ \times 50 \text{ cycles/degree} = 475 \text{ cycles}$$

Since we require at least two pixels per cycle:

$$475 \text{ cycles} \times 2 \text{ pixels/cycle} = 950 \text{ lines}$$

7.2.3 Brightness Adaptation

The vision system responds to a wide range of brightness levels. The response actually varies based on the average brightness observed, and is limited by the *dark threshold* and the *glare limit*. Light intensities below the dark threshold or above the glare limit are either too dark to see, or blinding. We cannot see across the entire range at any one time, but our system will adapt to existing light conditions. This is largely a result of the pupil in the eye, which acts as a diaphragm on the lens by controlling the amount it is open or closed, and thus controls the amount of light that can enter.

As we have seen, due to the function of the eye's sensors, subjective brightness is a logarithmic function of the light intensity incident on the eye. Figure 7.2.9 is a plot of the range and variation of the system. The vertical axis shows the entire range of subjective brightness over which the system responds, and the horizontal corresponds to the measured brightness. The horizontal axis is actually the log of the light intensity, so this results in an approximately linear response. A typical response curve for a specific lighting condition can be seen in the smaller curve plotted; any brightness levels below this curve will be seen as black. This small curve can be extended to higher levels (above the main curve), but if the lighting conditions change, the entire small curve will simply move upward.

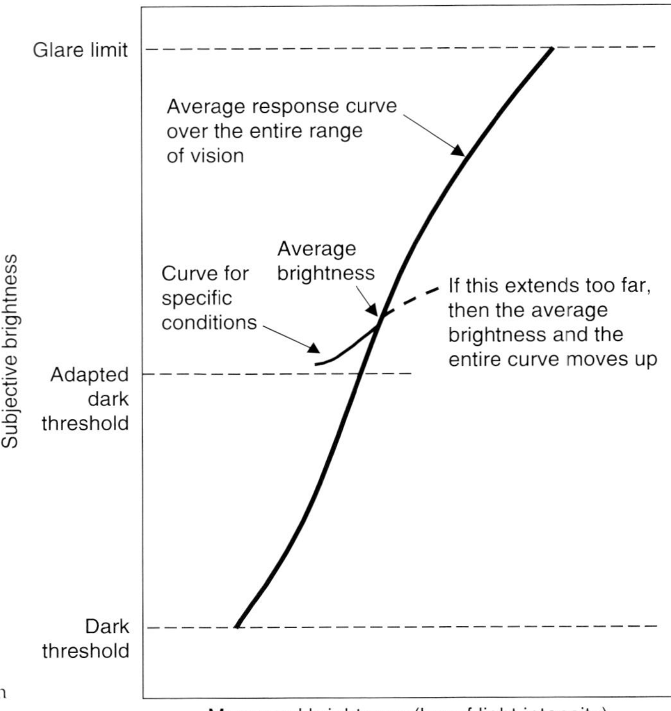

FIGURE 7.2.9
Brightness adaptation in the human visual system.

In images we observe many brightness levels, and the vision system can adapt to a wide range, as we have seen. However, it has been experimentally determined that we can only detect about 20 changes in brightness in a small area within a complex image. But, for an entire image, due to the brightness adaptation that our vision system exhibits, it has been determined that about 100 different gray levels are necessary to create a realistic image. For 100 gray levels in a digital image, we need at least 7 bits per pixel ($2^7 = 128$). If fewer gray levels are used, we observe *false contours* (bogus lines) resulting from gradually changing light intensity not being accurately represented, as in Figure 7.2.10.

7.2.4 Temporal Resolution

The *temporal resolution* of the human visual system deals with how we respond to visual information as a function of time. This is most useful when considering video and motion in images, where time can simply be thought of as an added dimension. In other words, temporal resolution deals with frame rates, or how slow can we update an image on a video screen or in a motion picture (movie) without seeing flashing between individual frames? Although we deal primarily with two-dimensional (row and column) stationary images in this book, a basic understanding of the temporal response of the human visual system is necessary to have a complete overview of human vision.

In Figure 7.2.11 we see a plot of temporal contrast sensitivity on the vertical axis, versus frequency (time-based frequency, not spatial frequency) on the horizontal. This is a plot of what is known as flicker sensitivity. *Flicker sensitivity* refers to our

(a)

(b)

FIGURE 7.2.10
False contouring. (a) Original image at 8 bits/pixel for 256 possible gray levels. (b) False contours can be seen by using only 3 bits/pixel for 8 possible gray levels, some sample false contour lines are marked by arrows.

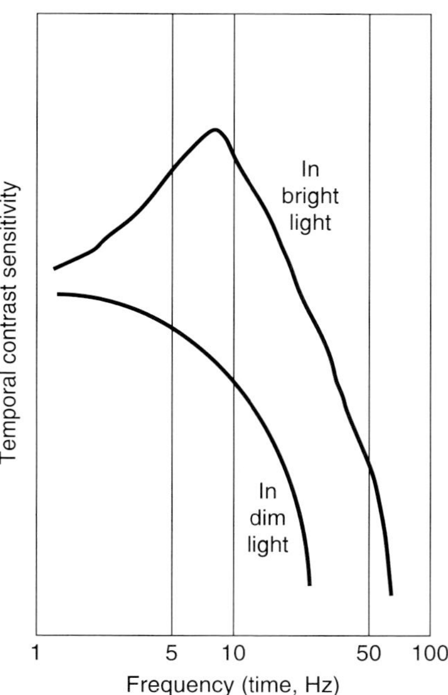

FIGURE 7.2.11
Temporal resolution.

ability to observe a flicker (flashing) in a video signal displayed on a monitor. The variables here are brightness and frequency. Two plots are shown to illustrate the variation in response due to image brightness. Here, as with the spatial frequency response, the brighter the display, the more sensitive we are to changes. It should be noted that the changes are relative, that is, a percentage change based on the average brightness.

The primary point here is that the human visual system has a *temporal cutoff frequency* of about 50 Hz (cycles per second), so we will not observe flicker if the image is refreshed at that rate or faster. Current video standards meet this criterion, and any new ones developed will need to meet or exceed it. The NTSC television standard used in the United States has a *field* rate of 60 fields per second, and the European standard, PAL, has a *field* rate of 50 fields per second. Note these television standards use interlaced video, to conserve bandwidth for transmission, while most computer monitors use noninterlaced video and operate at about 72–75 *frames* per second. A field rate of 60 per second provides a frame rate of 30 per second. According to our temporal cutoff frequency of 50 this should cause visible flicker. Why then does our television image not visibly flicker?—due to a phenomenon of the display elements called persistence. *Persistence* means the display elements will continue to emit light energy while the next alternating field is being displayed. So, even though a field is only one-half of the frame (every other line), the effective frame rate is approximately equal to the field rate.

7.2.5 Perception and Illusion

To fully understand our ability for visual perception, the current biological system model is limited. Our ability to see and to perceive visually involves more than simply applying the current physical model of the vision system to the arrangement of elements in the image. Some phenomena have been observed that are caused by the physical limitations of the visual system, such as spatial frequency resolution and brightness adaptation, while others are less well understood. Perception involves the brain as a processing unit, and how it functions is not fully known.

We saw that the neural system exhibits an effect called lateral inhibition (see Figure 7.2.5) that emphasizes edges in our visual field. One important visual phenomenon that can be at least partially attributed to lateral inhibition is called the *Mach Band effect*. This effect creates an optical illusion, as can be seen in Figure 7.2.12. Here we observe that when there is a sudden change in intensity, our vision system response overshoots the edge, thus creating a scalloped effect. This phenomenon has the effect of accentuating edges, and helps us to distinguish, and separate, objects within an image. This ability, combined with our brightness adaptation response, allows us to see outlines even in dimly lit areas.

Another phenomenon that shows that the perceived brightness of the human visual system is more than just a function of image brightness values is called *simultaneous contrast*. This means that the perceived brightness depends not only on the brightness levels, but also on the brightness levels of adjacent areas. Figure 7.2.13 illustrates this, where the center circle is the same gray level in all the images, but each has a different background gray level.

Other visual phenomena, commonly called *optical illusions*, are created when the brain completes missing spatial information or misinterprets objects' attributes. Similar to the simultaneous contrast phenomenon, the illusions occur as a result of the particular arrangement of the objects in the image. Figure 7.2.14 shows some of these illusions. These illusions emphasize the concept that visual perception depends not simply on individual objects, but also on the background and on how the objects are arranged. In other words, like most things in life, context has meaning and perception is relative, not absolute.

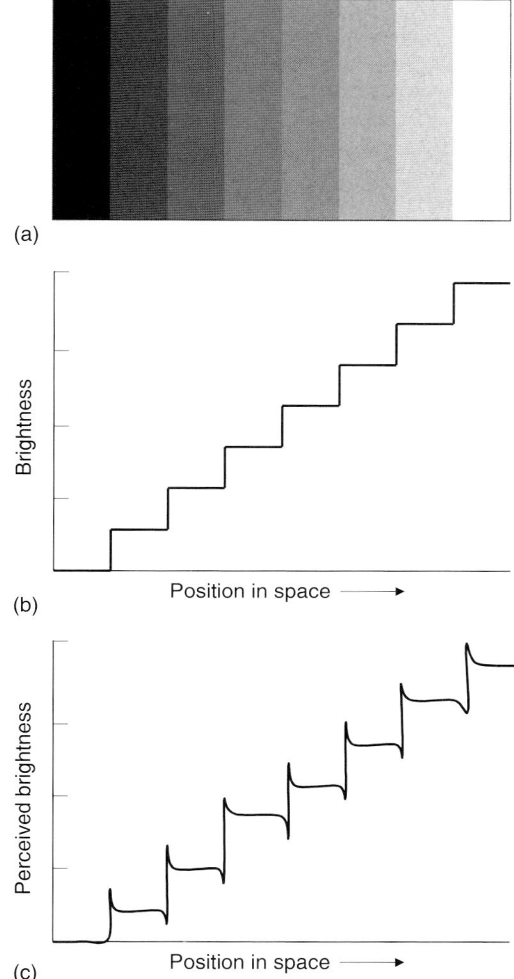

FIGURE 7.2.12
Mach band effect. (a) Image with uniformly distributed gray levels. (b) Actual brightness values. (c) Perceived brightness values due to the Mach band effect which causes overshoot at the edges, creating a scalloped effect. (*Note*: the amount of overshoot is exaggerated for clarity.)

7.3 Image Fidelity Criteria

To determine exactly what information is important, and to be able to measure image quality, we need to define image fidelity criteria. The information required is application specific, so the imaging specialist needs to be knowledgeable of the various types and approaches to measuring image quality.

Fidelity criteria can be divided into two classes: (1) objective fidelity criteria, and (2) subjective fidelity criteria. The *objective fidelity criteria* are borrowed from digital signal processing and information theory, and provide us with equations that can be used to measure the amount of error in a processed image by comparison to a known image. We will refer to the processed image as a reconstructed image—typically, one that can be created from a compressed data file or by using a restoration method. Thus, these measures are only applicable if an original or standard image is available for comparison. *Subjective fidelity criteria* require the definition of a qualitative scale to assess image quality. This scale can then be used by human test subjects to determine image fidelity.

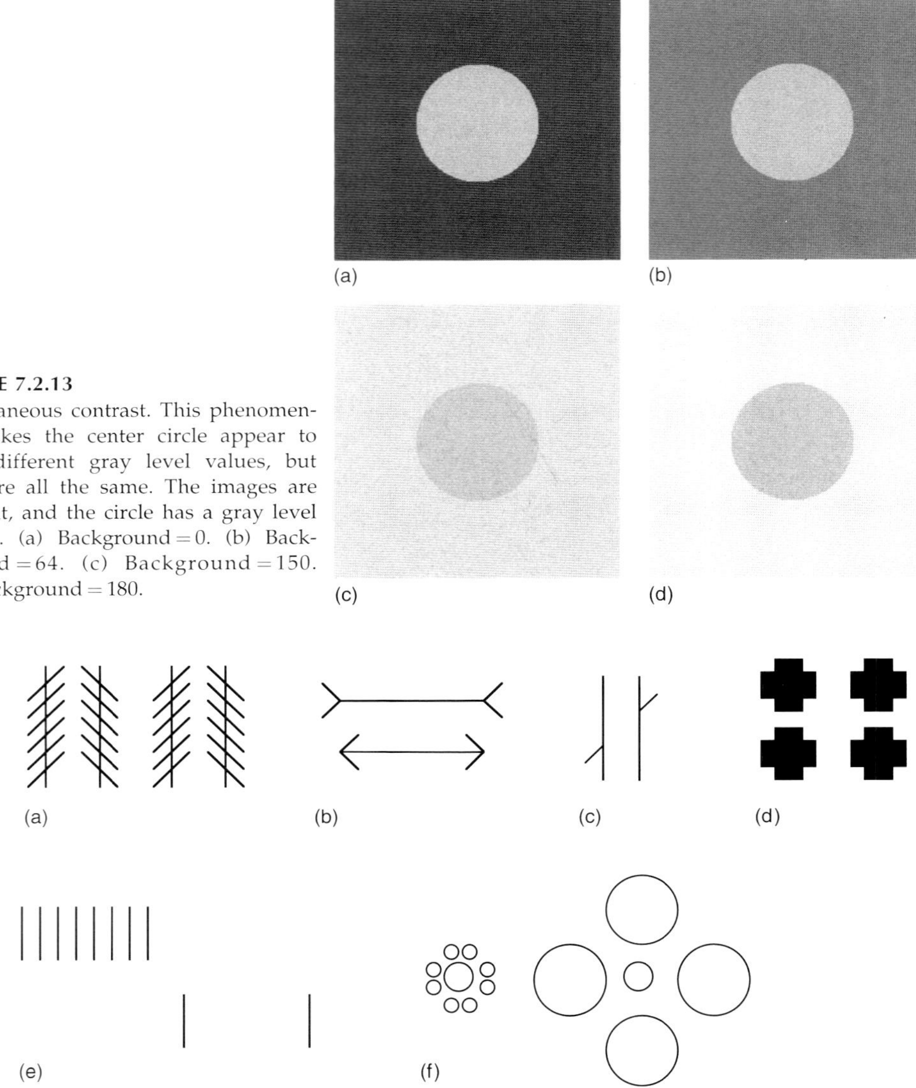

FIGURE 7.2.13
Simultaneous contrast. This phenomenon makes the center circle appear to have different gray level values, but they are all the same. The images are all 8-bit, and the circle has a gray level of 127. (a) Background = 0. (b) Background = 64. (c) Background = 150. (d) Background = 180.

FIGURE 7.2.14
Optical illusions. (a) Even though the vertical lines are parallel, they appear tilted. (b) The top line appears longer than the bottom one. (c) The two diagonal line segments appear not to be collinear. (d) Is this four black crosses, or connected white rectangles? (e) The outer two lines in the upper group appear to be farther apart than the two lines in the lower group. (f) The two center circles are the same size, but the one surrounded by larger circles appears smaller.

In order to provide unbiased results, evaluation with subjective measures requires careful selection of the test subjects and carefully designed evaluation experiments.

7.3.1 Objective Fidelity Measures

The objective criteria, although widely used, are not necessarily correlated with our perception of image quality. For instance, an image with a low error as determined by an

FIGURE 7.3.1
Objective fidelity measures are not always correlated with visual perception. The peak signal-to-noise ratio SNR_{PEAK} of an image with itself is theoretically infinite, so a high SNR_{PEAK} implies a good image and a lower SNR_{PEAK} implies an inferior image. (a) Original image. (b) Image of gaussian noise, the SNR_{PEAK} of this image with the original is 11.17. (c) Original image after edge detection and thresholding, $SNR_{PEAK} = 5.15$. (d) Original image after edge detection and contrast enhancement, $SNR_{PEAK} = 8.81$. With this measure, image (b) is better than (c) or (d) to represent the original!

objective measure may actually look much worse than an image with a high error metric. Consider Figure 7.3.1, according to an objective error metric image (b) better represents image (a) than images (c) and (d). However, the objective measures can be useful as relative measures in comparing differing versions of the same image. Figure 7.3.2 illustrates this by showing a series of four images which get progressively worse and have an objective measure that decreases accordingly.

Commonly used objective measures are the root-mean-square error, e_{RMS}, the root-mean-square signal-to-noise ratio, SNR_{RMS}, and the peak signal-to-noise ratio, SNR_{PEAK}. To understand these measures we first define the error between an original or standard pixel value, and the reconstructed pixel value as:

$$\text{error}(r, c) = \hat{I}(r, c) - I(r, c)$$

where

$$I(r, c) = \text{the original or standard image}$$

$$\hat{I}(r, c) = \text{the reconstructed image}$$

FIGURE 7.3.2
Objective fidelity measures can be useful. (a) Original image. We can see as the image gets visually worse, the peak SNR decreases, as expected. (b) Original image quantized to 16 gray levels using IGS. The peak SNR of it and original image is 35.01. (c) Original image with gaussian noise added with a variance of 200 and mean 0. Peak SNR of it and original image is 28.14. (d) Original image with gaussian noise added with a variance of 800 and mean 0. Peak SNR of it and original image is 22.73.

Next, we can define the total error in an $N \times N$ reconstructed image as:

$$\text{Total error} = \sum_{r=0}^{N-1} \sum_{c=0}^{N-1} \left[\hat{I}(r,c) - I(r,c)\right]$$

Typically, we do not want the positive and negative errors to cancel each other out, so we square the individual pixel error. The *root-mean-square error* is found by taking the square root ("root") of the error squared ("square") divided by the total number of pixels in the image ("mean"):

$$e_{\text{RMS}} = \sqrt{\frac{1}{N^2} \sum_{r=0}^{N-1} \sum_{c=0}^{N-1} \left[\hat{I}(r,c) - I(r,c)\right]^2}$$

If we consider the reconstructed image, $\hat{I}(r,c)$, to be the "signal," and the error to be "noise," we can define the *root-mean-square signal-to-noise ratio* as:

$$\text{SNR}_{\text{RMS}} = \sqrt{\frac{\sum_{r=0}^{N-1} \sum_{c=0}^{N-1} \left[\hat{I}(r,c)\right]^2}{\sum_{r=0}^{N-1} \sum_{c=0}^{N-1} \left[\hat{I}(r,c) - I(r,c)\right]^2}}$$

Another related metric, the *peak signal-to-noise ratio*, is defined as:

$$\mathrm{SNR}_{\mathrm{PEAK}} = 10\log_{10}\frac{(L-1)^2}{\frac{1}{N^2}\sum_{r=0}^{N-1}\sum_{c=0}^{N-1}\left[\hat{I}(r,c)-I(r,c)\right]^2}$$

where $L =$ the number of gray levels (e.g., for 8-bits $L = 256$).

These objective measures are often used in the research because they are easy to generate and seemingly unbiased, but remember that these metrics are not necessarily correlated to our perception of an image (see Figure 7.3.1).

7.3.2 Subjective Fidelity Measures

The subjective measures are better than the objective measures for image evaluation, if the goal is to achieve high-quality images as defined by our visual perception. To generate a subjective score for an image, or set of images, requires designing and performing experiments in which a group of people evaluate the images. The methodology used for subjective testing includes creating a database of images to be tested, gathering a group of people that are representative of the desired population, and then having all the test subjects evaluate the images according to a predefined scoring criterion. The results are then analyzed statistically, typically using the averages and standard deviations as metrics.

Subjective fidelity measures can be classified into three categories. The first type are referred to as *impairment tests*, where the test subjects score the images in terms of how bad they are. The second type are *quality tests*, where the test subjects rate the images in terms of how good they are. The third type are called *comparison tests*, where the images are evaluated on a side-by-side basis. The comparison type tests are considered to provide the most useful results, as they provide a relative measure, which is the easiest metric for most people to determine. Impairment and quality tests require an absolute measure, which is more difficult to determine in an unbiased fashion. In Table 7.1, are examples of internationally accepted scoring scales for these three types of subjective fidelity measures.

In the design of experiments that measure subjective fidelity care must be taken so that the experiments are reliable, robust and repeatable. The specific conditions must be carefully defined and controlled. The following exemplify the items that need specified: (1) the scoring scale to be used—it is best to have scoring sheets designed so the test subjects can easily rate the images with a minimum of distraction, (2) the display to be used, including the brightness, contrast, etc., settings, (3) the resolution setting for the display, (4) the lighting conditions in the room, (5) the distance the test

TABLE 7.1

Subjective Fidelity Scoring Scales

Impairment	Quality	Comparison
5—Imperceptible	A—Excellent	+2 much better
4—Perceptible, not annoying	B—Good	+1 better
3—Somewhat annoying	C—Fair	0 the same
2—Severely annoying	D—Poor	−1 worse
1—Unusable	E—Bad	−2 much worse

subjects are from the display screen, (6) the amount of time the images are to be displayed, (7) the number of test subjects participating in the experiment, (8) the type of people performing the experiment, that is, are they "average people off the street," or experts in a specific field, (9) the metrics used for the results, for example averages and standard deviations of all the scores. The details for the experiment will depend on the application and additional parameters may be required that are specific to the application.

7.4 Key Points

OVERVIEW FOR DIGITAL IMAGE PROCESSING

- The output of image processing is an image that has been enhanced, restored or compressed
- The output image is meant to be viewed by people, as opposed to computer vision/image analysis where the output is some form of higher level representation for a computer or analysis
- Image enhancement and restoration both involve techniques to make a better image
- Image compression involves development of techniques to make smaller files while still retaining high quality images
- Metrics to measure image quality must be defined to compare improvement methods
- The human visual system must be understood to help determine what makes a better image

HUMAN VISUAL PERCEPTION

The Human Visual System

- Vision is our most powerful sense
- Vision enables us to gather information and learn through observation
- Vision allows us to interact with our environment without physical contact
- The human visual system (HVS) has two primary components: the eye and the brain
- The eye and brain are connected by the optic nerve
- The eye is the receiving sensor and the brain is the processor
- The HVS works as follows: (1) light energy is focused by the lens of the eye onto the sensors on the retina at the back of the eye, (2) these sensors respond to this light energy by an electro-chemical reaction that sends an electrical signal down the optic nerve to the brain, (3) the brain uses these nerve signals to create neurological patterns that we perceive as images
- Visible light falls in the range of about 380 to 780 nm, but it is standard to use 400 to 700 nm since the response outside of this range is minimal
- It is typical to divide this range into three spectral bands corresponding to red (600 to 700 nm), green (500 to 600 nm) and blue (400 to 500 nm)

The eye—the receiving sensor with two types of light energy receptors, photoreceptors, which use an electro-chemical process to covert light energy into electrical energy or neural signals
Rods—sensors in the eye that are used for *scotopic* (night) vision; see only brightness (not color); on the order of 100 million; low resolution due to multiple rods connected to each nerve
Cones—sensor in the eye used for *photopic* (daylight) vision; they see color; on the order of 10 million; high resolution due to a separate nerve for each cone; three types for red, green, and blue
Retina—area on the back of the eye where the rods and cones are located
Fovea—area of highest concentration of cones, about 300,000 in a circular area with a 1.5 mm diameter
Blind spot—place on the retina where the optic nerve connects
Tristimulus curves—response curves of the cones, wavelength of light versus relative gain, allow calculation of relative RGB values as follows:

$$\mathrm{R} = K \int_{400}^{700} \mathrm{R}(\lambda) b(\lambda)\, \mathrm{d}\lambda$$

$$\mathrm{G} = K \int_{400}^{700} \mathrm{G}(\lambda) b(\lambda)\, \mathrm{d}\lambda$$

$$\mathrm{B} = K \int_{400}^{700} \mathrm{B}(\lambda) b(\lambda)\, \mathrm{d}\lambda$$

Where $b(\lambda)$ is the incident photon flux (as in Chapter 2), $R(\lambda)$, $G(\lambda)$, and $B(\lambda)$ are the tristimulus curves, and K is a constant that is dependent on the sensor's area, the length of the time the signal is measured, and the units used.
Cameras can be designed to mimic the HVS response by using sensors that have response curves like the tristimulus curves
Metamer(s)—two colors with different spectral distributions that appear the same, so the RGB values are identical
Neural system model (Figure 7.2.5)—logarithmic response of the rods and cones followed by a multiplicative weighting factor and then summed
Lateral inhibition—emphasizes edges to the HVS by performing a highpass filter using negative weights on adjacent sensor values, causes the *Mach Band effect*

Spatial Frequency Resolution

- Our spatial frequency resolution determines the amount of detail we can see
- It can be used to determine pixel size and quantify specifications for image sensors and displays
- It has been experimentally measured by displaying various spatial frequencies and varying the contrast until the viewer can see the separation
- It has been experimentally determined that we have the most sensitivity to middle frequencies (see Figure 7.2.8)

Resolution—has to do with the ability to separate adjacent pixels, if we can see two adjacent pixels as being separate then we can resolve the two

Spatial frequency—how fast a brightness signal changes in space, in general dependent on distance from the viewer

Cycles per degree—a metric that allows us to decouple viewer distance from the concept of spatial frequency

Cutoff frequency—50 cycles per degree

Brightness Adaptation

- The human visual system (HVS) responds to a wide range of brightness levels
- The HVS adapts to existing lighting conditions, which allows us to see over a small part of the overall range at any one time (see Figure 7.2.9)
- The adaptation is due to the pupil which acts as a diaphragm on the lens by controlling the amount of light that can enter
- Subjective brightness is a logarithmic function of the incident light energy
- About 100 brightness levels are needed in images

Dark threshold—below this brightness level all appears black to the HVS

Glare limit—above this brightness value is blinding, all we see is white

Temporal Resolution

- The *temporal resolution* of the human visual system deals with how we respond to visual information as a function of time
- Applicable to video and motion in images, where time can simply be thought of as an added dimension
- Temporal resolution deals with frame rates, or how slow can we update an image on a video screen or in a motion picture (movie) without seeing flashing between individual frames

Flicker sensitivity—our ability to observe a flicker (flashing) in a video signal displayed on a monitor

Temporal cutoff frequency—the minimum field/frame rate needed to avoid visible flicker, about 50 hertz (cycles per second)

Perception and Illusion

- To fully understand our ability to perceive visually, the current biological system model is limited
- Some phenomena are caused by physical system limits, such as spatial frequency resolution and brightness adaptation
- Perception involves the brain, which is not fully understood
- Perception is relative, not absolute, and depends on context

Mach Band effect—an effect caused by the lateral inhibition process inherent in the visual neural system which emphasizes edges in images (Figure 7.2.12)

Simultaneous contrast—a phenomenon of the HVS that causes perceived brightness to be dependent on adjacent brightness (Figure 7.2.13)

Optical illusion—created when the brain completes missing spatial information or misinterprets objects' attributes (Figure 7.2.14)

IMAGE FIDELITY CRITERIA

- Image fidelity criteria are necessary to determine exactly what information is important and to measure image quality
- They can be divided into two classes, objective fidelity criteria and subjective criteria

Objective fidelity criteria—equations that can be used to measure the amount of error in a processed, or reconstructed, image by comparison to a known image

Reconstructed image—one that can be created from a compressed data file or by using a restoration method

Subjective fidelity criteria—require the definition of a qualitative scale to assess image quality which is used by human test subjects to determine image fidelity

Objective Fidelity Measures

- Objective fidelity measures are not necessarily correlated to human visual perception (see Figure 7.3.1)
- They can be useful as a relative measure in comparing differing versions of the same image (see Figure 7.3.2)

Total error:

$$\text{Total error} = \sum_{r=0}^{N-1}\sum_{c=0}^{N-1}\left[\hat{I}(r,c) - I(r,c)\right]$$

where $I(r,c)$ = the original or standard image and $\hat{I}(r,c)$ = the reconstructed image

Root-mean-square error:

$$e_{\text{RMS}} = \sqrt{\frac{1}{N^2}\sum_{r=0}^{N-1}\sum_{c=0}^{N-1}\left[\hat{I}(r,c) - I(r,c)\right]^2}$$

Root-mean-square signal-to-noise ratio:

$$\text{SNR}_{\text{RMS}} = \sqrt{\frac{\sum_{r=0}^{N-1}\sum_{c=0}^{N-1}\left[\hat{I}(r,c)\right]^2}{\sum_{r=0}^{N-1}\sum_{c=0}^{N-1}\left[\hat{I}(r,c) - I(r,c)\right]^2}}$$

Peak signal-to-noise ratio:

$$\text{SNR}_{\text{PEAK}} = 10\log_{10}\frac{(L-1)^2}{\frac{1}{N^2}\sum_{r=0}^{N-1}\sum_{c=0}^{N-1}\left[\hat{I}(r,c) - I(r,c)\right]^2}$$

where L = the number of gray levels

Subjective Fidelity Measures

- The subjective measures more accurately reflect our visual perception than objective measures
- To generate a subjective score for an image requires designing and performing experiments in which a group of people evaluate the images according to a predefined scoring criterion
- Experimental design requires careful definition and controls so that the experiments are reliable, robust and repeatable
- The results are then analyzed statistically, typically using the averages and standard deviations as metrics
- Three categories: (1) impairment tests, (2) quality tests, (3) comparison tests

Impairment tests—test subjects rate images in terms of how bad they are

Quality tests—test subjects rate images in terms of how good they are

Comparison tests—test subjects evaluate images on a side-by-side basis

Comparison tests provide the most useful results due to relative measures providing the most consistent results from human test subjects.

7.5 References and Further Reading

Books that integrate computer imaging topics with human vision include [Nixon/Aguado 01], [Deutsch 93], [Arbib/Hanson 90], [Levine 85], and [Marr 82]. For more on color in computing systems as related to human vision, see [Giorgianni/Madden 98] and [Durrett 87]. An interesting book that describes our visual perception by exploring visual disorders is [Farah 04]. A comprehensive treatment of color science can be found in [Wyszecki/Stiles 82]. For more details on mathematical models for vision see [Mallot 01], [Levine 85], and [Marr 82]. [Sid-Ahmed 95] contains more details relating the human visual system to television signal processing. To delve deeper into video images, [Jack 01] is a useful reference, and [Poynton 03] contains more information for high definition television (HDTV). More details and examples of optical illusions and other perception based phenomena are found in [Levine 85] and [Marr 82]. Image fidelity information can be found in [Gonzalez/Woods 02], [Watson 93], and [Golding 78].

Arbib, M.A. and Hanson, A.R., editors, *Vision, Brain and Cooperative Computation*, Cambridge, MA: MIT Press, 1990.

Deutsch, S. and Deutsch, A., *Understanding the Nervous System—An Engineering Perspective*, NY: IEEE Press, 1993.

Durrett, H.J., editor, *Color and the Computer*, San Diego, CA: Academic Press, 1987.

Farah, M.J., *Visual Agnosia*, Cambridge, MA: MIT Press, 2004.

Giorgianni, E.J., Madden, T.E., *Digital Color Management*, Reading, MA: Addison Wesley, 1998.

Golding, L.S., Quality Assessment of Digital Television Signals, *SMPTE Journal*, Volume 87, March 1978, pp. 153–157.

Gonzalez, R.C., Woods, R.E., *Digital Image Processing*, Upper Saddle River, NJ: Prentice Hall, 2002.

Jack, K., *Video Demystified*, LLH Publications, 2001.

Levine, M.D., *Vision in Man and Machine*, NY: McGraw Hill, 1985.

Mallot, H.A., *Computational Vision*, Cambridge, MA: MIT Press, 2001.
Marr, D., *Vision*, Freeman and Company, 1982.
Nixon, M.S., Aguado, A.S., *Feature Extraction and Image Processing*, Woburn, MA: Newnes/Butterworth-Heinemann, 2001.
Poynton, C., *Digital Video and HDTV*, Morgan Kaufman, 2003.
Sid-Ahmed, M.A., *Image Processing: Theory, Algorithms, and Architectures*, Englewood Cliffs, NJ: Prentice Hall, 1995.
Watson, A.B. (ed.), *Digital Images and Human Vision*, Cambridge, MA: MIT Press, 1993.
Wyszecki, G., and Stiles, W. S., *Color Science: Concepts and Methods, Quantitative Data and Formulae*, New York: Wiley, 1982.

7.6 Exercises

1. (a) What is the distinction between image processing and computer vision/image analysis? (b) What are the three main areas of image processing?
2. Explain why we need to understand human visual perception for image processing applications.
3. (a) What are the two primary components of the human visual system? (b) How are these two components connected?
4. What are the three sequential processes describing the way the human visual system works?
5. (a) What is the range of visible light wavelengths? (b) What are the types of imaging sensors in the eye?
6. (a) How do we see color? (b) What are the tristimulus curves? (c) What part of the eye has the most spatial resolution? Why?
7. Assume that the tristimulus curves are approximated by bandpass filters with a flat response with a gain of one, using the red, green, and blue bands as shown in Figure 7.2.2. Find the RGB values for the following incident photon flux: (a) $b(\lambda) = 10$, (b) $b(\lambda) = 5\lambda$.
8. Using the same assumptions as in Exercise 7, and given the following spectral distribution, find the RGB values:

$$b(\lambda) = \begin{cases} 15 & \text{for } 400 \le \lambda \le 475 \\ 5 & \text{for } 510 \le \lambda \le 580 \\ 10 & \text{for } 650 \le \lambda \le 700 \\ 0 & \text{elsewhere} \end{cases}$$

9. (a) How is it possible for two colors to have different spectral distributions, but appear the same? (b) What is this called?
10. What is lateral inhibition and why is it important for human vision?
11. (a) How is spatial frequency measured to decouple viewer distance from the equation? (b) In these terms, what is the standard spatial cutoff frequency for the HVS? (c) How does average brightness affect spatial resolution? Why?
12. (a) Sketch an image which shows a horizontal square wave of frequency 2. (b) Sketch an image which shows a vertical square wave of frequency 3. (c) Given an image of a square wave at 84 cycles per degree, describe how it appears.

13. An inventor wants to build a weather radar instrument for private aircraft. Rather than using a traditional radial sweep display, the system electronically warps spatial information so that it can be displayed in the form of a $6'' \times 6''$ raster pattern (square grid). What maximum spatial resolution would you suggest (a power of 2, for digital display reasons) if the pilot's eyes are about $28''$ from the screen?
14. (a) If a video display is being designed operating at a 50 hertz frame rate, and 1024×1024 pixels, what maximum frequency will the video amplifier have to pass in order to utilize the full resolving capability of the human eye? (assume a minimum of two pixels per cycle, and ignore blanking intervals). (b) How close can the viewer sit without seeing discrete dots?—find this distance, D, as a function of x, where x is the width of the (square) pixel, i.e., $D = f(x)$.
15. Use CVIPtools to explore the spatial frequency resolution of your vision system. Use *Utilities->Create*. (a) Create 256×256 images of vertical *square* waves of frequencies 16, 32, 64, and 128. Hold out your thumb at arm's length (this is about one degree) and back away from the computer screen until the thumb covers one of the 256×256 images completely. Which of the images can you see all the lines and in which images do they blend together? (b) Repeat part (a), but create horizontal square waves. Are the results the same?
16. Use CVIPtools to explore the spatial frequency resolution of your vision system. Use *Utilities->Create*. (a) Create 256×256 images of vertical *sine* waves of frequencies 25, 50, 64, and 100. Back away from the screen until one of the 256×256 images is in about one degree of your field of view. Which of the images can you see the sine waves clearly and in which images do they blend together? (b) Repeat part (a), but with horizontal sine waves. Are the results the same?
17. Use CVIPtools to explore the spatial frequency resolution of your vision system. Use *Utilities->Create*. Create 256×256 images of checkerboard patterns with squares of size 16×16, 8×8, 4×4 and 2×2. Back away from the screen until one of the 256×256 images is in about one degree of your field of view. Which of the images can you see the squares clearly and in which images do they blend together?
18. (a) Why is subjective brightness a logarithmic function of light intensity on the eye? (b) How does the eye adapt to various lighting conditions? (c) Can we see over our entire brightness range at any one time? Why or why not?
19. (a) About how many brightness levels are required to create a realistic image for the HVS? (b) How many bits do we need? (c) What happens if we do not use enough bits for the brightness values?
20. Use CVIPtools to explore the number of brightness levels required for the HVS. (a) Open a monochrome 8-bit per pixel image. (b) Use *Utilities->Convert->Gray Level Quantization*, using the *Standard* option, to create images with 128, 64, 32, 16, 8, 4, and 2 gray levels. (c) Put these images side by side. How many gray levels do you need to avoid image artifacts? d) Perform (a)–(c) with a variety of monochrome images.
21. Use CVIPtools to explore the number of brightness levels required for the HVS. (a) Open a color 24-bit per pixel image. (b) Use *Utilities->Convert->Gray Level Quantization*, using the *Standard* option, to create images with 128, 64, 32, 16, 8, 4, and 2 gray levels per color band. (c) Put these images side by side. How many gray levels per band do you need to avoid image artifacts? (d) Perform (a)–(c) with a variety of color images.

22. (a) What is flicker sensitivity? (b) What is the temporal cutoff frequency for the HVS? (c) How does the average brightness affect the temporal cutoff frequency?
23. (a) Name the phenomenon in the neural system that helps to create the Mach Band effect. (b) Explain how this mechanism works.
24. (a) What is simultaneous contrast? (b) What are optical illusions? (c) Sketch one of the optical illusions and create a theory to explain the phenomenon.
25. Given the following 4-bit per pixel, 4×4 images, calculate: (a) root-mean-square-error, (b) root-mean-square signal-to-noise ratio, and (c) peak signal-to-noise ratio.

$$\text{Original image}\begin{bmatrix} 10 & 10 & 8 & 7 \\ 7 & 7 & 8 & 7 \\ 6 & 6 & 5 & 7 \\ 12 & 12 & 13 & 14 \end{bmatrix} \quad \text{Reconstructed image}\begin{bmatrix} 12 & 12 & 7 & 7 \\ 8 & 8 & 8 & 8 \\ 6 & 6 & 6 & 6 \\ 12 & 12 & 12 & 12 \end{bmatrix}$$

26. (a) Name the three types of subjective fidelity measures. (b) Which type do you think is the best? Why?
27. A photo developing studio has a new image enhancement technique, but they are uncertain about how to set one of the parameters to achieve results that will please the customers the most. In preliminary testing they have determined that the parameter should be set to 1.5 or 2.2. Design an experiment to measure image quality using a subjective fidelity measure to help them. Be sure to specify all the details.
28. A cable television company is trying to improve their customer service. To do that they need to determine when the average customer will call for service if the video signal (image) is slowly degrading due to water in the line. Design an experiment to measure image quality using a subjective fidelity measure to help them. Be sure to specify all the details.

7.6.1 Programming Exercise: Spatial Resolution

1. Write a function to create images of square waves. Let the user specify the image size, the orientation and frequency of the waves.
2. Write a function to create images of sine waves. Let the user specify the image size, the orientation and frequency of the waves.
3. Write a function to create images of checkerboards. Let the user specify the image size, and the height and width of the rectangles.
4. Use these functions to experimentally determine the horizontal and vertical spatial frequency cutoff of the HVS.

7.6.2 Programming Exercise: Brightness Adaptation

1. Write a function that takes an 8-bit per pixel monochrome image and allows the user to specify the number of bits per pixel in the output image: 1, 2, 3, 4, 5, 6 or 7.
2. Write a function that takes a 24-bit per pixel color (three band) image and allows the user to specify the number of bits per pixel per band in the output image: 1, 2, 3, 4, 5, 6 or 7.
3. Select a variety of images and use these functions to experimentally determine the number of bits needed by the HVS to avoid false contours.

7.6.3 Programming Exercise: Optical Illusions

1. Write a function to create images of the optical illusions shown in Figure 7.2.14. If desired, use the CVIPtools library functions in the Geometry library, such as *create_line* or *create_circle,* along with any needed logic or arithmetic functions in the ArithLogic library.
2. Show the images to your friends and family. Do they all see the illusions the in same way?

8

Image Enhancement

8.1 Introduction and Overview

Image enhancement techniques are employed to emphasize, sharpen, and/or smooth image features for display and analysis. *Image enhancement* is the process of applying these techniques to facilitate the development of a solution to a computer imaging problem. Consequently, the enhancement methods are application-specific and are often developed empirically. Figure 8.1.1 illustrates the importance of the application by the feedback loop from the output image back to the start of the enhancement process, and models the experimental nature of the development. In this figure we define the enhanced image as $E(r, c)$. The range of applications includes using enhancement techniques as preprocessing steps to ease the next processing step or as postprocessing steps to improve the visual perception of a processed image, or image enhancement may be an end in itself. Enhancement methods operate in the spatial domain, manipulating the pixel data, or in the frequency domain, by modifying the spectral components (Figure 8.1.2). Some enhancement algorithms use both the spatial and frequency domains.

The type of techniques include *point operations*, where each pixel is modified according to a particular equation that is not dependent on other pixel values, *mask operations*, where each pixel is modified according to the values in a small neighborhood (subimage), or *global operations*, where all the pixel values in the image are taken into consideration. Spatial domain processing methods include all three types, but frequency domain operations, by nature of the frequency (and sequency) transforms, are global operations. Of course, frequency domain operations can become "mask operations," based only on a local neighborhood, by performing the transform on small image blocks instead of the entire image.

Enhancement is used as a preprocessing step in some computer vision applications to ease the vision task, for example, to enhance the edges of an object to facilitate guidance of a robotic gripper. Enhancement is also used as a preprocessing step in applications where human viewing of an image is required before further processing. For example, in one application, high-speed film images had to be correlated with a computer simulated model of an aircraft. This process was labor intensive because the high speed film generated many images per second and difficult due to the fact that the images were all dark. This task was made considerably easier by enhancing the images before correlating them to the model, enabling the technician to process many more images in one session.

Image enhancement is used for postprocessing to generate a visually desirable image. For instance, we may perform image restoration to eliminate image distortion and find that the output image has lost most of its contrast. Here, we can apply some basic image enhancement method to restore the image contrast. Alternatively, after a compressed

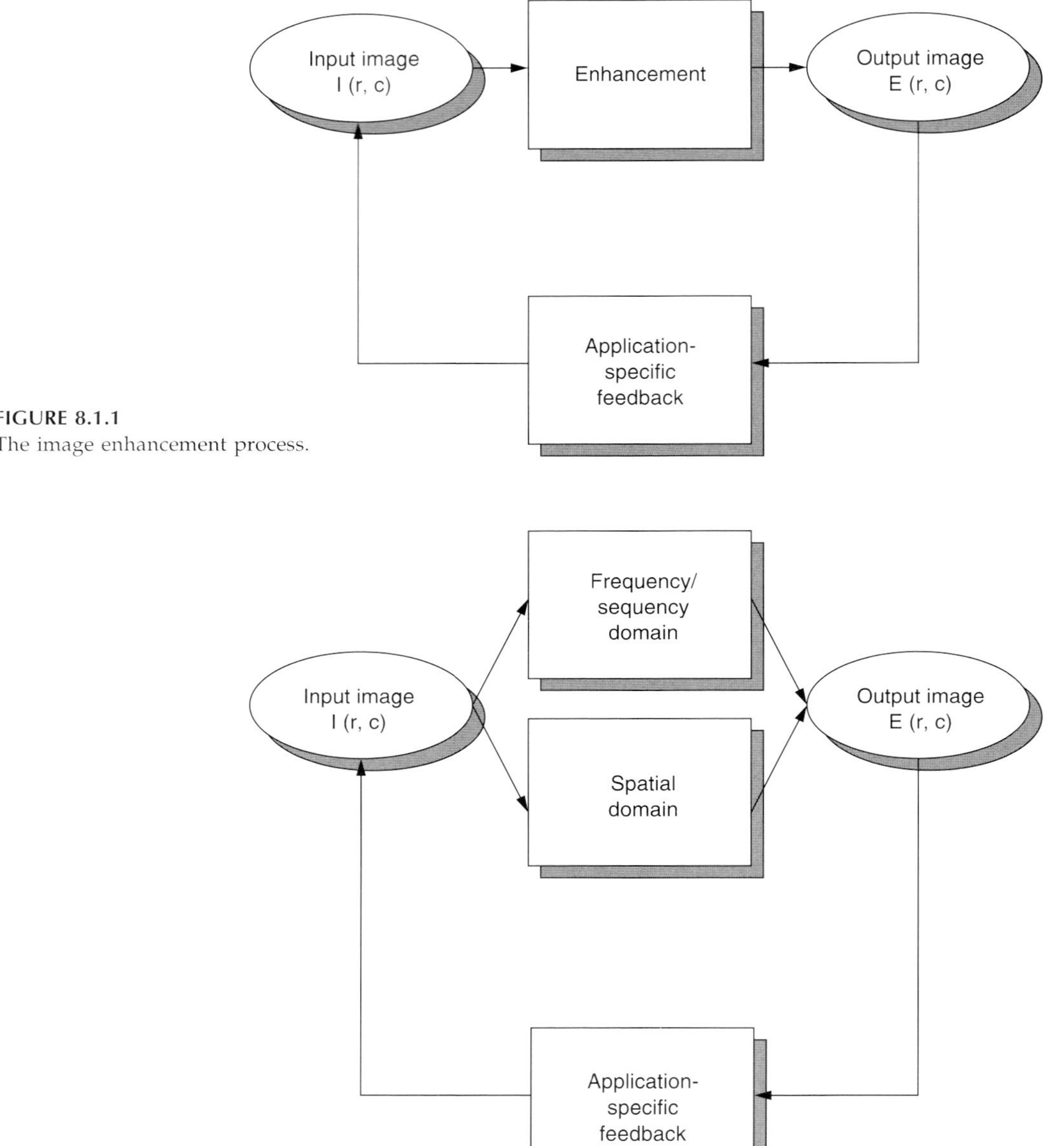

FIGURE 8.1.1
The image enhancement process.

FIGURE 8.1.2
Image enhancement.

image has been restored to its "original" state (decompressed), some post-processing enhancement may significantly improve the look of the image. For example, a block-based transform compression algorithm may generate an image with undesirable "blocky" artifacts, and post-processing it with a smoothing filter (lowpass or mean) will improve the appearance.

Overall, image enhancement methods are used to make images look better. What works for one application, may not be suitable for another application, so the development of enhancement methods require problem domain knowledge, as well as image enhancement expertise. Assessment of the success of an image enhancement algorithm is often "in the eye of the beholder," so image enhancement is as much an art as it is a science.

8.2 Gray Scale Modification

Gray scale modification, also called gray level scaling or gray level transformation, is the process of taking the original gray level values and changing them to improve the image. Typically, this relates to improving image contrast and brightness. Image contrast is a measure of the distribution and range of the gray levels—the difference between the brightest and darkest pixel values and how the intermediate values are arranged. Image brightness usually refers to the overall average, or mean, pixel value in the image. Depending on the application we may need to increase or decrease contrast, brightness or both.

8.2.1 Mapping Equations

One method to modify the gray levels in an image is by the use of a mapping equation. The *mapping equation* changes the pixel's (gray level) values based on a mathematical function that uses brightness values as input. The outputs of the equation are the enhanced pixel values. The mapping equation is typically, but not necessarily, linear; nonlinear equations can be modeled by piecewise linear models. The use of mapping equations to modify the gray scale belongs in the category of point operations, and typical applications include contrast enhancement and feature enhancement. The notation used for the mapping equation is as follows:

$$\text{Mapping Equation} \rightarrow E(r,c) = M[I(r,c)] \quad \text{where } M[\ ] \text{ is the mapping equation}$$

The primary operations applied to the gray scale of an image are to compress or stretch it. We typically compress gray level ranges that are of little interest to us, and stretch the gray level ranges where we desire more information. This is illustrated in Figure 8.2.1a, where the original image data is shown on the horizontal axis, and the modified values are shown on the vertical axis. The linear equations corresponding to the lines shown on the graph represent the mapping equations. If the slope of the line is between zero and one, this is called *gray level compression*, while if the slope is greater than one it is called *gray level stretching*. In Figure 8.2.1a, the range of gray level values from 28 to 75 are stretched, while the other gray values are left alone. The original and modified images are shown in Figure 8.2.1b,c, where we can see that stretching this range exposed previously hidden visual information. The following example shows how to find the mapping equation shown in Figure 8.2.1.

Example 8.2.1

For the ranges 0 to 28 and 75 to 255 the input equals the output. For the range 28 to 75, we want to stretch the range from 28 to 255. To do this we need a linear equation. If we use the standard form $y = mx + b$, where m is the slope, and b is the y-intercept, we can find the equation as follows (note that in this case y corresponds to the mapping equation $M[\ ]$, and x is the input image gray level (brightness) values $I(r,c)$):

1. We know two points on the line, (28, 28) and (75, 255), so:

$$m = \frac{y_1 - y_2}{x_1 - x_2} = \frac{255 - 28}{75 - 28} = \frac{227}{47} \approx 4.83$$

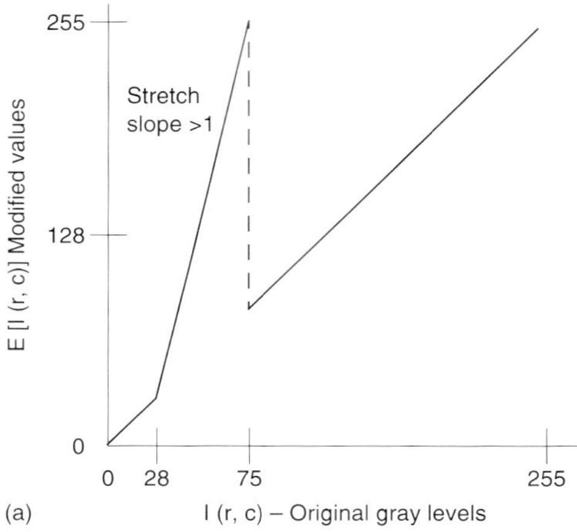

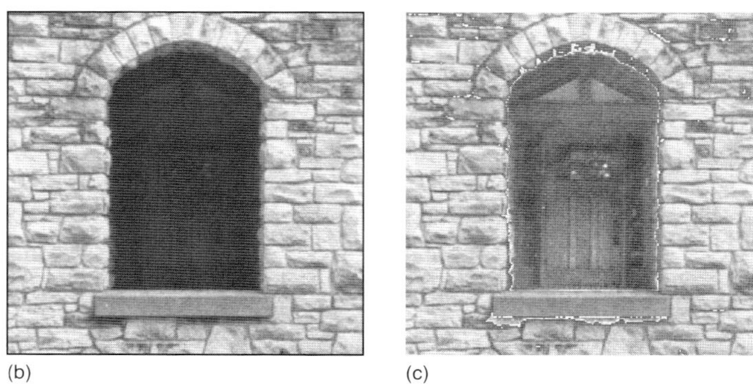

FIGURE 8.2.1
Gray scale modification. (a) Gray level stretching. (b) Original image. (c) Image after modification.

2. $y = 4.83x + b$. Putting in a point to solve for the y-intercept, b:

$$255 = 4.83(75) + b$$
$$b = -107.25$$

3. So the equation of the line for the range between 28 and 75 is:

$$M[I(r,c)] = 4.83[I(r,c)] - 107.25$$

4. Therefore:

$$M[I(r,c)] = \begin{cases} I(r,c) & \text{for } 0 < I(r,c) < 28 \\ (4.83[I(r,c)] - 107.25) & \text{for } 28 \leq I(r,c) \leq 75 \\ I(r,c) & \text{for } 75 < I(r,c) < 255 \end{cases}$$

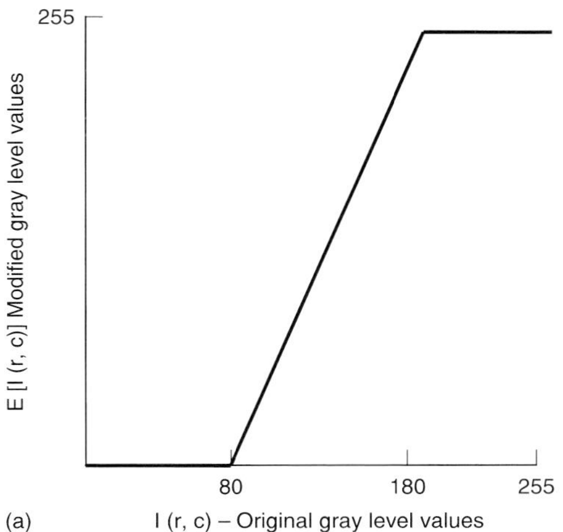

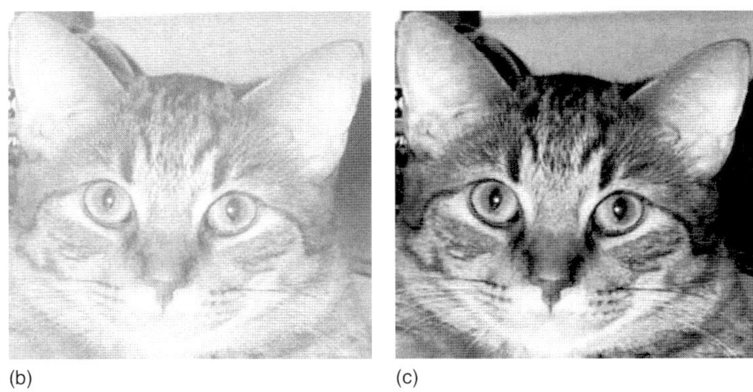

FIGURE 8.2.2
Gray-level stretching with clipping at both ends. (a) The mapping equation. (b) The original image. (c) The modified image with the stretch gray levels.

In some cases we may want to stretch a specific range of gray levels, while clipping the values at the low and high ends. Figure 8.2.2a illustrates a linear function to stretch the gray levels between 80 and 180, while clipping any values below 80 and any values above 180. The original and modified images are shown in Figure 8.2.2b,c, where we see the resulting enhanced image.

A *digital negative* can be created with a mapping equation as follows:

$$M[I(r,c)] = \mathrm{MAX} - I(r,c)$$

Where MAX is the maximum gray value. This is the equivalent of performing a logical NOT on the input image. This process of complementing an image can be useful as an enhancement technique. Because the eye responds logarithmically to brightness changes, details characterized by small brightness changes in the bright regions may not be visible. Complementing the image converts these small deviations in the bright regions to the dark regions, where they may be easier to detect. Partial complementing of an image also produces potentially useful results. An example would be to leave the lower half of the gray scale untouched while complementing the upper half. Dark regions in the original image are unaffected while bright regions are complemented.

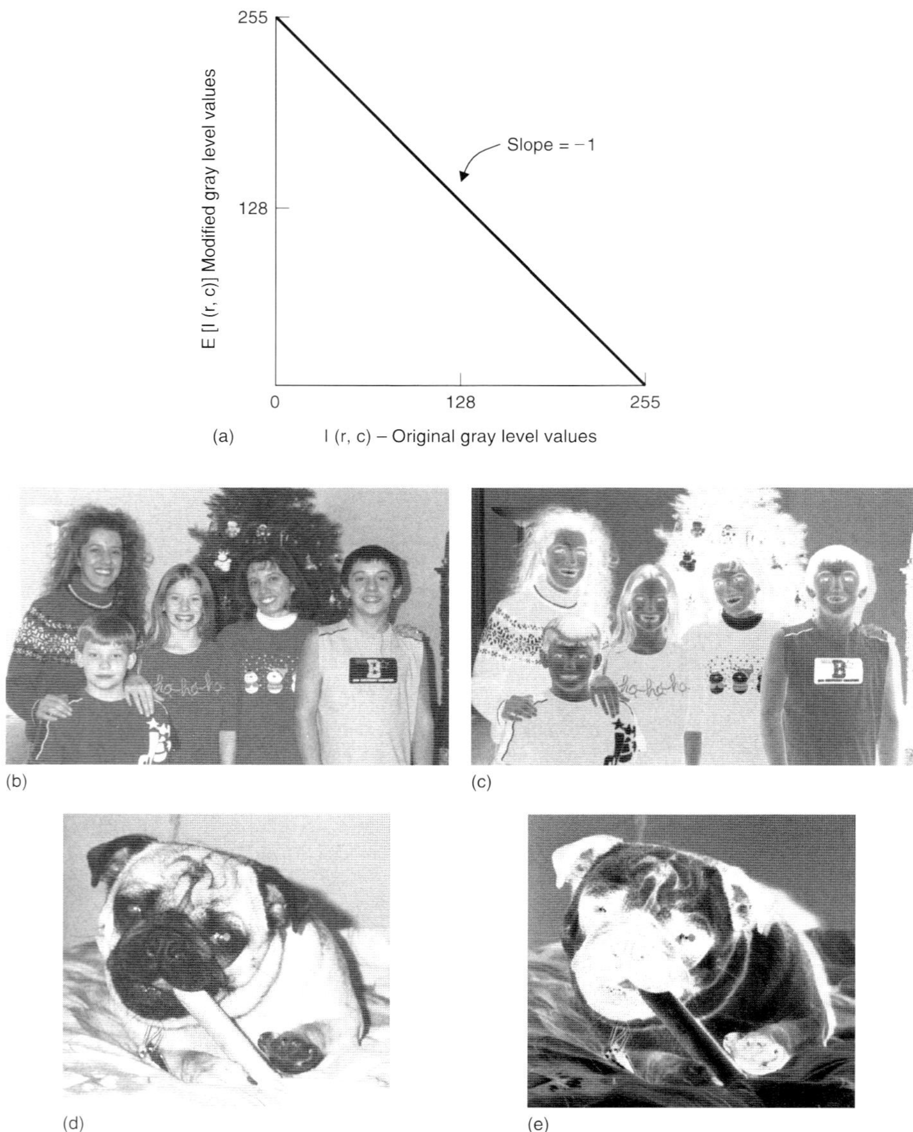

FIGURE 8.2.3
Digital negative. (a) Mapping equation. (b) Original image. (c) Negative image modified by the inverse mapping equation which is the equivalent of a logical NOT.

Figure 8.2.3a shows the mapping equation for creating an image complement, also called the inverse mapping equation, and Figure 8.2.3b,c and Figure 8.2.3d,e show examples of original images and their negatives. In some cases details will become more visible in the negative image; for example, the highlights in the faces in images b & c are more obvious in the negative. Also, in images, Figure 8.2.3d,e, the reflections in the dog's eyes are more apparent in the negative image (where they appear as dark spots), as well as details in the fur on the upper part of the dog's head.

Another type of mapping equation, used for feature extraction, is called *intensity level slicing*. Here we are selecting specific gray level values of interest, and mapping them to a specified (typically high/bright) value. For example, we may have an application where it has been empirically determined that the objects of interest are in the gray level range of 150 to 200. Using the mapping equations illustrated in Figure 8.2.4, we can generate the resultant images shown. The first operation shows a one-to-one mapping that outputs the input image unchanged. The next two illustrate intensity level slicing where a specific gray value range is "sliced" out to be highlighted. With this type of operation we can either leave the "background" gray level values unchanged (Figure 8.2.4c,d), or map them to black (Figure 8.2.4e,f). Note that they do not need to be turned black; any gray level value may be specified.

CVIPtools can be used to perform gray scale modification by using the *Enhancement-> Histograms* window, and selecting *Linear Modification* (see Figure 8.2.5). Currently, it allows for the modification of one contiguous range of gray values by specifying the *Start* and *End* value for the input range to be modified, and the *Initial value* and the *slope* of the mapping equation. The *Initial value* is the value that the *Start* value gets mapped to by the mapping equation. CVIPtools also has selections to *keep out of range data* or *set out of range data to 0*.

Example 8.2.2

To use CVIPtools to implement the modification equation for the digital negative shown in Figure 8.2.3a:

1. *Start* = 0
2. *End* = 255
3. *Initial value* = 255
4. *Slope* = −1
5. Since there is no out of range data, this selection is irrelevant, so click *APPLY*.

Example 8.2.3

To use CVIPtools to implement *intensity level slicing* by using the modification equation shown in Figure 8.2.4c:

1. *Start* = 150
2. *End* = 200
3. *Initial value* = 255
4. *Slope* = 0
5. Select *keep out of range data* and click *APPLY*.

Example 8.2.4

To use CVIPtools to implement *intensity level slicing* by using the modification equation shown in Figure 8.2.4e:

1. *Start* = 150
2. *End* = 200
3. *Initial value* = 255
4. *Slope* = 0
5. Select *set of out range data to 0* and click *APPLY*.

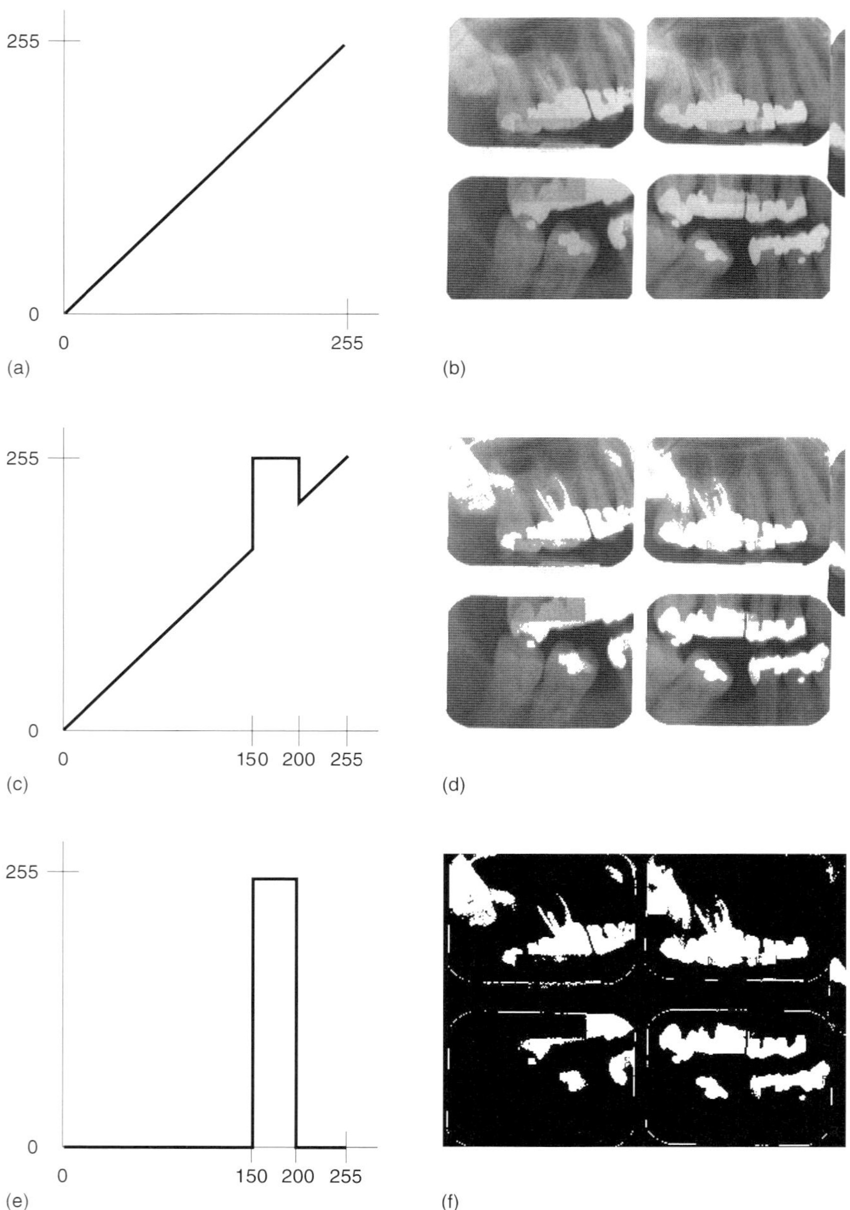

FIGURE 8.2.4
Intensity level slicing. (a) This operation returns the original gray levels. (b) Original image. (c) This operation intensifies the desired gray level range while not changing the other values. (d) Image sliced to emphasize gray values from 150 to 200; background unchanged. (e) This operation intensifies the desired gray level range while changing the other values to black. (f) Image sliced to emphasize gray values from 150 to 200; background changed to black.

To realize a piece-wise linear modification equation on various gray level ranges in the same image, simply perform each linear piece on the original image and select *set out of range data to 0* each time. Next, use a logical OR on all the output images to create the final image. An example of this for a two piece linear equation is shown in Figure 8.2.6, where the first piece of the equation stretches the gray levels from 0 to 80,

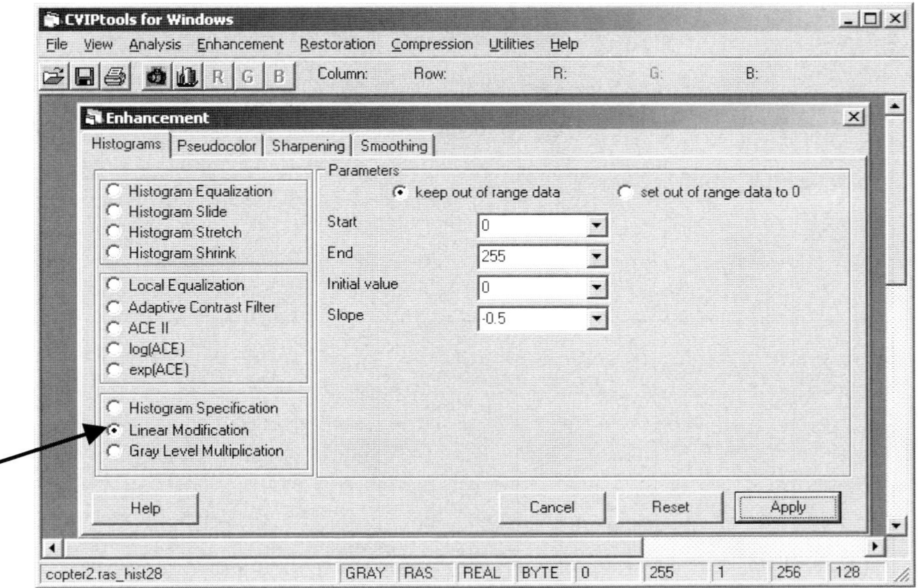

FIGURE 8.2.5
Gray scale modification with CVIPtools. Use the *Enhancement* window and select the *Histograms* tab. Next, select *Linear Modification* and set the parameters as desired.

and the second piece compresses the gray levels from 80 to 255. The step-by-step process for this is shown here:

Example 8.2.5

STEP 1: For the first piece:

1. *Start* = 0
2. *End* = 80
3. *Initial value* = 0
4. We can use the two endpoints on the line to find the slope (0, 0) and (80, 180), so:

$$\text{slope} = \frac{y_1 - y_2}{x_1 - x_2} = \frac{180 - 0}{80 - 0} = 2.25$$

5. Select *set out of range data to 0.*

STEP 2: For the second piece:

1. *Start* = 80
2. *End* = 255
3. *Initial value* = 180

4. We can use the two endpoints on the line to find the slope (80, 180) and (255, 255), so:

$$\text{slope} = \frac{y_1 - y_2}{x_1 - x_2} = \frac{255 - 180}{255 - 80} = \frac{75}{175} \approx 0.4286$$

5. Select *set out of range data to 0* and click *APPLY.*

STEP 3:

1. Select the output images from STEP 1 and STEP 2 and perform a logical OR.

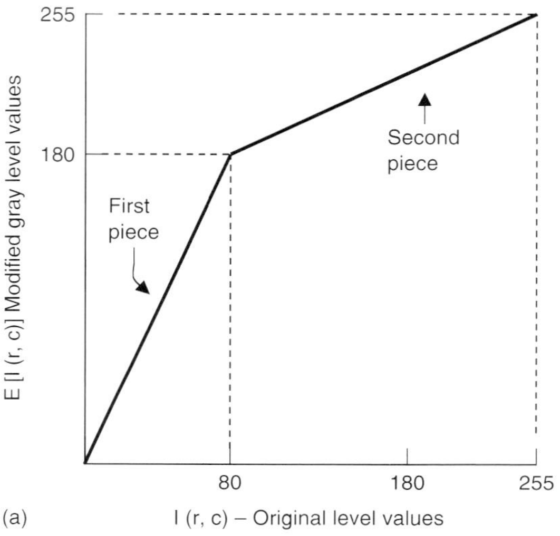

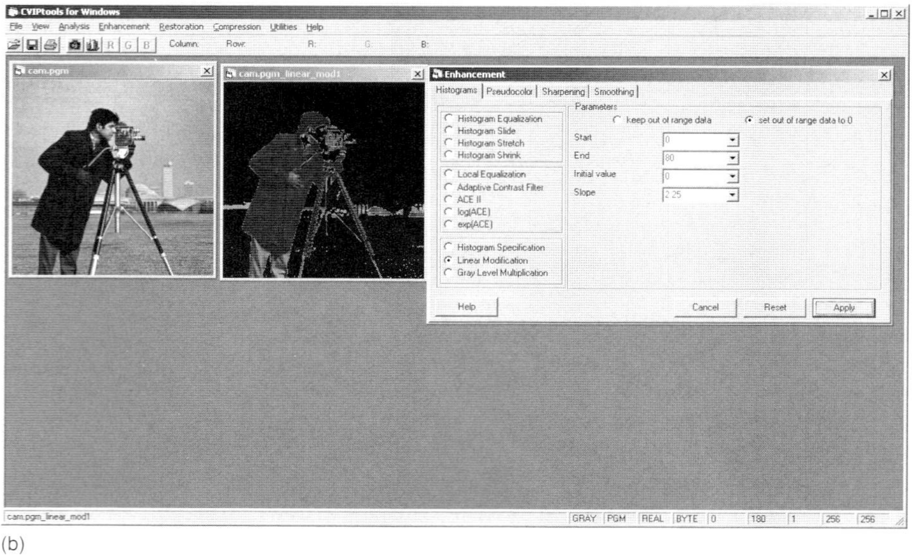

FIGURE 8.2.6
Performing a piece-wise linear modification with CIVPtools. (a) The mapping equation. (b) Screen shot of CVIPtools with the parameter set for the first piece, along with the input and output images.

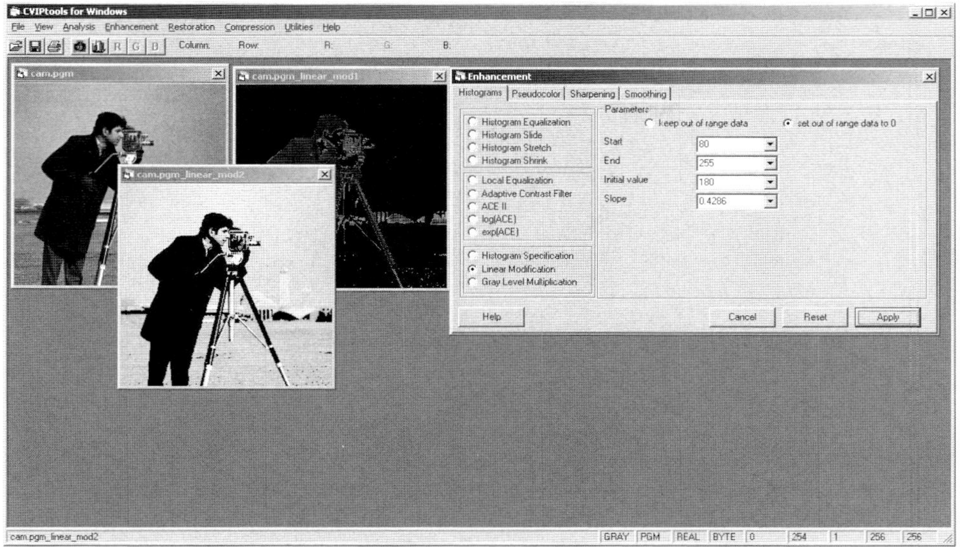

(c)

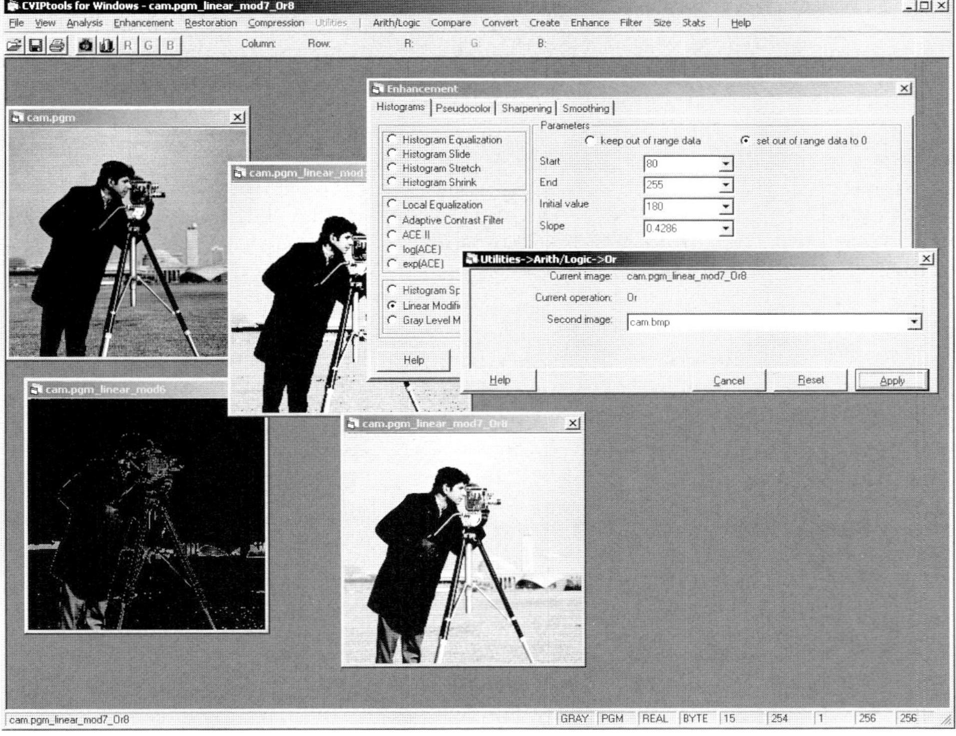

(d)

FIGURE 8.2.6 (Continued)
Performing a piece-wise linear modification with CIVPtools. (c) Screen shot of CVIPtools with parameters set for second piece, along with the input and output images. (d) Screen shot of CVIPtools with parameters set for the logical OR of the results from the first two linear modifications to create the final image.

FIGURE 8.2.7
Images from piece-wise linear modification. (a) Original image. (b) Output image after first piece modification. (c) Output image after second piece modification. (d) Final output image after (b) and (c) are OR'd together. Notice the improved detail in dark areas, such as the coat, glove, and pants, due to the gray level streteching of the low values.

Figure 8.2.7 shows the images in more detail from the piece-wise linear modification example. Here we see the improved detail in dark areas, such as the coat, glove and pants, due to the gray level stretching in to 0 to 80 range.

One of the commonly used nonlinear transforms is the logarithmic function that we used to display spectral images (see Section 5.2.4). This function is useful when the dynamic range of the input data is very large, and is also referred to as *range compression*. Another useful nonlinear transform is the *power-law transform*, where the mapping equations are of the following form:

$$E(r,c) = M[I(r,c)] = K_1[I(r,c)]^{\gamma} \quad \text{where } K_1 \text{ and } \gamma \text{ are positive constants}$$

Imaging equipment, such as cameras, displays and printers typically react according to a power-law equation. This means that their response is not linear; for values of γ greater than 1 images will appear darker, and for values less than 1 images will appear lighter. If the response function of the device is given by the above power-law transform equation, then it can be compensated for by application of a *gamma-correction* equation of the following form:

$$E(r,c) = M[I(r,c)] = K_2[I(r,c)]^{1/\gamma} \quad \text{where } K_2 \text{ and } \gamma \text{ are positive constants}$$

Gamma correction is important for proper display of images, whether on a computer monitor or on a printed page. To use CVIPtools for gamma correction, use the *histogram specification* selection, which is explored in the following section.

8.2.2 Histogram Modification

An alternate perspective to gray scale modification that performs a similar function is referred to as histogram modification. This approach will also lead to the use of a mapping equation, but instead of simply considering the gray levels, the histogram shape and range is the focus.

As previously discussed, the *gray level histogram* of an image is the distribution of the gray levels in an image. In general, a histogram with a small spread has low contrast, and a histogram with a wide spread has high contrast, while an image with its histogram clustered at the low end of the range is dark, and a histogram with the values clustered at the high end of the range corresponds to a bright image (as was shown in Figure 6.2.5). Examination of the histogram is one of the most useful tools for image enhancement, as it contains information about the gray level distribution in the image and makes it easy to see the modifications that may improve an image.

The histogram can be modified by a mapping function, which will stretch, shrink (compress), or slide the histogram. Histogram stretching and histogram shrinking are forms of gray scale modification, sometimes referred to as *histogram scaling*. In Figure 8.2.8 we see a graphical representation of histogram stretch, shrink and slide.

The mapping function for a histogram stretch can be found by the following equation:

$$\text{Stretch}\,(I(r,c)) = \left[\frac{I(r,c) - I(r,c)_{\text{MIN}}}{I(r,c)_{\text{MAX}} - I(r,c)_{\text{MIN}}}\right][\text{MAX} - \text{MIN}] + \text{MIN}$$

where $I(r,c)_{\text{MAX}}$ is the largest gray level value in the image $I(r,c)$, $I(r,c)_{\text{MIN}}$ is the smallest gray level value in $I(r,c)$ and MAX and MIN correspond to the maximum and minimum gray level values possible (for an 8-bit image these are 0 and 255).

This equation will take an image and stretch the histogram across the entire gray level range, which has the effect of increasing the contrast of a low contrast image (see Figure 8.2.9). If a stretch is desired over a smaller range, different MAX and MIN values can be specified. If most of the pixel values in an image fall within a small range, but a few outliers force the histogram to span the entire range, a pure histogram stretch will not improve the image. In this case it is useful to allow a small percentage of the pixel values to be clipped at the low and high end of the range (for an 8-bit image this means truncating at 0 and 255). Figure 8.2.10 shows an example of this where we see a definite improvement with the stretched and clipped histogram compared to the pure histogram stretch.

The opposite of a histogram stretch is a histogram shrink, which will decrease image contrast by compressing the gray levels. The mapping function for a histogram shrink can be found by the following equation:

$$\text{Shrink}\,(I(r,c)) = \left[\frac{\text{Shrink}_{\text{MAX}} - \text{Shrink}_{\text{MIN}}}{I(r,c)_{\text{MAX}} - I(r,c)_{\text{MIN}}}\right]\left[I(r,c) - I(r,c)_{\text{MIN}}\right] + \text{Shrink}_{\text{MIN}}$$

where $I(r,c)_{\text{MAX}}$ is the largest gray level value in the image $I(r,c)$, $I(r,c)_{\text{MIN}}$ is the smallest gray level value in $I(r,c)$, and $\text{Shrink}_{\text{MAX}}$ and $\text{Shrink}_{\text{MIN}}$ correspond to the maximum and minimum desired in the compressed histogram.

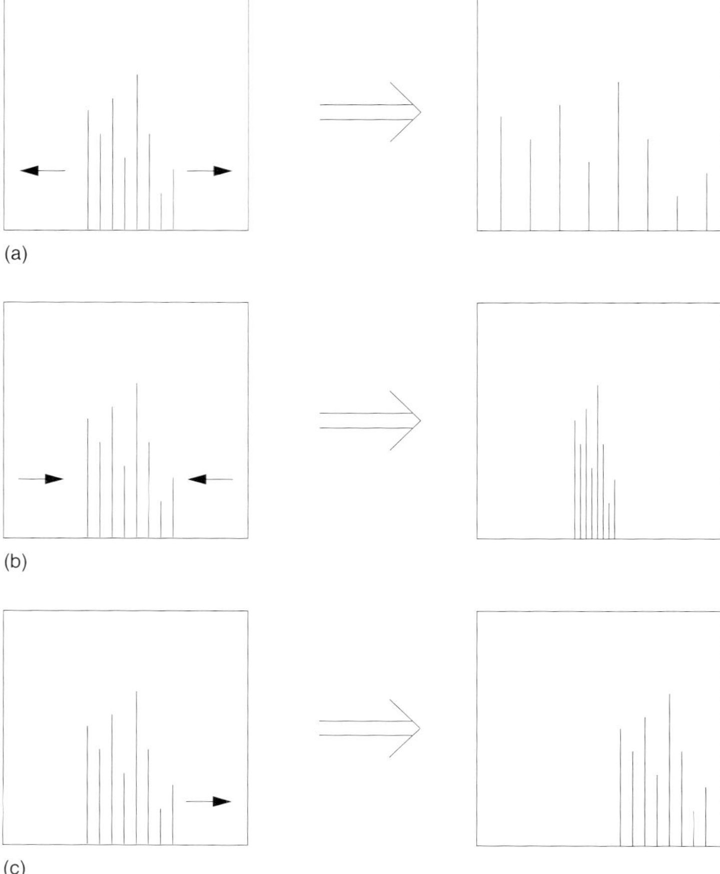

FIGURE 8.2.8
Histogram modification. (a) Histogram stretch. (b) Histogram shrink. (c) Histogram slide.

Figure 8.2.11 illustrates a histogram shrink procedure. In Figure 8.2.11a,b we see an original image and its histogram, and Figure 8.2.11c,d show the result of the histogram shrink. In general, this process produces an image of reduced contrast, and may not seem to be useful as an image enhancement tool. However, we will see (Section 8.3) an image sharpening algorithm (unsharp masking) that uses the histogram shrink process as part of an enhancement technique.

The histogram slide technique can be used to make an image either darker or lighter, but retain the relationship between gray level values. This can be accomplished by simply adding or subtracting a fixed number from all of the gray level values, as follows:

$$\text{Slide}\,(I(r,c)) = I(r,c) + \text{OFFSET}$$

where the OFFSET value is the amount to slide the histogram.

In this equation we assume that any values slid past the minimum and maximum values will be clipped to the respective minimum or maximum. A positive OFFSET value will increase the overall brightness, while a negative OFFSET will create a darker

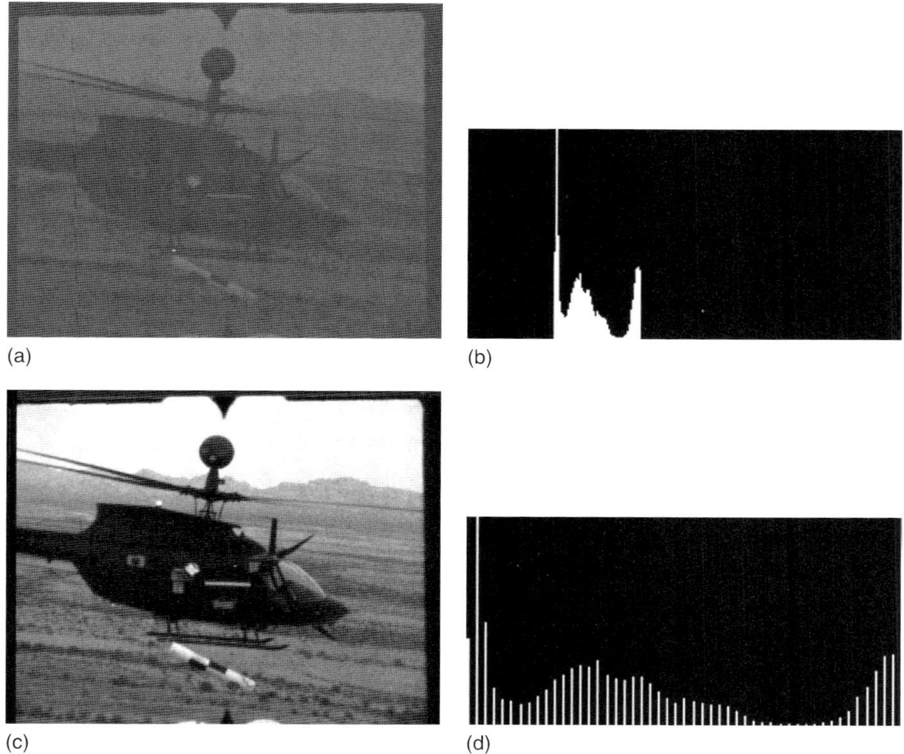

FIGURE 8.2.9
Histogram stretching. (a) Low-contrast image. (b) Histogram of image (a). Notice the tight cluster. (c) Image (a) after histogram stretch. (d) Histogram of image after stretch.

image. Figure 8.2.12a shows image that has been brightened by a histogram slide with a positive OFFSET value, and Figure 8.2.12c shows an image darkened by a negative OFFSET value.

Figure 8.2.13 shows the CVIPtools *Enhancement* window with the *Histograms* tab selected. In this figure the histogram stretch operation is selected, which displays the parameters for the operation in the right side of the window. Note that histogram slide or shrink can also be selected with the radiobuttons on the left, which will popup the appropriate parameters in the *Parameters* box on the right. After the operation is selected, and the parameters are set as desired, clicking on the APPLY button will perform the operation.

Histogram equalization is an effective technique for improving the appearance of a poor image. Its function is similar to that of a histogram stretch but often provides more visually pleasing results across a wider range of images. *Histogram equalization* is a technique where the histogram of the resultant image is as flat as possible (with histogram stretching the overall shape of the histogram remains the same). The theoretical basis for histogram equalization involves probability theory, where we treat the histogram as the probability distribution of the gray levels. This is reasonable, since the histogram is the distribution of the gray levels for a particular image.

The histogram equalization process for digital images consists of four steps: (1) find the running sum of the histogram values, (2) normalize the values from step (1) by dividing by the total number of pixels, (3) multiply the values from step (2) by the maximum gray level value and round, and (4) map the gray level values to the results

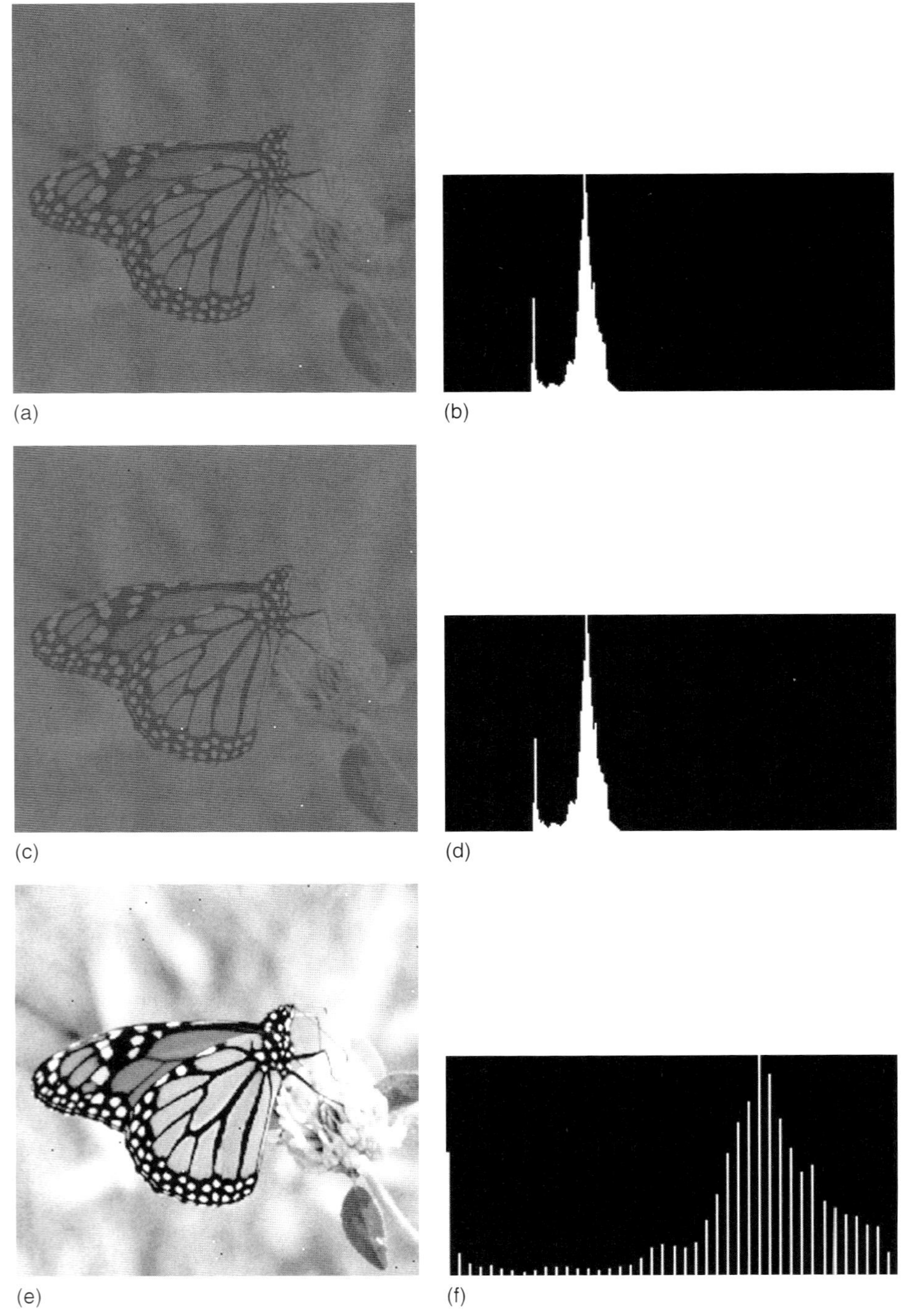

FIGURE 8.2.10
Histogram stretching with clipping. (a) Original image. (b) Histogram of original image. (c) Image after histogram stretching with out clipping. (d) Histogram of image (c). (e) Image after histogram stretching with clipping 1% of the values at the high and low ends. (f) Histogram of image (e).

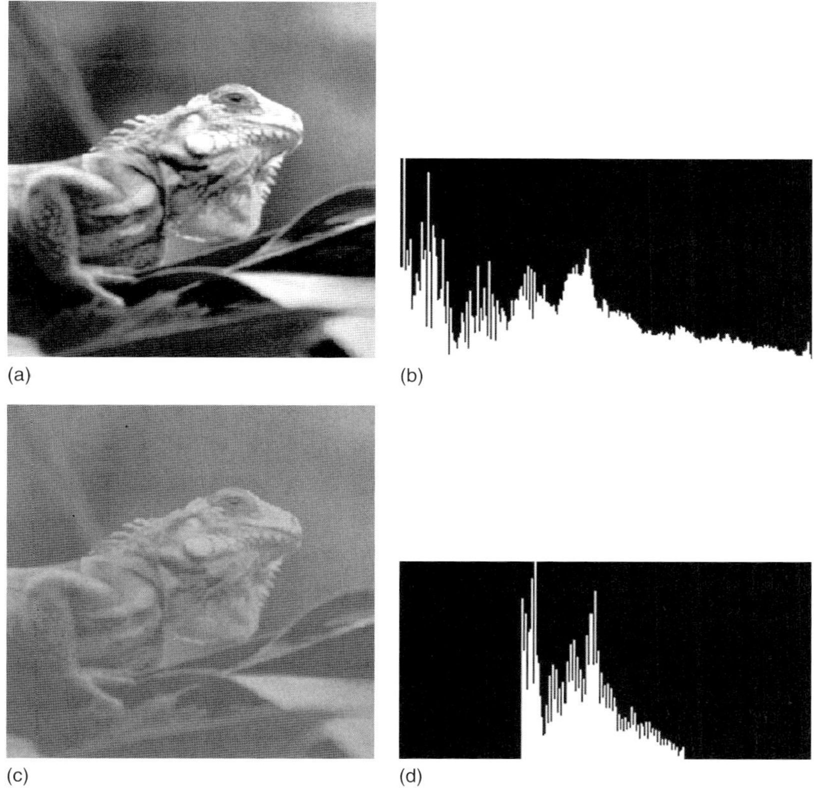

FIGURE 8.2.11
Histogram shrinking (a) Original image. (b) Histogram of image (a). (c) Image after shrinking the histogram to the range [75, 175]. (d) Histogram of image (c).

from step (3) using a one-to-one correspondence. An example will help to clarify this process:

Example 8.2.6

We have an image with 3-bits per pixel, so the possible range of values is 0 to 7. We have an image with the following histogram:

Gray level value	Number of pixels (histogram values)
0	10
1	8
2	9
3	2
4	14
5	1
6	5
7	2

STEP 1: Create a running sum of the histogram values. This means the first value is 10, the second is $10+8=18$, next $10+8+9=27$, and so on. Here we get 10, 18, 27, 29, 43, 44, 49, 51.

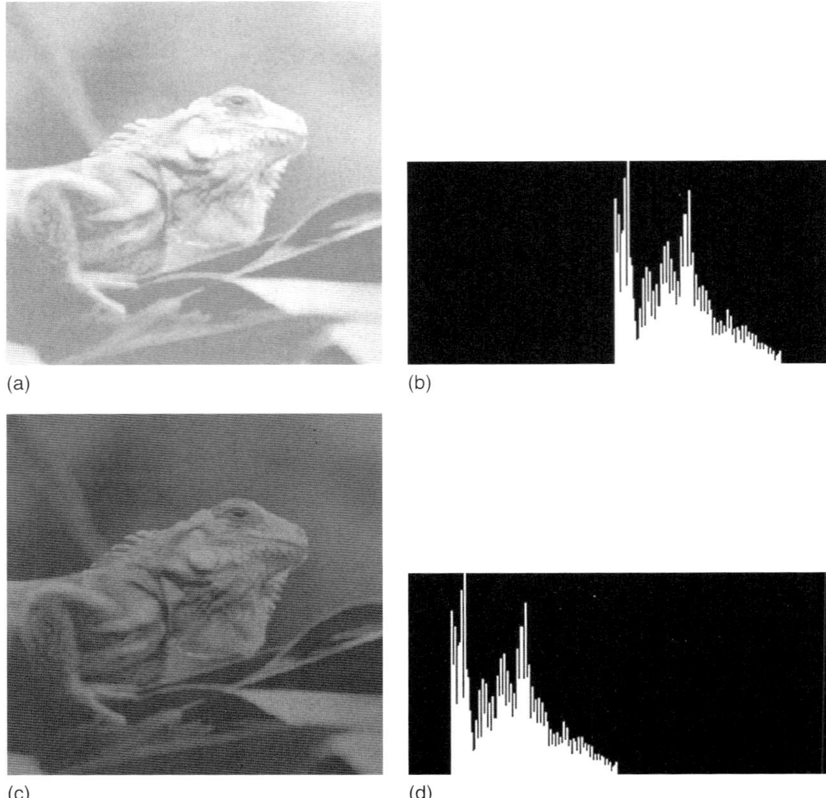

FIGURE 8.2.12
Histogram slide. The original image for these operations is the image from Figure 8.2.11c that had undergone a histogram shrink process. (a) The resultant image from sliding the histogram down up 50. (b) The histogram of image (a). (c) The resultant image from sliding the histogram down by 50. (d) The histogram of image (c).

STEP 2: Normalize by dividing by the total number of pixels. The total number of pixels is: $10+8+9+2+14+1+5+0=51$ (note this is the last number from Step 1), so we get: 10/51, 18/51, 27/51, 29/51, 43/51, 44/51, 49/51, 51/51.

STEP 3: Multiply these values by the maximum gray level values, in this case 7, and then round the result to the closest integer. After this is done we obtain: 1, 2, 4, 4, 6, 6, 7, 7.

STEP 4: Map the original values to the results from Step 3 by a one-to-one correspondence. This is done as follows:

Original gray level value	Histogram equalized values
0	1
1	2
2	4
3	4
4	6
5	6
6	7
7	7

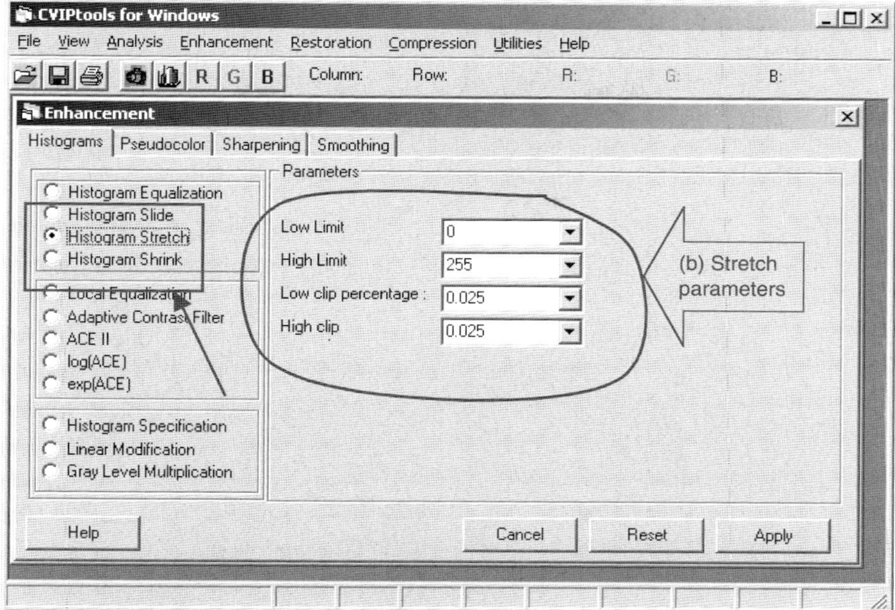

FIGURE 8.2.13
CVIPtools histogram stretch, shrink and slide. (a) Histogram stretch, histogram shrink, and histogram slide. (b) Parameters for histogram stretch, note that a percentage for clipping can be selected at the low and high ends.

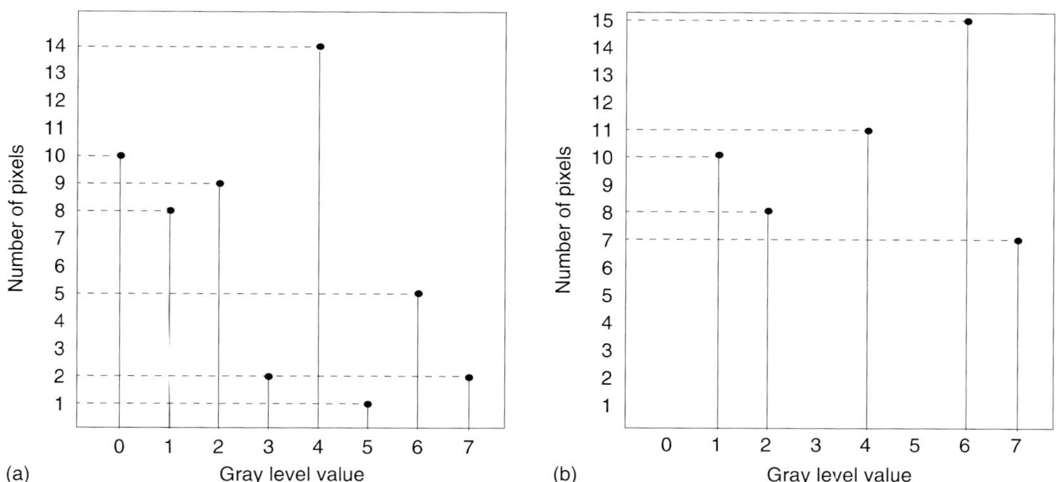

FIGURE 8.2.14
Histogram equalization. (a) Original histogram. (b) After histogram equalization.

All pixels in the original image with gray level 0 are set to 1, values of 1 are set to 2, 2 set to 4, 3 set to 4, and so on. After the histogram equalization values are calculated it can be implemented efficiently with a look-up-table (LUT), as discussed in Chapter 2 (see Figure 2.4.8). In Figure 8.2.14 we see the original histogram and the resulting histogram equalized histogram. Although the result is not flat, it is closer to being flat than the original histogram.

Histogram equalization of a digital image will not typically provide a histogram that is perfectly flat, but it will make it as flat as possible. For the equalized histogram to be completely flat, the pixels at a given gray level might need to be redistributed across more than one gray level. This could be done, but would greatly complicate the process, as some redistribution criteria would need to be defined. In most cases the visual gains achieved by doing this would be negligible, and could in some cases be negative. In practice, it is not done.

Figure 8.2.15 shows the result of histogram equalizing images of various average brightness and contrast. In Figure 8.2.15a,b histogram equalization was applied to a

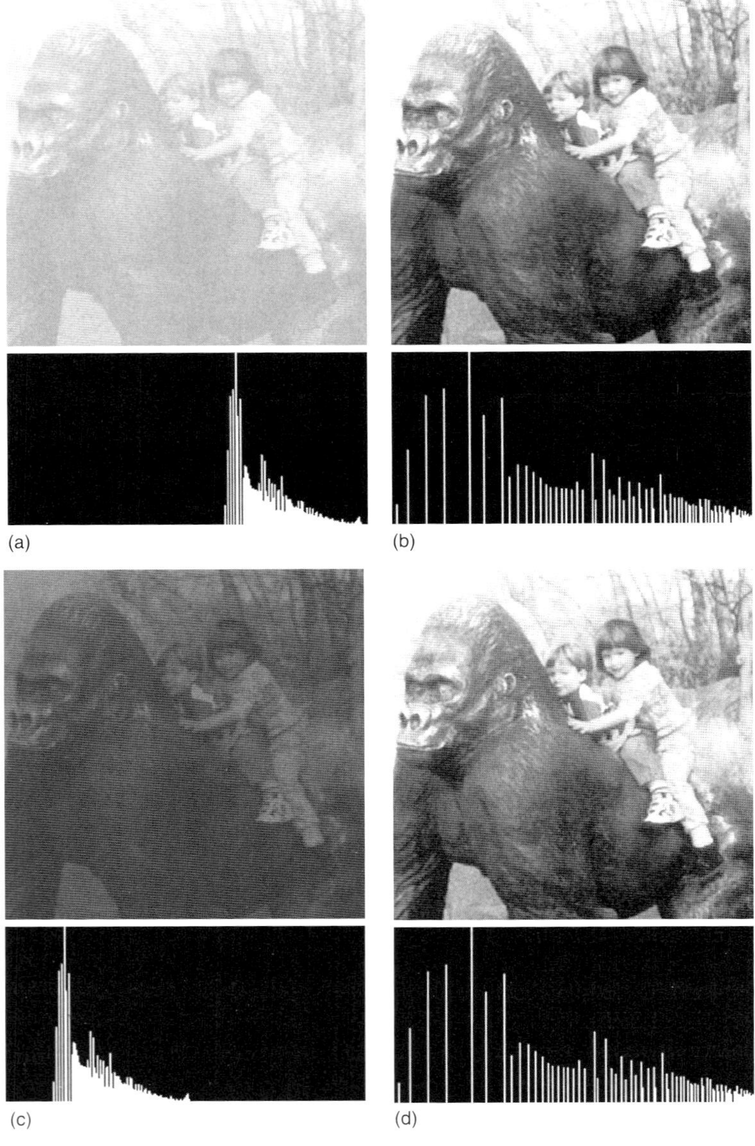

FIGURE 8.2.15
Histogram equalization examples. Images on the left (a, c, e, g) are input images of varying average brightness and contrast. Images on the right (b, d, f, h) are the resultant images after histogram equalization. The histograms of the images are directly below them. As can be seen, histogram equalization provides similar results regardless of the input image.

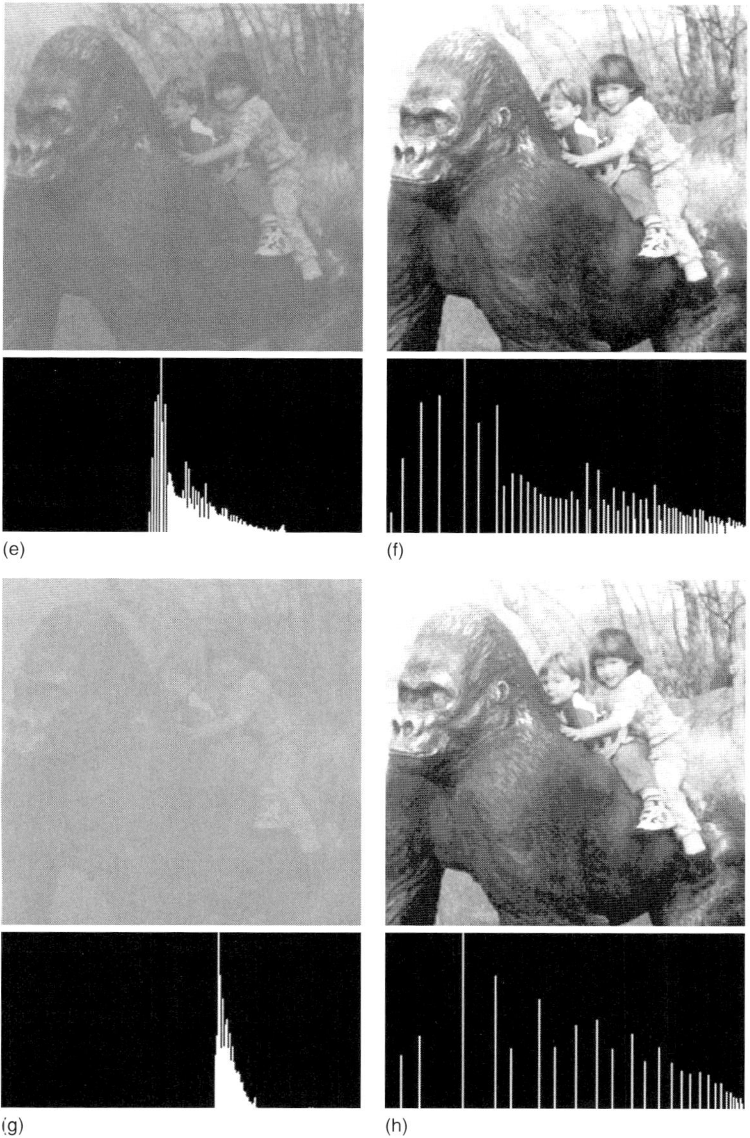

FIGURE 8.2.15 (Continued)
Histogram equalization examples.

bright image, and in Figure 8.2.15c,d to a dark image. In Figure 8.2.15e,f histogram equalization was applied to an image of medium average brightness, and Figure 8.2.15g,h shows an image of very low contrast. The results of this process are often very dramatic, as illustrated in this figure. This figure also shows that histogram equalization provides similar results regardless of the characteristics of the input image.

Histogram equalization may not always provide the desired effect, since its goal is fixed—to distribute the gray level values as evenly as possible. To allow for interactive histogram manipulation, the ability to specify the histogram is necessary. *Histogram specification* is the process of defining a histogram and modifying the histogram of the original image to match the histogram as specified. The key concept in the

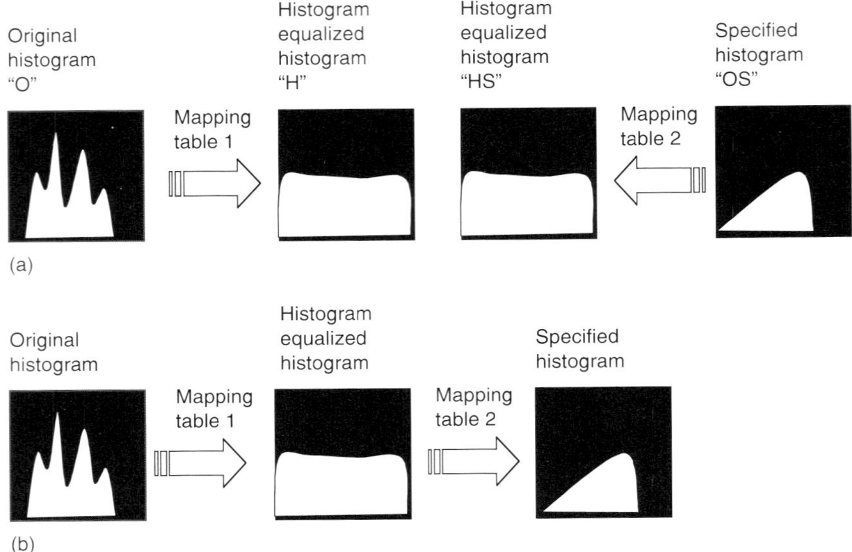

FIGURE 8.2.16
Histogram specification. This figure is a conceptual look at histogram specification. (a) Here we depict the histogram equalized versions of the original image histogram and the specified histogram. Now we have a common histogram for both, the histogram equalized version should both be approximately flat. (b) Now we can use the histogram equalization mapping tables to get from the original histogram to the specified histogram.

histogram specification process is to picture the original image being histogram equalized, and the specified histogram being histogram equalized. Now, we have a common point since the histogram equalization process results in a flat histogram, or in the case of digital images, a histogram that is as close to flat as possible. The process is illustrated in Figure 8.2.16.

This process can be implemented by: (1) specify the desired histogram, (2) find the mapping table to histogram equalize the image, Mapping Table 1, (3) find the mapping table to histogram equalize the values of the specified histogram, Mapping Table 2, (4) use mapping Tables 1 and 2 to find the mapping table to map the original values to the histogram equalized values and then to the specified histogram values, (5) use the table from step (4) to map the original values to the specified histogram values. This process is best illustrated by example:

Example 8.2.7

STEP 1: Specify the desired histogram:

Gray level value	Number of pixels in desired histogram
0	1
1	5
2	10
3	15
4	20
5	0
6	0
7	0

(a)

(b)

FIGURE 1.3.2
Noise removal. (a) Noisy image. (b) Noise removed with image restoration.

(a)

(b)

FIGURE 2.2.3
The reflectance function. Here we see that the way in which an object reflects the incident light, the reflectance function, has a major effect on how it appears in the resulting image. The reflectance function is an intrinsic property of the object and relates to both color and texture. (a) Monochrome image showing brightness only, the color determines how much light is reflected and the surface texture determines the angle at which the light is reflected. (b) Color image, the color determines which wavelengths are absorbed and which are reflected.

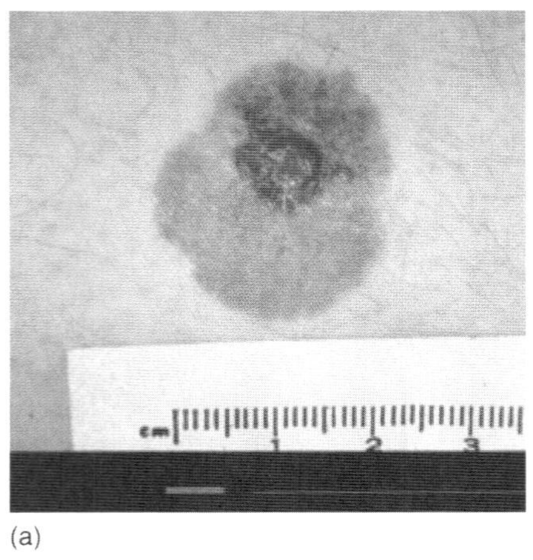

(a)

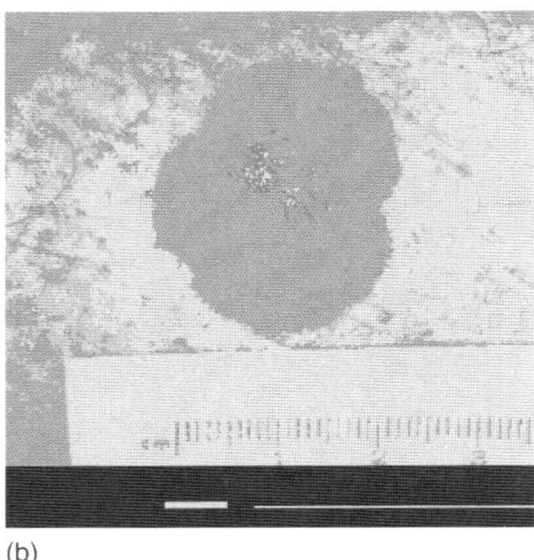

(b)

FIGURE 4.3.8
SCT/Center segmentation algorithm. (a) Original image. (b) SCT/Center segmentation of skin tumor using four colors.

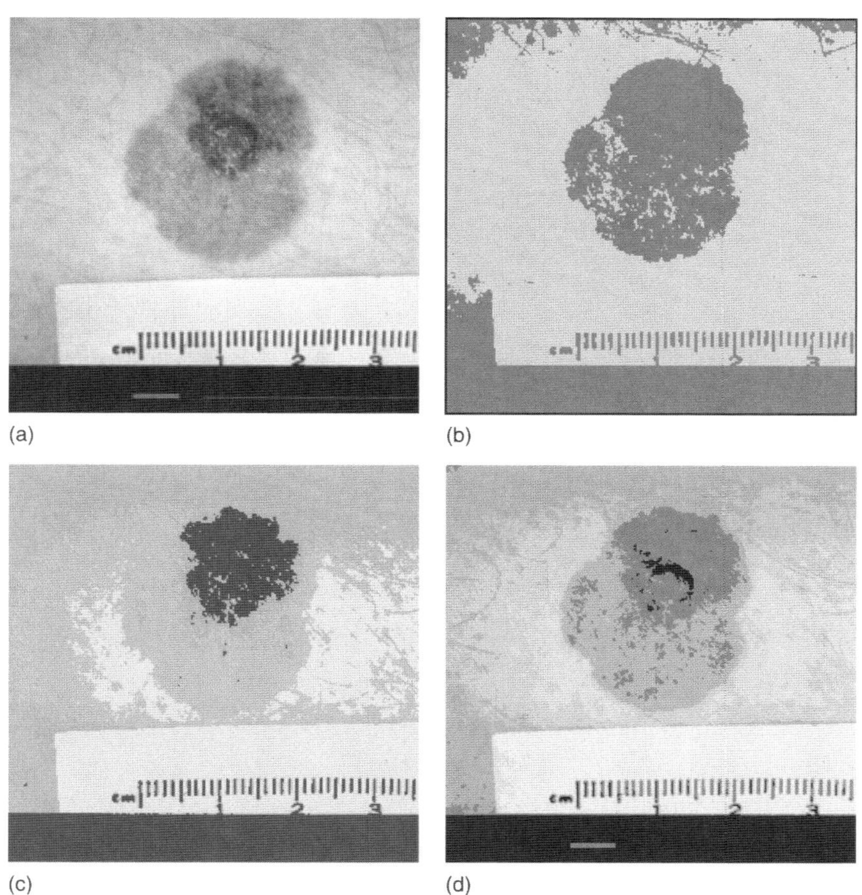

FIGURE 4.3.10
PCT/Median segmentation algorithm. (a) Original image. (b) PCT/Median segmented image with two colors. (c) PCT/Median segmented image with four colors. (d) PCT/Median segmented image with eight colors.

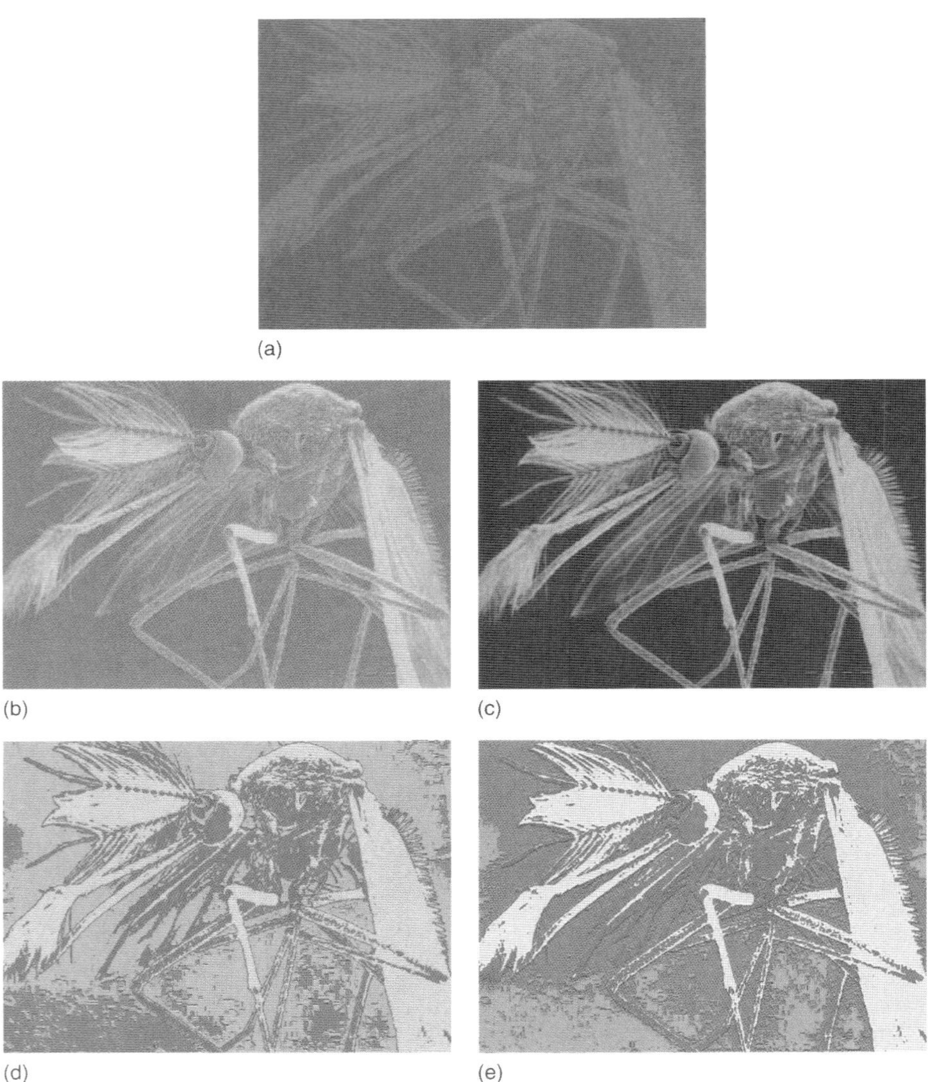

FIGURE 8.2.25
Image examples of pseudocolor in the spatial domain. (a) Original scanning electron image of a mosquito (photo courtesy of Sue Eder, SIUE). (b,c) Gray level mapping pseudcolor. (d,e) Intensity slicing pseudocolor.

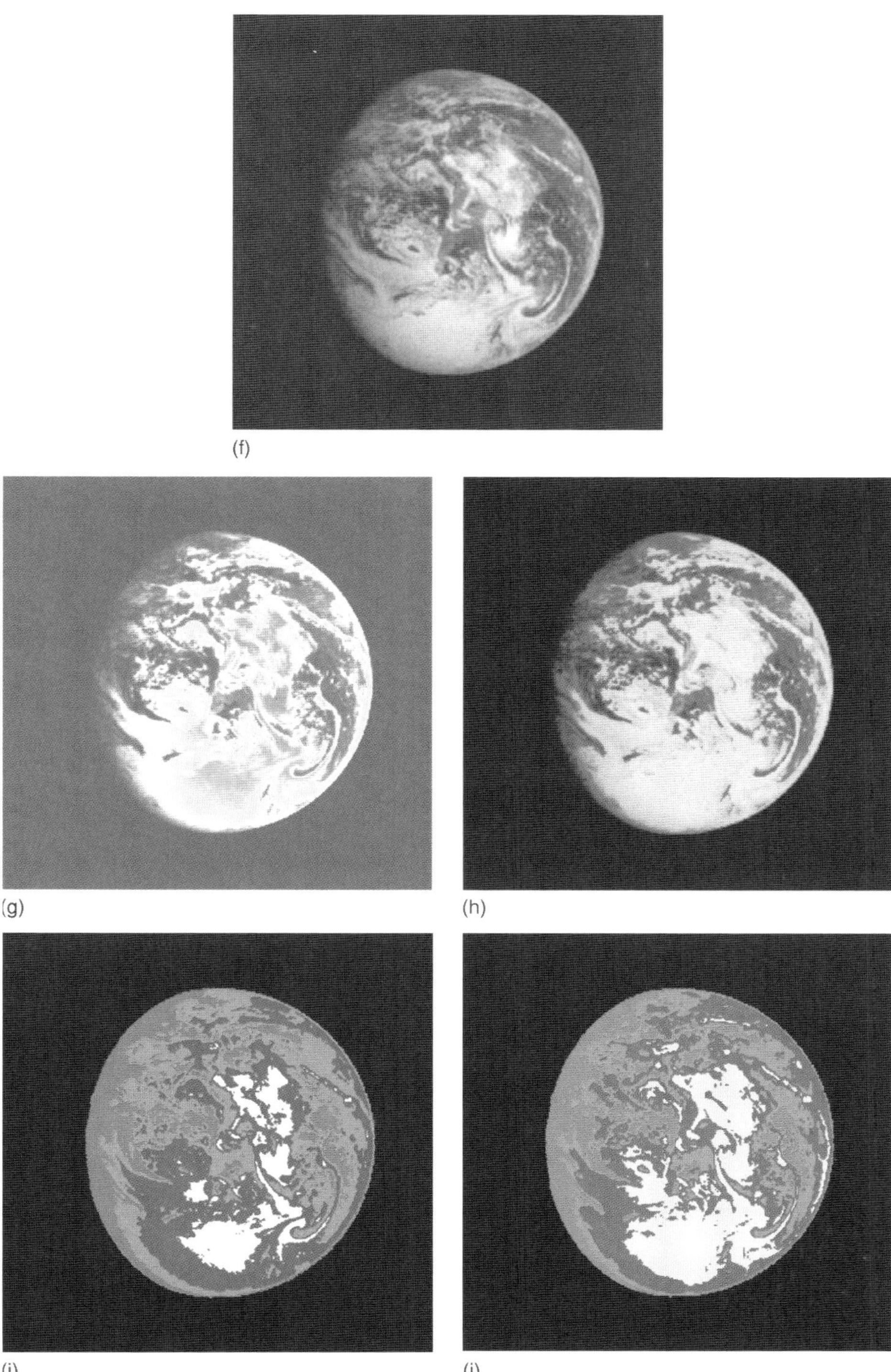

FIGURE 8.2.25 (Continued)
Image examples of pseudocolor in the spatial domain. (f) Original satellite image of the earth. (g,h) Gray level mapping pseudocolor. (i,j) Intensity slicing pseudocolor.

FIGURE 8.2.27
Frequency domain pseudo-color. (a) Original 256×256 image, so in the Fourier domain the highest frequency is 128. (b) Result with cutoff frequencies of 10 and 100, with the lowpass result mapped to red, bandpass to green, and highpass to blue. (c) Result with cutoff frequencies of 10 and 100, with the lowpass mapped to blue, bandpass to green, and highpass to red. (d) Result with cutoff frequencies of 5 and 50, with the lowpass mapped to red, bandpass to blue, and highpass to green.

FIGURE 8.2.28
Histogram equalization of color images. (a) Original poor contrast image. (b) Histogram equalization based on the red color band. (c) Histogram equalization based on the green color band. (d) Histogram equalization based on the blue color band. Note that in this case the red band gives the best results. This will depend on the image and the desired result.

(a)

(b)

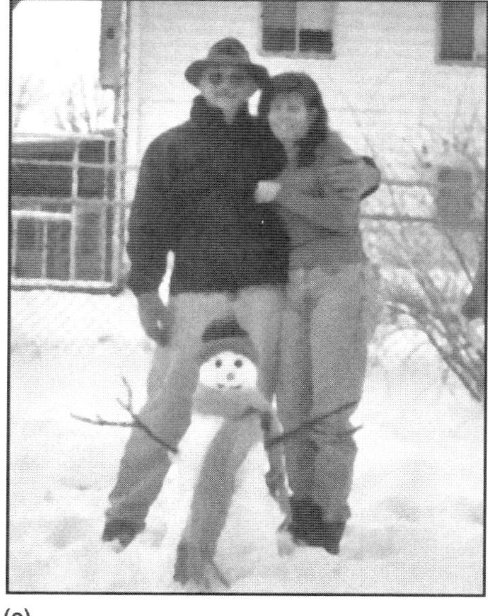

(c)

FIGURE 9.3.1
Median filter with color image. (a) Image with added salt-and-pepper noise, the probability for salt = probability for pepper = 0.08. (b) After median filtering with a 3×3 window, all the noise is not removed. (c) After median filtering with a 5×5 window, all the noise is removed, but the image is blurry acquiring the "painted" effect.

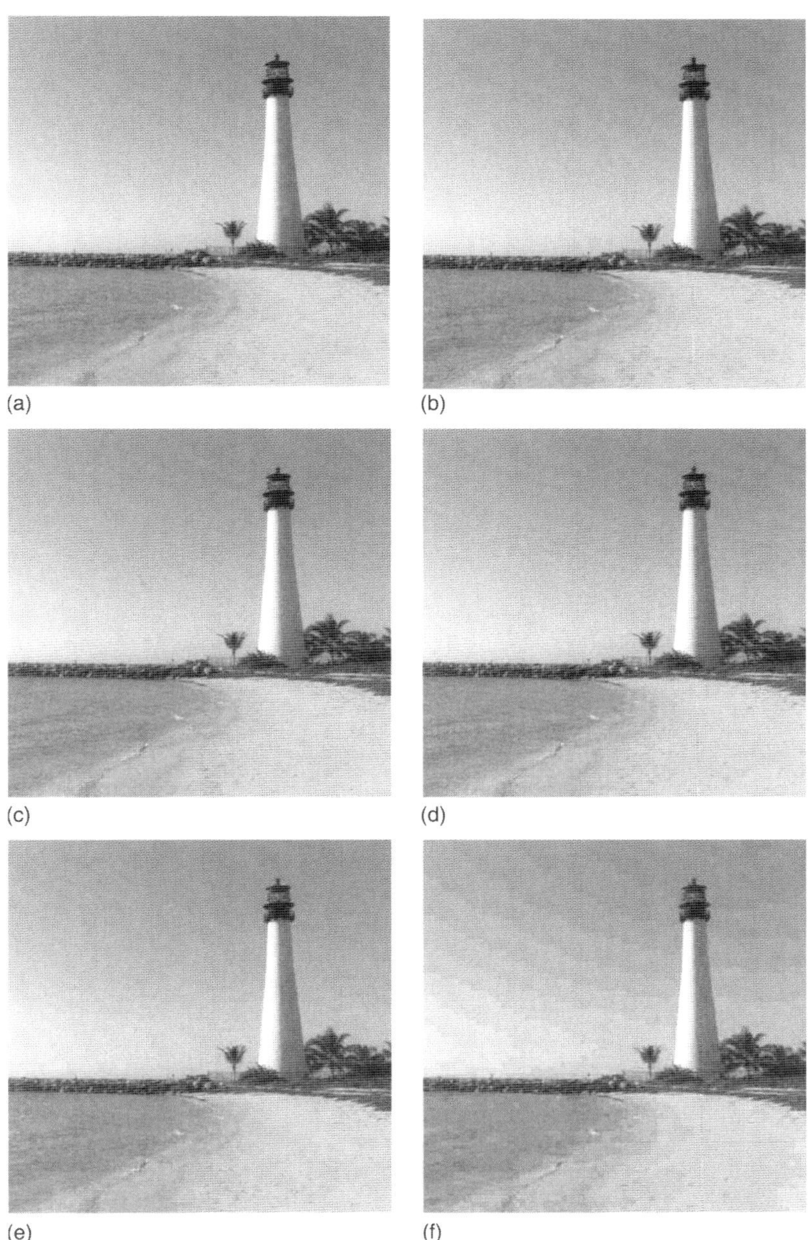

FIGURE 10.3.21
The original DCT-based JPEG algorithm applied to a color image. (a) The original image. (b) Compression ratio = 34.34. (c) Compression ratio = 57.62. (d) Compression ratio = 79.95. (e) Compression ratio = 131.03. (f) Compression ratio = 201.39.

FIGURE 10.3.24
The JPEG2000 algorithm applied to a color image. (a) The original image. (b) Compression ratio = 130, compare to Figure 10.3.21e. (c) Compression ratio = 200, compare to Figure 10.3.21f. (d) A 128×128 subimage cropped from the standard JPEG image and enlarged to 256×256 using zero-order hold. (e) A 128×128 subimage cropped from the JPEG2000 image and enlarged to 256×256 using zero-order hold. The JPEG2000 image is much smoother, even with the zero-order hold enlargement.

STEP 2: For this we will use the image and mapping table from the previous example, where the histogram equalization mapping table is given by:

Original gray level value—O	Histogram equalized values—H
0	1
1	2
2	4
3	4
4	6
5	6
6	7
7	7

STEP 3: Find the histogram equalization mapping table for the specified histogram:

Gray level value—OS	Histogram equalized values—HS
0	round(1/51)*7 = 0
1	round(6/51)*7 = 1
2	round(16/51)*7 = 2
3	round(31/51)*7 = 4
4	round(51/51)*7 = 7
5	round(51/51)*7 = 7
6	round(51/51)*7 = 7
7	round(51/51)*7 = 7

STEP 4: Use Mapping Tables 1 and 2 to find the final mapping table by mapping the values first to the histogram equalized values and then to the specified histogram values. Notice in Mapping Table 2, we switched the columns to match Figure 8.2.16a.

Mapping Table 1		Mapping Table 2		
O	H	HS	OS	M
0	1	0	0	1
1	2	1	1	2
2	4	2	2	3
3	4	4	3	3
4	6	7	4	4
5	6	7	5	4
6	7	7	6	4
7	7	7	7	4

The M column for this table is obtained by mapping the value in H to the closest value in HS, which brings the two histogram equalized mappings together, and then using the corresponding row in OS for the entry in M. For example, start with the first original gray level value, 0, which maps to a 1 in H. Now, to map with Table 2, we find the closest value in HS, which is 1. This 1 from HS maps to a 1 in OS, so we write a 1 for that entry in M. Another example, the original gray value 2 maps to the third entry in H which is a 4. Using Table 2, we find the closest value in HS, which is 4. This 4 from HS maps back to a 3 in OS, so we write a 3 for that entry in M. If we consider the gray value 4, it maps to 6 in H, we see the 6 must map to 7 (the closest value), but the 7 appears on rows 4, 5, 6, and 7. Which one do we select? It depends on what we want, picking the largest value will provide maximum contrast, but picking the smallest (closest) value will produce a more gradually changing image. Typically, the smallest is chosen, since we can always perform a histogram stretch or equalization on the output image, if we desire to maximize contrast.

STEP 5: Use the table from STEP 4 to perform the histogram specification mapping. For this all we need are columns O (or OS) and M:

O	M
0	1
1	2
2	3
3	3
4	4
5	4
6	4
7	4

Now, all the 0's get mapped to 1's, the 1's to 2's, the 3's to 3's and so on.

In practice, the desired histogram is often specified by a continuous (possibly nonlinear) function, for example a sine or a log function. To obtain the numbers for the specified histogram the function is sampled, the values are normalized to 1, and then multiplied by the total number of pixels in the image. Figure 8.2.17 shows the result of specifying an exponential and a log function for the histogram functions. Remember that this is not a gray level mapping function as was discussed before, but is the desired shape of the output histogram.

8.2.3 Adaptive Contrast Enhancement

Adaptive contrast enhancement refers to modification of the gray level values within an image based on some criterion that adjusts its parameters as local image characteristics change. Since the operation depends on other pixels in local areas, neighborhoods, it is primarily a mask type operation. Additionally, some adaptive contrast operators use global image statistics, hence are also global operations.

The simplest adaptive contrast enhancement method is to perform a histogram modification technique, but instead of doing it globally (on the entire image), applying it to the image on a block by block basis. In this case, the block size corresponds to the local neighborhood and the enhancement is adaptive because the output depends only on the local histogram. Thus, this technique is also called *local enhancement*.

In Figure 8.2.18 are the results of applying histogram equalization to various block sizes within an image, and these are contrasted with global histogram equalization. In Figure 8.2.18c,d we can see that this technique brings out minute details, is very sensitive to noise in the image, and the resulting image, although full of detail, is not very visually pleasing. As an enhancement method it is useful to multiply the image after

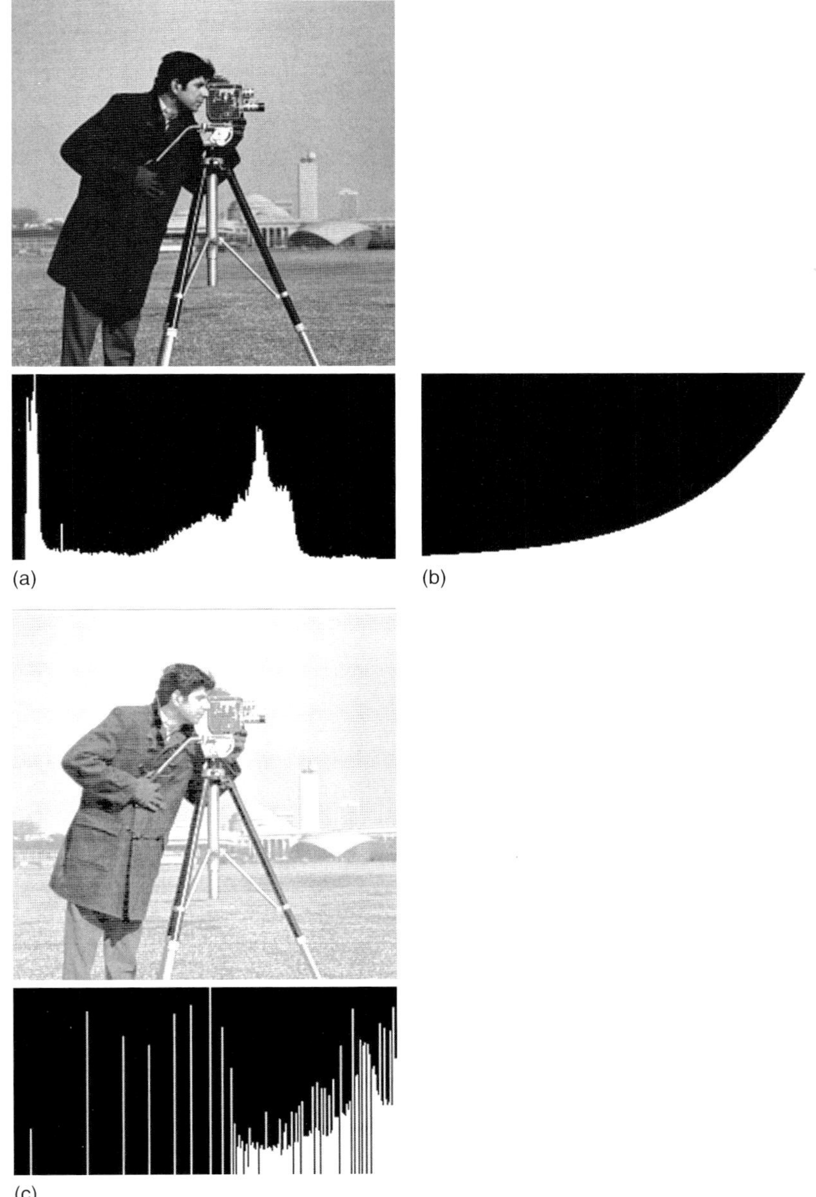

FIGURE 8.2.17
Histogram specification examples. (a) Original image and histogram. (b) Specified histogram, *exp(0.015*x)*. (c) The output image and its histogram.

local histogram equalization by a number less than one and then adding it back to the original, as shown in Figure 8.2.18e,f.

The *adaptive contrast enhancement* (ACE) filter is used with an image that appears to have uneven contrast, where we want to adjust the contrast differently in different regions of the image. It works by using both local and global image statistics to determine the amount of contrast adjustment required. This filter is adaptive in

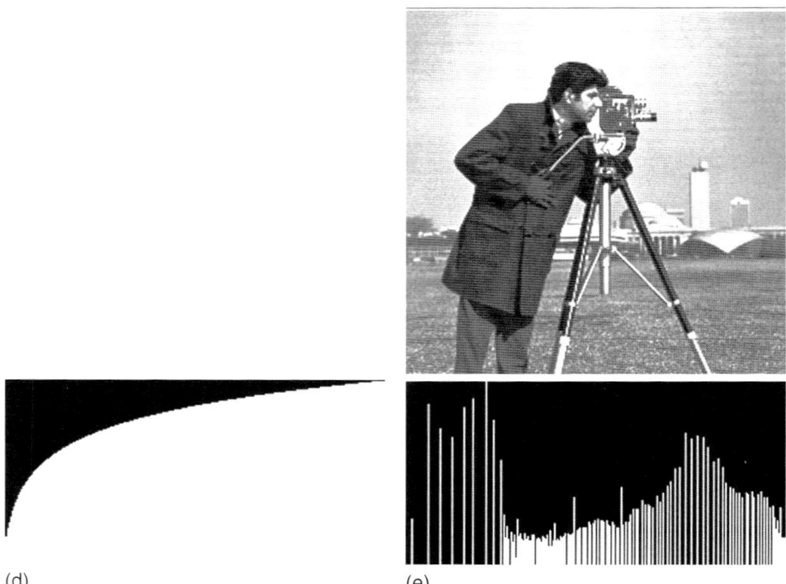

FIGURE 8.2.17 (Continued)
Histogram specification examples. (d) Specified histogram, *log(0.5*x + 2)*. (e) The output image and its histogram.

the sense that its behavior changes based on local image statistics, unlike the standard histogram modification techniques which use only global parameters and result in fixed gray level transformations. The image is processed using the sliding window concept (see Figure. 8.2.19), and the local image statistics are found by considering only the current window (subimage), and the global parameters are found by considering the entire image. It is defined as follows:

$$\text{ACE} \Rightarrow E(r,c) = k_1\left[\frac{m_{I(r,c)}}{\sigma_l(r,c)}\right][I(r,c) - m_l(r,c)] + k_2 m_l(r,c)$$

where

$m_{I(r,c)}$ = the mean (average) for the entire image $I(r,c)$
σ_l = local standard deviation the current $n \times n$ window $= \sqrt{\sum (I(r,c) - m_l)^2/(n^2 - 1)}$
m_l = local mean in current window
k_1 = local gain factor constant, between 0 and 1
k_2 = local mean constant, between 0 and 1

From the equation we can see that this filter subtracts the local mean from the original data, and weights the result by the local gain factor, $k_1[m_{I(r,c)}/\sigma_l(r,c)]$. This has the effect of intensifying local variations, and can be controlled by the constant, k_1. Areas of low contrast (low values of $\sigma_l(r,c)$) are boosted. In practice it may be helpful to set a minimum and maximum value for the local gain, this option is available in CVIPtools. After the local gain factor is multiplied by the difference between the current pixel and the local mean, the mean is then added back to the result, weighted by k_2, to restore the local average brightness. We can also see from this equation that negative values may result, so the output image will be remapped to put values within the standard gray level range.

FIGURE 8.2.18
Local histogram equalization. (a) Original image. (b) Image after global histogram equalization. (c) Image after local histogram equalization with a block size of 8×8. (d) Image after local histogram equalization with a block size of 16×16. (e) The image from (d) multiplied by 0.5 and added back to the original. (f) The image from (d) multiplied by 0.25 and added back to the original.

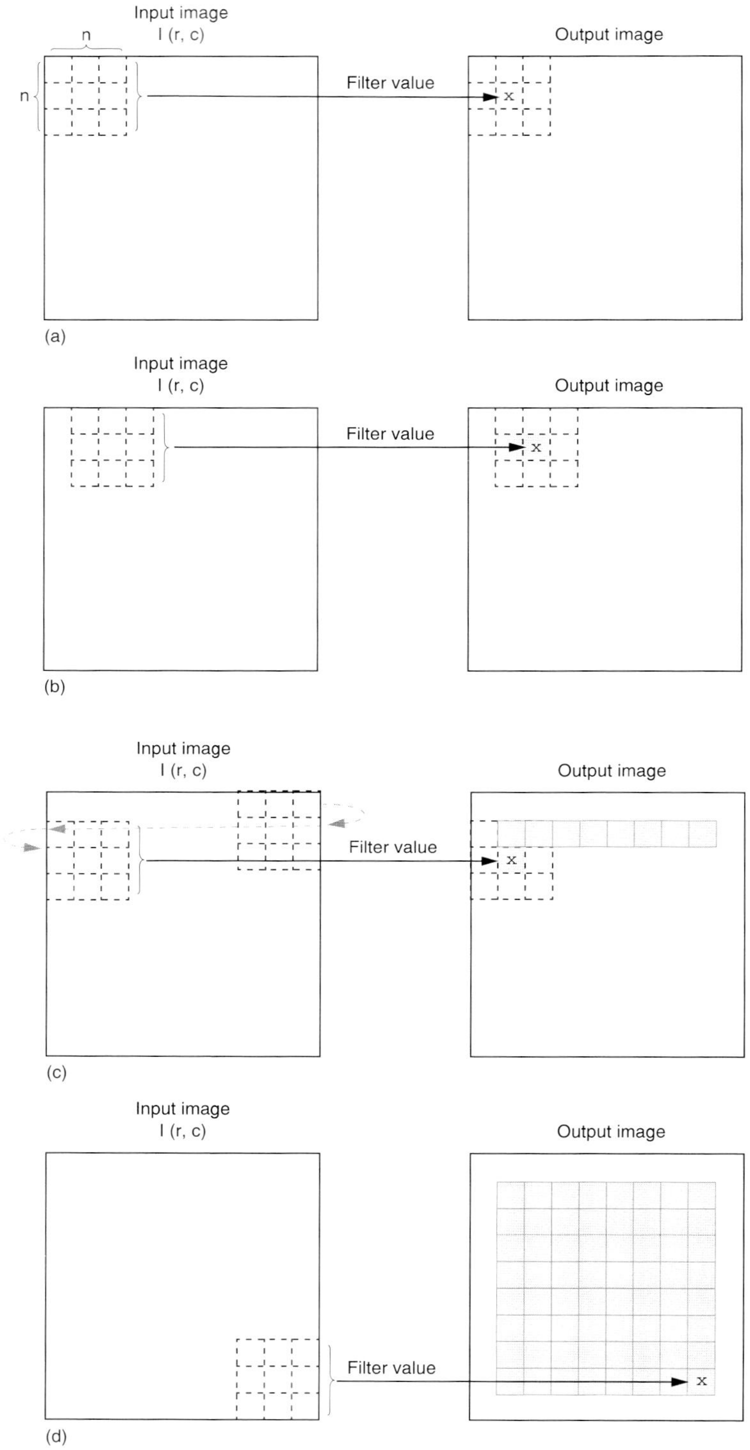

FIGURE 8.2.19
Filtering with a sliding window. (a) The input image is overlaid with an $n \times n$ window, and the filter value of the pixels covered by the window is placed in the output image at location x. (b) The window is moved one pixel to the right, and the filter value of the pixels now covered by the window is placed in the output image at location x. (c) When the end of a row is reached, the window is moved back to the left edge of the image and down one row. (d) The entire image has been processed. Note the unprocessed outer rows and columns.

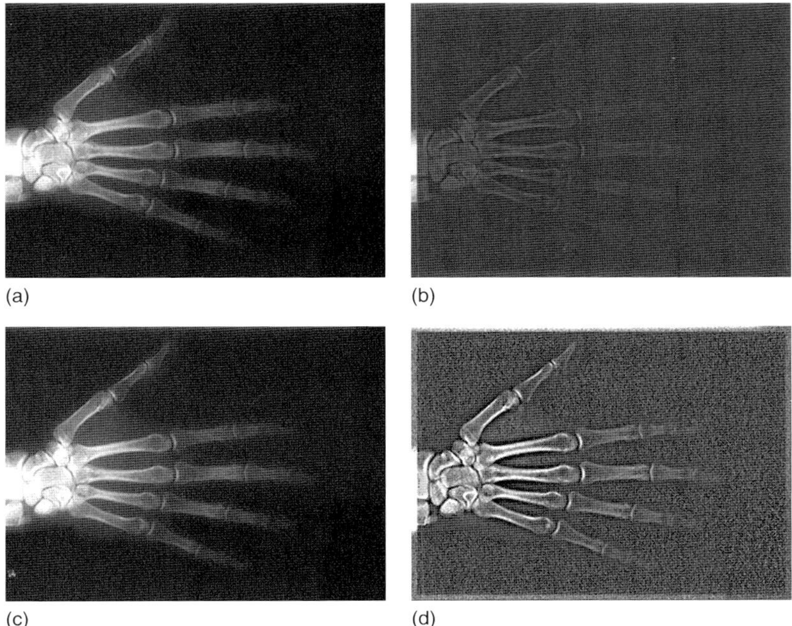

FIGURE 8.2.20
Adaptive contrast filter (ACE). (a) Original x-ray image. (b) Image after ACE filter (ignore outer rows and columns that are not processed). (c) Original image after a histogram stretch, note that the bright areas near the wrist are washed out. (d) ACE filter results followed by a histogram stretch, since the filter adapts to local image statistics contrast is enhanced in both bright and dark areas.

Figure 8.2.20 illustrates results from this filter with a local window size of 11, a local gain factor of 0.9, and a local mean factor of 0.1. The direct output from the ACE filter typically needs some form of postprocessing to improve the contrast—compare Figure 8.2.20b,d. Comparing Figure 8.2.20c,d we see that we retain more detail in both the bright and dark areas of the image, but note that most of the detail in the dark, background areas appears to be attributable to noise, suggesting that some noise removal could improve the results. These images illustrate the experimental nature of developing image enhancement algorithms and the fact that the algorithms tend to be application dependent.

A simplified variation of the ACE filter, we will call the ACE2 filter, is given by the following equation:

$$\text{ACE2} \Rightarrow E(r,c) = k_1[I(r,c) - m_l(r,c)] + k_2 m_l(r,c)$$

where

m_l = local mean in current window
k_1 = local gain factor constant, between 0 and 1
k_2 = local mean constant, between 0 and 1

This filter is less computationally intensive than the original ACE filter and provides similar results. Figure 8.2.21 shows the result of applying this filter. In this figure a local gain factor of 0.9 was used, a window size of 11, and a local mean factor of 0.1. Comparing Figure 8.2.21c,d we can see that since the filter adapts to local image statistics contrast is improved in both bright and dark areas of the image.

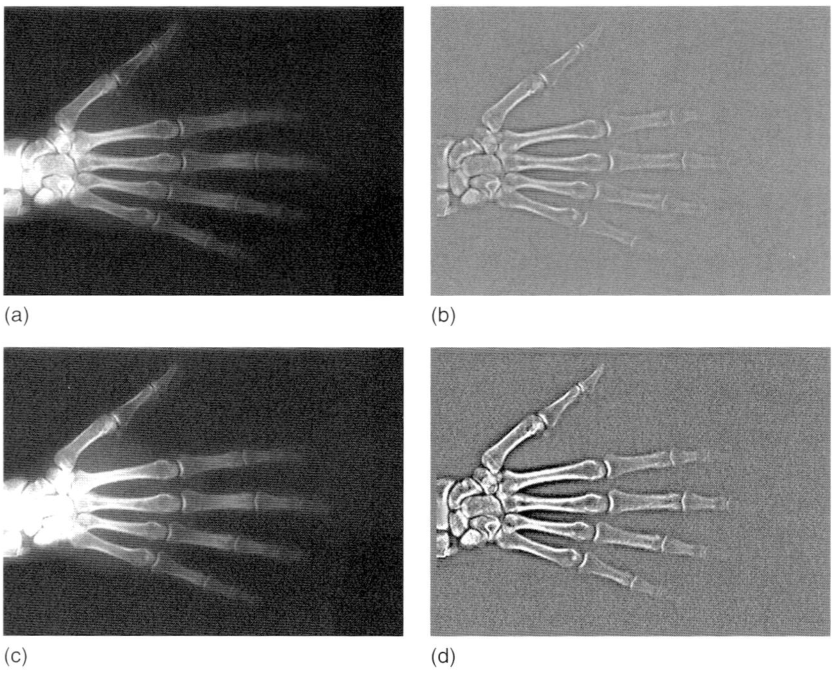

FIGURE 8.2.21
Adaptive contrast filter 2 (ACE2). (a) Original x-ray image. (b) Image after ACE2 filter. (c) Original image after a histogram stretch, note that the bright areas near the wrist are washed out. (d) ACE2 filter results followed by a histogram stretch, since the filter adapts to local image statistics contrast is enhanced in both bright and dark areas.

Other variations of the ACE filter include logarithmic and exponential ACE filters. The log-ACE filter equation is:

$$\text{Log-ACE} \Rightarrow E(r,c) = k_1\left[\ln\left(\tilde{I}(r,c)\right) - \ln(\tilde{m}_l(r,c))\right] + k_2\tilde{m}_l(r,c)$$

where

$\tilde{I}(r,c)$ = normalized complement of image = $1 - I(r,c)/\text{MAX}$, MAX is maximum gray value (e.g. 255)
$\tilde{m}_l$ = normalized complement of local mean = $1 - m_l(r,c)/\text{MAX}$
m_l = local mean in current window
k_1 = local gain factor constant, between 0 and 1
k_2 = local mean constant, between 0 and 1

Figure 8.2.22 compares various window sizes and postprocessing methods with the Log-ACE filter. The exponential ACE filter equation is as follows:

$$\text{Exp-ACE} \Rightarrow E(r,c) = \text{MAX} \times \left[\frac{I(r,c)}{\text{MAX}}\right]^{k_1} + \left[\frac{m_l(r,c)}{I(r,c)}\right]^{k_2}$$

where

m_l = local mean in current window
MAX = maximum gray value (e.g. 255)

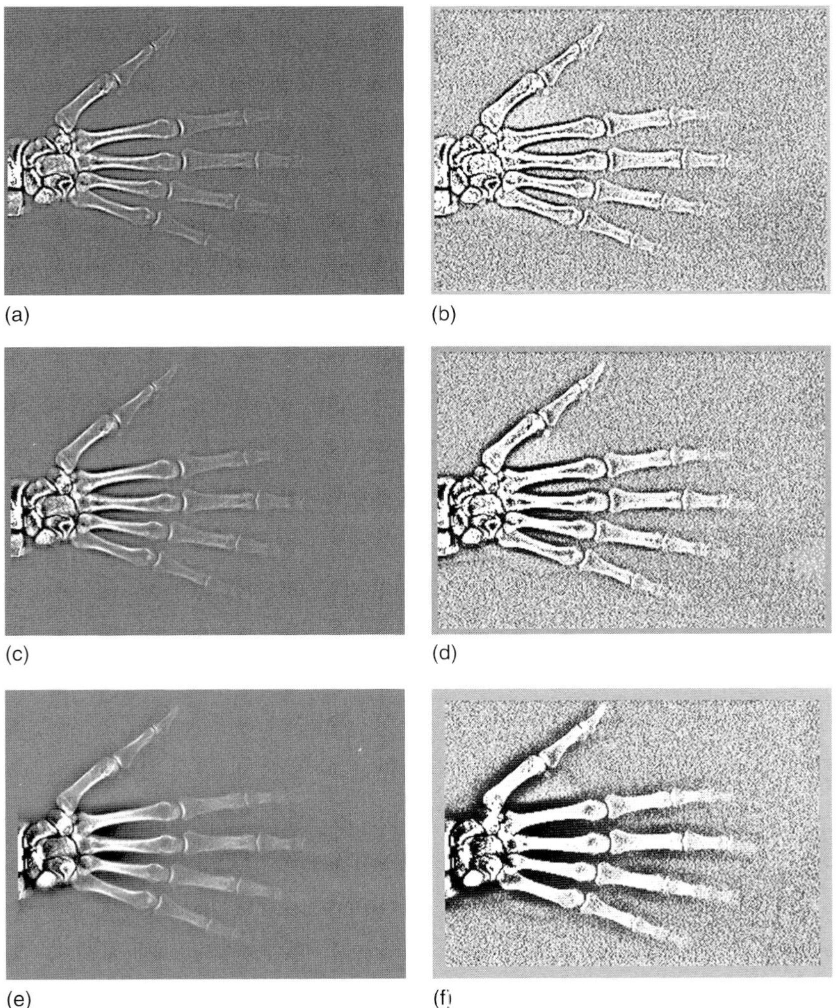

FIGURE 8.2.22
Logarithmic adaptive contrast filter (Log-ACE), varying the window size and the postprocessing method. The same original hand x-ray used in Figure 8.2.20a was the original for these results. The images on the left were postprocessed with a histogram stretch, and those on the right with a histogram equalization process. (a,b) Window size 7. (c,d) Window size 15. (e,f) Window size 21.

k_1 = local gain factor exponent
k_2 = local mean factor exponent

Figure 8.2.23 shows results of applying the Exp-ACE filter.

8.2.4 Color

One of the reasons that color is important for image enhancement is that the human visual system can perceive thousands of colors in a small spatial area, but only about 100 gray levels. Additionally, color contrast can be more dramatic than gray level contrast, and various colors have different degrees of psychological impact on the

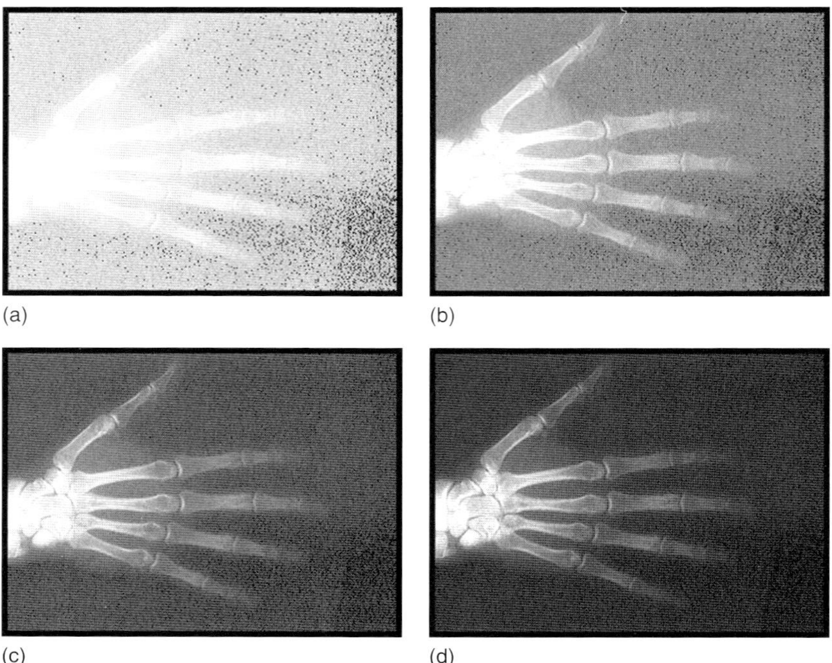

FIGURE 8.2.23
Exponential adaptive contrast filter (Exp-ACE), varying the local mean factor. The same original hand x-ray image used in Figure 8.2.20a was the original for these results. A window size of 11×11 and a local gain factor of 0.9 were used. (a) Local mean factor = 0.1. (b) local mean factor = 0.25. (c) local mean factor = 0.5. (d) local mean factor = 0.75. As the local mean factor is increased more of the original image is retained. Also note that no postprocessing was applied with the Exp-ACE filter.

observer. Taking advantage of these aspects of our visual perception to enhance gray level images we apply a technique called pseudocolor. *Pseudocolor* involves mapping the gray level values of a monochrome image to red, green, and blue values, creating a color image. The pseudocolor techniques can be applied in both the spatial and frequency domains. Pseudocolor is often applied to images where the relative values are important, but the specific representation is not—for example, satellite, microscopic, or x-ray images.

In the spatial domain a gray level to color transform is defined, which has three different mapping equations for each of the red, green, and blue color bands. The equations selected are application-specific, and are functions of the gray levels in the image, $I(r,c)$. So we have three equations, as follows:

$$I_{\mathrm{R}}(r,c) = \mathrm{R}[I(r,c)]$$

$$I_{\mathrm{G}}(r,c) = \mathrm{G}[I(r,c)]$$

$$I_{\mathrm{B}}(r,c) = \mathrm{B}[I(r,c)]$$

where R[ ], G[ ], and B[ ] are the mapping equations to map the gray levels to the red, green, and blue components. These equations can be linear or nonlinear. A simple example, called *intensity slicing*, splits the range of gray levels into separate colors.

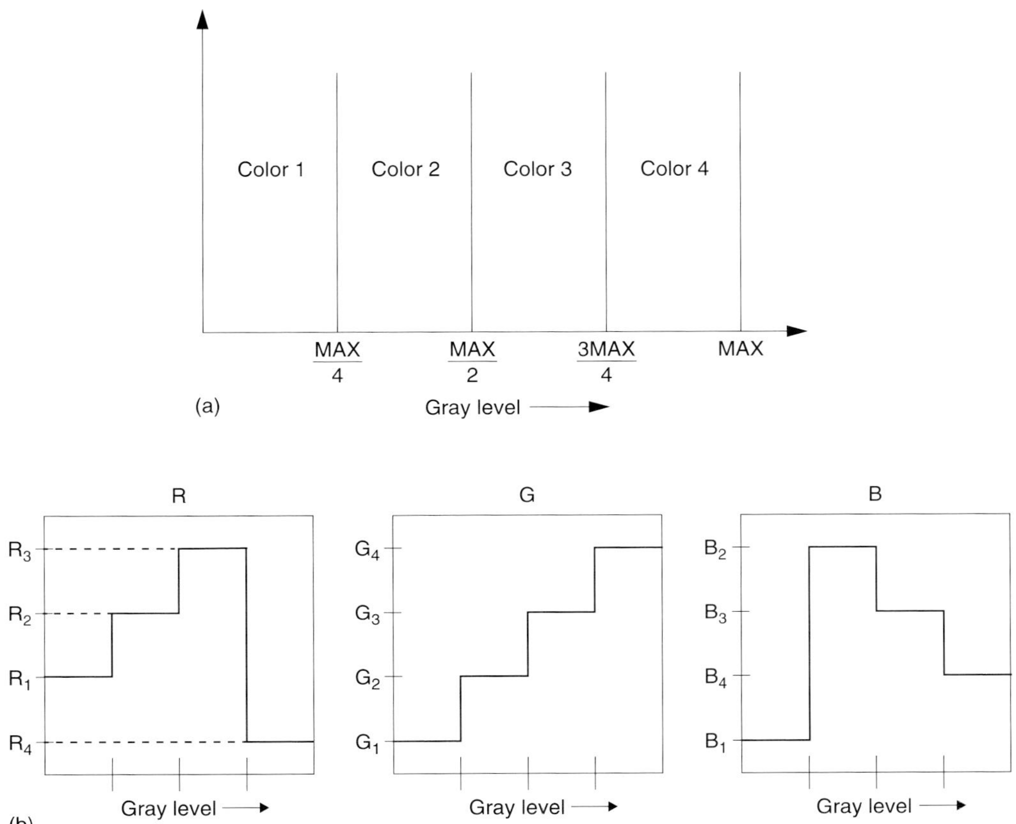

FIGURE 8.2.24
Pseudocolor in the spatial domain. (a) Intensity slicing. (b) Mapping equations.

For this, the gray levels that fall within a specified range are mapped to fixed RGB values (colors). Figure 8.2.24 illustrates the intensity slicing method for pseudocolor. Figure 8.2.24a shows the gray scale range evenly divided into four different colors. The colors in the first range, 0 to MAX/4, are mapped to Color_1, the second range, MAX/4 to MAX/2, are mapped to Color_2, and so on. If we define Color_i as (R_i, G_i, B_i), we obtain the mapping equations given in Figure 8.2.24b. In this case the equations are constants over specified ranges; however, they can be any type of equations.

Figure 8.2.25 shows two images and various results of spatial domain pseudocolor mapping. The original images are of a mosquito from a scanning electron microscope and a satellite image of the earth. As can be seen in these images, the specific colors chosen are application dependent, and the methods used may experimental or analytical. To use an analytical approach the methodology is to examine the gray level distribution in terms of objects of interest in the image and assign colors as desired. Figure 8.2.25 shows results from using the CVIPtools *Enhancement->Pseudocolor->Gray level mapping* and *Intensity slicing* functions. Gray level mapping allows the user to select the mapping equation from a set of standard mappings; while with Intensity slicing selection of up to four gray level ranges and four colors are allowed. For maximum flexibility, the user can use *Enhancement->Pseudocolor->Gray level mapping II*, which provides the user with a graphical interface to enter mapping equations.

In addition to operation in the spatial domain, we can perform pseudocolor in the frequency domain. This is typically accomplished by performing a Fourier transform on the image, and then applying a lowpass, bandpass, and highpass filter to the transformed data. These three filtered outputs are then inverse transformed and the individual outputs are used as the RGB components of the color image. A block diagram of the process is illustrated in Figure 8.2.26a. Typical postprocessing includes histogram equalization, but is application-dependent. Although these filters may be of any type, they are often chosen to cover the entire frequency domain by dividing it into three separate bands, corresponding to lowpass, bandpass, and highpass filters (Figure 8.2.26b). Figure 8.2.27 shows image examples of frequency domain pseudocolor mapping.

The pseudocolor techniques provide us with methods to change a gray scale image into a color image. Additionally, we may wish to apply some of the enhancement techniques, such as histogram modification, directly to color images. One method for

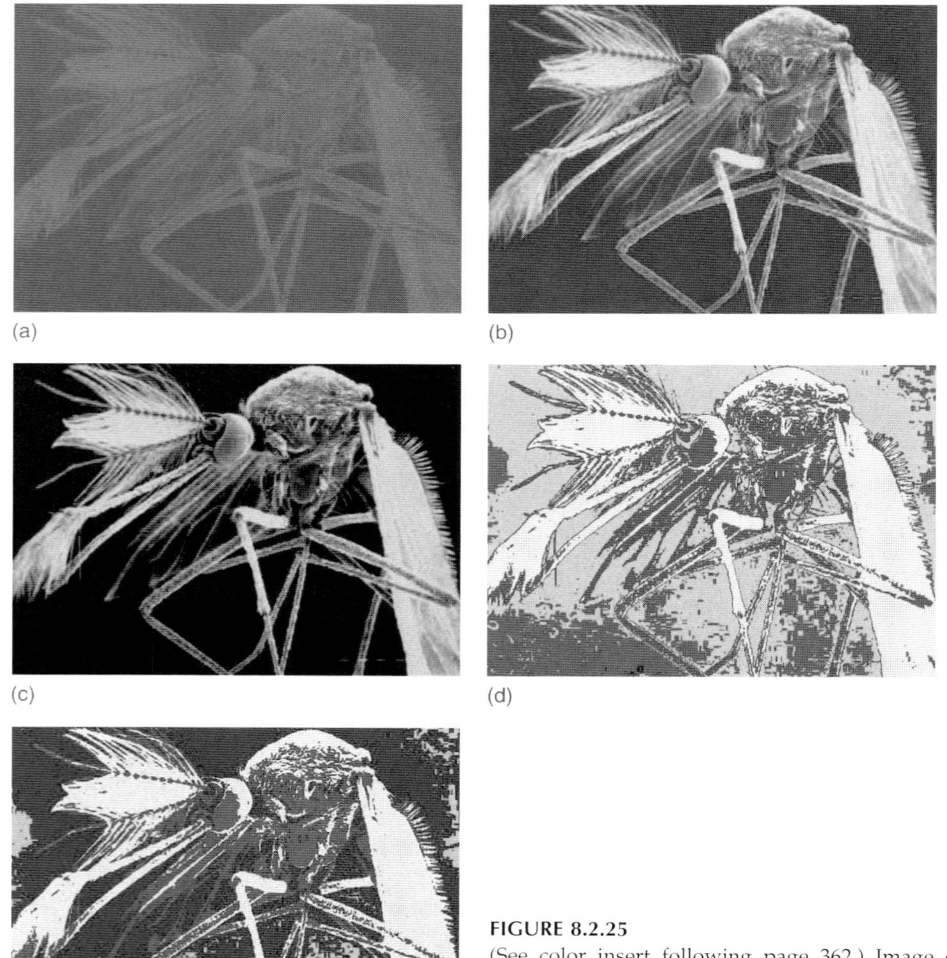

FIGURE 8.2.25
(See color insert following page 362.) Image examples of pseudocolor in the spatial domain. (a) Original scanning electron image of a mosquito (photo courtesy of Sue Eder, SIUE). (b,c) Gray level mapping pseudcolor. (d,e) Intensity slicing pseudocolor.

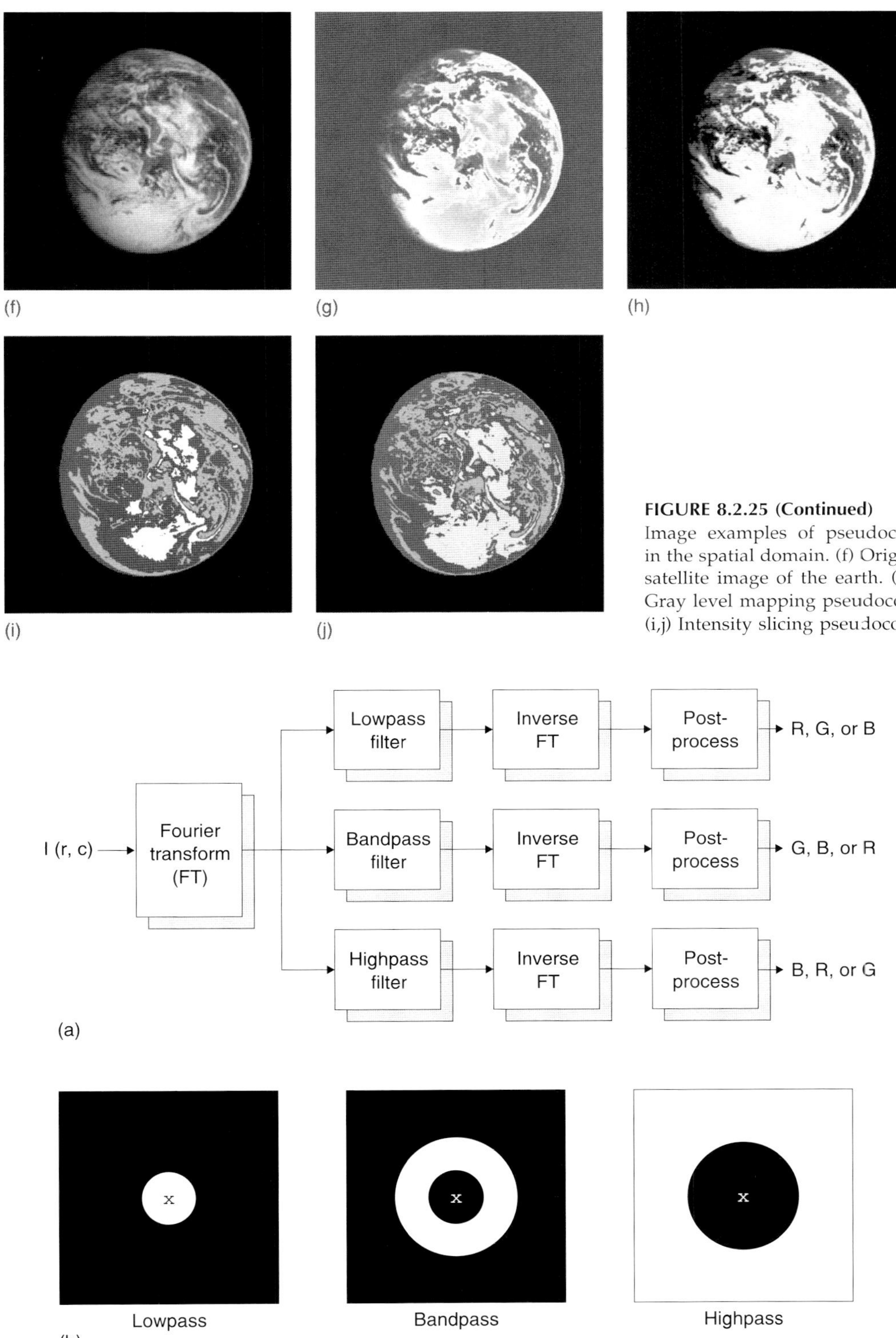

FIGURE 8.2.25 (Continued)
Image examples of pseudocolor in the spatial domain. (f) Original satellite image of the earth. (g,h) Gray level mapping pseudocolor. (i,j) Intensity slicing pseudocolor.

FIGURE 8.2.26
Pseudocolor in the frequency domain. (a) Block diagram of process. (b) Fourier filters (x = origin).

FIGURE 8.2.27
(See color insert following page 362.) Frequency domain pseudo-color. (a) Original 256 × 256 image, so in the Fourier domain the highest frequency is 128. (b) Result with cutoff frequencies of 10 and 100, with the lowpass result mapped to red, bandpass to green, and highpass to blue. (c) Result with cutoff frequencies of 10 and 100, with the lowpass mapped to blue, bandpass to green, and highpass to red. (d) Result with cutoff frequencies of 5 and 50, with the lowpass mapped to red, bandpass to blue, and highpass to green.

doing this is to treat color images as three band gray scale images. Thus, we can apply any and all of the gray scale modification techniques, including histogram modification, to color images by applying the method to each color band separately. The problem with this approach is that the colors will change, which is typically not the desired effect. We need to retain the relative color (the ratios between red, green, and blue for each pixel), in order to avoid color shifts.

The relative color can be retained by applying the gray scale modification technique to one of the color bands (red, green, or blue), and then using the ratios from the original image to find the other values (see Figure 8.2.28). Typically the most important color band is selected, and this choice is very much application-specific. This technique will not always provide us with the desired result, either. Often, we really want to apply the gray scale modification method to the image brightness only, even with color images. This is done by using the HSL transform, applying the gray scale modification technique to the brightness band only (L), and then performing the inverse HSL transform. This effect will be similar to application to gray scale images.

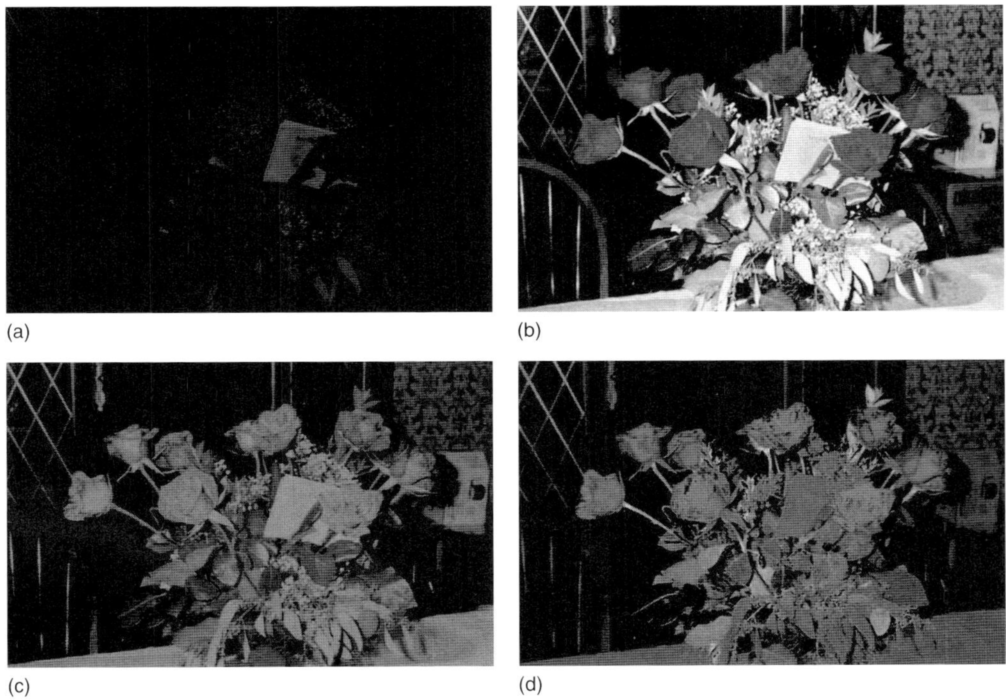

FIGURE 8.2.28
(See color insert following page 362.) Histogram equalization of color images. (a) Original poor contrast image. (b) Histogram equalization based on the red color band. (c) Histogram equalization based on the green color band. (d) Histogram equalization based on the blue color band. Note that in this case the red band gives the best results. This will depend on the image and the desired result.

8.3 Image Sharpening

Image sharpening deals with enhancing detail information in an image. The detail information is typically contained in the high spatial frequency components of the image, so most of the techniques contain some form of highpass filtering. The detail information includes edges, and, in general, correspond to image features that are small spatially. This information is visually important, because it delineates object and feature boundaries, and is important for textures in objects.

In the following sections, representative algorithms and techniques for image sharpening are discussed. Mask operations in the spatial domain and their equivalent global operations in the frequency domain are considered. Many image sharpening algorithms consist of three general steps: (1) extract high frequency information, (2) combine the high frequency image with the original image to emphasize image detail, and (3) maximizing image contrast via histogram manipulation.

8.3.1 Highpass Filtering

Filters that emphasize high frequency information have been introduced in Chapter 4 (spatial domain) and Chapter 5 (frequency domain). Here we will consider techniques

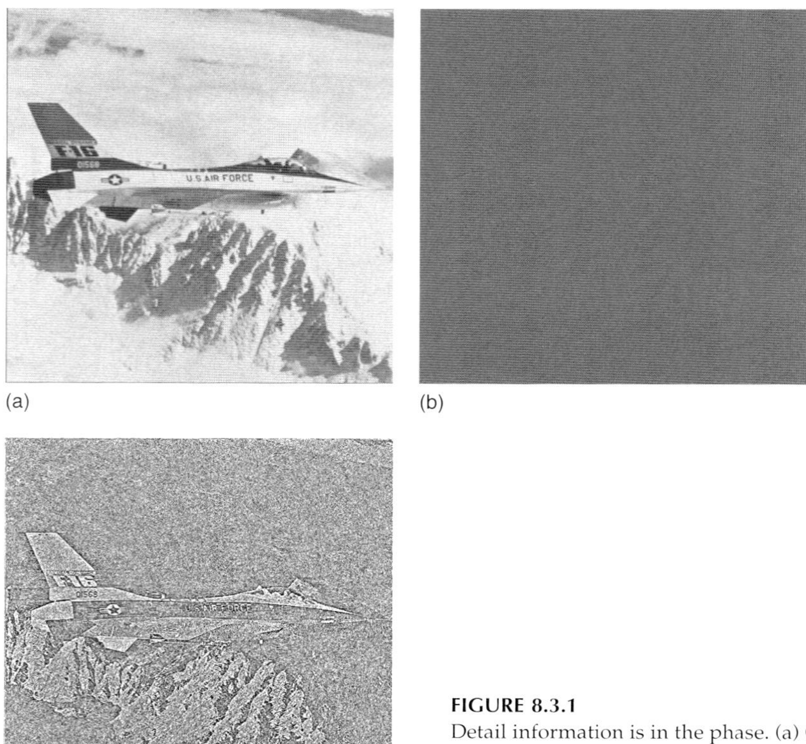

FIGURE 8.3.1
Detail information is in the phase. (a) Original image. (b) Results of performing a Fourier transform, normalizing the magnitudes to 1, and then performing the inverse transform—this provides results based on phase only. (c) Histogram equalized version of (b).

specifically for image sharpening. Highpass filtering for image enhancement typically requires some form of post-processing, such as histogram equalization, to create an acceptable image. Additionally, highpass filtering alone is seldom used for enhancement, but is often part of a more complex enhancement algorithm such as those discussed in later sections. Highpass filtering, in the form of edge detection, is often used in computer vision applications to delineate object outlines.

Edge detectors are spatial domain convolution mask approximations to the equivalent frequency domain filter. One method to find an approximate spatial convolution mask that minimizes mean square error is to use the Moore–Penrose generalized inverse matrix. This technique is beyond the scope of the discussion here, but more information can be found in the references. *Phase contrast filtering*, also discussed in the references, is similar to highpass filtering, but is based on the assumption that most visual information is in the phase (see Figure 8.3.1).

8.3.2 High Frequency Emphasis

As we have seen in Chapter 5, high frequency emphasis can be used to enhance details in an image (Figure 5.7.8). The highpass filter alone will accentuate edges in the image, but loses a large portion of the visual information by filtering out the low spatial frequency components. This problem is solved with the high frequency emphasis filter which boosts the high frequencies and retains some of the low frequency information (see Figures 5.7.7 and 5.7.8) by adding an offset to the filter function. When this is done, care must be taken to avoid overflow in the resulting image. The results from overflow will appear as noise, typically white and black points (depending on how the data conversion is handled). This problem can be avoided by careful use of proper data types, correct data conversion when necessary, and appropriate remapping of the data before display.

A similar result can be obtained in the spatial domain by using a high boost spatial filter. The high boost spatial filter mask is of the following form:

$$\begin{bmatrix} -1 & -1 & -1 \\ -1 & x & -1 \\ -1 & -1 & -1 \end{bmatrix}$$

This mask is convolved with the image, and the value of x determines the amount of low frequency information retained in the resulting image. A value of 8 will result in a highpass filter (the output image will contain only the edges), while larger values will retain more of the original image. If values of less than 8 are used for x, the resulting image will appear as a negative of the original. Figure 8.3.2 shows the results from using various values of x for high boost spatial filtering. The resultant images are also shown with a histogram stretch as a post-processing step for further enhancement.

As was done with the edge detection spatial masks, the high boost mask can be extended with -1's and a corresponding increase in the value of x. Larger masks will emphasize the edges more (make them wider), and help to mitigate the effects of any noise in the original image. For example a 5×5 version of this mask is:

$$\begin{bmatrix} -1 & -1 & -1 & -1 & -1 \\ -1 & -1 & -1 & -1 & -1 \\ -1 & -1 & x & -1 & -1 \\ -1 & -1 & -1 & -1 & -1 \\ -1 & -1 & -1 & -1 & -1 \end{bmatrix}$$

If we create an $N \times N$ mask, the value for x for a highpass filter is $N \times N - 1$, in this case 24 ($5 \times 5 - 1$). Note that other forms for the highboost spatial sharpening mask can be generated such as the following:

$$\begin{bmatrix} 1 & -2 & 1 \\ -2 & x & -2 \\ 1 & -2 & 1 \end{bmatrix} \begin{bmatrix} 0 & -1 & 0 \\ -1 & x & -1 \\ 0 & -1 & 0 \end{bmatrix}$$

8.3.3 Directional Difference Filters

Directional difference filters are similar to the spatial domain high boost filter, but emphasize the edges in a specific direction. These filters are also called *emboss filters*, due to the effect they create on the output image. The filter masks are of the following form:

$$\begin{bmatrix} 0 & +1 & 0 \\ 0 & 0 & 0 \\ 0 & -1 & 0 \end{bmatrix} \quad \begin{bmatrix} +1 & 0 & 0 \\ 0 & 0 & 0 \\ 0 & 0 & -1 \end{bmatrix} \quad \begin{bmatrix} 0 & 0 & 0 \\ +1 & 0 & -1 \\ 0 & 0 & 0 \end{bmatrix} \quad \begin{bmatrix} 0 & 0 & -1 \\ 0 & 0 & 0 \\ +1 & 0 & 0 \end{bmatrix}$$

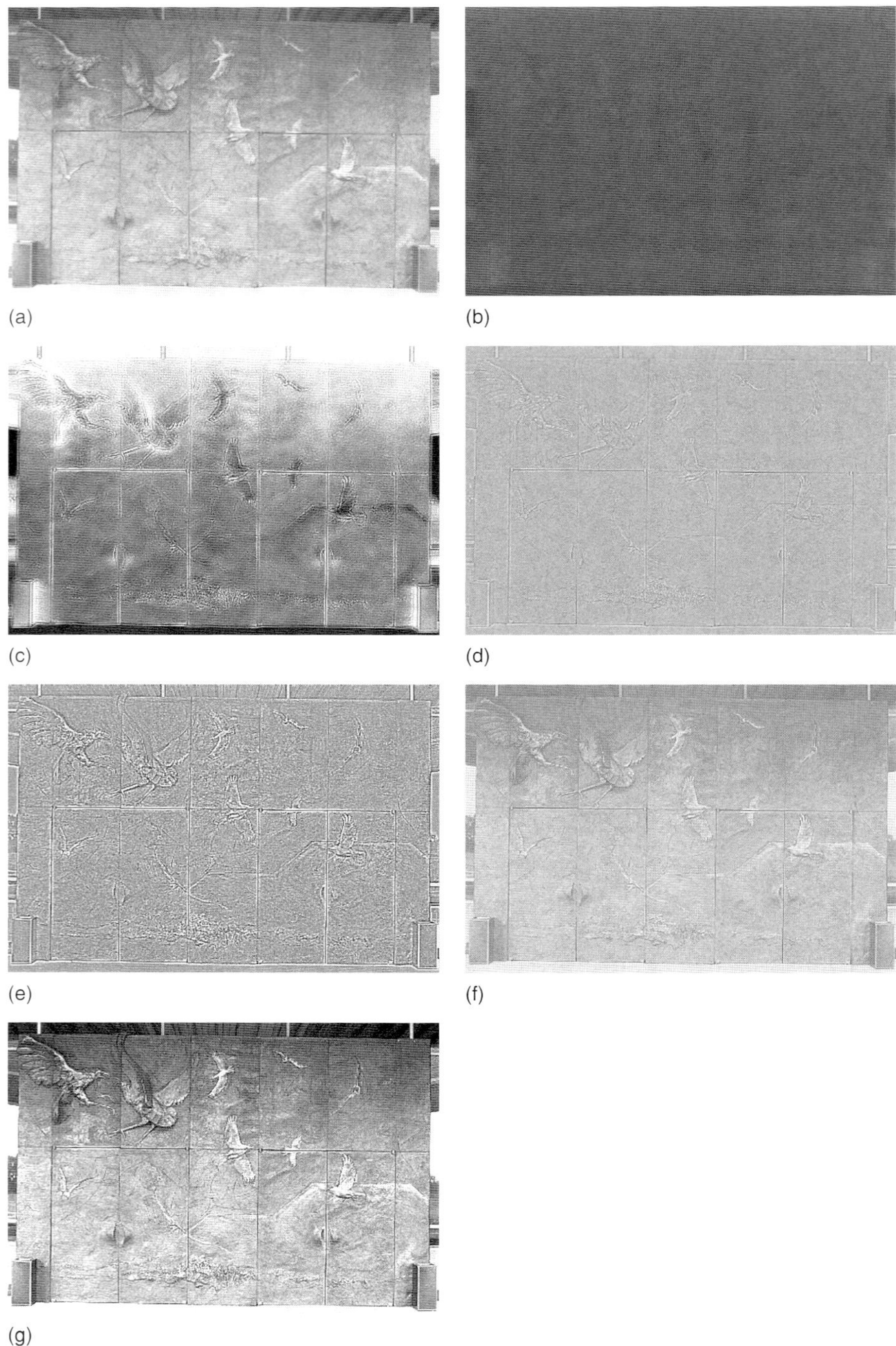

FIGURE 8.3.2
High boost spatial filtering (a) Original image. (b) results of performing a highboost spatial filter with a 3×3 mask and $x = 6$. (c) Histogram stretched version of (b). Note the image is a negative of the original. (d) Results of performing a highboost spatial filter with a 3×3 mask and $x = 8$. (e) Histogram stretched version of (d), note the image contains edge information only. (f) Results of performing a highboost spatial filter with a 3×3 mask and $x = 12$. (g) Histogram stretched version of (f).

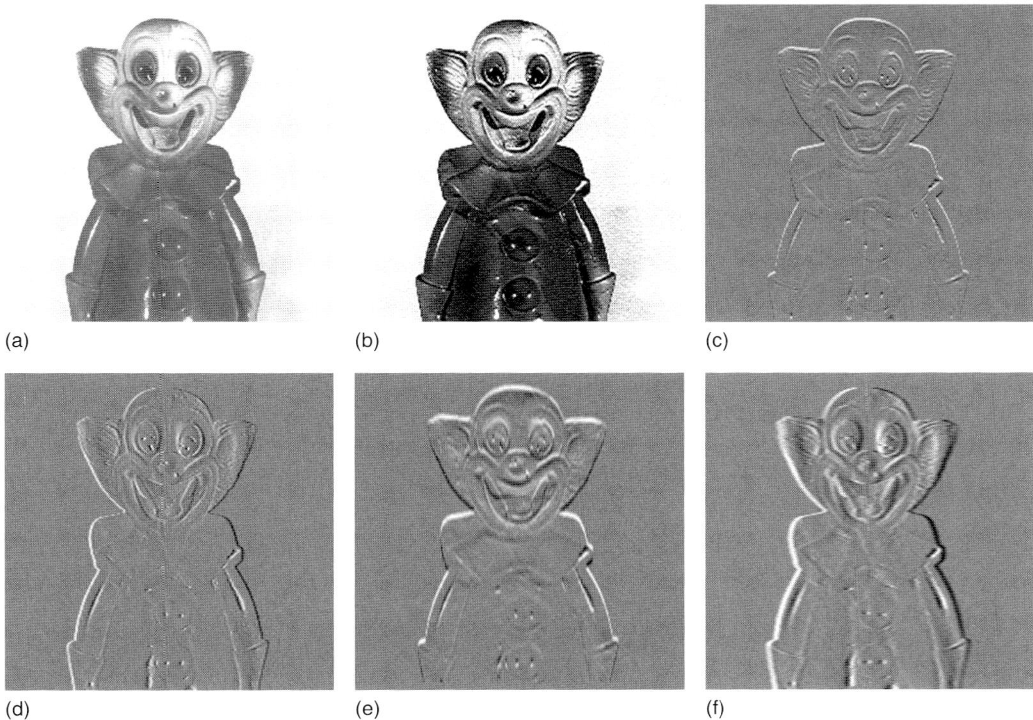

FIGURE 8.3.3
Directional difference filters. (a) Original image. (b) Image sharpened by adding the difference filter result to the original image, followed by a histogram stretch. (c) 3×3 filter result with the +1 and −1 in the vertical direction, which emphasizes horizontal lines. (d) 3×3 filter result with the +1 and −1 in the horizontal direction, which emphasizes vertical lines. (e) 7×7 filter result with the +1 and −1's in the vertical direction, which emphasizes horizontal lines. (f) 7×7 filter result with the +1 and −1's in the horizontal direction, which emphasizes vertical lines.

where a different directional mask can be created by rotating the outer +1 and −1. Larger masks can be generated by extending the +1 and −1's as follows:

$$\begin{bmatrix} 0 & 0 & +1 & 0 & 0 \\ 0 & 0 & +1 & 0 & 0 \\ 0 & 0 & 0 & 0 & 0 \\ 0 & 0 & -1 & 0 & 0 \\ 0 & 0 & -1 & 0 & 0 \end{bmatrix} \quad \begin{bmatrix} +1 & 0 & 0 & 0 & 0 \\ 0 & +1 & 0 & 0 & 0 \\ 0 & 0 & 0 & 0 & 0 \\ 0 & 0 & 0 & -1 & 0 \\ 0 & 0 & 0 & 0 & -1 \end{bmatrix} \quad \begin{bmatrix} 0 & 0 & 0 & 0 & 0 \\ 0 & 0 & 0 & 0 & 0 \\ +1 & +1 & 0 & -1 & -1 \\ 0 & 0 & 0 & 0 & 0 \\ 0 & 0 & 0 & 0 & 0 \end{bmatrix}$$

Increasing the mask size will create wider edges in the resultant image. Figure 8.3.3 illustrates the results of using these filters. In CVIPtools these filters are part of *Utilities->Filter*.

8.3.4 Homomorphic Filtering

The digital images we process are created from optical images. Optical images consist of two primary components, the lighting component and the reflectance component. The lighting component results from the lighting conditions present when the image is captured, and can change as the lighting conditions change. The reflectance component results from the way the objects in the image reflect light and are determined by the intrinsic properties of the object itself, which (normally) do not change. In many applications it is useful to enhance the reflectance component, while reducing the contribution from the lighting component. *Homomorphic filtering* is a frequency domain filtering process that compresses the brightness (from the lighting conditions), while enhancing the contrast (from the reflectance).

The image model for homomorphic filters is as follows:

$$I(r,c) = L(r,c)R(r,c)$$

where $L(r,c)$ represents the contribution of the lighting conditions and $R(r,c)$ represents the contribution of the reflectance properties of the objects.

The homomorphic filtering process assumes that $L(r,c)$ consists of primarily slow spatial changes (low spatial frequencies), and is responsible for the overall range of the brightness in the image. The assumptions for $R(r,c)$ are that it consists primarily of high spatial frequency information, which is especially true at object boundaries, and it is responsible for the local contrast (the spread of the brightness range within a small spatial area). These simplifying assumptions are valid for many types of real images.

The homomorphic filtering process consists of five steps: (1) a natural log transform (base e), (2) the Fourier transform, (3) Filtering, (4) the inverse Fourier transform, and (5) the inverse log function—the exponential. This process is illustrated in a block diagram in Figure 8.3.4. The first step allows us to decouple the $L(r,c)$ and $R(r,c)$ components, since the logarithm function changes a product into a sum. Step 2 puts the image into the frequency domain, so that we can perform the filtering in Step 3. Next, Steps 4 and 5 do the inverse transforms from Steps 1 and 2, to get our image data back into the spatial domain. The only factor left to be considered is the filter function, $H(u,v)$.

The typical filter for the homomorphic filtering process is shown in Figure 8.3.5. Here we see that we can specify three parameters—the high frequency gain, the low frequency gain, and the cutoff frequency. Typically the high frequency gain is greater than 1, and the low frequency gain is less than 1. This provides us with the desired effect of boosting the $R(r,c)$ components, while reducing the $L(r,c)$ components. The selection of the cutoff frequency is highly application-specific, and needs to be chosen so that no important information is lost. In practice the values for all three parameters are often determined empirically.

Figure 8.3.6 shows results from application of homomorphic filtering to a poor image. In this case, the homomorphic filter returns an image of low contrast, so the contrast is enhanced by a histogram stretch procedure. A comparison is made between the homomorphic filter followed by a histogram stretch (Figure 8.3.6b) and simply stretching the original image's histogram (Figure 8.3.6c). We see that the homomorphic filter provides an image with greater visual detail, where it is much easier to see the pilot in the helicopter.

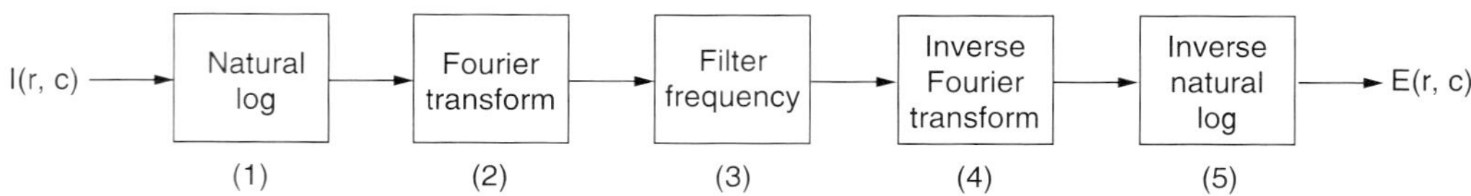

FIGURE 8.3.4
The homomorphic filtering process.

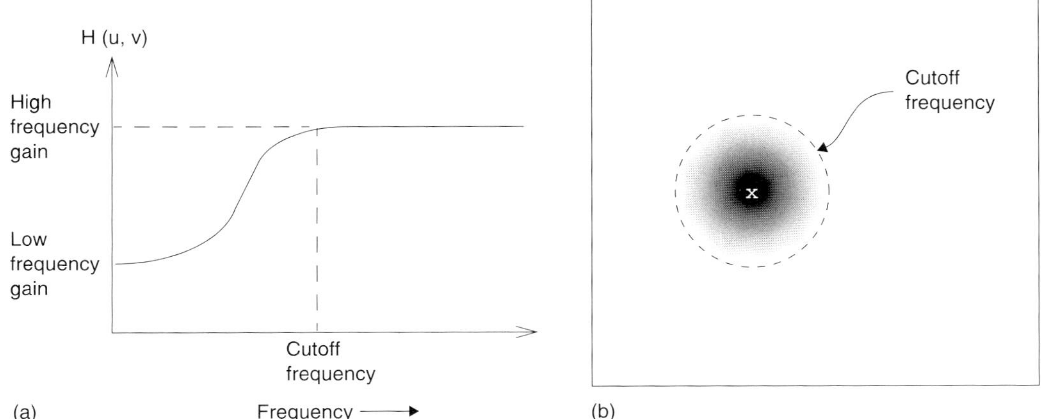

FIGURE 8.3.5
(a) Cross-section of homomorphic filter, $H(u, v)$. (b) 2-D filter diagram (x = origin).

FIGURE 8.3.6
Homomorphic filtering example. (a) Original image. (b) Result of homomorphic filter followed by histogram stretching: upper gain = 1.2, lower gain = 0.5, cutoff frequency = 16. (c) Result of histogram stretch without the use of homomorphic filtering.

8.3.5 Unsharp Masking

The unsharp masking algorithm has been used for many years by photographers to enhance images. It sharpens the image by subtracting a blurred (lowpass) version of the original image. This was accomplished during film development by superimposing

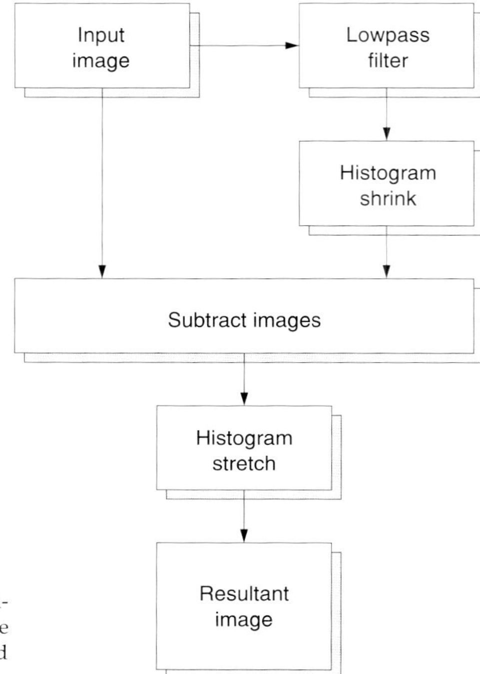

FIGURE 8.3.7
Unsharp masking enhancement algorithm flowchart. Unsharp masking subtracts a blurred (lowpassed) version of the original, which is similar to adding an edge-enhanced (highpassed) version.

a blurred negative onto the corresponding positive film to produce a sharper result. The process is similar to *adding* a detail enhanced (highpass) version of the image to the original. To improve image contrast we have included histogram modification as part of our unsharp masking enhancement algorithm.

A flowchart for this process is shown in Figure 8.3.7. Here we see that the original image is lowpass filtered, followed by a histogram shrink to the lowpass filtered image. The resultant image from these two operations is then subtracted from the original image, and the result of this operation undergoes a histogram stretch to restore the image contrast. This process works because subtracting a slowly changing edge (the lowpass filtered image) from faster changing edges (in the original), has the visual effect of causing overshoot and undershoot at the edges, which has the effect of emphasizing the edges. By the scaling the lowpassed image with a histogram shrink we can control the amount of edge emphasis desired. In Figure 8.3.8, we see results of application of the unsharp masking algorithm with different ranges for the histogram shrink process. Here we see that as the range for the histogram shrink is increased, the resulting image has a greater edge emphasis.

8.3.6 Edge Detector-Based Sharpening Algorithms

The following two algorithms are implemented in CVIPtools as Sharpening Algorithm I and II. They are both based on using edge detectors to enhance the edges, followed by contrast enhancement via histogram modification. Sharpening Algorithm I works as follows: (1) performs a Laplacian edge detection on the original image, (2) performs

FIGURE 8.3.8
Unsharp masking. (a) Original image. (b) Unsharp masking with lower limit = 0, upper = 100, with 2% low and high clipping. (c) Unsharp masking with lower limit = 0, upper = 150, with 2% low and high clipping. (d) Unsharp masking with lower limit = 0, upper = 200, with 2% low and high clipping.

a Sobel edge detection on the original image, (3) multiplies the resulting images from the Laplacian and the Sobel, (4) adds the product from (3) to the original image, (5) histogram stretches the result from (4).

In CVIPtools various options are available for this algorithm. With the *Intermediate Image Remapping* option you can remap the product (from Step 3) to BYTE range (0–255) before you add it to the original image. This has the effect of improving image contrast, but with reduced sharpening effect. Two different Laplacian masks can be selected, but Mask II tends to give more raggedy edges. Choosing a larger mask size for the Sobel will tend to brighten the major edges, but will also tend to smear them out slightly. The final two selections, "Low (High) clip percent," determine how much of the histogram is clipped during the final histogram stretch. Figure 8.3.9 shows results from this algorithm.

Sharpening Algorithm II is simpler, faster and provides similar results, but with less flexibility. It works as follows: (1) performs a histogram shrink to the range [10, 100], (2) performs a Roberts edge detection, (3) optionally adds the result to the original image, (4) remaps to BYTE data type, and (5) performs a histogram equalization. To add the result to the original image the user selects the appropriate box. Figure 8.3.10 shows results from this algorithm.

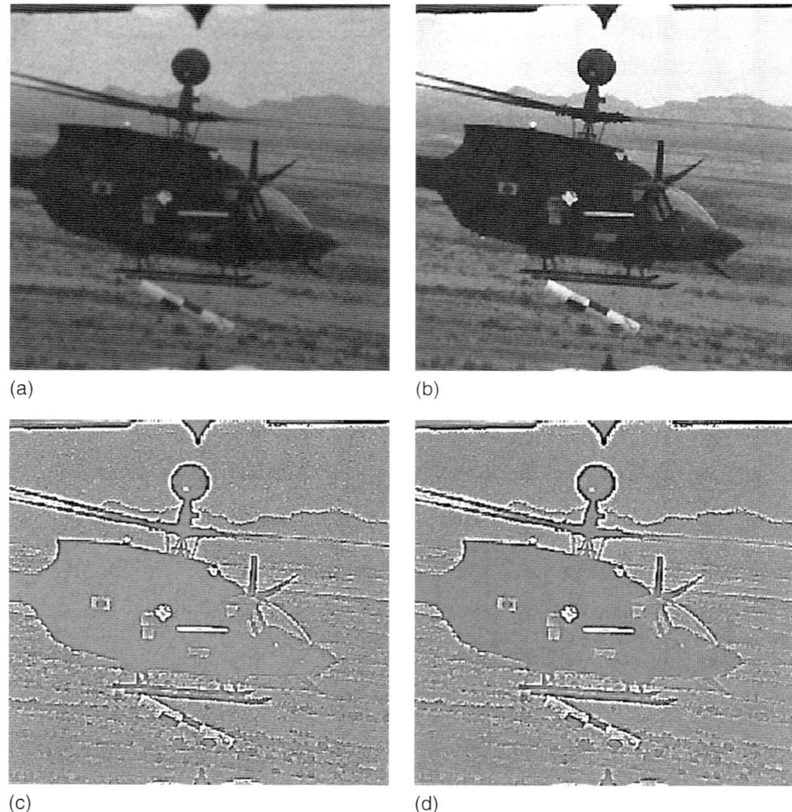

(a) (b) (c) (d)

FIGURE 8.3.9
Sharpening Algorithm I. (a) Original image. (b) Parameters: Laplacian Mask I, 3 × 3 Sobel, 0.005 clipping on both low and high end for the histogram stretch, intermediate image is remapped, and result is added to original image. (c) Parameters: same as (b) except intermediate image is not remapped, and 0.03 is used for clipping on the histogram stretch. (d) Parameters: same as (c), but with a 7 × 7 Sobel.

(a) (b)

FIGURE 8.3.10
Sharpening Algorithm II. (a) Original image. (b) Results of Sharpening Algorithm II. Note that some of the image artifacts, especially in the sky, become quite prominent.

8.4 Image Smoothing

Image smoothing is used for two primary purposes: to give an image a softer or special effect, or to mitigate noise effects. Chapter 9 discusses smoothing filters in more depth for handling noise in images; for image enhancement we will focus on basic smoothing methods for creating a softer effect. Image smoothing is accomplished in the spatial domain by considering a pixel and its neighbors and eliminating any extreme values with median filters or by averaging with neighboring pixels with mean filters. In the frequency domain, image smoothing is accomplished by some form of lowpass filtering. Since the high spatial frequencies contain the detail, including edge information, the elimination of this information via lowpass filtering will provide a smoother image. Any fast or sharp transitions in the image brightness will be filtered out, thus providing the desired effect.

8.4.1 Frequency Domain Lowpass Filtering

In the frequency domain, lowpass filtering can be done as discussed in Chapter 5. An ideal filter can cause undesirable artifacts, while the Butterworth filter does not. Lowpass filtering creates an image with a smooth appearance since it suppresses any rapidly changing brightness values in the original image. The lowpass filters do this by attenuating high spatial frequency information, which corresponds to the rapid changes (edges). The amount of information suppressed is determined by the cutoff frequency of the filter.

8.4.2 Convolution Mask Lowpass Filtering

As was seen in Chapter 5, the convolution theorem allows us to use filter masks in the spatial domain to perform filtering. Given a frequency domain filter specification, an equivalent convolution mask can be approximated with the Moore–Penrose generalized inverse (see references). For lowpass filtering these masks are typically some form of average (mean) filters. The coefficients for these filter masks are all positive, unlike the highpass filters where the center is surrounded by negative coefficients. Here are some common spatial convolution masks for lowpass filtering, where the first two filters are standard arithmetic mean filters, and the last two masks are approximations to Gaussian filters:

$$\begin{bmatrix} 1 & 1 & 1 \\ 1 & 1 & 1 \\ 1 & 1 & 1 \end{bmatrix} \quad \begin{bmatrix} 1 & 1 & 1 \\ 1 & 2 & 1 \\ 1 & 1 & 1 \end{bmatrix} \quad \begin{bmatrix} 2 & 1 & 2 \\ 1 & 4 & 1 \\ 2 & 1 & 2 \end{bmatrix} \quad \begin{bmatrix} 1 & 2 & 1 \\ 2 & 4 & 2 \\ 1 & 2 & 1 \end{bmatrix}$$

As is seen, the coefficients in the mask may be biased, that is, they may not all be 1's. This is typically for application specific reasons. For example, we may want to weight the center pixel, or the diagonal pixels, more heavily than the other pixels. Note also that these types of masks are often multiplied by $1/N$, where N is the sum of the mask coefficients. As examples, the first mask is multiplied by 1/9, the second by 1/10, and so on. This is the equivalent of linearly remapping the image data (typically to BYTE) after the convolution operation.

FIGURE 8.4.1
Mean filters (3 × 3). These filters create a soft effect in images. (a) Original image. (b) Arithmetic filter. (c) Midpoint filter. (d) Gaussian filter. (e) Contra-harmonic, order = +1. (f) Y_p mean, order = +1.

The use of mean filters to eliminate noise in images is discussed in Chapter 9, where we will see that the various types of mean filters are most effective with different types of noise. For image smoothing, the results from most of the spatial mean filters are visually similar—the image is smoothed out providing a softer visual effect (see Figure 8.4.1). We can use a larger mask size for a greater smoothing effect. In Figure 8.4.2 we see the results of using an arithmetic filter and various mask sizes. We can see that as the mask size increases, the amount of smoothing increases, and at some point the smoothing becomes a noticeable blurring. By comparing the same size arithmetic mean and Gaussian filters, we see that the Gaussian creates a more natural effect and less noticeable blurring as the mask size increases.

8.4.3 Median Filtering

A median filter (Chapters 3 and 9) can also be used to create a similar smoothing effect, but with large mask sizes it creates an almost painted (and blurred) look. In Figure 8.4.3 we see the results from applying a median filter with various mask sizes.

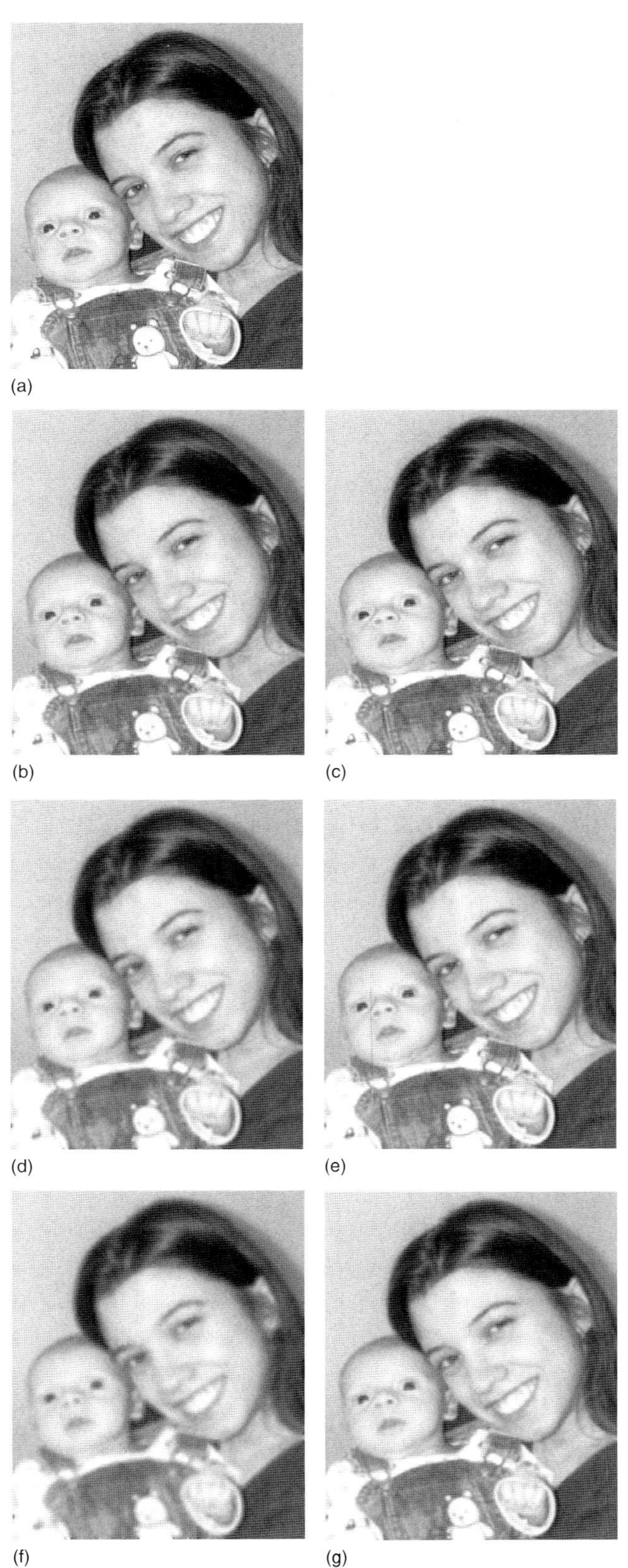

FIGURE 8.4.2
Image smoothing with arithmetic mean and Gaussian filters by varying the mask size. (a) Original image. (b) Arithmetic mean, 3×3. (c) Gaussian, 3×3. (d) Arithmetic mean, 5×5. (e) Gaussian, 5×5, (f) Arithmetic mean, 7×7. (g) Gaussian, 7×7.

(a) (b) (c)

(d) (e) (f)

FIGURE 8.4.3
Image smoothing with a median filter. (a) Original image. (b) 3×3 median filter. (c) 5×5 median filter. (d) 7×7 median filter. (e) 9×9 median filter. (f) 11×11 median filter. Note that as ther filter size increases above a 5×5 the resulting image acquires a painted effect.

We can see that details smaller than the mask size are eliminated, and as the mask gets larger the image begins to take on an artificial look. With a large mask the median filter will take a long time to process, so in practice a fast algorithm or a pseudo-median filter may be used.

Fast algorithms for median filtering operate by efficiently maintaining the sorting of the data as we move across the image. An alternative to using a fast algorithm is the *pseudomedian filter*, which approximates the operation of a median filter, but is simpler and more computationally efficient. The pseudomedian is defined as follows:

$$\text{PMED}(S_L) = (1/2)\text{MAXIMIN}(S_L) + (1/2)\text{MINIMAX}(S_L)$$

where S_L denotes a sequence of elements $s_1, s_2, \ldots, s_L$ and where for $M = (L+1)/2$:

$$\text{MAXIMIN}(S_L) = \text{MAX}\big[[\text{MIN}(s_1, \ldots, s_M)], [\text{MIN}(s_2, \ldots, s_{M+1})], \ldots, [\text{MIN}(s_{LM+1}, \ldots, s_L)]\big]$$

$$\text{MINIMAX}(S_L) = \text{MIN}\big[[\text{MAX}(s_1, \ldots, s_M)], [\text{MAX}(s_2, \ldots, s_{M+1})], \ldots, [\text{MAX}(s_{LM+1}, \ldots, s_L)]\big]$$

For example, the pseudomedian of length five is defined as:

$$\text{PMED}(a,b,c,d,e) = (1/2)\text{MAX}[\text{MIN}(a,b,c), \text{MIN}(b,c,d), \text{MIN}(c,d,e)] + (1/2)\,\text{MIN}[\text{MAX}(a,b,c), \text{MAX}(b,c,d), \text{MAX}(c,d,e)]$$

The MIN followed by MAX contributions of the first part of the equation always result in the actual median or a value smaller, while the MAX followed by the MIN contributions result in the actual median or a value larger. The average of the two contributions tends to cancel out the biases, thus creating a valid approximation to the median filtering process.

8.5 Key Points

OVERVIEW: IMAGE ENHANCEMENT

- Image enhancement techniques are employed to emphasize, sharpen, and/or smooth image features for display and analysis
- *Image enhancement* is the process of applying these techniques to facilitate the development of a solution to a computer imaging problem
- Enhancement methods are application-specific and often developed empirically
- We define the enhanced image as $E(r, c)$, and application feedback is an important part of the process (see Figure 8.1.1)
- Enhancement methods operate in both the spatial and frequency/sequency domains (see Figure 8.1.2)
- Three types of operation: (1) point operations—each pixel is modified by an equation that is not dependent on other pixel values, (2) mask operations—each pixel is modified based on its values as well as its neighbors, (3) global operations—all pixel values in the image are needed
- Enhancement techniques are used in both preprocessing, to ease the further imaging tasks, and in post-processing to create a more visually desirable image
- Enhancement methods are also used as an end in itself to improve images for viewing

GRAY SCALE MODIFICATION

- Gray scale modification is also called gray level scaling or gray level modification
- Gray scale modification is the process of taking the original gray level values and changing them to improve the image
- Typically this is performed to improve image contrast and/or brightness
- Image contrast is a measure of the distribution and range of the gray levels, while image brightness usually refers to the overall average, or mean, pixel value in the image

MAPPING EQUATIONS

- Mapping equations provide a general technique for modifying the gray level values in an image
- The *mapping equation* changes the pixel's (gray level) values based on a mathematical function that uses brightness values as input and outputs the enhanced pixel values
- The mapping equation is typically, but not necessarily, linear
- Nonlinear equations can be mapped by piecewise linear equations
- The use of mapping equations to modify the gray scale belong in the category of point operations
- Notation: Mapping Equation$\rightarrow E(r,c) = M[I(r,c)]$ where $M[\,]$ is the mapping equation
- A linear mapping equation will perform *gray level stretching* if the slope is greater than one, and *gray level compression* if the slope is less than one
- A digital negative can be created with:

$$M[I(r,c)] = \text{MAX} - I(r,c)$$

  where MAX is the maximum gray value
- *Intensity level slicing* is a particular kind of gray level mapping where specific ranges on the input image are mapped to easily seen values in the output image, such as white = 255
- In CVIPtools gray scale modification is done with *Enhancement->Histograms->Linear Modification*; *Enhancement->Pseudocolor->Gray level mapping* can also be used in conjunction with *Utilities->Create->Extract Band*
- If the dynamic range on the input data is very large, log transforms may be used, such as with Fourier spectral data (see Section 5.2.4)
- Imaging equipment, such as cameras, displays and printers typically react according to a *power-law equation* of the following form:

$$E(r,c) = M[I(r,c)] = K_1[I(r,c)]^{\gamma}$$

  where K_1 and γ are positive constants
- A device with a response of the power-law transform, can be compensated for by application of a *gamma-correction* equation of the following form:

$$E(r,c) = M[I(r,c)] = K_2[I(r,c)]^{1/\gamma}$$

  where K_2 and γ are positive constants
- Gamma correction is important for proper display of images, whether on a computer monitor or on a printed page

Histogram Modification

- Histogram modification performs a function similar to gray level mapping, but works by consideration of the histogram shape and spread
- The basic operations are: stretch, to increase contrast, shrink, to reduce contrast, and slide, to change average brightness

- Histogram shrinking and stretching are also called *histogram scaling*
- Histogram shrink:

$$\text{Shrink}\,(I(r,c)) = \left[\frac{\text{Shrink}_{\text{MAX}} - \text{Shrink}_{\text{MIN}}}{I(r,c)_{\text{MAX}} - I(r,c)_{\text{MIN}}}\right]\left[I(r,c) - I(r,c)_{\text{MIN}}\right] + \text{Shrink}_{\text{MIN}}$$

- Histogram stretch:

$$\text{Stretch}\ \ (I(r,c)) = \left[\frac{I(r,c) - I(r,c)_{\text{MIN}}}{I(r,c)_{\text{MAX}} - I(r,c)_{\text{MIN}}}\right][\text{MAX} - \text{MIN}] + \text{MIN}$$

- Histogram slide: Slide $(I(r,c)) = I(r,c) + \text{OFFSET}$
- *Histogram equalization* is a technique where the histogram of the resultant image is as flat as possible (see example and Figures 8.2.14 and 8.2.15)
- The histogram equalization process for digital images consists of four basic steps: (1) find the running sum of the histogram values, (2) normalize the values from Step (1) by dividing by the total number of pixels, (3) multiply the values from Step (2) by the maximum gray level value and round, and (4) map the gray level values to the results from Step (3) using a one-to-one correspondence
- *Histogram specification* is the process of modifying the histogram of an image to match a specified histogram
- The histogram specification process is implemented by the following steps (see example and Figures 8.2.16 and 8.2.17): (1) specify the desired histogram, (2) find the mapping table to histogram equalize the image, Mapping Table 1, (3) find the mapping table to histogram equalize the values of the specified histogram, Mapping Table 2, (4) use mapping Tables 1 and 2 to find the mapping table to map the original values to the histogram equalized values and then to the specified histogram values, (5) use the table from Step (4) to map the original values to the specified histogram values

Adaptive Contrast Enhancement

- *Adaptive contrast enhancement* is modification of the gray level values based on some criterion that adjusts its parameters as local image characteristics change
- It is primarily a mask type operation, but some adaptive contrast operators use global image statistics, hence are also global operations
- The simplest method is to perform histogram equalization on a block-by-block basis, also called local enhancement (see Figure 8.2.18)
- The *Adaptive Contrast Enhancement* (ACE) filters use local image statistics to improve images with uneven contrast:

$$\text{ACE} \Rightarrow E(r,c) = k_1\left[\frac{m_{I(r,c)}}{\sigma_l(r,c)}\right][I(r,c) - m_l(r,c)] + k_2 m_l(r,c)$$

where
$m_{I(r,c)}$ = the mean (average) for the entire image $I(r,c)$
σ_l = local standard deviation in the current window (see Figure 8.2.20)
m_l = local mean in current window

k_1 = local gain factor constant, between 0 and 1
k_2 = local mean constant, between 0 and 1

The ACE2 filter is less computationally intensive than the original ACE filter and provides similar results (see Figure 8.2.21):

$$\text{ACE} \Rightarrow E(r,c) = k_1[I(r,c) - m_l(r,c)] + k_2 m_l(r,c)$$

Other variations include the log and exp ACE filters (see Figures 8.2.22 and 8.2.23)

$$\text{Log-ACE} \Rightarrow E(r,c) = k_1\left[\ln\left(\tilde{I}(r,c)\right) - \ln(\tilde{m}_l(r,c))\right] + k_2\tilde{m}_l(r,c)$$

where
$\tilde{I}(r,c)$ = normalized complement of image = $1 - I(r,c)/\text{MAX}$, MAX is maximum gray value (e.g. 255)
$\tilde{m}_l$ = normalized complement of local mean = $1 - m_l\,(r,c)/\text{MAX}$
m_l = local mean in current window
k_1 = local gain factor constant, between 0 and 1
k_2 = local mean constant, between 0 and 1

$$\text{Exp-ACE} \Rightarrow E(r,c) = \text{MAX} \times \left[\frac{I(r,c)}{\text{MAX}}\right]^{k_1} + \left[\frac{m_l(r,c)}{I(r,c)}\right]^{k_2}$$

where
m_l = local mean in current window
MAX = maximum gray value (e.g. 255)
k_1 = local gain factor exponent
k_2 = local mean factor exponent

Color

- The human visual system can perceive thousands of colors in a small spatial area, but only about 100 gray levels
- Color contrast can be more dramatic than gray level contrast, and various colors have different degrees of psychological impact on the observer
- *Pseudocolor* involves mapping the gray level values of a monochrome image to red, green, and blue values, creating a color image
- Pseudocolor techniques can be applied in both the spatial and frequency domains
- Spatial domain mapping equations:

$$I_R(r,c) = R[I(r,c)]$$

$$I_G(r,c) = G[I(r,c)]$$

$$I_B(r,c) = B[I(r,c)]$$

- *Intensity slicing* is a simple method of spatial domain pseudocolor, where each gray level range is mapped to a specific color (see Figures 8.2.24 and 8.2.25)
- Pseudocolor in the frequency domain is performed by: (1) Fourier transform, (2) lowpass, bandpass, and highpass filter, (3) inverse Fourier transform the three filter results, (4) use the three resulting images for the red, green, and blue color bands (see Figures 8.2.26 and 8.2.27)

- Histogram modification can be performed on color images, but doing it on each color band separately may give an undesirable output—the colors will shift
- To perform histogram modification on color images: (1) retain the RGB ratios and perform the modification on one band only, then use the ratios to get the other two bands' values, or (2) perform a color transform, such as HSL, do the modification on the lightness (brightness band), then do the inverse color transform

IMAGE SHARPENING

- Image sharpening deals with enhancing detail information in an image, typically edges and textures
- Detail information is typically in the high spatial frequency information, so these methods include some form of highpass filtering

Highpass Filtering

- Highpass filters are often an important part of multi-step sharpening algorithms
- Spatial domain highpass filters in the form of convolution masks can be approximated by the Moore–Penrose generalized inverse matrix (see references)
- Phase contrast filtering is a technique with results similar to highpass filtering and is based on the idea that most detail information is in the phase of the Fourier transform (see references and Figure 8.3.1)

High Frequency Emphasis

- A high frequency emphasis (HFE) filter is essentially a highpass filter with an offset in the filter function to boost high frequencies and retain some of the low frequency information
- A spatial domain high boost filter (Figure 8.3.2) provides results similar to the frequency domain HFE filter, and is of the following form:

$$\begin{bmatrix} -1 & -1 & -1 \\ -1 & x & -1 \\ -1 & -1 & -1 \end{bmatrix} \qquad \begin{bmatrix} -1 & -1 & -1 & -1 & -1 \\ -1 & -1 & -1 & -1 & -1 \\ -1 & -1 & x & -1 & -1 \\ -1 & -1 & -1 & -1 & -1 \\ -1 & -1 & -1 & -1 & -1 \end{bmatrix}$$

For a high boost $N \times N$ mask, x should be N^2 or greater, or an image negative will result

Directional Difference Filters

- *Directional difference filters,* also called *emboss filters,* are similar to the spatial domain high boost filter, but emphasize the edges in a specific direction (Figure 8.3.3)
- The filter masks are of the following form:

$$\begin{bmatrix} 0 & +1 & 0 \\ 0 & 0 & 0 \\ 0 & -1 & 0 \end{bmatrix} \qquad \begin{bmatrix} +1 & 0 & 0 \\ 0 & 0 & 0 \\ 0 & 0 & -1 \end{bmatrix} \qquad \begin{bmatrix} 0 & 0 & 0 \\ +1 & 0 & -1 \\ 0 & 0 & 0 \end{bmatrix} \qquad \begin{bmatrix} 0 & 0 & -1 \\ 0 & 0 & 0 \\ +1 & 0 & 0 \end{bmatrix}$$

Homomorphic Filtering

- *Homomorphic filtering* is a frequency domain filtering process that compresses the brightness, while enhancing the contrast
- It is based on modeling the image as a product of the lighting and reflectance properties of the objects
- It assumes that the lighting components in the image are primarily low spatial frequency
- The homomorphic filtering process (Figure 8.3.4): (1) a natural log transform (base e), (2) the Fourier transform, (3) filtering, (4) the inverse Fourier transform, and (5) the inverse log function—the exponential
- The filter used in Step (3) of the process is essentially a high frequency emphasis filter that allows for the specification of both the high and low frequency gain (Figure 8.3.5)

Unsharp Masking

- The *unsharp masking algorithm* (Figure 8.3.7): (1) lowpass filter original, (2) shrink histogram of lowpassed image, (3) subtract result from original, (4) stretch histogram
- The unsharp masking algorithm (results shown in Figure 8.3.8) sharpens the image by subtracting a blurred (lowpass) version of the original image

Edge-Detector-Based Sharpening Algorithms

- Sharpening Algorithm I: (1) performs a Laplacian edge detection on the original image, (2) performs a Sobel edge detection on the original image, (3) multiplies the resulting images from the Laplacian and the Sobel, (4) adds the product from Step (3) to the original image, (5) histogram stretches the result from Step (4). (Figure 8.3.9)
- Sharpening Algorithm II: (1) performs a histogram shrink to the range [10, 100], (2) performs a Roberts edge detection, (3) optionally adds the result to the original image, (4) remaps to BYTE data type, (5) performs a histogram equalization (Figure 8.3.10)

IMAGE SMOOTHING

- Image smoothing for enhancement is to give the image a softer look
- Image smoothing is also used for noise mitigation (see Chapter 9)

Frequency Domain Lowpass Filtering

- Frequency domain lowpass filters smooth images by attenuating high frequency components that correspond to rapidly changing brightness values
- Ideal filters cause undesirable artifacts, but a Butterworth filter does not (see Chapter 5)

Convolution Mask Lowpass Filtering

- The masks are spatial domain approximations to frequency domain filters (see Moore–Penrose matrix reference)
- The masks are typically some form of an averaging filter, such as arithmetic or Gaussian approximations, such as:

$$\begin{bmatrix} 1 & 1 & 1 \\ 1 & 1 & 1 \\ 1 & 1 & 1 \end{bmatrix} \quad \begin{bmatrix} 1 & 1 & 1 \\ 1 & 2 & 1 \\ 1 & 1 & 1 \end{bmatrix} \quad \begin{bmatrix} 2 & 1 & 2 \\ 1 & 4 & 1 \\ 2 & 1 & 2 \end{bmatrix} \quad \begin{bmatrix} 1 & 2 & 1 \\ 2 & 4 & 2 \\ 1 & 2 & 1 \end{bmatrix}$$

- Lowpass filters tend to blur an image, which creates a smooth or softer effect (Figure 8.4.1)
- Increasing the mask size increases the blur amount (Figure 8.4.2)
- Other mean filters are explored in more detail in Chapter 9

Median Filtering

- Median filters (Chapters 3 and 9) create a smoothing effect, and large mask sizes create a painted effect (Figure 8.4.3)
- The median filter is computationally intensive, especially for large mask sizes, so a *pseudomedian filter* approximation can be used:

$$\text{PMED}(S_L) = (1/2)\text{MAXIMIN}(S_L) + (1/2)\text{MINIMAX}(S_L)$$

where S_L denotes a sequence of elements $s_1, s_2, \ldots, s_L$ and where for $M = (L+1)/2$

$$\text{MAXIMIN}(S_L) = \text{MAX}\Big[[\text{MIN}(s_1, \ldots, s_M)], [\text{MIN}(s_2, \ldots, s_{M+1})], \ldots, [\text{MIN}(s_{LM+1}, \ldots, s_L)]\Big]$$

$$\text{MINIMAX}(S_L) = \text{MIN}\Big[[\text{MAX}(s_1, \ldots, s_M)], [\text{MAX}(s_2, \ldots, s_{M+1})], \ldots, [\text{MAX}(s_{LM+1}, \ldots, s_L)]\Big]$$

8.6 References and Further Reading

References that contain major chapters on image enhancement include: [Gonzalez/Woods 02], [Shapiro/Stockman 01], [Russ 99], [Jahne 97], [Pratt 95], [Baxes 94], [Lim 90], and [Jain 89]. Gray scale modification is discussed in [Gonzalez/Woods 02], [Rosenfeld/Kak 82], and [Jain 89]. More on gamma correction can be found in [Gonzalez/Woods 92/02], [Watt/Polocarpo 98], and [Giorgianni/Madden 98]. A more complete theoretical treatment of histogram modification is given in [Gonzalez/Woods 02], [Castelman 96], [Pratt 91], [Banks 90], [Jain 89], and [Rosenfeld/Kak 82]. A useful adaptive histogram modification technique is discussed in [Pratt 91]. For details on the ACE2 filter see [Lee 80], and for more information on the log and exponential ACE filters see [Deng/Cahill/Tobin 95]. A conceptual perspective to gray level transforms and histogram modification is provided in [Baxes 94], while [Jain/Kasturi/Schnuck 95], [Myler/Weeks 93], and [Sid-Ahmed 95] provide a practical treatment. Adaptive filters are discussed in [Gonzalez/Woods 02]. Pseudocolor is discussed in [Gonzalez/Woods 02], [Pratt 91], and [Jain 89]. [Lim 90], [Schalkoff 89], and [Gonzalez/Woods 02] provide different perspectives to unsharp masking. Phase contrast filtering is discussed in [Sid-Ahmed 95]. Details on use of the Moore–Penrose matrix for generating convolution masks based on frequency domain filter models can be found in [Gonzalez/Woods 92]. Various image sharpening and smoothing methods are discussed in all the following references. Two excellent handbooks that contain practical information for application-based image enhancement are [Russ 99] and [Jahne 97].

Banks, S., *Signal Processing, Image Processing and Pattern Recognition*, Upper Saddle River, NJ: Prentice Hall, 1990.

Baxes, G.A., *Digital Image Processing: Principles and Applications*, NY: Wiley, 1994.

Bracewell, R.N., *Two-Dimensional Imaging*, Upper Saddle River, NJ: Prentice Hall, 1995.

Castleman, K.R., *Digital Image Processing*, Upper Saddle River, NJ: Prentice Hall, 1996.

Deng, G., Cahill, L.W., Tobin, G.R., The study of logarithmic image processing model and its application to image enhancement, *IEEE Transaction on Image Processing*, 4, pp. 506–511, 1995.

Galbiati, L.J., *Machine Vision and Digital Image Processing Fundamentals*, Upper Saddle River, NJ: Prentice Hall, 1990.

Giorgianni, E.J., Madden, T.E., *Digital Color Management: Encoding Solutions*, Reading, MA: Addison Wesley, 1998.

Gonzalez, R.C., Woods, R.E., *Digital Image Processing*, Reading, MA: Addison-Wesley, 1992.

Gonzalez, R.C., Woods, R.E., *Digital Image Processing*, Upper Saddle River, NJ: Prentice Hall, 2002.

Jahne, B., *Practical Handbook on Image Processing for Scientific Applications*, Boca Raton, FL: CRC Press, 1997.

Jain, A.K., *Fundamentals of Digital Image Processing*, Upper Saddle River, NJ: Prentice Hall, 1989.

Jain, R., Kasturi, R., Schnuck, B.G., *Machine Vision*, NY: McGraw Hill, 1995.

Lee, J.S., Digital image enhancement and noise filtering by use of local statistics, *IEEE Trans. on Pattern Analysis and Machine Intelligence*, vol. 2, pp. 165–168, 1980.

Lim, J.S., *Two-Dimensional Signal and Image Processing*, Upper Saddle River, NJ: Prentice Hall 1990.

Myler, H.R., Weeks, A.R., *Computer Imaging Recipes in C*, Upper Saddle River, NJ: Prentice Hall, 1993.

Pratt, W.K., *Digital Image Processing*, NY: Wiley, 1991.

Rosenfeld, A., Kak, A.C., *Digital Picture Processing*, San Diego, CA: Academic Press, 1982.

Russ, J.C., *The Image Processing Handbook*, Boca Raton, FL: CRC Press, 1999.

Schalkoff, R.J., *Digital Image Processing and Computer Vision*, NY: Wiley, 1989.

Shapiro, L., Stockman, G., *Computer Vision*, Upper Saddle River, NJ: Prentice Hall, 2001.

Sid-Ahmed, M.A., *Image Processing: Theory, Algorithms, and Architectures*, Upper Saddle River, NJ: Prentice Hall, 1995.

Watt, A., Policarpo, F., *The Computer Image*, New York, NY: Addison-Wesley, 1998.

8.7 Exercises

1. (a) Image enhancement can be performed in two domains, what are they? (b) Why is application specific feedback important in image enhancement?
2. Name and define the three types of techniques used in enhancement.
3. (a) To what type of technique does gray scale modification belong? (b) What is a mapping equation? (c) For a linear mapping equation, how does the slope of the line affect the results? (d) Explain how gray scale modification works.

4. For an 8-bit image, sketch a mapping equation that will provide more image detail in the range 50 to 100, while leaving the other values unchanged.
5. For an 8-bit image, find the mapping equation to stretch the original image range [0 to 50] to [0 to 150], while leaving the other values unchanged.
6. For an 8-bit image, sketch the mapping equation that will stretch the range [100 to 200] over the entire range, with clipping at both the low and high ends.
7. For an 8-bit image, sketch the mapping equation that will create a digital negative.
8. (a) Using intensity level slicing on an 8-bit image, sketch the mapping equation that will turn values between 35 and 50 white, while leaving other values unchanged, b) sketch the mapping equation that will turn values between 50 and 75 white, while making other values black.
9. (a) Sketch a mapping equation that will stretch the range $[0, 50]$ to $[0, 150]$ and compress the range $[50, 255]$ to $[150, 255]$, b) Use CVIPtools to implement the mapping equation.
10. (a) Why is gamma correction used? (b) Explain the equations for gamma correction.
11. (a) Why is range compression used? (b) What is an example of an application for a logarithmic mapping equation?
12. (a) Given a low contrast 8-bit image with a gray level range from 50 to 125, what is the equation to stretch a histogram over the entire 8-bit range? (b) What is the equation to shrink the histogram to the range $[25, 50]$, (c) What is the equation to slide the histogram up by 100?
13. In what case do we need to clip when performing a histogram stretch? Explain.
14. Given an image with 3-bits per pixel, with the following histogram:

Gray level	Number of pixels
0	0
1	5
2	10
3	15
4	8
5	5
6	0
7	0

Find the histogram mapping table and the resulting histogram after histogram equalization.

15. Given an image with 3-bits per pixel, with the following histogram:

Gray level	Number of Pixels
0	0
1	1
2	5
3	3
4	2
5	12
6	2
7	0

Find the histogram mapping table and the resulting histogram after histogram equalization.

16. Given the following tables of an image histogram and a specified histogram, find the mapping tables and the resulting histogram after histogram specification process is performed.

Image Histogram

Gray value	Number of pixels
0	5
1	5
2	0
3	0
4	5
5	11
6	3
7	6

Specified Histogram

Gray value	Number of pixels
0	0
1	0
2	5
3	10
4	10
5	5
6	5
7	0

17. Normally histogram equalization is a global process. Explain how can be used for local enhancement.

18. Use CVIPtools to explore the standard histogram operations on a gray level image. Select a monochrome image of your choice and do the following: (a) Display the histogram by selecting *File->Show Histogram*, or by clicking the histogram icon which looks like a tiny bar graph. (b) Perform a histogram equalization and display the histogram. (c) Perform a histogram slide up by 50 and down by 50 and display the histograms, verify the results are correct, (d) Perform a histogram stretch without clipping (set to 0), and with 0.025 clipping on both ends, display the histograms—are they correct? (e) Perform a histogram shrink to the range [1, 100], display the histogram and verify it is correct.

19. Use CVIPtools to explore the standard histogram operations on a color image. Select a color image of your choice and do the following: (a) Display the histogram by selecting *File->Show Histogram*, or by clicking the histogram icon which looks like a tiny bar graph. (b) Use *Enhancement->Histograms->Histogram Equalization* to perform a histogram equalization four times, each time selecting a different band to use (*Value*, *Red*, *Green*, and *Blue*), and display the histograms. What band does the parameter selection *Value* use? Explain. (c) Perform a histogram slide up by 50 and down by 50 and display the histograms, verify the results are correct. (d) Perform a histogram stretch without clipping (set to 0), and

with 0.025 clipping on both ends, display the histograms—are they correct? (e) Perform a histogram shrink to the range [1, 100], display the histogram and verify it is correct.

20. Use CVIPtools to explore histogram specification. Select a monochrome of your choice and do the following: (a) Display the histogram by selecting *File->Show Histogram*, or by clicking the histogram icon which looks like a tiny bar graph. (b) Use the default *sin(0.025*x)* for the *Formula*, and look at the histograms—does the output look like the specified histogram, why or why not? (c) Change the *Formula* to *sin(0.25*x)*, and look at the histograms—does the output look like the specified histogram, why or why not? (d) Change the *Formula* to *sin(0.005*x)*, and look at the histograms—does the output look like the specified histogram, why or why not? (e) Change the *Formula* to *ramp(2.0x + 5)*, and look at the histograms—does the output look like the specified histogram, why or why not? Does it look like any of the other specified histograms? (f) Experiment with the other formulas, especially the *log* and *exp*. After your experimentation can you draw any general conclusions regarding histogram specification? (g) Select a color image of your choice and repeat (a)–(f). With the color image experiment with using different *Formulas* for each band.
21. Given the following 5×5 subimage (using 5×5 as the window size) from an image with 3 bits per pixel and average gray value of 6, find the resulting value for the center pixel by letting $k_1 = 0.8$ and $k_2 = 0.2$ and applying the following filters: (a) ACE, (2) ACE2, (3) log-ACE, (4) exp-ACE.

$$\begin{bmatrix} 5 & 5 & 5 & 5 & 5 \\ 3 & 3 & 1 & 1 & 1 \\ 3 & 3 & 1 & 1 & 1 \\ 3 & 3 & 1 & 1 & 1 \\ 5 & 5 & 5 & 5 & 5 \end{bmatrix}$$

22. Use CVIPtools to explore the ACE filters. Open the image in Figure 8.2.20a and use the ACE filters to create the images in Figures 8.2.20, through 8.2.23.
23. (a) Why use pseudocolor? (b) List the two domains in which pseudocolor is performed and describe a method in each.
24. Given the following 4-bit per pixel image, create a pseudocolor image by applying intensity slicing. Divide the gray level range into four equal regions and map the low range to bright red, the next bright green, the next bright blue, and the next to bright yellow (red + green). Express the image as a 5×5 matrix with a triple at each pixel locations for the RGB values.

$$\begin{bmatrix} 6 & 7 & 7 & 7 & 5 \\ 3 & 3 & 1 & 1 & 1 \\ 2 & 2 & 11 & 12 & 12 \\ 13 & 13 & 0 & 0 & 0 \\ 8 & 9 & 9 & 10 & 15 \end{bmatrix}$$

25. Use CVIPtools to explore pseudocolor. Select a monochrome image of your choice and: (a) Select *Gray level mapping* and perform the operation with the default values, (b) change the shape of the mapping equations so they are all the same—how does the image appear? Change to a different shape, but make them all the same, how does the image appear now? (c) Select *Intensity Slicing* and apply it with the default parameter values. How do the colors compare to the results from (a)? (d) Change the input ranges so that the entire 0–255 range is not covered and apply the operation both with and without Set *Out of Range Data to 0* selected. Are the results what you expected? Can you think of an application where this parameter is useful? (e) Experiment with changing the output colors. What colors do you add to create yellow? Purple? Cyan? After your experimentation can you draw any general conclusions?
26. Use CVIPtools to explore pseudocolor in the frequency domain. (a) Select *Frequency domain mapping* and apply with the default values—what colors are most prominent? (b) Change the colors from RGB order to BRG order. How does this affect the colors you see in the output image? (c) Experiment with changing the cutoff frequencies. After your experimentation can you draw any general conclusions?
27. Use CVIPtools to explore pseudocolor with *Enhancement->Pseudocolor->Gray level mapping II*. This function provides a graphical interface and more options than the one explored in Exercise 25. Select a monochrome image of your choice and: (a) Select *Gray level mapping II* and click on the *Custom Remap Curve* button, which will display a new window for you to enter your mapping curves. Select the red band and use the left mouse button to input new points, and the right mouse button to delete points. The data points can also be dragged with the left mouse button. Points can also be entered manually by inputting the (X, Y) values and clicking the *Add* button. Next, create curves for the green and blue bands, then select *All*, which will show you the mapping curves. Select the interpolation method desired and use the *Save* button to save your mapping file. Note that the default directory for the mapping files is in $CVIPtools\bin\remap. Next press *APPLY* on the enhancement window to perform the pseudocolor operation. (b) View the histogram for your pseudocolor image, can you see any correlation between it and the mapping equations? (c) Experiment with creating different mapping equations and viewing the output images and their histograms. In general, can you see any correlation between the histograms and the mapping equations? (d) Apply the mapping files you have created to other images with the *Load Mapping File* option. Compare the results from using the same mapping file to different images. Are they similar? Why or why not?
28. Many image sharpening algorithms consist of three basic steps, what are they? Provide an example operation for each step.
29. Use CVIPtools to explore the high boost spatial filter. Open the image in Figure 8.3.2a and create the images in Figure 8.3.2b–g.
30. Use CVIPtools to develop your own image sharpening algorithm. Select an image that you want to sharpen (if you cannot find one, then use a good image and slightly blur it with *Utilities->Filter->Specify a Blur*). Be sure to examine the histograms of your output images during development. (a) Use *Analysis->Transforms* to extract a phase only image. Develop your own sharpening algorithm by using this image and the original image. (b) Use *Analysis->Edge/Line Detection* to generate edge only images. Develop your own sharpening algorithm by using

these images and the original image. (c) Use *Utilities->Filter->Difference Filter* to generate images. Develop your own sharpening algorithm by using these images and the original image. (d) After your algorithm development can you draw any general conclusions?

31. (a) What is the image model used for homomorphic filtering? (b) List the steps in the homomorphic filtering process. (c) Explain how the filter shown in Figure 8.3.5 relates to the model defined in (a).
32. Use CVIPtools to explore homomorphic filtering. Open the image in Figure 8.3.6a and create the images in Figure 8.3.6b,c. Experiment with varying the parameters. Can you obtain better results than shown in the figure?
33. (a) Explain the historical reasons underlying the development of the unsharp masking algorithm. (b) Describe and explain the steps in the algorithm.
34. Use CVIPtools to explore unsharp masking. Open the image in Figure 8.3.8a and create the images in Figure 8.3.8b–d. Experiment with varying the parameters. After your experimentation can you draw any general conclusions?
35. Use CVIPtools to explore Sharpening Algorithms I and II. Open the image in Figure 8.3.9a. (a) Create the images in Figure 8.3.9b–d. (b) Create the image in Figure 8.3.10b. (c) Use these two algorithms and experiment with varying the parameters. (d) Based on what you have learned develop your own sharpening algorithm. How do your results compare to the results from Sharpening Algorithms I and II and/or the algorithms you developed in Exercise 30?
36. (a) List two reasons for image smoothing. (b) In general, how is image smoothing accomplished?
37. (a) Describe convolution masks used for image smoothing. In general, what can be said about the mask coefficients? (b) What is the primary difference in the results from an arithmetic mean compared to a Gaussian spatial filter? (c) With arithmetic mean filters the results can be normalized by dividing by the sum of the mask coefficients. What is another method to accomplish this? (d) What happens as the filter mask size is increased?
38. (a) How would you describe an image that has been smoothed by a median filter with a relatively large mask? (b) Since median filtering is computationally intensive, what is an alternative that is more efficient, but gives similar results?
39. Explain why the results are different if we use an FFT and a DCT for lowpass frequency domain smoothing, even though we use the same cutoff frequency for the filter.
40. Use CVIPtools to explore image smoothing. (a) Use the FFT smoothing and the mean filter. Experiment with varying the parameters until the output images look similar. Go to *Analysis->Transforms* and perform an FFT on the similar looking output images. Do the spectra look similar? Why or why not? (b) Perform the operations in (a), but use the Y_p-mean and the midpoint filter. (c) Perform the operations in (a), but use the Gaussian and the contra-harmonic filter. (d) After your experimentation can you draw any general conclusions?

8.7.1 Programming Exercise: Digital Negative

1. Write a function that implements the inverse mapping equation to create a digital negative of an image.

2. Write a function to create a partial complement of an image. Have the user input the gray level value which is the lower limit at which the complement is performed. The value will be between 0 and 255, with 255 having no effect, and 0 complementing the entire image.

8.7.2 Programming Exercise: Piecewise Gray Level Mapping

1. Write a function to implement piecewise gray level mapping. Have the user input up to three linear mapping equations. The user will specify: (a) the input data range, (b) the initial value for the output data, (c) the slope of the linear mapping equation(s). Any out-of-range values (any values not in the ranges specified by the user) will be unchanged.
2. Modify the function to allow the user to specify the method for handling out-of-range data. Let the user select: (a) leave the data unchanged, or (b) set the data to a user specified value.

8.7.3 Programming Exercise: Gamma Correction

1. Write a function to perform gamma correction. Let the user specify: (a) gamma, γ, and (b) the constant, K_2.
2. Experiment by using the function on various images and displaying them on different monitors, and printing them on different printers. Also, when experimenting with the monitors, change the lighting conditions in the room and observe how this affects the results.

8.7.4 Programming Exercise: Histogram Modification

1. Write a function to implement a histogram stretch/shrink. Clip if the numbers go out of BYTE range.
2. Write a function to perform a histogram slide on an image. Have your program find the maximum and minimum gray level values in the image and calculate the largest value of a left or right slide that is possible before gray level saturation (clipping) occurs. Warn the user of this, but let them clip if desired.
3. Enhance your histogram stretch to allow for a specified percentage of pixels to be clipped at either the low end (set to zero) or high end (set to 255), or both.
4. Modify the histogram stretch to allow for out-of-range results, followed by a remap (see the *Mapping* library). A data type other than BYTE will be needed (see the *Image* library).

8.7.5 Programming Exercise: Histogram Equalization

1. Incorporate the CVIPtools function *histeq*, in the *Histogram* library, into your CVIPlab program.
2. Write your own function to perform histogram equalization. Compare this to using the CVIPtools function, *histeq*.
3. Incorporate the CVIPtools function *local_histeq* into your CVIPlab. Compare the results of using this to the *histeq* function.

8.7.6 Programming Exercise: ACE Filters

1. Write a function to implement the ACE2 filter. Let the user specify the window size and the values for k_1 and k_2. Compare the results from this function to those obtained with CVIPtools.
2. Write a function to implement the ACE filter. Let the user specify the window size and the values for k_1 and k_2. Compare the results from this function to those obtained with CVIPtools.
3. Write function to implement the exp-ACE filter. Let the user specify the window size and the values for k_1 and k_2. Compare the results from this function to those obtained with CVIPtools.
4. Incorporate the CVIPtools function *log_ace_filter* (in the *SpatialFilter* library) into your CVIPlab program. Verify that the function is working properly by comparing the results to those you obtain with CVIPtools.

8.7.7 Programming Exercise: Pseudocolor

1. Write a function to perform intensity slicing pseudocolor. Let the user select values for four different input data ranges, as well as the RGB values for the four output colors. Do not modify any out of range values (values not included in the four specified ranges).
2. Modify the function so the user can select: (a) to not modify out-of-range data, or (b) to set the out-of-range data to a user specified value.
3. Incorporate the CVIPtools function *pseudocol_freq* (in the *Color* library) into your CVIPlab program. This function will perform frequency domain pseudocolor by using the Fourier transform and lowpass, bandpass, and highpass filters.

8.7.8 Programming Exercise: Basic Enhancement Convolution Masks

1. Write a program to implement spatial convolution masks. Let the user select from one of the following masks:
   Lowpass filter masks (smoothing):

$$\frac{1}{9}\begin{bmatrix} 1 & 1 & 1 \\ 1 & 1 & 1 \\ 1 & 1 & 1 \end{bmatrix} \qquad \frac{1}{10}\begin{bmatrix} 1 & 1 & 1 \\ 1 & 2 & 1 \\ 1 & 1 & 1 \end{bmatrix} \qquad \frac{1}{16}\begin{bmatrix} 1 & 2 & 1 \\ 2 & 4 & 2 \\ 1 & 2 & 1 \end{bmatrix}$$

   Highboost filter masks (sharpening):

$$\begin{bmatrix} -1 & -1 & -1 \\ -1 & 9 & -1 \\ -1 & -1 & -1 \end{bmatrix} \qquad \begin{bmatrix} 1 & -1 & 1 \\ -2 & 5 & -2 \\ 1 & -2 & 1 \end{bmatrix} \qquad \begin{bmatrix} 0 & -1 & 0 \\ -1 & 5 & -1 \\ 0 & -1 & 0 \end{bmatrix}$$

2. Modify the function to allow the user to input the coefficients for a 3×3 mask.
3. Experiment with using the smoothing and sharpening masks. Try images with and without added noise.

4. Compare the results of using your spatial masks to frequency domain filtering using CVIPtools. Examine and compare the spectra of the resulting images.
5. Modify the function to handle larger masks. Expand the above masks as described in this chapter and in Chapter 4.

8.7.9 Programming Exercise: Unsharp Masking

1. Write a function to perform unsharp masking enhancement. Use the flow chart given in Figure 8.3.7. Use any functions that you have already written—lowpass filtering (via spatial convolution masks), subtraction, histogram shrink, and stretch. Use a 3×3 arithmetic mean spatial mask for the lowpass filter, and a shrink range of 0 to 100.
2. Modify the function to allow the user to select at least three different lowpass filter masks, and to select the histogram shrink range.
3. Modify the function so it will automatically select the shrink range based on the histogram of the image after lowpass filtering. Allow the user to specify the percentage for the shrink.

8.7.10 Programming Exercise: Sharpening Algorithms

1. Take the algorithms that you developed in Exercise 30 and/or Exercise 35 and implement them in your CVIPlab program.
2. Compare the results from your program to those you obtained by using CVIPtools. If there are any differences in the results, can you explain them? Pay careful attention to data types and remapping as you do the comparison.

8.7.11 Programming Exercise: Median Filtering

1. Write a median filtering function; allow the user to enter the window (mask) size. Compare the median filter results to lowpass filter results for image smoothing. Use the *Utilities->Compare* options.
2. Incorporate the CVIPtools function *median_filter,* in the *SpatialFilter* library, into your CVIPlab program. Is it faster or slower than your median filtering function?
3. Use CVIPtools to compare the spectra of the median filtered images to images that have been lowpass filtered in both the spatial and frequency domain.

9

Image Restoration

9.1 Introduction and Overview

Image restoration methods are used to improve the appearance of an image by application of a restoration process that uses a mathematical model for image degradation. The modeling of the degradation process differentiates restoration from enhancement where no such model is required. Examples of the types of degradation considered include blurring caused by motion or atmospheric disturbance, geometric distortion caused by imperfect lenses, superimposed interference patterns caused by mechanical systems, and noise from electronic sources. It is assumed that the degradation model is known, or can be estimated. The idea is to model the degradation process and then apply the inverse process to restore the original image. In general, image restoration is more of an art than a science; the restoration process relies on the experience of the individual to successfully model the degradation process. In this chapter we will consider the various types of degradation that can be modeled, and discuss the various techniques available to restore the image. The types of degradation models include both spatial and frequency domain considerations.

In practice the degradation process model is often not known and must be experimentally determined or estimated. Any available information regarding the images and the systems used to acquire and process them is helpful. This information, combined with the developer's experience, can be applied to find a solution for the specific application. A general block diagram for the image restoration process is provided in Figure 9.1.1. Here we see that sample degraded images and knowledge of the image acquisition process are inputs to the development of a degradation model. After this model has been developed the next step is the formulation of the inverse process. This inverse degradation process is then applied to the degraded image, $d(r,c)$, which results in the output image, $\hat{I}(r,c)$. This output image, $\hat{I}(r,c)$, is the restored image which represents an estimate of the original image, $I(r,c)$. Once the estimated image has been created, any knowledge gained by observation and analysis of this image is used as additional input for the further development of the degradation model. This process continues until satisfactory results are achieved. With this perspective, we can define *image restoration* as the process of finding an approximation to the degradation process and finding the appropriate inverse process to estimate the original image.

9.1.1 System Model

The degradation process model consists of two parts, the degradation function and the noise function. The general model in the spatial domain is as follows:

$$d(r,c) = h(r,c) * I(r,c) + n(r,c)$$

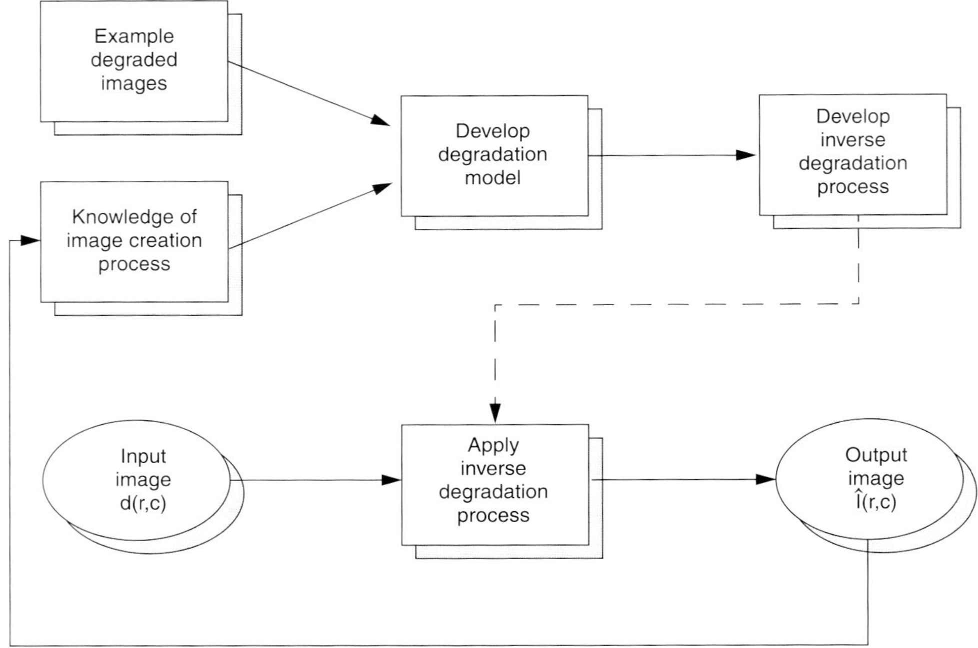

FIGURE 9.1.1
Image restoration process.

where the $*$ denotes the convolution process

$$\begin{aligned} d(r,c) &= \text{degraded image} \\ h(r,c) &= \text{degradation function} \\ I(r,c) &= \text{original image} \\ n(r,c) &= \text{additive noise function} \end{aligned}$$

Because convolution in the spatial domain is equivalent to multiplication in the frequency domain, the frequency domain model is:

$$D(u,v) = H(u,v)I(u,v) + N(u,v)$$

where:

$$\begin{aligned} D(u,v) &= \text{Fourier transform of the degraded image} \\ H(u,v) &= \text{Fourier transform of the degradation function} \\ I(u,v) &= \text{Fourier transform of the original image} \\ N(u,v) &= \text{Fourier transform of the additive noise function} \end{aligned}$$

Based on our definition of the image restoration process, and the preceding model, we can see that what needs to be done is to find the degradation function, $h(r,c)$ (or its frequency domain representation $H(u,v)$), and the noise model, $n(r,c)$ (or $N(u,v)$). Note that other models can be defined; specifically a multiplicative noise model where the noise function is not added to the image but is multiplied by the image. To handle this case, we typically take the logarithm of the degraded image, thus decoupling the noise and image functions into an additive process (see Chapter 8 on homomorphic filtering).

9.2 Noise Models

What is noise? *Noise* is any undesired information that contaminates an image. Noise appears in images from a variety of sources. The digital image acquisition process, which converts an optical image into a continuous electrical signal that is then sampled, is the primary process by which noise appears in digital images. At every step in the process there are fluctuations caused by natural phenomena which add a random value to the exact brightness value for a given pixel. The noise inherent in the electronics is also affected by environmental conditions such as temperature, and may vary during the acquisition of an image database. Other types of noise, such as periodic noise, may be introduced during the acquisition process as a result of the physical systems involved.

9.2.1 Noise Histograms

The noise models in this section consider the noise a random variable with a probability density function (PDF) that describes its shape and distribution. The actual distribution of noise in a specific image is the histogram of the noise. In other words, the histogram is a specific example of the theoretical model or PDF of the noise. To make the histogram look more exactly like the theoretical model, many example images of the noise could be created and then averaged.

In typical images the noise can be modeled with either a Gaussian ("normal"), uniform or salt-and-pepper ("impulse") distribution. The shape of the distribution of these noise types as a function of gray level can be modeled as a histogram and can be seen in Figure 9.2.1. In Figure 9.2.1a we see the bell-shaped curve of the Gaussian noise distribution, which can be analytically described by:

$$\text{HISTOGRAM}_{\text{Gaussian}} = \frac{1}{\sqrt{2\pi\sigma^2}} e^{-(g-m)^2/2\sigma^2}$$

where:

$$g = \text{gray level}$$
$$m = \text{mean (average)}$$
$$\sigma = \text{standard deviation } (\sigma^2 = \text{variance})$$

About 70% of all the values fall within the range from one standard deviation (σ) below the mean (m) to one above, and about 95% fall within two standard deviations. Theoretically, this equation defines values from $-\infty$ to $+\infty$, but since the actual gray levels are only defined over a finite range, the number of pixels at the lower and upper values will be higher than this equation predicts. This is due to the fact that all the noise values below the minimum will be clipped to the minimum, and those above the maximum will be clipped at the maximum value. This is a factor that must be considered with all theoretical noise models, when applied to a fixed, discrete range such

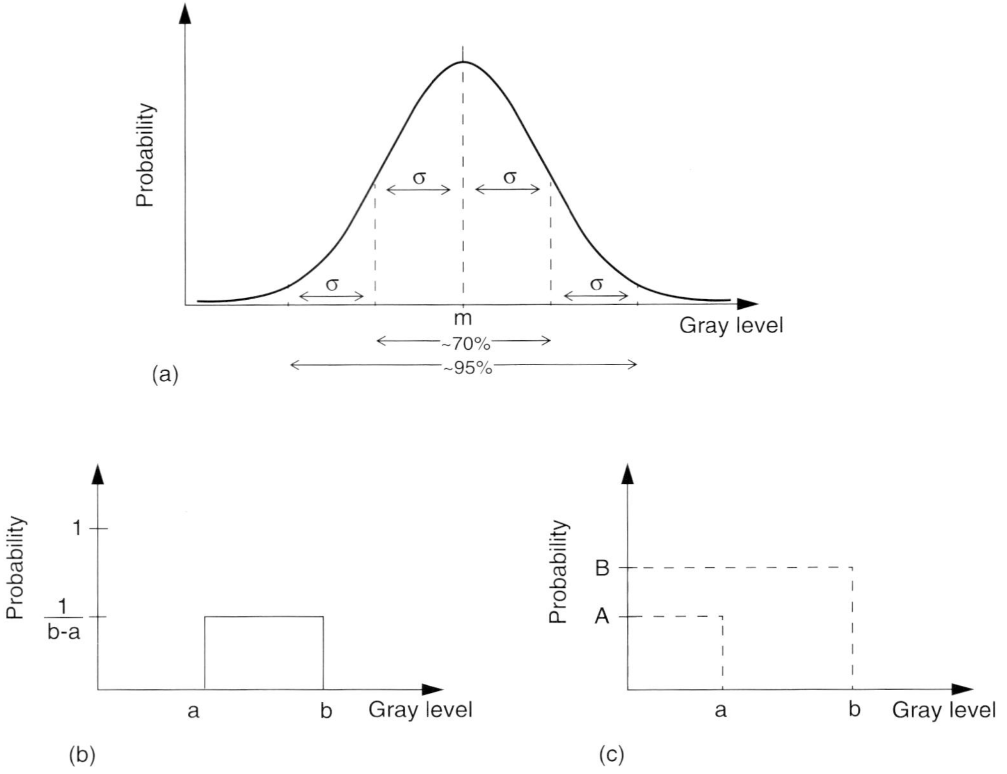

FIGURE 9.2.1
Gaussian, uniform and salt-and-pepper noise distribution. (a) Gaussian noise. (b) Uniform noise. (c) Salt and pepper noise.

as with digital images (e.g., 0 to 255). In Figure 9.2.1b is the uniform distribution:

$$\text{HISTOGRAM}_{\text{Uniform}} = \begin{cases} \dfrac{1}{b-a} & \text{for } a \le g \le b \\ 0 & \text{elsewhere} \end{cases}$$

$$\text{mean} = \frac{a+b}{2}$$

$$\text{variance} = \frac{(b-a)^2}{12}$$

With the uniform distribution, the gray level values of the noise are evenly distributed across a specific range, which may be the entire range (0 to 255 for 8-bits), or a smaller portion of the entire range. In Figure 9.2.1c is the salt-and-pepper distribution:

$$\text{HISTOGRAM}_{\text{Salt and Pepper}} = \begin{cases} A & \text{for } g = a \text{ (``pepper'')} \\ B & \text{for } g = b \text{ ( ``salt'')} \end{cases}$$

In the salt-and-pepper noise model there are only two possible values, a and b, and the probability of each is typically less than 0.2—with numbers greater than this the

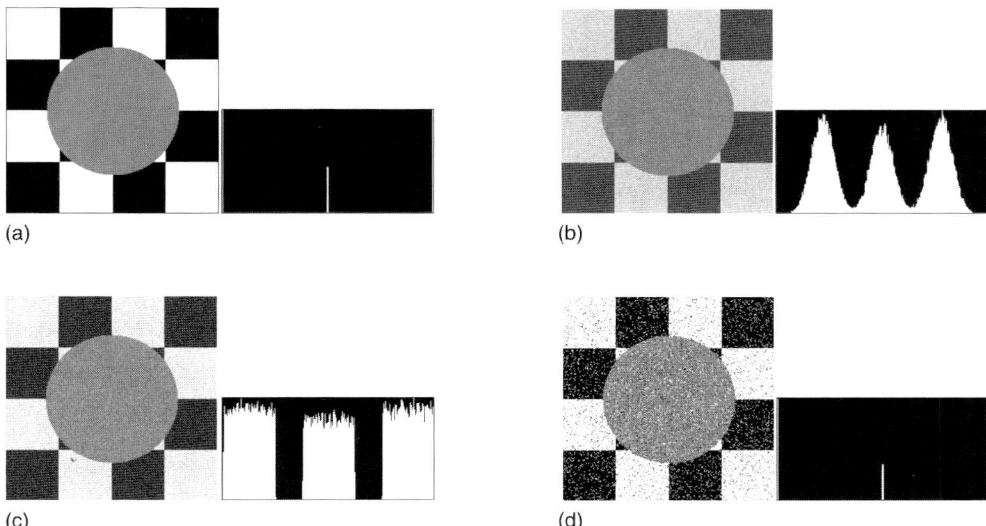

FIGURE 9.2.2
Gaussian, uniform and salt-and-pepper noise. (a) Original image without noise, and its histogram. (b) Image with added Gaussian noise with mean = 0 and variance = 600, and its histogram. (c) Image with added uniform noise with mean = 0 and variance = 600, and its histogram. (d) Image with added salt-and-pepper noise with the probability of each 0.08, and its histogram.

noise will swamp out the image. For an 8-bit image, the typical value for pepper-noise is 0, and 255 for salt-noise.

The Gaussian model is most often used for natural noise processes, such as those occurring from electronic noise in the image acquisition system. The random electron fluctuations within resistive materials in sensor amplifiers or photodetectors results in thermal noise, which is the most common cause. This electronic noise is most problematic with poor lighting conditions or very high temperatures. The Gaussian model is also valid for film grain noise, if photographic film is part of the imaging process.

The salt-and-pepper type noise (also called impulse noise, shot noise or spike noise) is typically caused by malfunctioning pixel elements in the camera sensors, faulty memory locations, or timing errors in the digitization process. Uniform noise is useful, since it can be used to generate any other type of noise distribution, and is often used to degrade images for the evaluation of image restoration algorithms since it provides the most unbiased or neutral noise model. In Figure 9.2.2, we see examples of these three types of noise added to images, along with their histograms. Visually, the Gaussian and uniform noisy images appear similar, but the image with added salt-and-pepper noise is very distinctive.

Radar range and velocity images typically contain noise that can be modeled by the Rayleigh distribution, defined by:

$$\text{HISTOGRAM}_{\text{Rayleigh}} = \frac{2g}{\alpha} e^{-g^2/\alpha}$$

where:

$$\text{mean} = \sqrt{\frac{\pi\alpha}{4}}$$

$$\text{variance} = \frac{\alpha(4 - \pi)}{4}$$

Negative exponential noise occurs in laser-based images, and if this type of image is lowpass filtered the noise can be modeled as gamma noise. The equation for negative exponential noise (assuming g and α are both positive):

$$\text{HISTOGRAM}_{\text{Negative Exponential}} = \frac{e^{-g/\alpha}}{\alpha}$$

where:

$$\text{mean} = \alpha$$

$$\text{variance} = \alpha^2$$

The equation for gamma noise:

$$\text{HISTOGRAM}_{\text{Gamma}} = \frac{g^{\alpha-1}}{(\alpha-1)!a^{\alpha}} e^{-g/a}$$

where:

$$\text{mean} = a\alpha$$

$$\text{variance} = a^2\alpha$$

The histograms (distributions) for these can be seen in Figure 9.2.3. The Rayleigh distribution peaks at $\sqrt{\alpha/2}$, and negative exponential noise is actually gamma noise with the peak moved to the origin ($\alpha = 1$). Many of the types of noise that occur in natural phenomena can be modeled as some form of exponential noise such as those described here. Figure 9.2.4 shows images with these types of noise added, along with their histograms.

In addition to the noise distribution, another important consideration is the spatial frequency content of the noise. Typically the noise is treated as *white noise*, which assumes a constant spectral content. This means that, unlike regular images, there are equal amounts of low, medium and high frequency content in the noise image. As we have seen, in most real images the spatial frequency energy is concentrated in the low frequencies. Therefore, in an image with added noise, much of the high frequency content is due to noise. This information will be useful in the development of models for noise removal. In Figure 9.2.5 are real images and their Fourier spectra, and noise images and their Fourier spectra. Here we can see that the noise images appear to have a much more evenly distributed spectrum than the real images.

9.2.2 Periodic Noise

Periodic noise in images is typically caused by electrical and/or mechanical systems. This type of noise can be identified in the frequency domain as impulses corresponding to sinusoidal interference (see Figure 9.2.6). During image acquisition mechanical jitter or vibration can result in this type of noise appearing in the image. The vibration can be caused by motors or engines, wind or seas, depending on the location of the image sensing device. Electrical interference in the system may also result in additive sinusoidal noise corrupting the image during acquisition. If this type of noise can be isolated it can be removed with bandreject and notch filters as is shown in Section 9.5.6.

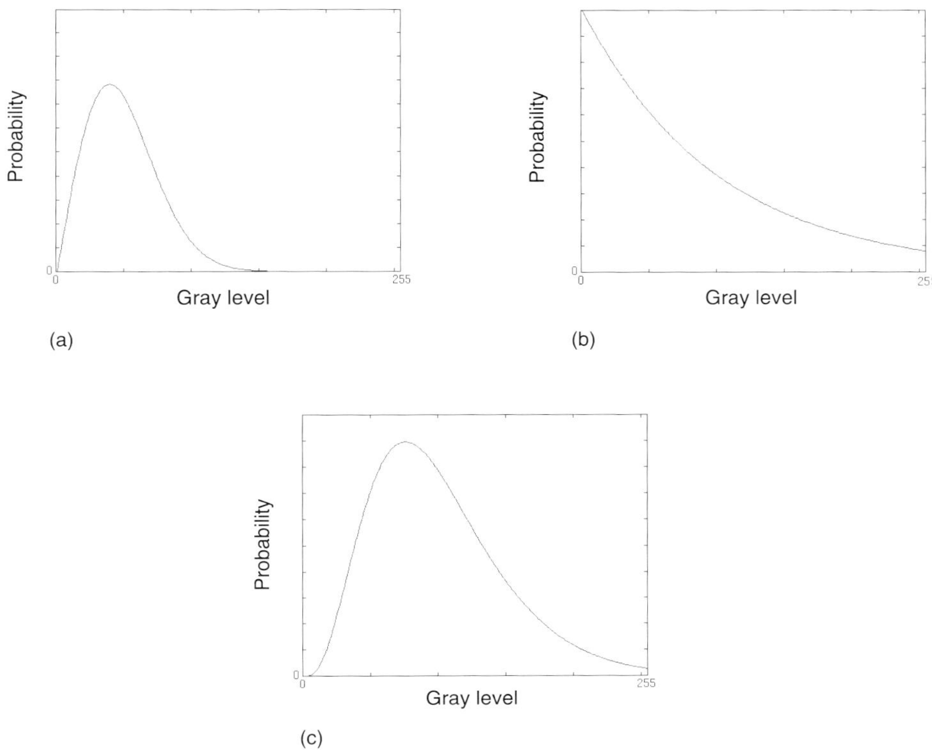

FIGURE 9.2.3
Rayleigh, negative exponential and gamma noise distributions. (a) Rayleigh distribution. (b) Negative exponential distribution. (c) Gamma distribution.

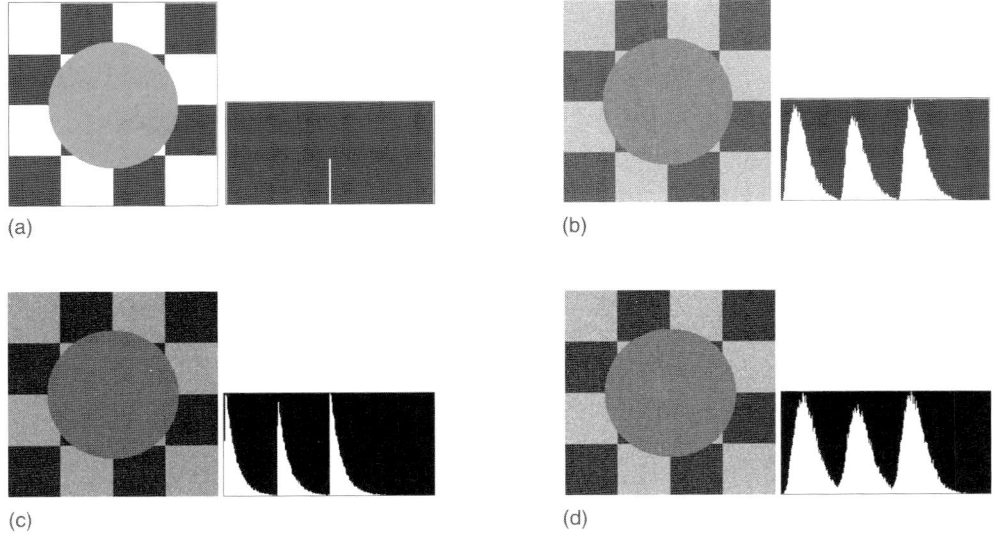

FIGURE 9.2.4
Rayleigh, negative exponential and gamma noise. (a) Original image without noise, and its histogram. (b) Image with added Rayleigh noise with variance = 600, and its histogram. (c) Image with added negative exponential noise with variance = 600, and its histogram. (d) Image with added gamma noise with variance = 600 and $\alpha = 6$, and its histogram.

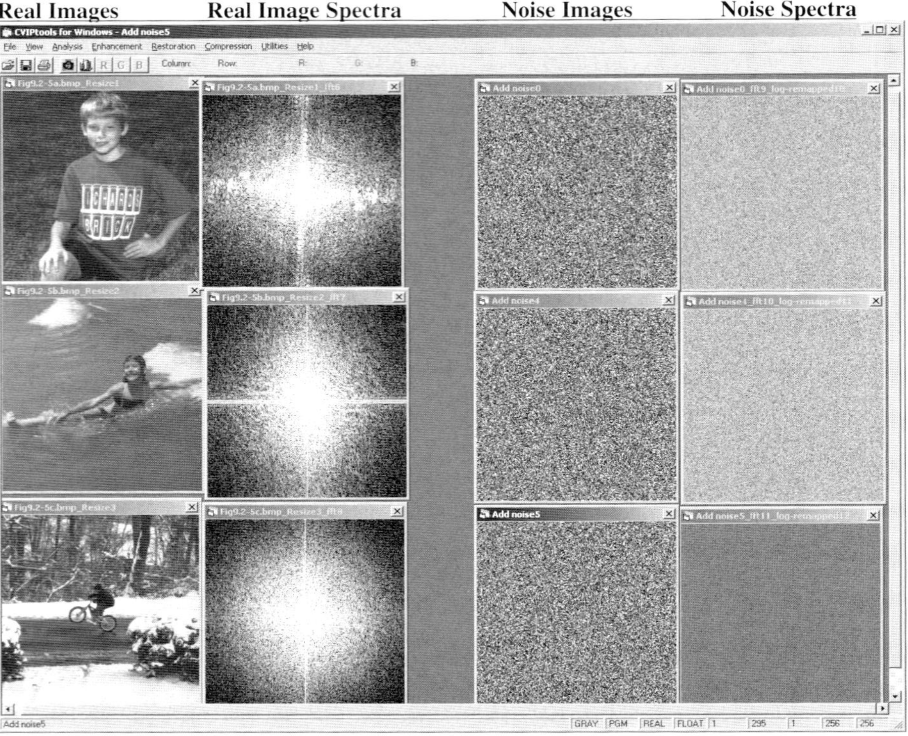

FIGURE 9.2.5
Fourier spectra of real images and Fourier spectra of noise images. On the left are three real images and their Fourier spectra. On the right are three noise only images and their Fourier spectra. Note that in real images the energy is concentrated in the low frequency areas, but in the noise images it is fairly evenly distributed.

9.2.3 Estimation of Noise

There are various approaches to determining the type of noise that has corrupted an image. Ideally, we want to find an image (or subimage) that contains only noise, and then we can use its histogram for the noise model. For example, if we have access to the system that generated the images, noise images can be acquired by aiming the imaging device (e.g., camera) at a blank wall—the resulting image will contain only an average (D.C.) value as a result of the lighting conditions and any fluctuations will be from noise.

If we cannot find "noise-only" images, a portion of the image is selected that has a known histogram, and that knowledge is used to determine the noise characteristics. This may be a subimage of constant value (Figure 9.2.7) or a well-defined line—any portion of the image where we know what to expect in the histogram. We can then subtract the known values from the histogram, and what is left is our noise model. We can then compare this noise model to the models described here and select the best match. In order to develop a valid model with any of these approaches, many such images (or subimages) need to be evaluated. For more information on the theoretical approach to noise modeling see references on digital signal processing, statistical or stochastic processes, and communications theory. In practice, the best model is often determined empirically.

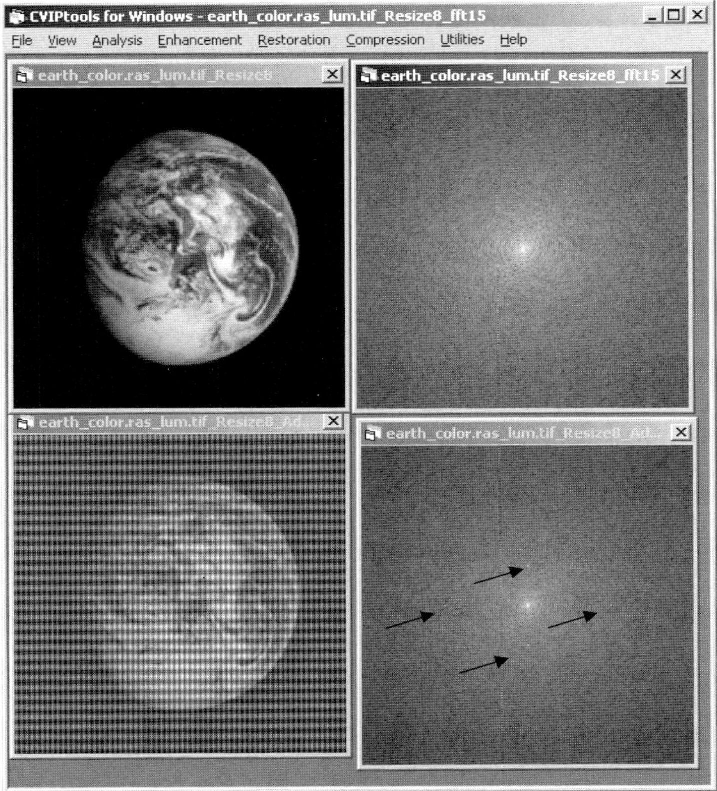

FIGURE 9.2.6
Image corrupted by periodic noise. On the top are the original image and its spectrum; under it are the image with additive sinusoidal noise, and its spectrum. Note the four impulses corresponding to the noise appearing as white dots—two on the vertical axis and two on the horizontal axis.

9.3 Noise Removal Using Spatial Filters

Spatial filters can be effectively used to remove various types of noise in digital images. These spatial filters typically operate on small neighborhoods, 3×3 to 11×11, and some can be implemented as convolution masks. For this section, we will use the degradation model defined in Section 9.1.1, with the assumption that $h(r, c)$ causes no degradation, so the only corruption to the image is caused by additive noise, as follows:

$$d(r, c) = I(r, c) + n(r, c)$$

where,

$$d(r, c) = \text{degraded image}$$

$$I(r, c) = \text{original image}$$

$$n(r, c) = \text{additive noise function}$$

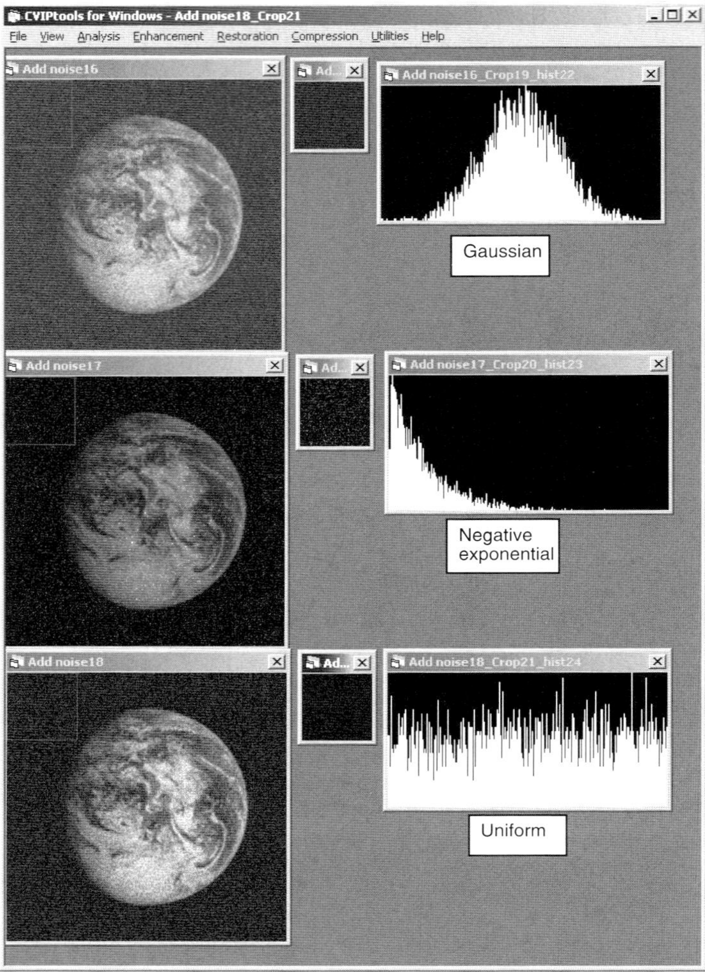

FIGURE 9.2.7
Estimating the noise with crop and histogram. On the left are three images with different noise types added. The upper left corner is cropped from the image and is shown in the middle. The histogram for the cropped subimage is shown on the right. Although the noise images look similar, the histograms are quite distinctive—Gaussian, negative exponential and uniform.

The two primary categories of spatial filters for noise removal are order filters and mean filters. The *order filters* are implemented by arranging the neighborhood pixels in order from smallest to largest gray level value, and using this ordering to select the "correct" value, while the *mean filters* determine, in one sense or another, an average value. The mean filters work best with Gaussian or uniform noise, and the order filters work best with salt-and-pepper, negative exponential, or Rayleigh noise.

The mean filters have the disadvantage of blurring the image edges, or details; they are essentially lowpass filters. As we have seen, much of the high frequency energy in noisy images is from the noise itself, so it is reasonable that a lowpass filter can be used to mitigate noise effects. We have also seen that order filters such as the median can be used to smooth images, thereby attenuating high frequency energy. However, the order filters are nonlinear, so their results are sometimes unpredictable.

In general, there is a tradeoff between preservation of image detail and noise elimination. To understand this concept consider an extreme case where the entire image is replaced with the average value of the image. In one sense, we have eliminated any noise present in the image, but we have also lost all the information in the image. Practical mean and order filters also lose information in their quest for noise elimination, and the trick is to minimize this information loss while maximizing noise removal. Ideally, a filter that adapts to the underlying pixel values is desired. A filter that changes its behavior based on the gray level characteristics (statistics) of a neighborhood is called an *adaptive filter,* and these filters are effective for use in many practical applications.

9.3.1 Order Filters

Order filters are based on a specific type of image statistics called order statistics. Typically, these filters operate on small subimages, *windows,* and replace the center pixel value (similar to the convolution process). *Order statistics* is a technique that arranges all the pixels in sequential order, based on gray level value. The placement of the value within this ordered set is referred as the *rank*. Given an $N \times N$ window, W, the pixel values can be ordered from smallest to largest, as follows:

$$I_1 \leq I_2 \leq I_3 \leq \cdots \leq I_{N^2}$$

where $\{I_1, I_2, I_3 \ldots, I_{N^2}\}$, are the Intensity (gray level) values of the subset of pixels in the image, that are in the $N \times N$ window, W (that is, $(r, c) \in W$).

Example 9.3.1

Given the following 3×3 subimage:

$$\begin{bmatrix} 110 & 110 & 114 \\ 100 & 104 & 104 \\ 95 & 88 & 85 \end{bmatrix}$$

The result from applying order statistics to arrange them is:

$$\{85, 88, 95, 100, 104, 104, 110, 110, 114\}$$

The most useful of the order filters is the median filter. The *median filter* selects the middle pixel value from the ordered set. In the above example the median filter selects the value 104, since there are 4 values above it and 4 values below it. The median filtering operation is performed on an image by application of the sliding window concept (see Figure 8.2.19), similar to what is done with convolution. Note that with this technique the outer $[(N+1)/2] - 1$ rows and columns are not replaced. In practice this is usually not a problem due to the fact that the images are much larger than the masks, and these "wasted" rows and columns are often filled with zeros (or cropped off the image). For example, with a 3×3 mask, we lose one outer row and column, a 5×5 loses two rows and columns—this is not usually significant for a typical 256×256 or 512×512 image. Results from using the median filter for salt-and-pepper (impulse) noise are shown in Figure 9.3.1.

The maximum and minimum filters are two order filters that can be used for elimination of salt *or* pepper (impulse) noise. The *maximum filter* selects the largest value within an ordered window of pixel values, so is effective at removing pepper-type (low values) noise. The *minimum filter* selects the smallest value and works when the noise is primarily of the salt-type (high values). In Figure 9.3.2a,b, the application of a minimum

(a)

(b)

(c)

FIGURE 9.3.1
(See color insert following page 362.) Median filter with color image. (a) Image with added salt-and-pepper noise, the probability for salt = probability for pepper = 0.08. (b) After median filtering with a 3 × 3 window, all the noise is not removed. (c) After median filtering with a 5 × 5 window, all the noise is removed, but the image is blurry acquiring the "painted" effect.

(a)

(b)

(c)

(d)

FIGURE 9.3.2
Minimum and maximum filters. (a) Image with "salt" noise; probability of salt = 0.04. (b) Result of minimum filtering image (a); mask size = 3 × 3. (c) Image with "pepper" noise; probability of pepper = 0.04. (d) Result of maximum filtering image (c); mask size = 3 × 3.

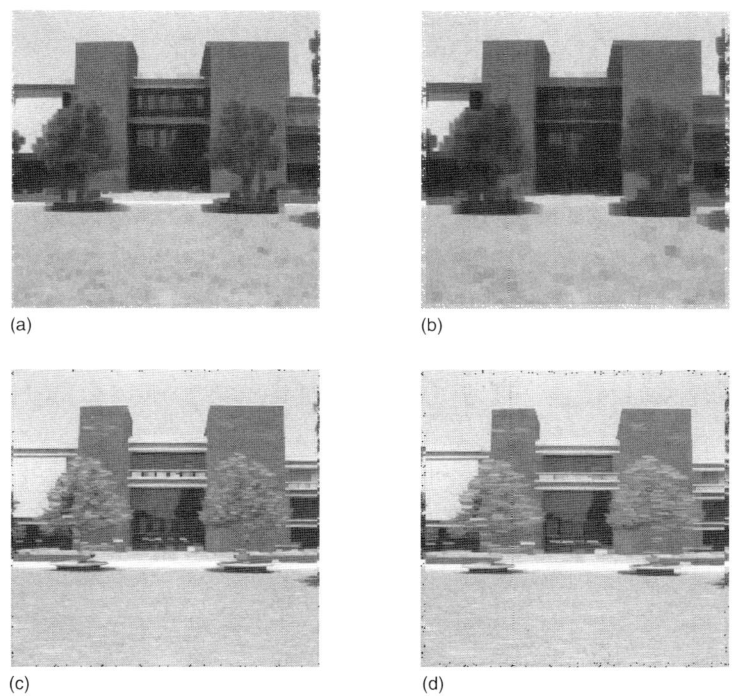

FIGURE 9.3.3
Various window sizes for maximum and minimum filters. (a) Result of minimum filtering image 9.3.2a; mask size = 5×5. (b) Result of minimum filtering image 9.3.2a; mask size = 9×9. (c) Result of maximum filtering image 9.3.2c; mask size = 5×5. (d) Result of maximum filtering image 9.3.2c; mask size = 9×9.

filter to an image contaminated with salt-type noise is shown, and in Figure 9.3.2c,d a maximum filter is applied to an image corrupted with pepper-type noise is shown. Here we see that these filters are effective for removing these types of noise, while still retaining essential image information. As the size of the window gets bigger, the more information loss occurs; with windows larger than about 5×5 the image acquires an artificial, "painted," effect (Figure 9.3.3), similar to the median filter.

In a manner similar to the median, minimum, and maximum filter, order filters can be defined to select a specific pixel rank within the ordered set. For example, we may find for certain types of pepper noise selecting the second highest value works better than selecting the maximum value. This type of ordered selection is very sensitive to the type of images and their use—it is application specific. Another example might be selecting the third value from the lowest, and using a larger window, for specific types of salt noise. It should be noted that, in general, a minimum or low rank filter will tend to darken an image and a maximum or high rank filter will tend to brighten an image—this effect is especially noticeable in areas of fine detail and high contrast.

The final two order filters are the midpoint and alpha-trimmed mean filters. They are actually both order and mean filters since they rely on ordering the pixel values, but are then calculated by an averaging process. The *midpoint filter* is the average of the maximum and minimum within the window, as follows:

$$\text{Ordered set} \rightarrow I_1 \leq I_2 \leq I_3 \leq \cdots \leq I_{N^2}$$

$$\text{Midpoint} = \frac{I_1 + I_{N^2}}{2}$$

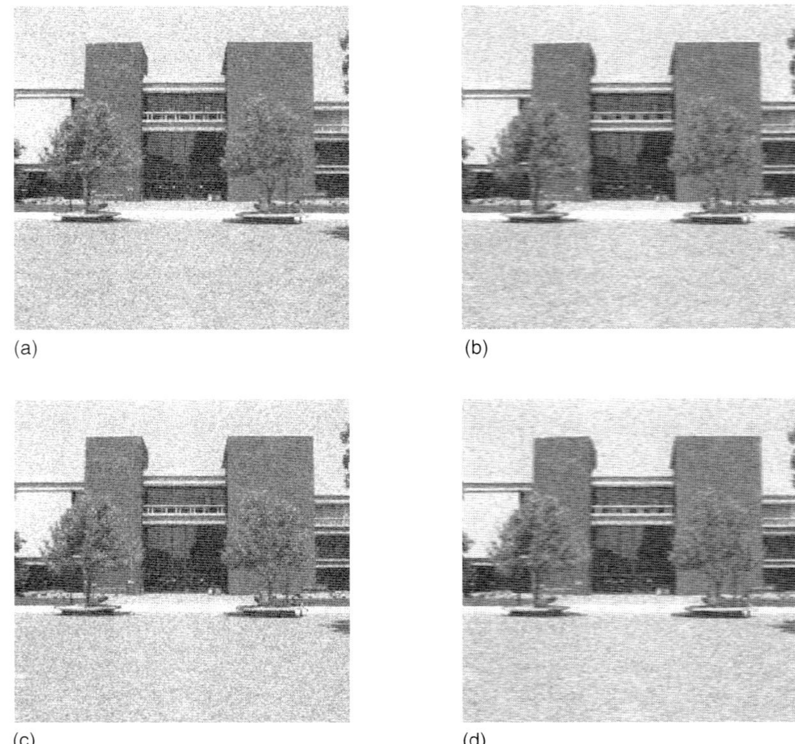

FIGURE 9.3.4
Midpoint filter. (a) Image with Gaussian noise, variance = 300, mean = 0. (b) Result of midpoint filter mask size = 3. (c) Image with uniform noise, variance = 300, mean = 0. (d) Result of midpoint filter mask size = 3.

The midpoint filter is most useful for Gaussian and uniform noise, as illustrated in Figure 9.3.4.

The *alpha-trimmed mean* is the average of the pixel values within the window, but with some of the endpoint ranked values excluded. It is defined as follows:

$$\text{Ordered set} \rightarrow I_1 \leq I_2 \leq \cdots \leq I_N{}^2$$

$$\text{Alpha-trimmed mean} = \frac{1}{N^2 - 2T} \sum_{i=T+1}^{N^2-T} I_i$$

where T is the number of pixel values excluded at each end of the ordered set, and can range from 0 to $(N^2 - 1)/2$.

The alpha-trimmed mean filter ranges from a mean to median filter, depending on the value selected for the T parameter. For example, if $T = 0$, the equation reduces to finding the average gray level value in the window, which is an arithmetic mean filter. If $T = (N^2 - 1)/2$, the equation becomes a median filter. This filter is useful for images containing multiple types of noise, for example Gaussian and salt-and-pepper noise. In Figure 9.3.5 are the results of applying this filter to an image with both Gaussian and salt-and-pepper noise.

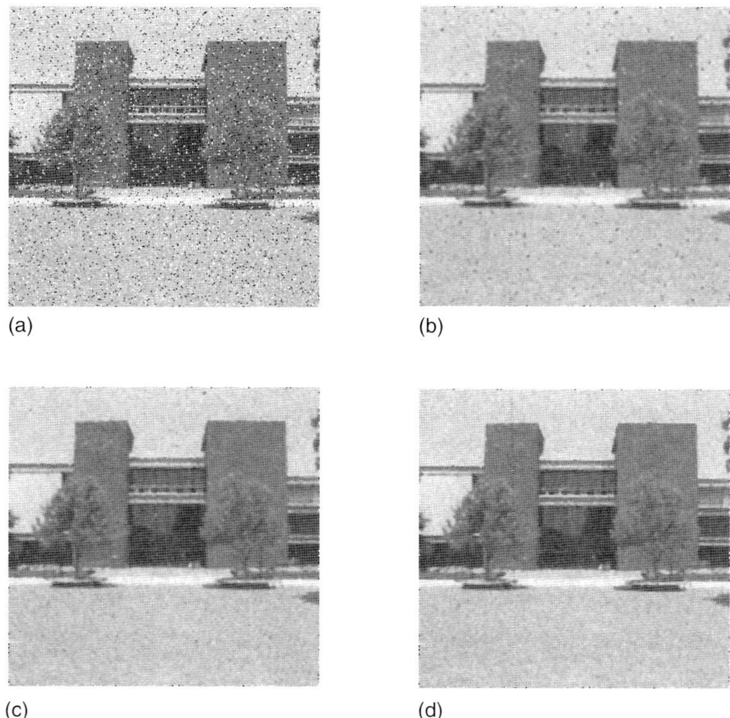

FIGURE 9.3.5
Alpha-trimmed mean. This filter can vary between a mean filter and a median filter. (a) Image with added noise: zero-mean Gaussian noise with a variance of 200, and salt-and-pepper noise with probability of each = 0.03. (b) Result of alpha-trimmed mean filter, mask size = 3×3, $T = 1$. (c) Result of alpha-trimmed mean filter, mask size = 3×3, $T = 2$. (d) Result of alpha-trimmed mean filter, mask size = 3×3, $T = 4$. As the T parameter increases the filter becomes more like a median filter, so becomes more effective at removing the salt-and pepper noise.

9.3.2 Mean Filters

The mean filters function by finding some form of an average within the $N \times N$ window, using the sliding window concept to process the entire image. The most basic of these filters is the *arithmetic mean filter*, which finds the arithmetic average of the pixel values in the window, as follows:

$$\text{Arithmetic mean} = \frac{1}{N^2} \sum_{(r,c) \in W} d(r,c)$$

where N^2 = the number of pixels in the $N \times N$ window, W.

The arithmetic mean filter smooths out local variations within an image, so it is essentially a lowpass filter. It can be implemented with a convolution mask where all the mask coefficients are $1/N^2$. This filter will tend to blur an image, while mitigating the noise effects. In Figure 9.3.6 are the results of an arithmetic mean applied to images with various types of noise. It can be seen that the larger the mask size, the more pronounced the blurring effect. This type of filter works best with Gaussian, gamma and uniform noise.

In Figure 9.3.7, the Fourier spectra for images before and after adding noise and using a mean filter are shown. Here we see that the mean filter removes some of the high

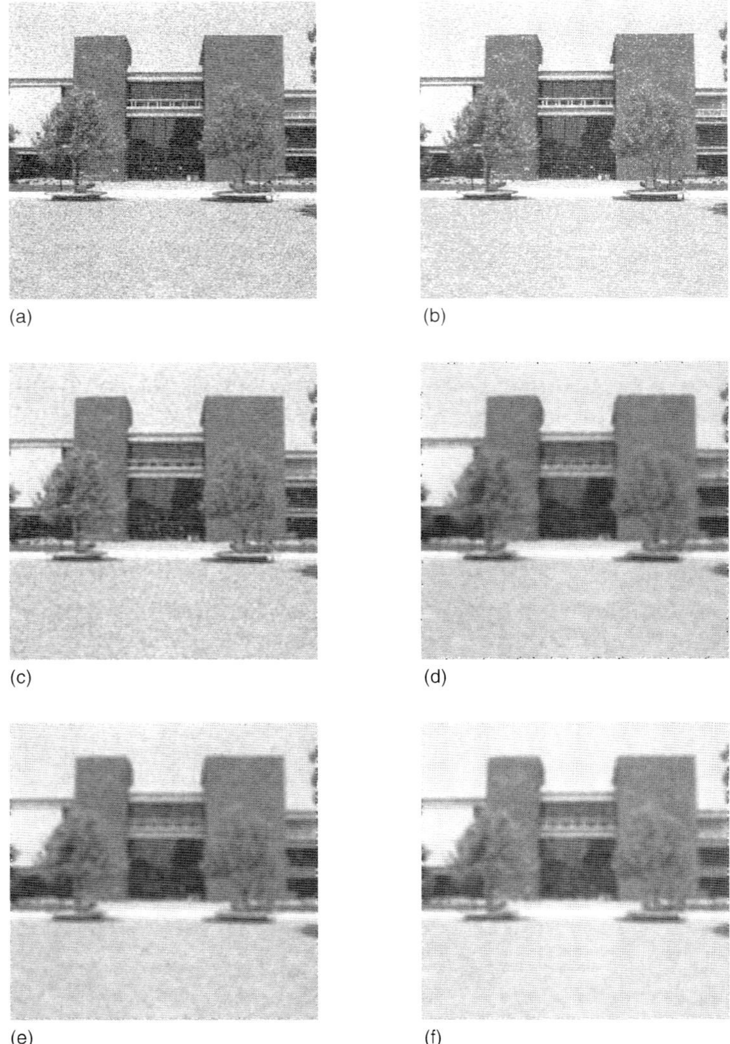

FIGURE 9.3.6
Arithmetic mean filter. (a) Image with Gaussian noise variance = 300, mean = 0. (b) Image with gamma noise variance = 300, alpha = 1. (c) Result of arithmetic mean filter, mask size = 3, on image with Gaussian noise. (d) Result of arithmetic mean filter, mask size = 3, on image with gamma noise. (e) Result of arithmetic mean filter, mask size = 5, on image with Gaussian noise. (f) Result of arithmetic mean filter, mask size = 5, on image with gamma noise.

frequency energy caused by the added noise. We also can see artifacts caused by the finite size of the convolution mask, combined with the fact that at the edge of the mask the coefficients drop off abruptly from $1/N^2$ to zero. These artifacts can be reduced by using coefficients with a Gaussian distribution (as described in Chapter 8), and results are shown in Figure 9.3.8. Here we also see that the blur is reduced with a Gaussian filter, compared to an arithmetic mean filter.

The images in these two figures illustrate that the theoretical mitigation of noise effects, in this case by reducing the high frequency energy with an arithmetic mean filter, does not necessarily create what we would consider a "good" image. The undesirable blurring effect, which reduces image details, is not as pronounced with some of the other

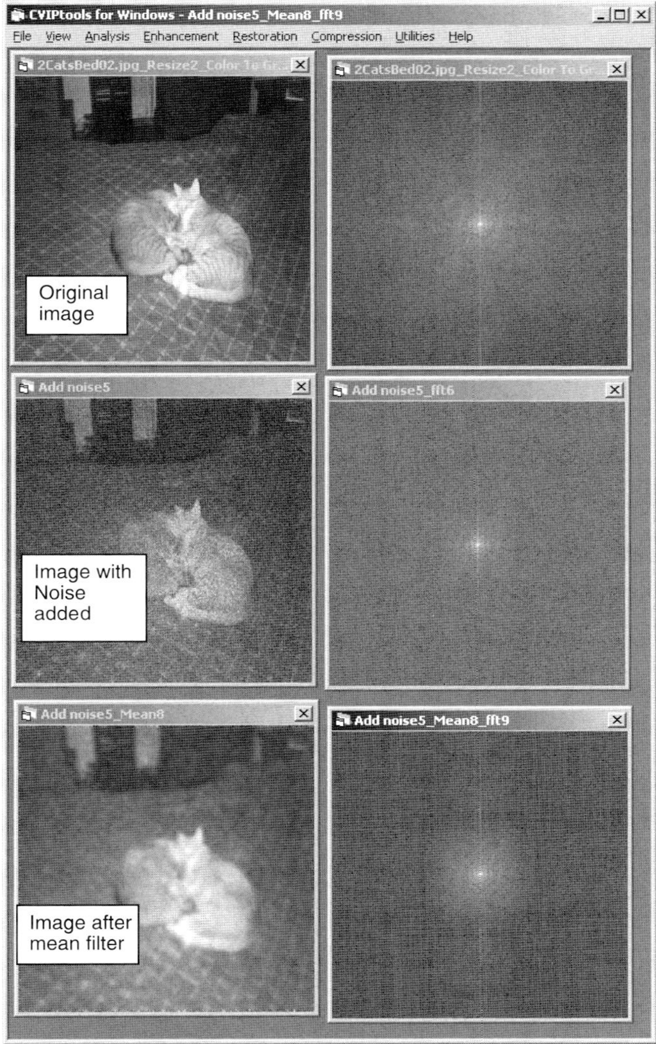

FIGURE 9.3.7
Fourier spectra from using mean filter to reduce Gaussian noise effects. Comparing the spectrum of the original image and the spectrum of the image with added noise, we can see the increase in high frequency energy—the brightness is more uniform throughout the spectrum after adding noise. After mean filtering, the increased noise energy in the high frequencies has been reduced and the energy along the vertical axis has been restored. Artifacts as a result of the finite size of the convolution mask are also visible.

mean filters. The following mean filters, and the adaptive filters, are designed to minimize this loss of detail information.

The *contra-harmonic mean filter* works well for images containing salt OR pepper type noise, depending on the filter order, R:

$$\text{Contra-harmonic mean} = \frac{\sum_{(r,c)\in W} d(r,c)^{R+1}}{\sum_{(r,c)\in W} d(r,c)^{R}}$$

where W is the $N \times N$ window under consideration.

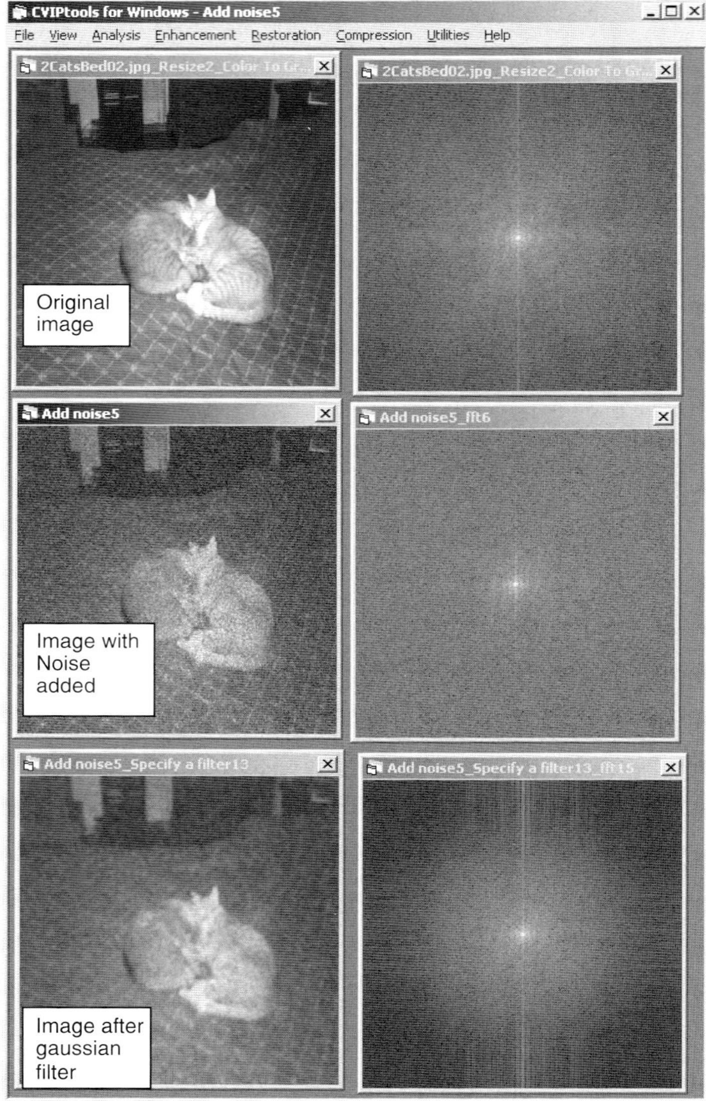

FIGURE 9.3.8
Fourier spectra from using Gaussian filter to reduce Gaussian noise effects. Comparing this figure to 9.3.7, we can see that using a convolution mask with coefficients that change gradually will reduce the artifacts that result from the finite size of the convolution mask.

For negative values of R, it eliminates salt-type noise, while for positive values of R, it eliminates pepper-type noise. This is shown in Figure 9.3.9.

The *geometric mean filter* works best with Gaussian noise, and retains detail information better than an arithmetic mean filter. It is defined as the product of the pixel values within the window, raised to the $1/N^2$ power:

$$\text{Geometric mean} = \prod_{(r,c)\in W} [d(r,c)]^{1/N^2}$$

FIGURE 9.3.9
Contra-harmonic mean filter. (a) Image with salt noise, probability = 0.04. (b) Result of contra-harmonic mean filter, mask size = 3, order = −3. (c) Image with pepper noise, probability = 0.04. (d) Result of contra-harmonic mean filter, mask size = 3, order = +3.

In Figure 9.3.10 are the results of applying this filter to images with Gaussian (Figure 9.3.10a,b) and pepper noise (Figure 9.3.10c,d). As shown in Figure 9.3.10d, this filter is ineffective in the presence of pepper noise—with zero (or very low) values present in the window, the equation returns a zero (or very small) number.

The *harmonic mean filter* also fails with pepper noise, but works well for salt noise. It is defined as follows:

$$\text{Harmonic mean} = \frac{N^2}{\sum\limits_{(r,c)\in W} 1/d(r,c)}$$

This filter also works with Gaussian noise, retaining detail information better than the arithmetic mean filter. In Figure 9.3.11 are the results from applying the harmonic mean filter to an image with Gaussian noise (Figure 9.3.11a,b), and to an image corrupted with salt noise (Figure 9.3.11c,d).

The Y_p *mean filter* is defined as follows:

$$Y_p \text{ mean} = \left[\sum_{(r,c)\in W} \frac{d(r,c)^P}{N^2} \right]^{1/P}$$

FIGURE 9.3.10
Geometric mean filter. (a) Image with Gaussian noise, variance = 300, mean = 0. (b) Result of geometric mean filter, mask size = 3, on image with Gaussian noise. (c) Image with pepper noise, probability = 0.04. (d) Result of geometric mean filter, mask size = 3, on image with pepper noise.

This filter removes salt noise for negative values of P, and pepper noise for positive values of P. Figure 9.3.12 illustrates the use of the Y_p filter.

9.3.3 Adaptive Filters

The previously described filters are not adaptive because their basic behavior does not change as the image is processed, even though their output depends on the underlying pixel values. Some, such as the alpha-trimmed mean, can vary between a mean and median filter, but this change in filter behavior is fixed for a given value of the T parameter—the behavior does not change during processing. However, an *adaptive filter* alters its basic behavior depending on the underlying pixel values, which allows it to retain image detail while still removing noise. The typical criteria for determining filter behavior involve some measure of local brightness and contrast. In this section we will explore two adaptive filters; one that uses standard statistical measures and one that uses order statistics for measuring local variation.

The minimum mean-squared error (MMSE) filter is a good example of an adaptive filter, which exhibits varying behavior based on local image statistics. By using the first and second order statistics, the mean and the variance, we get a measure of local variation. The mean measures the average local brightness, and the variance measures local contrast. These two metrics, combined with some knowledge of the noise

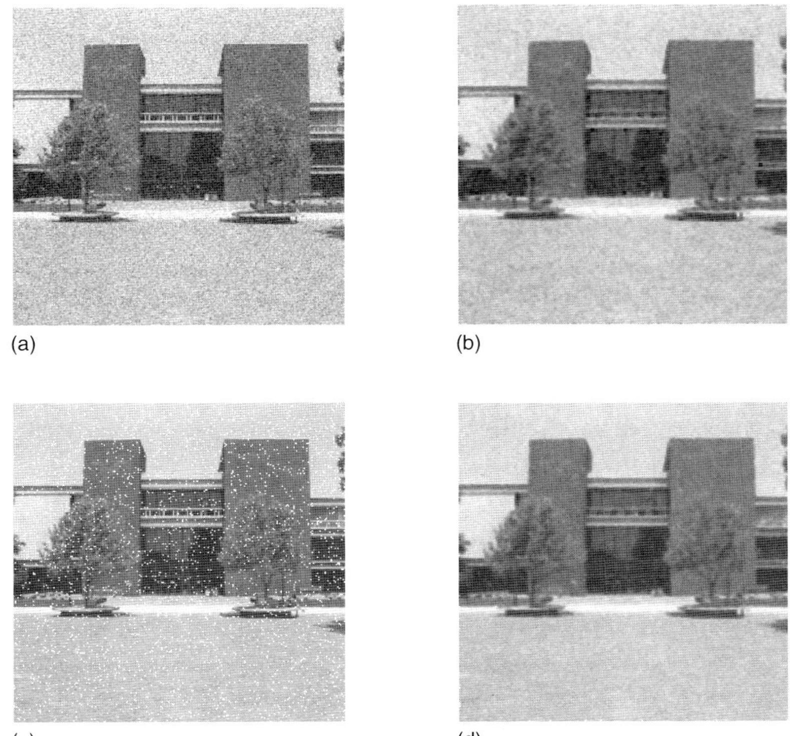

FIGURE 9.3.11
Harmonic mean filter. (a) Image with Gaussian noise, variance = 300, mean = 0. (b) Result of harmonic mean filter, mask size = 3, on image with Gaussian noise. (c) Image with salt noise, probability = 0.04. (d) Result of harmonic mean filter, mask size = 3, on image with salt noise.

variance, are used to determine filter behavior. The MMSE filter works best with Gaussian or uniform noise and is defined as follows:

$$\text{MMSE} = d(r, c) - \frac{\sigma_n^2}{\sigma_l^2}[d(r, c) - m_l(r, c)]$$

where:

$$\sigma_n^2 = \text{noise variance}$$

$$W \rightarrow \text{the current } N \times N \text{ window centered at } d(r, c)$$

$$\sigma_l^2 = \text{local variance} = \frac{\sum_{I(r,c) \in W} (I(r, c) - m_l)^2}{N^2 - 1}$$

$$m_l = \text{local mean} = \frac{1}{N^2} \sum_{I(r,c) \in W} I(r, c)$$

With no noise in the image, the noise variance equals zero, and this equation will return the original unfiltered image. In background regions of the image, areas of

FIGURE 9.3.12
Y_p mean filter. (a) Image with salt noise, probability = 0.04. (b) Result of Y_p mean filter, mask size = 3, order = −3 on image with salt noise. (c) Image with pepper noise, probability = 0.04. (d) Result of Y_P mean filter, mask size = 3, order = +3 on image with pepper noise.

fairly constant value in the original uncorrupted image, the noise variance will equal the local variance, and the equation reduces to the mean filter. In areas of the image where the local variance is much greater then the noise variance, the filter returns a value close to the unfiltered image data. This is desired since high local variance implies high detail (edges), and an adaptive filter tries to preserve the original image detail.

In general, the MMSE filter returns a value that consists of the unfiltered image data, $d(r, c)$, with some of the original value subtracted out and some of the local mean added. The amount of the original and local mean used to modify the original are weighted by the noise-to-local-variance ratio, σ_n^2/σ_l^2. As this ratio increases, implying primarily noise in the window, the filter returns primarily the local average. As this ratio goes down, implying high local detail, the filter returns more of the original unfiltered image. By operating in this manner, the MMSE filter adapts itself to the local image statistics, preserving image details while removing noise. Figure 9.3.13 illustrates the use of the MMSE filter on an image with added Gaussian noise. Here we specify the window (kernel) size and the noise variance to be used.

Another useful adaptive filter is the *adaptive median filter*. The primary strength of the adaptive median filter is the removal of salt-and-pepper noise, but it also attempts to smooth other types of noise and to avoid the distortion of small image structures

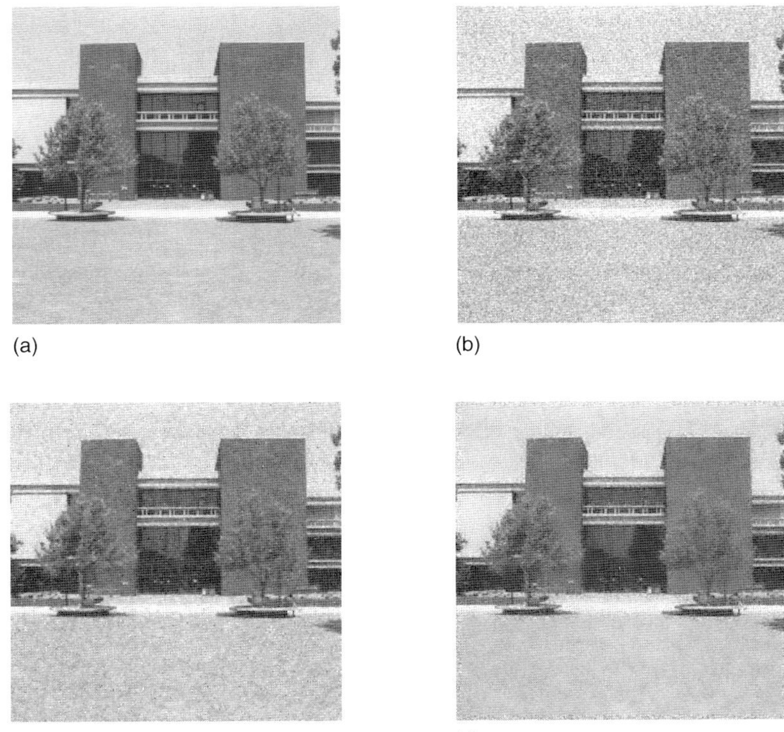

FIGURE 9.3.13
MMSE filter. (a) Original image. (b) Image with Gaussian noise variance = 300, mean = 0. (c) Result of MMSE filter kernel size = 3, noise variance = 300. (d) Result of MMSE filter kernel size = 9, noise variance = 300.

seen with the standard median filter (see Figures 8.4.3 and 9.3.1). This filter is algorithmic in nature and has a variable window size that increases until a certain criterion is met. To describe the algorithm for the adaptive median filter we need to define the following:

$$d(r,c) = \text{the degraded image}$$
$$W \rightarrow \text{the current } N \times N \text{ window centered at } d(r,c)$$
$$W_{max} = \text{maximum window size}$$
$$g_{min} = \text{minimum gray level in the window}, W$$
$$g_{max} = \text{maximum gray level in } W$$
$$g_{med} = \text{median gray level in } W$$

The adaptive median filter algorithm is defined using two levels, as follows:

Level 1:

If $(g_{max} < g_{med} < g_{min})$
Then go to Level 2

Else increase window size, $N = N + 2$
If (window size) $\leq W_{max}$
Then go to Level 1
Else output $= d(r, c)$

Level 2:
If $(g_{max} < d(r, c) < g_{min})$
Then output $= d(r, c)$
Else output $= g_{med}$

The function of *Level 1* is to determine if the standard median filter output is impulse noise. If it equals the MAX or MIN it might be impulse noise, so we increase the window size and try again. If it is not, we go to *Level 2* and test to see if the current pixel is impulse noise. If it is, we output the median value, if not we output the current value. This will tend to preserve edges. An example, using salt noise (255), will help to illustrate this.

Example 9.3.2

Given the following two subimages, with salt noise:

$$\textit{Image 1}: \begin{bmatrix} 10 & 9 & 11 \\ 11 & 255 & 10 \\ 12 & 13 & 9 \end{bmatrix} \qquad \textit{Image 2}: \begin{bmatrix} 255 & 9 & 255 \\ 11 & 255 & 10 \\ 255 & 13 & 255 \end{bmatrix}$$

Sorting the values for *Image 1*: [255 13 12 11 11 10 10 9 9],

Level 1: $(g_{max} < g_{med} = 11 < g_{min})$ is TRUE, so go to *Level 2*

Level 2: $(g_{max} < d(r, c) = 255 < g_{min})$ is FALSE, so output $= 11$

In this case we eliminated the salt noise successfully

Sorting the values for *Image 2*: [255 255 255 255 255 13 11 10 9],

Level 1: $(g_{max} < g_{med} = 255 < g_{min})$ is FALSE, so increase window size. Note that in this case, the standard median would not remove this noise pixel.

Increase the window size to a 5×5 and we find:

$$\begin{bmatrix} 11 & 12 & 11 & 10 & 10 \\ 9 & 255 & 9 & 255 & 10 \\ 255 & 11 & 255 & 10 & 11 \\ 12 & 255 & 13 & 255 & 13 \\ 13 & 13 & 12 & 11 & 12 \end{bmatrix}$$

Now the median value is between the MAX and MIN, so we fall through to Level 2 and output the median value, thus eliminating the salt noise even in a sea of noise.

Another aspect of the adaptive median filter is that it will preserve image details better than the standard median filter. This is shown in the following example.

Example 9.3.3

Given the following subimage:

$$\begin{bmatrix} 177 & 255 & 180 \\ 180 & 189 & 170 \\ 192 & 196 & 180 \end{bmatrix}$$

Sorting the values for *Image 1*: [255 196 192 189 180 180 180 177 170],

Level 1: $(g_{\max} < g_{\text{med}} = 180 < g_{\min})$ is TRUE, so go to *Level 2*

Level 2: $(g_{\max} < d(r,c) = 189 < g_{\min})$ is TRUE, so output = 189

In this case we have retained image detail, whereas the standard median would have output the median, 180, thus losing detailed information.

Figure 9.3.14 compares a standard median filter with the adaptive median filter. Here we see that the adaptive median retains image detail much better than the standard filter, while removing most of the noise. Look carefully at the leaves on the trees, and the window frames and balcony railing that disappear with the standard median filter. Of course there is a cost—increased computational complexity and processing time.

9.4 The Degradation Function

Any measurement of a quantity that varies in time, such as brightness, has some inherent degradation in the form of blurring since it requires a finite amount of time to measure. During that time the signal will fluctuate and the imaging device will typically include all the values during the measured time interval in the final result. The physics of the device may integrate the signal, add the values or average them over the measurement interval. Additional blurring may occur due to imperfect lenses, motion of the object or imaging device, and spatial quantization.

These types of degradations can be either spatially-invariant or spatially-variant. A *spatially-invariant* degradation affects all pixels in the image the same, the pixel's location does not affect the distortion. Examples include poor lens focus and motion of the camera. *Spatially-variant* degradations are dependent on spatial location and are more difficult to model. Examples include imperfections in a lens or movement of individual objects in the scene. Although more difficult to model than degradations that do not change with location, they can often be modeled as being spatially-invariant over small regions. Additionally, image degradation functions can be considered to be linear or nonlinear; and we will only consider linear degradation functions.

9.4.1 The Spatial Domain—The Point Spread Function

The model presented includes a degradation function, $h(r,c)$, and additive noise $n(r,c)$. Noise models have been discussed and here we will focus on $h(r,c)$, the degradation function. If we assume no additive noise, we have the following model:

$$d(r,c) = h(r,c) * I(r,c)$$

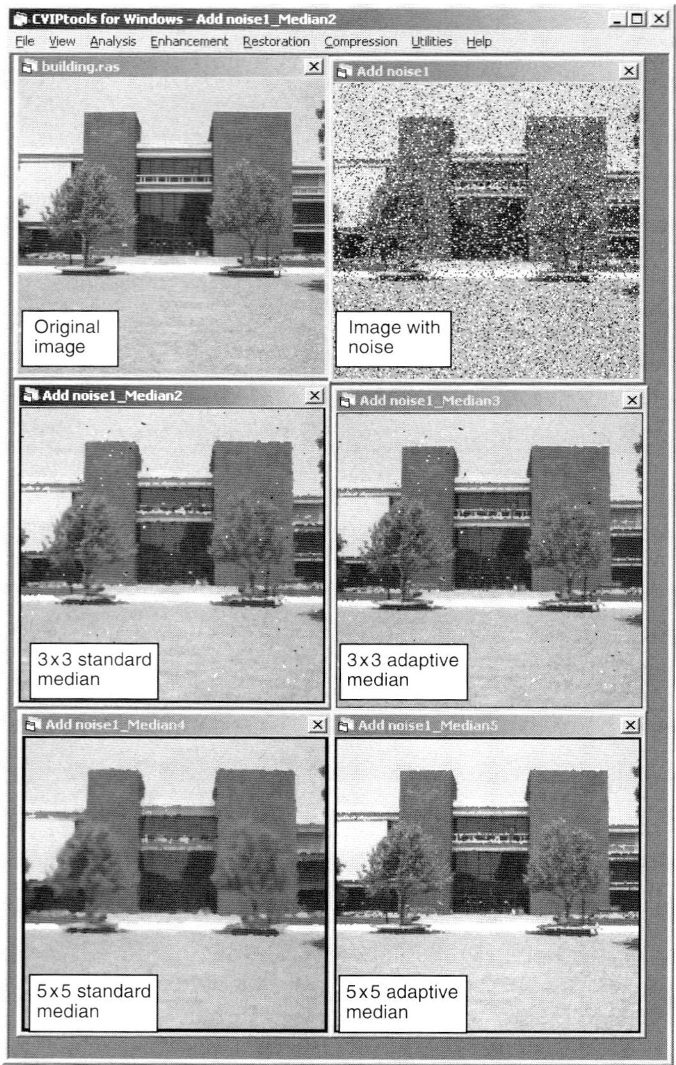

FIGURE 9.3.14
Adaptive median filter compared to standard median filter. The adaptive median filter removes noise while preserving image detail much better than the standard median filter.

where the $*$ denotes the convolution process

$$d(r, c) = \text{degraded image}$$
$$h(r, c) = \text{degradation function}$$
$$I(r, c) = \text{original image}$$

For those familiar with linear system theory, $h(r,c)$ is the two dimensional equivalent of the impulse response. Remember that the impulse response is the output of a linear system when the input is a single, narrow (ideally zero width) pulse, and that it completely characterizes the system. With a linear system, if we know the impulse response we know the response of the system to any complex signal—it is simply the signal convolved with the impulse response.

The response of an imaging system, $h(r,c)$, is called the *point spread function* (PSF). As the name implies the PSF is the blur function, or spread function, for a single point of light and describes what happens when it passes through a system. Theoretically, the PSF of a *linear, spatially-invariant* (shift invariant) system can be empirically determined by imaging a single point of light, which is the two-dimensional equivalent of an impulse signal. Also, PSFs can be derived or developed for various types of image degradation such as motion blur or atmospheric turbulence.

A simple model for the PSF of motion blur can be developed as follows. Consider that an object is moving while a camera shutter is open. During that time the object will appear across several of the imaging elements or pixels. Figure 9.4.1 shows two models for motion blur along the column axis. Figure 9.4.1a shows an image meant to model a point, which has been enlarged for ease of viewing (it is actually a circle of radius 4 created with CVIPtools—note the distortion due to spatial quantization). In Figure 9.4.1b,c we see the result from uniform blur, and the model for the PSF. Since it takes a finite amount of time for the camera's shutter to close, the PSF model in Figure 9.4.1e more accurately models a mechanical system where the gain slowly decreases over time.

In CVIPtools the PSF blur models can be simulated with *Utilities->Filter->Specify a Blur*. The types of blur can be linear in one direction—horizontal, vertical, or diagonal; or it can be two-dimensional. Two-dimensional masks can have circular or rectangle symmetry; both have the same blur in all directions. The PSF mask is used to model

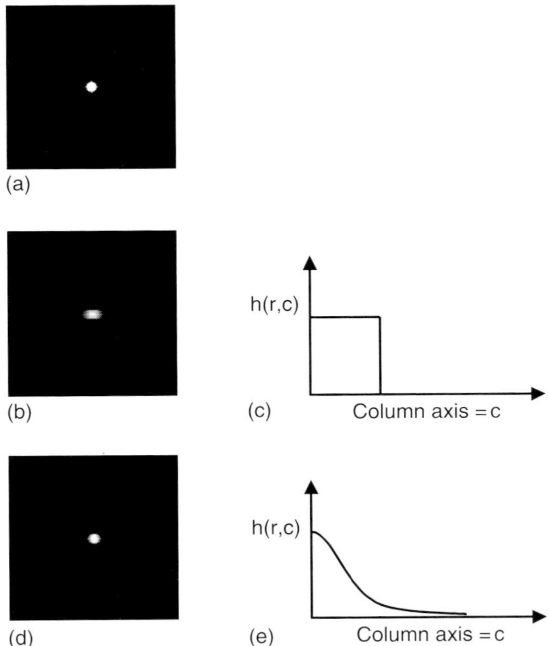

FIGURE 9.4.1
Modeling the point spread function (PSF) for motion blur. (a) A digital image of a point, enlarged for ease of viewing, also note distortion due to spatial quantization. (b) Image after uniform motion blur along the column axis. (c) PSF model for uniform motion blur. (d) Image after Gaussian motion blur along the column axis. (e) PSF model for Gaussian motion blur, which models a mechanical shutter closing or the speed decreasing. In practice this is often approximated by a linear function.

$h(r,c)$, typical models are shown in Figure 9.4.2. The non-zero terms, designated by x's in the figure, can have a uniform, Gaussian, or centered-weighted distribution.

In Figure 9.4.3, are shown representative values for the various types of blur filter masks. Once a blur filter mask has been selected, the mask must be padded with zeros up to the size of the image, before the Fourier transform of $h(r,c)$ is determined.

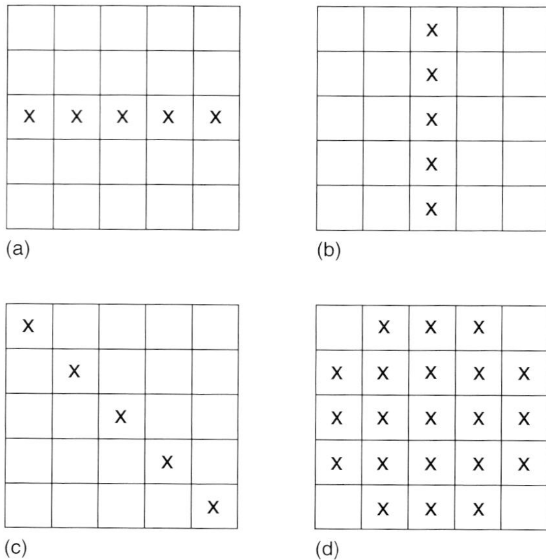

FIGURE 9.4.2
Blur (PSF) masks. (a) Horizontal PSF mask; x = non-zero term. (b) Vertical PSF mask; x = non-zero term. (c) Diagonal PSF mask; x = non-zero term. (d) Circular PSF mask; x = non-zero term.

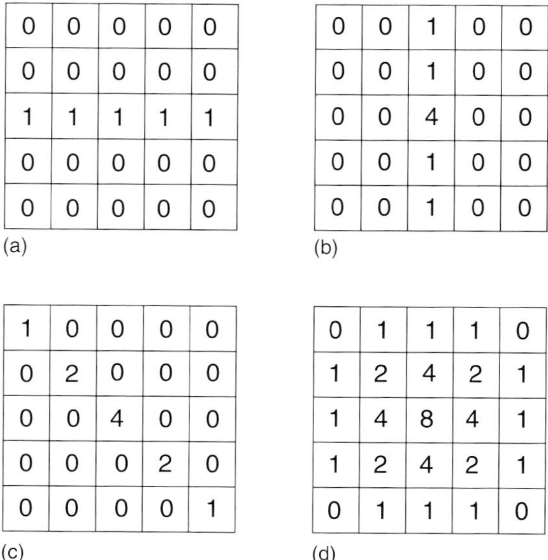

FIGURE 9.4.3
Typical blur mask coefficients. (a) Horizontal PSF mask with uniform blur. (b) Vertical PSF mask with center-weighting. (c) Diagonal PSF mask with an approximation to a Gaussian distribution. (d) Circular PSF mask with an approximation to a Gaussian distribution.

Once the blur mask has been zero-padded, this extended $h(r, c)$ needs to be shifted, with wrap-around, so that the center of the original blur mask is at the $(0, 0)$ point in the image, that is the upper left corner. In other words, the coefficients will appear in the corners, based on Fourier symmetry. If this is not done phase shifts will occur in the output image.

After the PSF is determined, a model is developed to reverse the process. The application of the inverse model is called *deconvolution*, because we are trying to undo the convolution process that degraded (blurred) the image. If we do not know the exact degradation function, which is typically true in practice, then the process is referred to as *blind deconvolution* and the PSF must be estimated. This inverse process is often performed in the frequency domain.

9.4.2 The Frequency Domain—The Modulation/Optical Transfer Function

The Fourier transform of the degradation function, $H(u, v)$, is also referred to as the modulation transfer function (MTF), or the optical transfer function (OTF). The MTF typically refers to the transfer function of the system, while the OTF may refer to the transfer function of a lens or, in general, the optics in the system. Like the PSF, the MTF is used to completely characterize a linear, spatially-invariant system.

Using our previously defined model and assuming no additive noise, the frequency domain model for the degradation process is as follows:

$$D(u, v) = H(u, v)I(u, v)$$

where:

$$\begin{aligned} D(u, v) &= \text{Fourier transform of the degraded image} \\ H(u, v) &= \text{Fourier transform of the degradation function} \\ I(u, v) &= \text{Fourier transform of the original image} \end{aligned}$$

Examples of specific mathematical models that have been developed include motion blur, poor lens focus, atmospheric turbulence and CCD interactions. Motion blur occurs if the object or camera moves during image acquisition. Atmospheric turbulence degradation occurs during imaging of astronomical objects. Blurring also occurs from the spatial quantization inherent in a CCD array, and the interaction between adjacent CCD elements. Given knowledge of the image acquisition process, the appropriate model can be applied.

For an image that is acquired with a camera with a mechanical shutter, $H(u, v)$ due to motion blur along the column axis is as follows:

$$H(u, v) = \frac{\sin(\pi STv)}{\pi Sv}$$

where:

$$\begin{aligned} S &= \text{the constant speed in the direction of the column axis} \\ T &= \text{the time interval the shutter is open} \end{aligned}$$

Note that this is a *sinc* function, $(\sin x/x)$, which is the Fourier transform of a rectangle function, such as the PSF of the motion blur in Figure 9.4.1b, c.

9.4.3 Estimation of the Degradation Function

The degradation function can be estimated primarily by combinations of these three methods: (1) image analysis, (2) experimentation, and (3) mathematical modeling. The degraded image can be analyzed by examination of a known point or line in an image. If the width of the point or line is known, we can estimate the PSF by measuring the width of the known feature in the blurred image. This will give us some idea of how wide the PSF blur mask should be. If the imaging system is available these points or lines can be found by the use of test charts. Some images may contain valid point sources; such as the stars in astronomical images. Lines can be found by analyzing the images; for example, edges of buildings or object borders. We can also analyze the gray level distribution of the blur and attempt to model it mathematically to determine the proper PSF mask coefficients.

As we have seen, the degradation function or PSF can be found experimentally, *if the system is available and the conditions under which the image was acquired have not changed*—all we need to do is to send a point of light through the system and see what comes out. The output is the PSF, in this case $h(r, c)$. However, it is not always practical to implement a point source of light, and a more reliable method is to use sinusoidal inputs at many different spatial frequencies to determine the MTF, $H(u, v)$. For many applications the system that created the images may not be available, or the conditions under which the image was acquired are unknown.

Mathematical models are often used to gain insight into image degradation. One example is the motion blur model previously discussed. Another example is the atmospheric turbulence degradation model used in astronomy and remote sensing:

$$H(u, v) = e^{-k(u^2+v^2)^{5/6}}$$

where k is an experimentally determined constant.

The constant is related to the severity of the turbulence (larger k, more turbulence), usually related to the amount of temperature variation in the atmosphere. The temperature variation causes image distortion such as is seen in the desert or around a heat source. The exponent 5/6 can be replaced with 1 to create a simpler model. The simple model is a Gaussian lowpass filter which blurs the image as we have seen.

For further study, details of mathematical models for other degradation functions can be found in the references. Mathematical models, image analysis, and lots of experimentation, combined with the experience and intuition of the expert are often necessary to estimate the degradation function successfully.

9.5 Frequency Domain Filters

Frequency domain filtering operates by using the Fourier transform representation of images. This representation consists of information about the spatial frequency content of the image, also referred to as the spectrum of the image. In Figure 9.5.1 is the general model for frequency domain filtering. The Fourier transform is performed on three spatial domain functions: (1) the degraded image, $d(r, c)$, (2) the degradation function, $h(r, c)$, and (3) the noise model, $n(r, c)$. Next, the frequency

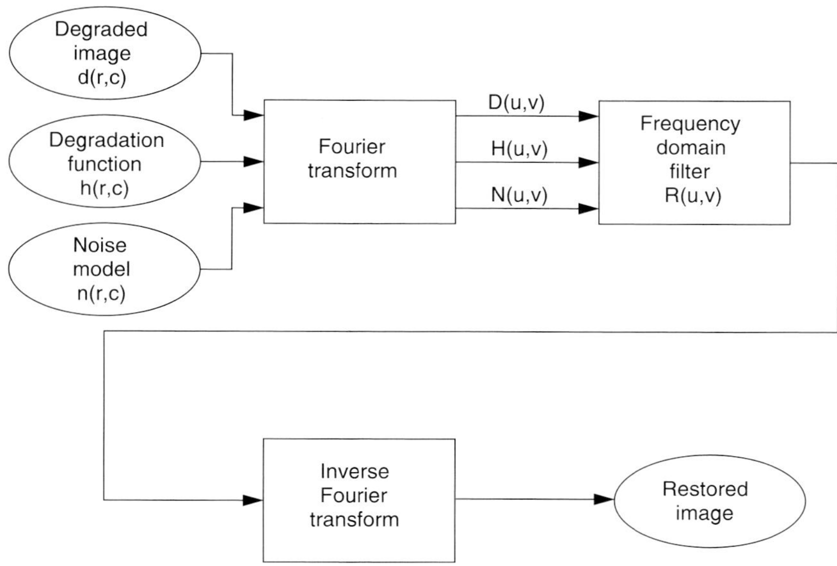

FIGURE 9.5.1
Frequency domain filtering.

domain filter is applied to the Fourier transform outputs, $N(u,v)$, $D(u,v)$, and $H(u,v)$. The output of the filter operation undergoes an inverse Fourier transform to give the restored image.

The frequency domain filters incorporate information regarding the noise and the PSF into their model, and are based on the mathematical model provided in Section 9.1.1:

$$D(u,v) = H(u,v)I(u,v) + N(u,v)$$

where:

$$\begin{aligned} D(u,v) &= \text{Fourier transform of the degraded image} \\ H(u,v) &= \text{Fourier transform of the degradation function} \\ I(u,v) &= \text{Fourier transform of the original image} \\ N(u,v) &= \text{Fourier transform of the additive noise function} \end{aligned}$$

In order to obtain the restored image, the general form is as follows:

$$\hat{I}(r,c) = F^{-1}\left[\hat{I}(u,v)\right] = F^{-1}\left[R_{\text{type}}(u,v)D(u,v)\right]$$

where:

$$\begin{aligned} \hat{I}(r,c) &= \text{the restored image, an approximation to } I(r,c) \\ F^{-1}[\,] &= \text{the inverse Fourier transform} \\ R_{\text{type}}(u,v) &= \text{the restoration (frequency domain) filter, the subscript} \\ &\quad\ \text{defines the type of filter} \end{aligned}$$

The filters discussed here include the inverse filter, the classical Wiener filter, the parametric Wiener filter, the power spectrum equalization filter, the constrained least squares filter, the geometric mean filter, bandpass, bandreject, and notch filters. A general mathematical model using the geometric mean filter is provided, and from this model many of the other filters can be generated.

Many of these filters are based on the assumption that the noise and image signals are *stationary*. This means that the spatial frequency energy content does not vary across the image, which is usually not valid for most real images, but is usually acceptable for noise images (as shown in Figure 9.2.5). For an image to be a stationary signal, if we measure the power spectrum over a small area it should be approximately equal to the power spectrum of the entire image. Most images are highly nonstationary; for example, some areas in the image may appear fairly constant and thus have mostly low frequency energy; whereas object boundaries and textured objects will have a lot of energy in high frequencies. Adaptive filtering techniques are discussed for managing this problem.

9.5.1 Inverse Filter

The inverse filter uses the foregoing model, with the added assumption of no noise ($N(u,v) = 0$). If this is the case, the Fourier transform of the degraded image is:

$$D(u,v) = H(u,v)I(u,v) + 0$$

So, the Fourier transform of the original image can be found as follows:

$$I(u,v) = \frac{D(u,v)}{H(u,v)} = D(u,v)\frac{1}{H(u,v)}$$

Using the previously defined notation for the restoration filters:

$$\text{Inverse filter} = R_{\text{inv}}(u,v) = \frac{1}{H(u,v)}$$

To find the original image we take the inverse Fourier transform of $I(u,v)$:

$$I(r,c) = F^{-1}[I(u,v)] = F^{-1}\left[\frac{D(u,v)}{H(u,v)}\right] = F^{-1}\left[D(u,v)\frac{1}{H(u,v)}\right]$$

where $F^{-1}[\,]$ represents the inverse Fourier transform.

The equation implies that the original, undegraded image can be obtained by multiplying the Fourier transform of the degraded image, $D(u,v)$, by $1/H(u,v)$, and then inverse Fourier transforming the result. Thus, the restoration filter applied is $1/H(u,v)$, the *inverse filter*. Note that this inversion is a point-by-point inversion, *not* a matrix inversion.

Example 9.5.1

$$H(u,v) = \begin{bmatrix} 50 & 50 & 25 \\ 20 & 20 & 20 \\ 20 & 35 & 22 \end{bmatrix} \qquad \frac{1}{H(u,v)} = \begin{bmatrix} \frac{1}{50} & \frac{1}{50} & \frac{1}{25} \\ \frac{1}{20} & \frac{1}{20} & \frac{1}{20} \\ \frac{1}{20} & \frac{1}{35} & \frac{1}{22} \end{bmatrix}$$

To find $1/H(u, v)$, we take each term separately and divide it into 1.

Unfortunately, in practice, there are complications that arise when this technique is applied. If there are any points in $H(u,v)$ that are zero, we face a mathematical dilemma—division by zero. If the assumption of no noise is correct, then the degraded image transform, $D(u,v)$, will also have corresponding zeros and we are left with an indeterminate ratio, 0/0. If the assumption is incorrect, and the image has been corrupted by additive noise, then the zeros will not coincide, and the image restored by the inverse filter will be obscured by the contribution of the noise terms. This can be seen by considering the following equation:

$$D(u,v) = H(u,v)I(u,v) + N(u,v)$$

Then, when we apply the inverse filter, we obtain:

$$\hat{I}(u,v) = \frac{D(u,v)}{H(u,v)} = \frac{H(u,v)I(u,v)}{H(u,v)} + \frac{N(u,v)}{H(u,v)}$$

$$= I(u,v) + \frac{N(u,v)}{H(u,v)}$$

As the values in $H(u,v)$ become very small, the second term becomes very large, and it overshadows the $I(u,v)$ term, which is the original image we are trying to recover.

One method to deal with this problem is to limit the restoration to a specific radius about the origin in the spectrum, called the restoration cutoff frequency. For spectral components beyond this radius, we can set the filter gain to 0, so $\hat{I}(u,v) = D(u,v) \times 0 = 0$. This is the equivalent of an ideal lowpass filter, which may result in blurring and ringing. In practice, the selection of the cutoff frequency must be experimentally determined, and is highly application specific. In Figure 9.5.2, we see the result of application of the inverse filter to an image blurred by an 11×11 Gaussian convolution mask. Here we see that selection of a cutoff frequency that is too low may provide poor results, and with a cutoff frequency too high the resulting image is overwhelmed by noise effects.

With some types of degradation, the function $H(u,v)$ falls off quickly as we move away from the origin in the spectrum. In this case we may want to set the filter gain to 1 for frequencies beyond the restoration cutoff. Another possibility is to model a Butterworth filter, or something between the extremes of setting the gain to 0 or 1. In practice a similar result can be achieved by limiting the gain of the filter to some maximum value.

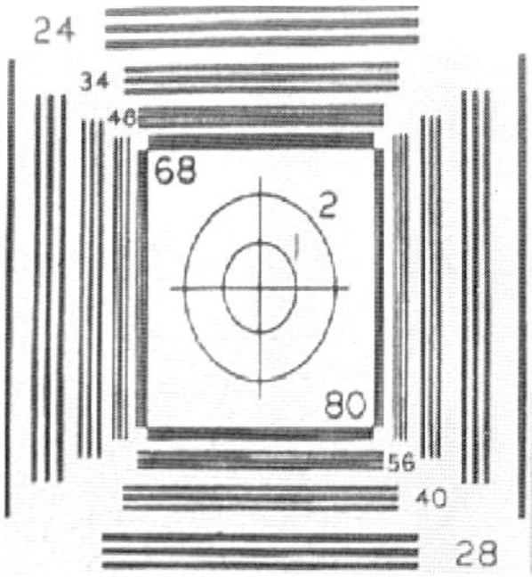

(a)

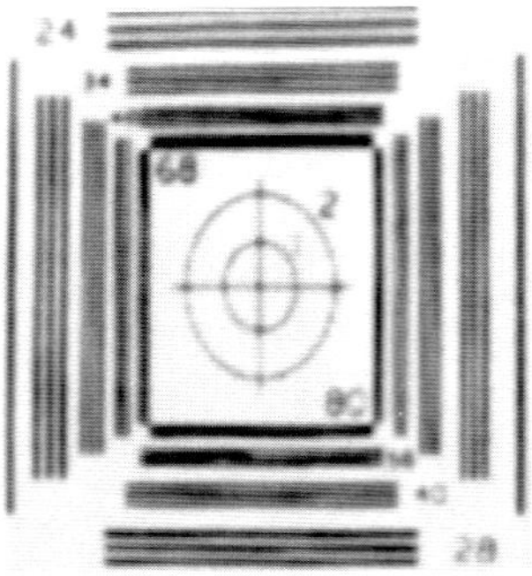

(b)

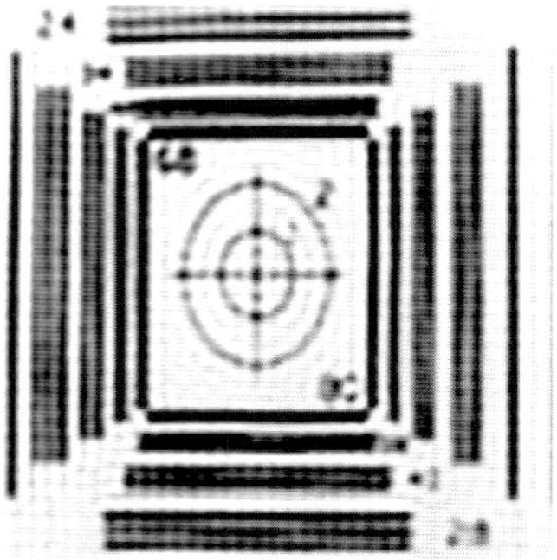

(c)

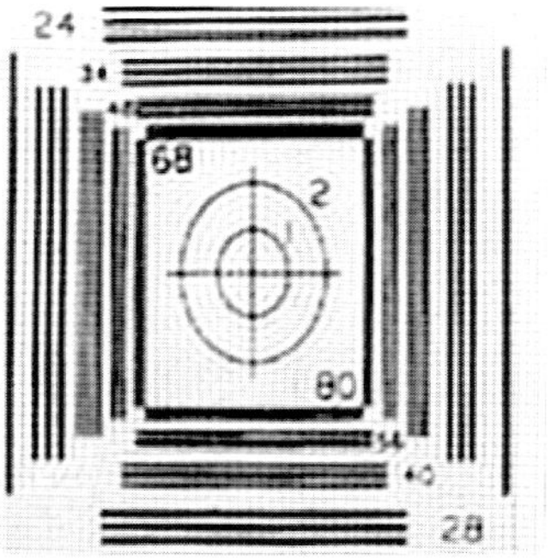

(d)

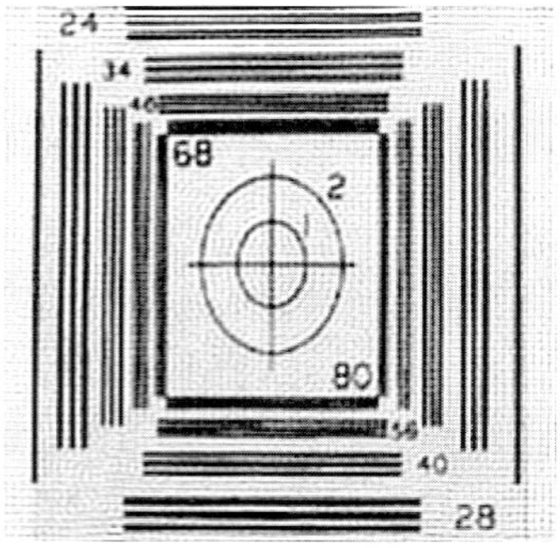

(e)

(f)

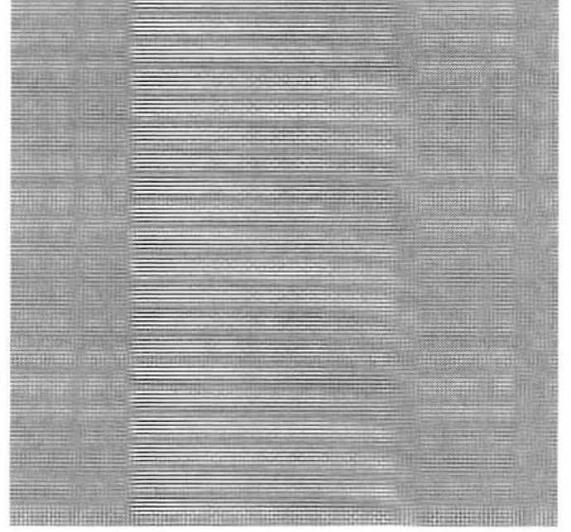

(g)

FIGURE 9.5.2
Inverse filter. (a) Original image. (b) Image blurred with an 11×11 Gaussian convolution mask. (c) Inverse filter, with cutoff frequency = 40, histogram stretched with 3% low and high clipping to show detail. (d) Inverse filter, with cutoff frequency = 60, histogram stretched. (e) Inverse filter, with cutoff frequency = 80, histogram stretched. (f) Inverse filter, with cutoff frequency = 100, histrogram stretched. (g) Inverse filter, with cutoff frequency = 120, histrogram stretched.

A related method is to use the *pseudoinverse filter*, which handles zeros in $H(u,v)$ as follows:

$$R_{\mathrm{PI}}(u,v) = \begin{cases} \dfrac{1}{H(u,v)} & \text{for } H \neq 0 \\ 0 & \text{for } H = 0 \end{cases}$$

This will provide a mathematically stable filter that will not blow up, approach infinity, as $H(u,v)$ approaches zero. In practice the filter gain is set to zero whenever the magnitude of $H(u,v)$ goes below a specified threshold.

9.5.2 Wiener Filter

The Wiener filter, also called a minimum mean-square estimator (developed by Norbert Wiener in 1942), alleviates some of the difficulties inherent in inverse filtering by attempting to model the error in the restored image through the use of statistical methods. Once the error is modeled, the average error is mathematically minimized, thus the term *minimum mean square estimator*. The resulting equation is the Wiener filter:

$$R_W(u,v) = \frac{H^*(u,v)}{\left|H(u,v)\right|^2 + \left[\dfrac{S_n(u,v)}{S_I(u,v)}\right]}$$

where:

$$H^*(u,v) = \text{complex conjugate of } H(u,v)$$

$$S_n(u,v) = \left|N(u,v)\right|^2 = \text{power spectrum of the noise}$$

$$S_I(u,v) = \left|I(u,v)\right|^2 = \text{power spectrum of the original image}$$

$$\left[\frac{S_n(u,v)}{S_I(u,v)}\right] = \text{power spectrum ratio}$$

This equation assumes a square image of size $N \times N$. The complex conjugate can be found by negating the imaginary part of a complex number. Note that the power spectrum ratio is the related to the signal-to-noise ratio inverted. Other practical considerations are discussed in Section 9.5.7. Examining this equation will provide us with some understanding of how it works.

If we assume that the noise term, $S_n(u,v)$, is zero, this equation reduces to an inverse filter, since $\left|H(u,v)\right|^2 = H^*(u,v)H(u,v)$. As the noise term increases, the denominator of the Wiener filter increases, thus decreasing the value of $R_W(u,v)$. Thus, as the contribution of the noise increases, the filter gain decreases. This seems reasonable—in portions of the spectrum uncontaminated by noise we have an inverse filter, whereas in portions of the spectrum heavily corrupted by noise, the filter attenuates the signal (see Figure 9.5.3), with the amount of attenuation being determined by the ratio of the noise spectrum to the uncorrupted image spectrum.

The Wiener filter is applied by multiplying it by the Fourier transform of the degraded image and the restored image is obtained by taking the inverse Fourier transform of the

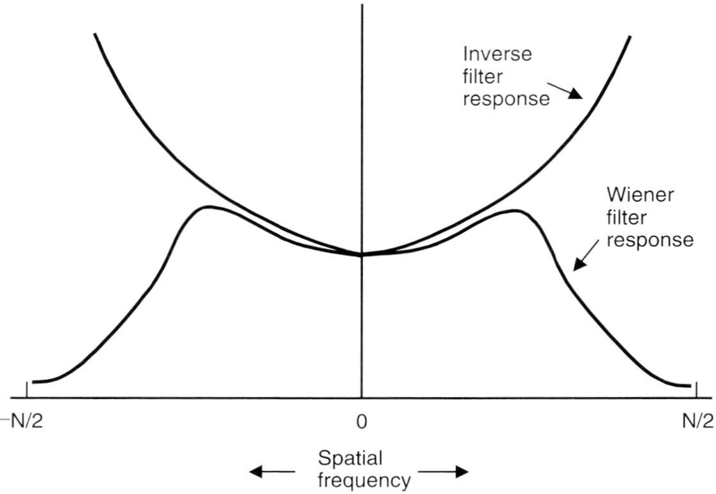

FIGURE 9.5.3
Wiener filter compared to inverse filter response. This plot shows that the Wiener filter gain falls off at high frequencies where the noise tends to dominate the image. It also shows a standard inverse filter response which will amplify noise at high frequencies as seen in the images of Figure 9.5.2f, g. An $N \times N$ image is assumed.

result, as follows:

$$\hat{I}(r,c) = F^{-1}\left[\hat{I}(u,v)\right] = F^{-1}[R_W(u,v)D(u,v)]$$

Figure 9.5.4 compares the inverse filter and the Wiener filter. The filters are applied to images which have been blurred and then had various amounts of Gaussian noise added. With small amounts of noise, the inverse filter works adequately, but when the noise level is increased, the Wiener filter results are obviously superior.

In practical applications the original, uncorrupted, image is not typically available, so the power spectrum ratio is replaced by a parameter, K, whose optimal value must be experimentally determined:

$$R_W(u,v) = \frac{H^*(u,v)}{|H(u,v)|^2 + K}$$

This form of the Wiener filter equation we call the *Practical Wiener*. Examining this equation and using our knowledge that the noise power spectrum is typically flat, white noise, it may seem that the parameter, K, should also be a function of frequency that makes the gain of $R_W(u,v)$ decrease at high frequencies.

9.5.3 Constrained Least Squares Filter

The constrained least squares (CLS) filter provides an alternate to the practical Wiener filter by replacing the power spectrum ratio with a function that varies with frequency. This filter was initially developed to eliminate some of the artifacts caused by Wiener filters. This is done by including a smoothing criterion in the filter derivation, so that the result will not have undesirable oscillations (these appear as "waves" in the image),

(a)

(b)

(c)

(d)

(e)

(f)

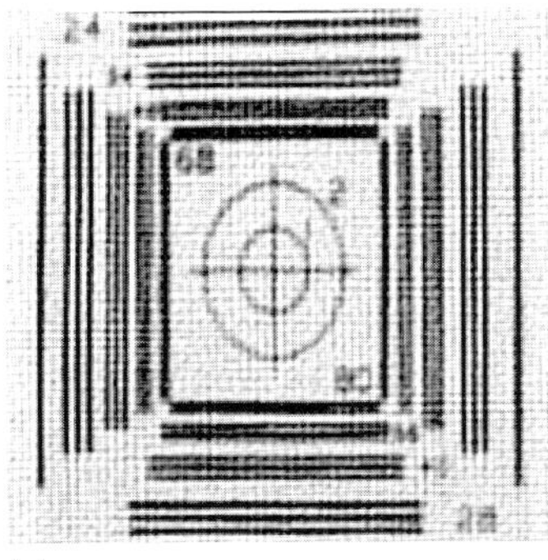

(g)

FIGURE 9.5.4
Wiener filter. (a) Image blurred with an 11×11 Gaussian convolution mask. (b) Image with Gaussian noise, variance = 5, mean = 0. (c) Inverse filter, with cutoff frequency = 80, histogram stretched with 3% low and high clipping to show detail. (d) Wiener filter, with cutoff frequency = 80, histogram stretched. (e) Image with Gaussian noise, variance = 200, mean = 0. (f) Inverse filter, with cutoff frequency = 80, histogram stretched. (g) Wiener filter, with cutoff frequency = 80, histogram stretched.

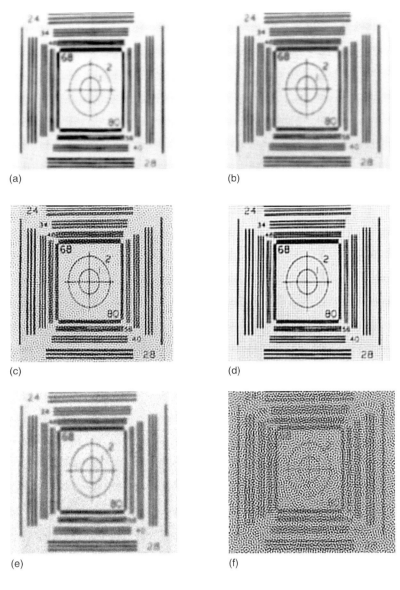

as sometimes occurs with other frequency domain filters. The constrained least squares filter is given by:

$$R_{\mathrm{CLS}}(u,v) = \frac{H^*(u,v)}{|H(u,v)|^2 + \gamma\,|P(u,v)|^2}$$

where:

$$\gamma = \text{adjustment factor}$$

$$P(u,v) = \text{the Fourier transform of smoothness criterion function}$$

Note that this filter is the same as the Wiener filter, but with the noise-to-signal power spectrum ratio replaced by the smoothing criterion function. Also, it is the same as the practical Wiener with the K parameter replaced by the smoothing criterion function. The adjustment factor's value is experimentally determined, and is application dependent. A standard function to use for $p(r,c)$ (the inverse Fourier transform of $P(u,v)$) is the Laplacian filter mask, as follows:

$$p(r,c) = \begin{bmatrix} 0 & -1 & 0 \\ -1 & 4 & -1 \\ 0 & -1 & 0 \end{bmatrix}$$

This corresponds to a highpass filter, but since it appears in the denominator of the filter function, it acts as a lowpass filter. Remember that before $P(u,v)$ is calculated, the $p(r,c)$ function must be extended with zeros (zero-padded) to the same size as the image. Figure 9.5.5 shows results of applying this filter.

The constrained least squares filter is applied by multiplying it by the Fourier transform of the degraded image, and the restored image is obtained by taking the inverse Fourier transform of the result, as follows:

$$\hat{I}(r,c) = F^{-1}\left[\hat{I}(u,v)\right] = F^{-1}[R_{\mathrm{CLS}}(u,v)D(u,v)]$$

9.5.4 Geometric Mean Filters

The geometric mean filter equation provides us with a general form for many of the frequency domain restoration filters. It is defined as follows:

$$R_{\mathrm{GM}}(u,v) = \left[\frac{H^*(u,v)}{|H(u,v)|^2}\right]^{\alpha} \left[\frac{H^*(u,v)}{|H(u,v)|^2 + \gamma\left[\frac{S_n(u,v)}{S_{\mathrm{I}}(u,v)}\right]}\right]^{1-\alpha}$$

The terms are as previously defined, with γ and α being positive real constants. If $\alpha = 1/2$ and $\gamma = 1$, this filter is called a *power spectrum equalization filter.* If $\alpha = 1/2$, then this filter is an average between the inverse filter and the Wiener filter, hence the term geometric mean, although it is standard to refer to the general form of the equation as *geometric mean filter(s).*

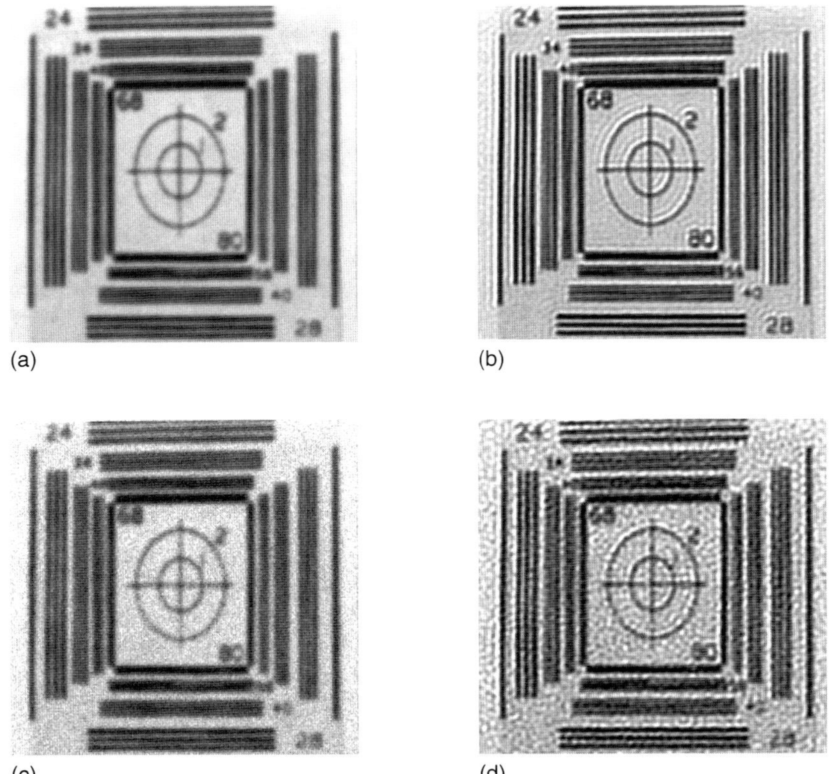

FIGURE 9.5.5
Constrained least squares filter. (a) Blurred image with added Gaussian noise, mean = 0, variance = 5. (b) Result of CLS filter on image (a). (c) Blurred image with added Gaussian noise, mean = 0, variance = 200. (d) Result of CLS filter on image (c).

The geometric mean filter is applied by multiplying it by the Fourier transform of the degraded image, and the restored image is obtained by taking the inverse Fourier transform of the result, as follows:

$$\hat{I}(r,c) = F^{-1}\left[\hat{I}(u,v)\right] = F^{-1}[R_{\mathrm{GM}}(u,v)D(u,v)]$$

If $\alpha = 0$, this filter is called a *parametric Wiener filter*. The equation reduces to the Wiener filter equation, but with γ included as an adjustment parameter:

$$R_{\mathrm{PW}}(u,v) = \frac{H^{*}(u,v)}{\left|H(u,v)\right|^{2} + \gamma\left[\dfrac{S_{n}(u,v)}{S_{\mathrm{I}}(u,v)}\right]}$$

When $\gamma = 1$, this filter becomes a standard Wiener filter, and when $\gamma = 0$, this filter becomes the inverse filter. As γ is adjusted, the results vary between these two filters, with larger values providing more of the Wiener filtering effect.

The parametric Wiener filter is applied by multiplying it by the Fourier transform of the degraded image, and the restored image is obtained by taking the inverse Fourier transform of the result, as follows:

$$\hat{I}(r,c) = F^{-1}\left[\hat{I}(u,v)\right] = F^{-1}[R_{\text{PW}}(u,v)D(u,v)]$$

In general, the frequency domain filters discussed to this point work well for small amounts of blurring and moderate amounts of additive noise. The inverse filter is inadequate with too much noise, and the Wiener filter has the tendency to cause undesirable artifacts in the resultant image. The constrained least squares filter helps to minimize the Wiener-type artifact, and the parametric Wiener and the geometric mean provide additional parameters, which can be adjusted for application-specific needs.

9.5.5 Adaptive Filtering

The frequency domain filters discussed thus far, which are generalized in the geometric mean filter model, are spatially invariant filters. They do not change their characteristics based on spatial location, so are applied to the entire image in the same manner. Additionally, they are all derived based on the assumption that the signals are stationary. As previously mentioned, an image that satisfies the stationary criteria has similar spectral distributions across all subimages, obviously an invalid assumption due to varying objects and textures in the image. Fine textures have more high frequency energy than coarse textures, and regions with edges have more high frequency energy content than smooth, constant regions. In general, images are highly nonstationary. The degradations that occur may also vary from one image region to the next, they may be spatially-variant.

Given these violations to the assumptions under which the filter models were developed, it is not surprising that the results are suboptimal. The concept underlying adaptive filtering is that the filter will adapt to the local image characteristics, as was discussed with the adaptive spatial filters. In the spatial domain the adaptive filters changed at each pixel based on local image characteristics, a form of *pixel-by-pixel* processing. This is impractical in the frequency domain due to its computational complexity.

Block-by-block or *subimage-by-subimage* processing is more practical in the frequency domain. With this approach the image is divided into blocks, typically between 8×8 and 32×32 pixels, and then the results are combined. This is equivalent to processing each block with its own spatially invariant filter, such as those given by the geometric mean filter model. In this case, the parameters are tuned to the characteristics of each block. The underlying idea is that better results will be obtained by adapting to local image characteristics than by using a fixed filter for the entire image. Of course implementing an adaptive filter has an associated cost of increased computational complexity when compared to use of a fixed filter.

One problem with this approach is caused by the filter characteristics changing at block boundaries. This may cause artificial brightness changes at block boundaries, which creates the *blocking effect* or *blocking artifact*. This effect appears in images as false lines between image blocks. In some image restoration applications this may not be as severe a problem as it is in image compression applications, and more about it is discussed in Chapter 10. Post-processing the object boundaries with lowpass filters can help to mitigate these effects. Another method is to overlap the subimages by using a window function, which allows neighboring subimages to slowly merge instead of abruptly change at the boundaries.

To apply adaptive filtering in the frequency domain on the block-by-block basis, the task is the same as application to an entire image. However, in this case, each subimage (block) is treated as an image and an optimal processing filter must be determined for each block by using the previously described methods. Computational intelligence based methods, using techniques such as neural networks and genetic algorithms, can be applied to develop a system that will learn to adjust its operation to adapt to the image. More information on adaptive filtering in the frequency domain can be explored with the references.

9.5.6 Bandpass, Bandreject and Notch Filters

The shape of the bandpass, bandreject and notch filters is shown in Figure 5.7.9. These filters are useful in analyzing and restoring images that require the removal of periodic noise patterns such as discussed in Section 9.2.2 (see Figure 9.2.6). The bandreject and notch filters will eliminate or attenuate the noise, while the bandpass is useful for analyzing the noise pattern itself.

The *notch filter* is a special form of a bandreject filter; instead of eliminating an entire ring of frequencies in the spectrum, it only "notches" out selected frequencies. This type of filter is most useful for an image that has been corrupted with a sinusoidal interference pattern. This type of image degradation is often seen in poor broadcast television images, and is also a common artifact in images that have been obtained where the imaging device resides on some type of vibrating mechanical system—for example, a ship or a satellite.

For this type of image degradation, the spectrum will reveal the problem. Figure 9.5.6b,d shows the type of spectrum that result from the sinusoidal interference. Bright spots in the spectrum corresponding to the interference can be seen. In Figure 9.5.6e,f, the restored image and the spectrum are shown. The portions of the spectrum that were causing the interference have been removed, effectively eliminating the interference pattern and noticeably improving the appearance of the image.

The bandreject filter is useful when the interfering periodic noise is at a fixed frequency, but of varying orientation. For example, in Figure 9.5.7 we have an image that has both horizontal and vertical sinsuiodal interference at a spatial frequency of 32 added to the original 256×256 image. In this figure we see the corrupted image and its spectrum, and the resulting image and its spectrum after performing an ideal bandreject filter with cutoff of frequencies of 31 and 33. If we instead apply a bandpass filter we can extract the noise interference pattern itself (Figure 9.5.8), thus providing a useful tool to analyze noise patterns. (Remember that in CVIPtools the transform must be performed first, followed by the filter operation. Additionally, as part of the filter operation, the inverse transform is automatically performed.)

With natural images the interference patterns are not as simple and clean as those we obtain by creating sinusoidal images and adding them using CVIPtools. Real world images typically have more complex interference patterns, and analysis of the spectrum coupled with experimentation is required to achieve satisfactory results. Figure 9.5.9 shows a much more complex interference pattern along with the results of the first step in attempting to restore the image. Further processing to obtain a better image is left as an exercise for the reader (Exercise 27). The basic technique involves successive application of bandreject and notch filters to remove prominent spikes in the spectrum. A more sophisticated method involves extracting the interference patterns and

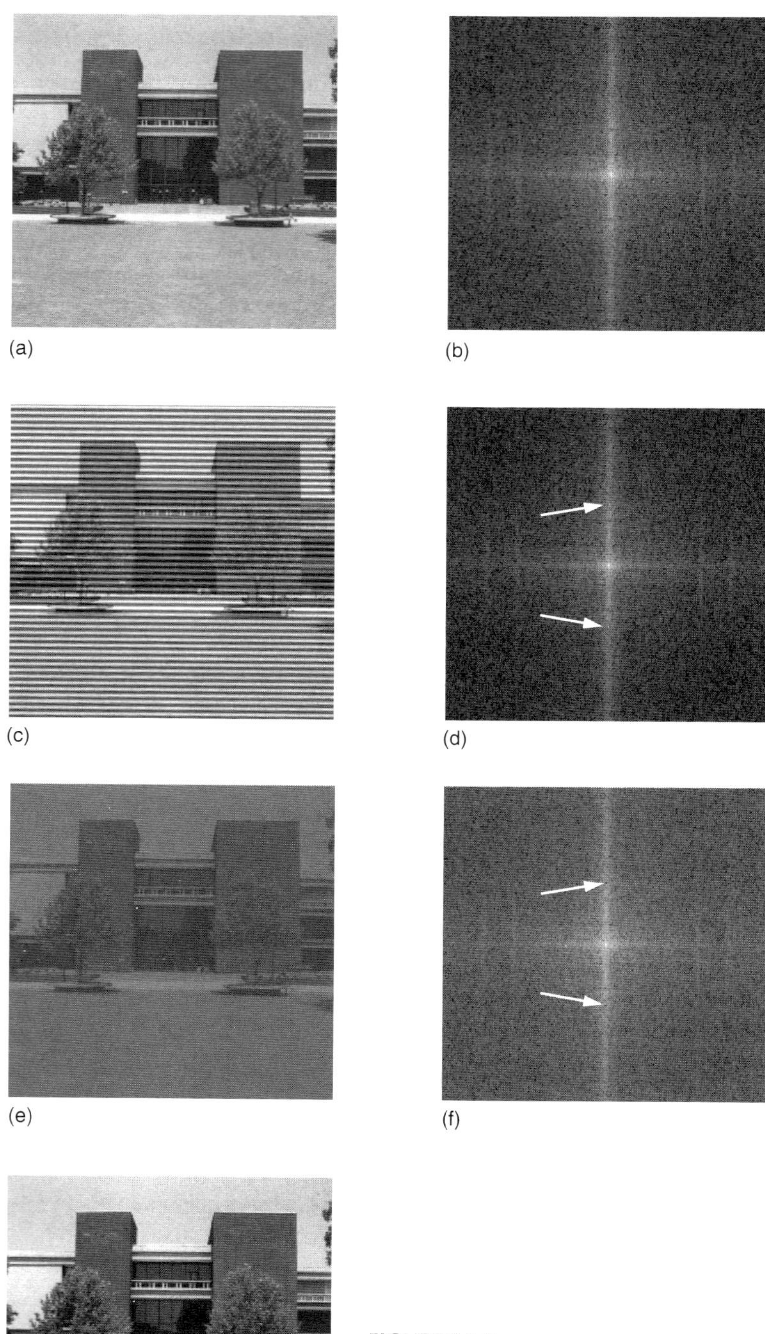

FIGURE 9.5.6
Notch filter. (a) Original image. (b) Spectrum of original Image. (c) Image corrupted with sinusoidal noise. (d) Spectrum of corrupted image; arrows point to contribution from interference. (e) Image restored by notch filtering. (f) Spectrum of filtered image; arrows point to masked sinusoidal contribution. (g) Image further enhanced with histogram techniques.

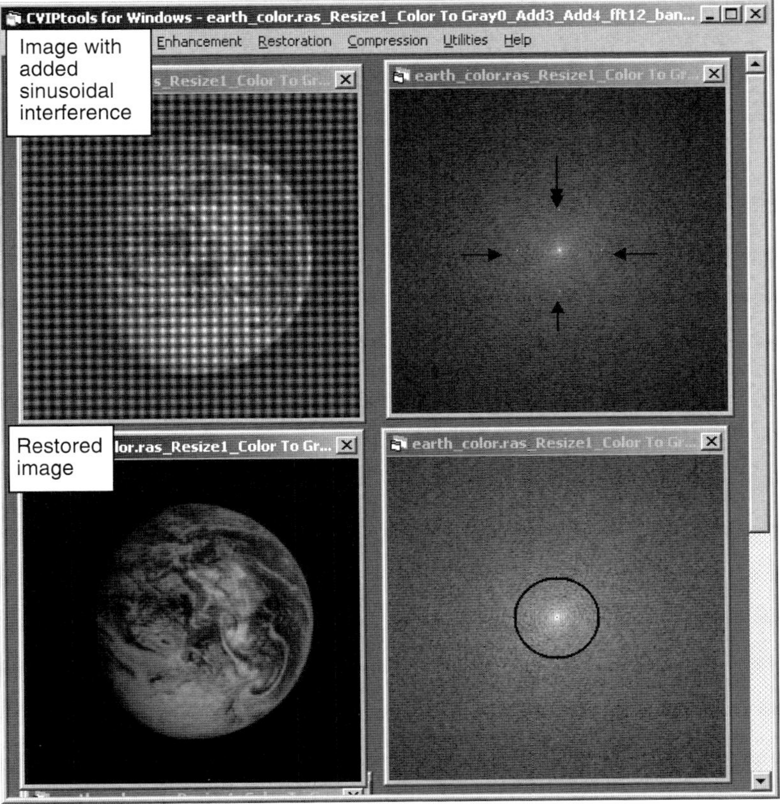

FIGURE 9.5.7
Bandreject filter for removal of periodic noise. This image has horizontal and vertical sinusoidal waves added to the original image. The original image is 256×256 pixels, and the noise frequency is 32. Here we see the image with added noise and its spectrum, and the restored image and its spectrum after applications of the bandreject filter.

then subtracting a variable, weighted amount of the noise pattern from the original, degraded image. In this case, the weight can be based on local image statistics; details of the procedure can be explored in the references.

9.5.7 Practical Considerations

Using the Fourier transform as defined in Chapter 5, care must be taken when implementing the frequency domain filters. It is common practice to define the 2-D Fourier transform with a constant, $1/N$, in both the forward and inverse directions (as in Chapter 5), when it actually has a $1/N^2$ term in the forward direction only. This is done for symmetry, and it has no adverse effect on the Fourier transform pair, since this is a linear process. However, it may affect the outcome of the frequency domain filters.

To avoid problems, the simplest method is to multiply each Fourier transformed image by $1/N$, perform the filter calculations, multiply by the degraded image, and then multiply the result by N before passing it to the inverse Fourier transform. Note that for the power spectral density ratios, the division by N is not required since any constant

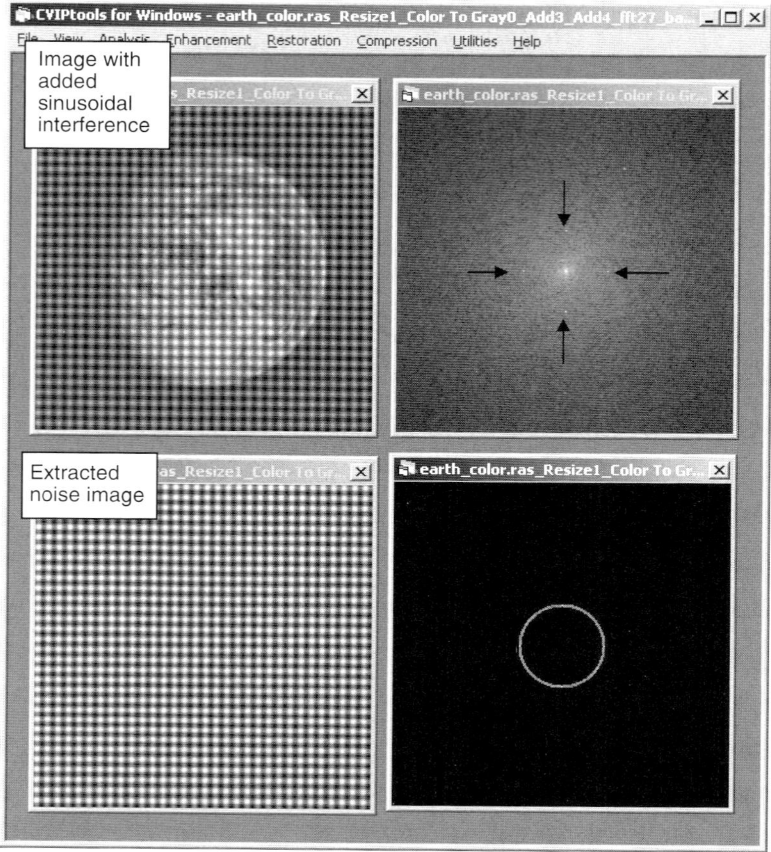

FIGURE 9.5.8
Bandpass filter for extraction of periodic noise patterns. This image has horizontal and vertical sinusoidal waves added to the original image. The original image is 256 × 256 pixels, and the noise frequency is 32. Here we see the image with added noise and its spectrum, and the extracted noise image and its spectrum after application of the bandpass filter.

multipliers will cancel when the ratio is taken. There are many different ways to deal with this problem, but it must be considered and dealt with appropriately or the results can be incorrect.

Care must be taken so that the degradation image, $h(r,c)$, and the noise image, $n(r,c)$, model the degradation process correctly. For example, most images are of type BYTE and thus have a range of 0 to 255. Typically the degradation image magnitude should be normalized, and the noise image should be mapped to the range of the added noise. Although we have discussed dealing with zeros in $H(u, v)$, it is also helpful to limit the gain of the restoration filter so that very small values in denominator will not overwhelm the resulting image. When the image has been restored, simple post-processing image enhancement methods, such as histogram equalization or a histogram stretch, can dramatically improve the visual results (compare Figures 9.5.6e,g).

Other methods of image restoration are explored in the references, including advanced preprocessing and postprocessing techniques to improve the results of the filters given here. Although many of these other methods are more complex, the improvements achieved are often minimal, or only applicable to a limited domain. Often image

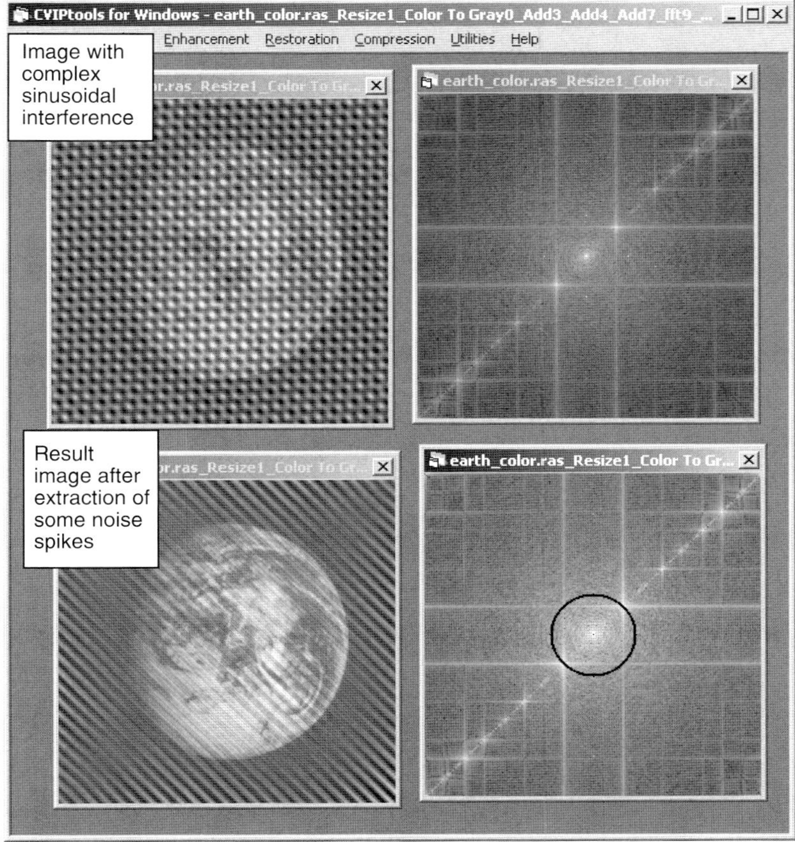

FIGURE 9.5.9
A more realistic complex noise pattern caused by periodic interference. Here we see the image with a more complex pattern of periodic noise, which is more like that found in natural images. We can see in its spectrum multiple periodic spikes and lines. The result after extraction of some of the primary noise by a bandreject filter is shown.

restoration requires a combination of techniques and, as with many computer imaging tasks, require application-domain specific information.

9.6 Geometric Transforms

The previous sections in this chapter have all been about modifying the brightness values to restore a degraded image. It was assumed that the pixel *location* was correct. This final section in the image restoration chapter has to do with images that have been spatially, or geometrically, distorted. In this section we will only consider two-dimensional distortion, which is adequate for most digital images. To restore images that have undergone geometric distortion requires the application of geometric transforms.

Geometric transforms are used to modify the location of pixel values within an image, typically to correct images that have been spatially warped. These methods are often referred to as *rubber-sheet transforms*, because the image is modeled as a sheet of rubber and stretched and shrunk, or otherwise manipulated, as required to correct for any spatial distortion. This type of distortion can be caused by defective optics in an image acquisition system, distortion in image display devices, or 2-D imaging of 3-D surfaces. The methods are used in map making, image registration, image morphing, and other applications requiring spatial modification. It should be noted that the geometric transforms can also be used in image warping where the goal is to take a "good" image and distort it spatially.

The simplest geometric transforms—translate, rotate, zoom and shrink—have already been discussed in Chapter 3. These transforms are limited to moving the pixels within an image in a fixed, regular manner, and do not really distort the image, but merely move pixel values. The more sophisticated geometric transforms, such as those discussed here, require two steps: (1) spatial transform and (2) gray level interpolation. The model used for the geometric transforms is seen in Figure 9.6.1. The spatial transform provides the location of the output pixel, and the gray level interpolation is necessary since pixel row and column coordinates provided by the spatial transform are not necessarily integers. The image is processed one pixel at a time, until the entire image has been transformed.

9.6.1 Spatial Transforms

Spatial transforms are used to map the input image location to a location in the output image; it defines how the pixel values in the output image are to be arranged. This process can be modeled as in Figure 9.6.2, where the original, undistorted image, is $I(r, c)$, and the distorted (or degraded) image is $d(\hat{r}, \hat{c})$. The distorted image coordinates can be defined by the two equations:

$\hat{r} = \hat{R}(r, c)$, defines the row coordinate for the distorted image

$\hat{c} = \hat{C}(r, c)$, defines the column coordinate for the distorted image

The primary idea presented here is to find a mathematical model for the geometric distortion process, specifically the two equations $\hat{R}(r, c)$ and $\hat{C}(r, c)$ and then apply the inverse process to find the restored image.

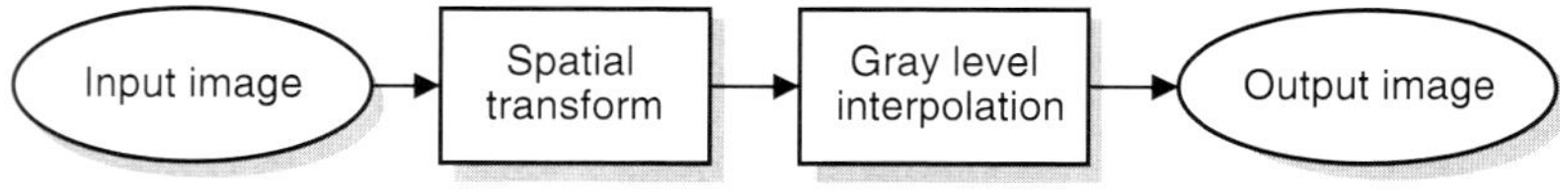

FIGURE 9.6.1
Geometric transforms.

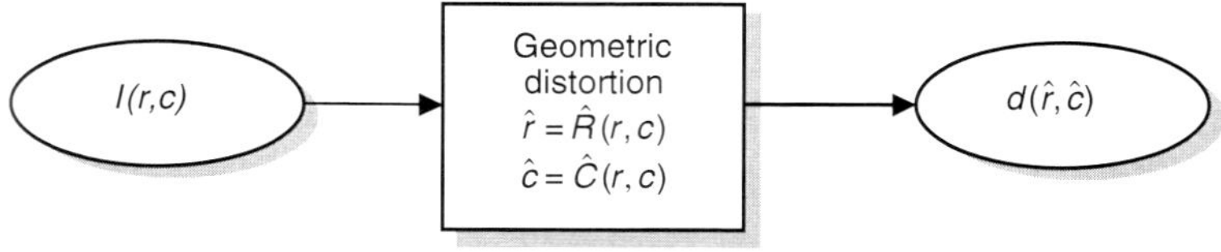

FIGURE 9.6.2
Spatial transforms.

The type of distortion considered may vary across the image, so different equations for different portions of the image are often required. To determine the necessary equations, we need to identify a set of points in the original image that match points in the distorted image. These sets of points are called *tiepoints*, and are used to define the equations $\hat{R}(r, c)$ and $\hat{C}(r, c)$. The form of these equations is typically bilinear, although higher-order polynomials can be used. The higher-order polynomials are much more computationally intensive, and there is no guarantee of better results—in some cases, the results may be worse (although it is wise to remember that we are dealing with subjective analysis regarding better or worse, and that image restoration is more of an art than a science).

The method to restore a geometrically distorted image consists of three steps: (1) Define quadrilaterals (four-sided polygons) with known, or best-guessed tiepoints for the entire image, (2) Find the equations $\hat{R}(r, c)$ and $\hat{C}(r, c)$ for each set of tiepoints, and (3) remap all the pixels within each quadrilateral subimage using the equations corresponding to those tiepoints.

Figure 9.6.3 illustrates Step 1. The two images are divided into subimages, defined by the tiepoints (Figure 9.6.3a). Figure 9.6.3b shows the center subimage from both the distorted and the original images, and the corresponding tiepoints. The four corners are the tiepoints for this subimage, and provide us with four pixels whose location is known in both images.

In Step 2, using a bilinear model for the mapping equations, these four points are used to generate the equations:

$$\hat{R}(r, c) = k_1 r + k_2 c + k_3 rc + k_4 = \hat{r}$$

$$\hat{C}(r, c) = k_5 r + k_6 c + k_7 rc + k_8 = \hat{c}$$

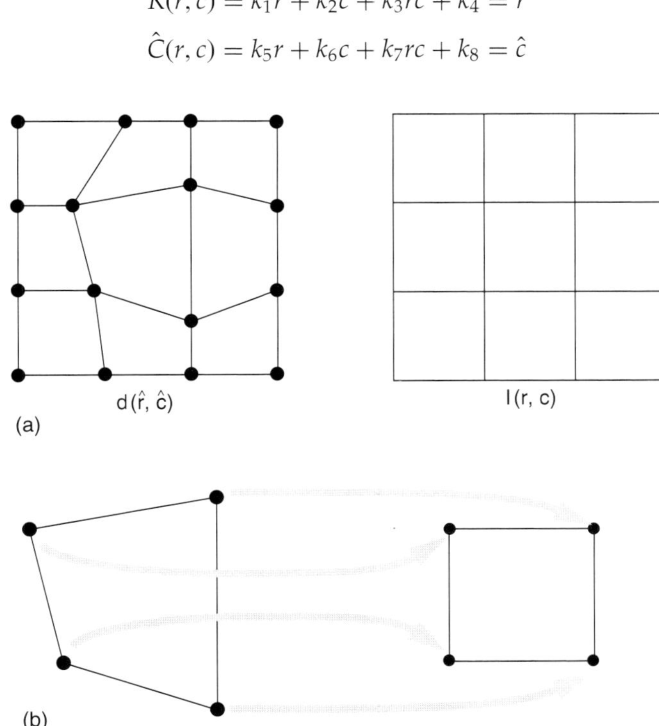

FIGURE 9.6.3
Restoring geometric distortion. (a) Images divided into quadrilateral subimages. (b) Center subimage showing corresponding points, also known as tie-points.

The k_i values are constants to be determined by solving the eight simultaneous equations. Since we have defined four tiepoints, we have eight equations where r, c, $\hat{r}$ and $\hat{c}$ are known (two for each point, one mapping the row coordinate and one mapping the column coordinate). Now we can solve the eight equations for the eight unknowns, and we have the necessary equations for the coordinate mapping in Step 3.

Step 3 involves application of the mapping equations, $\hat{R}(r,c)$ and $\hat{C}(r,c)$, to all the (r,c) pairs in the corresponding quadrilateral in $I(r,c)$. For example:

Example 9.6.1

Assume we have found the following mapping equations:

$$\hat{R}(r,c) = 5r + 3c + 3rc + 2 = \hat{r}$$
$$\hat{C}(r,c) = 1r + 1c + 2rc + 0 = \hat{c}$$

To find $I(2,3)$, substitute $(r,c) = (2,3)$ into the above equations and we find:

$$\hat{R}(r,c) = 5(2) + 3(3) + 3(2)(3) + 2 = 39$$
$$\hat{C}(r,c) = 1(2) + 1(3) + 2(2)(3) + 0 = 17$$

Now, we let $I(2,3) = d(39,27)$.

Assuming all the pixel value mappings worked out as well as the example, we could recover our original image, $I(r,c)$, exactly. However, in practice the k_i values are not likely to cooperate and be integers. The following example illustrates this.

Example 9.6.2

Assume we have found the following mapping equations:

$$\hat{R}(r,c) = 4.5r + 3c + 3.5rc + 2.4 = \hat{r}$$
$$\hat{C}(r,c) = 1.6r + 1c + 2.4rc + 0 = \hat{c}$$

To find $I(2,3)$, substitute $(r,c) = (2,3)$ into the above equations and we find:

$$\hat{R}(r,c) = 4.5(2) + 3(3) + 3.5(2)(3) + 2.4 = 41.4$$
$$\hat{C}(r,c) = 1.6(2) + 1(3) + 2.4(2)(3) + 0 = 20.6$$

Now, we want to set $I(2,3) = d(41.4, 20.6)$.

The difficulty in the above example arises when we try to determine the value of $d(41.4, 20.6)$. Since the digital images are defined only at the integer values for (r,c), gray interpolation must be performed. In this case, we define $\hat{I}(r,c)$ as an estimate to the original image $I(r,c)$ to represent the restored image.

9.6.2 Gray Level Interpolation

The simplest method of gray level interpolation is the *nearest neighbor method*, where the pixel is assigned the value of the closest pixel in the distorted image. In the above example the value of $\hat{I}(2,3)$ is set to the value of $d(41,21)$, the row and column values determined by rounding (0.5 and above is rounded up to the next highest integer) the $\hat{r}$ and $\hat{c}$ result. This method is similar to the zero-order hold described in Section 3.2.1 for image enlargement. This method does not necessarily provide optimal results, but has the advantage of being easy to implement and computationally fast. With the nearest neighbor approach, object edges will tend to appear jagged or blocky.

Alternatively, we can use a more advanced method to interpolate the value. In general these methods will be more computationally intensive, but will provide more visually pleasing results. Figure 9.6.4 illustrates how this is done. The four surrounding pixel values in the distorted image are used to estimate the desired value, and this estimated value is used in the restored image. This can be done in a variety of ways. The easiest method is to find a *neighborhood average*. This can be done one-dimensionally, using the adjacent rows or columns, or it can be done two-dimensionally using all four neighbors. The selection is application-specific, but in general the 2-D average of the four neighbors will provide a better output image. The results are typically rounded to the nearest integer for most images. The neighborhood average method will provide smoother object edges, but the result will be slightly blurry.

To achieve better results, a technique similar to the method used to find the spatial coordinates can be applied. This technique uses *bilinear interpolation* and is done with the following equation:

$$g(\hat{r},\hat{c}) = k_1\hat{r} + k_2\hat{c} + k_3\hat{r}\hat{c} + k_4$$

where:

$$g(\hat{r},\hat{c}) = \text{the gray level interpolating equation.}$$

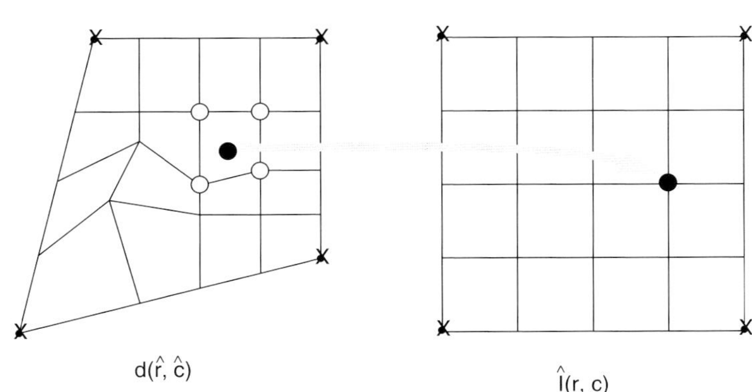

FIGURE 9.6.4
Gray level interpolation.

Note that these constants, k_i, are different than the constants used in the spatial mapping equations. The four unknown constants are found by using the four surrounding points shown in Figure 9.6.4. The values for row and column, $(\hat{r}, \hat{c})$, and the gray level values at each point are used.

Example 9.6.3

Suppose the four surrounding points are as follows:

$$d(\hat{r}, \hat{c}) \rightarrow d(1,2) = 50, \quad d(1,3) = 55, \quad d(2,2) = 44, \quad d(2,3) = 48$$

Then we define the following four equations, and solve for the constants, k_i:

$$50 = k_1(1) + k_2(2) + k_3(1)(2) + k_4$$
$$55 = k_1(1) + k_2(3) + k_3(1)(3) + k_4$$
$$44 = k_1(2) + k_2(2) + k_3(2)(2) + k_4$$
$$48 = k_1(2) + k_2(3) + k_3(2)(3) + k_4$$

Solving these equations simultaneously gives us:

$$k_1 = -4, \quad k_2 = 6, \quad k_3 = -1, \quad k_4 = 44$$
$$\therefore \; g(\hat{r}, \hat{c}) = -4\hat{r} + 6\hat{c} - \hat{r}\hat{c} + 44$$

After the equation, $g(\hat{r}, \hat{c})$, is found, the interpolated value can be determined. To do this we insert the noninteger values for row and column into the gray level interpolating equation, and the resulting $g(\hat{r}, \hat{c})$ value is the interpolated gray level value.

Example 9.6.4

The preceding example assumes that the row and column coordinates are between rows 1 and 2, and column 2 and 3; for example $\hat{r} = 1.3$ and $\hat{c} = 2.6$. Applying these values to the preceding gray level interpolating equation, we obtain:

$$g(1.3, 2.6) = -4(1.3) + 6(2.6) - (1.3)(2.6) + 44 = 51.02 \approx 51$$

The gray level value of 51 (or 51.02 can be used if the image is of FLOAT data type) is then inserted into the restored image at the row and column location used to generate $\hat{r} = 1.3$ and $\hat{c} = 2.6$ from the mapping equations.

Figure 9.6.5 illustrates geometric restoration and compares the three gray level interpolation methods discussed.

For applications requiring even higher quality results, such as medical imaging or computer-aided design (CAD) graphics, more mathematically complex methods can be used. For example, *cubic convolution interpolation* will fit a smooth surface over a larger group of pixels to provide a reasonably optimal gray level value at any point on the surface. The added computational complexity is not necessary for many computer imaging applications, where the results from bilinear interpolation are usually adequate. Details on this and other more sophisticated methods can be found in the references.

9.6.3 The Geometric Restoration Procedure

Now that all the required tools have been discussed, using tiepoints for spatial transformation and three methods for gray level interpolation, we present the complete procedure in more detail for restoring an image that has undergone geometric distortion. The procedure is as follows:

1. Find tiepoints throughout the image mapping the distorted image, $d(\hat{r}, \hat{c})$ to the restored image, $\hat{I}(r, c)$. $\hat{I}(r, c)$ is the estimate to the original, undistorted image $I(r, c)$.
2. For each quadrilateral find the equations for $\hat{R}(r, c)$ and $\hat{C}(r, c)$.
3. For each value of (r, c) in $\hat{I}(r, c)$, apply the equation pair, $\hat{R}(r, c)$ and $\hat{C}(r, c)$, corresponding to the mapped quadrilateral to find $(\hat{r}, \hat{c})$.

(a)

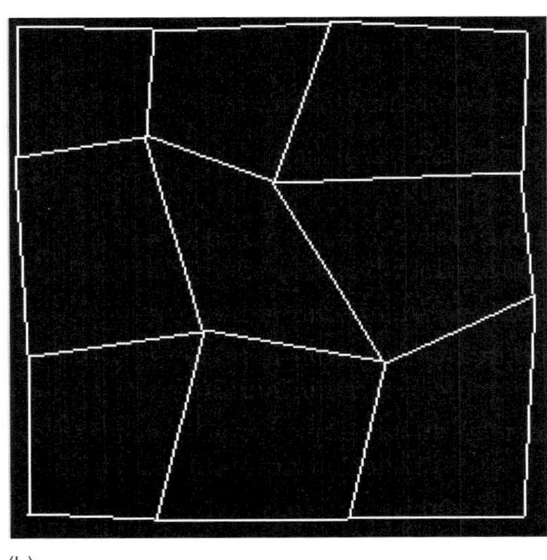
(b)

(c)

(d)

FIGURE 9.6.5
Geometric restoration example. (a) Original image. (b) A mesh defined by 16 tiepoints will be used to first distort and then restore the image. (c) The original image has been distorted using the bilinear interpolation method. (d) Restoration by the nearest neighbor method shows the blocky effect that occurs at edges.

(e)

(f)

FIGURE 9.6.5 (Continued)
Geometric restoration example. (e) Restoration with neighborhood averaging interpolation provides smoother edges than with the nearest neighbor method, but also blurs the image. (f) Restoration by bilinear interpolation provides optimal results. Note that some distortion occurs at the boundaries of the mesh quadrilaterals.

4. Perform the selected method of gray level interpolation using the values for $(\hat{r}, \hat{c})$ found in Step 3 to find the gray value for $\hat{I}(r, c)$.
5. Continue Step 3 and Step 4 until all values for $\hat{I}(r, c)$ are found and we have our restored image.

Many variations of this method are possible. For example, the shape of the tiepoint polygons need not be quadrilaterals. They could be triangles, or they could vary throughout the image depending on the needs of the application. Other gray level interpolation methods are possible, and the three presented here are representative and commonly used. Gray level interpolation is really simply a mathematical estimation problem, and numerous techniques may be used.

In cases where images of the same scene are taken from multiple views and we want to match them, the geometric transformation process to map the images to a common coordinate system is called *image registration*. Image registration methods employ the techniques discussed here and in some cases may require more complex three-dimensional transformations. These transformations, along with the myriad of other methods available for geometric restoration, can be explored further in the references.

9.6.4 Geometric Restoration with CVIPtools

To perform geometric restoration or distortion in CVIPtools select the *Restoration-> Geometric Transforms* window. To create *mesh* files, which are used to warp and restore images, select the *Enter a new mesh file* option. The mesh file contains the coordinates of the tiepoints in the image. The user can select the number of tiepoints, which determines the size of the distortion grid, and then enter the points on an image by holding the Alt key on the keyboard and left-clicking the mouse at the point of interest.

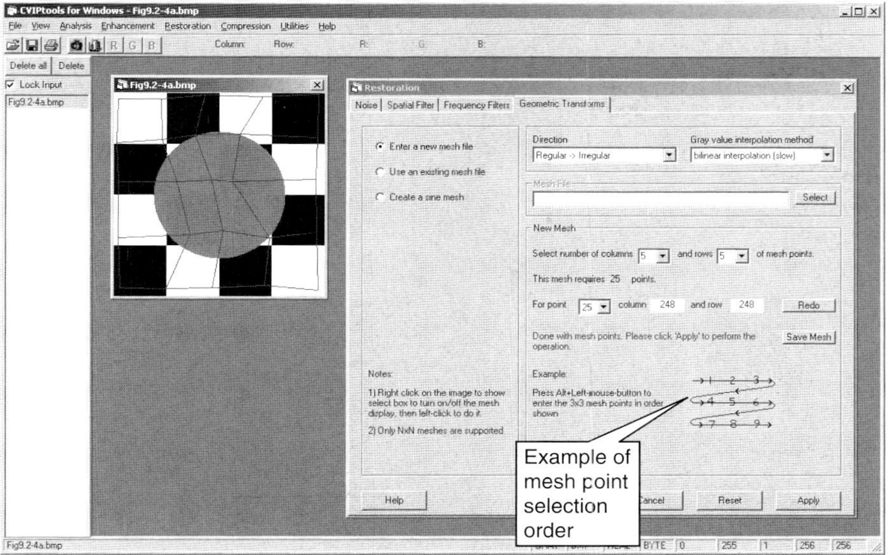

FIGURE 9.6.6
CVIPtools geometric restoration window. This figure shows CVIPtools after creation of a 5×5 mesh on a test image. The mesh is created on the currect image by selecting the *Enter a new mesh file* option, selecting the mesh size, here 5×5, and then holding the Alt key on the keyboard and clicking the left mouse button at the desired points. The order of the point selection is left to right and top to bottom, as the 3×3 example on the Restoration window shows. As the points are selected, the column and row coordinates appear in the window. If a mistake is made during the mesh point selection process, you can start over by clicking the *Redo* button.

CVIPtools will automatically connect the points and display them as an overlay on the image (see Figure 9.6.6). The mesh overlay can be displayed or removed with the right mouse button followed by a left-click. After a mesh has been created it can be saved as a mesh file with the *Save Mesh* button. After a mesh file has been saved it can be loaded and used to warp or restore an image with the *Use an existing mesh file* option.

After the mesh file has been created or loaded, the user selects the direction for the mapping. The choices are from a regular to an irregular grid, *Regular->Irregular*, or from an irregular to a regular grid, *Irregular->Regular*. In a regular grid, the quadrilaterals are all square, whereas in an irregular grid the quadrilaterals are warped. This is best shown by example, see Figure 9.6.7. In Figure 9.6.7c imagine a regular grid, like the background checkerboard, being overlaid on the original image and then being warped to the mesh overlay shown in Figure 9.6.7b. In Figure 9.6.7e imagine the distorted mesh being overlaid on the original image, and then it being stretched to a regular grid.

After the mapping direction is selected the user can select one of the three gray level interpolation methods: (1) nearest neighbor, (2) bilinear interpolation, or (3) neighborhood average. To restore an image that has been warped or distorted, simply select the distorted image and reverse the direction, as is shown in Figure 9.6.7(d) and (f). Note that artifacts occur at the grid boundaries, and we see in Figure 9.6.7f that pixels beyond the boundaries of the mesh (compare Figure 9.6.7(b) and (f) at the edges), cannot be recovered.

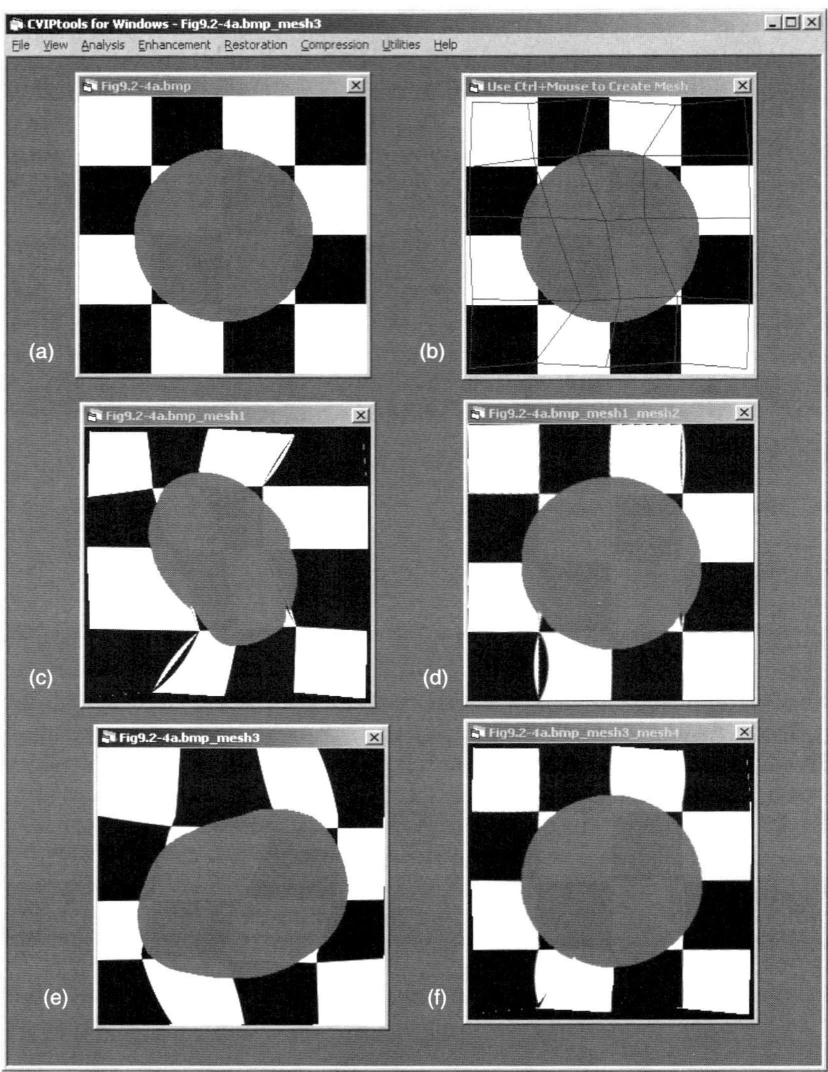

FIGURE 9.6.7
Regular and irregular image mappings in CVIPtools. (a) Original image. (b) Image with the distortion mesh overlay. (c) Image distorted by *Regular->Irregular* mapping. (d) Image (c) restored by *Irregular->Regular* mapping. (e) Image distorted by *Irregular->Regular* mapping. (f) Image (e) restored by *Regular->Irregular* mapping.

9.7 Key Points

INTRODUCTION AND OVERVIEW

- *Image restoration* is the process of finding an approximation to the degradation process and finding the appropriate inverse process to estimate the original image (Figure 9.1.1)
- Restoration differs from enhancement because it uses a mathematical model for image degradation

- Examples of the types of degradation include blurring caused by motion or atmospheric disturbance, geometric distortion caused by imperfect lenses, superimposed interference patterns caused by mechanical systems, and noise from electronic sources
- The types of degradation models include both spatial and frequency domain considerations
- In general image restoration is more of an art than a science

System Model

- Spatial domain:

$$d(r,c) = h(r,c) * I(r,c) + n(r,c)$$

where the $*$ denotes the convolution process:

$$d(r,c) = \text{degraded image}$$

$$h(r,c) = \text{degradation function}$$

$$I(r,c) = \text{original image}$$

$$n(r,c) = \text{additive noise function}$$

- Because convolution in the spatial domain is equivalent to multiplication in the frequency domain, the frequency domain model is:

$$D(u,v) = H(u,v)I(u,v) + N(u,v)$$

where:

$$D(u,v) = \text{Fourier transform of the degraded image}$$

$$H(u,v) = \text{Fourier transform of the degradation function}$$

$$I(u,v) = \text{Fourier transform of the original image}$$

$$N(u,v) = \text{Fourier transform of the additive noise function}$$

- Alternately, a multiplicative noise model can be defined, where we take the logarithm of the degraded image to decouple noise and image

NOISE MODELS

- *Noise* is any undesired information that contaminates an image
- Noise appears from the digital image acquisition process, where fluctuations caused by natural phenomena add a random value to the exact brightness value for a given pixel
- Noise in electronics is affected by environmental conditions such as temperature, which varies over time
- Other types of noise, such as periodic noise, may be introduced during the acquisition process as a result of the physical systems involved

Noise Histograms

- We consider noise to be a random variable with a probability density function (PDF) to describe its shape and distribution
- The histogram of a noise image approximates the PDF
- Typical image noise models are uniform, Gaussian, and salt-and-pepper (impulse)
- Gaussian model is valid for random electron fluctuations and film grain noise
- Electronic noise is problematic with poor lighting or high temperatures:

$$\text{HISTOGRAM}_{\text{Gaussian}} = \frac{1}{\sqrt{2\pi\sigma^2}}\,\mathrm{e}^{-(g-m)^2/2\sigma^2}$$

where:

$$g = \text{gray level}$$

$$m = \text{mean (average)}$$

$$\sigma = \text{standard deviation}\,(\sigma^2 = \text{variance})$$

- Uniform PDF can be used to generate any other noise model
- Uniform noise is used to evaluate image restoration algorithms, as it is the most unbiased:

$$\text{HISTOGRAM}_{\text{Uniform}} = \begin{cases} \dfrac{1}{b-a} & \text{for } a \le g \le b \\ 0 & \text{elsewhere} \end{cases}$$

- Salt-and-pepper noise, also called shot, spike, or impulse noise is typically caused by faulty electronics:

$$\text{HISTOGRAM}_{\text{Salt and Pepper}} = \begin{cases} A & \text{for } g = a \text{ (``pepper'')} \\ B & \text{for } g = b \text{ (``salt'')} \end{cases}$$

- Radar range and velocity image noise is modeled by Rayleigh noise

$$\text{HISTOGRAM}_{\text{Rayleigh}} = \frac{2g}{\alpha}\mathrm{e}^{-g^2/\alpha}$$

- Negative exponential noise occurs in laser-based images:

$$\text{HISTOGRAM}_{\text{Negative Exponential}} = \frac{\mathrm{e}^{-g/\alpha}}{\alpha}$$

- In lowpass filtered laser-based images the noise can be modeled as gamma noise:

$$\text{HISTOGRAM}_{\text{Gamma}} = \frac{g^{\alpha-1}}{(\alpha-1)!a^{\alpha}}\,\mathrm{e}^{-g/a}$$

Periodic Noise

- Periodic noise in images is typically caused by electrical and/or mechanical systems, such as engine vibration or electrical system interference during image acquisition

- It appears in the frequency domain as impulses corresponding to sinusoidal interference (see Figure 9.2.6)
- It can be removed with bandreject and notch filters (Section 9.5.6)

Estimation of Noise

- Find an image (or subimage) that contains only noise, and use its histogram for the noise model
- Noise only images can be acquired by aiming the imaging device (e.g., camera) at a blank wall
- If we cannot find "noise-only" images, select a portion of the image with a known histogram, subtract the known values from the histogram, only noise is left
- To develop a valid model many such subimages need to be evaluated

NOISE REMOVAL USING SPATIAL FILTERS

- Spatial filters typically operate on small neighborhoods, 3×3 to 11×11
- The degradation model used for this section assumes that $h(r, c)$ causes no degradation:

$$d(r, c) = I(r, c) + n(r, c)$$

- The two primary categories of spatial filters for noise removal are order filters and mean filters
- *Order filters* work by ordering the pixels based on brightness values and using the ordered list to select the "correct" value
- Order filters work best with salt-and-pepper, negative exponential, or Rayleigh noise
- The order filters are nonlinear, so their results are sometimes unpredictable
- *Mean filters* measure some form of average value
- Mean filters work best with Gaussian or uniform noise
- Mean filters have the disadvantage of blurring the image details, they are lowpass filters
- A tradeoff exists between preservation of image detail and noise elimination
- Adaptive filters change their behavior based on local gray level characteristics (statistics), so are more effective at preserving detail while removing noise

Order Filters

- Order filters use a technique called *order statistics* that arranges all the pixels in sequential order, based on gray level value (pixel brightness)
- The placement of the value within this ordered set is referred as the *rank*
- These filters operate on small subimages, *windows*, and replace the center pixel
- The *median* filter selects the middle value from the ordered set
- Median filters work well with salt-and-pepper noise
- The *maximum* filter selects the largest value, works with pepper noise
- The *minimum* filter selects the smallest value, works with salt noise

- The *midpoint* filter selects the average of the minimum and maximum value, useful for Gaussian and uniform noise
- The *alpha-trimmed mean* is the average of the ordered set, with some end-point values excluded
- The alpha-trimmed mean varies from a mean to a median filter, so is useful for images containing multiple noise types:

$$\text{Ordered set} \rightarrow I_1 \leq I_2 \leq \cdots \leq I_{N^2}$$

$$\text{Alpha-trimmed mean} = \frac{1}{N^2 - 2T} \sum_{i=T+1}^{N^2-T} I_i$$

where T is the number of pixel values excluded at each end of the ordered set, and can range from 0 to $(N^2 - 1)/2$.

Mean Filters

- Mean filters function by finding some form of an average with an $N \times N$ window
- The *arithmetic mean filter* will blur an image and is useful for Gaussian, gamma, and uniform noise:

$$\text{Arithmetic mean} = \frac{1}{N^2} \sum_{(r,c) \in W} d(r,c)$$

where N^2 = the number of pixels in the $N \times N$ *window*, W.

- The *contra-harmonic mean filter*, works well for images containing salt (for negative R) or pepper (for positive R) noise:

$$\text{Contra-harmonic mean} = \frac{\sum_{(r,c) \in W} d(r,c)^{R+1}}{\sum_{(r,c) \in W} d(r,c)^{R}}$$

- The *geometric mean filter* works best with Gaussian noise, and retains detailed information better than an arithmetic mean filter:

$$\text{Geometric mean} = \prod_{(r,c) \in W} [d(r,c)]^{1/N^2}$$

- The *harmonic mean filter* works well for salt and Gaussian noise:

$$\text{Harmonic mean} = \frac{N^2}{\sum_{(r,c) \in W} \frac{1}{d(r,c)}}$$

- The Y_p *mean filter*, removes salt noise for negative values of P, and pepper noise for positive values of P:

$$Y_p \text{ mean} = \left[\sum_{(r,c) \in W} \frac{d(r,c)^P}{N^2} \right]^{1/P}$$

Adaptive Filters

- *Adaptive filters* alter their behavior based on local statistical measures, which allows them to retain image detail while still removing noise
- These local measures can be based on order statistics or standard statistics
- These local measures typically involve image brightness and contrast
- The minimum-mean-squared error (MMSE) filter works best for Gaussian or uniform noise:

$$\text{MMSE} = d(r, c) - \frac{\sigma_n^2}{\sigma_l^2}[d(r, c) - m_l(r, c)]$$

where:

σ_n^2 = noise variance

σ_l^2 = local variance (in the window under consideration)

m_l = local mean (average in the window under consideration)

- The noise to local variance ratio, σ_n^2/σ_l^2, controls the amounts of the degraded image retained and mean added
- As this ratio increases the filter returns primarily the local average
- As this ratio goes down, implying high local detail, the filter returns more of the original unfiltered image
- The *adaptive median filter* is algorithmic in nature and retains detail much better than the standard median
- The adaptive median filter algorithm:

$d(r, c)$ = the degraded image

$W \rightarrow$ the current $N \times N$ window centered at $d(r, c)$

$W_{\max}$ = maximum window size

$g_{\min}$ = minimum gray level in the window, W

$g_{\max}$ = maximum gray level in W

g_{med} = median gray level in W

Level 1:
If ($g_{\max} < g_{\text{med}} < g_{\min}$)
Then go to Level 2
Else increase window size, $N = N + 2$
If (window size) $\leq W_{\max}$
Then go to Level 1
Else output $= d(r, c)$

Level 2:
If ($g_{\max} < d(r, c) < g_{\min}$)
Then output $= d(r, c)$
Else output $= g_{\text{med}}$

THE DEGRADATION FUNCTION

- Degradation occurs in the form of blurring due to the signal fluctuating during the measured time interval, imperfect lenses, motion of the object or imaging device, and spatial quantization
- The degradation is either spatially-invariant or spatially-variant
- *Spatially-invariant* degradation affects all pixels in the image the same
- Examples of spatially-invariant degradation includes poor lens focus and camera motion
- *Spatially-variant* degradations are dependent on spatial location and are more difficult to model
- Examples of spatially-variant degradations include imperfections in a lens or object motion
- Spatially-variant degradations can often be modeled as being spatially-invariant over small regions
- Image degradation functions can be considered to be linear or nonlinear

Point Spread Function

- Assuming no additive noise:

$$d(r, c) = h(r, c) * I(r, c)$$

where the $*$ denotes the convolution process
- $h(r, c)$ is called the *point spread function* (PSF), or the blur function
- The PSF of a *linear, spatially-invariant* (shift invariant) system can be empirically determined by imaging a single point of light
- The PSF completely characterizes a linear, spatially-invariant system
- The PSF for motion blur is a rectangular or Gaussian function (see Figure 9.4.1)
- PSF blur models can be simulated in CVIPtools with *Utilities->Filter->Specify a Blur*

Modulation/Optical Transfer Function

- Assuming no additive noise:

$$D(u, v) = H(u, v)I(u, v)$$

- The Fourier transform of the degradation function, $H(u, v)$, is also referred to as the modulation transfer function (MTF), or the optical transfer function (OTF)
- The MTF typically refers to the transfer function of the system
- The OTF refers to the transfer function of the optics in the system
- The MTF completely characterizes a linear, spatially-invariant system
- Motion blur along the column axis from a camera with a mechanical shutter:

$$H(u, v) = \frac{\sin(\pi STv)}{\pi Sv}$$

where:

$S =$ the constant speed in the direction of the column axis

$T =$ the time interval the shutter is open

Estimation of the Degradation Function

- The degradation function can be estimated primarily by combinations of: (1) image analysis, (2) experimentation, and (3) mathematical modeling
- *Image analysis:* examine a known point or line in an image, and estimate the PSF by measuring the width and distribution of the known feature in the blurred image
- *Experimentation:* (1) The PSF can be found by imaging a point of light, *if the system is available and the conditions under which the image was acquired have not changed*, (2) A more reliable method is to use sinusoidal inputs at many different spatial frequencies to find the MTF, $H(u,v)$
- *Mathematical modeling examples*: (1) the motion blur model:

$$H(u,v) = \frac{\sin(\pi STv)}{\pi Sv}$$

(2) atmospheric turbulence degradation model used in astronomy and remote sensing:

$$H(u,v) = e^{-k(u^2+v^2)^{5/6}}$$

where k is an experimentally determined constant

- Mathematical models, image analysis, and lots of experimentation, combined with the experience and intuition of the expert are often necessary to estimate the degradation function successfully.

FREQUENCY DOMAIN FILTERS

- The frequency domain filters are based on the mathematical model provided in Section 9.1.1:

$$D(u,v) = H(u,v)I(u,v) + N(u,v)$$

- Application of the restoration filter is as follows:

$$\hat{I}(r,c) = F^{-1}\left[\hat{I}(u,v)\right] = F^{-1}\left[R_{\text{type}}(u,v)D(u,v)\right]$$

where:

$\hat{I}(r,c) =$ the restored image, an approximation to $I(r,c)$

$F^{-1}[\,] =$ the inverse Fourier transform

$R_{\text{type}}(u,v) =$ the Restoration (frequency domain) filter, the subscript defines the type of filter

- These filters assume the image and noise functions are *stationary,* which means spatial frequency content is fairly constant across the entire image
- The noise function can be assumed stationary (Figure 9.2.5)
- The stationary assumption is invalid for most real images, and adaptive filtering can help manage this problem

Inverse Filter

- The inverse filter model assumes no noise:

$$D(u, v) = H(u, v)I(u, v) + 0$$

- Inverse filter $= R_{\text{inv}}(u, v) = \dfrac{1}{H(u, v)}$

- In practice the assumption of no noise is usually invalid and the noise term will obscure the image at high frequencies
- The faulty noise assumption can be handled by: (1) limit the filter radius, (2) limit the filter gain, (3) use the pseudo-inverse filter

$$\text{Pseudo-inverse filter} = R_{\text{PI}}(u, v) = \begin{cases} \dfrac{1}{H(u, v)} & \text{for } H \neq 0 \\ 0 & \text{for } H = 0 \end{cases}$$

- In practice the gain of the pseudoinverse is set to 0 when the magnitude of $H(u, v)$ goes below a user specified threshold

Wiener Filter

- The Wiener is a minimum mean-square estimator which theoretically minimizes average error
- Wiener filter equation for an $N \times N$ image:

$$R_W(u, v) = \frac{H^*(u, v)}{|H(u, v)|^2 + \left[\dfrac{S_n(u, v)}{S_I(u, v)}\right]}$$

where:

$$H^*(u, v) = \text{complex conjugate of } H(u, v)$$

$$S_n(u, v) = |N(u, v)|^2 = \text{power spectrum of the noise}$$

$$S_I(u, v) = |I(u, v)|^2 = \text{power spectrum of the original image}$$

- The Wiener filter response is reduced at high frequencies compared to an inverse filter (see Figure 9.5.3) due to the power spectrum ratio, $\left[\frac{S_n(u, v)}{S_I(u, v)}\right]$
- The *Practical Wiener* filter replaces the power spectrum ratio with an experimentally determined constant, because in real applications $S_I(u, v)$ cannot be determined:

$$R_W(u,v) = \frac{H^*(u,v)}{|H(u,v)|^2 + K}$$

Constrained Least Squares Filter

- Instead of a constant for the power spectrum ratio, as in the Practical Wiener, the constrained least squares (CLS) filter has a function that varies with frequency
- The CLS filter includes a smoothing criterion function in its development to eliminate artifacts from filters such as Wiener filters:

$$R_{\text{CLS}}(u,v) = \frac{H^*(u,v)}{|H(u,v)|^2 + \gamma\,|P(u,v)|^2}$$

where:

$\gamma =$ adjustment factor

$P(u,v) =$ the Fourier transform of smoothness criterion function

- Referring to Figure 9.5.3 we want $P(u, v)$ to make $R_{\text{CLS}}(u, v)$ act as a lowpass filter, so it is actually a highpass in the denominator

Geometric Mean Filters

- The geometric mean filter equation provides a general form for the frequency domain restoration filters:

$$R_{\text{GM}}(u,v) = \left[\frac{H^*(u,v)}{|H(u,v)|^2}\right]^{\alpha} \left[\frac{H^*(u,v)}{|H(u,v)|^2 + \gamma\left[\frac{S_n(u,v)}{S_I(u,v)}\right]}\right]^{1-\alpha}$$

with α and γ being positive real constants

- For $\alpha = 1/2$ and $\gamma = 1$, this filter is called a *power spectrum equalization filter*
- For $\alpha = 0$, this filter is called a *parametric Wiener (PW) filter*:

$$R_{\text{PW}}(u,v) = \frac{H^*(u,v)}{|H(u,v)|^2 + \gamma\left[\frac{S_n(u,v)}{S_I(u,v)}\right]}$$

- For $\gamma = 1$, the PW is a standard Wiener filter
- For $\gamma = 0$, the PW is the inverse filter
- With the PW, as γ is adjusted, the results vary between the inverse and Wiener filters, with larger values providing more of the Wiener filtering effect

Adaptive Filtering

- The development of the preceding filters assumes stationary signals, images are highly nonstationary

- Block-by-block processing can be performed where the filter parameters are adjusted based on the block (subimage) characteristics, creating an adaptive filtering process
- Use of an adaptive filter will provide better results than a fixed filter, at the cost of increased computationally complexity
- Blocking artifacts may occur at block boundaries, which can be handled by post-processing lowpass filters, or overlapping blocks using window functions
- Computational intelligence methods, such as neural networks and genetic algorithms, can be applied to develop adaptive filters

Bandpass, Bandreject and Notch Filters

- These filters are used to analyze and restore images containing periodic noise
- A notch filter is a special form of a bandreject filter that only filters out specific frequencies
- The procedure for elimination of periodic noise involves examining the spectrum for spikes and artifacts, then removing them

GEOMETRIC TRANSFORMS

- Geometric transforms are by their very nature spatially-variant
- They are known as rubber-sheet transforms where the image is deformed or warped as if on a sheet of rubber
- They are used to restore a spatially distorted image, or to warp an image
- They require two steps: (1) spatial transform, (2) gray level interpolation
- The spatial transform provides location of the output pixel
- Gray level interpolation is needed due to the non-integer coordinates supplied by the spatial transform

Spatial Transforms

- *Spatial transforms* are used to map the input image location to a location in the output image
- The spatial transform requires two equations to map the distorted image, $d(\hat{r}, \hat{c})$, to the undistorted image $I(r, c)$:

  $\hat{r} = \hat{R}(r, c)$, defines the row coordinate for the distorted image

  $\hat{c} = \hat{C}(r, c)$, defines the column coordinate for the distorted image

- To find these equations requires known matching points, called *tiepoints*, in both the original image and the distorted image
- The method to restore a geometrically distorted image consists of three steps: (1) define quadrilaterals with known tiepoints for the entire image, (2) find the equations $\hat{R}(r, c)$ and $\hat{C}(r, c)$ for each set of tiepoints, and (3) remap all the pixels within each quadrilateral subimage using the equations corresponding to those tiepoints.

- A bilinear model for the mapping equations can be used for Step 2:

$$\hat{R}(r,c) = k_1 r + k_2 c + k_3 rc + k_4 = \hat{r}$$
$$\hat{C}(r,c) = k_5 r + k_6 c + k_7 rc + k_8 = \hat{c}$$

- The k_i values are constants to be determined by solving the eight simultaneous equations where r, c, $\hat{r}$ and $\hat{c}$ are known
- Non-integer results for $\hat{r}$ and $\hat{c}$ requires gray level interpolation

Gray Level Interpolation

- Three methods: (1) nearest neighbor, (2) neighborhood average, (3) bilinear interpolation
- Nearest neighbor finds the closest value; it is fast, easy, but provides ragged edges
- Neighborhood average is of medium complexity, reasonably fast, provides smooth but blurred edges
- Bilinear interpolation is the most complex, slowest, but has the best results (see Figure 9.6.5)
- Bilinear interpolation is accomplished by solving this equation using the four surrounding points, and plugging in the non-integer values for $\hat{r}$ and $\hat{c}$:

$$g(\hat{r},\hat{c}) = k_1\hat{r} + k_2\hat{c} + k_3\hat{r}\hat{c} + k_4$$

The Geometric Restoration Procedure

1. Find tiepoints throughout the image mapping the distorted image, $d(\hat{r},\hat{c})$ to the restored image, $\hat{I}(r,c)$. $\hat{I}(r,c)$ is the estimate to the original, undistorted image $I(r,c)$.
2. For each quadrilateral find the equations for $\hat{R}(r,c)$ and $\hat{C}(r,c)$
3. For each value of (r,c) in $\hat{I}(r,c)$, apply the equation pair, $\hat{R}(r,c)$ and $\hat{C}(r,c)$, corresponding to the mapped quadrilateral to find $(\hat{r},\hat{c})$.
4. Perform the selected method of gray level interpolation using the values for $(\hat{r},\hat{c})$ found in Step 3 to find the gray value for $\hat{I}(r,c)$.
5. Continue Step 3 and Step 4 until all values for $\hat{I}(r,c)$ are found and we have our restored image.

9.8 References and Further Reading

The first complete text on image restoration is [Andrews/Hunt 77], and this book provides a solid foundation for the work that has been done since. For more background and theory on image restoration see [Bates/McDonnell 89] and [Sezan/Tekalp 96]. More information can also be found in the chapters on this topic in [Gonzalez/Woods 02], [Sonka/Hlavac/Boyle 99], [Castleman 96], [Sid-Ahmed 95], [Bracewell 95], [Pratt 91], [Lim 90], [Banks 90], [Jain 89], and [Rosenfeld/Kak 82].

For the section on noise, the references [Gonzalez/Woods 02], [Castleman 96], [Myler/Weeks 93], [Pratt 91], [Kennedy/Neville 86], [Peebles 87] and [Andrews/Hunt 77] were consulted. More information regarding spatial filters can be found in [Petrou/Bosdogianni 99], [Tekalp 95], [Myler/Weeks 93], [Lu/Antoniou 92], [Pitas/Venetsanopoulis 90], and [Haykin 91]. Details on the adaptive median algorithm are found in [Gonzalez 02] and [Hwang/Haddad 95]. A detailed model for photodetector and film grain noise is found in [Pratt 91]. For general information on linear systems theory and digital signal processing theory see [Oppenheim/Schafer 89], [Stanley/Dougherty 84] and [Tretter 76]. For more information on noise and estimation theory see [VanTrees 68]. For more information on statistical or stochastic processes see [Gray/Davidsson 86] and [Peebles 87].

Consulted references for frequency domain filters include [Gonzalez/Woods 02], [Sonka/Hlavac/Boyle 99], [Jansson 97], [Castleman 96], [Bracewell 95], [Sid-Ahmed 95], [Pratt 91], [Lim 90], [Jain 89], [Bates/McDonnell 89], and [Rosenfeld/Kak 82]. [Gonzalez/Woods 02] and [Pratt 91] have more information on estimation of degradation functions. [Castleman 96] provides practical approaches to estimating the degradation function for image blurring, and [Sid-Ahmed 95] has an algorithm and code to estimate image blur. [Sonka/Hlavac/Boyle 99], [Jahne 97], [Pratt 91], and [Jain 89] provide more details on mathematical models for degradation functions. [Andrews/Hunt 77] provide more information on various PSFs, including spatially variant types, which are not discussed here. [Russ 99] provides practical information and examples of removal of image blurring. [Bates/McDonnell 89] provide advanced methods of preprocessing to improve the results of these filters. Much of the seminal work in communications theory, for example work on the Wiener filter, can be found in [Sloane/Wyner 93].

An excellent text that provides in depth coverage of adaptive restoration, via neural networks, fuzzy set theory and genetic algorithms, is [Perry/Wong/Guan 02]. For a discussion of adaptive algorithms based on specific conditions see [Lim 90]. A procedure for determining the variable weights in restoring an image with a complex periodic interference pattern (Section 9.5.6) can be found in [Gonzalez/Woods 02].

The geometric transforms sources include [Gonzalez/Woods 02], [Castleman 96], and [Pratt 91]. For more in depth information on geometric transforms see [Sonka/Hlavac/Boyle 99] and [Jahne 97]. For more information on sophisticated interpolation techniques see [Watt/Policarpo 98] and [Hill 90].

Andrews, H.C., Hunt, *Digital Image Restoration*, Upper Saddle River, NJ: Prentice Hall, 1977.

Banks, S., *Signal Processing, Image Processing and Pattern Recognition*, Upper Saddle River, NJ: Prentice Hall, 1990.

Bates, R.H., McDonnell, *Image Restoration and Reconstruction*, Oxford, UK: Oxford University Press, 1989.

Bracewell, R.N., *Two-Dimensional Imaging*, Upper Saddle River, NJ: Prentice Hall, 1995.

Castleman, K.R., *Digital Image Processing*, Upper Saddle River, NJ: Prentice Hall, 1996.

Gonzalez, R.C., Woods, R.E., *Digital Image Processing*, Upper Saddle River, NJ: Prentice Hall, 2002.

Gray, R.M., Davisson, L.D., *Random Processes: A Mathematical Approach for Engineers*, Upper Saddle River, NJ: Prentice Hall, 1986.

Hwang, H., Haddad, R.A., Adaptive median filters: new algorithms and results, *IEEE Transaction on Image Processing*, pp. 499–501, April 1995.

Haykin, S., *Adaptive Filter Theory*, Upper Saddle River, NJ: Prentice Hall, 1991.

Hill, F.S., *Computer Graphics*, NY: Macmillan, 1990.

Jahne, B., *Practical Handbook on Image Processing for Scientific Applications*, Boca Raton, FL: CRC Press, 1997.

Jain, A.K., *Fundamentals of Digital Image Processing*, Upper Saddle River, NJ: Prentice Hall, 1989.

Jansson, P.A., Editor, *Deconvolution of Images and Spectra*, 2nd Edition, NY: Academic Press, 1997.

Kennedy, J.B., Neville, A.M., *Basic Statistical Methods for Engineers and Scientists*, NY: Harper and Row 1986.

Lim, J.S., *Two-Dimensional Signal and Image Processing*, Upper Saddle River, NJ: Prentice Hall, 1990.

Lu, W.S., Antoniou, A., *Two-Dimensional Filters*, NY: Marcel Dekker, 1992.

Myler, H.R., Weeks, A.R., *Computer Imaging Recipies in C*, Upper Saddle River, NJ: Prentice Hall, 1993.

Myler, H.R., Weeks, A.R., *The Pocket Handbook of Image Processing Algorithms in C*, Upper Saddle River, NJ: Prentice Hall, 1993.
Oppenheim, A.V., Schafer, R.W., *Discrete-Time Signal Proccessing*, Upper Saddle River, NJ: Prentice Hall, 1989.
Peebles, P.Z., *Probability, Random Variables, and Random Signal Principles*, NY: McGraw-Hill, 1987.
Perry, S.W., Wong, H., Guan, L., *Adaptive Image Processing: A Computational Intelligence Perspective*, NY: CRC Press, 2002.
Petrou, M., Bosdogianni, P., *Image Processing: The Fundamentals*, West Sussex, England: John Wiley & Sons Ltd, 1999.
Pitas, I., Venetsanopoulos, *Nonlinear Digital Filters*, Boston, MA: Kluwer Academic, 1990.
Pratt, W.K., *Digital Image Processing*, NY: Wiley, 1991.
Rosenfeld, A., Kak, A.C., *Digital Picture Processing*, San Diego, CA: Academic Press, 1982.
Russ, J.C., *The Image Processing Handbook*, Boca Raton, FL: CRC Press, 1999.
Sezan, I., Tekalp, A.M., *Image Restoration*, Upper Saddle River, NJ: Prentice Hall, 1996.
Sid-Ahmed, M.A., *Image Processing: Theory, Algorithms, and Architectures*, NY: McGraw Hill, 1995.
Sloane, N.J.A., Wyner, A.D., editors, *Claude Elwood Shannon, Collected Papers*, NY: IEEE Press, 1993.
Sonka, M., Hlavac, V., Boyle, R., *Image Processing, Analysis and Machine Vision*, Pacific Grove, CA: Brooks/Cole Publishing Company, 1999.
Stanley, W.D., Dougherty G.R., Dougherty, R., *Digital Signal Processing*, Reston, VA: Reston Publishing, Prentice Hall, 1984.
Tekalp, A.M, *Digital Video Processing*, Upper Saddle River, NJ: Prentice Hall, 1995.
Tretter, S.A., *Introduction to Discrete-Time Signal Processing*, NY: Wiley, 1976.
Watt, A., Policarpo, F., *The Computer Image*, NY: Addison-Wesley, 1998.

9.9 Exercises

1. (a) How do image restoration and image enhancement differ? (b) How are restoration and image enhancement alike?
2. List four examples of causes of image degradation.
3. (a) Sketch a block diagram of the image restoration process. (b) Briefly discuss each block.
4. (a) What is the equation in the spatial domain for the degradation process model? (b) In the frequency domain? (c) Is this the only possible model? Discuss.
5. (a) Define noise in images. (b) List sources of noise in images. Discuss. (c) List names of the mathematical models for the primary types of noise in images, and describe a cause for each. (d) What type of noise occurs in radar images? (e) What type of noise occurs in laser images? (f) What is periodic noise?
6. Given noisy images, how can we estimate the noise model?
7. Use CVIPtools to explore noise histograms. Select *Restoration->Noise* and an image of your choice and do the following. (a) Add zero-mean gaussian noise with a variance of 25 to the image and view the histogram of the image with and without noise, place the histogram next to the image on the screen. (b) Add zero-mean gaussian noise with a variance of 800 to the image and view the histogram, compare all three histograms, what do you observe? (c) Find an area in your image that is fairly constant in the original and crop it out of the two images with added noise, display the histograms. Can you tell the noise type? (d) Select the *Use black image* option to create a noise only image and compare its histogram to the histograms of the cropped sections (of image plus noise) (e) Repeat (a)–(d) with negative exponential. (f) Repeat (a)–(d) with uniform noise.

8. (a) What are three types of spatial filters for noise removal? (b) List an advantage and disadvantage of each type. (c) What type works best and why?
9. Apply the following filters to the 3×3 (window size is 3×3) subimages below, and find the output for each. (a) median, (b) maximum, (c) minimum, (d) midpoint, (e) alpha-trimmed mean with $T = 2$.

$$\text{Subimage 1} \begin{bmatrix} 255 & 118 & 112 \\ 122 & 121 & 111 \\ 120 & 119 & 112 \end{bmatrix} \quad \text{Subimage 2} \begin{bmatrix} 100 & 100 & 98 \\ 0 & 99 & 96 \\ 100 & 99 & 93 \end{bmatrix}$$

$$\text{Subimage 3} \begin{bmatrix} 10 & 11 & 10 \\ 12 & 12 & 11 \\ 9 & 10 & 9 \end{bmatrix}$$

10. Use CVIPtools to explore the order filters, *Restoration->Spatial Filter*, for noise removal. Experiment by adding Gaussian and salt-and-pepper noise to images. Refer to Section 9.3.1 and verify the claims made regarding the types of noise each filter will handle.
11. Apply the following filters to the 3×3 (window size is 3×3) subimages below, and find the output for each: (a) arithmetic mean, (b) contra-harmonic mean with $R = -2$, (c) contra-harmonic mean with $R = +2$, (d) geometric mean, (e) harmonic mean, (f) Y_p mean with $P = -1$, (g) Y_p mean with $P = +2$.

$$\text{Subimage 1} \begin{bmatrix} 255 & 118 & 112 \\ 122 & 121 & 111 \\ 120 & 119 & 112 \end{bmatrix} \quad \text{Subimage 2} \begin{bmatrix} 100 & 100 & 98 \\ 0 & 99 & 96 \\ 100 & 99 & 93 \end{bmatrix}$$

$$\text{Subimage 3} \begin{bmatrix} 10 & 11 & 10 \\ 12 & 12 & 11 \\ 9 & 10 & 9 \end{bmatrix}$$

12. Use CVIPtools to explore the mean filters, *Restoration->Spatial Filter*, for noise removal. Experiment by adding Gaussian, gamma, uniform, and salt *or* pepper noise to images. Refer to Section 9.3.2 and verify the claims made regarding the types of noise each filter will deal with. Note that "dealing with" the noise does not always create a "good" image. Use *Analysis->Transforms* to compare the spectra of the images before adding the noise, with the added noise and after filtering. Even if the images do not look "good," can you see some mitigation of the noise effects in the spectra?
13. Apply the spatial domain MMSE filter to the center pixel of the following subimages (window size of 3×3), using a noise variance of 100:

$$\text{Subimage 1} \begin{bmatrix} 255 & 118 & 112 \\ 122 & 121 & 111 \\ 120 & 119 & 112 \end{bmatrix} \quad \text{Subimage 2} \begin{bmatrix} 100 & 100 & 98 \\ 0 & 99 & 96 \\ 100 & 99 & 93 \end{bmatrix}$$

$$\text{Subimage 3} \begin{bmatrix} 10 & 11 & 10 \\ 12 & 12 & 11 \\ 9 & 10 & 9 \end{bmatrix}$$

14. (a) What are the advantages of the adaptive median algorithm compared to the standard median filter? (b) What the disadvantages?
15. (a) List and explain two examples of spatially-invariant degradations. (b) List and explain two examples of spatially-variant degradations.
16. (a) What does *PSF* stand for and what does it mean? (b) For a *PSF* to completely characterize an imaging system, what are the constraints on the system? (c) Describe two models for motion blur. d) What are *deconvolution* and *blind deconvolution*?
17. (a) What do *OTF* and *MTF* stand for and what do they mean? (b) What does it mean for an image to be stationary? (c) Are images typically stationary? Explain.
18. (a) At high frequencies noise may obscure the image signal when using an inverse filter. Explain why. (b) What can be done to help solve this problem?
19. (a) What is the power spectrum ratio and how is it related to the signal-to-noise ratio? (b) Sketch the filter response of the Wiener and the inverse filter and explain why the Wiener works better in the presence of noise. (c) Why use a constant in place of the power spectrum ratio in the Wiener filter?
20. Use CVIPtools to explore the inverse and the Wiener filter. *Use Restoration-> Frequency Filters* and *Utilities->Create->Add Noise* and *Utilities->Filter->Specify a Blur*. Select a square image that is an even power of 2, for example 256×256 or 512×512. (a) Blur the image to simulate motion blur along the column axis, using the following parameters: 7×7 mask, horizontal line for blur shape, blur method constant, and weight $= 1.0$. Compare results of using the Wiener and the inverse filters on the blurry image. For the degradation function, select *Specify a function* and set the parameters the same as with the blur. For the Wiener use *Utilities->Create->Black Image* to create a black image to be used as the noise image (since we did not add noise). Experiment with setting the *cutoff frequency* and the *gain limit* to obtain good results. (b) Use the blurry image and add zero-mean Gaussian noise with a variance of 100. Also, with *Utilities-> Create->Noise* select *Use a black image* to create the noise image. Compare results of using the Wiener and the inverse filters. For the degradation function, select *Specify a function* and set the parameters the same as with the blur. For the Wiener use the noise image you created. Experiment with setting the *cutoff frequency* and the *gain limit* to obtain good results. Compare your results to part (a). (c) Repeat (b) but add noise with a variance of 800. How do the results compare?
21. (a) Explain why the CLS filter may give better results than the Wiener filter. (b) Use CVIPtools and apply the CLS filter to the images you created in Exercise 20. (c) Compare the results of the CLS and Wiener filter. Did it perform as you expected? Why or why not?
22. Repeat Exercise 20, but for the blur use a Gaussian circle to simulate a poorly focused lens.
23. (a) Write the equation for the geometric mean filter. What are the values for α and γ to create (b) power spectrum equalization filter?, (c) parametric Wiener filter?, (c) standard Wiener filter?, (d) inverse filter?
24. Repeat Exercise 20 with the practical Wiener and the parametric Wiener filters. Can you get results as good, or better, than with the standard Wiener? Why or why not?

25. Repeat Exercise 20 with the geometric mean and the power spectrum equalization filters. With the geometric mean try various values for α and γ. Can you get results as good, or better, than with the standard Wiener filter? Why or why not?
26. (a) What type of processing is normally done for adaptive filtering in the frequency domain? (b) What is the blocking effect and what causes it? (c) What are two methods for dealing with the blocking effect?
27. Use CVIPtools to explore bandpass, bandreject and notch filters. (a) Use image Fig9.5-9.bmp and try to do a better restoration than in the figure. (b) Select an image of your choice and experiment with adding sine wave and cosine wave images at various frequencies to the image. Use *Utilities->Create* and *Utilities->Arith/Logic*. Next, use notch and bandreject filters to try to remove the periodic noise. Use the bandpass filter to extract noise only images and examine the spectrum of these images to gain insight into the process.
28. (a) What are the two steps in geometric transforms? (b) Explain why the second step is necessary. (c) What are the advantages and disadvantages of the three types of gray level interpolation?
29. Given the following 16×16 distorted image, $d(\hat{r}, \hat{c})$, and the mapping equations, $\hat{R}(r, c)$and $\hat{C}(r, c)$, restore the 3×3 subimage where the row and column coordinates are between 0 and 2. That is, find the subimage represented by the x's.

$$\textit{subimage } \hat{I}(r,c) \left\{ \begin{array}{c|ccc} \cdot & 0 & 1 & 2 \\ 0 & x & x & x \\ 1 & x & x & x \\ 2 & x & x & x \end{array} \right.$$

$$\hat{R}(r, c) = 2r + 1c + 2rc + 1 = \hat{r}$$

$$\hat{C}(r, c) = 1r + 1c + 2rc + 0 = \hat{c}$$

$$d(\hat{r}, \hat{c}) = \begin{bmatrix} 5 & 4 & 6 & 7 & 8 & 5 & 5 & 5 & 6 & 6 & 6 & 6 & 6 & 6 & 6 & 7 \\ 9 & 9 & 9 & 9 & 9 & 9 & 6 & 6 & 6 & 6 & 6 & 6 & 6 & 6 & 6 & 6 \\ 6 & 6 & 6 & 5 & 4 & 3 & 2 & 3 & 4 & 5 & 4 & 3 & 2 & 6 & 2 & 2 \\ 2 & 2 & 2 & 2 & 2 & 2 & 2 & 2 & 2 & 2 & 2 & 2 & 2 & 2 & 2 & 2 \\ 2 & 2 & 2 & 2 & 2 & 2 & 2 & 2 & 2 & 2 & 2 & 2 & 2 & 2 & 2 & 2 \\ 2 & 2 & 2 & 5 & 5 & 5 & 5 & 5 & 5 & 5 & 5 & 5 & 5 & 5 & 5 & 5 \\ 5 & 5 & 5 & 7 & 7 & 7 & 7 & 7 & 7 & 7 & 7 & 7 & 7 & 8 & 8 & 8 \\ 8 & 7 & 6 & 8 & 5 & 7 & 4 & 5 & 6 & 3 & 4 & 5 & 4 & 6 & 7 & 8 \\ 9 & 8 & 7 & 8 & 7 & 8 & 7 & 8 & 7 & 8 & 7 & 9 & 9 & 9 & 9 & 9 \\ 9 & 9 & 9 & 9 & 9 & 9 & 9 & 9 & 9 & 9 & 9 & 4 & 4 & 4 & 4 & 4 \\ 4 & 4 & 4 & 4 & 4 & 4 & 4 & 4 & 4 & 2 & 2 & 2 & 2 & 2 & 2 & 2 \\ 2 & 2 & 2 & 2 & 1 & 1 & 1 & 1 & 1 & 1 & 1 & 1 & 1 & 1 & 1 & 1 \\ 1 & 1 & 1 & 1 & 6 & 5 & 4 & 3 & 2 & 3 & 4 & 5 & 6 & 5 & 6 & 5 \\ 8 & 9 & 0 & 8 & 7 & 6 & 0 & 0 & 0 & 0 & 0 & 0 & 0 & 0 & 0 & 0 \\ 0 & 0 & 0 & 7 & 7 & 7 & 7 & 7 & 6 & 6 & 6 & 6 & 6 & 6 & 6 & 6 \\ 6 & 6 & 6 & 6 & 6 & 6 & 6 & 6 & 6 & 5 & 5 & 4 & 4 & 4 & 4 & 4 \end{bmatrix}$$

30. Given the following 16×16 distorted image, $d(\hat{r}, \hat{c})$, and the mapping equations, $\hat{R}(r, c)$ and $\hat{C}(r, c)$, restore the 3×3 subimage where the row and column coordinates are between 0 and 2. That is, find the subimage represented by the x's. Use (a) nearest neighbor, (b) neighborhood average using the four edge neighbors (horizontal and vertical)

$$\text{subimage } \hat{I}(r,c) \left\{ \begin{array}{c|ccc} \cdot & 0 & 1 & 2 \\ \hline 0 & x & x & x \\ 1 & x & x & x \\ 2 & x & x & x \end{array} \right.$$

$$\hat{R}(r,c) = 1.3r + 1c + 2rc + 1.2 = \hat{r}$$

$$\hat{C}(r,c) = 0.75r + 1c + 1.8rc + 2.1 = \hat{c}$$

$$d(\hat{r},\hat{c}) = \begin{bmatrix} 5 & 4 & 6 & 7 & 8 & 5 & 5 & 5 & 6 & 6 & 6 & 6 & 6 & 6 & 6 & 7 \\ 9 & 9 & 9 & 9 & 9 & 9 & 6 & 6 & 6 & 6 & 6 & 6 & 6 & 6 & 6 & 6 \\ 6 & 6 & 6 & 5 & 4 & 3 & 2 & 3 & 4 & 5 & 4 & 3 & 2 & 6 & 2 & 2 \\ 2 & 2 & 2 & 2 & 2 & 2 & 2 & 2 & 2 & 2 & 2 & 2 & 2 & 2 & 2 & 2 \\ 2 & 2 & 2 & 2 & 2 & 2 & 2 & 2 & 2 & 2 & 2 & 2 & 2 & 2 & 2 & 2 \\ 2 & 2 & 2 & 5 & 5 & 5 & 5 & 5 & 5 & 5 & 5 & 5 & 5 & 5 & 5 & 5 \\ 5 & 5 & 5 & 7 & 7 & 7 & 7 & 7 & 7 & 7 & 7 & 7 & 7 & 8 & 8 & 8 \\ 8 & 7 & 6 & 8 & 5 & 7 & 4 & 5 & 6 & 3 & 4 & 5 & 4 & 6 & 7 & 8 \\ 9 & 8 & 7 & 8 & 7 & 8 & 7 & 8 & 7 & 8 & 7 & 9 & 9 & 9 & 9 & 9 \\ 9 & 9 & 9 & 9 & 9 & 9 & 9 & 9 & 9 & 9 & 9 & 4 & 4 & 4 & 4 & 4 \\ 4 & 4 & 4 & 4 & 4 & 4 & 4 & 4 & 4 & 2 & 2 & 2 & 2 & 2 & 2 & 2 \\ 2 & 2 & 2 & 2 & 1 & 1 & 1 & 1 & 1 & 1 & 1 & 1 & 1 & 1 & 1 & 1 \\ 1 & 1 & 1 & 1 & 6 & 5 & 4 & 3 & 2 & 3 & 4 & 5 & 6 & 5 & 6 & 5 \\ 8 & 9 & 0 & 8 & 7 & 6 & 0 & 0 & 0 & 0 & 0 & 0 & 0 & 0 & 0 & 0 \\ 0 & 0 & 0 & 7 & 7 & 7 & 7 & 7 & 6 & 6 & 6 & 6 & 6 & 6 & 6 & 6 \\ 6 & 6 & 6 & 6 & 6 & 6 & 6 & 6 & 6 & 5 & 5 & 4 & 4 & 4 & 4 & 4 \end{bmatrix}$$

31. Given that we have found the mapping equations for a quadrilateral and determined the following corresponding pixel coordinates:

(r, c) coordinates for the restored image	**Corresponding ($\hat{r}, \hat{c}$) coordinates found by the mapping equations**
(4, 3)	(1.2, 2.1)
(4, 4)	(1.3, 2.6)
(5, 3)	(1.8, 2.2)
(5, 4)	(1.9, 2.8)

(a) Use the following degraded image and find the bilinear interpolation equation needed to find the x's in the following restored image. (b) Apply the equation and find the values for the x's:

$$\hat{I}(r,c) = \begin{bmatrix} 22 & 22 & 24 & 23 & 34 & 35 & 35 & 36 \\ 23 & 22 & 23 & 24 & 55 & 55 & 56 & 57 \\ 23 & 44 & 45 & 48 & 49 & 55 & 56 & 57 \\ 22 & 46 & 49 & 48 & 51 & 51 & 52 & 52 \\ 23 & 49 & 48 & x & x & 48 & 49 & 49 \\ 49 & 49 & 49 & x & x & 49 & 48 & 49 \\ 50 & 50 & 47 & 48 & 49 & 50 & 50 & 50 \\ 50 & 50 & 51 & 50 & 51 & 51 & 51 & 50 \end{bmatrix} \quad d(\hat{r},\hat{c}) = \begin{bmatrix} 50 & 50 & 51 & 52 & 55 & 55 & 55 & 55 \\ 52 & 51 & 50 & 55 & 56 & 56 & 56 & 55 \\ 48 & 46 & 44 & 48 & 48 & 47 & 48 & 44 \\ 42 & 42 & 40 & 40 & 40 & 41 & 42 & 43 \\ 48 & 49 & 51 & 50 & 55 & 56 & 56 & 56 \\ 49 & 46 & 46 & 44 & 48 & 47 & 55 & 55 \\ 50 & 50 & 50 & 45 & 47 & 46 & 54 & 54 \\ 50 & 50 & 50 & 50 & 51 & 52 & 53 & 54 \end{bmatrix}$$

32. Use CVIPtools, *Restoration->Geometric Transforms*, to explore geometric transforms and restoration. Select an image to distort and restore. (a) Create a 5×5 warping mesh by selecting *Enter a new mesh file*, inputting 5 for the *Number of rows* and 5 for the *Number of columns*. Next, input *Regular->Irregular* for the *Direction* and *bilinear interpolation* for the *Gray value interpolation method*. Now select the points for the mesh with the mouse by holding the Alt key on the keyboard and clicking the left mouse button. (b) Click *Apply* to distort the image, (c) Restore the image by selection *Irregular->Regular* for the direction. Do this three times, each time selecting a different *Gray level interpolation method*. Compare the results with each of the methods, which one is best? What type of artifacts do you observe? On your computer is the relative speed of each method noticeable?
33. Use CVIPtools, *Restoration->Geometric Transforms*, to explore geometric transforms and restoration. Select an image to distort and restore. (a) Create a 7×7 warping mesh by selecting *Enter a new mesh file*, inputting 7 for the *Number of rows* and 7 for the *Number of columns*. Next, input *Irregular->Regular* for the *Direction* and *bilinear interpolation* for the *Gray value interpolation method*. Now select the points for the mesh with the mouse by holding the *control* key on the keyboard and clicking the left mouse button. (b) Click *Apply* to distort the image. (c) Restore the image by selection *Regular->Irregular* for the direction. Do this three times, each time selecting a different *Gray level interpolation method*. Compare the results with each of the methods, which one is best? What type of artifacts do you observe? On your computer is the relative speed of each method noticeable?

9.9.1 Programming Exercise: Noise

1. Write a function to create an image with Gaussian noise. Let the user specify the mean and the variance.
2. Modify the function to allow the user to create a noise only image or to add the noise to an image. For a noise only image let the user specify the image size. Also, let the user specify if the image is to be remapped to BYTE or left as FLOAT data type.

3. Incorporate the six CVIPtools noise functions into your CVIPlab program (see the *Noise* library). Allow the user to create a noise only image or to add the noise to an image. For a noise only image let the user specify the image size. Also, let the user specify if the image is to be remapped to BYTE or left as FLOAT data type. To create a noise only image you will need to create an image filled with zeros to be passed to the CVIPtools functions.

9.9.2 Programming Exercise: Order Filters

1. Write functions to perform median, maximum, and minimum filters. Let the user specify the mask size.
2. Write a function to implement the adaptive median filter algorithm. Let the user input the maximum window size. Compare the results to the standard median filter.
3. Incorporate the CVIPtools functions *alpha_filter* (alpha-trimmed mean filter) and *midpoint_filter* from the *SpatialFilter* library into your CVIPlab program.

9.9.3 Programming Exercise: Mean Filters

1. Write functions to perform arithmetic and geometric mean filters. Let the user specify the mask size.
2. Incorporate the CVIPtools functions *contra_filter* (contra-hramonic mean filter), *harmonic_filter* (harmonic mean filter) and from the *SpatialFilter* library into your CVIPlab program.

9.9.4 Programming Exercise: MMSE Filter

1. Write a function to implement the adaptive MMSE spatial filter. Let the user input the noise variance and the window size.
2. Incorporate the CVIPtools function *mmse_filter* from *the SpatialFilter Library* into your CVIPlab program. Compare the results from the CVIPtools function to the one you wrote. Are they the same? Why or why not?

9.9.5 Programming Exercise: Frequency Domain Filters

1. Write a function to implement the Inverse filter in the frequency domain. Inputs to the function are: (1) the degraded image, (2) an image with the PSF, (3) a maximum cutoff frequency. Use the CVIPtools functions *fft_transform* and *ifft_transform* from the *Transform* library to perform the Fourier transform. *Note:* PSF images can be created by creating a circle with a zero radius and blurring it, and translating it to the corners.
2. Modify your function implement a pseudoinverse filter, where the user can specify the minimum threshold for setting the filter gain to zero.
3. Incorporate the CVIPtools function *wiener* (Wiener filter) from the *Transform-Filter* library into your CVIPlab program. Compare results of its use in your CVIPlab to using it in CVIPtools. Are the results the same? Why or why not?

9.9.6 Programming Exercise: Geometric Transforms

1. Write a program to perform geometric distortion and restoration using the method outlined in this chapter. Limit it to a single mesh and use nearest neighbor gray level interpolation. Have the user input the parameters from the keyboard: (a) the image file name, (b) the (r, c) coordinates for the four mesh points, (c) to warp (regular to irregular) or restore (irregular to regular) the image.
2. Modify your program to perform neighborhood average gray level interpolation. Let the user select all four neighbors or vertical or horizontal neighbors only for the averaging process.
3. Incorporate the CVIPtools function *mesh_warping* from the *Geometry* library into your CVIPlab program. This is the function used in CVIPtools *Restoration-> Geometric Transforms*. Other associated functions include: *display_mesh, keyboard_to_mesh, mesh_to_file, bilinear_interp,* and *solve_c.*

10

Image Compression

10.1 Introduction and Overview

The field of image compression continues to grow at a rapid pace. As we look to the future, the need to store and transmit images will only continue to increase faster than the available capability to process all the data. Even with the rapid growth in computer power and the increase in Internet bandwidth, the ability to process and transmit the desired amount of image data continues to be problematic. Additionally, advances in video technology and the corresponding growth in the multi-media market, including high-definition television, are creating a demand for new, better, and faster image compression algorithms.

Applications that require image compression are many and varied. Use of images and graphics in business documents is rapidly increasing—from product catalogs to stock reports, and these documents are often stored in databases and transmitted over the Internet. Many organizations are making their entire libraries available on the Internet; from the U.S. Library of Congress to professional organizations such as the IEEE and ACM. Satellite images are collected and transmitted daily, for weather, political, environmental, and sociological uses, and can include 30 or more spectral bands imaged at very high resolution. Use of medical imaging modalities continues to grow and the effective management of image databases is essential to the practice of medicine. Additionally, these images often need to be transmitted for the increasing number of telemedicine applications. Hospitals are archiving enormous amounts of medical image data daily, businesses and governments are using teleconferencing at an ever increasing pace, and broadcast television standards are evolving to require higher resolution images. These applications, along with many others, are helping to push image compression to the forefront of the image processing field.

Compression algorithm development starts with applications to two-dimensional (2-D) still images. Because video and television signals consist of consecutive frames of 2-D image data, the development of compression methods for 2-D still data is of paramount importance. After the 2-D methods are developed, they are often extended to video (motion imaging). Here, we will focus on image compression of single frames of image data.

What is image compression? *Image compression* involves reducing the size of image data files, while retaining necessary information. Image segmentation methods, which are primarily a data reduction process, were explored in Chapter 4 and can be used for compression. However, segmentation methods tend to reduce too much of the data and are only useful in a limited number of applications. One of the key aspects of a good compression scheme involves the second part of the definition—*retaining necessary information*. What information is necessary?—as usual, it is application specific.

The reduced file created by the compression process is called the *compressed file* and is used to reconstruct the image, resulting in the *decompressed image*. The original image, before any compression is performed, is called the *uncompressed image* file. The ratio of the original, uncompressed image file and the compressed file is referred to as the *compression ratio*. The compression ratio is denoted by:

$$\text{Compression Ratio} = \frac{\text{Uncompressed file size}}{\text{Compressed file size}} = \frac{\text{SIZE}_\text{U}}{\text{SIZE}_\text{C}};$$

$$\text{Often written as} \rightarrow \text{SIZE}_\text{U} : \text{SIZE}_\text{C}$$

Example 10.1.1

The original image is 256×256 pixels, single-band (grayscale), 8-bits per pixel. This file is 65,536 bytes (64k). After compression the image file is 6554 bytes. The compression ratio is: $\text{SIZE}_\text{U}/\text{SIZE}_\text{C} = 65{,}536/6554 = 9.999 \approx 10$. This can also be written as 10:1.

This is called a "10 to 1 compression", a "10 times compression", or can be stated as "compressing the image to 1/10 its original size". Another way to state the compression is to use the terminology of *bits per pixel*. For an $N \times N$ image:

$$\text{Bits per pixel} = \frac{\text{Number of bits}}{\text{Number of pixels}} = \frac{(8)(\text{Number of bytes})}{N \times N}$$

Example 10.1.2

Using the preceding example, with a compression ratio of 65,536/6554 bytes, we want to express this as bits per pixel. This is done by first finding the number of pixels in the image: $256 \times 256 = 65{,}536$ pixels. We then find the number of bits in the compressed image file: (6554 bytes)(8 bits/byte) = 52,432 bits. Now we can find the bits per pixel by taking the ratio: 52,432/65,536 = 0.8 bits/pixel.

The reduction in file size is necessary to meet the bandwidth requirements for many transmission systems, and for the storage requirements in computer databases. The amount of data required for digital images is enormous. For example, a single 512×512, 8-bit, image requires 2,097,152 bits for storage. If we wanted to transmit this image over the Internet with a standard dialup connection, it would take minutes for transmission—too long for most people to wait.

Example 10.1.3

To transmit an RGB (color) 512×512, 24-bit (8-bit per pixel per color) image via modem at 56 kbaud (kilo-bits per second), it would take about:

$$\frac{(512 \times 512 \text{ pixels})(24 \text{ bits/pixel})}{(56 \times 1024 \text{ bits/second})} \approx 109 \text{ seconds} \approx 1.8 \text{ minutes}$$

This number is based on the actual transmission rate being the maximum, which is typically not the case due to Internet traffic, overhead bits, and transmission errors.

Additionally, considering that a web page might contain more than one of these images, the time it takes is simply too long. For high quality images the required resolution can be much higher than the previous example. For example, a 35 mm photograph has an effective resolution of about three or four thousand pixels in each dimension.

Example 10.1.4

To transmit a digitized color 35 mm slide scanned at 4000×3000 pixels, and 24-bits, at 56 kbaud would take about:

$$\frac{(4000 \times 3000 \text{ pixels})(24 \text{ bits/pixel})}{(56 \times 1024 \text{ bits/sec})} \approx 5022 \text{ sec} \approx 83 \text{ min, too long to wait!}$$

Of course fast Internet connections are becoming more and more prevalent via cable modems from the cable company and DSL (digital subscriber lines) connections from the phone company. With DSL current speeds range from 144 kbaud to 1.5 mega-bits-per-second (Mbps), and cable modems can provide data rates up to 3 Mbps. The next example applies the maximum rate to the preceding example.

Example 10.1.5

To transmit a digitized color 35 mm slide scanned at 4000×3000 pixels, and 24-bits, at 3 Mbps would take about:

$$\frac{(4000 \times 3000 \text{ pixels})(24 \text{ bits/pixel})}{(3 \times 1024 \times 1024 \text{ bits/sec})} \approx 91 \text{ seconds} \approx 1.5 \text{ minutes}$$

This is a great improvement over the 81 minutes, but is still longer than we care to wait, especially if the web page contains multiple images. Consider the transmission of video images, where we need multiple frames per second. Consider just one second of video data that has been digitized at 640×480 pixels per frame, and requiring 15 frames per second for interlaced video.

Example 10.1.6

To transmit one second of interlaced video that has been digitized at 640×480 pixels:

$$\frac{(640 \times 480 \times 15 \text{ frames/sec})(24 \text{ bits/pixel})}{(3 \times 1024 \times 1024 \text{ bits/sec})} \approx 35 \text{ seconds}$$

Waiting 35 seconds for one second's worth of video is not exactly real time! Even attempting to transmit uncompressed video over the highest speed Internet connection is impractical. The preceding examples only begin to approach the capability needed for image storage and transmission in the 21st century. For example, a Landsat satellite transmits images that are typically 250 megabytes—6100×6100 pixels in seven spectral bands. The Japanese Advanced Earth Observing Satellite (ADEOS) transmits image data at the rate of 120 Mbps. Even with high speed connections, applications such as high definition television, real-time teleconferencing, and transmission of multiband high resolution satellite images, leads us to the conclusion that image compression is not only desirable but a necessity.

As previously mentioned, the key to a successful compression scheme comes with the second part of the definition of image compression—*retaining necessary information*. To understand this we must differentiate between data and information. For digital images, *data* refers to the pixel gray level values that correspond to the brightness of a pixel at a point in space. *Information* is an interpretation of the data in a meaningful way. Data are used to convey information, much like the way the alphabet is used to convey information via words. Information is an elusive concept; it can be application specific. For example, in a binary image that contains text only, the necessary information may only involve the text being readable, whereas for a medical image the necessary information may be every minute detail in the original image.

There are two primary types of image compression methods—those that preserve the data, and those that allow some loss of data. The first type are called *lossless methods*, since no data are lost and the original image can be recreated exactly from the compressed data. For complex images these methods are limited to compressing the image file to about one-half to one-third its original size (2 : 1 to 3 : 1), often the achievable compression is much less. For simple images such as text-only images, lossless methods may achieve much higher compression. The second type of compression methods are called *lossy*, since they allow a loss in the actual image data, so the original uncompressed image *cannot* be created *exactly* from the compressed file. For complex images these techniques can achieve compression ratios of about 10 to 50, and still retain high-quality visual information. For simple images, or lower quality results, compression ratios as high as 100 to 200 can be attained.

Compression algorithms are developed by taking advantage of the redundancy that is inherent in image data. Four primary types of redundancy can be found in images: (1) coding, (2) interpixel, (3) interband, and (4) psychovisual redundancy. *Coding redundancy* occurs when the data used to represent the image is not utilized in an optimal manner. For example, if we have an 8-bit per pixel image which allows 256 gray level values, but the actual image contains only 16 gray level values, this is a sub-optimal coding—only 4-bits per pixel are actually needed. *Interpixel redundancy* occurs because adjacent pixels tend to be highly correlated. This is a result of the fact that in most images the brightness levels do not change rapidly, but change gradually, so adjacent pixel values tend to be relatively close to each other in value (for video, or motion images, this concept can be extended to include *inter-frame redundancy*, redundancy between frames of image data). *Interband redundancy* occurs in color images due to the correlation between bands within an image—if we extract the red, green, and blue bands we can see that they look similar. The fourth type, *psychovisual redundancy*, refers to the fact that some information is more important to the human visual system than other types of information. For example, we can only perceive spatial frequencies below about 50 cycles per degree (see Section 7.2), so any higher spatial frequency information is of little interest to us.

The key in image compression algorithm development is to determine the minimal data required to retain the necessary information. The compression is achieved by taking advantage of the redundancy that exists in images. If the redundancies are removed prior to compression, for example with a decorrelation process, a more effective compression can be achieved. To help determine which information can be removed and which information is important, the image fidelity criteria as defined in Chapter 7 are used. These measures provide metrics for determining image quality. In the case of image compression, the "reconstructed image" discussed in Chapter 7 refers to the decompressed image. It should be noted that the information required is application specific, and that, with lossless schemes, there is no need for a fidelity criteria.

Most of the compressed images shown in this chapter were generated with CVIPtools, which consists of code that has been developed for educational and research purposes. The compressed images shown are not necessarily representative of the best commercial applications that use the techniques described, because the commercial compression algorithms are often *combinations* of the techniques described herein. Additionally, commercial applications have been developed much more extensively, and may provide better compression ratios and better resulting images than those shown here. In this chapter we chose to show results from each individual technique to illustrate how they work and what we can expect from them. The final two sections briefly outline some of the more sophisticated commercial algorithms, such as the relatively new JPEG2000, which combine many of the separate techniques described.

10.1.1 Compression System Model

The compression system model consists of two parts: the compressor, and the decompressor. The *compressor* consists of a preprocessing stage and encoding stage, whereas the *decompressor* consists of a decoding stage followed by a postprocessing stage (Figure 10.1.1). Before encoding, preprocessing is performed to prepare the image for the encoding process, and consists of any number of operations that are application specific. After the compressed file has been decoded, postprocessing can be performed to eliminate some of the potentially undesirable artifacts brought about by the compression process. Often, many practical compression algorithms are a combination of a number of different individual compression techniques.

The compressor can be further broken down into stages as illustrated in Figure 10.1.2. The first stage in preprocessing is data reduction. Here, the image data can be reduced by gray level and/or spatial quantization, or can undergo any desired image improvement (for example, noise removal) process. The second step in preprocessing is the mapping process, which maps the original image data into another mathematical space where it is easier to compress the data. Next, as part of the encoding process, is the quantization stage, which takes the potentially continuous data from the mapping stage and puts it in discrete form. The final stage of encoding involves coding the resulting data, which maps the discrete data from the quantizer onto a code in an optimal manner.

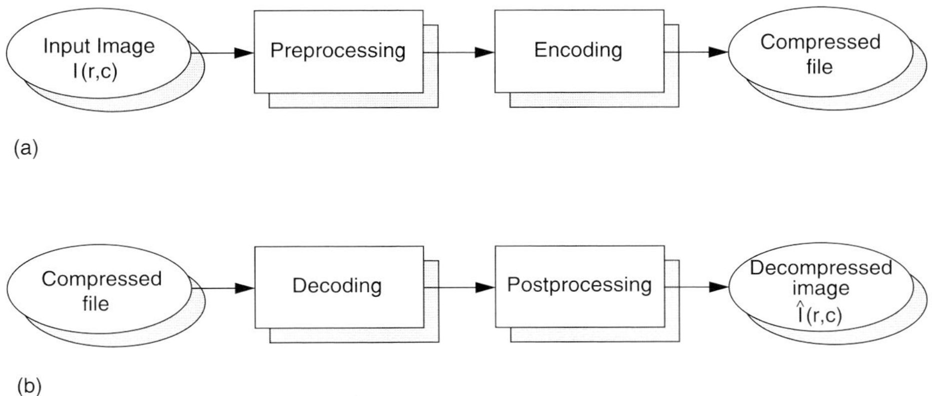

FIGURE 10.1.1
Compression system model. (a) Compression. (b) Decompression.

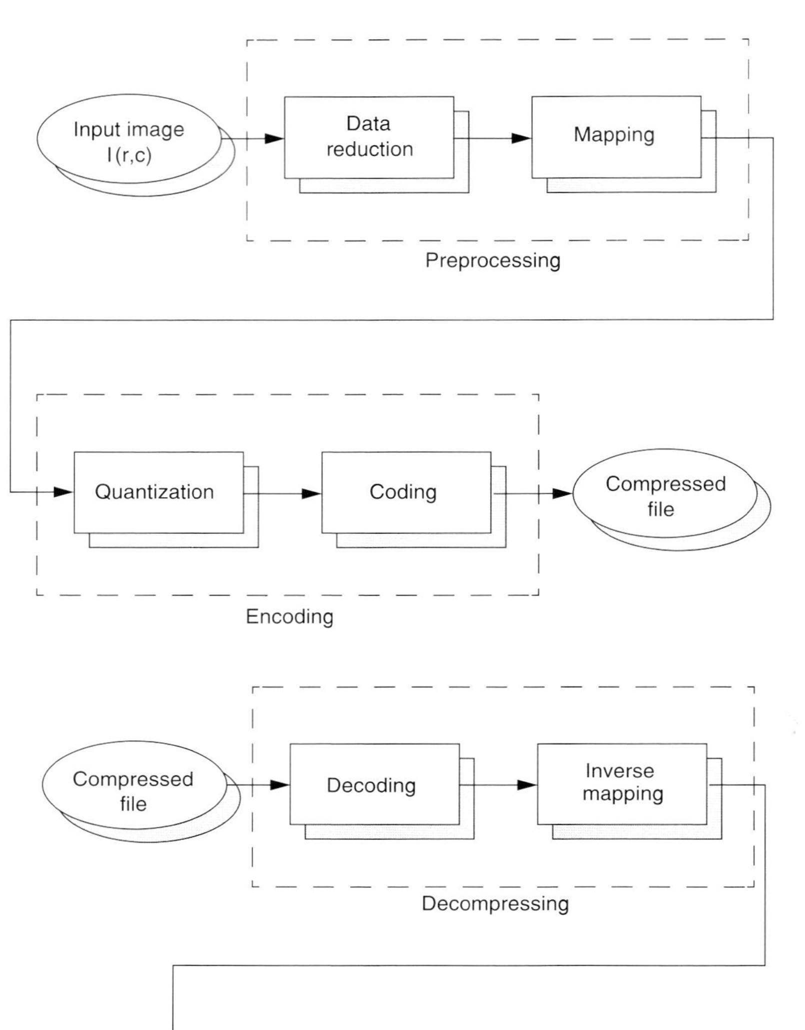

FIGURE 10.1.2
The compressor.

FIGURE 10.1.3
The decompressor.

A compression algorithm may consist of all the stages, or it may consist of only one or two of the stages.

The decompressor can be further broken down into the stages shown in Figure 10.1.3. Here the decoding process is divided into two stages. The first, the decoding stage, takes the compressed file and reverses the original coding by mapping the codes to the original, quantized values. Next, these values are processed by a stage that performs an inverse mapping to reverse the original mapping process. Finally, the image may be postprocessed to enhance the look of the final image. In some cases this may be done to reverse any preprocessing, for example, enlarging an image that was shrunk in the data reduction process. In other cases the postprocessing may simply enhance the image to ameliorate any artifacts from the compression process itself.

The development of a compression algorithm is highly application specific. During the preprocessing stage of compression, processes such as enhancement, noise removal, or quantization are applied. The goal of preprocessing is to prepare the image for the encoding process by eliminating any irrelevant information, where *irrelevant* is defined by the application. For example, many images that are for viewing purposes only can be preprocessed by eliminating the lower bit planes, without losing any useful information. In Figure 10.1.4 are shown the eight bit-planes corresponding to an 8-bit image. Each bit plane is shown as an image by using white if the corresponding bit is a 1, and black if the bit is a 0. Here we see that the lower bit planes contain little information, and can be eliminated with no significant information loss.

FIGURE 10.1.4
Bit plane images. (a) Original, 8-bit image. (b) Bit plane 7, the most significant bit. (c) Bit plane 6. (d) Bit plane 5. (e) Bit plane 4. (f) Bit plane 3.

(g)

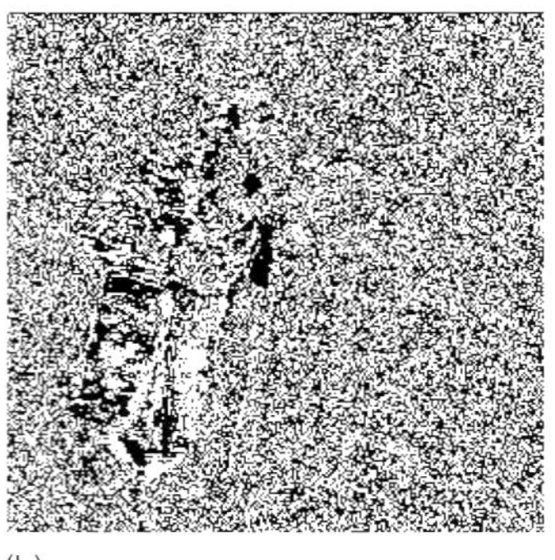
(h)

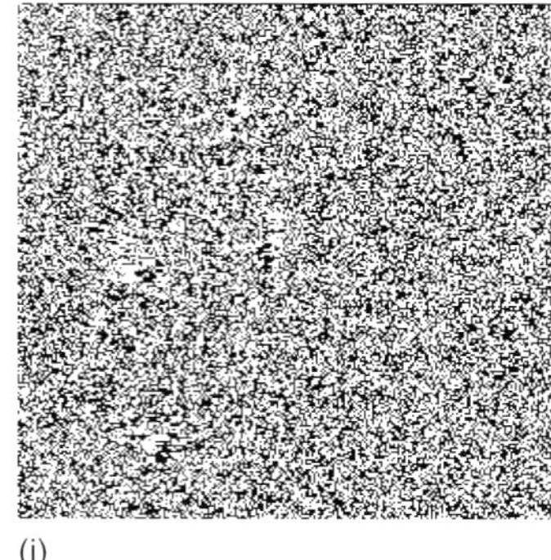
(i)

FIGURE 10.1.4 (Continued)
Bit plane images. (g) Bit plane 2. (h) Bit plane 1. (i) Bit plane 0, the least significant bit.

The mapping process is important because image data tends to be highly correlated. What this means is that there is a lot of redundant information in the data itself. Specifically, if the value of one pixel is known, it is highly likely that the adjacent pixel value is similar. By finding a mapping equation that decorrelates the data this type of data redundancy can be removed. One method to do this is to find the difference between adjacent pixels and encode these values, this is called *differential coding*. Second, the principal components transform can be used, which provides a theoretically optimal decorrelation (see Figure 5.6.1). Color transforms are used to decorrelate data between image bands. Also, the spectral domain is used for image compression, so this first stage may include mapping into the frequency or sequency domain where the energy in the image is compacted into primarily the lower frequency/sequency components. These methods are all *reversible*, that is information preserving, although all mapping methods are not reversible. The concept of reversibility is important to a compression method.

Depending on the mapping equation used, quantization may be necessary to convert the data into digital form (BYTE data type). This is because many of these mapping methods will result in floating point data which requires multiple bytes for representation—not very efficient if our goal is data reduction. There are two ways to do the quantization: uniform quantization, or nonuniform quantization. In *uniform quantization* all the quanta, or subdivisions into which the range is divided, are of equal width. In *nonuniform quantization*, these quantization bins are not all of equal width (as shown in Figure 3.2.18). Often, nonuniform quantization bins are designed to take advantage of the response of the human visual system. For example, very high brightness levels appear the same—white—so wider quantization bins may be used over this range. In the spectral domain, the higher frequencies may also be quantized with wider bins because we are more sensitive to lower and midrange spatial frequencies and most images have little energy at high frequencies. The concept of nonunform quantization bin sizes is also described as a *variable bit rate*, since the wider quantization bins imply fewer bits to encode, while the smaller bins need more bits. It is important to note that the quantization process is not reversible, so some information may be lost during quantization. Additionally, since it is not a reversible process, the inverse process does not exist, so it does not appear in the decompression model (Figure 10.1.3).

The coding stage of any image compression algorithm is very important. The coder provides a one-to-one mapping, each input is mapped to a unique output by the coder, so it is a reversible process. The code can be an *equal length code*, where all the code words are the same size, or an *unequal length code* with variable length code words. In most cases, an unequal length code is the most efficient for data compression, but requires more overhead in the coding and decoding stages. Many of the lossless methods described here are primarily efficient coding techniques.

10.2 Lossless Compression Methods

Lossless compression methods are necessary in some imaging applications. For example, with medical images, legal requirements often mandate that any archived medical images are stored without any data loss. In general, any images that are to be used in a court of law will be suspect if a lossy compression technique has been applied. Many of the lossless techniques were developed for nonimage data, and, consequently are not optimal for image compression. In general, the lossless techniques alone provide marginal compression of complex image data, often in the range of only a 10% reduction in file size. However, lossless compression techniques may be used for both preprocessing and postprocessing in image compression algorithms to obtain the extra 10% compression. Additionally, for simple images the lossless techniques can provide substantial compression.

The underlying theory for lossless compression (also called *data compaction*) comes from the area of communications and information theory, with a mathematical basis in probability theory. One of the most important concepts used here is the idea of information content and randomness in data. Using information theory, an event that is less likely to occur is said to contain more information than an event that is likely to occur. For example, consider the following statements:

1. The earth will continue to revolve around the sun.
2. An earthquake will occur tomorrow.
3. A matter transporter will be invented in the next 10 years.

Which statement, in the sense stated above, has the most information? Statement 1 contains relatively little information, because this is an event that we all know will occur—it has a probability approaching 100% (we hope!). Statement 2 contains more information, because the event "earthquake will occur" has a probability less than 100%. Statement 3 contains the most information, because it is a highly unlikely event. This perspective on information is the *information theoretic definition* and should not be confused with our working definition that requires information in images to be useful, not simply novel. This brief background is provided to help explain some of the following concepts.

An important concept here is the idea of measuring the average information in an image, referred to as the *entropy*. The entropy for an $N \times N$ image can be calculated by this equation:

$$\text{Entropy} = -\sum_{i=0}^{L-1} p_i \log_2(p_i) \quad \text{(in bits/pixel)}$$

where

$$p_i = \text{the probability of the } i\text{th gray level} = n_k/N^2$$
$$n_k = \text{the total number of pixels with gray value } k$$
$$L = \text{the total number of gray levels (e.g. 256 for 8-bits)}$$

This measure provides us with a theoretical minimum for the average number of bits per pixel that could be used to code the image. This number is theoretically optimal, and can be used as a metric for judging the success of a coding scheme.

Example 10.2.1

Let $L = 8$, meaning there are 3 bits/pixel in the original image. Now, let's say the number of pixels at each gray level value is equal (they have the same probability), that is:

$$p_0 = p_1 = \cdots = p_7 = \frac{1}{8}$$

Now, we can calculate the entropy as follows:

$$\text{Entropy} = -\sum_{i=0}^{7} p_i \log_2(p_i) = -\sum_{i=0}^{7} \frac{1}{8} \log_2\left(\frac{1}{8}\right) = 3$$

This tells us that the theoretical minimum for lossless coding for this image is 3 bits per pixel. In other words, there is no code that will provide better results than the one currently used (called the natural code, since $000_2 = 0$, $001_2 = 1$, $010_2 = 2, \ldots, 111_2 = 7$). This example illustrates that the image with the most random distribution of gray levels, a uniform distribution, has the highest entropy.

Example 10.2.2

Let $= 8$, thus we have a natural code with 3 bits per pixel in the original image. Now lets say that the entire image has a gray level of 2, so:

$$p_2 = 1 \quad \text{and} \quad p_0 = p_1 = p_3 = p_4 = p_5 = p_6 = p_7 = 0$$

And the entropy is:

$$\text{Entropy} = -\sum_{i=0}^{7} p_i \log_2(p_i) = -(1)\log_2(1) + 0 + \cdots + 0 = 0$$

This tells us the theoretical minimum for coding this image is 0 bits per pixel. Why is this?—Because the gray level value is known to be 2. To code the entire image we need only one value, this is called the certain event, it has a probability of 1.

The two preceding examples illustrate the range of the entropy:

$$0 \leq \text{ Entropy } \leq \log_2(L)$$

The examples also illustrate the information theory perspective regarding information and randomness. The more randomness that exists in an image, the more evenly distributed the gray levels, and more bits per pixel are required to represent the data (see Figure 10.2.1). This also correlates to information—more randomness implies each individual value is less likely, which means more information is contained in each pixel value, so we need more bits to code each pixel value. This also provides us with one of

FIGURE 10.2.1
Entropy. Entropy as measured in bit per pixel (bpp) for the following images: (a) Original image, entropy = 7.032 bpp. (b) Image after local histogram equalization, block size 4, entropy = 4.348 bpp. (c) Image after binary threshold, entropy = 0.976 bpp. (d) Circle with a radius of 32, entropy = 0.283 bpp. (e) Circle with a radius of 64, entropy = 0.716 bpp. (f) Circle with a radius of 32, and a linear blur radius of 64, entropy = 2.030 bpp.

the key concepts in coding theory: we want to assign a fewer number of bits to code more likely events. Intuitively, this makes sense. Given an image to code, a minimum overall file size will be achieved if a smaller number of bits is used to code the most frequent gray levels.

The entropy measure also provides us with a metric to evaluate coder performance. We can measure the average number of bits per pixel (*L*ength) in a coder by the following:

$$L_{\text{ave}} = \sum_{i=0}^{L-1} l_i p_i$$

where

$$l_i = \text{length in bits of the code for } i\text{th gray level}$$
$$p_i = \text{histogram-probability of } i\text{th gray level}$$

This can then be compared to the entropy, which provides the theoretical minimum. The closer L_{ave} is to the entropy, the better the coder.

10.2.1 Huffman Coding

The Huffman code, developed by D. Huffman in 1952, is a minimum length code. This means that given the statistical distribution of the gray levels (the histogram), the Huffman algorithm will generate a code that is as close as possible to the minimum bound, the entropy. This method results in an *unequal* (or *variable*) *length code*, where the size of the code words can vary. For complex images, Huffman coding alone will typically reduce the file by 10% to 50% (1.1 : 1 to 1.5 : 1), but this ratio can be improved to 2 : 1 or 3 : 1 by preprocessing for irrelevant information removal.

The Huffman algorithm can be described in five steps:

1. Find the gray level probabilities for the image by finding the histogram
2. Order the input probabilities (histogram magnitudes) from smallest to largest
3. Combine the smallest two by addition
4. GOTO Step 2, until only two probabilities are left
5. By working backward along the tree, generate code by alternating assignment of 0 and 1

This procedure is best illustrated by example.

Example 10.2.3

We have an image with 2-bits per pixel, giving four possible gray levels. The image is 10 rows by 10 columns. In Step 1 we find the histogram for the image. This is shown in Figure 10.2.2a, where we see that gray level 0 has 20 pixels, gray level 1 has 30 pixels, gray level 2 has 10 pixels, and gray level 3 has 40 pixels with the value. These are converted into probabilities by normalizing to the total number of pixels in the image. Next, in Step 2, the probabilities are ordered as in Figure 10.2.2b. For Step 3, we combine the smallest two by addition. Step 4 repeats Steps 2 and 3, where we reorder (if necessary) and add the two smallest probabilities as in Figure 10.2.2d. This step is repeated until only two values remain. Since we have only two left in our

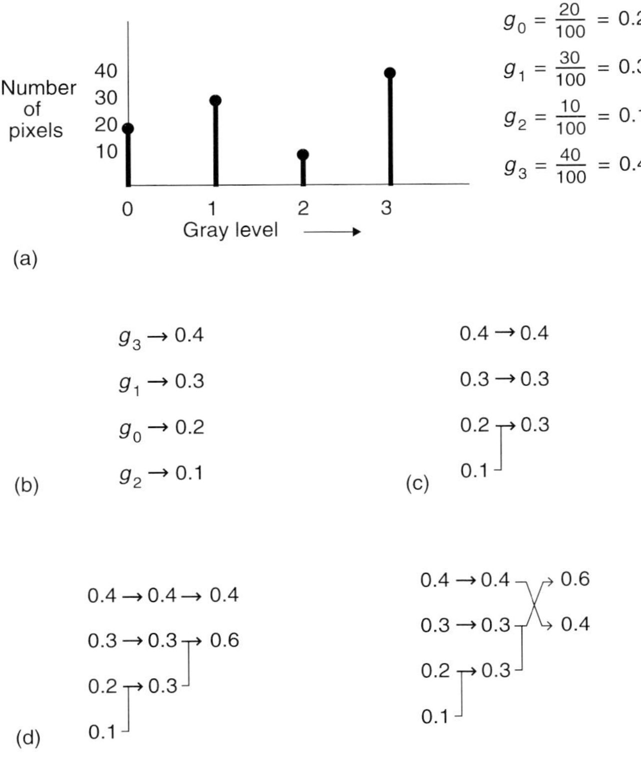

FIGURE 10.2.2
Huffman coding example. (a) Step 1: Histogram. (b) Step 2: Order. (c) Step 3: Add. (d) Step 4: Reorder and add until only two values remain.

example, we can continue to Step 5 where the actual code assignment is made. The code assignment is shown in Figure 10.2.3. We start on the right-hand side of this tree and assign 0's and 1's, working our way back to the original probabilities. Figure 10.2.3a shows the first assignment of 0 and 1. A 0 is assigned to the 0.6 branch, and a 1 to the 0.4 branch. In Figure 10.2.3b, the assigned 0 and 1 are brought back along the tree, and wherever a branch occurs the code is put on both branches. Now (Figure 10.2.3c), we assign the 0 and 1 to the branches labeled 0.3, appending to the existing code. Finally (Figure 10.2.3d), the codes are brought back one more level, and where the branch splits another assignment of 0 and 1 occurs (at the 0.1 and 0.2 branch). Now we have the Huffman code for this image in Table 10.1.

Note that two of the gray levels now have 3 bits assigned to represent them, but one gray level only has 1 bit assigned to represent it. The gray level represented by 1 bit, g_3, is the most likely to occur (40% of the time) and thus has the *least information in the information theoretic sense*. Remember that we learned from information theory that symbols with less information require fewer bits to represent them. The original image had an average of 2 bits/pixel, let us examine the entropy in bits per pixel, and average bit length for the Huffman coded image file.

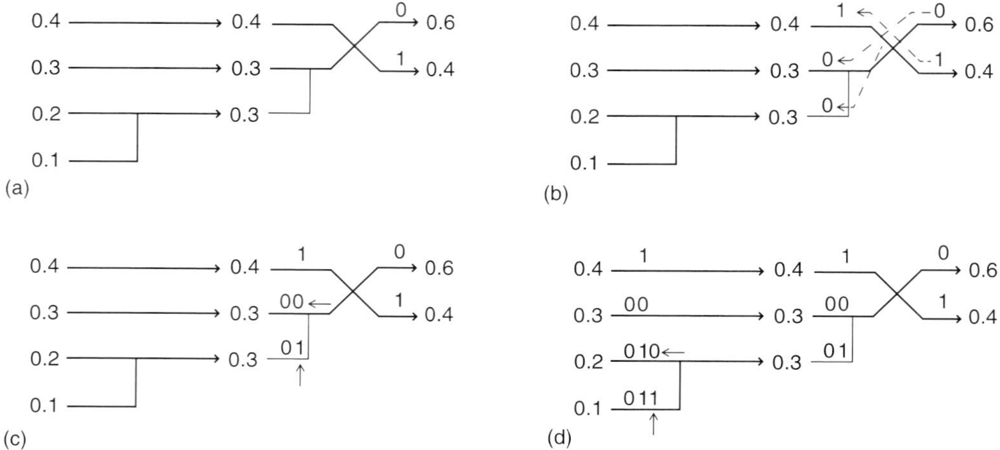

FIGURE 10.2.3
Huffman coding example, Step 5. (a) Assign 0 and 1 to the right-most probabilities. (b) Bring 0 and 1 back along the tree. (c) Append 0 and 1 to previously-added branches. (d) Repeat the process until the original branch is labeled.

TABLE 10.1

Original Gray Level (Natural Code)	Probability	Huffman Code
g_0: 00_2	0.2	010_2
g_1: 01_2	0.3	00_2
g_2: 10_2	0.1	011_2
g_3: 11_2	0.4	1_2

Example 10.2.4

$$\text{Entropy} = -\sum_{i=0}^{i} p_i \log_2(p_i)$$

$$= -[(0.2)\log_2(0.2) + (0.3)\log_2(0.3)] + (0.1)\log_2(0.1) + (0.4)\log_2(0.4)$$

$$\approx 1.846 \text{ bits/pixel}$$

(*Note:* $\log_2(x)$ can be found by taking $\log_{10}(x)$ and multiplying by 3.322)

$$L_{\text{ave}} = \sum_{i=0}^{L-1} l_i p_i$$

$$= 3(0.2) + 2(0.3) + 3(0.1) + 1(0.4)$$

$$= 1.9 \text{ bits/pixel} \quad \text{(average length with Huffman code)}$$

In the example, we observe a 2.0 : 1.9 compression, which is about a 1.05 compression ratio, providing about 5% compression. From the example we can see that the Huffman code is highly dependent on the histogram, so any preprocessing to simplify the histogram will help improve the compression ratio.

10.2.2 Run-Length Coding

Run-length coding (RLC) is an image compression method that works by counting the number of adjacent pixels with the same gray level value. This count, called the *run-length*, is then encoded and stored. Here we will explore several methods of run-length encoding—basic methods used primarily for binary (two-valued) images, and extended versions for gray scale images. We will also briefly discuss RLC standards.

Basic RLC is used primarily for binary images, but can work with complex images that have been preprocessed by thresholding to reduce the number of gray levels to two. There are various ways to implement basic RLC, and the first step is to define the required parameters. We can either use horizontal RLC, counting along the rows, or vertical RLC, counting along the columns. In basic horizontal RLC, the number of bits used for the encoding depends on the number of pixels in a row. If the row has 2^n pixels, then the required number of bits is n, so that a run that is the length of the entire row can be encoded.

Example 10.2.5

A 256×256 image requires 8-bits, since $2^8 = 256$.

Example 10.2.6

A 512×512 image requires 9-bits, since $2^9 = 512$.

The next step is to define a convention for the first RLC number in a row—does it represent a run of 0's or 1's? Defining the convention for the first RLC number to represent 0's, we can look at the following example.

Example 10.2.7

The image is an 8×8 binary image, which requires 3 bits for each run-length coded word. In the actual image file are stored 1's and 0's, although upon display the 1's become 255 (white) and the 0's are 0 (black). To apply RLC to this image, using horizontal RLC:

$$\begin{bmatrix} 0 & 0 & 0 & 0 & 0 & 0 & 0 & 0 \\ 1 & 1 & 1 & 1 & 0 & 0 & 0 & 0 \\ 0 & 1 & 1 & 0 & 0 & 0 & 0 & 0 \\ 0 & 1 & 1 & 1 & 1 & 1 & 0 & 0 \\ 0 & 1 & 1 & 1 & 0 & 0 & 1 & 0 \\ 0 & 0 & 1 & 0 & 0 & 1 & 1 & 0 \\ 1 & 1 & 1 & 1 & 0 & 1 & 0 & 0 \\ 0 & 0 & 0 & 0 & 0 & 0 & 0 & 0 \end{bmatrix}$$

The RLC numbers are:

First row: 8
Second row: 0, 4, 4
Third row: 1, 2, 5
Fourth row: 1, 5, 2
Fifth row: 1, 3, 2, 1, 1
Sixth row: 2, 1, 2, 2, 1
Seventh row: 0, 4, 1, 1, 2
Eighth row: 8

Note that in the second and seventh rows, the first RLC number is 0, since we are using the convention that the first number corresponds to the number of zeros in a run.

This basic method can be extended to gray level images by using a technique called bit-plane RLC. *Bitplane-RLC* works by applying basic RLC to each bit-plane independently. In Figure 10.2.4 the concept of bit-planes is illustrated. For each binary digit in the gray level value, an image plane is created, and this image plane (a string of 0's and 1's) is then encoded using RLC. Typical compression ratios of 0.5 to 1.2 are achieved with complex 8-bit monochrome images; so, without further processing, this is not a good compression technique for complex images. Bitplane-RLC is most useful for simple images, such as graphics files, where much higher compression ratios are achieved. The compression results using this method can be improved by preprocessing to reduce the number of gray levels, but then the compression is *not lossless*.

With lossless bitplane RLC we can improve the compression results by taking our original pixel data (in *natural code*) and mapping it to a *Gray code* (named after Frank Gray), where adjacent numbers differ in only one bit. Because adjacent pixel values are highly correlated, adjacent pixel values tend to be relatively close in gray level value, and this can be problematic for RLC.

Example 10.2.8

In Figure 10.2.5 is shown the 4-bit Gray code and the natural binary code. The Gray code, by definition, only has one bit changing in adjacent codes. However, in, for example, the 7 to 8 transition with the natural code, all 4 bits change:

Natural Code	Gray Code
0 1 1 1	0 1 0 0
↓ ↓ ↓ ↓	↓ ↓ ↓ ↓
1 0 0 0	1 1 0 0

When a situation such as the above example occurs, each bitplane experiences a transition, which adds a code for the run in each bitplane. However, with the Gray code, only one bitplane experiences the transition, so it only adds one extra code word. By preprocessing with a Gray code we can achieve about a 10% to 15% increase in compression with bitplane-RLC for typical images.

Another way to extend basic RLC to gray level images is to include the gray level of a particular run as part of the code. Here, instead of a single value for a run, two parameters are used to characterize the run. The pair (G, L) correspond to the gray level

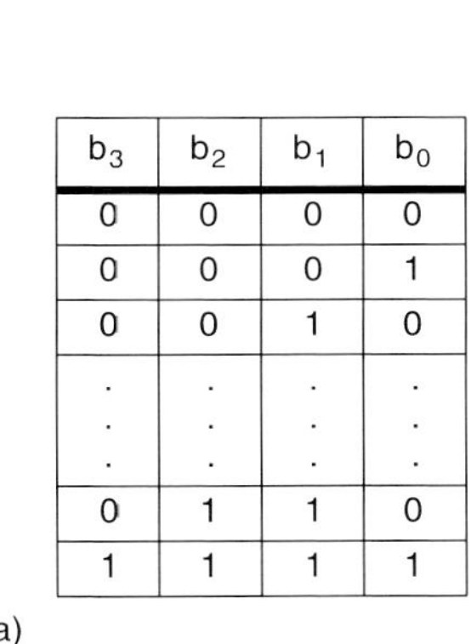

b_3	b_2	b_1	b_0
0	0	0	0
0	0	0	1
0	0	1	0
⋮	⋮	⋮	⋮
0	1	1	0
1	1	1	1

(a)

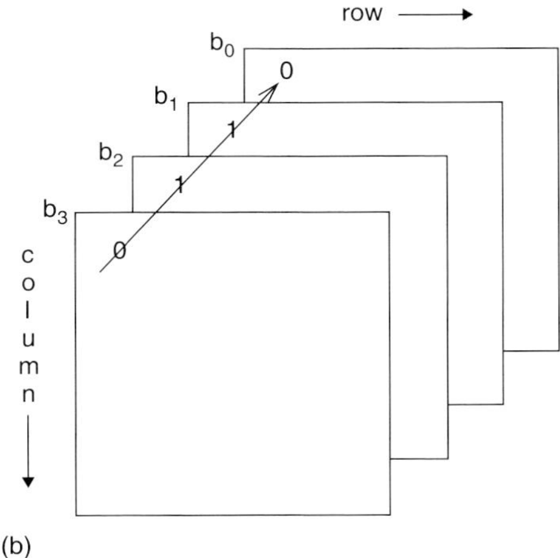

(b)

FIGURE 10.2.4
Bit plane run-length coding. (a) 4 bits/pixel designation. (b) Bit-planes $b \rightarrow \begin{matrix} b3 & b2 & b1 & b0 \\ 0 & 1 & 1 & 0 \end{matrix}$.

Decimal	4-bit Natural Code	4-bit Gray Code
0	0000	0000
1	0001	0001
2	0010	0011
3	0011	0010
4	0100	0110
5	0101	0111
6	0110	0101
7	0111	0100
8	1000	1100
9	1001	1101
10	1010	1111
11	1011	1110
12	1100	1010
13	1101	1011
14	1110	1001
15	1111	1000

(a)

7: 0 1 1 1
↓ ↓ ↓ ↓ ↓
8: 1 0 0 0

(b)

FIGURE 10.2.5
Gray code. (a) Gray code versus natural code. (b) The natural code transition of 7 to 8 changes all 4 bits.

value, G, and the run length, L. This technique is only effective with images containing a small number of gray levels.

Example 10.2.9

Given the following 8×8, 4-bit image:

$$\begin{bmatrix} 10 & 10 & 10 & 10 & 10 & 10 & 10 & 10 \\ 10 & 10 & 10 & 10 & 10 & 12 & 12 & 12 \\ 10 & 10 & 10 & 10 & 10 & 12 & 12 & 12 \\ 0 & 0 & 0 & 10 & 10 & 10 & 0 & 0 \\ 5 & 5 & 5 & 0 & 0 & 0 & 0 & 0 \\ 5 & 5 & 5 & 10 & 10 & 9 & 9 & 10 \\ 5 & 5 & 5 & 4 & 4 & 4 & 0 & 0 \\ 0 & 0 & 0 & 0 & 0 & 0 & 0 & 0 \end{bmatrix}$$

The corresponding gray levels pairs are as follows:

First row: 10, 8
Second row: 10, 5   12, 3
Third row: 10, 5   12, 3
Fourth row: 0, 3   10, 3   0, 2
Fifth row: 5, 3   0, 5
Sixth row: 5, 3   10, 2   9, 2   10, 1
Seventh row: 5, 3   4, 3   0, 2
Eighth row: 0, 8

These numbers are then stored in the RLC compressed file as:

10, 8, 10, 5, 12, 3, 10, 5, 12, 3, 0, 3, 10, 3, 0, 2, 5,
3, 0, 5, 5, 3, 10, 2, 9, 2, 10, 1, 5, 3, 4, 3, 0, 2, 0, 8

The decompression process requires the number of pixels in a row, and the type of encoding used.

Standards for RLC have been defined by the International Telecommunications Union-Radio (ITU-R, previously CCIR). These standards, initially defined for use with FAX transmissions, have become popular for binary image compression. They use horizontal RLC, but postprocess the resulting RLC with a Huffman encoding scheme. Newer versions of this standard also utilize a two-dimensional technique where the current line is encoded based on a previous line. This additional processing helps to reduce the file size. These encoding methods provide compression ratios of about 15 to 20 for typical documents.

10.2.3 Lempel–Ziv–Welch Coding

The Lempel–Ziv–Welch (LZW) coding algorithm works by encoding strings of data. For images, these strings of data correspond to sequences of pixel values. It works by creating a string table that contains the strings and their corresponding codes. The string table is updated as the file is read, with new codes being inserted whenever a new string is encountered. If a string is encountered that is already in the table, the corresponding code for that string is put into the compressed file.

LZW coding uses code words with more bits than the original data. For example, with 8-bit image data, an LZW coding method could employ 10-bit words. The corresponding string table would then have $2^{10} = 1024$ entries. This table consists of the original 256 entries, corresponding to the original 8-bit data, and allows 768 other entries for string codes. The string codes are assigned during the compression process, but the actual string table is not stored with the compressed data. During decompression the information in the string table is extracted from the compressed data itself.

For the GIF (and TIFF) image file format the LZW algorithm is specified, but there has been some controversy over this, since the algorithm is patented (by Unisys Corporation under patent #4,558,302). Since these image formats are widely used, other methods similar in nature to the LZW algorithm have been developed to be used with these, or similar, image file formats. Similar versions of this algorithm include the *adaptive Lempel–Ziv*, used in the UNIX compress function, and the *Lempel–Ziv 77* algorithm used in the UNIX gzip function.

10.2.4 Arithmetic Coding

In arithmetic coding there is not a direct correspondence between the code and the individual pixel values. *Arithmetic coding* transforms input data into a single floating point number between 0 and 1. As each input symbol (in this case, pixel value) is read, the precision required for this number becomes greater. Because images are very large and the precision of digital computers finite, an entire image must be divided into small subimages to be encoded.

Arithmetic coding uses the probability distribution of the data (histogram), so it can theoretically achieve the maximum compression specified by the entropy. It works by successively subdividing the interval between 0 and 1, based on the placement of the current pixel value in the probability distribution. This is best illustrated by example.

Example 10.2.10

Given a 16×16, 2-bit image with the histogram shown in Figure 10.2.6a, we can define an arithmetic coding probability table shown in Figure 10.2.6b. The probability values are the ratio of the specific gray level value to the total number of pixels in the image (in this case $16 \times 16 = 256$). The initial subinterval specifies how the 0 to 1 interval is divided based on the distribution, where the width of the subinterval is equal to the probability, and the subinterval starts where the previous one stops. In Figure 10.2.6c, the actual arithmetic coding process is illustrated, with an example pixel value sequence of: 0, 0, 3, 1. Starting on the left, the initial 0 to 1 interval is subdivided, based on the probability distribution. Next, the first pixel value '0' is coded by extracting the subinterval corresponding to the '0' and subdividing it again, based on the same relative distribution. This process is repeated for each pixel value in the sequence until a final interval is determined, in this case from 58/1024 to 62/1024, or 0.056640625 to 0.060546875. Any value within this subinterval, such as 0.057 or 0.060, can be used to represent this sequence of gray level values.

In practice, this technique may be used as part of an image compression scheme, but is impractical to use alone. It is one of the options available in the JPEG standard.

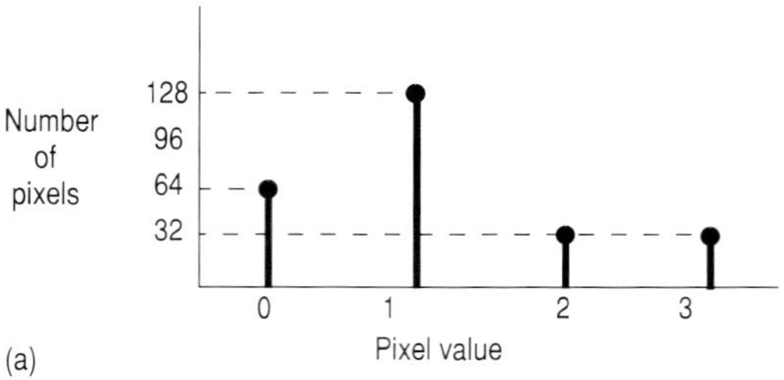

Pixel value	Probability	Initial sub-interval
0	64/256 = 1/4	0 - 1/4
1	128/256 = 1/2	1/4 - 3/4
2	32/256 = 1/8	3/4 - 7/8
3	32/256 = 1/8	7/8 - 1

(b)

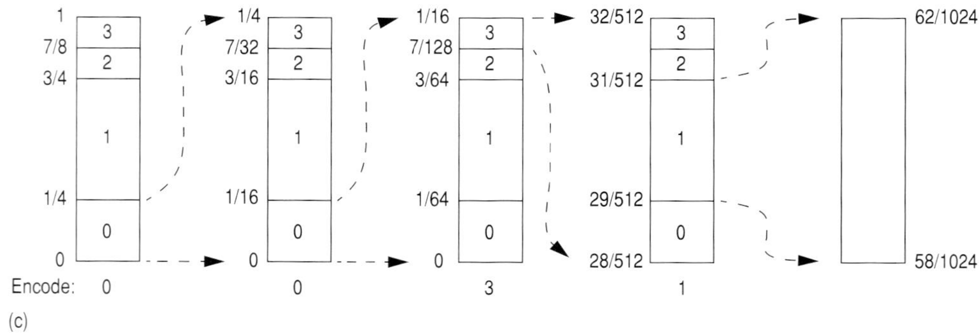

FIGURE 10.2.6
Arithmetic coding. (a) Histogram. (b) Probability table. (c) Coding process for "0, 0, 3, 1".

10.3 Lossy Compression Methods

In order to achieve high compression ratios with complex images, lossy compression methods are required. Lossy compression provides tradeoffs between image quality and degree of compression, which allows the compression algorithm to be customized to the application. With some of the more advanced methods, images can be compressed 10 to 20 times with virtually no visible information loss, and 30 to 50 times with minimal degradation (see Figure 10.3.1). Newer techniques, such as JPEG2000, can achieve reasonably good image quality with compression ratios as high as 100 to 200 (see figure 10.3.24). Image enhancement and restoration techniques can be combined with lossy compression schemes to improve the appearance of the decompressed image.

The lossy compression methods discussed are representative of the available tools for compression algorithm development and provide a wide variety of compression ratios and image quality. Many of the methods have adjustable parameters to allow the user to select the desired compression ratio and image fidelity. In general, a higher

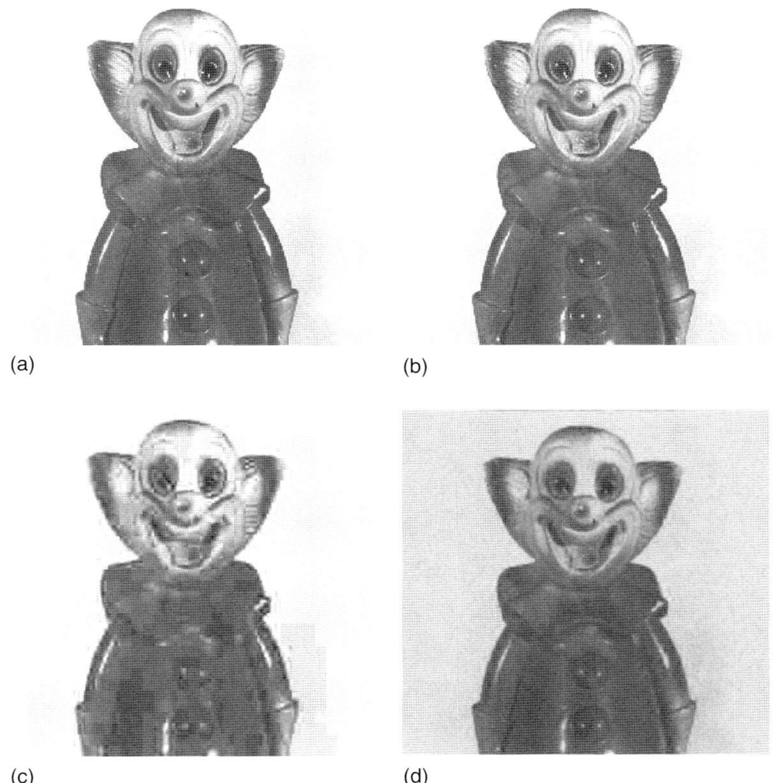

FIGURE 10.3.1
Lossy image compression. (a) Original image. (b) JPEG compression, 10:1 ratio. (c) JPEG compression, 48:1 ratio. (d) Wavelet/vector quantization compression, 36:1 ratio.

compression ratio results in a poorer image, but the results are highly image dependent. A technique that works well for one application may not be suitable for another.

Lossy compression is performed in both the spatial and transform domains. We will explore methods that utilize each of these domains, and some that use both. In the spatial domain we will discuss gray-level run length coding (GLRLC), block truncation coding (BTC), vector quantization (VQ), differential predictive coding (DPC), and fractal-based methods. In the transform domain we will discuss filtering, zonal coding, threshold coding, and the JPEG (Joint Photographer's Expert Group) algorithms. We will also look at techniques for combining these methods into hybrid compression algorithms which use both the spatial and transform domains.

10.3.1 Gray-Level Run-Length Coding

In Section 10.2 on lossless compression we discussed methods of extending basic run-length coding (RLC) to gray level images, by using bit-plane coding. The RLC technique can also be used for lossy image compression, by reducing the number of gray levels, and then applying standard RLC techniques. As with the lossless techniques, preprocessing by Gray code mapping will improve the compression ratio. Figure 10.3.2 shows results with this method and also lists the compression ratio with and without Gray code preprocessing.

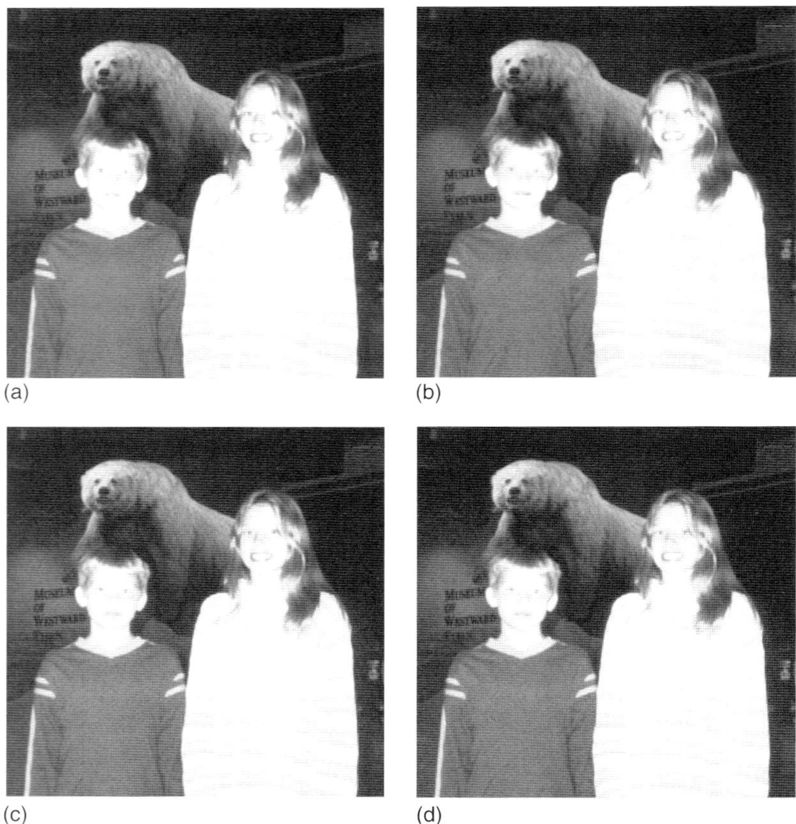

FIGURE 10.3.2
Lossy bitplane run length coding. Note that no compression occurs until reduction of 5 bits/pixel. (a) Original image, 8 bits/pixel, 256 gray levels. (b) Image after reduction to 7 bits/pixel, 128 gray levels, compression ratio 0.55, with Gray Code preprocessing 0.66. (c) Image after reduction to 6 bits/pixel, 64 gray levels, compression ratio 0.77, with Gray code preprocessing 0.97. (d) Image after reduction to 5 bits/pixel, 32 gray levels, compression ratio 1.20, with Gray code preprocessing 1.60.

A more sophisticated RLC algorithm for encoding gray level images is called the *dynamic window-based RLC*. This algorithm relaxes the criterion of the runs being the same value and allows for the runs to fall within a gray level range, called the *dynamic window range*. This range is dynamic because it starts out larger than the actual gray level window range, and maximum and minimum values are narrowed down to the actual range as each pixel value is encountered. This process continues until a pixel is found out of the actual range. The image is encoded with two values, one for the run length and one to approximate the gray level value of the run. This approximation can simply be the average of all the gray level values in the run, or a more complex method may be used to calculate the representative value.

Example 10.3.1

Given the following pixel values in sequence:

65 67 66 64 63 68 70

and a window range of 5.

FIGURE 10.3.2 (Continued)
Lossy bitplane run length coding. Note that no compression occurs until reduction of 5 bits/pixel. (e) Image after reduction to 4 bits/pixel, 16 gray levels, compression ratio 2.17, with Gray code preprocessing 2.79. (f) Image after reduction to 3 bits/pixel, 8 gray levels, compression ratio 4.86, with Gray code preprocessing 5.82. (g) Image after reduction to 2 bits/pixel, 4 gray levels, compression ratio 13.18, with Gray code preprocessing 15.44. (h) Image after reduction to 1 bit/pixel, 2 gray levels, compression ration 44.46, with Gray code preprocessing 44.46.

The first value is called the reference value (in this case = 65). A dynamic window range is then defined that has:

$$\text{MINIMUM} = \text{ reference} - (\text{ window length} - 1)$$

and

$$\text{MAXIMUM} = \text{ references} + (\text{ windowlength} - 1)$$

In this case the dynamic window is: $[65 - (5 - 1)]$ to $[65 + (5 - 1]) = 61$ to 69.
The next value encountered, 67, is used to adjust this range. The range based on this value alone is from 63 to 71. The new dynamic range is based on the intersection of the range from this new value with the previous range, so the new range is 63 to 69. This process continues until the value of 68 is encountered. At this point the range has been narrowed down to 63 to 67, so the 68 is out of range. This run is then encoded as:

$$\text{RUN LENGTH} = 5$$
$$\text{GRAY LEVEL} = (65 + 67 + 66 + 64 + 63)/5 = 65$$

In Figure 10.3.3 are results of the dynamic window-based RLC, where the average was used as the representative value. This particular algorithm also uses some preprocessing to allow for the run-length mapping to be coded so that a run can be any length and is not constrained by the length of a row (see reference for details).

10.3.2 Block Truncation Coding

Block truncation coding (BTC) works by dividing the image into small subimages and then reducing the number of gray levels within each block. This reduction is performed by a quantizer that adapts to the local image statistics. The levels for the quantizer are chosen to minimize a specified error criterion, and then all the pixel values within each block are mapped to the quantized levels. The necessary information to decompress the image is then encoded and stored. Many different BTC algorithms have been defined by using various types of quantization and error criteria, as well as various preprocessing and postprocessing methods. The more sophisticated algorithms provide better results, but with a corresponding increase in computational complexity.

The basic form of BTC divides the image into $n \times n$ blocks and codes each block using a two-level quantizer. The two levels are selected so that the mean and variance of the gray levels within the block are preserved. Each pixel value within the block is then compared with a threshold, typically the block mean, and then is assigned to one of the two levels.

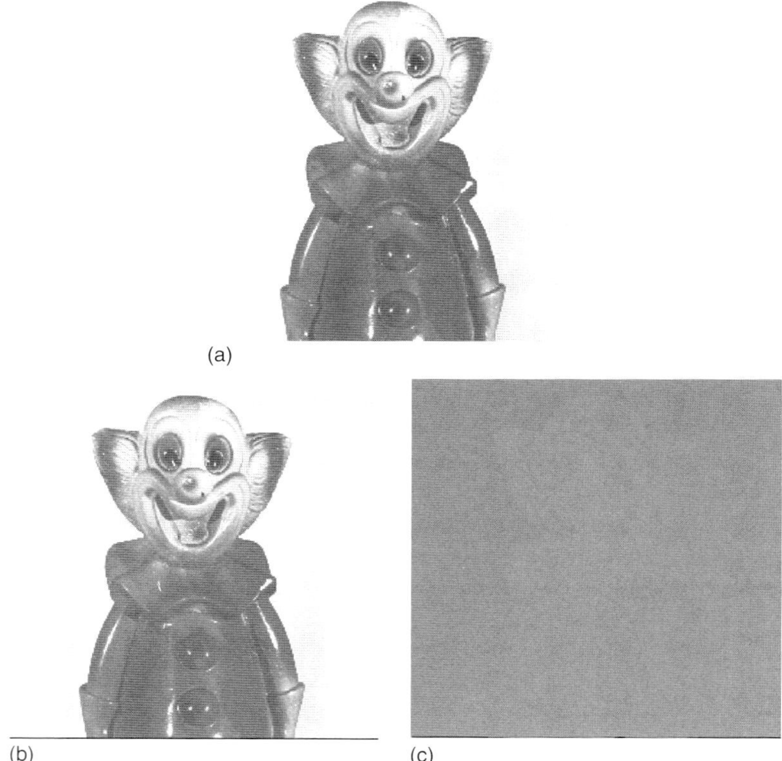

FIGURE 10.3.3
Dynamic window range RLC. (a) Original image. (b) Window length = 10, compression = 4 : 1. (c) Error image of (b), multiplied to show detail.

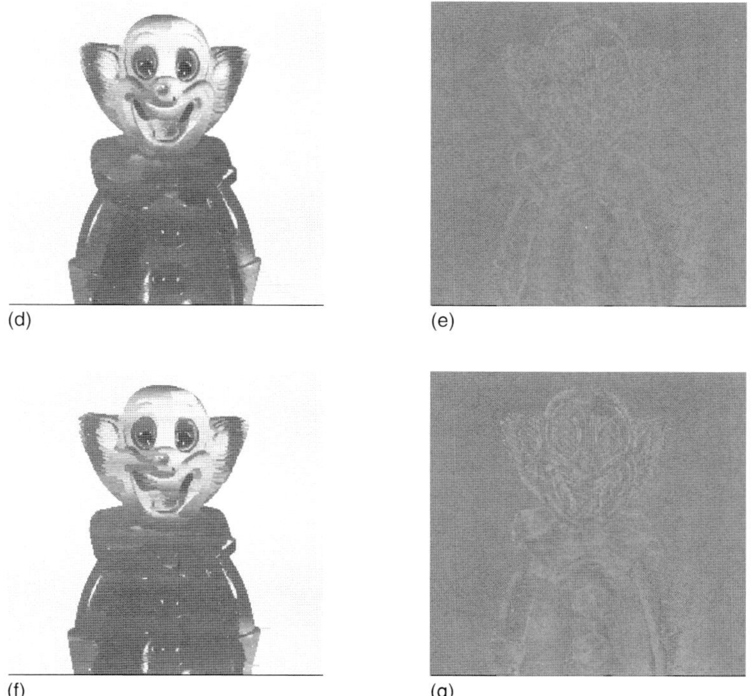

FIGURE 10.3.3 (Continued)
Dynamic window range RLC. (d) Window length = 35, compression = 8:1. (e) Error image of (d), multiplied to show detail. (f) Window length = 60, compression = 12.6:1. (g) Error image of (f), multiplied to show detail.

If it is above the mean it is assigned the high level code, if it is below the mean, it is assigned the low level code. If we call the high value H and the low value L, we can find these values via the following equations:

$$H = m_b + \sigma_b \sqrt{\frac{n^2 - q}{q}}$$

$$L = m_b - \sigma_b \sqrt{\frac{q}{n^2 - q}}$$

where the block size is $n \times n$

$$b = \text{the current block}$$

$$m_b = \text{the block mean} = \frac{1}{n^2} \sum_{I(r,c) \in b} I(r,c)$$

$$\sigma_b = \text{the block variance} = \sqrt{\frac{1}{n^2} \sum_{I(r,c) \in b} [I(r,c)]^2 - m_b^2}$$

$$q = \text{the number of values in the block} \geq m_b$$

Letting $n = 4$, we can do the following analysis: After the H and L values are found, the 4×4 block is encoded with four bytes: two bytes to store the two levels, H and L, and two

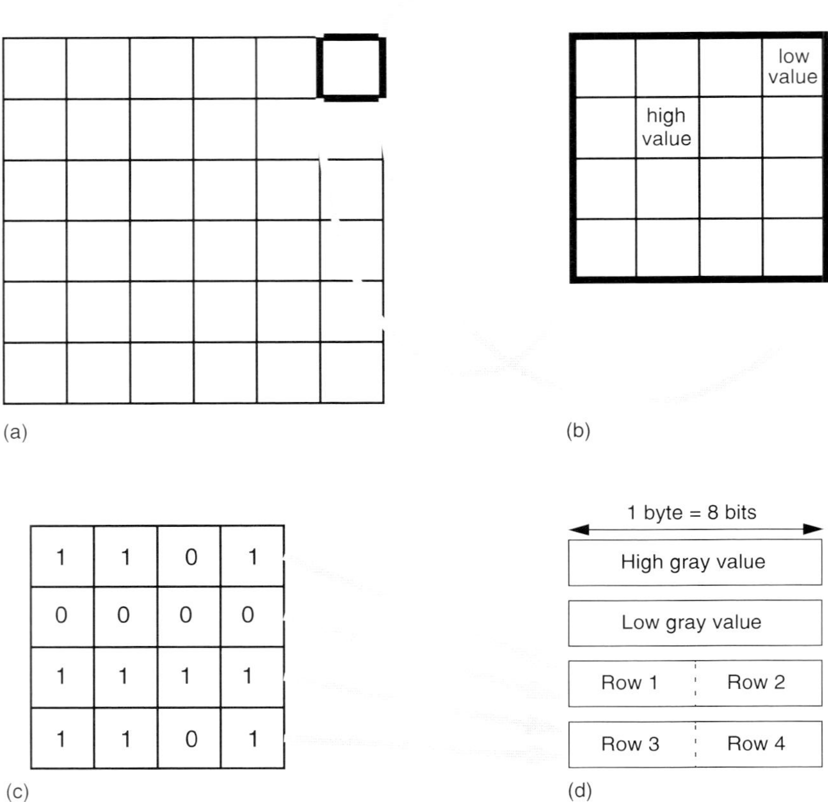

FIGURE 10.3.4
Basic block truncation coding. (a) Divide image into 4×4 blocks. (b) Find high and low values for blocks. (c) Assign a "0" to each pixel less than the mean, "1" to each pixel greater than the mean. (d) Encode 4×4 block with 4 bytes.

bytes to store a bit string of 1's and 0's corresponding to the high and low codes for that particular block. This is illustrated in Figure 10.3.4, where we see the bit string for the 4×4 block is packed into two bytes. Since the original 4×4 subimage has 16 bytes, and the resulting code has 4 bytes (two for the high and low values, and two for the bit string), this provides a 16:4 or 4:1 compression.

The following example illustrates finding the values for a specific block using basic BTC.

Example 10.3.2

Given the following 4×4 subimage, apply basic BTC and find the resulting values.

$$\begin{bmatrix} 12 & 16 & 15 & 17 \\ 13 & 16 & 17 & 17 \\ 4 & 4 & 35 & 35 \\ 42 & 42 & 12 & 12 \end{bmatrix}$$

$$m_b = \frac{1}{n^2}\sum_{I(r,c)\in b} I(r,c) = \frac{1}{16}[12+16+15+17+13+16+17+17+4+4+35$$
$$+35+42+42+12+12]$$
$$= 19.3125$$

$$\sigma_b = \sqrt{\frac{1}{n^2}\sum_{I(r,c)\in b}[I(r,c)]^2 - m_b^2}$$
$$= \sqrt{\frac{1}{16}[12^2+16^2+15^2+17^2+13^2+16^2+17^2+17^2+4^2+4^2+35^2+35^2+42^2+42^2+12^2+12^2]-(19.3125)^2}$$
$$\approx 11.85$$

There are 4 pixel values greater than the mean, so $q = 4$.

$$H = m_b + \sigma_b\sqrt{\frac{n^2-q}{q}} = 19.3125 + 11.85\sqrt{\frac{16-4}{4}} \approx 40$$

$$L = m_b - \sigma_b\sqrt{\frac{q}{n^2-q}} = 19.3125 - 11.85\sqrt{\frac{4}{16-4}} \approx 13$$

Now, find the bit string by using 0 for values less than the mean and 1 for values greater than the mean:

$$\begin{bmatrix} 12 & 16 & 15 & 17 \\ 13 & 16 & 17 & 17 \\ 4 & 4 & 35 & 35 \\ 42 & 42 & 12 & 12 \end{bmatrix} \Rightarrow \begin{bmatrix} 0 & 0 & 0 & 0 \\ 0 & 0 & 0 & 0 \\ 0 & 0 & 1 & 1 \\ 1 & 1 & 0 & 0 \end{bmatrix} \Rightarrow (0000000000111100_2)$$

The high value, H, and the low value, L, will be stored along with the bit string. The subimage, when decompressed will be:

$$\begin{bmatrix} 13 & 13 & 13 & 13 \\ 13 & 13 & 13 & 13 \\ 13 & 13 & 40 & 40 \\ 40 & 40 & 13 & 13 \end{bmatrix}$$

Although the results of this algorithm are image dependent, it tends to produce images with blocky effects as shown in Figure 10.3.5b. These artifacts can be smoothed by applying enhancement techniques such as median and average (lowpass) filters (Figure 10.3.5d,e).

More advanced BTC algorithms can be explored in the references, and Figure 10.3.6 illustrates the multilevel BTC algorithm which uses a four-level quantizer. This algorithm allows for varying the block size, and a larger block size should provide higher compression, but with a corresponding decrease in image quality. With this particular implementation, we get the decreasing image quality, but the compression ratio is fixed.

(a)

(b)

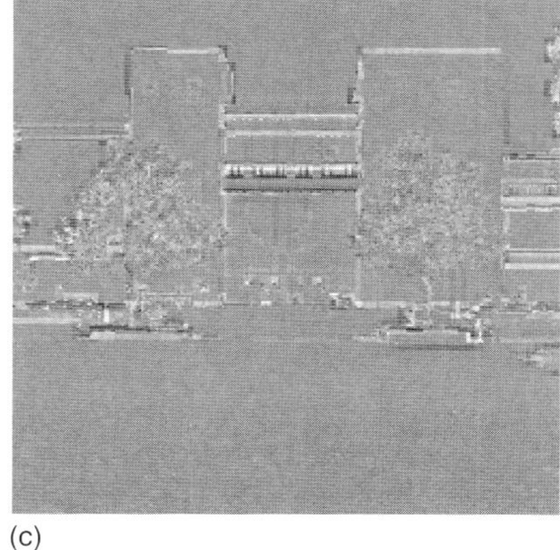

(c)

(d)

(e)

FIGURE 10.3.5
Block truncation coding (BTC). (a) Original image. (b) Block truncation coded image, compression = 4 : 1. (c) Error image of (b), histogram stretched to show detail. (d) Image (b) post-processed with a 3×3 median filter. (e) Image (b) post-processed with a 3×3 averaging filter.

10.3.3 Vector Quantization

Vector quantization is the process of mapping a vector that can have many values to a vector that has a smaller (quantized) number of values. For image compression, the vector corresponds to a small subimage, or block.

Example 10.3.3

Given the following 4 × 4 subimage:

$$\begin{bmatrix} 65 & 70 & 71 & 75 \\ 71 & 70 & 71 & 81 \\ 81 & 80 & 81 & 82 \\ 90 & 90 & 91 & 92 \end{bmatrix}$$

This can be re-arranged into a 1-D vector by putting the rows adjacent as follows:

[row1 row2 row3 row4] = [65 70 71 75 71 70 71 81 81 80 81 82 90 90 91 92]

(a)

(b)

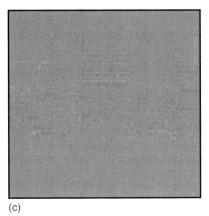

(c)

FIGURE 10.3.6
Multilevel block truncation coding. (a) Original image. (b) Multilevel BTC, block size = 8 × 8, compression = 3.99 : 1. (c) Error image of (b), multiplied to show detail.

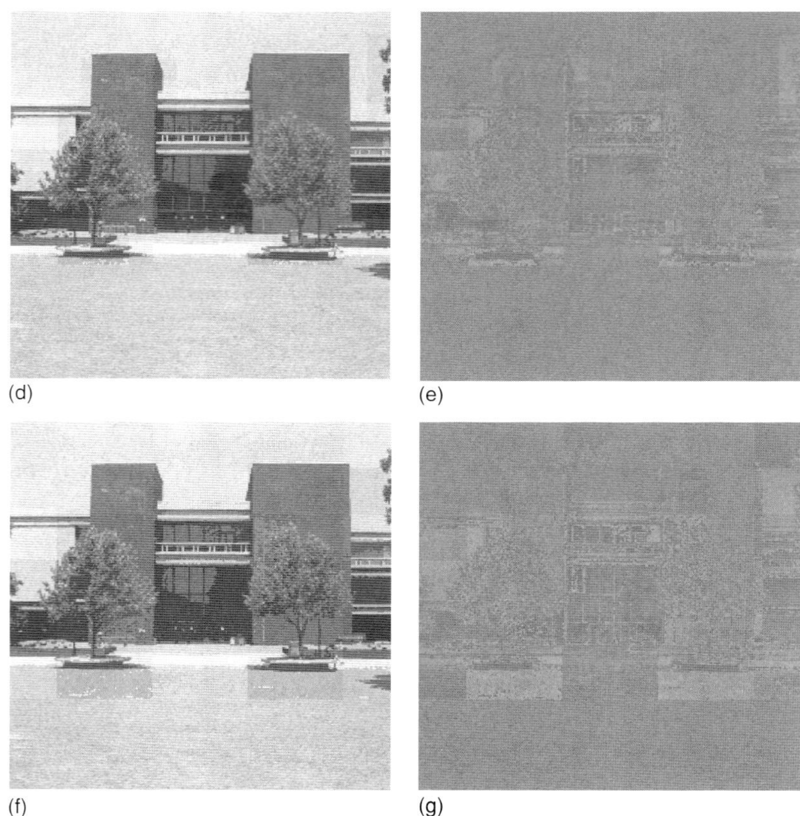

FIGURE 10.3.6 (Continued)
Multilevel block truncation coding. (d) Multilevel BTC, block size = 16 × 16, compression = 3.99 : 1. (e) Error image of (d) multiplied to show detail. (f) Multilevel BTC, block size = 32 × 32, compression = 3.99 : 1. (g) Error image of (f) multiplied to show detail.

The previous types of quantization have to do with taking a single value and reducing the number of bits used to represent that value—this is called *scalar quantization*, and is most easily achieved by rounding or truncation. Information theory (Shannon's rate distortion theory) tells us that better compression can always be achieved by vector quantization than with scalar quantization. Vector quantization treats the entire subimage (vector) as a single entity and quantizes it by reducing the total number of bits required to represent the subimage. This is done by utilizing a *codebook*, which stores a fixed set of vectors, and then coding the subimage by using the index (address) into the codebook.

Example 10.3.4

Given an 8-bit, 256 × 256 image, we devise a vector quantization scheme that will encode each 4 × 4 block with one of the vectors in a codebook of 256 entries. We determine that we want to encode a specific subimage with vector number 122 in the codebook. For this subimage we then store the number 122 as the index into the codebook. Then when the image is decompressed, the vector at the 122 address in the codebook, is used for that particular subimage. This is illustrated in Figure 10.3.7. This will require 1 byte (8-bits) to be stored for each 4 × 4 block, providing a data reduction of 16 bytes for a 4 × 4 block to 1 byte, or 16 : 1.

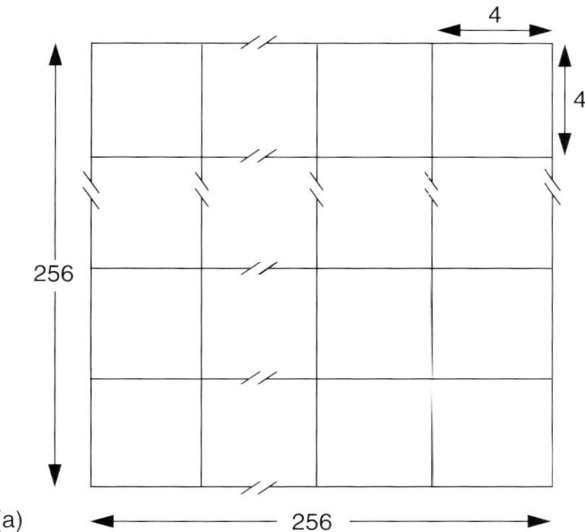

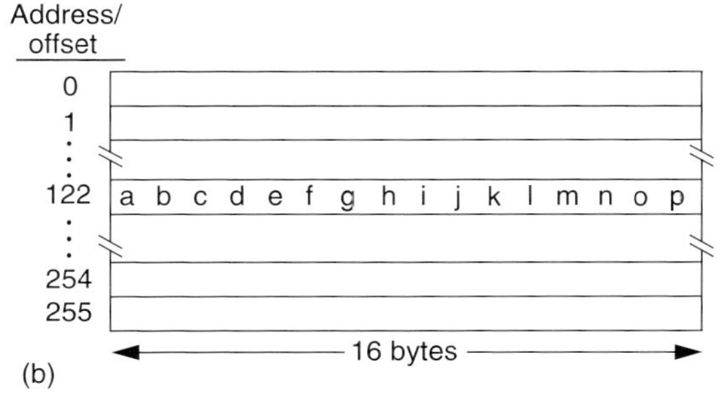

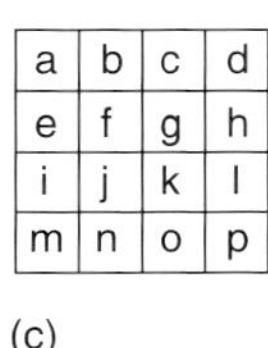

FIGURE 10.3.7
Quantizing with a codebook. (a) Original 256×256 image divided into 4×4 blocks. (b) Codebook with 256 16-byte entries. (c) A subimage decompressed with vector #122.

In the example we achieved a 16:1 compression, but note that this assumes that the codebook is not stored with the compressed file. However, the codebook will need to be stored unless a generic codebook is devised which could be used for a particular type of image—then we need only store the name of that particular codebook file. In the general case, better results will be obtained with a codebook that is designed for a particular image.

Example 10.3.5

If we include the codebook in the compressed file from the previous example, the compression ratio will not be quite as good. For every 4×4 block we will have 1 byte. This gives us:

$$\left(\frac{256 \text{ pixels}}{4 \text{ pixels/block}}\right)\left(\frac{256 \text{ pixels}}{4 \text{ pixels/block}}\right) = 4096 \text{ blocks}$$

At 1 byte for each 4×4 block, this give us 4096 bytes for the codebook addresses. Now we also include the size of the codebook, 256×16:

$$4096 + (256)(16) = 8192 \text{ bytes for the coded file}$$

The original 8-bit, 256×256 image contained:

$$(256)(256) = 65,536 \text{ bytes}$$

Thus, we obtain a compression of:

$$\frac{65,536}{8192} = 8 \rightarrow 8:1 \text{ compression}$$

In this case, including the codebook cut the compression in half, from 16:1 to 8:1.

Now, how do we decide which vectors will be stored in the codebook? This is typically done by a training algorithm that finds a set of vectors that best represent the blocks in the image. This set of vectors is determined by optimizing some error criterion, where the error is defined as the sum of the vector distances between the original subimages and the resulting decompressed subimages. The standard method is to use the Linde–Buzo–Gray (LBG) algorithm; one implementation of the LBG algorithm for codebook generation is as follows:

Step 1. Given an arbitrary codebook, encode each input vector according to the nearest-neighbor criterion. Use a distance metric to compare all the input vectors to the encoded vectors, and then sum these errors (distances) to provide a distortion measure. If the distortion is small enough (less than a predefined threshold), quit. If not, go to Step 2.

Step 2. For each codebook entry, compute the euclidean centroid of all the input vectors encoded into that specific codebook vector.

Step 3. Use the computed centroids as the new codebook, and go to Step 1.

The LBG algorithm, also called the K-means or the clustering algorithm, along with other iterative codebook design algorithms do not, in general, yield globally optimum codes. These algorithms will converge to a local minimum in the error (distortion) space. Theoretically, to improve the codebook, the algorithm is repeated with different initial random codebooks and the one codebook that minimizes distortion is chosen. However, the LBG algorithm will typically yield "good" codes if the initial codebook is carefully chosen. One simple method to find a good initial codebook is to subdivide the vector space and find the centroid for the sample vectors within each division. These centroids are then used as the initial codebook. Alternately, a subset of the training vectors, preferably spread across the vector space, can be randomly selected and used to initialize the codebook.

The primary advantage of vector quantization is simple and fast decompression, but with the high cost of complex compression. The compression process consists of generating the codebook, which can be computationally expensive, especially if careful attention is given to creating an optimal codebook. The decompression process requires the use of the codebook to recreate the image, which is easily implemented with a look-up table (LUT). This type of compression is useful for applications where the images are compressed once and decompressed many times, such as images on an Internet site. Real-time applications, such as video conferencing, need a compression scheme that is fast for both compression and decompression.

Vector quantization can be applied in both the spatial and spectral domains. In Figure 10.3.8 we see VQ applied in the spatial domain using 4×4 subimages, and varying the codebook size. As the codebook size is increased the image quality improves and

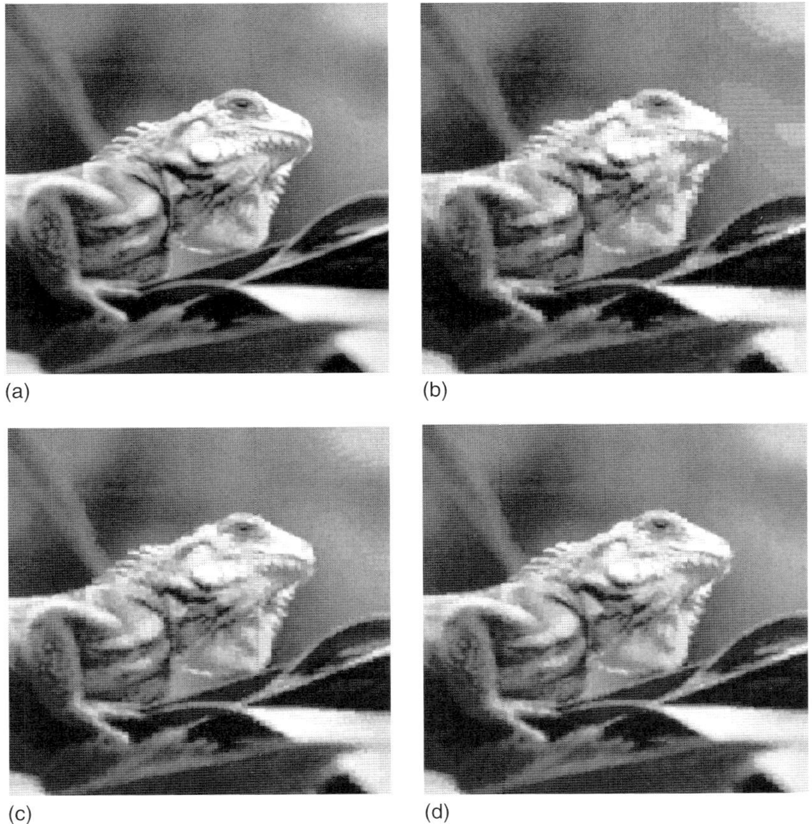

FIGURE 10.3.8
Vector quantization in the spatial domain. (a) Original image. (b) VQ with 4×4 vectors, and a codebook of 128 entries, compression ration = 11.49. (c) VQ with 4×4 vectors, and a codebook of 256 entries, compression ration = 7.93. (d) VQ with 4×4 vectors, and a codebook of 512 entries, compression ration = 5.09.

the compression ratio decreases. Figure 10.3.9 shows results from using VQ in the transform domain, which is explored more in Sections 10.3.6 and 10.3.7.

10.3.4 Differential Predictive Coding

Differential predictive coding works by predicting the next pixel value based on the previous values, and encoding the difference between the predicted value and the actual value (for analog signals, this is also called Differential Pulse Code Modulation or DPCM). This technique takes advantage of the fact that adjacent pixels are highly correlated (gray level values are close), except at object boundaries. This correlation makes it easy to predict the next pixel value based on previous pixel values, and we need only encode the difference between the estimate and the actual value. Typically the difference, or error, will be small which minimizes the number of bits required for compressed file. This error is then quantized, to further reduce the data and to optimize visual results, and can then be coded.

A block diagram of this process is shown in Figure 10.3.10, where we can see the predictor must be in the feedback loop so that it matches the decompression system. The system must be initialized by retaining the first value(s) without any

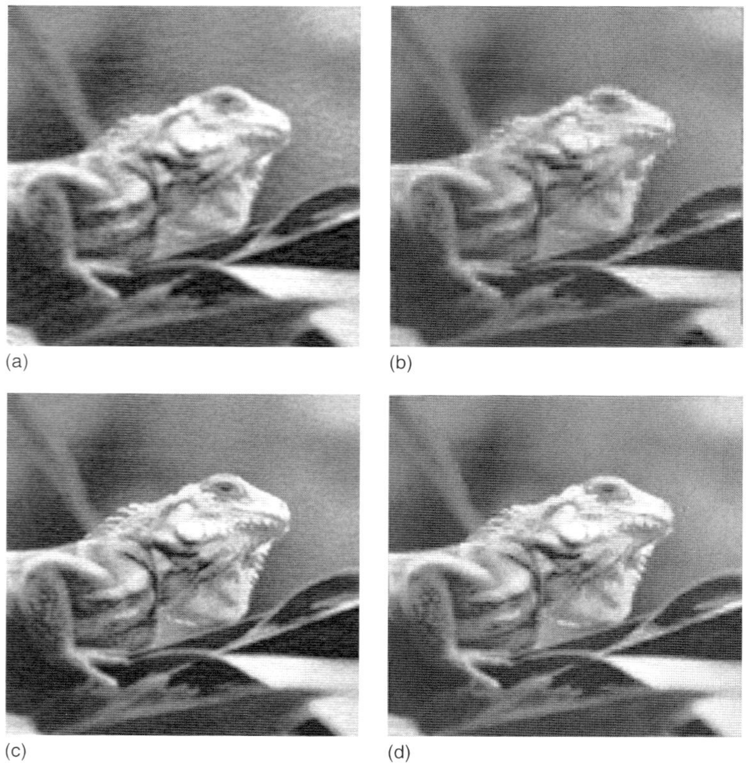

FIGURE 10.3.9
Vector quantization in the transform domain. The original image is the image in Figure 10.3.8a. (a) VQ with the discrete cosine transform, compression = 9.21. (b) VQ with the wavelet transform, compression ration = 9.21. (c) VQ with the discrete cosine transform, compression ratio = 3.44. (d) VQ with the wavelet transform, compression ration = 3.44.

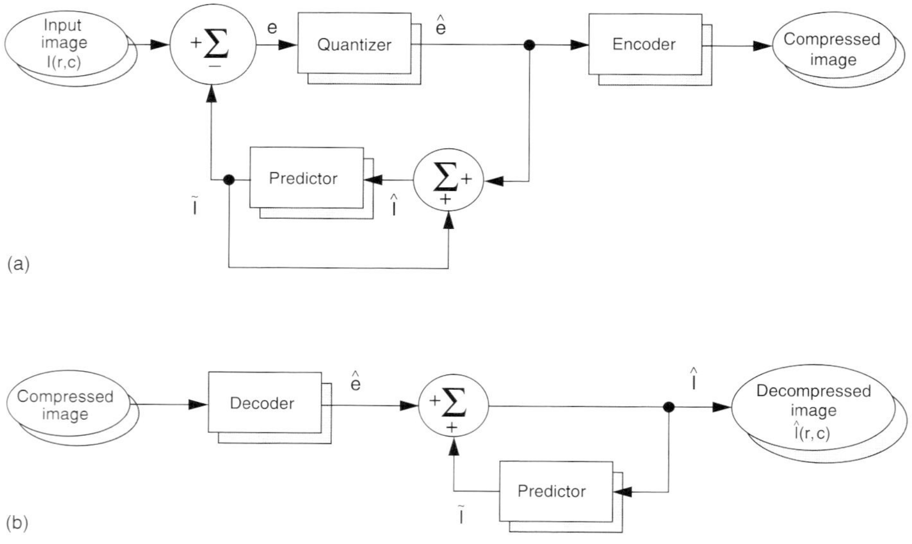

FIGURE 10.3.10
Differential predictive coding (DPC). (a) Compression. (b) Decompression.

compression, in order to calculate the first prediction. From the block diagram, we have the following:

$$\tilde{I} = \text{the predicted next pixel value}$$
$$\hat{I} = \text{the reconstructed pixel velue}$$
$$e = I - \tilde{I} = \text{error}$$
$$\hat{e} = \hat{I} - \tilde{I} = \text{quantized error}$$

The prediction equation is typically a function of the previous pixel(s), and can also include global or application-specific information.

The theoretically optimum predictor that uses only the previous value is based on minimizing mean squared error between the original and the decompressed image, and is given by:

$$\tilde{I}(r, c+1) = \rho\hat{I}(r, c) + (1 - \rho)\bar{I}(r, c)$$
$$\text{where: } \bar{I}(r, c) = \text{the average value for the image}$$
$$\rho = \text{the normalized correlation between pixel values}$$

For most images ρ is between 0.85 and 0.95. Once the next pixel value has been predicted, the error is calculated:

$$e(r, c+1) = I(r, c+1) - \tilde{I}(r, c+1)$$

This error signal is then quantized, such that:

$$\hat{e}(r, c+1\ ) = \tilde{I}(r, c+1) - \tilde{I}(r, c+1)$$

This quantized error can then be encoded using a lossless encoder, such as a Huffman coder. It should be noted that it is important that the predictor uses the same values during both compression and decompression; specifically the reconstructed values and not the original values (see Figure 10.3.10). In Figure 10.3.11 we see the results from using the original image values in the prediction, compared to using the reconstructed (decompressed) pixel values in the predictor. With these examples the quantization used was simply truncation ("clipping").

The prediction equation can be one-dimensional or two-dimensional, that is, it can be based on previous values in the current row only, or on previous rows also (see Figure 10.3.12). The following prediction equations are typical examples of those used in practice, with the first being one-dimensional and the next two being two-dimensional:

$$\tilde{I}(r, c+1) = 0.97\hat{I}(r, c)$$
$$\tilde{I}(r, c+1) = 0.49\hat{I}(r, c) + 0.49\hat{I}(r-1, c+1)$$
$$\tilde{I}(r, c+1) = 0.74\hat{I}(r, c) + 0.74\hat{I}(r-1, c+1) - 0.49\hat{I}(r-1, c)$$

Using more of the previous values in the predictor increases the complexity of the computations for both compression and decompression, and it has been determined that using more than three of the previous values provides no significant improvement in the resulting image.

The results of DPC can be improved by using an optimal quantizer, such as the Lloyd–Max quantizer, instead of simply truncating the resulting error. The Lloyd–Max quantizer

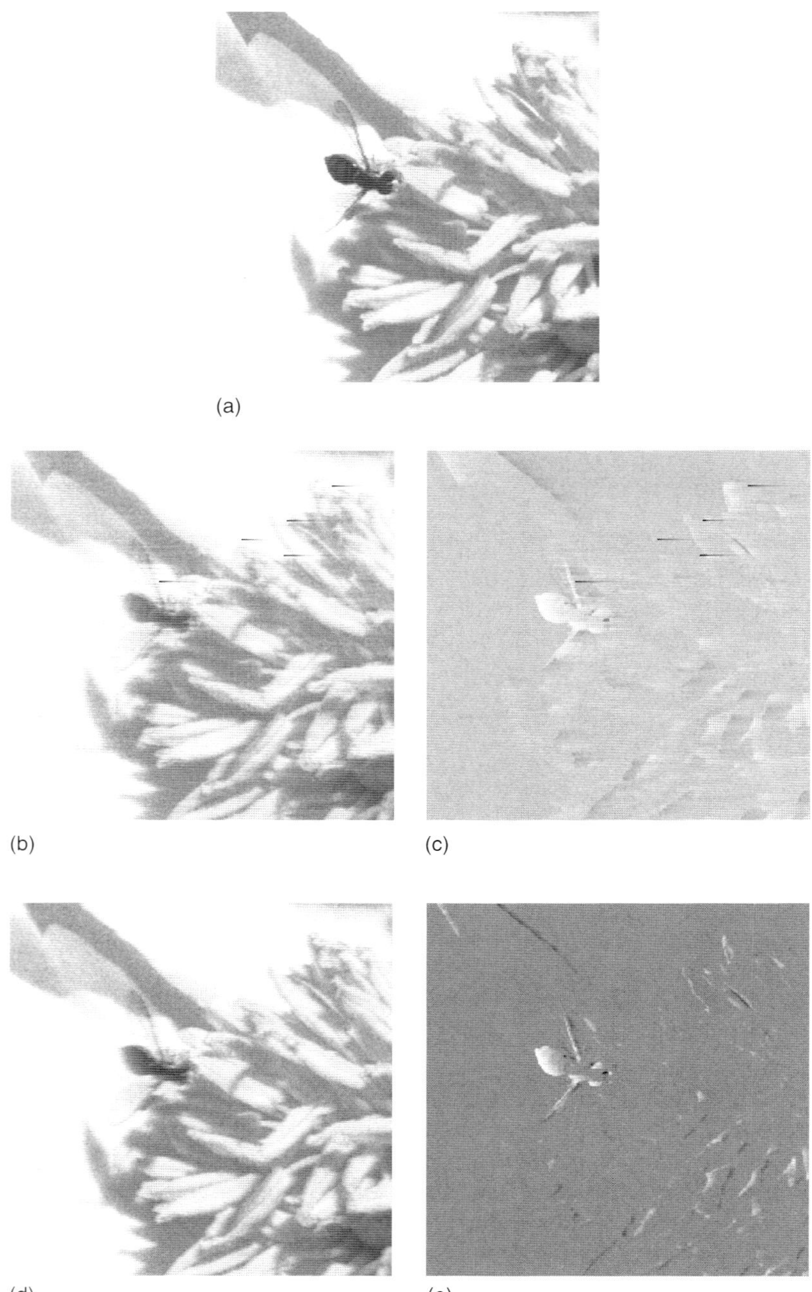

FIGURE 10.3.11
DPC example. (a) Original image. (b) DPC using original values in predictor, clipping to the maximum, 5 bits/pixel, normalized correlation 0.90. (c) Error image of (b), multiplied to show detail. (d) DPC using reconstructed values in predictor, clipping to the maximum, 5 bits/pixel, normalized correlation 0.90. (e) Error image of (d) multiplied to show detail.

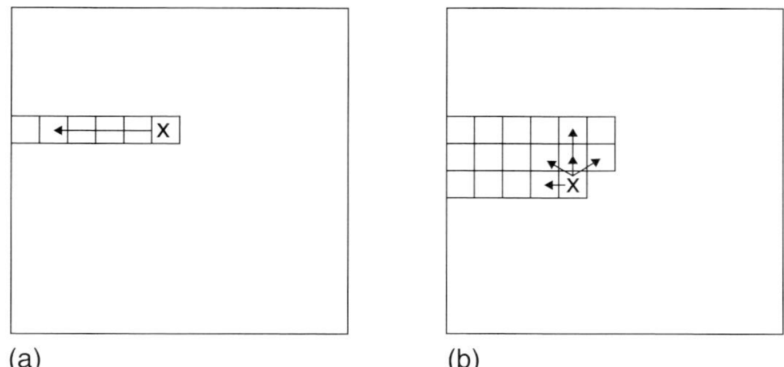

FIGURE 10.3.12
DPC predictor dimensions. (a) One-dimensional predictor, based on current row only x = current pixel. (b) Two-dimensional predictor is based on current and previous row or rows.

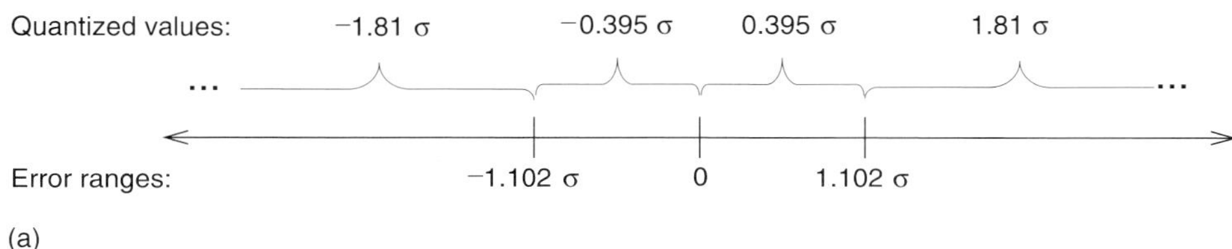

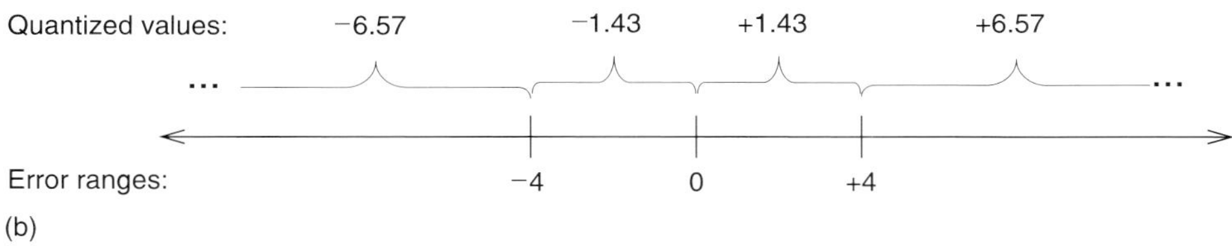

FIGURE 10.3.13
Lloyd–Max quantizer. (a) 2 bit Lloyd–Max quantizer with Laplacian error distribution. (b) An example for $\sigma = 3.63$. The ± 1.43 is typically rounded to ± 1, and the ± 6.57 is rounded to ± 7.

assumes a specific distribution for the prediction error. Assuming a 2-bit code for the error, and a Laplacian distribution for the error, the Lloyd-Max quantizer is defined as follows (see Figure 10.3.13):

Error Range		Quantized Value
$0 \leq e < 1.102\sigma$	$\rightarrow$	$+0.395\sigma$
$1.102\sigma \leq e < \infty$	$\rightarrow$	$+1.81\sigma$
$-1.102\sigma \leq e < 0$	$\rightarrow$	-0.395σ
$-1.102\sigma \leq e < -\infty$	$\rightarrow$	-1.81σ

where σ = the standard deviation of the the error distribution.

Tables for the coefficients for n-bit codes can be found in the references. For most images, the standard deviation, σ, for the error signal is between 3 and 15. After the data is quantized it can be further compressed with a lossless coder such as Huffman or

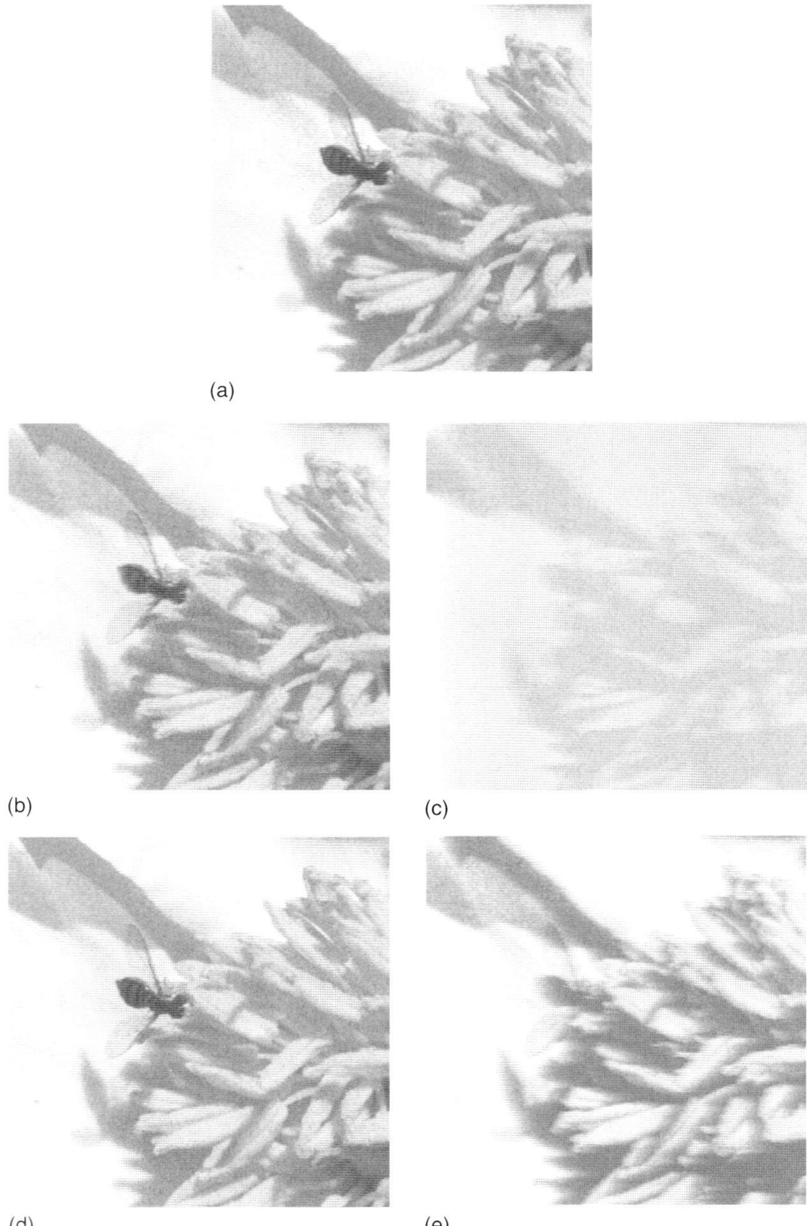

FIGURE 10.3.14
DPC quantization. (a) Original image. (b) Lloyd–Max quantizer, using 2 bits/pixel, normalized correlation = 0.90, with standard deviation = 10. (c) Truncation quantizer, using 2 bits/pixel, normalized correlation = 0.90. (d) Lloyd–Max quantizer, using 4 bits/pixel, normalized correlation = 0.90, with standard deviation = 10. (e) Truncation quantizer, using 4 bits/pixel, normalized correlation = 0.90.

arithmetic coding. In Figure 10.3.14 is a comparison of using the Lloyd–Max quantizer and truncation as a quantization method. Figure 10.3.15 shows the error images and decompressed images using different bit rates for DPC compression (with Lloyd–Max quantization and a one-dimensional predictor).

10.3.5 Model-Based and Fractal Compression

Model-based or *intelligent* compression works by finding models for objects within the image and using model parameters for the compressed file. The techniques used are similar to computer vision methods where the goal is to find descriptions of the objects in the image. The objects are often defined by lines or shapes (boundaries), so a Hough transform may be used, while the object interiors can be defined by statistical texture modeling. Methods have also been developed that use texture modeling in the wavelet domain. The model-based methods can achieve very high compression ratios, but the decompressed images often have an artificial look to them. Fractal methods are an example of model-based compression techniques.

Fractal image compression is based on the idea that if an image is divided into subimages, many of the subimages will be self-similar. This means that some parts of the image look like other parts of the image. More specifically, *self-similar* means that one subimage can be represented as a skewed, stretched, rotated, scaled, and/or translated version of another subimage. Treating the image as a geometric plane, these mathematical operations (skew, stretch, scale, rotate, translate) are called *affine transformations* and can be represented by the following general equations:

$$r' = k_1 r + k_2 c + k_3$$
$$c' = k_4 r + k_5 c + k_6$$

where r' and c' are the new cordinates, and k_i are constants.

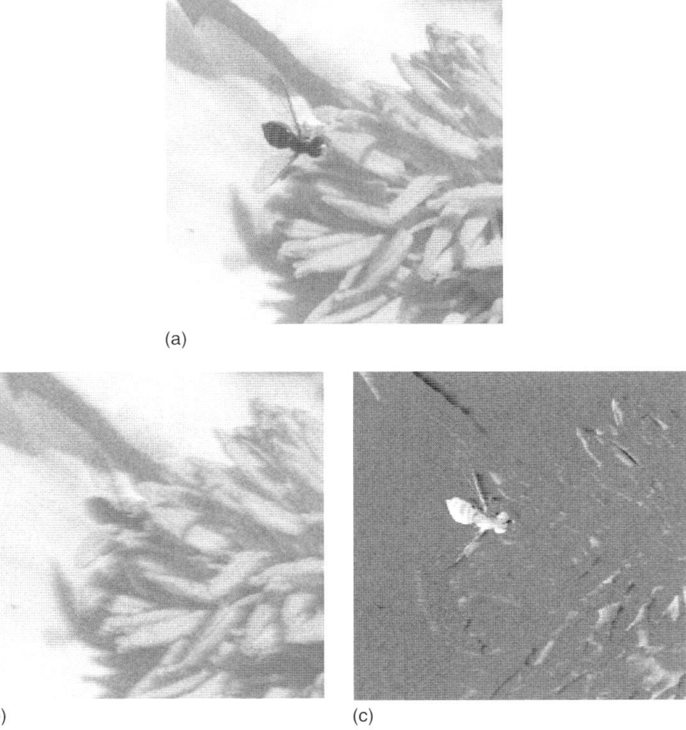

FIGURE 10.3.15
DPC quantization. (a) Original image. (b) Lloyd–Max quantizer, using 1 bit/pixel, normalized correlation = 0.90, with standard deviation = 10. (c) Error image for (b).

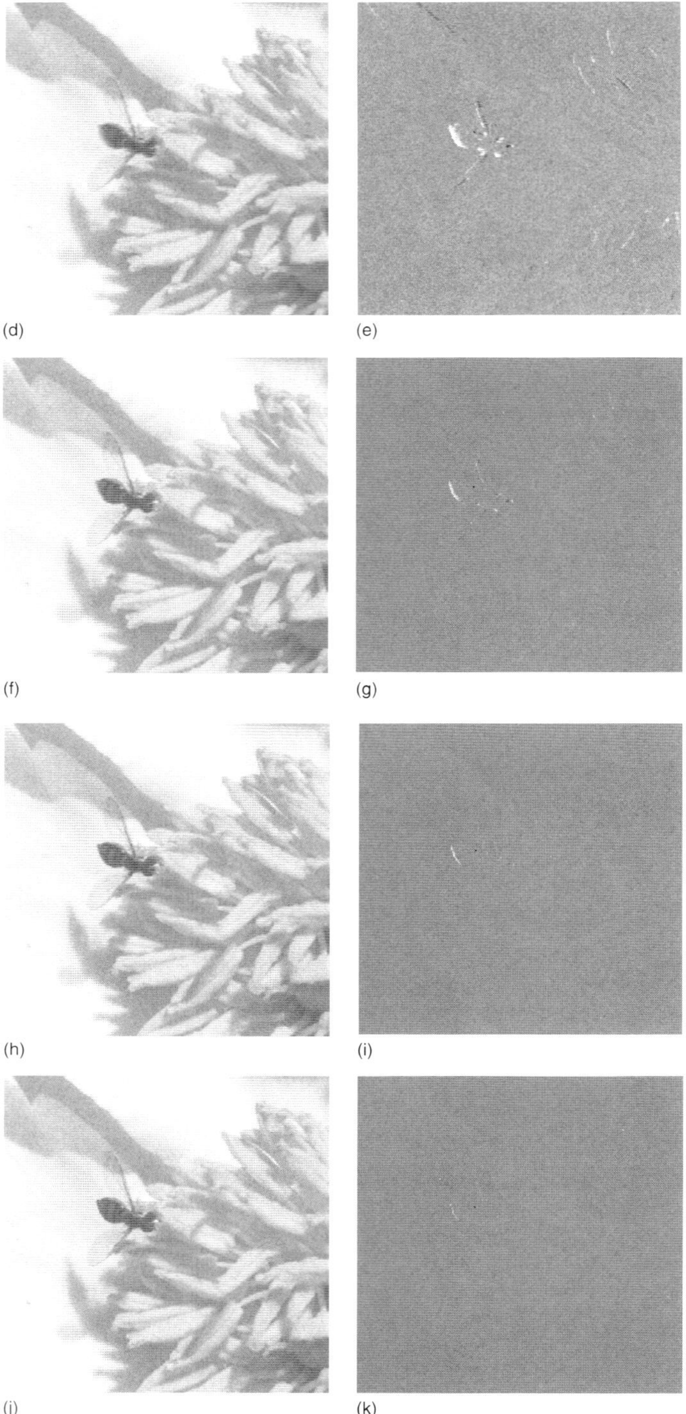

FIGURE 10.3.15 (Continued)
DPC quantization. (d) Lloyd–Max quantizer, using 2 bits/pixel, normalized correlation = 0.90, with standard deviation = 10. (e) Error image for (d). (f) Lloyd–Max quantizer, using 3 bits/pixel, normalized correlation = 0.90, with standard deviation = 10. (g) Error image for (f). (h) Lloyd–Max quantizer, using 4 bits/pixel, normalized correlation = 0.90, with standard deviation = 10. (i) Error image for (h). (j) Lloyd–Max quantizer, using 5 bits/pixel, normalized correlation = 0.90, with standard deviation = 10. (k) Error image for (j).

Fractal compression is somewhat like vector quantization, except that the subimages, or blocks, can vary in size and shape. The idea is to find a good set of basis images, or fractals, that can undergo affine transformations, and then be assembled into a good representation of the image. After this has been done we only need to store the fractals (basis images), and the necessary affine transformation coefficients in the compressed file.

Fractal compression can provide high quality images and very high compression rates, but often at a very high cost. The quality of the resulting decompressed image is directly related to the amount of time taken in generating the fractal compressed image. If the compression is done offline, one time, and the images are to be used many times, it may be worth the cost. For example, thousands of images have been compressed and stored with fractals in the popular Microsoft Encarta encyclopedia. Another advantage of fractals is that they can be magnified as much as is desired, so one fractal compressed image file can be used for any resolution or size of image.

To apply fractal compression, the image is first divided into non-overlapping regions that completely cover the image, called *domains*. Then, regions of various size and shape are chosen for the basis images, called the *range* regions. The range regions are typically larger than the domain regions, can be overlapping and do not cover the entire image. The goal is to find the set affine transformations to best match the range regions to the domain regions. The methods used to find the best range regions for the image, as well as the best transformations, are many and varied and can be explored in the references.

Figure 10.3.16 illustrates some examples of images that have been compressed with fractal techniques. The first three are examples of the familiar cameraman image at various compression ratios. The error images are shown with the compressed images, where we can see that most of the distortion occurs at object edges and areas of high image detail, as with most compression methods (note that the background gray in the error images represents zero—no error). If we look closely at Figure 10.3.16c,d we can see that the fractal compression scheme determined that the sky area directly above the tall tower looks similar to the camera tripod! In Figure 10.3.16g,h we show a checkerboard that has been compressed more than 500 times it original file size, with only minor errors at the edges.0

10.3.6 Transform Coding

Transform coding, is a form of block coding done in the transform domain. The image is divided into blocks, or subimages, and the transform is calculated for each block. Any of the previously defined transforms can be used, frequency (e.g., Fourier) or sequency (e.g., Walsh/Hadamard), but during development of the original JPEG algorithm it was determined that the discrete cosine transform (DCT) is optimal for most images. The newer JPEG2000 algorithms use the wavelet transform, which has been found to provide even better compression, and wavelet compression is explored in the next section.

After the transform has been calculated, the transform coefficients are quantized and coded. The primary reason this method is effective is because the frequency/sequency transform of images is very efficient at putting most of the information into relatively few coefficients, so many of the high frequency coefficients can be quantized to 0 (eliminated completely). This type of transform is really just a special type of mapping that uses spatial frequency concepts as a basis for the mapping. Remember that for image compression the main reason for mapping the original data into another mathematical

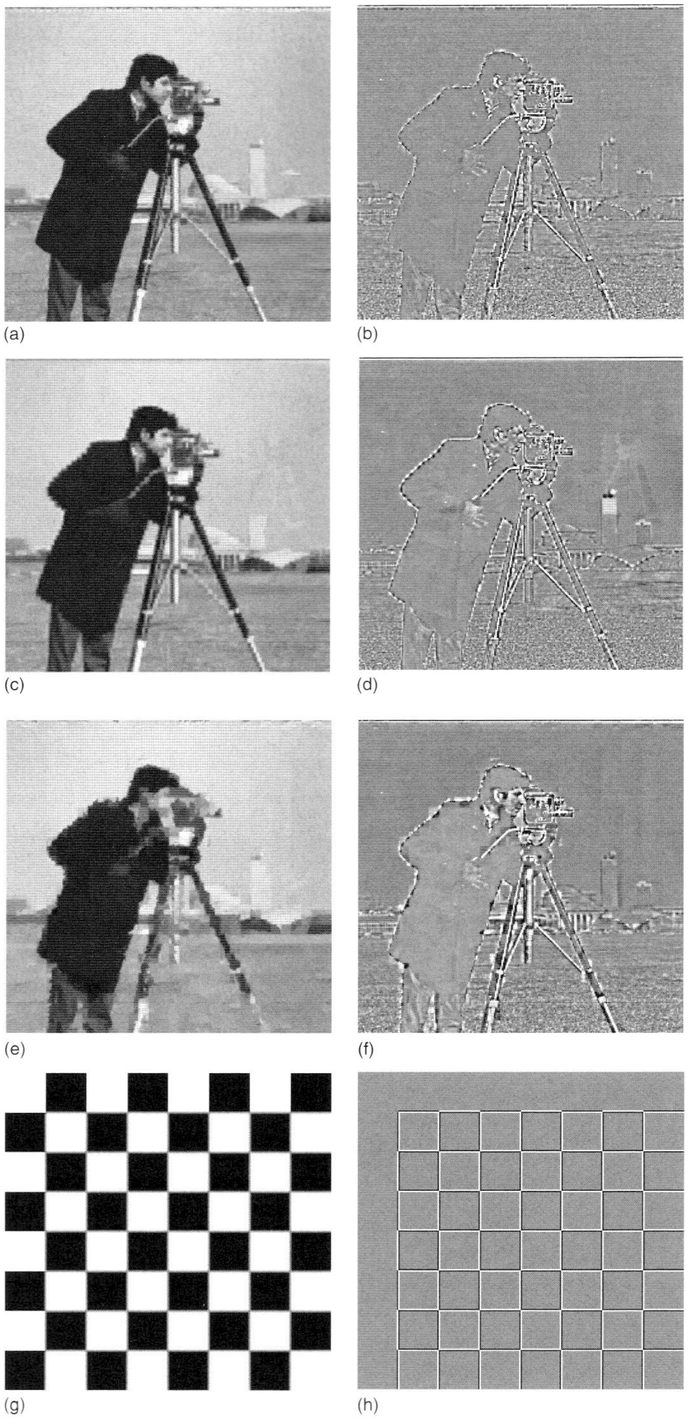

FIGURE 10.3.16
Fractal compression. (a) Our familiar cameraman image compressed with fractal encoding, compression ration = 9.19. (b) Error image for (a). (c) Compression ratio = 15.65. (d) Error image for (c). (e) Compression ratio = 34.06. (f) Error image for (e). (g) A checkerboard, compression ratio = 564.97. (h) Error image for (g). *Note*: Error images have been remapped for display so the background gray corresponds to zero, then they were enhanced by a histogram stretch to show detail.

space is to pack the information (or energy) into as few coefficients as possible. We have seen that with these transforms most of the energy in images in contained in the lower frequency terms.

The simplest form of transform coding is achieved by filtering—we can simply eliminate some of the high frequency coefficients. This alone will not provide much compression, since the transform data are typically floating point and thus 4 or 8 bytes per pixel (compared to the original pixel data at 1 byte per pixel), so quantization and coding is applied to the reduced data. One aspect of quantization includes a process called *bit allocation,* which determines the number of bits to be used to code each coefficient based on its importance. Typically, more bits are used for lower frequency components where the energy is concentrated for most images, resulting in a *variable bit rate* or *nonuniform quantization*. Using more bits provides more resolution—one bit has only two possible values, 8 bits have 256 possible values.

Example 10.3.6

We have decided to use transform coding with a DCT on an image by dividing it into 4×4 blocks. The selected bit allocation can be represented by the following mask:

$$\begin{bmatrix} 8 & 6 & 4 & 1 \\ 6 & 4 & 1 & 0 \\ 4 & 1 & 0 & 0 \\ 1 & 0 & 0 & 0 \end{bmatrix}$$

Where the numbers in the mask are the number of bits used to represent the corresponding transform coefficients. The upper left corner corresponds to the zero-frequency coefficient, DC or average value, and the frequency increases to the right and down. This allows the lower frequencies less quantization (more resolution), since they have more bits allocated to represent them. Because the number of bits varies with different coefficients, we have a *variable bit rate.* Additionally, more bits means finer quantization and fewer bits means coarser quantization resulting in *nonuniform quantization.*

Next a quantization scheme, such as Lloyd–Max quantization is applied. Since the zero-frequency coefficient for real images contains a large portion of the energy in the image and is always positive, it is typically treated differently than the higher frequency coefficients. Often this term is not quantized at all, or the differential between blocks is encoded. After they have been quantized, the coefficients can be coded using, for example, a Huffman or arithmetic coding method.

In addition to simple filtering, two particular types of transform coding have been widely explored: zonal and threshold coding. These two vary in the method they use for selecting the transform coefficients to retain (using ideal filters for transform coding selects the coefficients based on their location in the transform domain). *Zonal coding* involves selecting specific coefficients based on maximal variance, while *threshold coding* selects the coefficients above a specific value. In zonal coding, a zonal mask is determined for the entire image by finding the variance for each frequency component. This variance is calculated by using each subimage within the image as a separate sample and then finding the variance within this group of subimages (see Figure 10.3.17). The *zonal mask* is a bitmap of 1's and 0's, where the 1's correspond to the coefficients to retain, and the 0's to the ones to eliminate. Because the zonal

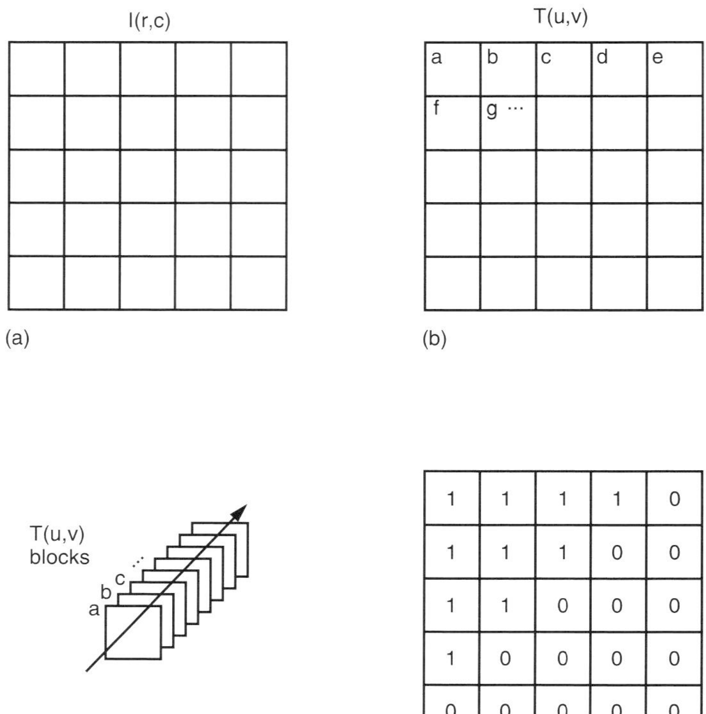

FIGURE 10.3.17
Zonal coding. (a) Divide the image into blocks. (b) Apply the transform to each block. (c) Treating each transform block from $T(u, v)$ as a separate sample, calculate the variance for each frequency component. Retain only the components with variance above a specified threshold. (d) Generate zonal masks; 1 = retain, 0 = eliminate. A typical mask is shown.

mask applies to the *entire image,* only one mask is required. In threshold coding a different threshold mask is required for each block, which increases file size as well as algorithmic complexity.

In practice, the zonal mask is often predetermined because the low frequency terms tend to contain the most information, and hence exhibit the most variance. In this case we select a fixed mask of a given shape and desired compression ratio, which streamlines the compression process. This saves the overhead involved in calculating the variance of each group of subimages for compression and also eases the decompression process. Typical masks may be square, triangular, or circular and the cutoff frequency is determined by the compression ratio. Figure 10.3.18 shows results of zonal compression with the cosine and Walsh transforms. In Figure 10.3.18d we see the blocking artifact at the block boundaries (the original image is 384×256, and the block size is 64×64). Comparing the DCT and the Walsh results we observe that the DCT provides more visually pleasing results, although obvious blurring occurs at the higher compression ratio.

One of the most commonly used image compression standards is primarily a form of transform coding. The Joint Photographic Expert Group (JPEG) met initially in 1987 under the auspices of the International Standards Organization (ISO) to devise an optimal still image compression standard. The result was a family of image compression methods

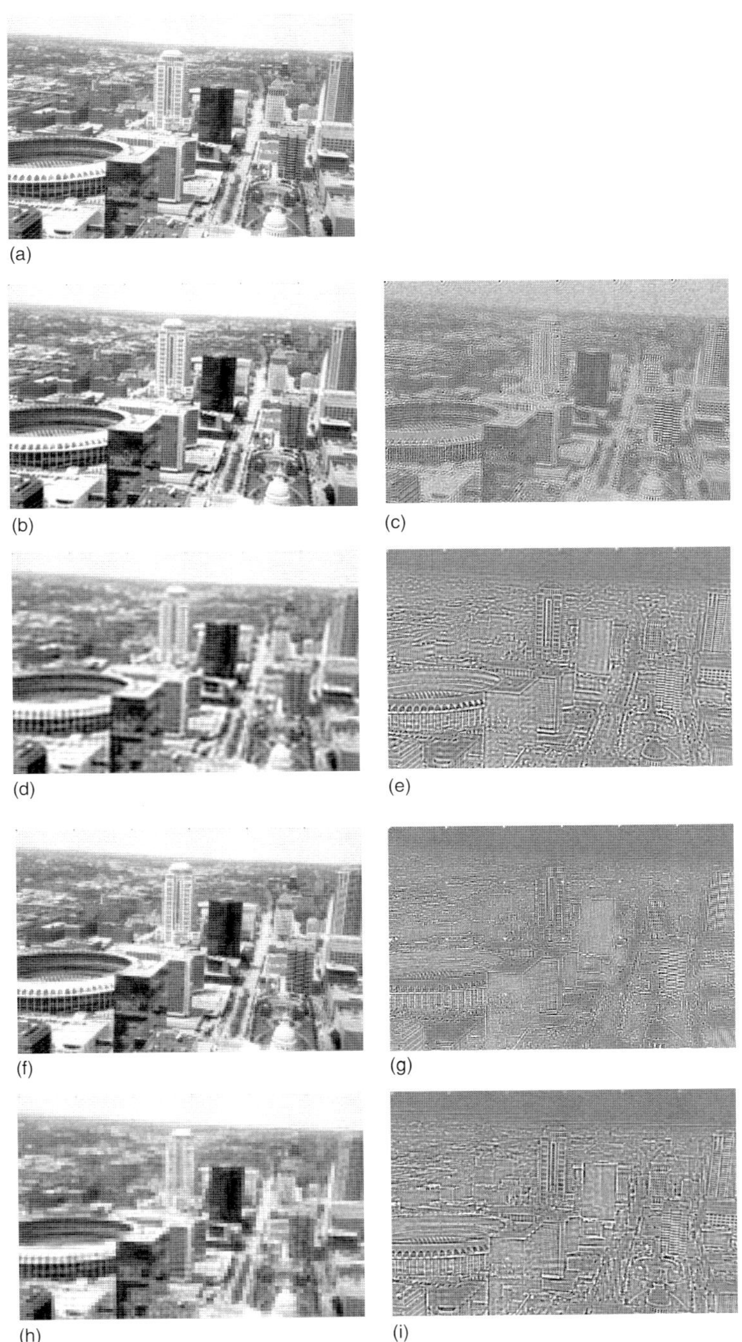

FIGURE 10.3.18
Zonal compression with DCT and Walsh transforms. A block size of 64×64 was used, a circular zonal mask, and DC coefficients were not quantized. (a) Original image, a view of St. Louis, Missouri, from the Gateway Arch. (b) Results from using the DCT with a compression ratio = 4.27. (c) Error image comparing the original and (b), histogram stretched to show detail. (d) Results from using the DCT with a compression ratio = 14.94. (e) Error image comparing the original and (d), histogram stretched to show detail. (f) Results from using the Walsh transform (WHT) with a compression ratio = 4.27. (g) Error image comparing the original and (f), histogram stretched to show detail. (h) Results from using the WHT with a compression ratio = 14.94. (i) Error image comparing the original and (h), histogram stretched to show detail.

for still images. The original JPEG standard uses the DCT and 8×8 pixel blocks as the basis for compression. Before computing the DCT, the pixel values are level shifted so that they are centered at zero.

Example 10.3.7

A typical 8-bit image has a range of gray levels of 0 to 255. Level shifting this range to be centered at zero involves subtracting 128 from each pixel value, so the resulting range is from -128 to 127.

After level shifting, the DCT is computed. Next, the DCT coefficients are quantized by dividing by the values in a quantization table and then truncated. For color signals JPEG transforms the RGB components into the *YCrCb* color space, and subsamples the two color difference signals (*Cr* and *Cb*), since we perceive more detail in the luminance (brightness) than in the color information. Once the coefficients are quantized, they are coded using a Huffman code. The zero-frequency coefficient (DC term) is differentially encoded relative to the previous block.

The quantization tables used by JPEG are shown in Figure 10.3.19. These 8×8 matrices correspond to the DCT coefficients with the zero frequency term in the upper left, and increasing in frequency to the right and down. The DCT coefficients are quantized by dividing by the numbers in the quantization tables, and then truncating. After this process many of the high frequency coefficients are zero, which greatly helps the compression process. As can be seen by comparing the two tables, the human visual is much more sensitive to changes in the luminance component (*Y* in the color space transform) than in the color (chrominance, *Cr* and *Cb*) components. These quantization tables were experimentally determined by JPEG to take advantage of the human visual system's response to spatial frequency which peaks around 4 or 5 cycles per degree (see Figure 7.2.8).

In Figure 10.3.20 are results from the original JPEG compression algorithm applied to a monochrome image. We can begin to see the block artifacts from the JPEG 8×8 DCT block in Figure 10.3.20d, with a compression ratio of 20:1. JPEG can achieve much higher compression ratios with color images, as shown in Figure 10.3.21. Here the

Mask	1	2	3	4	5	6	7	8
1	16	11	10	16	24	40	51	61
2	12	12	14	19	26	58	60	55
3	14	13	16	24	40	57	69	56
4	14	17	22	29	51	87	80	62
5	18	22	37	56	68	109	103	77
6	24	35	55	64	81	104	113	92
7	49	64	78	87	103	121	120	101
8	72	92	95	98	112	100	103	99

(a)

Mask	1	2	3	4	5	6	7	8
1	17	18	24	47	99	99	99	99
2	18	21	26	66	99	99	99	99
3	24	26	56	99	99	99	99	99
4	47	66	99	99	99	99	99	99
5	99	99	99	99	99	99	99	99
6	99	99	99	99	99	99	99	99
7	99	99	99	99	99	99	99	99
8	99	99	99	99	99	99	99	99

(b)

FIGURE 10.3.19
Original JPEG DCT coefficient quantization tables. (a) Luminance (brightness) quantization table corresponding to the *Y* component of the color transform. (b) Chrominance (color) quantization table corresponding to the *Cr* and *Cb* components of the color transform. The zero frequency (DC) term is in the upper left corner, at (1, 1), and the frequencies increase to the right and down. As the quantization coefficient increases, the quantization gets coarser. For many high frequency terms the quantization coefficient is large enough that, after dividing and truncating, they are zero.

block artifacts are not visible until we reach compression ratios more than 50 : 1, as seen Figure 10.3.21c,d, and become most noticeable above 100 : 1 as shown in Figure 10.3.21 e,f. The newer JPEG2000 compression algorithm is based on the wavelet transform, which we have seen retains both spatial and frequency information, so it is included in the next section on hybrid transforms.

10.3.7 Hybrid and Wavelet Methods

Hybrid compression methods use both the spatial domain and the transform domain. For example, the original image (spatial domain) can be differentially mapped, and then this differential image can be transform coded. Alternately, a one-dimensional transform can be performed on the rows, and this transformed data can undergo differential predictive coding along the columns. These methods are often used for compression of analog video signals. For digital images these techniques can be applied to blocks (subimages), as well as rows or columns. Vector quantization is often combined with these methods to achieve higher compression ratios.

(a)

(b)

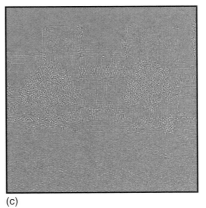

(c)

FIGURE 10.3.20
The original DCT-based JPEG algorithm. (a) Original image. (b) JPEG compression = 10 : 1. (c) Error image for (b), multiplied by 8 to show detail.

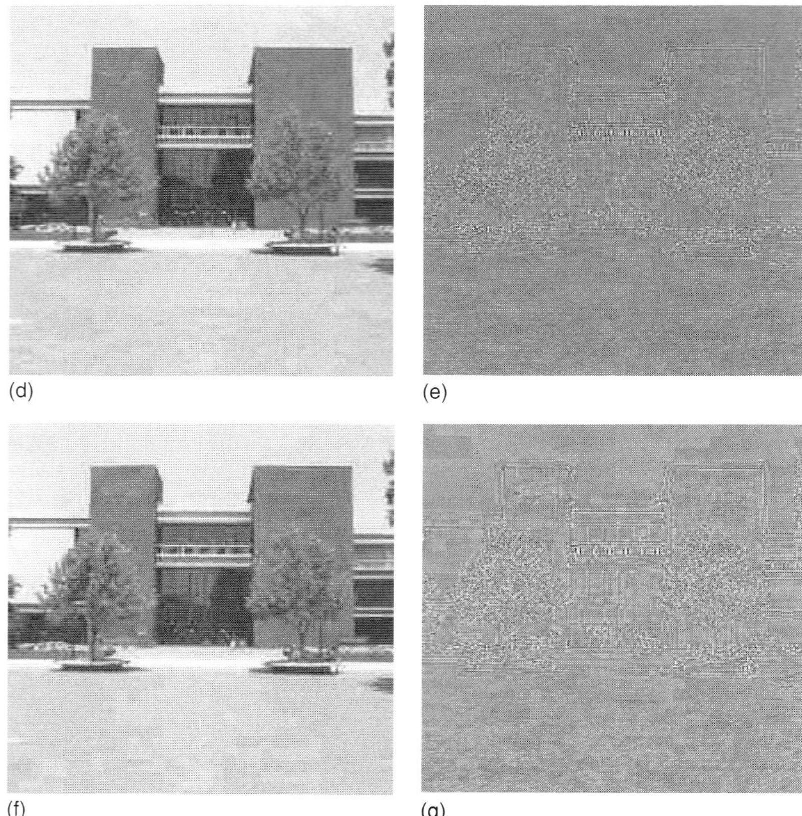

FIGURE 10.3.20 (Continued)
The original DCT-based JPEG algorithm. (d) JPEG compression = 20 : 1. (e) Error image for (d), multiplied by 8 to show detail. (f) JPEG compression = 30 : 1. (g) Error image for (f), multiplied by 8 to show detail.

The wavelet transform, which localizes information in both the spatial and frequency domain, is used in newer hybrid compression methods. For example, the JPEG2000 standard is based on the wavelet transform. The wavelet transform provides superior performance to the DCT-based techniques, and also is useful in progressive transmission for Internet and database use. Progressive transmission allows low quality images to appear quickly and then gradually improve over time as more detail information is transmitted or retrieved. With this approach the user need not wait for an entire high quality image before they decide to view it or move on.

The wavelet transform combined with vector quantization has led to the development of experimental compression algorithms, which can be explored in CVIPtools. The general algorithm is as follows:

1. Perform the wavelet transform on the image by using convolution masks (described in Section 5.8).
2. Number the different wavelet bands from 0 to $N-1$, where N is the total number of wavelet bands, and 0 is the lowest frequency (in both horizontal and vertical directions) band (see examples in Figure 10.3.22).

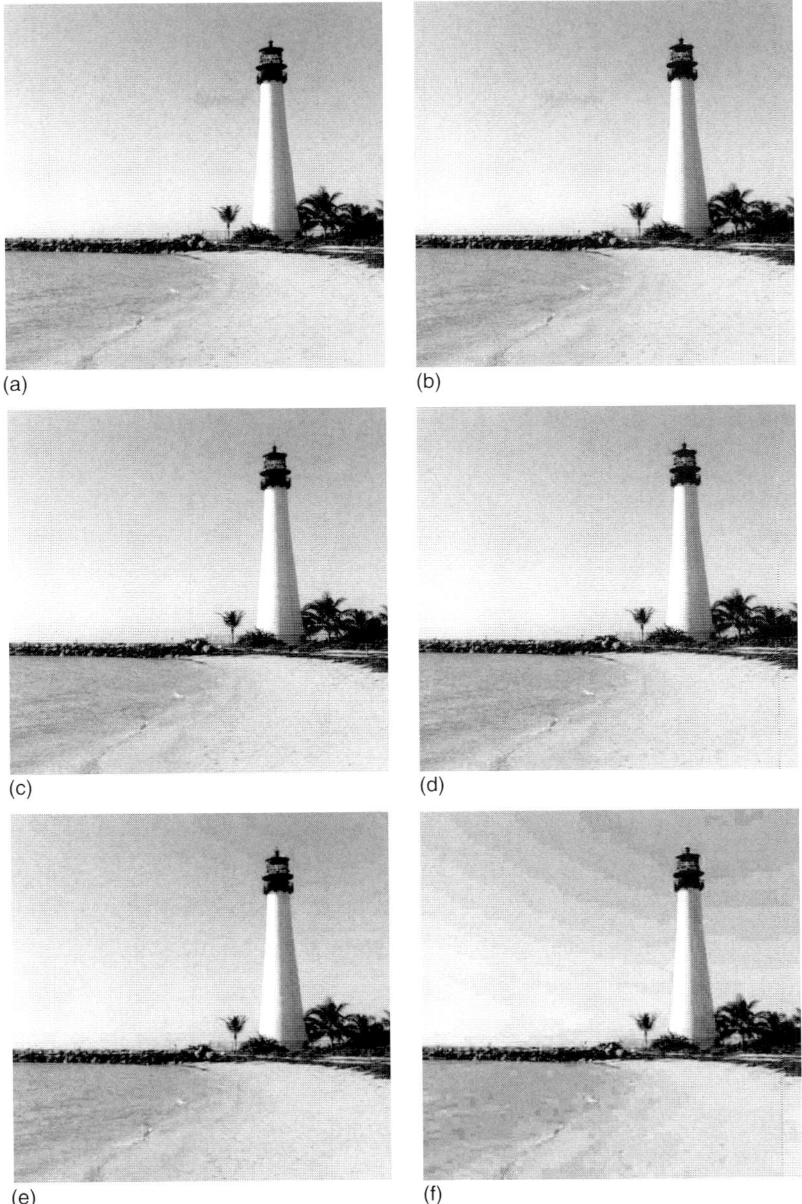

FIGURE 10.3.21
(See color insert following page 362.) The original DCT-based JPEG algorithm applied to a color image. (a) The original image. (b) Compression ratio = 34.34. (c) Compression ration = 57.62. (d) Compression ratio = 79.95. (e) Compression ratio = 131.03. (f) Compression ratio = 201.39.

3. Scalar quantize the 0 band linearly to 8 bits.
4. Vector quantize the middle bands using a small block size (e.g., 2×2). Decrease the codebook size as the band number increases.
5. Eliminate the highest frequency bands.

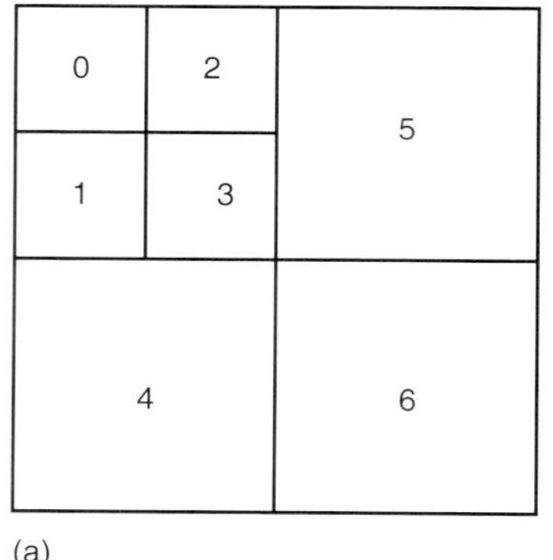

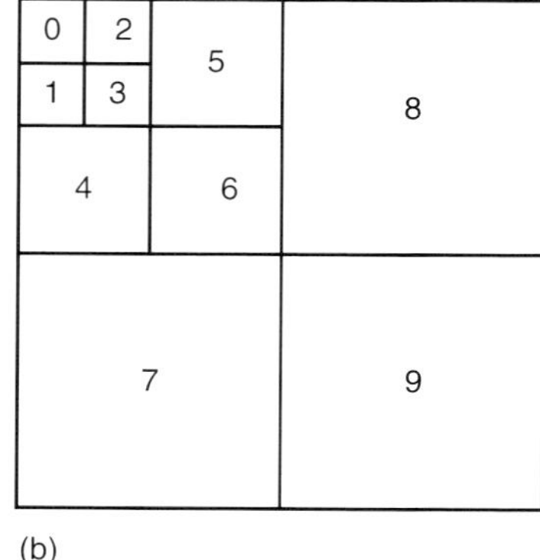

FIGURE 10.3.22
Wavelet/vector quantization compression. (a) Numbering for 7 bands. (b) Numbering for 10 bands.

TABLE 10.2

	WVQ2		WVQ3		WVQ4 (PCT)	
Band Number	**Blocksize**	**# of bits**	**Blocksize**	**# of bits**	**Blocksize**	**# of bits**
0	(scalar)	8	(scalar)	8	(scalar)	8
1, 2	2 × 2	8	2 × 2	8	2 × 2	8
3	2 × 2	6	2 × 2	6	2 × 2	6
4, 5	2 × 2	5	2 × 2	5	2 × 2	5
6	2 × 2	5	2 × 4	4	2 × 4	5
7, 8	2 × 4	4	X	0	2 × 4	5
9	X	0	X	0	X	0

The example algorithms shown here utilize 10-band wavelet decomposition (Figure 10.3.22b), with the Daubecies 4 element basis vectors, in combination with the vector quantization technique. They are called Wavelet/Vector Quantization followed by a number (WVQ#); specifically WVQ2, WVQ3, and WVQ4. In addition, one algorithm (WVQ4) employs the PCT for preprocessing, before subsampling the second and third PCT bands by a factor of 2 : 1 in the horizontal and vertical direction. Table 10.2 contains the details of the parameters for each of these algorithms.

This table lists the wavelet band numbers versus the three WVQ algorithms. For each WVQ algorithm, we have a blocksize, which corresponds to the vector size, and the number of bits, which, for vector quantization, corresponds to the codebook size. The lowest wavelet band is coded linearly using 8-bit scalar quantization. Vector quantization is used for bands 1–8, where the number of bits per vector defines the size of the codebook. For example, if we use 8 bits per vector for the codebook, this corresponds to a codebook of 256 (2^8) entries. The highest band is completely eliminated (0 bits are used to code them) in WVQ2 and WVQ4, while the highest three bands are eliminated in WVQ3. For WVQ2 and WVQ3, each of the red, green, and blue color planes are individually encoded using the parameters in the table. For WVQ4, however, the PCT is used to preprocess the image and the parameters in the table are used only for the PCT color band; the other two color bands, which contain much less information, are first subsampled and then encoded with a small number of bits. Figure 10.3.23 shows results of these algorithms.

The JPEG2000 standard is also based on the wavelet transform. It provides high quality images at very high compression ratios. Many software companies currently support it, and plugins are available for major imaging software. The committee that developed the standard had these goals for JPEG2000: (1) to provide better compression than the DCT-based JPEG algorithm, (2) to allow for progressive transmission of high quality images, (3) to be able to compress binary and continuous tone images by allowing 1 to 16 bits for image components, (4) to allow random access to subimages, (5) to be robust to transmission errors, and (6) to allow for sequentially image encoding.

The JPEG2000 compression method begins by level shifting the data to center it at zero, followed by an optional transform to decorrelate the data, such as a color transform for color images. The one-dimensional wavelet transform is applied to the rows and columns, and the coefficients are quantized based on the image size and number of wavelet bands utilized. These quantized coefficients are then arithmetically coded (see Figure 10.2.6) on a bitplane basis (see Figure 10.2.4). The JPEG2000 standard has many different options, and uses many of the techniques explored in the chapter. Figure 10.3.24 shows results from using JPEG2000 to compress a color image at high compression ratios, compare these to the images in Figure 10.3.21e,f.

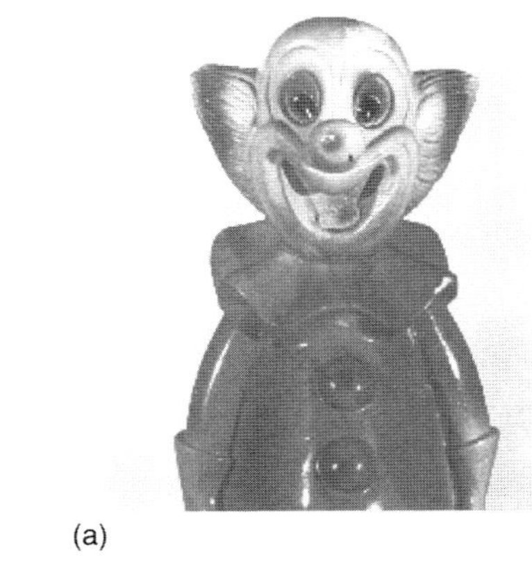
(a)

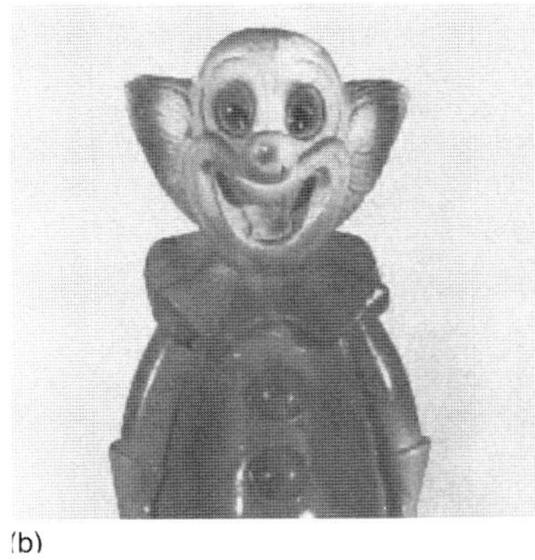
(b)

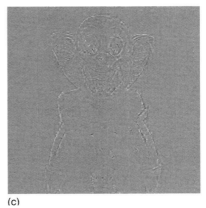

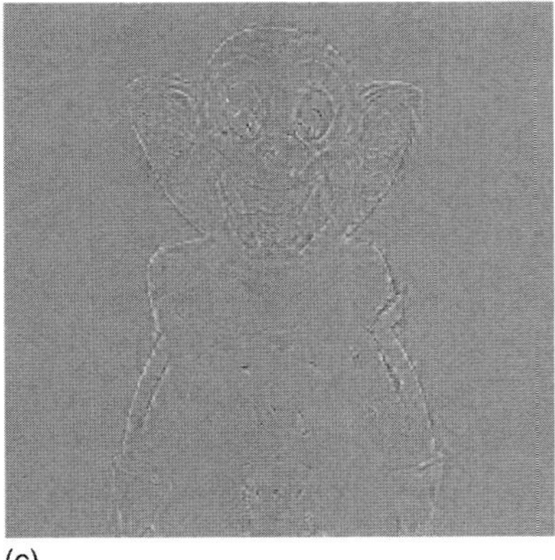
(c)

FIGURE 10.3.23
Wavelet/vector quantization (WVQ) compression example. (a) Original image. (b) WVQ2 compression ratio 10:1. (c) Error of image (b).

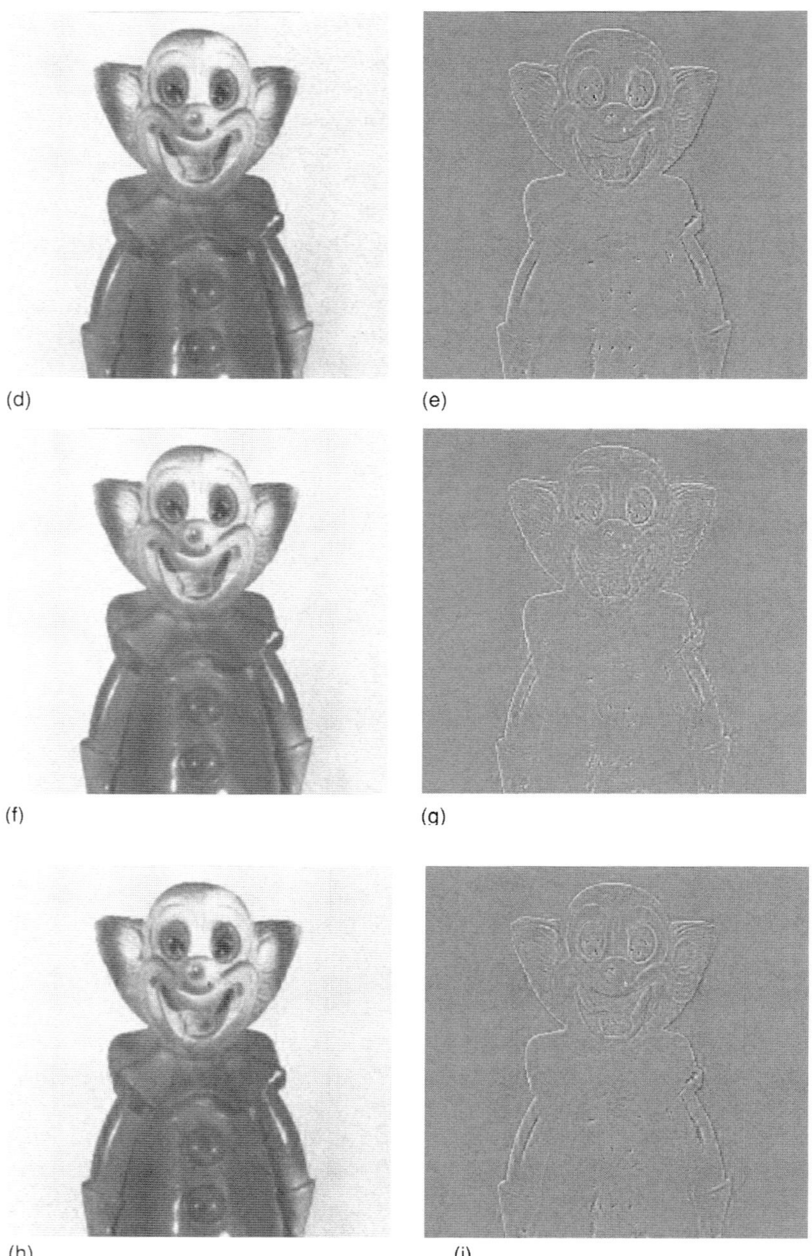

FIGURE 10.3.23 (Continued)
Wavelet/vector quantization (WVQ) compression example. (d) WVQ3 compression ratio 15 : 1. (e) Error of image (d). (f) WVQ4 compression ratio 33 : 1. (g) Error of image (f). (h) WVQ4 compression ratio 36 : 1. (i) Error of image (h).

FIGURE 10.3.24
(See color insert following page 362.) The JPEG2000 algorithm applied to a color image. (a) The original image. (b) Compression ratio = 130, compare to Figure 10.3.21e. (c) Compression ratio = 200, compare to Figure 10.3.21f. (d) A 128 × 128 subimage cropped from the standard JPEG image and enlarged to 256 × 256 using zero-order hold. (e) A 128 × 128 subimage cropped from the JPEG2000 image and enlarged to 256 × 256 using zero-order hold. The JPEG2000 image is much smoother, even with the zero-order hold enlargement.

10.4 Key Points

OVERVIEW: IMAGE COMPRESSION

- *Image compression* involves reducing the size of image data files, while retaining necessary information
- The field is continuing to grow due to demands from many different applications such as the Internet, businesses, high-definition television, satellite imaging, and medical imaging
- The original image before compression is the *uncompressed* image, the file created by the compression process is the *compressed* file, and the image recreated from the compressed file is the *decompressed* image
- The compression ratio is the size of the uncompressed file to the size of the compressed file, and can also be measured in bits per pixel (bpp)
- *Data* refers to the pixel gray level values and *information* is an interpretation of the data in a meaningful way
- *Lossless* compression methods allow for the exact recreation of the original image data, and can compress complex images to a maximum $\frac{1}{2}$ to $\frac{1}{3}$ the original size, corresponding to 2:1 to 3:1 compression ratios
- *Lossy* compression methods lose data, but can compress complex images 10:1 to 50:1 and retain high quality, and 100 to 200 times for lower quality, but acceptable, images
- Compression algorithms work by removing image redundancy
- Four types of redundancy in images: (1) coding, use of inefficient codes, (2) interpixel, adjacent pixels highly correlated, (3) interband redundancy, color images have high correlation between bands within an image; the red, green, and blue bands look similar, (4) psychovisual, human visual system has limitations
- Fidelity criteria (Chapter 7) are used to measure and compare compression algorithms

COMPRESSION SYSTEM MODEL

- The compression system model consists of two parts, the compressor and decompressor
- *Compressor*: (1) data reduction, (2) mapping, (3) quantization, (4) coding
- *Data reduction*: removal of application-specific irrelevant information by gray level and/or spatial quantization, noise removal, enhancement
- *Mapping*: to compact and decorrelate the data, using differential coding, frequency/sequency transforms, color transforms, and/or principal components transform
- *Quantization*: required because the mapping process often results in floating point data, can be uniform, equal subdivisions, or nonuniform, unequal subdivisions also called a variable bit rate; quantization is not reversible
- *Coding*: it is a reversible one-to-one mapping, can be an equal length code or unequal (variable) length code; unequal length codes are usually more efficient but more complex

- *Decompressor*: (1) decoding, reverse the coding process, (2) inverse mapping, reverses the mapping process, (3) postprocessing, to enhance the decompressed image

LOSSLESS COMPRESSION METHODS

- No data loss, necessary for some applications such as medical or legal images
- Often may only get a 10% reduction in file size, maximum of 2:1 or 3:1 for complex images
- Lossy compression is also called *data compaction*
- Information theory defines information based on the probability of an event, knowledge of an unlikely event has more information than knowledge of a likely event
- Entropy measures information from an information theoretic perspective, and provides the minimum lower bound for a coder:

$$\text{Entropy} = -\sum_{i=0}^{L-1} p_i \log_2(p_i) \quad \text{(in bits/pixel)}$$

where: $p_i =$ the probability of the ith gray level $= n_k/N^2$

$n_k =$ the total number of pixels with gray value k

$L =$ the total number of gray levels (e.g., 256 for 8-bits)

- The average length of a code should be close to the entropy for a good code:

$$L_{\text{ave}} = \sum_{i=0}^{L-1} l_i p_i$$

where: $l_i =$ length in bits of the code for ith gray level

$p_i =$ histogram - probability of ith gray level

Huffman Coding

- The Huffman code is a variable (unequal) length code
- It will generate a code close to the entropy which is the minimum lower bound
- To generate a Huffman code requires five steps: (1) find the gray level probabilities for the image by finding the histogram, (2) order the input probabilities (histogram magnitudes) from smallest to largest, (3) combine the smallest two by addition, (4) GOTO Step 2, until only two probabilities are left, (5) by working backward along the tree, generate code by alternating assignment of 0 and 1

Run-Length Coding

- Run-length coding (RLC) works by counting adjacent pixels with the same value; the count is called the *run-length*
- RLC works best for binary, two-valued, images
- Gray or color images can use RLC by storing the run length and the gray value(s)

- Alternately, extension to gray or color images is bitplane RLC, which applies RLC to each bitplane separately
- Preprocessing with a Gray code can improve compression by bitplane RLC by typically 10% to 15%
- Standards developed for RLC by ITU-R provide compression ratios of 15 to 20 for typical documents

Lempel–Ziv–Welch Coding

- Lempel–Ziv–Welch (LZW) methods work by encoding strings of data
- During compression it creates a string table containing strings and the corresponding codes
- The string table is not stored but extracted from the compressed data itself
- It uses code words with more bits than the original data, extra bits used as index for string entries in the string table
- This algorithm and its variations are used in many image files formats such as GIF and TIFF

Arithmetic Coding

- *Arithmetic coding* transforms input data into a single floating point number between 0 and 1
- An entire image must be divided into small subimages to be encoded
- The precision required for this number becomes greater as the subimage size increases
- Arithmetic coding uses the histogram as a probability model, so can theoretically achieve the maximum compression specified by the entropy
- Arithmetic coding is part of the JPEG standard

Lossy Compression Methods

- *Lossy compression methods* lose image data, so cannot recreate the original image *exactly*
- Lossy compression is necessary to achieve high compression ratios with complex images
- Images can be compressed 10 to 50 times with minimal visible information loss
- Newer methods, such as JPEG2000, can achieve reasonably good image quality with compression ratios as high as 100 to 200 for color images

Gray-Level Run Length Coding

- One method is to reduce the number of bits per pixel and then apply standard RLC with bit-plane coding
- Compression ratio can be improved by preprocessing with Gray code
- A more sophisticated method is dynamic window-based RLC
- *Dynamic window-based RLC* works by allowing a range of gray levels to be replaced by a representative value, the gray level range adapts to the data

Block Truncation Coding

- Block truncation coding (BTC) works by dividing the image into blocks and reducing the number of gray levels in a block
- The gray levels are reduced by a quantizer that adapts to local statistics
- Basic BTC divides the image into 4×4 blocks, and outputs high (H) and low (L) values for the block, and a bit-string to encode which pixels are H and L

$$H = m_b + \sigma_b \sqrt{\frac{n^2 - q}{q}}$$

$$L = m_b - \sigma_b \sqrt{\frac{q}{n^2 - q}}$$

- Many other BTC algorithms have been developed, including 4-level and multilevel quantizers

Vector Quantization

- *Vector quantization* (VQ) maps a vector that can have many values to a vector that has a smaller (quantized) number of values
- VQ can be applied in both the spectral or spatial domains
- With images the vector is a subimage or a block within the image
- Information theory tells us that better compression can be achieved with vector quantization than with scalar quantization (rounding or truncating individual values)
- Vector quantization is achieved by use of a *codebook* that contains a fixed set of vectors, and storing the index (address) into the codebook
- The standard algorithm to generate the codebook is the Linde–Buzo–Gray (*LBG*) algorithm, also called the *K-means* or the *clustering* algorithm:

Step 1. Given an arbitrary codebook, encode each input vector according to the nearest-neighbor criterion. Use a distance metric to compare all the input vectors to the encoded vectors, and then sum these errors (distances) to provide a distortion measure. If the distortion is small enough (less than a predefined threshold), quit. If not, go to Step 2.

Step 2. For each codebook entry, compute the euclidean centroid of all the input vectors encoded into that specific codebook vector.

Step 3. Use the computed centroids as the new codebook, and go to Step 1.

- Generating a good code with the LBG algorithm requires careful choice of the initial codebook
- Initial codebook can be chosen by: (1) randomly selecting vectors in the training set, preferably spread across the vector space, or (2) subdividing the vector space and finding the centroid of the training vectors in each division
- Vector quantization has simple decompression, but complex compression

Differential Predictive Coding

- *Differential predictive coding (DPC)* predicts the next pixel value based on previous values, and encodes the difference between predicted and actual value (the error signal)
- The compression results from the error signal being small, because prediction is easy as adjacent pixels are highly correlated
- The algorithm is lossy due to quantization (see block diagram in Figure 10.3.10)
- The theoretically optimal predictor is:

$$\tilde{I}(r, c+1) = \rho\hat{I}(r, c) + (1-\rho)\bar{I}(r, c)$$

where: $\bar{I}(r, c) =$ the average value for the image

$\rho =$ the normalized correlation between pixel values

- The normalized correlation, ρ, is between 0.85 and 0.95 for most images
- DPC results can be improved by use of a Lloyd–Max quantizer
- After quantization, a lossless coder, such as Huffman or arithmetic, can be applied

Model-Based and Fractal Compression

- *Model-based* or *intelligent* compression works modeling objects in the image and storing object descriptions
- Fractal methods are an example of model-based techniques
- Fractal compression is based on the idea that various regions in the image are *self-similar,* which means that one subimage can be represented as a skewed, stretched, rotated, scaled, and/or translated version of another subimage
- The mathematical operations, skew, stretch, scale, rotate, and translate, are called *affine transformations* and can be represented by the following general equations:

$$r' = k_1 r + k_2 c + k_3$$

$$c' = k_4 r + k_5 c + k_6$$

where r' and c' are the new coordinates, and k_i are constants

- Fractal image compression divides an image into subimages, and selects some to serve as models called *range* regions to map to the *domain* regions which represent the entire image
- These range regions are the fractals, which are like basis images, that can undergo affine transformations and be assembled into a good representation of the image
- The compressed file stores the fractals and the necessary affine transformation coefficients
- Model-based methods can provide high compression ratios, but have complex and costly compression methods

Transform Coding

- Transform coding works by dividing the image into blocks, performing the transform, and then quantizing and coding the coefficients
- Filtering is the simplest form of transform coding
- To maximize compression, *bit allocation* is applied to quantize the coefficients, frequencies of more importance are allocated more bits, thus are more finely quantized
- *Zonal coding* sets a threshold on the variance for each component and generates a *zonal mask* to determine which coefficients to retain
- In practice, a fixed mask may be used with zonal coding since low frequency terms are typically the ones with high variance
- These fixed masks are typically square, circular or triangular with the cutoff frequency determined by the desired compression ratio
- *Threshold coding* sets a threshold on the magnitude of the coefficients in each block, but requires the added overhead of a bit mask for each block
- The original JPEG standard is based on transform coding using the DCT with 8×8 blocks
- Original JPEG algorithm: (1) level shift data to center at zero, (2) transform *RGB* data into *YCrCb* color space, for color images, (3) compute DCT for 8×8 blocks, (4) quantize coefficients using tables, (5) encode with Huffman, (6) differentially encode the DC coefficients between blocks

Hybrid and Wavelet Methods

- Hybrid methods use both the spatial and spectral domains
- Algorithms exist that combine differential coding and spectral transforms for analog video compression
- The wavelet transform, which localizes information in both the spectral and spatial domains, combined with vector quantization can provide good results
- General wavelet/vector quantization (WVQ) algorithm: (1) Perform the wavelet transform on the image by using convolution masks. (2) Number the different wavelet bands from 0 to $N-1$, where N is the total number of wavelet bands, and 0 is the lowest frequency (in both horizontal and vertical directions) band. (3) Scalar quantize the 0 band linearly to 8 bits. (4) Vector quantize the middle bands using a small block size (e.g., 2×2). Decrease the codebook size as the band number increases. (5) Eliminate the highest frequency bands
- The JPEG2000 standard is based on the wavelet transform, and is finding widespread support
- Committee goals for JPEG2000: (1) to provide better compression than the DCT-based JPEG algorithm, (2) to allow for progressive transmission of high quality images, (3) to be able to compress binary and continuous tone images by allowing 1 to 16 bits for image components, (4) to allow random access to subimages, (5) to be robust to transmission errors, and (6) to allow for sequential image encoding
- The JPEG2000 compression method: (1) Level shift the data to center it at zero. (2) An optional transform to decorrelate the data, such as a color transform for color images. (3) The one-dimensional wavelet transform is applied to the rows

and columns. (4) The coefficients are quantized based on the image size and number of wavelet bands utilized. (5) The coefficients are then arithmetically coded on a bitplane basis

10.5 References and Further Reading

The compression system model presented is based on the model in [Gonzalez/Woods 02]. More details and examples on redundancy in images can be found in [Gonzalez/Woods 02], and coding irrelevancy in [Castelman 96]. Fundamentals of information and coding theory is in [Tranter/Ziemer 85] and [Gonzalez/Woods 02]. The Huffman coding technique is found in [Huffman 52], [Gonzalez/Woods 02], [Netravali/Haskell 88], [Rosenfeld/Kak 82], [Jain 89], and [Sid-Ahmed 95]. More information on run length coding, including two-dimensional methods can be found in [Gonzalez/Woods 02], [Tekalp 95], [Jain 89], and [Hunter/Robinson 80]. Details on LZW coding are contained in [Ziv/Lempel 77] and [Welch 84]. The arithmetic coding method can be found in [Gonzalez/Woods 02].

The dynamic window-based RLC algorithmic details are in [Kumaran/Umbaugh 95]. Block truncation coding is explored in [Wu/Coll 93], [Rabbini/Jones 91], [Delp/Mitchel 79], and [Rosenfeld/Kak 82].The references for vector quantization include [Tekalp 95], [Rabbini/Jones 91], [Netravali/Haskell 88], [Linde/Buzo/Gray 80]. Differential predictive techniques are explored in [Gonzalez/Woods 02], [Jain 89], [Netravali/Haskell 88], and [Rosenfeld/Kak 82]. More on transform coding can be found in [Gonzalez/Woods 02], [Sid-Ahmed 95], [Jain 89], [Netravali/Haskell 88], and [Rosenfeld/Kak 82].

Model-based compression using texture and the wavelet transform is described in [Ryan/Sanders/Fisher/Iverson 96]. More information on the wavelet/vector quantization compression algorithms can be found in [Guo/Umbaugh/Cheng 01], [Kjoelen/Umbaugh/Zuke 98], and [Kjoelen 95]. Wavelet compression references include [Taubman/Marcellin 02], [Gonzalez/Woods 02], and [Welstead 99].

For an overview of JPEG see [Gonzalez/Woods 02] and [Sonka 99]. For more information on fractal-based image compression see [Welstead 99], [Watt/Policarpo 98], and [Fisher 95]. For details on JPEG2000 see [Taubman/Marcellin 02]. More information on compression of video (motion) images can be found in [Poynton 03], [Orzessek/Sommer 98], [Tekalp 95], [Sid-Ahmed 95], [Clarke 95], [Kuo 95], [Bhaskaran/Konstantinides 95], and [Netravali/Haskell 88].

Bhaskaran, V., Konstantinides, K., *Image and Video Compression Standards: Algorithms and Architectures*, Boston: Kluwer Academic Publishers, 1995.

Castleman, K.R., *Digital Image Processing*, Englewood Cliffs, NJ: Prentice Hall, 1996.

Clarke, R.J., *Digital Compression of Still Images and Video*, San Diego, CA: Academic Press, 1995.

Delp, E.J., Mitchell, O.R., *Image Compression Using Block Truncation Coding, IEEE Transactions on Communications*, Vol. 27, No. 9, pp. 1335–1342, September 1979.

Fisher. Y., editor, *Fractal Image Compression: Theory and Application*, NY: Springer-Verlag, 1995.

Gonzalez, R.C., Woods, R.E., *Digital Image Processing*, Upper Saddle River, NJ: Prentice Hall, 2002.

Guo, L., Umbaugh, S., Cheng, Y., *Compression of Color Skin Tumor Images with Vector Quantization, IEEE Engineering in Medicine and Biology Magazine*, Vol. 20, No. 6, Nov/Dec 2001, pp. 152–164.

Huffman, D.A., A Method for the Reconstruction of Minimum Redunancy Codes, *Proceedings of the IRE*, Volume 40, Number 10, pp. 1098–1101, 1952.

Hunter, R., Robinson, A.H., International Digital Facsimile Coding Standards, *Proceedings of the IEEE*, Vol. 68, No. 7, pp. 854–867, 1980.

Jain, A.K., *Fundamentals of Digital Image Processing*, Englewood Cliffs, NJ: Prentice Hall, 1989.

Kjoelen, A., Umbaugh, S. E, Zuke, M., Compression of Skin Tumor Images, *IEEE Engineering in Medicine and Biology Magazine*, Vol. 17, No. 3, May/June 1998, pp.73–80.

Kjoelen, A., *Wavelet Based Compression of Skin Tumor Images*, Master's Thesis in Electrical Engineering, Southern Illinois University at Edwardsville, 1995.

Kou, W., *Digital Image Compression: Algorithms and Standards*, Boston: Kluwer Academic Publishers, 1995.

Kumaran, M., Umbaugh, S.E., A Dynamic Window-Based Runlength Coding Algorithm Applied to Gray-Level Images, *Graphical Models and Image Processing*, Vol. 57, No. 4, pp. 267–282, July 1995.

Linde, Y., Buzo, A., Gray, R.M., An Algorithm for Vector Quantizer Design, *IEEE Transactions on Communications*, Vol. 28, No. 1, pp. 84–89, January 1980.

Netravali, A.N., Haskell, B.G., *Digital Pictures: Representation and Compression*, NY: Plenum Press, 1988.

Orzessek, M., Sommer, P., *ATM and MPEG-2: Integrating Digital Video into Broadband Networks*, Upper Saddle River, NJ: Prentice Hall PTR, 1998.

Poynton, C., *Digital Video and HDTV Algorithms and Interfaces*, Morgan Kaufmann, 2003.

Rabbani, M., Jones, P.W., *Digital Image Compression Techniques*, SPIE—International Society for Optical Engineeering, 1991.

Rosenfeld, A., Kak, A.C., *Digital Picture Processing*, San Diego, CA: AcademicPress, 1982.

Ryan, T.W, Sanders L.D., Fisher, H.D., Iverson, A.E., Image Compression by Texture Modeling in the Wavelet Domain, *IEEE Transactions on Image Processing*, Vol. 5, No. 1, pp. 26–36, January 1996.

Sid-Ahmed, M.A. *Image Processing: Theory, Algorithms, and Architectures*, NY: McGraw Hill, 1995.

Sonka, M., Hlavac, V., Boyle, R., *Image Processing, Analysis and Machine Vision*, Pacific Grove, CA: Brooks/Cole Publishing Company, 1999.

Taubman, D.S., Marcellin, M.W., *JPEG2000: Image Compression Fundamentals, Standards and Practice*, Norwell, MA: Kluwer Academic Publishers, 2002.

Tekalp, A.M., *Digital Video Processing*, Upper Saddle River, NJ: Prentice Hall, 1995.

Tranter, W.H, Ziemer, R.E., *Principles of Communications: Systems, Modulation, and Noise*, Boston: Houghton Mifflin Company, 1985.

Watt, A., Policarpo, F., *The Computer Image*, New York, NY: Addison-Wesley, 1998.

Welch, T.A., A Technique for High-Performance Data Compression, *IEEE Computer*, Vol. 17, No. 6, pp. 8–19, 1984.

Welstead, S., *Fractal and Wavelet Image Compression Techniques*, Bellingham, WA: SPIE Press, 1999.

Wu, Y., Coll, D.C., Multilevel Block Truncation Coding Using a Minimax Error Criterion for High-Fidelity Compression of Digital Images, *IEEE Transactions on Communications*, Vol. 41, No. 8, August 1993.

Ziv, J., Lempel, J., A Universal Algorithm for Sequential Data Compression, *IEEE Transactions on Information Theory*, Vol. 24, No. 5, pp. 530–537, 1977.

10.6 Exercises

1. (a) Define image compression. (b) List three reasons why image compression is important.
2. (a) What is the term for an image before compression? (b) after compression? (c) after reconstruction? (d) If a color (three-band RGB) image that is 512×512 pixels with 8-bits per pixel per band is compressed with a compression ratio of 16, what size is the compressed data? (e) If the same image is compressed at 0.5 bits per pixel per band, what size is the compressed data?

3. (a) About how long would it take to transmit a 640 × 480 uncompressed color image that has 8-bits per color band, using a 56 kb modem? (b) Using a 3 Mbs cable connection? (c) How long to receive a satellite image that is 6000 × 6000, seven spectral bands with 16 bits per band, with a 3 Mbs cable modem?
4. (a) Discuss the difference between data and information (*not* the information theoretic definition) in images. (b) Describe an example to illustrate. (c) Name and define the two types of compression methods and typical compression ratios for them.
5. (a) Name and define the four types of redundancy in images. (b) Give examples of each. (c) How is redundancy in images related to image compression? (c) How do we measure the quality of a decompressed image?
6. (a) Sketch a block diagram of the compression system model. (b) Briefly discuss each block.
7. (a) Describe three methods to decorrelate image data. (b) What does it mean for a process to be reversible? (c) Are the methods you discussed in (a) reversible?
8. (a) Describe two ways to perform quantization. (b) Is quantization reversible? (c) Describe two general types of codes. (d) Is the coding process reversible?
9. (a) Why would we use lossless compression, and what is another term for it? (b) What is the information theoretic definition of information and how is it different from our standard definition of information in images?
10. (a) Given a 3-bit per pixel image with the following histogram, find the entropy:

Gray Value	0	1	2	3	4	5	6	7
Number of Pixels	324	100	212	194	4	66	50	74

(b) What are the minimum and maximum possible values of entropy for an image of this type? (c) Is it possible to devise a code for the image that will provide an average number of bits per pixel that is less than the entropy?

11. Use CVIPtools to explore entropy of simple images. For all the images use the default size of 256 × 256. (a) Create two checkerboard images with different size cells with *Utilities->Create->Checkerboard*. Use *Utilities->Stats->Image Statistics* to find the entropy of each (*Note*: you may need to ***double*** click *Apply* on the statistics window to be sure you have the current image information). Explain your results. (b) Create an ellipse that is 128 × 64 and another that is 32 × 64 and find the entropy of each. Explain your results. (c) Create a circle with radius of 32. Create a second circle with a radius of 32 but check the *Blur radius* box, and set it to 64. Find the entropy of each and explain the results. (d) Repeat parts (a)–(c), but invert each image using the NOT operator. Are the results what you expected? Explain. (e) Create a rectangle image that is half white and half black. For example, start at (0, 0) with a width of 128 and height of 256. Next, create an image that is 25% white and 75% black, and then invert it with the NOT operator. Find the entropy of all three images. Explain your results.

12. (a) Find the Huffman code for a 2-bit per pixel image with the following histogram:

Gray Level	0	1	2	3
Number of Pixels	400	200	300	100

(b) Find the entropy for the image, and the average number of bits per pixel with the code you obtained in (a). Do you think this is a good code? Why?

13. (a) Find the run length code of the following 1 bpp image using horizontal RLC and the conventions as defined in this chapter:

$$\begin{bmatrix} 0 & 1 & 1 & 0 & 0 & 1 & 0 & 0 \\ 1 & 1 & 1 & 1 & 0 & 0 & 0 & 0 \\ 1 & 0 & 1 & 0 & 0 & 0 & 1 & 1 \\ 0 & 1 & 1 & 1 & 1 & 1 & 0 & 0 \\ 0 & 1 & 1 & 1 & 0 & 0 & 1 & 0 \\ 1 & 1 & 1 & 0 & 0 & 0 & 1 & 0 \\ 1 & 1 & 1 & 1 & 1 & 1 & 1 & 0 \\ 0 & 1 & 0 & 0 & 0 & 0 & 0 & 0 \end{bmatrix}$$

14. (a) Using conventions defined in this chapter and horizontal RLC, find the RLC for each bit plane for the following 2 bpp image (numbers are base 10):

$$\begin{bmatrix} 3 & 3 & 1 & 1 \\ 2 & 1 & 1 & 0 \\ 2 & 1 & 0 & 0 \\ 1 & 1 & 0 & 0 \end{bmatrix}$$

(b) Apply bitplane RLC to the following 3 bpp image:

$$\begin{bmatrix} 6 & 6 & 7 & 7 \\ 6 & 6 & 6 & 6 \\ 4 & 3 & 4 & 3 \\ 1 & 1 & 0 & 0 \end{bmatrix}$$

(c) Explain why a Gray code may improve the compression with bitplane RLC. In the image from part (b), which row will benefit the most from using a gray code? Why?

15. (a) If we use LZW coding with 8 bpp image data, and used 12 bits for the code words, how many entries can be in the string table? (b) How many string codes can we have?

16. Given the following histogram, find an arithmetic code for this row of a 2 bpp image:

$$\begin{bmatrix} 3 & 3 & 2 & 1 \end{bmatrix}$$

Gray Level	0	1	2	3
Number of Pixels	20	40	10	30

17. Use CVIPtools to explore lossy RLC with the *Compression->Lossy* window. Select the *View->Debug Information* option on the main window. (a) Select an image of your choice and apply *Bitplane Run Length Coding*. To start, check all the bits so they are all retained and record the compression ratio. Next, starting at bit 0, uncheck one bit at a time and record the compression ratio each time while lining up the output images. How many bits are used when the compression ratio first becomes greater than 1? How many bits are used when compression artifacts become visibly noticeable? (b) Preprocess your original image with a Gray code conversion, and perform the same experiment as in (a). Did the compression results improve? How much was the average increase or decrease for the compression ratio? (c) Apply *Dynamic windows-based RLC* to your test image, using window sizes of 2, 5, 10, 20, and 50. Record the compression ratio each time. Compare images with the same compression ratio from (a) and (b) to those you created here. With which method do you get the best results? (d) Preprocess your original image with a Gray code conversion, and perform the same experiment as in (c). Did the compression results improve? How much was the average increase or decrease for the compression ratio? Explain your results.
18. Given the following 4×4 subimage, apply basic BTC and find the resulting values.

$$\begin{bmatrix} 22 & 21 & 15 & 17 \\ 20 & 21 & 17 & 17 \\ 5 & 5 & 40 & 40 \\ 42 & 41 & 19 & 20 \end{bmatrix}$$

19. Use CVIPtools to explore block truncation coding (BTC) with the *Compression-> Lossy* window. Select the *View->CVIP Function Information* option on the main window. (a) Select an image of your choice and apply BTC and vary the block size using, 4, 8, 16 and 32. Record the compression ratio each time. (b) Next, select *Utilities-> Compare* and compare the original image to each of the four compressed images using *RMS error* and *SNR* and record the results. Now compare the compression ratio, RMS errors and the SNRs. Do you see any relationship as the block size increases? Explain.
20. (a) Describe vector quantization (VQ) and scalar quantization. (b) In general, which technique do you think will give better image compression results? Why? (c) What is a codebook and how is it used in image compression? (d) Describe an algorithm for codebook generation, (e) Describe a method to initialize the codebook and explain why this is important. (f) With VQ is it faster to compress or decompress images? Explain.
21. Use CVIPtools to explore vector quantization in the spatial domain. Select the *View->CVIP Function Information* option on the main window. (a) Select a 256×256 image of your choice and apply *Vector Quantization* (VQ) using *Vector Height* = 2, *Vector Width* = 2, *Error Threshold* = 20, select *Generate codebook; Save codebook with coding result*, and vary the codebook size with 512, 256, 128, and 64. Record the compression ratios and align the images from largest codebook to smallest. (b) Repeat (a) with 4×4 blocks. (c) Repeat (a) with 8×8 blocks. (d) Compare the images with the same size codebook across the three different block sizes. Which block size provides better looking images? Does a lower compression ratio always imply a better image? Explain.

22. Use CVIPtools to explore vector quantization in the transform domain. Select the *View->CVIP Function Information* option on the main window. a) Select a monochrome (gray) 256 × 256 image of your choice and apply *XVQ* (Xform *VQ*). Set *Data type* = BYTE, *Remap type* = LINEAR, Select *DCT* and create images with *VQ1, VQ2, VQ3,* and *VQ4*. Record the compression ratios and compare the images visually and by finding the SNR of the compressed and original image(s). Which images look better, and which have the better SNR? (b) Repeat (a) with a color image and compare *DCT* and *PCT_DCT* results. (c) Repeat (a) with a color image and compare *WAVLET* and *PCT_WAVLET* results.
23. (a) Why is it easy to generate a good prediction equation for use in differential predictive coding (DPC)? (b) What exactly is the error signal in DPC and why is it quantized? (c) What is the maximum number of previous values useful for prediction? Why?
24. Use CVIPtools to explore DPC. Select the *View->CVIP Function Information* option on the main window. (a) Select an image of your choice and apply *DPC* using the default parameters. Vary the number of bits and observe the compression ratio and image quality. Are these two correlated? (b) Use the default parameters by clicking on the *Reset* button, but vary the *Predictor* parameter. Which one provides better results? Why? (c) Next, vary the type of *Quantizer* and observe the compression ratio and image quality. Are these two correlated?
25. (a) Explain how model-based compression works. (b) What does *self-similar* mean and how is it related to model-based compression? (c) What is the name and form of the equation(s) for transforming a self-similar subimage into another subimage? (d) Discuss the advantages and disadvantages of fractal image compression. (e) How do the terms *domain* and *range* apply to fractal image compression?
26. (a) What are the differences between *zonal* and *threshold* coding? (b) What is meant by *bit allocation*? (c) In practice, the implementation of zonal coding may differ from its formal definition. How and why?
27. Use CVIPtools to explore zonal coding with *Compression->Lossy->Zonal*. Select the *View->CVIP Function Information* option on the main window. (a) Select a 256 × 256 image of your choice and apply Zonal compression with the following parameters: *Blocksize* = 32, *Transform* = DCT, *Compression ratio* = 5, *Mask shape* = triangle, *DC quantize* unchecked. Next, change the transform to FFT and apply. What difference do you observe between the DCT and FFT results? Explain. (b) Apply Zonal compression with the following parameters: *Blocksize* = 64, *Transform* = DCT, *Compression ratio* = 10, *Mask shape* = circle, *DC quantize* unchecked. Next, change the transform to Walsh and apply. What difference do you observe between the DCT and Walsh results? Explain. (c) Use *Utilities->Compare* to find the RMS error and the SNR between the DCT zonal compressed and the original image and the Walsh zonal compressed image and the original. Which image is better according to these metrics? Do you agree with the results?
28. Use CVIPtools to explore JPEG compression with *Compression->Lossy-JPEG*. Select the *View->CVIP Function Information* option on the main window. (a) Select a 256 × 256 monochrome image of your choice and apply JPEG with the default parameters, but vary the *Quality* parameter. Set *Quailty* to 5, 10, 20, 50, 70, and 90 and record the compression ratio for each. What is the quality factor

and compression ratio when you first cannot see artifacts that make the image look bad? (b) Select a 256×256 color image of your choice and apply JPEG with the default parameters, but vary the *Quality* parameter. Set *Quailty* to 5, 10, 20, 50, 70, and 90 and record the compression ratio for each. What is the quality factor and compression ratio when you first cannot see artifacts that make the image look bad? (c) Are the answers to (b) and (c) the same? Explain.

10.6.1 Programming Exercise: Signal-to-Noise Ratio and Root-Mean-Square Error Metrics

1. Write a function to compare two images using the peak SNR and RMS error metrics (see Chapter 7). Compare your results to those obtained with CVIPtools. Are they the same? Why or why not?
2. Write a function to find the RMS SNR. How do the results compare to the peak SNR metric?

10.6.2 Programming Exercise: Huffman Coding

1. Write a function to implement Huffman coding with a single band (monochrome) image. Compare your compression ratios to those in CVIPtools. Are your results the same, better or worse? Can you explain why?
2. Extend your function to work with color images. Compare your compression ratios to those in CVIPtools. Are your results the same, better or worse? Can you explain why?
3. Incorporate the CVIPtools library C functions *huf_compress* and *huf_decompress* from the compression library into your CVIPlab. Test it and compare it to your Huffman function.

10.6.3 Programming Exercise: Run-Length Coding

1. Write a function to implement run-length coding of binary images.
2. Extend the function to perform bitplane RLC on gray images.
3. Modify the function so the user can select which bit planes to retain.
4. Incorporate the CVIPtools C functions *glr_compress* and *glr_decompress* into your CVIPlab program. Test it and compare results to those obtained with CVIPtools.

10.6.4 Programming Exercise: Block Truncation Coding

1. Write a function to implement basic BTC using a block size of 4×4.
2. Extend your function to allow the user to specify the block size.
3. Compare your results to those obtained with CVIPtools.

10.6.5 Programming Exercise: Differential Predictive Coding

1. Incorporate the CVIPtools C functions *dpc_compress* and *dpc_decompress* into your CVIPlab program. Test it and compare results to those obtained with CVIPtools.

10.6.6 Programming Exercise: Zonal Coding

1. Incorporate the CVIPtools C function *zon_compress* and *zon_decompress* into your CVIPlab program. Note: these functions include the DC term in the remapping, so it is quantized along with the higher frequency terms. Test it and compare results to those obtained with CVIPtools.
2. Incorporate the CVIPtools C function *zon2_compress* and *zon2_decompress* into your CVIPlab program. Note: these functions do not quantize the DC terms. Test it and compare results to those obtained with CVIPtools.

Section IV

Programming with CVIPtools

11

CVIPlab

11.1 Introduction to CVIPlab

The CVIPlab program was created to allow for experimentation with the CVIPtools functions outside of the CVIPtools environment. It is essentially a prototype program containing a sample CVIP function and a simple menu-driven user interface to ease program use. By following the format of this prototype function, and using library function prototypes (Chapter 12), and the *Help* pages (with CVIPtools) the user can implement any algorithms developed in the CVIPtools environment in their own standalone program. Additionally, the user can incorporate any of their own C or C++ functions into this program. To make it easy for those who are not experienced programmers, we have provided files and a brief tutorial for programming with Microsoft's Visual C++ 6.0 compiler. However, any compiler can be used by the experienced programmer.

In addition to the CVIPtools libraries, the CVIPlab program requires these three files: CVIPlab.c, threshold_lab.c, and CVIPlab.h. The CVIPlab.c file contains the main CVIPlab program, the threshold.c file contains a sample function, and the CVIPlab.h is a header file for function declarations. Additionally, for programming in Visual C++ 6.0, these files are required: CVIPlab_Project.rc, CVIPlab_Project.h, CVIPlab_Project.cpp, resource.h, Stdafx.cpp, Stdafx.h, CVIPlab_Project.dsw and CVIPlab_Project.dsp. CVIPlab_Project.h and CVIPlab_Project.cpp are generated by the project wizard in Visual C++ to support the empty console project. Stdafx.cpp and Stdafx.h are generated by the Visual C++ project wizard to support pre-compiled header files; resource.h and CVIPlab_Project.rc are to define resources used by the project.

CVIPlab.c contains a list of header files to include, function declarations, and three functions: *main*, *input*, and *threshold_setup*. The *main* function is declared as *void*, which means it does not return anything, and contains code for the menu-driven user interface for CVIPlab. The *input* function returns an image pointer and illustrates how to read an image file into a CVIPtools image structure, and display the resulting image. The *threshold_setup* function accepts an image pointer as input, gets the threshold value from the user, passes these to the *threshold_lab* function and returns an image pointer to the resultant image. The actual processing, in this case performing a threshold operation on an image, is done by the *threshold_lab* function which is contained in the file threshold_lab.c. By studying these functions the reader can see how to access and process image files using some of the CVIPtools library functions.

The CVIPlab.c program is commented to describe the details more completely, and is included here:

```
/* ===========================================================
  *
  *  Computer Vision and Image Processing Lab - Dr. Scott Umbaugh SIUE
  * ===========================================================
  *       File Name: CVIPlab.c
  *         Description: This is the skeleton program for the Computer Vision
  *                      and Image Processing Labs
  *  Initial Coding Date: April 23, 1996
  *     Last update Date: June 26, 2004
  *         Portability: Standard (ANSI) C
  *           Credit(s): Scott Umbaugh, Zhen Li, Kun Luo, Dejun Zhang
  *                      Southern Illinois University at Edwardsville

  *********************************************************************/
  /*
  ** include header files
  */

  #include  "CVIPtoolkit.h"
  #include  "CVIPconvert.h"
  #include  "CVIPdef.h"
  #include  "CVIPimage.h"
  #include  "CVIPlab.h"

  #define  CASE_MAX 10

  /* Put the command here, as VIDEO_APP, to run your image acquisition
  application program */
  #define VIDEO_APP "explorer.exe"

  /*
  ** function declarations
  */

  Image *threshold_Setup(Image *inputImage);
  Image *input();
  /*
  ** start main funct
  */
  void main_cviplab(){
      IMAGE_FORMAT      format;         /* the input image format */
      Image             *cvipImage;     /* pointer to the CVIP Image structure */
      Image             *cvipImage1;    /* pointer to the CVIP Image structure */
      char              *outputfile;    /* output file name */
      int               choice;
      CVIP_BOOLEAN      done=CVIP_NO;

      print_CVIP("\n\n\n\n***************************************");
      print_CVIP("**************************** ");
      print_CVIP("\n*\t\t Computer Vision and Image Processing Lab\t *");
      print_CVIP("\n*\t\t\t <Your Name Here> \t\t *");
      print_CVIP("\n******************************************");
      print_CVIP("*************************\n\n\n");
```

```
    while(!done) {
      print_CVIP("\t\t0.\tExit \n");
      print_CVIP("\t\t1.\tGrab and Snap an Image \n");
      print_CVIP("\t\t2.\tThreshold Operation \n");
      print_CVIP("\n\nCVIPlab>>");

      /*
      ** obtain an integer between 0 and CASE_MAX from the user
      */
      choice=getInt_CVIP(10, 0, CASE_MAX);

      switch(choice) {
      case 0:
           done=CVIP_YES;
           break;

      case 1:
           if (ShellExecute(NULL,"Open", VIDEO_APP,NULL,NULL, SW_SHOW)<=32)
           print_CVIP("Error while running Video Program");
           break;

      case 2:
           /*Get the input image */
           cvipImage=input();
           if(cvipImage==NULL)
           {
                error_CVIP("main", "could not read input image");
                break;
           }
           /* calls the threshold function */
           cvipImage=threshold_Setup(cvipImage);
           if (!cvipImage)
           {
               error_CVIP("main", "threshold fails");
               break;
           }
           /*
           ** display the resultant image
           */
           view_Image(cvipImage,"");
           delete_Image(cvipImage);

           break;

      default:
           print_CVIP("Sorry ! You Entered a wrong choice ");
           break;
           }
      }
}
/*** end of the function main*/

/*
** The following function reads in the image file specified by the user,
```

```
** stores the data and other image info. in a CVIPtools Image structure,
** and displays the image.
*/

Image* input(){
      char    *inputfile;
      Image    *cvipImage;
      /*
      ** get the name of the file and stores it in the string 'inputfile '
      */
      print_CVIP("\n\t\tEnter the Input File Name: ");
      inputfile=getString_CVIP();
      /*
      ** creates the CVIPtools Image structure from the input file
      */
      cvipImage=read_Image(inputfile, 1);
      if(cvipImage==NULL) {
          error_CVIP("init_Image", "could not read image file");
          free(inputfile);
          return NULL;
      }

      /*** display the source image*/
      view_Image(cvipImage,inputfile);

      /*
      **IMPORTANT: free the dynamic allocated memory when it is not needed
      */
      free(inputfile);

      return cvipImage;
}

/*
** The following setup function asks the threshold value from the user. After
** it gets the threshold value, it will call the threshold_Image() function.
*/
Image *threshold_Setup(Image *inputImage){
      unsigned int    threshval;    /* Threshold value */
/*
** Gets a value between 0 and 255 for threshold
*/

    print_CVIP("\n\t\tEnter the threshold value: ");
    threshval=getInt_CVIP(10, 0, 255);

    return threshold_lab(inputImage, threshval);
}
```

The following is the threshold function contained in the threshold_lab.c file. Note that it is a good idea in the programming exercises to append all your file and function names with something, such as _lab or your initials, to avoid compilation and linker naming conflicts.

```
/********************************************************************
* =================================================================
*
*  Computer Vision and Image Processing Lab - Dr. Scott Umbaugh SIUE
*
* =================================================================
*
*            File Name: threshold_lab.c
*          Description: it contains the function to threshold BYTE images
*  Initial Coding Date: April 23, 1996
*          Portability: Standard (ANSI) C
*            Credit(s): Zhen Li & Kun Luo
*                       Southern Illinois University at Edwardsville
*
** Copyright (c) 1995, 1996, SIUE - Scott Umbaugh, Kun Luo, Yansheng Wei
*********************************************************************/

/*** include header files*/

#include "CVIPtoolkit.h"
#include "CVIPconvert.h"
#include "CVIPdef.h"
#include "CVIPimage.h"
#include "CVIPlab.h"

/*
** The following function will compare the actual gray level of the input image
** with the threshold limit. If the gray-level value is greater than the
** threshold limit then the gray level is set to 255 (WHITE_LAB) else to
** 0 (BLACK_LAB). Note that the '_LAB' or '_lab' is appended to names used
** in CVIPlab to avoid naming conflicts with existing constant and function
** (e.g. threshold_lab) names.
*/

#define     WHITE_LAB  255
#define     BLACK_LAB  0

Image *threshold_lab(Image *inputImage, unsigned int threshval){
  byte            **image;       /* 2-d matrix data pointer */
  unsigned int    r,             /* row index */
                  c,             /* column index */
                  bands;         /* band index */
  unsigned int    no_of_rows,    /* number of rows in image */
                  no_of_cols,    /* number of columns in image */
                  no_of_bands;   /* number of image bands */
  /*
  ** Gets the number of image bands (planes)
  */
  no_of_bands = getNoOfBands_Image(inputImage);

  /*** Gets the number of rows in the input image*/

  no_of_rows =getNoOfRows_Image(inputImage);
```

```
  /*
  ** Gets the number of columns in the input image
  */
  no_of_cols=getNoOfCols_Image(inputImage);

  /*
  ** Compares the pixel value at the location (r,c)
  ** with the threshold value. If it is greater than
  ** the threshold value it writes 255 at the location
  ** else it writes 0. Note that this assumes the input
  ** image is of data type BYTE.
  */
  for(bands=0; bands<no_of_bands; bands++) {
    /*
    ** reference each band of image data in 2-D matrix form;
    ** which is used for reading and writing the pixel values
    */
    image=getData_Image(inputImage, bands);
    for(r=0; r<no_of_rows; r++) {
      for(c=0; c<no_of_cols; c++) {
        if(image[r][c] > (byte) threshval)
                image[r][c]=WHITE_LAB;
        else
          image[r][c]=BLACK_LAB;
        }
    }
  }
  return inputImage;
}
/*** end of function threshold_lab*/
```

11.2 Toolkits, Toolboxes and Application Libraries

All of the functions in the CVIPtools program are accessible to those programming with CVIPlab. The functions are arranged in a hierarchical grouping of libraries, with the Toolkit Libraries at the lowest level, the Toolbox Libraries at the next level, and the Application Libraries at the highest level, as illustrated in Figure 11.2.1. This hierarchical grouping is devised such that each successive level can use the building blocks (functions) available to it from the previous level(s).

The *Toolkit Libraries* contain low level functions, such as input/output functions, matrix manipulation functions, and memory management functions. The *Toolbox Libraries* are the primary libraries for use in application development; they contain the functions that are available from the GUI in CVIPtools, such as the many analysis, enhancement, restoration, compression, and utility functions. At the highest level, the *Application Libraries*, are the libraries generated by those using the CVIPlab environment to develop computer imaging applications. In some cases, useful functions are modified and extracted from an application and put into a Toolbox Library. Chapter 12 contains function prototypes for all Toolbox library functions, and some of the commonly used Toolkit functions. For more details and examples see

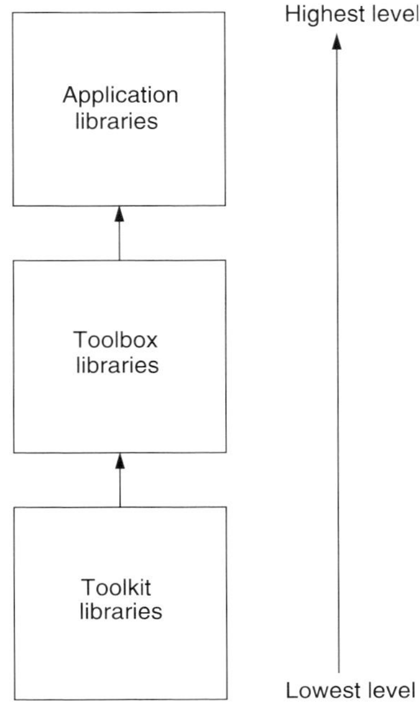

FIGURE 11.2.1
CVIPtools libraries.

the *Help pages* with CVIPtools, and for a quick look at all the available library functions see Appendix C.

11.3 Compiling and Linking CVIPlab

11.3.1 How to Build the CVIPlab Project with Microsoft's Visual C++ 6.0

1. Install CVIPtools, which includes the CVIPlab package, from the CD (see Appendix A).
2. Choose the desired location for the installation—in this guide, we use C:\CVIPtools4.3\CVIPlab as the working folder.
3. Run Microsoft's Visual C++ 6.0.
4. Open CVIPlab project file named *CVIPlab_Project.dsw* in C:\cviplab as shown in Figure 11.3.1.
5. Build the project by selecting *Build* → *Build CVIPlab_Project.exe* as shown in Figure 11.3.2.
6. Activate the output window by selecting *Output* from the *View* menu, or press Alt+2, (if it is not shown). *CVIPlab_Project.exe* should be compiled with 0 errors as in Figure 11.3.3a, and the executable file is located in C:\CVIPtools4.3\CVIPlab\ Debug. It should be noted that it is not unusual to get warning messages during compilation. These warning messages should be investigated as they may indicate poor programming practices that can cause problems. In this case, the last three warning messages are due to variables that are not referenced, meaning

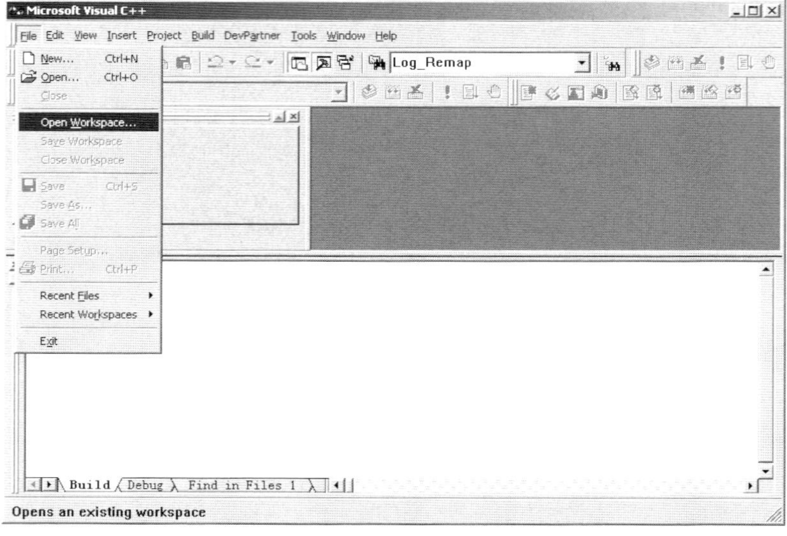

(a)

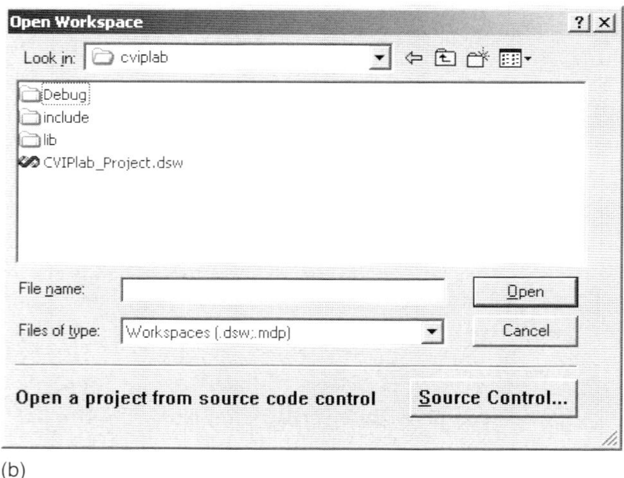

(b)

FIGURE 11.3.1
Opening CVIPlab_Project.dsw. (a) Select *Open Workspace*. (b) Select the file CVIPlab_Project.dsw.

they are not currently used in the program. Here, these variables are included for future use so we will not worry about them.

7. Press F5 to run the program with the debugger on; alternately, Cntrl-F5 will run it without the debugger. Select "2" and enter the file name for an image in the directory, or enter the full path name for an image elsewhere—here we used cam.pgm. Enter a threshold value to perform the threshold operation, as shown in Figure 11.3.3b. If you see this, you have compiled and run CVIPlab successfully!
8. If the computer you are using has video capture capability, you can add this to your CVIPlab as follows: Open file cviplab.c, go to the *#define VIDEO_APP* line directly after the include header section, modify the video/image capture string to the executable you plan to use (see Figure 11.3.4).

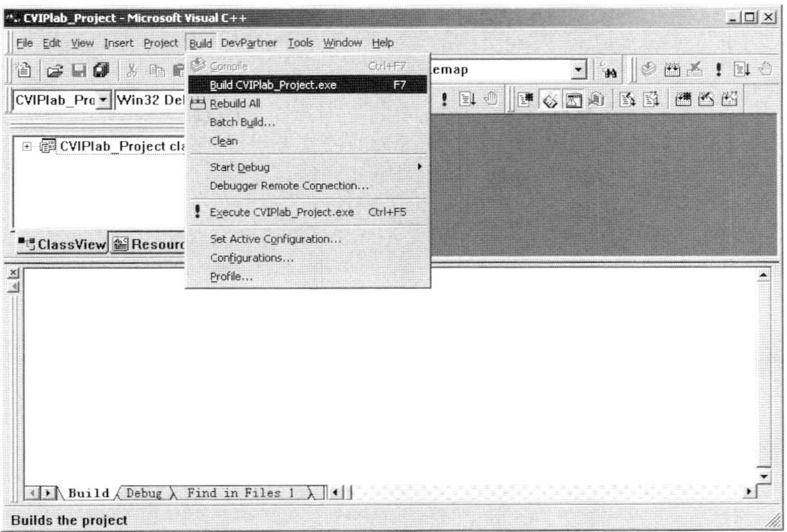

FIGURE 11.3.2
Building the project.

9. Run CVIPlab project by pressing F5. Select 1, and your video/image capture application will run.

11.3.2 The Mechanics of Adding a Function with Microsoft's Visual C++ 6.0

The following guide provides a step-by-step process for those unfamiliar with Visual C++. For those familiar with Visual C++, or those planning to use another compiler, skip to the next section, which provides details for adding a CVIP function to the CVIPlab menu. To add a function using Visual C++:

1. First create a new file by selecting *File->New* menu.
2. Select C++ source file, and input *test_new_file.c* as file name, as shown in Figure 11.3.5a.
3. Click the OK button and enter the text below in *test_new_file.c*, as shown in Figure 11.3.5b.

```
    int test_function(int i)
{
    return i+1;
}
```

4. Right click on *test_new_file.c* (located under CVIPlab_Project\Source Files) and select *Settings*. In the category combo box, select *Precompiled Headers* as shown in Figure 11.3.6a. Check *Not using precompiled headers*, and click OK shown in Figure 11.3.6b.
5. Press F7, or use the *Build* menu, to compile the project again. There should no error in the output box.
6. Double click on CVIPlab.h and CVIPlab.c to open them.

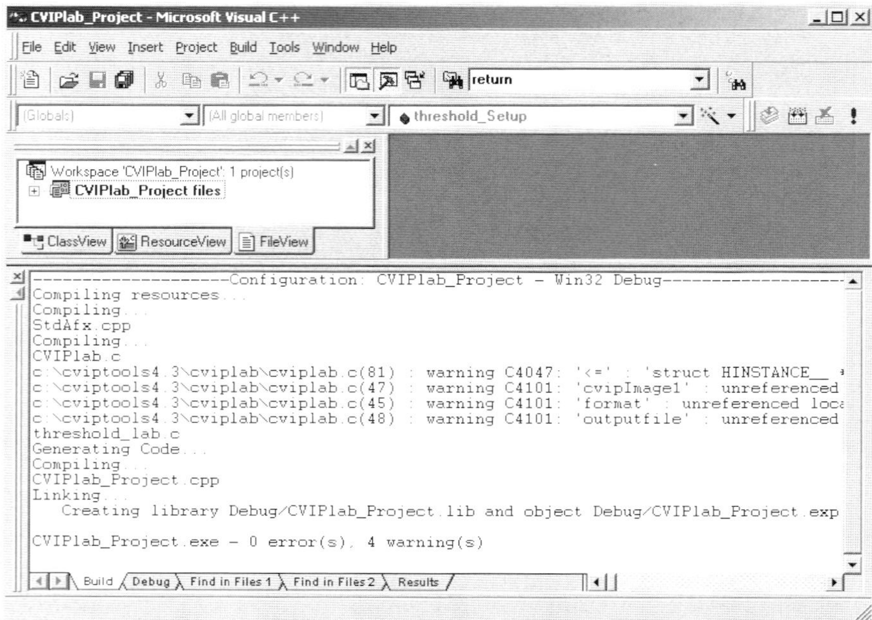

(a)

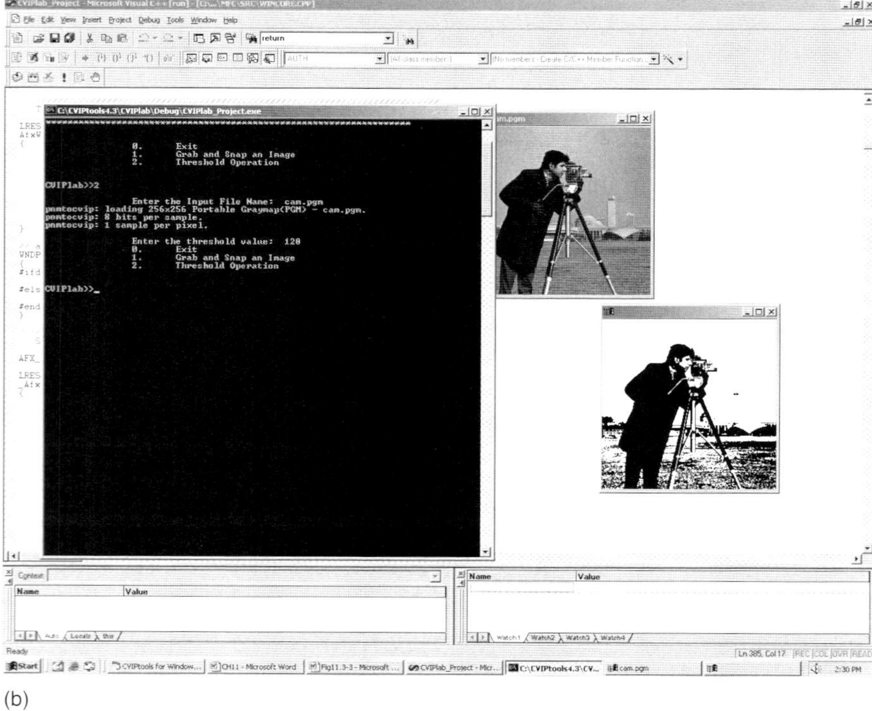

(b)

FIGURE 11.3.3
Compiling and running CVIPlab. (a) Screen after compilation with no errors. (b) Screen after running CVIPlab and performing a threshold operation on cam.pgm.

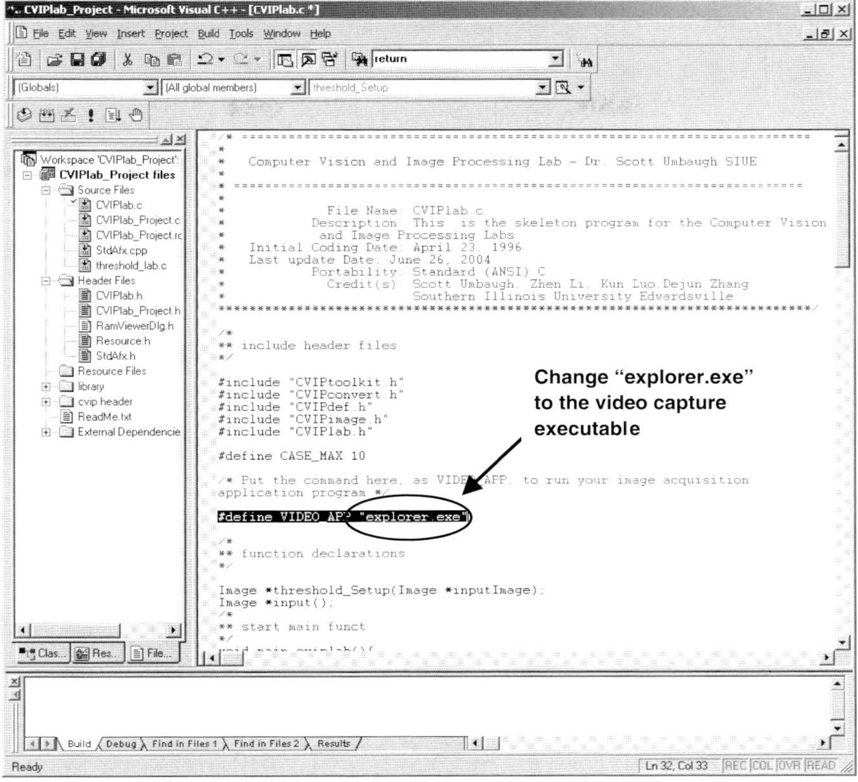

FIGURE 11.3.4
Adding your image capture program to CVIPlab.

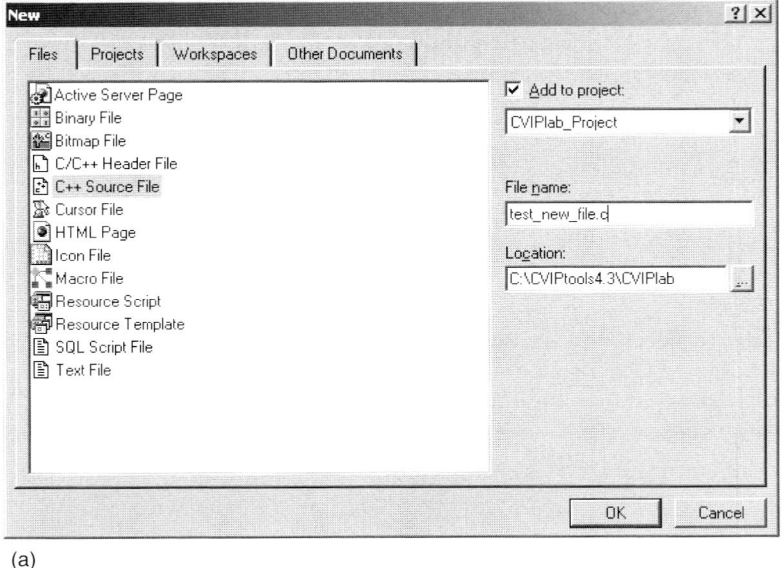

FIGURE 11.3.5
Adding a new function to CVIPlab. (a) Select *C++ Source File*.

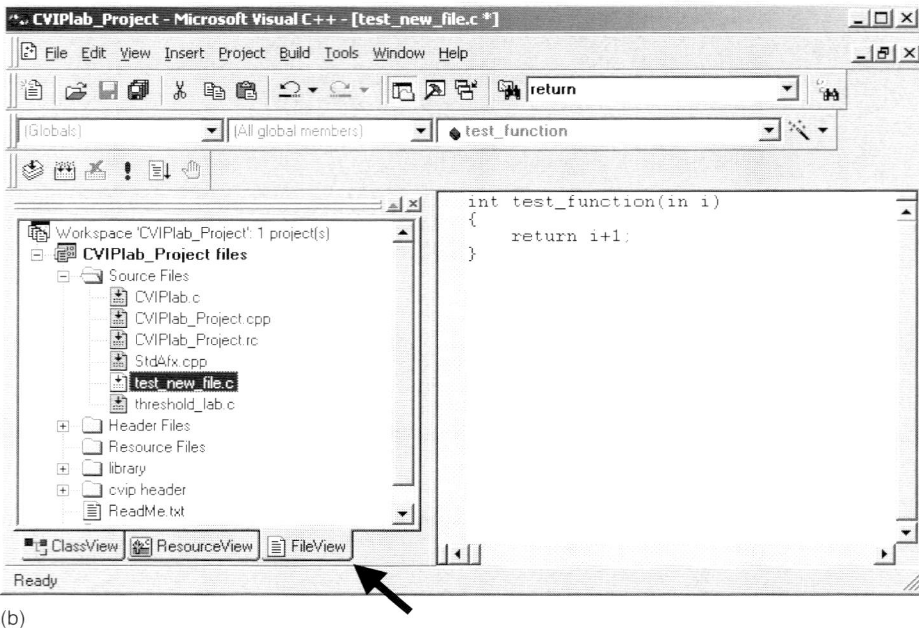

FIGURE 11.3.5 (Continued)
Adding a new function to CVIPlab. (b) The file name, *test_new file.c*, appears in the project file list—if the file names do not appear, click the *File View* tab below.

7. Find this line in the CVIPlab.h file:
   ```
   extern Image *threshold_lab(Image *imageP, unsigned int level)
   ```
   And directly after it add a new line (see Figure 11.3.7a):
   ```
   extern int test_function(int i);
   ```
8. Build the project again. It should pass the build without any error (Warning messages are OK).
9. Call this function in the main function of CVIPlab.c by inserting the following line:
   ```
   print_CVIP("test new function, return value is %d\n", rdquo;test_-
   function(1));
   ```
   after the function declarations; as shown in Figure 11.3.7b.
10. Press F7 to build the project. There should be no errors (again, you can ignore warnings).
11. Run the project by pressing F5. You should see:
    ```
    test new function, return value is 2
    ```
    in the second line of the DOS console window as shown in Figure 11.3.8. If you see this, you have added a new function successfully in the CVIPlab project.
12. Exit the compiler without saving these changes.

11.3.3 Using CVIPlab in the Programming Exercises

The previous section outlines the mechanics of adding a function with Visual C++. To follow the existing format of the program and organization of the files with any compiler

(a)

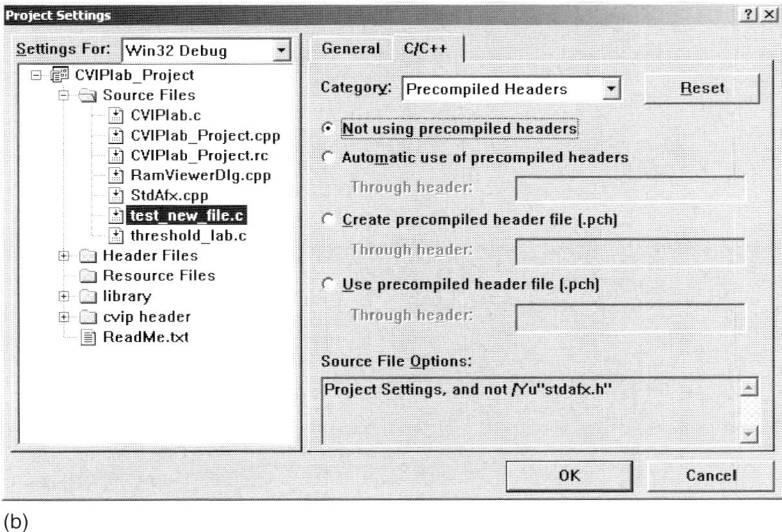

(b)

FIGURE 11.3.6
Settings for header files. (a) Select *Precompiled Headers*. (b) Check *Not using precompiled headers*.

(including Visual C++), do the following:

1. Create a file similar to threshold_lab.c for the *new_function*. The easiest method is to select the threshold.c file and perform a *Save* as the new_function.c. Next, edit the header to change the file name, description, modify the date, add your name, and change the old comments and the function name. The last step is to modify the code inside the band, row, and column *for* loop to perform the new function (see Figure 11.3.9).
2. Add the new function to the CVIPlab menu as shown in Figure 11.3.10a. Next, add a case statement for the function as shown in Figure 11.3.10b. The case statement

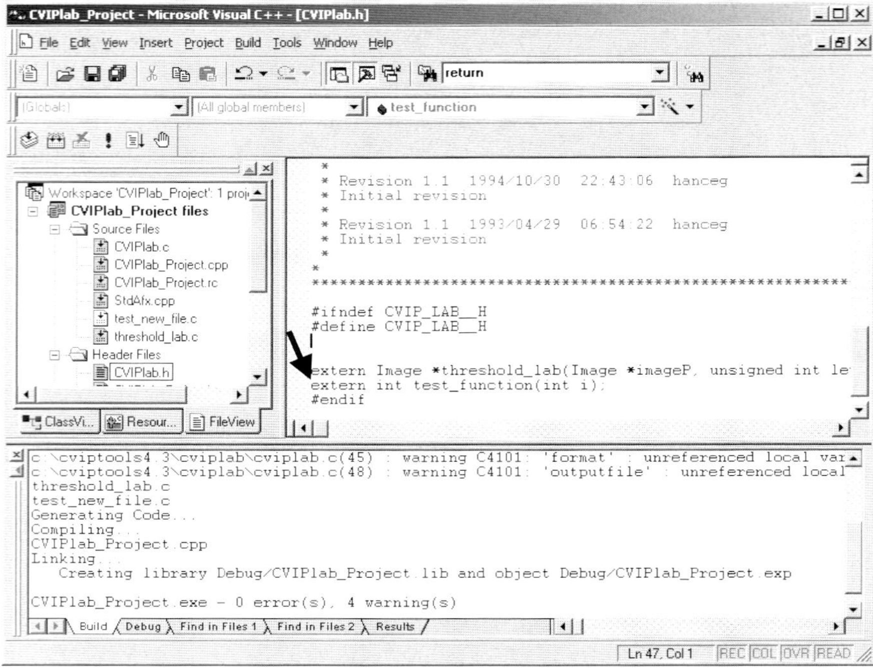

(a)

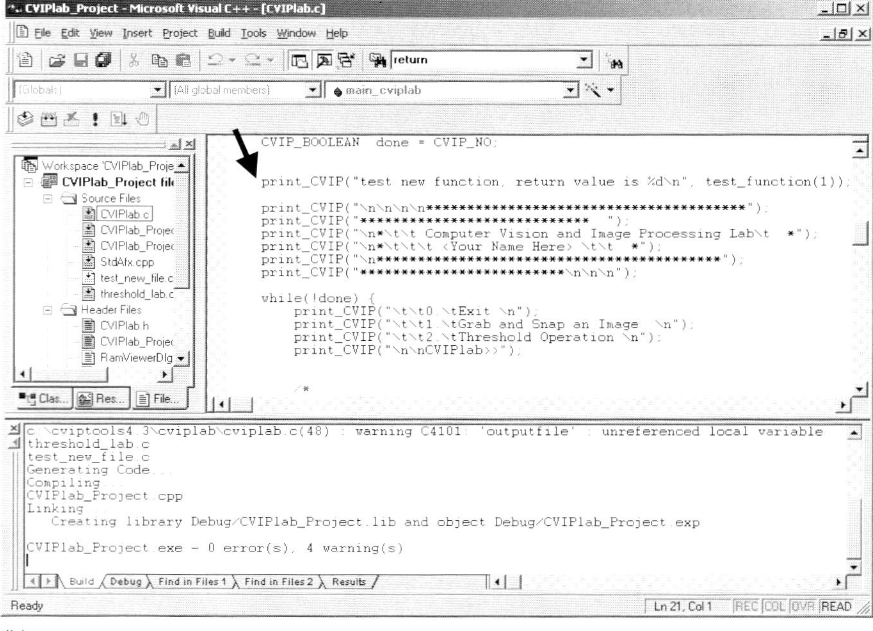

(b)

FIGURE 11.3.7
Editing CVIPlab.h and CVIPlab.c to add a new function. (a) Declare the new function in the header file CVIPlab.h. (b) Call the new function in CVIPlab.c.

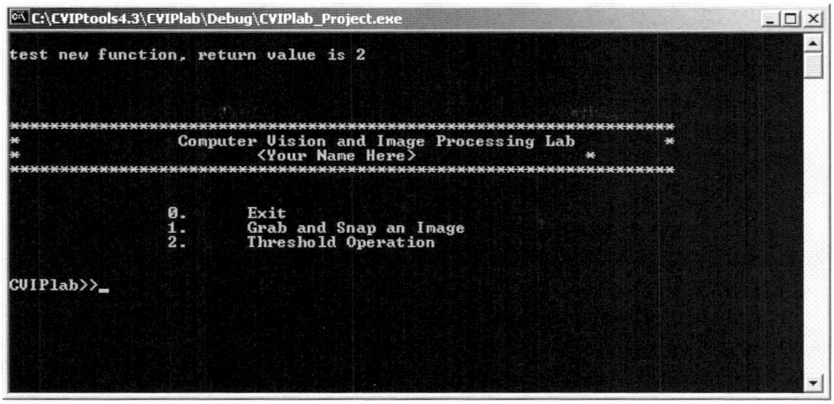

```
C:\CVIPtools4.3\CVIPlab\Debug\CVIPlab_Project.exe

test new function, return value is 2

*************************************************************************
*                 Computer Vision and Image Processing Lab              *
*                          <Your Name Here>                   *
*************************************************************************

                0.      Exit
                1.      Grab and Snap an Image
                2.      Threshold Operation

CVIPlab>>_
```

FIGURE 11.3.8
Output from your new added function, test_function.

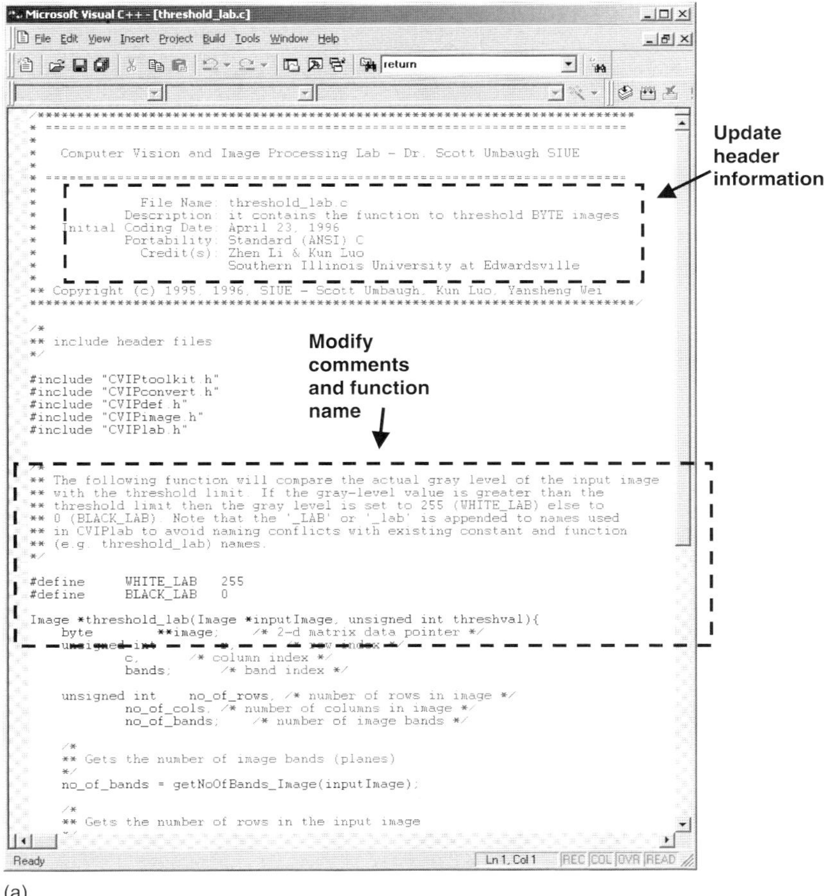

FIGURE 11.3.9
Create your new function by using the threshold function as a prototype. (a) Edit the header to change the file name, description, modify the date, add your name, and change the old comments and the function name.

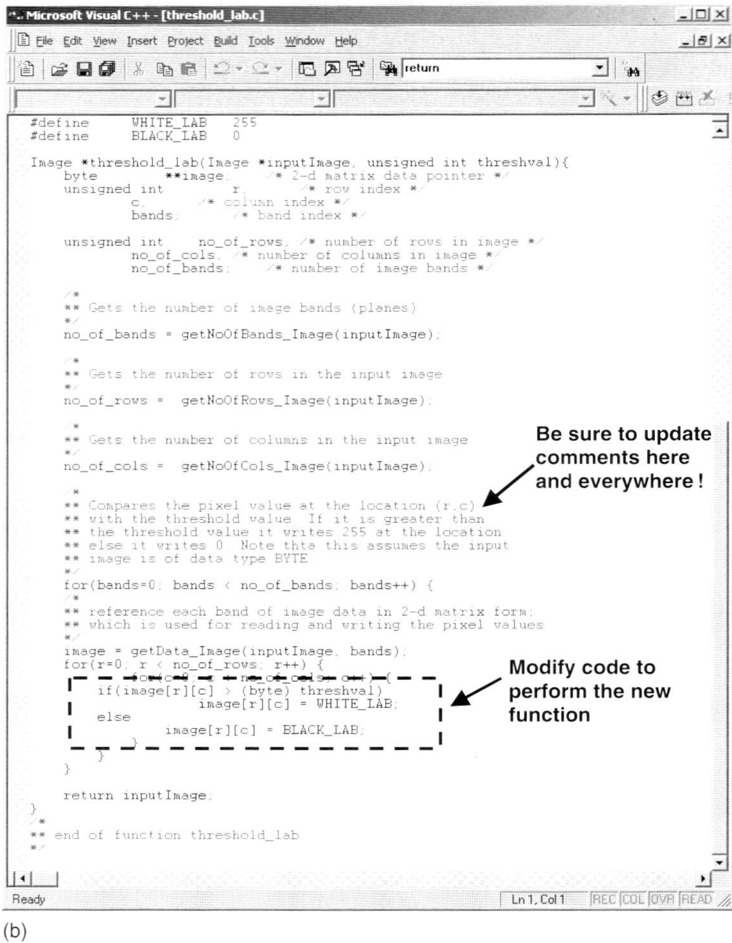

(b)

FIGURE 11.3.9 (Continued)
Create your new function by using the threshold function as a prototype. (b) The last step is to modify the code inside the band, row, and column *for* loop to perform the new function.

code for case 2 can be copied and used by modifying *threshold_Setup* to *new_function_Setup*.

3. Add the *new_function_Setup* to CVIPlab.c, similar to *threshold_Setup*.
4. Add the function prototype to the CVIPlab.h header file:
   extern Image **new_function*(new_function parameters . . .)

11.4 Image Data and File Structures

Details of the image data structures used in CVIPtools and CVIPlab are contained in this section. The CVIPlab programmer who is using the CVIPtools library functions typically does not need to understand all the details for using them. The library functions provide the user with a higher level interface so they can focus on learning about computer imaging. However, a basic understanding of the underlying data structures is necessary and useful for understanding problems that arise during development.

The data and file structures of interest are those that are required to process images. In traditional structured programming, a system can be modeled as a hierarchical set of functional modules, where the modules at one level are built of lower level modules. Similarly, the information used in CVIPlab, which consists primarily of image data, can use this hierarchical model. In this case we have a five-tiered model with the pixel data at the bottom, the vector data structure at the next level, the matrix data structure at the next level, image data structures next, and finally the image files at the top level. Figure 11.4.1a shows a triangle to illustrate this model, since it is naturally larger at the lower levels—it takes many pixels to make up a vector, many vectors to make an image, and so on.

In Figure 11.4.1b we see that a vector can be used to represent one row or column of an image, and the 2-D image data itself can be modeled by a matrix. The image data structure (Figure 11.4.1c) consists of a header that contains information about the type of image (see Section 2.4), followed by a matrix for each band of image data values. When the image data structure is written to a disk file, it is translated into the specified file format (for example, PostScript, TIFF, Sun Raster files, PPM, etc.). CVIPtools has its

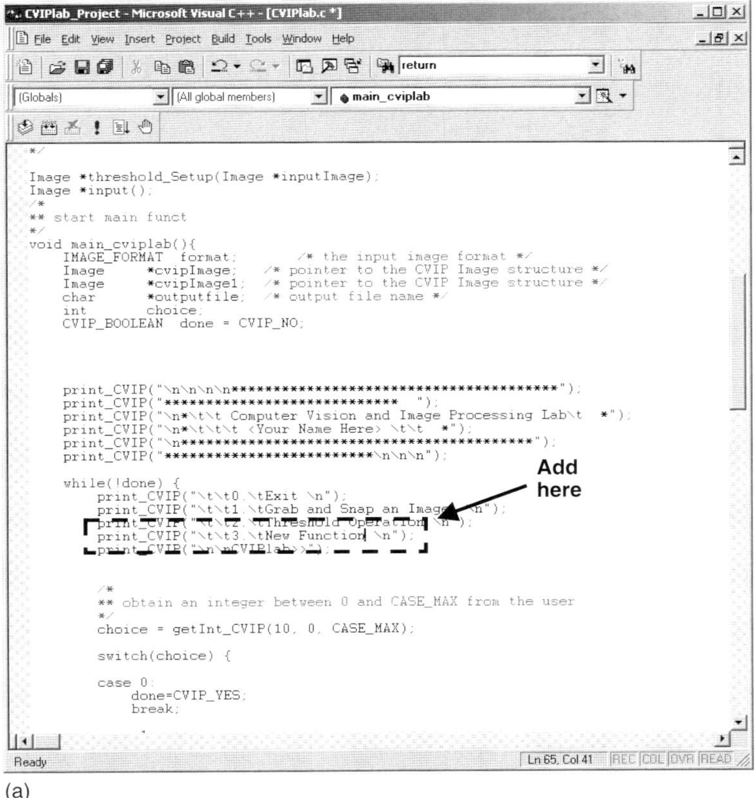

(a)

FIGURE 11.3.10
Add the case statement for the new function to CVIPlab. (a) Add your new function to the menu.

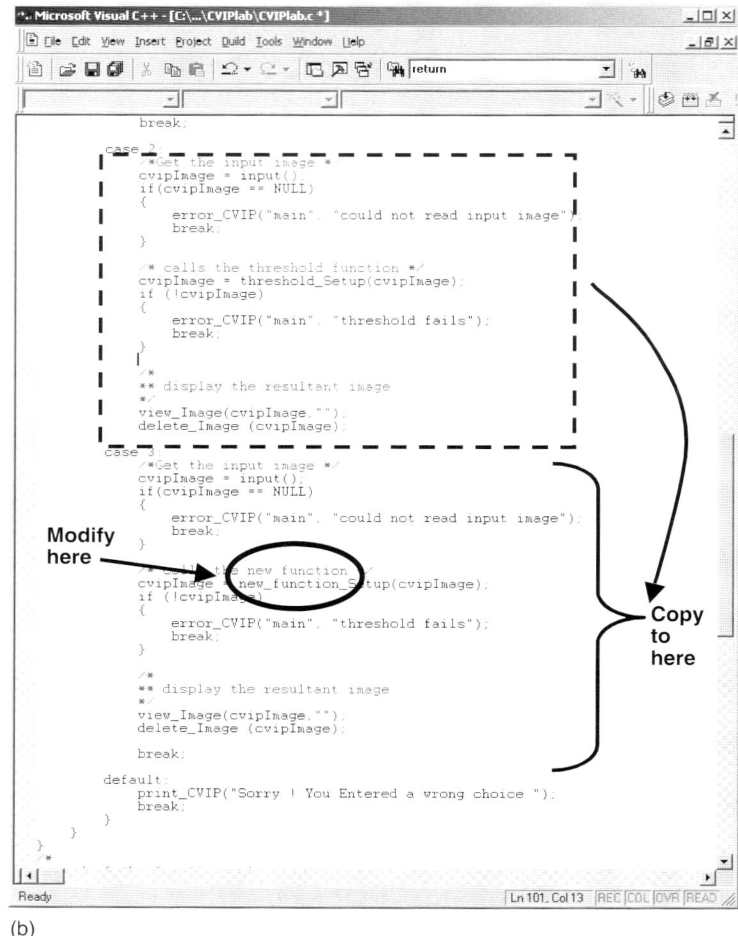

(b)

FIGURE 11.3.10 (Continued)
Add the case statement for the new function to CVIPlab. (b) Simply copy the statement from Case 2, and change the function name and update the comment.

own image file format, the Visualization in Image Processing (VIP) format, and also supports other standard file formats. Since most standard image files format assume 8-bit data, the VIP format is required for floating point data, complex data, as well as CVIPtools specific information.

The vector data structure can be defined by declaring an array in C of a given type, or by assigning a pointer and allocating a contiguous block of memory for the vector. A *pointer* is simply the address of the memory location where the data resides. In Figure 11.4.2 we see an illustration of a vector; the pointer to the vector is actually the address of the first element in the vector. For images, each element of the vector represents one pixel value, and the entire vector represents one row or column. The Vector library contains these functions.

The matrix structure is at the level above vectors. A matrix can be viewed as a one-dimensional vector, with M multiplied by N elements that has been mapped into a matrix with M rows and N columns. This is illustrated in Figure 11.4.3, where we see how

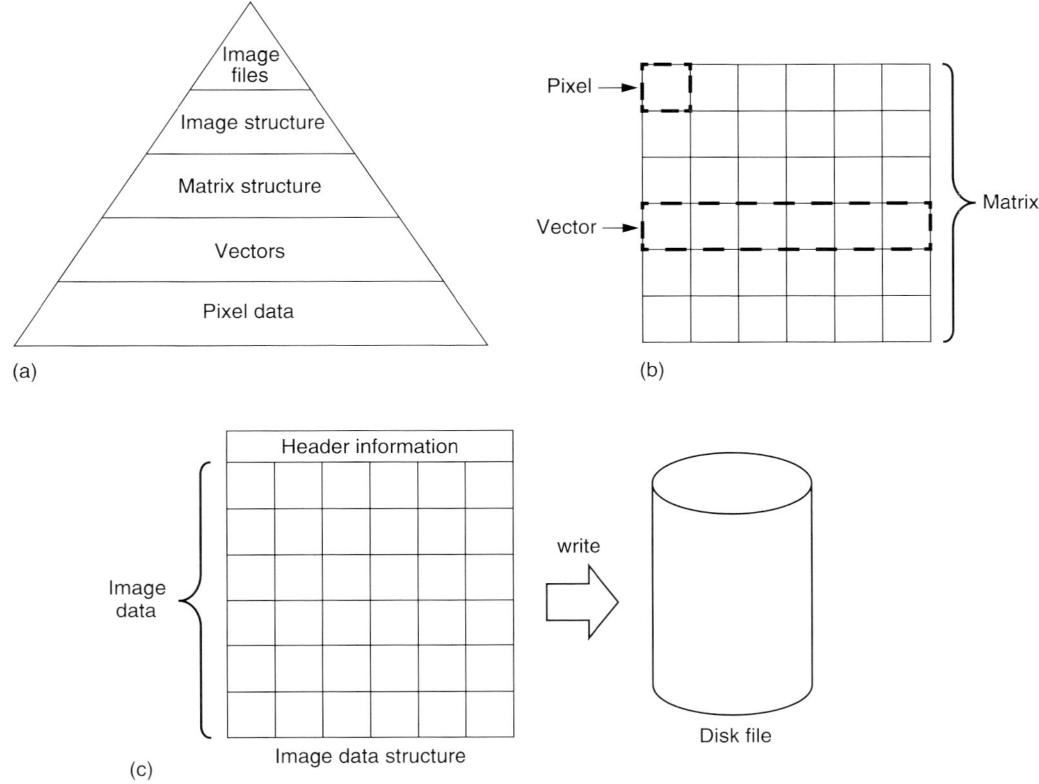

FIGURE 11.4.1
Image data and file structures. (a) Hierarchical model. (b) Image data representation. (c) Image data structure and disk file.

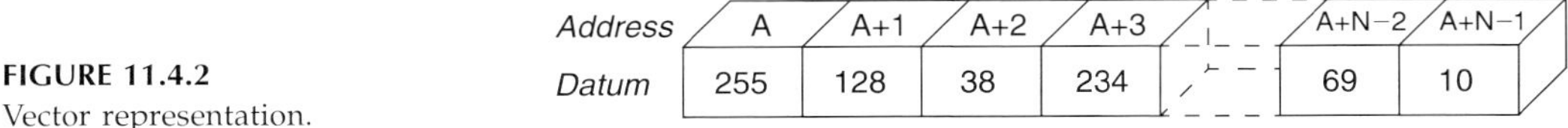

FIGURE 11.4.2
Vector representation.

a one-dimensional array can be mapped to a two-dimensional matrix via a pointer map. The matrix data structure is defined as follows:

```
typedef enum {CVIP_BYTE, CVIP_SHORT, CVIP_INTEGER, CVIP_FLOAT,
CVIP_DOUBLE} CVIP_TYPE;
  typedef enum {REAL, COMPLEX} FORMAT;
  typedef struct {
    CVIP_TYPE data_type;
    FORMAT data_format;
    unsigned int rows;
    unsigned int cols;
    void **rptr;     /*real data pointer*/
    void **iptr;     /*imaginary data pointer*/
  } Matrix;
```

The *data_type* field defines the type of data, such as BYTE or FLOAT, that is stored in the matrix. The *data_format* field describes whether the matrix elements are real or

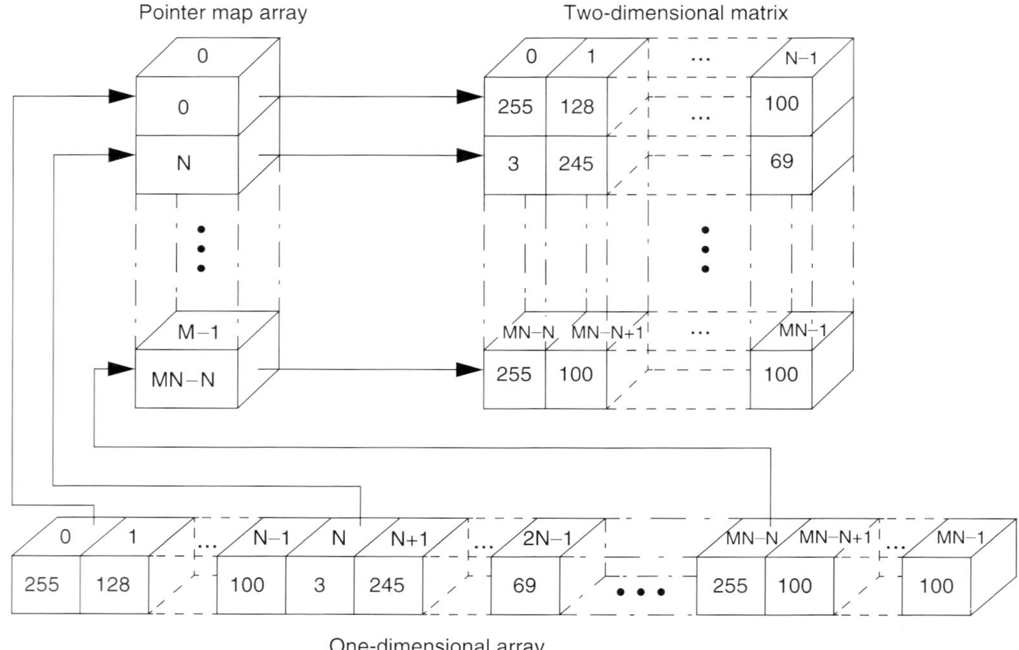

FIGURE 11.4.3
Matrices and pointers.

complex. The next two fields, *rows* and *cols*, contains the number of rows and columns in the matrix, and the last two, ***rptr* and ***iptr*, are two-dimensional pointers to the matrix elements (if the data_format is REAL, then the imaginary pointer is a null pointer). The Matrix library contains these functions, and the associated memory allocation and deallocation functions are called *new_Matrix* and *delete_Matrix*, respectively. The data type for the real and imaginary pointers is passed as a parameter to the function that creates and allocates memory for a matrix, the *new_Matrix* function. Once the matrix has been setup with *new_Matrix*, the data is accessed as a two-dimensional array by assigning a pointer with the *getData_Matrix* function; note that care must be taken to cast it to the appropriate data type.

The image structure is the primary data structure used for processing of digital images. It is at the level above the matrix data structure, since it consists of a matrix and additional information. The image data structure is defined as follows:

```
typedef enum {PBM, PGM, PPM, EPS, TIF, GIF, RAS, ITX,IRIS, CCC, BIN, VIP,
GLR, BTC, BRC, HUF, ZVL, ARITH, BTC2, BTC3, DPC, ZON, ZON2, SAFVR, JPG, WVQ,
FRA, VQ, XVQ, TRCM, PS} IMAGE_FORMAT;
  typedef enum {BINARY, GRAY_SCALE, RGB, HSL, HSV, SCT, CCT, LUV, LAB, XYZ}
  COLOR_FORMAT;
  typedef struct {
    IMAGE_FORMAT image_format;
    COLOR_FORMAT color_space;
    int bands;
    Matrix **image_ptr;
    HISTORY story;
  } IMAGE;
```

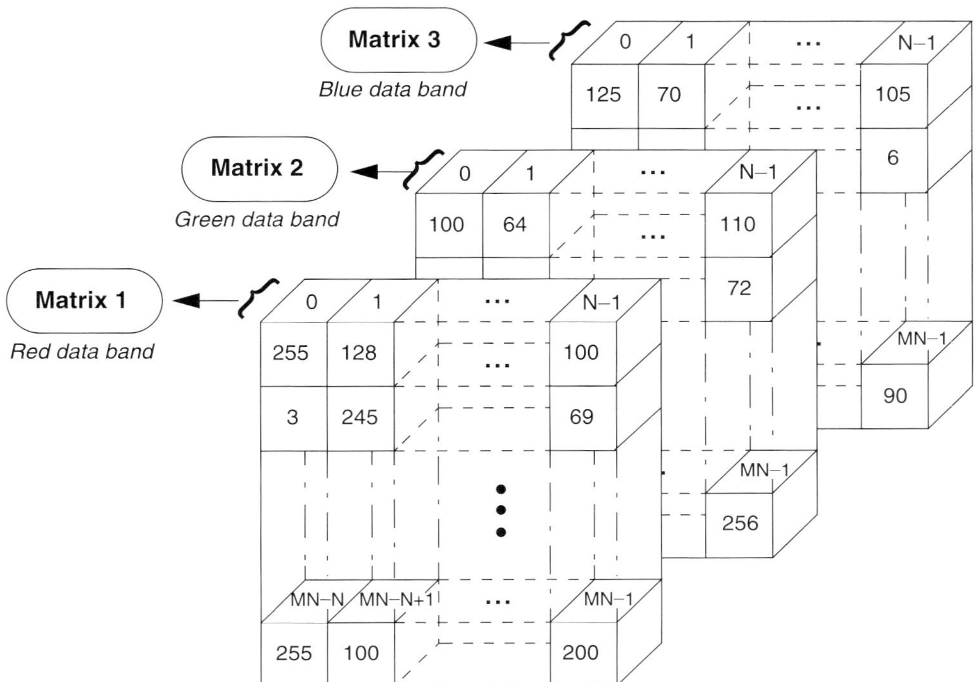

Representation of image data. In addition to the image data shown here, the image structure contains header information

FIGURE 11.4.4
Image data.

The first field, *image_format*, contains the file type of the original image. Note, however, that the image format does not necessarily tell us anything about the actual data in the image queue, especially after it has been processed. The second field, *color_space*, determines if the image is binary (two-valued), gray scale (typically 8-bit), or color (typically three-plane, 24-bit, RGB). If it is a color image, then this field is updated when a color space conversion is performed. The third field, *bands*, contains the number of bands in the image; for example a color image has three bands, and a gray scale image has one band. The next field, ***image_ptr*, is a pointer to an array of pointers to matrix data structures, where each matrix contains one band of pixel data (see Figure 11.4.4). The last field is for history information, and is used by the CVIPtools software to keep track of certain functions, such as transforms, that have been performed on an image in the CVIPtools image queue.

The history field in the image structure, *story*, is a pointer to a history data structure. The history data structure consists of packets of history information, where each *packet* contains information from a particular function. The history data structure is defined as follows:

```
typedef struct packet PACKET;
    struct packet {
      CVIP_TYPE *dtype;
      unsigned int dsize;
      void **dptr;
    };
```

```
typedef struct history *HISTORY;
  struct history {
    PROGRAMS ftag;
    PACKET *packetP;
    HISTORY next;
  };
```

Functions relating to the history are in the Image library, also see CVIPhistory.h.

At the highest level is the image file. The image file can be any of the types previously described as supported by CVIPtools. If the file is an 8-bit per pixel image file, which is typical, then CVIPtools may need to remap the data before it is written to a file. For example, if the range of the data is too large for 8-bits, if the data is in floating point format, or if it contains negative numbers, then the data in the image queue must be remapped before it can be written in 8-bit format. This is done automatically, if required. Keep in mind that any image that you see displayed is in the remapped format, so it will automatically be saved as what you see. However, in some cases, this may not be what is desired—we may want to retain the data as it is in the queue. To do this, the image must be saved in the CVIPtools image file format, the VIP format. The VIP file format allows the image data structure to be written to disk, and consists of an image header and the image data structure. The VIP structure is as follows:

```
VIP                -3 bytes (the ASCII letters "ViP")
  COMPRESS           -1 byte, ON or OFF, depending on whether the data is
  compressed)
  IMAGE_FORMAT       -1 byte, (defined in image data structure)
  COLOR_FORMAT       -1 byte, (e.g. BINARY, GRAYSCALE, RGB, LUV)
  CVIP_TYPE          -1 byte, data type, (e.g., CVIP_SHORT, CVIP_BYTE)
  NO_OF_BANDS        -1 byte, (1 for gray-level, 3 for color, other numbers
  allowed too)
  NO_OF_COLS         -2 bytes,
  NO_OF_ROWS         -2 bytes,
  FORMAT             -1 byte (REAL or COMPLEX)
  SIZEOF HISTORY     -4 bytes, (Size of history information, in bytes)
  HISTORY            -variable size (history information)
  RAW DATA           -variable size
```

If the file is a CVIPtools compressed image file, the first field in the RAW DATA corresponds to the type of compression that was performed (such as, btc, vq, etc.). Note that if the actual data stored in a VIP image file is examined, the number of bytes stored may vary from the above. This is due to the fact that we use the standard XDR (External Data Representation) functions to write the VIP files. By using these functions we assure file portability across computer platforms, but it results in most data types smaller than 32 bits (4 bytes) being written to the file in a standard 32-bit format. For example, in the Windows operating system the actual number of bytes stored in the image file is as follows:

```
VIP                -3 bytes (the ASCII letters "ViP")
  COMPRESS           -4 bytes, ON or OFF, depending on whether the data is
  compressed)
  IMAGE_FORMAT       -4 bytes, (defined in image data structure)
  COLOR_FORMAT       -4 bytes, (e.g., BINARY, GRAYSCALE, RGB, LUV)
```

```
CVIP_TYPE          -4 bytes, data type, (e.g. CVIP_SHORT, CVIP_BYTE)
NO_OF_BANDS        -4 bytes, (1 for gray-level, 3 for color, other numbers
allowed too)
NO_OF_COLS         -4 bytes,
NO_OF_ROWS         -4 bytes,
FORMAT             -4 bytes (REAL or COMPLEX)
SIZEOF HISTORY     -4 bytes, (Size of history information, in bytes)
HISTORY            - variable size (history information)
RAW DATA           - variable size
```

To use the VIP image file format, simply use the *read_Image* and *write_Image* functions contained in the Conversion library. These file read/write functions are in the Conversion library since a large portion of their functionality is to convert file types to and from the CVIPtools image data structure. These read/write functions will read/write any of the image file formats supported by CVIPtools, and require the programmer to deal with only one data structure, the image data structure.

11.5 CVIP Projects

11.5.1 Computer Vision Projects

The following process can be followed to streamline project development:

1. Use CVIPtools to explore algorithm development, using primarily the *Analysis* window for these types of projects.
2. First, apply segmentation by using the *Segmentation* window and/or edge detection.
3. Next, apply morphological filtering to reduce and solidify spatial objects (at bottom of *Segmentation* window).
4. Next, use the *Features* window to perform feature extraction.
5. Examine the feature file to come up with a classification algorithm. Or, if available, use the feature files as input to pattern classification or neural network software.
6. Code the algorithm developed into your CVIPlab program, as follows:

   (a) Find the function name that corresponds to the CVIPtools function (see Chapter 12 and the Appendices).
   (b) Use the Help with CVIPtools to see an example of how the function is used in a C program.
   (c) Do (a) and (b) for all functions needed.
   (d) Code the feature extraction algorithm with these functions in your CVIPlab program.
   (e) Code your classification algorithm into your CVIPlab program, or get the C function from your pattern classification software.

1. Test your program on real images input from an imaging device.

Project Topics

1. Implement a program for the recognition of geometric shapes; for example circles, squares, rectangles, and triangles. Make it robust so that it can determine if a shape is unknown. Images can be created with CVIPtools, but they should be blurred and noise added to them for realism. Alternately, the images can be captured with your image acquisition system by drawing the shapes, or by capturing images of real objects.
2. Experiment with the classification of tools; for example screwdrivers, wrenches, and hammers. Find features that will differentiate the classes of tools. Design a method for system calibration, and then identify various sizes of nuts and bolts, as well as number of threads per inch.
3. Implement a program to identify of different coins and bills. Make it robust so that it cannot be fooled by counterfeits.
4. Design a program to read bar codes. Experiment with different types of codes. Experiment with different methods to bring images into the system. In general, scanners will be easier to work with than cameras—verify this.
5. Implement a program to perform character recognition. Start with a subset, such as the numbers 0–9, and then expand using what you have learned. Experiment with different fonts, printers and lighting conditions.
6. Imagine we are designing a robotic system to put dishes in the dishwasher after a meal. Collect an image database of cups, glasses, bowls, plates, forks, knives, and spoons. Develop an algorithm to identify the objects.

11.5.2 Digital Image Processing Projects

The following process can be followed to streamline project development:

1. Use CVIPtools to explore algorithm development. Explore various options, sequences of operations, various parameter values, and so on, until desired results are achieved. Be creative, the beauty of CVIPtools is that anything can be tried and tested in a matter of seconds, and the results from different parameter values can easily be compared side by side.
2. Code the algorithm developed into the CVIPlab program, as follows:
   (a) Find the function name that corresponds to the CVIPtools function (see Chapter 12 and the Appendices).
   (b) Use the Help with CVIPtools to see an example of how the function is used in a C program.
   (c) Do (a) and (b) for all functions needed.
3. Put the functions into the CVIPlab program.
4. Write the necessary drivers to use the functions to implement the algorithm developed.
5. Test your program on images suitable for the application.

Project Topics

1. Find images that you want to process and improve. These may be images of poor contrast, blurred images, etc. Examples: personal photos taken by "Uncle Bob,"

medical images such as x-ray images, "UFO" images. Use CVIPtools to explore image enhancement.

2. Incorporate the CVIPtools function *gray_linear* (in the Histogram library) into your CVIPlab. Use this function to implement gray-level mapping pseudocolor. Apply it to x-ray images.
3. Implement a program for frequency domain pseudocolor enhancement. Experiment with the FFT, DCT, Haar, Walsh and Wavelet transforms. Apply it to ultrasound images.
4. Find images that have been degraded, for which a model is available or can be developed for the degradation. For example, satellite images, medical images, images from news stories (e.g., the JFK assassination). Use CVIPtools to explore image restoration.
5. Define a specific image domain of interest. Collect a number of these types of images. Explore image compression with CVIPtools. Design and perform subjective tests and compare the results to objective measures available in CVIPtools such as signal-to-noise ratio and RMS error.
6. Collect an image database of text images. These may be scans of typed material or images of license plates. Use *Utilities->Filter->Specify a Blur* to blur the images. Then add noise with *Utilities->Create->Add Noise* to add noise to the blurred images. Next, experiment with *Restoration->Frequency Filters* to restore images.

12

CVIPtools C Function Libraries

12.1 Introduction and Overview

This chapter contains a brief description of each of the CVIPtools Toolbox libraries, and prototypes for all the functions. Some of the commonly used Toolkit functions from the *Band*, *Mapping*, and *Image* libraries are also included. The libraries are in alphabetical order, as are all the functions contained in each library. Additionally, information about related functions are included to ease the function search process. This information will facilitate the use of these functions in the CVIPlab program, or any other C program.

In general, many functions return pointers to CVIPtools Image structures (IMAGE or Image are both valid designations). If the return value is NULL, an error has occurred. The general philosophy regarding memory management is that *whoever has control is responsible*. This means that any parameters passed to a function will be either used for return data or freed. It also means that if a programmer wants to retain a data structure, they should pass a *copy* of it to any CVIPtools function. By clearly following this simple rule, memory leaks can be avoided. For details of the operation of a specific function, see the *Help* pages in the CVIPtools. For a quick look at the function list see Appendix C.

12.2 Arithmetic and Logic Library—ArithLogic.lib

Functions for the application of arithmetic and logic operations to images are contained in this library. These functions require one or two Image pointers as input and all return an Image pointer. Related functions, specifically multiplication functions that perform gray-level mapping, are in the library *Histogram.lib*.

ARITHLOGIC LIBRARY FUNCTION PROTOTYPES

Image ***add_Image**(Image *inputIMAGE1, Image *inputIMAGE2)
<inputIMAGE1>– pointer to an image
<inputIMAGE2>– pointer to an image

Image ***and_Image**(Image *inputIMAGE1, Image *inputIMAGE2)
<inputIMAGE1>–pointer to an image
<inputIMAGE2>–pointer to an image

Image ***divide_Image***(Image *inputIMAGE1, Image *inputIMAGE2)
<inputIMAGE1> – pointer to an image
<inputIMAGE2> – pointer to an image

Image ***multiply_Image***(Image *inputIMAGE1, Image *inputIMAGE2)
<inputIMAGE1> – pointer to an image
<inputIMAGE2> – pointer to an image

Image ***not_Image***(Image *inputIMAGE)
<inputIMAGE> – pointer to an image

Image ***or_Image***(Image *inputIMAGE1, Image *inputIMAGE2)
<inputIMAGE1> – pointer to an image
<inputIMAGE2> – pointer to an image

Image ***subtract_Image***(Image *inputIMAGE1, Image *inputIMAGE2)
<inputIMAGE1> – pointer to an image
<inputIMAGE2> – pointer to an image

Image ***xor_Image***(Image *inputIMAGE1, Image *inputIMAGE2)
<inputIMAGE1> – pointer to an image
<inputIMAGE2> – pointer to an image

12.3 Band Image Library—Band.lib

Although *Band.lib* is a Toolkit library, consisting of lower level functions, these two are of particular utility so they are listed here. These functions allow for processing of individual bands of multiband images.

Image ***assemble_bands***(Image **inImgs, int noimgs);
<inImgs> – Pointer to array of image pointers
<noimgs> – number of image pointers contained in the array.

Image ***extract_band***(Image *inImg, int bandno);
<inImgs> – Pointer to image
<noimgs> – band number to be extracted from the image.

12.4 Color Image Library—Color.lib

The color library, *Color.lib*, primarily contains functions that modify color image information by a color transform. This includes principal components, luminance and various color space transforms. In addition, the frequency domain pseudocolor is contained here, but the gray level mapping pseudocolor that appears in CVIPtools uses the gray-level linear transform contained in *Histogram.lib*.

COLOR LIBRARY FUNCTION PROTOTYPES

Image ***colorxform**(const Image *rgbIMAGE, COLOR_FORMAT newcspace, float *norm, float *refwhite, int dir)
<rgbIMAGE> – pointer to an image (data type equal to or less precise than type CVIP_FLOAT)
<newcspace> – desired color space, one of: RGB, HSL, HSV, SCT, CCT, LUV, LAB, XYZ
<norm> – pointer to a normalization vector
<refwhite> – pointer to reference white values (for LUV and LAB only)
<dir> – direction of transform (1 = > (RGB->newcspace) else (newcspace->RGB)

Image ***luminance_Image**(Image *inIm)
<inIm> – pointer to an image

Image ***lum_average**(Image *input_Image)
<input_Image> – pointer to an image

Image ***ipct**(Image *imgP, CVIP_BOOLEAN is_mask, float *maskP)
<imgP> – pointer to an image
<is_mask> – whether to ignore a background color (CVIP_YES or CVIP_NO)
<maskP> – background color to ignore

Image ***pct**(Image *imgP, CVIP_BOOLEAN is_mask, float *maskP)
<imgP> – pointer to an image
<is_mask> – whether to ignore a background color (CVIP_YES or CVIP_NO)
<maskP> – background color to ignore

Image ***pct_color**(Image *imgP, CVIP_BOOLEAN is_mask, float *maskP, int choice)
<imgP> – pointer to Image structure
<is_mask> – whether to ignore a background color (CVIP_YES or CVIP_NO)
<maskP> – background color to ignore
<choice> – 1 = perform PCT, 2 = perform IPCT

Image ***pseudocol_freq**(Image *grayIMAGE,
int inner, int outer, int blow, int bband, int bhigh)
<grayIMAGE> – input gray image
<inner> – low cutoff frequency
<outer> – high cutoff frequency
<blow> – map lowpass results to band # (R = 0, G = 1, B = 2)
<bband> – map bandpass results to band # (R = 0, G = 1, B = 2)
<bhigh> – map highpass results to band # (R = 0, G = 1, B = 2)
(note: blow ! = bband ! = bhigh)

12.5 Compression Library—Compression.lib

The image compression library contains functions that compress and decompress images, as well as associated functions. The compression functions write the compressed data file to disk, and return a 0 upon successful completion and a −1 if an error occurs. The compressed data file is either in CVIPtools VIP format or a standard compression file format such as JPEG. The decompression functions take file names as input, and output Image pointers to the decompressed image. Two utility functions, *rms_error* and *srn*, which return the root-mean-square error and signal-to-noise ratio are included.

COMPRESSION LIBRARY FUNCTION PROTOTYPES

int **bit_compress**(Image *inputIMAGE, char *filename, byte sect)
<inputIMAGE> – pointer to the image
<filename> – pointer to a character array
<sect> – bitmask of planes to retain

Image ***bit_decompress**(char *filename)
<filename> – pointer to a character string containing the file name

Image ***bit_planeadd**(char *filename)
<filename> – pointer to a character array

int **btc_compress**(Image *inputIMAGE, char *filename)
<inputIMAGE> – pointer to the image

Image ***btc_decompress**(char *filename)
<filename> – pointer to a character string containing the file name

int **btc2_compress**(Image *inputIMAGE, char *filename, int blocksize)
<inputIMAGE> – pointer to an image
<filename> – pointer to character string containing the output file name
<blocksize> – blocksize

Image ***btc2_decompress**(char *fiilename)
<filename> – pointer to character string containing filename

int **btc3_compress**(Image *inputIMAGE, char *filename, int blocksize)
<inputIMAGE> – pointer to an image
<filename> – pointer to character string containing output filename
<blocksize> – blocksize

Image ***btc3_decompress**(char *filename)
<filename> – pointer to a character string containing the file name

int **dpc_compress**(Image *inputIMAGE, char *filename, float ratio, int bit_length, int clipping, int direction, int origin)
<inputIMAGE> – pointer to an image
<filename> – pointer to character string containing output filename
<ratio> – the correlation factor
<bit_length> – number of bits for compression (1 to 8)
<clipping> – clip to maximum value (1), otherwise 0
<direction> – scan image horizontally (0) or vertically (1)
<origin> – use original (1) or reconstructed (0) values

Image ***dpc_decompress**(char *filename)
<filename> – pointer to a character string containing the name of the compressed file

int **frac_compress**(Image *inputImage, char *filename, double tol, int min_part1, int max_part1, int dom_type1, int dom_step1, char c1, char c2, int s_bits1, int o_bits1)
<inputImage> – pointer to an Image structure
<filename> – character array
<tol> – tolerance value
<min_part1> – recursion size min.
<max_part1> – recursion size max.
<dom_type1> – domain type
<dom_step1> – domain step
<c1> – character(y/n) for searching 24 domain classes
<c2> – character(y/n) for searching 3 domain classes
<s_bits1> – scaling bits
<o_bits1> – offset bits

Image ***frac_decompress**(char *filename)
<filename> – name of the compressed file

int **glr_compress**(Image *inputIMAGE, char *filename, int win)
<inputIMAGE> – pointer to an image
<filename> – pointer to character string containing output filename
<win> – size of window (1–128)

Image ***glr_decompress**(char *filename)
<filename> – name of the compressed file

int **huf_compress**(Image *inputIMAGE, char *filename)
<inputIMAGE> – pointer to the image
<filename> – pointer to character string containing output filename

Image ***huf_decompress**(char *filename)
<filename> – pointer to character string containing compressed filename

int **jpg_compress**(Image *cvipImage, char *filename, int quality, CVIP_BOOLEAN grayscale, CVIP_BOOLEAN optimize, int smooth, CVIP_BOOLEAN verbose, char *qtablesFile)
<cvipImage> – pointer to the image
<filename> – pointer to character string containing filename
<quality> – quality factor, determines amount of compression
<grayscale> – output image grayscale only (CVIP_YES or CVIP_NO)?
<optimize> – fast or slower (better results) (CVIP_YES or CVIP_NO)?
<smooth> – smooth out artifacts, (CVIP_YES or CVIP_NO)?
<verbose> – text messages during compression (CVIP_YES or CVIP_NO)?
<qtablesFile> – pointer to file containing user sepcified quantization tables (NULL pointer will use default tables)

Image ***jpg_decompress**(char *filename, int colors, CVIP_BOOLEAN blocksmooth, CVIP_BOOLEAN grayscale, CVIP_BOOLEAN nodither, CVIP_BOOLEAN verbose)
<filename> – pointer to character string containing compressed filename
<colors> – number of colors to use
<blocksmooth> – postprocess to improve visual results for block artifacts (CVIP_YES or CVIP_NO)?

<grayscale> – output image grayscale (CVIP_YES or CVIP_NO)?
<nodither> – use no dithering on the output mage (CVIP_YES or CVIP_NO)?
<verbose> – text messages during compression (CVIP_YES or CVIP_NO)?

float ***rms_error**(Image *im1, Image *im2)
<im1> – Pointer to Image
<im2> – Pointer to Image

float ***snr**(Image *im1, Image *im2)
<im1> – Pointer to Image
<im2> – Pointer to Image

int **vq_compress** (Image *inputImage, char *outfile_name, int cdbook_in_file, int fixed_codebook, float in_error_thres, char *cdbook_file, int in_no_of_entries, int in_row_vector, int in_col_vector, XFORM_FMT xform)
<inputImage> – pointer to an Image structure
<outfile_name> – output file name
<cdbook_in_file> – codebook file writing control
<fixed_codebook> – codebook file reading control
<in_error_thres> – distortion control
<cdbook_file> – the codebook file name
<in_no_of_entries> – total number of vectors in the code-book
<in_row_vector> – total number of rows in a vector
<in_col_vector> – total number of cols in a vector
<xform> – what kind of transform and compression ratio are used

Image ***vq_decompress**(char *filename)
<filename> – pointer to a character string containing the compressed file name

Image ***xvq_compress**(Image *image, int xform, int scheme, char *filename, int file_type, int remap_type, int dc)
<image> – input image
<xform> – transform domain: 1 – DCT, 2 – PCT_DCT, 3 – WAVELET, 4 – PCT_WAVELET
<scheme> – VQ schemes: 1 – 8 – compression schemes in the domain mentioned above; 9 – customize the vector sets, but this can only be used in CVIPtools; 10 – VQ in spatial domain.
<filename> – temporary file name
<filetype> – the file type you want to save as: 1 – CVIP_FLOAT, 2 – CVIP_BYTE, 3 – CVIP_SHORT
<remap_type> – the remap method that you should choose when you save file as CVIP_BYTE data type or CVIP_SHORT data type: 1 – linear remap, 2 – log remap
<dc> – indicates whether you want to quantize DC term when doing vector quantization (only valid when you want to save file as CVIP_BYTE data type or CVIP_SHORT data type): 0 – quantize DC term, 1 – separate DC term and keep it in history

Image ***xvq_decompress**(char *filename)
<filename> – character array

int **zon_compress**(Image *inputIMAGE, char *filename, int block_size, int choice, int mask_type, float compress_ratio)
<inputIMAGE> – pointer to an Image

<filename> – pointer to character string containing output filename
<block_size> – a power of 2; kernel size is <block_size>^2
<choice> – transform to use: 1 = FFT 2 = DCT 3 = Walsh 4 = Hadamard
<mask_type> – type of kernel to use: 1 = triangle 2 = square 3 = circle
<compress_ratio> – compression ratio, from 1.0 (min) to (block_size*block_size/4) (max) for all kinds of transforms

Image ***zon_decompress**(char *filename)
<filename> – pointer to a character string containing the compressed file name

int **zon2_compress**(Image *inputIMAGE, char *filename, int block_size, int choice, int mask_type, float compress_ratio)
<inputIMAGE> – pointer to an Image
<filename> – pointer to character string containing output filename
<block_size> – a power of 2; kernel size is <block_size>^2
<choice> – transform to use: 1 = FFT 2 = DCT 3 = Walsh 4 = Hadamard
<mask_type> – type of kernel to use: 1 = triangle 2 = square 3 = circle
<compress_ratio> – compression ratio, from 1.0 (min) to (block_size*block_size/4) (max) for all kinds of transforms

Image ***zon2_decompress**(char *filename)
<filename> – pointer to a character string containing the compressed file name

int **zvl_compress**(Image *inputIMAGE, char *filename)
<inputIMAGE> – pointer to an image
<filename> – pointer to character string containing output filename

Image ***zvl_decompress**(char *filename)
<filename> – pointer to character string containing the compressed file name

12.6 Conversion Library—Conversion.lib

The conversion library contains all the functions that convert the various image file types to the CVIPtools Image structure, and back from the Image structure to the file type. However, the programmer does not need to use these functions directly, since the higher level read and write image functions (*read_Image* and *write_Image*) take care of any required overhead. The function that converts between gray code and natural binary code and a halftoning function are also in this library.

CONVERSION LIBRARY FUNCTION PROTOTYPES

Image ***bintocvip**(char *raw_image, FILE *inputfile, int data_bands, COLOR_ORDER color_order, INTERLEAVE_SCHEME interleaved, int height, int width, CVIP_BOOLEAN verbose)

Image ***bmptocvip**(char *name, FILE *in, int imageNumber, int showmessage)

Image ***ccctocvip**(char *prog_name, FILE *cccfile, int verbose)

Image ***CVIPhalftone**(Image *cvip_IMAGE, int halftone, int maxval, float fthreshval, CVIP_BOOLEAN retain_image, CVIP_BOOLEAN verbose)
<cvip_IMAGE> – pointer to input image
<halftone> – indicates method used to convert from grayscale to binary. (one of QT_FS, QT_THRESH, QT_DITHER8, QT_CLUSTER3, QT_CLUSTER4, QT_CLUSTER8)
<maxval> – specifies maximum range of input image (usually 255)
<fthreshval> – threshold value (for QT_THRESH) between [0.0 1.0].
<retain_image> – retain image after writing
<verbose> – shall I be verbose (CVIP_YES or CVIP_NO)?

void **cviptobin**(Image *raw_IMAGE, char *raw_image, FILE *outputfile, COLOR_ORDER color_order, INTERLEAVE_SCHEME interleaved, CVIP_BOOLEAN verbose)

void **cviptobmp**(Image *raw_Image, char *raw_image, FILE *outputfile, CVIP_BOOLEAN verbose)

void **cviptoccc**(Image *cvip_IMAGE, char *ccc_name, FILE *cccfile, int maxcolor, int dermvis, int verbose)

void **cviptoeps**(Image *cvip_IMAGE, char *eps_name, FILE *outputfile, float scale_x, float scale_y, CVIP_BOOLEAN verbose)

void **cviptogif**(Image *gif_IMAGE, char *gif_name, FILE *outfp, int interlace, int verbose)

void **cviptoitex**(Image *cvip_IMAGE, char *cvip_name, FILE *outputfile, char *image_comment, int verbose)

void **cviptoiris**(Image *cvipIMAGE, char *f_name, FILE *fp, int prt_type, int verb)

int **cviptojpg**(Image *cvipImage, char *filename, int quality, CVIP_BOOLEAN grayscale, CVIP_BOOLEAN optimize, int smooth, CVIP_BOOLEAN verbose, char *qtablesFile)

void **cviptopnm**(Image *cvip_IMAGE, char *pnm_name, FILE *outfp, int verbose)

void **cviptoras**(Image *ras_IMAGE,char *ras_name,FILE *outfp, int pr_type, int verbose)

void **cviptotiff**(Image *cvip_IMAGE, char *tiff_name, unsigned short compression, unsigned short fillorder, long g3options, unsigned short predictor, long rowsperstrip, int verbose)

CVIP_BOOLEAN **cviptovip**(Image *cvipIMAGE, char *filename, FILE *file, CVIP_BOOLEAN save_history, CVIP_BOOLEAN is_compressed, CVIP_BOOLEAN verbose)

Image ***epstocvip**(char *eps_image, FILE *inputfile, CVIP_BOOLEAN verbose)

Image ***giftocvip**(char *name, FILE *in, int imageNumber, int showmessage)

Image ***gray_binary**(Image *inputIMAGE, int direction)
<inputIMAGE> – pointer to an Image
<direction> – direction (0 = gray->binary 1 = binary->gray)

Image ***iristocvip**(char *f_name, FILE *fp, int format, int verb)

Image ***itextocvip**(char *itex_image, FILE *inputfile, CVIP_BOOLEAN verbose)

Image ***jpegtocvip**(char *filename, int quality, CVIP_BOOLEAN grayscale, CVIP_BOOLEAN optimize, int smooth, CVIP_BOOLEAN verbose, char *qtablesFile)

Image ***pnmtocvip**(char *pnm_file, FILE *ifp, int format, int verbose)

Image ***rastocvip**(char *rasterfile, FILE *ifp, int verbose)

Image ***read_Image**(char *filename, IMAGE_FORMAT format, int showmessages)
<filename> – pointer to an character string containing the file name
<format> – IMAGE_FORMAT, not used, retained for historical compatability
<showmessages> – shall I be verbose?

Image ***tifftocvip**(char *tiff_file, int verbose)

Image ***viptocvip**(char *filename, FILE *file, CVIP_BOOLEAN verbose)

int **write_Image**(Image *cvip_IMAGE, char *filename, CVIP_BOOLEAN retain_image, CVIP_BOOLEAN set_up, IMAGE_FORMAT new_format, CVIP_BOOLEAN showmessages)
<cvip_IMAGE> – pointer to valid CVIP Image structure
<filename> – pointer to an character string containing the file name
<retain_image> – retain image after writing (CVIP_YES or CVIP_NO)?
<set_up> – run setup (CVIP_YES or CVIP_NO)?
<new_format> – enumeration constant specifying the format of the file to be read
<showmessages> – shall I be verbose (CVIP_YES or CVIP_NO)?

12.7 Display Library—Display.lib

The display library contains functions relating to image display and viewing. The *view_Image* function provides the interface for image viewing which is most accessible and flexible for the CVIPlab programmer. The core display functions are in the display.c file (under~CVIPC\Display) and used by both CVIPlab (in *view_Image*) and CVIPtools.

void **view_Image**(Image *inputIMAGE, char *imagename)
<inputIMAGE> – pointer to the input Image structure
<imagename> – character string as the image name in display window

12.8 Feature Extraction Library—Feature.lib

The feature extraction library contains the functions that extract binary (object), histogram, texture, and spectral features from images. All of the feature functions require a labeled image, generated by the *label* function, and any spatial coordinate in the object of interest as input parameters. In addition, the spectral, texture, and histogram features require the original image as input. These three functions, called *spectral_feature*, *texture*, and *hist_feature* use the object selected in the labeled image as a mask on the original image, so that only the selected object is included in the calculations. Features can be extracted from any of the bands of a multiband image by first using the function *extract_band* in the Band library (in CVIPtools, this is done automatically for a multiband image).

FEATURE LIBRARY FUNCTION PROTOTYPES

long **area**(Image *labeledIMAGE, int r, int c)
<labeledIMAGE> – Pointer to the labeled image
<r> – row coordinate of a point on the labled image
<c> – column coordinate of a point on the labled image

double **aspect**(Image *labeledIMAGE, int r, int c)
<labeledIMAGE> – Pointer to the labeled image
<r> – row coordinate of the point on the labled image
<c> – column coordinate of the point on the labled image

int ***centroid**(Image *labeledIMAGE, int r, int c)
<labeledIMAGE> – pointer to a labeled image
<r> – row coordinate of a point on the labled image
<c> – column coordinate of a point on the labled image

int **euler**(Image *labeledIMAGE, int r, int c)
<labeledIMAGE> – pointer to a labeled image
<r> – row coordinate of a point on the labeled image
<c> – column coordinate of a point on the labeled image

double ***hist_feature**(Image *originalIMAGE,IMAGE *labeledIMAGE, int r, int c)
<originalIMAGE> – Pointer to the original image
<labeledIMAGE> – Pointer to the labeled image
<r> – row coordinate of a point on the labled image
<c> – column coordinate of a point on the labled image
Note: Returns 5 histogram features – mean, standard deviation, skew, energy and entropy via a pointer to double, its value is equal to the initial address of a one-dimensional array, which contains the five histogram features for each band. If the original image is a color image, the first five values are for band 0, the next five data are for band 1, and so on.

double **irregular**(Image *labeledIMAGE, int r, int c)
<labeledIMAGE> – pointer to a labeled image
<r> – row coordinate of a point on a labeled image
<c> – column coordinate of a point on a labeled image

Image ***label**(const Image *imageP)
<imageP> – pointer to an Image

double **orientation**(Image *labeledIMAGE, int r, int c)
<labeledIMAGE> – pointer to a labeled image
<r> – row coordinate of a point on a labeled image
<c> – column coordinate of a point on a labeled image

int **perimeter**(Image *labeledIMAGE, int r, int c)
<labeledIMAGE> – pointer to a labeled image
<r> – row coordinate of a point on the labled image
<c> – column coordinate of a point on the labled image

int ***projection**(Image *labeledIMAGE, int r, int c, int height, int width)
<labeledIMAGE> – pointer to a labeled image
<r> – row coordinate of a point on the labeled image
<c> – column coordinate of a point on the labeled image
<height> – image height after the object of interest is normalized
<width> – image width after the object of interest is normalized

double ***rst_invariant**(Image *label_image, int row, int col)
<label_image> – pointer to a labeled Image structure
<row> – a row coordinate within the object of interest
<column> – a column coordinate within the object of interest

POWER ***spectral_feature**(Image *originalIMAGE,IMAGE *labeledIMAGE, int no_of_rings, int no_of_sectors, int r, int c)
<originalIMAGE> – pointer to the original image
<labeledIMAGE> – Pointer to the labeled image
<no_of_rings> – number of rings
<no_of_sectors> – number of sectors
<r> – row coordinate of a point on the labled image
<c> – column coordinate of a point on the labled image
POWER data structure:

```
typedef struct
{
int       no_of_sectors;
int       no_of_bands;
int       imagebands;
double    *dc;
double    *sector;
double    *band;
} POWER;
```

TEXTURE ***texture**(const Image *ImgP, const Image *segP, int band, int r, int c, long int hex_equiv, int distance)
<ImgP> – pointer to source Image structure
<segP> – pointer to labeled Image structure
<band> – the band of the source image to be worked on
<r> – the row co-ordinate of the object
<c> – the column co-ordinate of the object
<hex_equiv> – the hex equivalent of the Texture feature map
<distance> – the pixel distance to calculate the co-occurence matrix

TEXTURE data structure:

```
typedef struct {
/* [0] -> 0 degree, [1] -> 45 degree, [2] -> 90 degree, [3] -> 135 degree, [4] -> average,
[5] -> range (max – min) */
    float ASM[6];            /* (1) Angular Second Moment */
    float contrast[6];       /* (2) Contrast */
    float correlation[6];    /* (3) Correlation */
    float variance[6];       /* (4) Variance */
    float IDM[6];            /* (5) Inverse Difference Moment */
    float sum_avg[6];        /* (6) Sum Average */
    float sum_var[6];        /* (7) Sum Variance */
    float sum_entropy[6];    /* (8) Sum Entropy */
    float entropy[6];        /* (9) Entropy */
    float diff_var[6];       /* (10) Difference Variance */
    float diff_entropy[6];   /* (11) Difference Entropy */
    float meas_corr1[6];     /* (12) Measure of Correlation 1 */
    float meas_corr2[6];     /* (13) Measure of Correlation 2 */
    float max_corr_coef[6];  /* (14) Maximal Correlation Coefficient */
    } TEXTURE;
```

double **thinness**(Image *labeledIMAGE, int r, int c)
<labeledIMAGE> – pointer to a labeled image
<r> – row coordinate of a point on a labeled image
<c> – column coordinate of a point on a labeled image

12.9 Geometry Library—Geometry.lib

The geometry library contains all functions relating to changing image size and orientation, as well as functions that create images of geometric shapes and sinusiodal waves. These functions all return Image structures, except *mesh_to_file* and *display_mesh* which are used as utility functions by the image warping and restoration function, *mesh_warping*.

GEOMETRY LIBRARY FUNCTION PROTOTYPES

Image ***bilinear_interp**(Image *inImg, float factor)
<inImg> – pointer to an image
<factor> – factor > 1 to enlarge, factor < 1 to shrink

Image ***copy_paste**(Image *srcImg, Image *destImg, unsigned start_r, unsigned start_c, unsigned height, unsigned width, unsigned dest_r, unsigned dest_c, CVIP_BOOLEAN transparent);
<srcImg> – source image to copy the subimage
<destImg> – destination image for pasting
<start_r> – row value of the upper-left corner of the subimage on scrImg
<start_c> – column value of the upper-left corner of the subimage on srcImg
<height> – height of desired subimage
<width> – width of desired subimage
<dest_r> – row value of the upper-left corner of the destImg area to paste the subimage

<dest_c> – column value of the upper-left corner of the destImg area to paste the subimage
<transparent> – whether the paste is transparent or not

Image ***create_black**(int width, int height)
<width> – desired image width
<height> – desired image height

Image ***create_circle**(int im_width, int im_height, int center_c, int center_r, int radius)
<im_width> – image width, number of columns
<im_height> – image height, number of rows
<center_c> – circle center column coordinate
<center_r> – circle center rwo coordinate
<radius> – radius of circle

Image ***create_checkboard**(int im_width, int im_height, int first_r, int first_c, int block_c, int block_r)
<im_width> – image width, number of columns
<im_height> – image height, number of rows
<first_c> – first column of checkerboard
<first_r> – first row fo checkerboard
<block_c> – width of checkerboard blocks
<block_r> – height of checkerboard blocks

Image ***create_cosine**(int img_size, int frequency, int choice)
<img_size> – number of rows (and columns) in new image
<frequency> – sine wave frequency
<choice> – enter 1 for horizontal, 2 for vertical cosine wave

Image ***create_degenerate_circle**(int im_width, int im_height, int Center_c, int center_r, int radius1, int radius2)
<im_width> – image width, number of columns
<im_height> – image height, number of rows
<center_c> – circle center column coordinate
<center_r> – circle center row coordinate
<radius1> – radius of circle
<radius2> – radius of the blur circle (Blur radius)
Note: The blur type is linear.

Image ***create_ellipse**(int width, int height, int centerrow, int centercol, int hor_length, int ver_length)
<width> – image width, number of columns
<height> – image height, number of rows
<centerrow> – row coordinate for the center of the ellipse
<centercol> – column coordinate of center of the ellipse
<hor_length> – length of the horizontal of the ellipse
<ver_length> – length of the vertical axis of the ellipse

Image ***create_line**(int im_width, int im_height, int start_c, int start_r, int end_c, int end_r)
<im_width> – image width, number of columns
<im_height> – image height, number of rows
<start_c> – first row coordinate of line

<start_r> – first row coordinate of line
<end_c> – last column of line
<end_r> – last row of line

Image ***create_rectangle**(int im_width, int im_height, int start_c, int start_r, int rect_width, int rect_height)
<im_width> – image width, number of columns
<im_height> – image height, number of rows
<start_c> – first column for rectangle
<start_r> – first row for rectangle
<rect_width> – width of rectangle
<rect_height> – rectangle height

Image ***create_sine**(int img_size, int frequency, int choice)
<img_size> – number of rows (and columns) in new image
<frequency> – sine wave frequency
<choice> – enter 1 for horizontal, 2 for vertical sine wave

Image ***create_squarewave**(int img_size, int frequency, int choice)
<img_size> – number of rows (and columns) in new image
<frequency> – sine wave frequency
<choice> – enter 1 for horizontal, 2 for vertical square wave

Image ***crop**(Image *imgP, unsigned row_offset, unsigned col_offset, unsigned rows, unsigned cols)
<imgP> – pointer to an image
<row_offset> – row coordinate of upper-left corner
<col_offset> – column coordinate of upper-left corner
<rows> – height of desired subimage
<cols> – width of desired subimage

int **display_mesh**(Image *inputImage, struct mesh *inmesh)

Image ***enlarge**(Image *cvipIMAGE, int row, int col)
<cvipIMAGE> – pointer to an image
<row> – number of rows for enlarged image
<column> – number of columns for enlarged image

void **mesh_to_file**(struct mesh *mesh_matrix, char *mesh_file);

Image ***mesh_warping**(Image *inputImage, struct mesh *inmesh, int method);
<inputImage> – pinter to Image structure
<inmesh> – mesh structure
<method> – Method used for gray level interlopolation: 1-nearest neighbor, 2-bilinear interpolation, 3-neighborhood average
Mesh data structure:

```
struct mesh_node {
     int x;
     int y;
  };
struct mesh {
     int width;
```

```
    int height;
    struct mesh_node** nodes;
  };
```

Image *__object_crop__ (Image *imgP, int no_of_coords, int *rcList, int format, int Rvalue, int Gvalue, int Bvalue)
<imgP> – pointer to the input image structure
<no_of_coords> – the number of (row, col)coordinates in rclist
<rcList> – pointer to an array of alternating [0] = row, [1] = column positions that create a pixel position in the image
<format> – 1 – crop rectangle containing border, 2 – border mask and 3 – border image
<Rvalue> – red value for border color
<Gvalue> – green value for border color
<Bvalue> – blue value for border color

Image *__rotate__(Image *input_IMAGE, float degree)
<input> – pointer to an image
<degree> – amount to rotate image (1–360)

Image *__shrink__(Image *input_IMAGE, float factor)
<input_Image> – pointer to an image
<factor> – scaling factor (0.1–1.0)
int __solve_c__(struct mesh_node intie[4], struct mesh_node outtie[4], float *c)
<intie> – input tie points
<outtie> – output tie points
<c> – pointer to result array

Image *__spatial_quant__(Image *cvipIMAGE, int row, int col, int method)
<cvipIMAGE> – pointer to an image
<row> – number of rows for reduced image
<column> – number of columns for reduced image
<method> – reduction method to use where: 1 = average, 2 = median, 3 = decimation

Image *__translate__(Image *cvipIMAGE, CVIP_BOOLEAN do_warp, int r_off, int c_off, int r_mount, int c_mount, int r_slide, int c_slide, float fill_out)
<cvipIMAGE> – pointer to an image
<do_warp> – wrap image during translation if CVIP_YES
<r_off> – row # of upper-left pixel in area to move
<c_off> – column # of upper-left pixel in area to move
<r_mount> – height of area to move
<c_mount> – width of area to move
<r_slide> – distance to slide vertically
<c_slide> – distance to slide horizontally
<fill_out> – value to fill vacated area in cut-and-paste

Image *__zoom__(Image *input_IMAGE, int quadrant, int r, int c, int width, int height, float temp_factor)
<input_IMAGE> – pointer to an Image
<quadrant> – 1 = UL, 2 = UR, 3 = LL, 4 = LR, 5 = ALL, 6 = Specify (x, y), dx, dy
<r> – column coordinate of area's upper-left corner
<c> – row coordinate of area's upper-left corner

<width> – width of area to enlarge
<height> – height of area to enlarge
<temp_factor> – degree of enlargement

12.10 Histogram Library—Histogram.lib

The histogram library, *Histogram.lib*, contains functions relating to modifying the image by histogram manipulation or gray-level mapping. The histogram shrink operation in CVIPtools is performed with the *remap_Image* function (see *Mapping.lib*).

HISTOGRAM LIBRARY FUNCTION PROTOTYPES

float ****define_histogram**(int bands, int mode, char **eq)
<bands> – number of bands in the image
<mode> – prompt the user for input (mode = 1), or use <eq> (mode = 0)
<eq> – string for mapping equation

float ****get_histogram**(Image *inputP)
<input_image> – pointer to an Image structure from which a histogram is obtained

Image ***get_histogram_Image**(Image *inputP)
<input_image> – pointer to an Image structure from which a histogram is obtained

Image ***gray_linear**(Image *inputIMAGE, double start, double end, double s_gray, double slope, int change)
<inputIMAGE> – pointer to an Image
<start> – initial gray level to modify
<end> – final gray level to modify
<s_gray> – new initial gray level
<slope> – slope of modifying line
<change> – 0 = change out-of-range pixels to black, 1 = don't modify out-of-range pixel values

Image ***gray_multiply**(Image *input, float ratio)
<input> – pointer to an Image
<ratio> – multiplier

Image ***gray_multiply2**(Image *input, float ratio)
<input> – pointer to an Image
<ratio> – multiplier

Image ***histeq**(Image *in, int band)
<in> – pointer to an image
<band> – which band (0, 1, or 2) to operate on; use 0 for gray

void **histogram_show**(float **histogram)
<histogram> – a 2-D array containing a histogram for each image band

Image ***hist_spec**(Image *imageP, int mode, char **input)
<imageP> – pointer to an image

<mode> – prompt the user for input (mode = 1), or use <eq> (mode = 0)
<input> – a 2-D string array for mapping equation for each image band

Image ***histogram_spec**(Image *imageP, float **histogram)
<imageP> – pointer to the input Image structure;
<histogram> – the specified histogram

Image ***hist_slide**(Image *input, int slide)
<input> – pointer to an image
<slide> – amount of histogram slide

Image ***hist_stretch**(Image *inputIMAGE, int low_limit, int high_limit, float low_clip, float high_clip)
<inputIMAGE> – pointer to an Image
<low_limit> – lower limit for stretch
<high_limit> – high limit for stretch
<low_clip> – percentage of low values to clip before stretching
<high_clip> – percentage of high values to clip before stretching

Image ***local_histeq**(Image *in, int size, int mb)
<in> – pointer to an Image structure
<size> – desired blocksize
<mb> – RGB band on which to calculate histogram (0,1,2)

Image ***make_histogram**(float **histogram, IMAGE_FORMAT image_format, COLOR_FORMAT color_format);
<histogram> – a 2-D float array of the histogram data;
<image_format> – the Image format of the resulting image;
<color_format> – the Color format of the resulting image

void **showMax_histogram**(float **histogram, char *title)
<histogram> – pointer to a histogram pointer
<title> – name given to histogram image

12.11 Image Library—Image.lib

Although the image library is a Toolkit library, consisting of lower level functions, the commonly used functions are included here for reference. For complete details, see the *Help* pages in CVIPtools.

int **cast_Image**(Image *src, CVIP_TYPE dtype)
<src> – pointer to Image structure
<type> – new data type

Image ***duplicate_Image**(const Image *a)
<a> – pointer to Image structure

void **delete_Image**(Image *A)
<A> – pointer to Image structure

unsigned **getNoOfBands_Image**(Image *image)
<image> – pointer to an Image

unsigned **getNoOfCols_Image**(Image *image)
<image> – pointer to an Image

unsigned **getNoOfRows_Image**(Image *image)
<image> – pointer to an Image

CVIP_TYPE **getDataType_Image**(Image *image)
<image> – pointer to an Image

Image ***new_Image**(IMAGE_FORMAT image_format, COLOR_FORMAT color_space, int bands, int heigth, int width, CVIP_TYPE data_type, FORMAT data_format)
<image_format> – original file format of image
<color_space> – current color space of image
<bands> – number of spectral bands
<height> – height of image (no. of rows)
<width> – width of image (no. of cols)
<data_type> – current data type of image
<data_format> – specifies real or complex data

12.12 Data Mapping Library—Mapping.lib

Although *Mapping.lib* is a Toolkit library, consisting of lower level functions, the commonly used functions are included here for reference. These functions are used primarily for remapping images for display purposes. For complete details, see the *Help* pages in CVIPtools.

Image ***condRemap_Image**(const Image *imageP, CVIP_TYPE dtype, unsigned dmin, unsigned dmax)
<imageP> – pointer to an Image
<dtype> – datatype of data to be mapped
<dmin> – minimum value for range
<dmax> – maximum value for range

Image ***logMap_Image**(Image *image, int band)
<image> – pointer to Image structure.
<band> – the band to do log mapping: $-1=$ all bands, $0=$ 1st band, $1=$ 2nd band, etc.

Image ***remap_Image**(const Image *imageP, CVIP_TYPE dtype, unsigned dmin, unsigned dmax)
<imageP> – pointer to an Image
<dtype> – datatype of data to be mapped
<dmin> – minimum value for range
<dmax> – maximum value for range

12.13 Morphological Library—Morphological.lib

The morphological library, *Morphological.lib*, contains all functions relating to image morphology. All the functions return Image structures and will operate on binary images. All the functions but *morphIterMod_Image* and *morpho*, which implement the iterative method described in Chapter 4, will accept gray-level images as input. Note that these functions can be used on multiband images with the use of the *extract_band* and *assemble_bands* functions in the Band library.

These functions are available in two forms: those that allow the user to set up the matrix structure for the morphological kernel, and those that only require an integer to specify one of the predefined kernels. The first type, with an **_Image** extension appended to the function name, are more flexible but more difficult to use. The second type, without the **_Image** extension, are easier to use because they require parameters like the morphological functions in CVIPtools.

MORPHOLOGICAL LIBRARY FUNCTION PROTOTYPES

Image ***MorphClose_Image**(Image *imageP, Matrix *kernelP, CVIP_BOOLEAN user_org, int row, int col)
<inputIMAGE> – pointer to an image
<kernelP> – a pointer to a Matrix structure
<user_org> – define center of kernel
<row> – user-defined row of kernel center
<col> – user-defined column of kernel center

Image ***MorphClose**(Image *inputIMAGE, int k_type, int ksize, int height, int width)
<inputIMAGE> – pointer to an Input
<k_type> – kernel type (1 = disk 2 = square 3 = rectangle 4 = cross)
<ksize> – size of the kernal (height and width of mask)
<height/thickness> – for square, rectangle/cross
<width/size> – for rectangle/cross

Image ***MorphDilate_Image**(Image *imageP, Matrix *kernelP, CVIP_BOOLEAN user_org, int row, int col)
<inputIMAGE> – pointer to an image
<kernelP> – a pointer to a Matrix structure
<user_org> – define center of kernel
<row> – user-defined row of kernel center
<col> – user-defined column of kernel center

Image ***MorphDilate**(Image *inputIMAGE, int k_type, int ksize, int height, int width)
<inputIMAGE> – pointer to an Input image structure
<k_type> – kernel type (1 = disk 2 = square 3 = rectangle 4 = cross)
<ksize> – size of the kernal (height and width of mask)
<height/thickness> – for square, rectangle/cross
<width/size> – for rectangle/cross

Image ***MorphErode_Image**(Image *imageP, Matrix *kernelP, CVIP_BOOLEAN user_ org, int row, int col)
<inputIMAGE> – pointer to an image

<kernelP> – a pointer to a Matrix structure
<user_org> – define center of kernel
<row> – user-defined row of kernel center
<col> – user-defined column of kernel center

Image ***MorphErode**(Image *inputIMAGE, int k_type, int ksize, int height, int width)
<inputIMAGE> – pointer to an Input
<k_type> – kernel type (1 = disk 2 = square 3 = rectangle 4 = cross)
<ksize> – size of the kernal (height and width of mask)
<height/thickness> – for square, rectangle/cross
<width/size> – for rectangle/cross

Image ***morphIterMod_Image**(Image *binImage, const Matrix **surMATS, CVIP_BOO LEAN(*const boolFUNC)(CVIP_BOOLEAN a, CVIP_BOOLEAN b), int no_of_sur, int connectedness, int no_of_iter, int f)
<binImage> – pointer to an image (binary image)
<surMATS> – pointer to set S (surrounds) for which a_ij = 1
<boolFUNC> – pointer to Boolean function of form L(a, b) (c_ij = L(a_ij, b_ij))
<no_of_sur> – number of surrounds
<connectedness> – the connectivity scheme being used; one of the constants: FOUR, EIGHT, SIX_NWSE, SIX_NESW.
<no_of_iter> – number of iterations to perform
<f> – number of subfields into which the image tesselation will be divided

Image ***morpho**(const Image *binImage, const char *surround_str, CVIP_BOOLEAN rotate, int boolFUNC, int connectedness, unsigned no_of_iter, int fields)
<binImage> – pointer to Image structure (binary image)
<surround_str> – pointer to a string holding the set of surrounds, such as "1, 7, 8."
<boolFUNC> – integer number for the Boolean function (1–6): 1: 0, 2: !a, 3: ab, 4: a plus;b, 5: a^b, 6: (!a)b
<rotate> – rotate or not (CVIP_YES, CVIP_NO)
<no_of_sur> – number of surrounds
<connectedness> – the connectivity scheme being used (FOUR, EIGHT, SIX_NWSE, or SIX_NESW)
<no_of_iter> – number of iterations to perform
<fields> – number of subfields into which the image tesselation will be divided

Image ***MorphOpen_Image**(Image *imageP, Matrix *kernelP, CVIP_BOOLEAN user_org, int row, int col)
<inputIMAGE> – pointer to an image
<kernelP> – a pointer to a Matrix structure
<user_org> – define center of kernel
<row> – user-defined row of kernel center
<col> – user-defined column of kernel center

Image ***MorphOpen**(Image *inputIMAGE, int k_type, int ksize, int height, int width)
<inputIMAGE> – pointer to an input IMAGE structure
<k_type> – kernel type (1 = disk 2 = square 3 = rectangle 4 = cross)
<ksize> – size of the kernal (height and width of mask)
<height/thickness> – for square, rectangle/cross
<width/size> – for rectangle/cross

12.14 Noise Library—Noise.lib

The noise library, *Noise.lib*, contains all functions that add noise to an image. The amount of noise added to the image, which will determine the signal-to-noise ratio, can be controlled through the variance parameter. The larger the variance, the more noise will be added. Note that a noise-alone image can be created by adding noise to an all black image (an all black image can be created using the function *create_black* in *Geometry.lib*).

NOISE LIBRARY FUNCTION PROTOTYPES

Image ***gamma_noise**(Image *imageP, float *var, int *alpha)
<imageP> – pointer to an image structure
<var> – variance of the noise distribution
<alpha> – alpha parameter for gamma distribution

Image ***gaussian_noise**(Image *imageP, float *var, float *mean)
<imageP> – pointer to an image structure
<var> – variance of the noise distribution
<mean> – mean or average value for distribution

Image ***neg_exp_noise**(Image *imageP, float *var)
<imageP> – pointer to an image structure
<var> – variance of the noise distribution

Image ***rayleigh_noise**(Image *imageP, float *var)
<imageP> – pointer to an image structure
<var> – variance of the noise distribution

Image ***speckle_noise**(Image *imageP, float *psalt, float *ppepper)
<imageP> – pointer to an image structure
<var> – variance of the noise distribution
<psalt> – probability of salt noise (high gray level = 255)
<ppepper> – probability of pepper noise (low gray level = 0)

Image ***uniform_noise**(Image *imageP, float *var, float *mean)
<imageP> – pointer to an image structure
<var> – variance of the noise distribution
<mean> – mean or average value of distribution

12.15 Segmentation Library—Segmentation.lib

The segmentation library, *Segmentation.lib*, contains all functions that perform image segmentation. These functions all require an input Image structure, and any parameters for the specific algorithm; they return the segmented images as Image structures.

SEGMENTATION LIBRARY FUNCTION PROTOTYPES

Image ***fuzzyc_segment**(Image *srcIMAGE, float variance)
<srcIMAGE> – pointer to an image
<variance> – value for Gaussian kernal variance

Image ***gray_quant_segment**(Image *cvipIMAGE, int num_bits)
<cvipIMAGE> – pointer to an image
<num_bits> – number of gray levels desired (2, 4, 8, . . ., 128)

Image ***hist_thresh_segment**(Image *imgP)
<imgP> – pointer to an image sturcture

Image ***igs_segment**(Image *inputIMAGE, int gray_level)
<inputIMAGE> – input image pointer
<gray_level> – the number of gray levels desired (2, 4, 8, . . ., 256)

Image ***median_cut_segment**(Image *imgP, int newcolors, CVIP_BOOLEAN is_bg, Color bg)
<impP> – pointer to an image
<newcolors> – desired number of colors
<is_bg> – is background color?
<bg> – background color
Color data structure:

```
struct ColorType {
   byte r, g, b;
   };
   typedef struct ColorType Color;
```

Image ***multi_resolution_segment**(Image *imgP, unsigned int choice, void *parameters, CVIP_BOOLEAN Run_PCT)
<imgP> – pointer to source Image structure
<level> – the level to begin procedure
<choice> – Predicate test chosen: (1) pure uniformity; (2) local mean vs. global; (3) local std. deviation vs. global mean; (4) Number of pixels within 2 times standard deviation; (5) Weighted gray level distance test; (6) Texture Homogeneity Test
<parameters> – Cutoff value usage determined by predicate test
<Run_PCT> – Choice to run PCT on color images

Image ***pct_median_segment**(Image *imgP, unsigned colors)
<impP> – pointer to an image
<colors> – desired number of colors

Image ***sct_split_segment**(Image *inP, int A_split, B_split)
<imgP> – a pointer to an image structure
<A_split> – number of colors to divide along angle A
<B_split> – number of colors to divide along angle B

Image ***split_merge_segment**(Image *imgP, unsigned int level, unsigned int choice, void *parameters, CVIP_BOOLEAN Run_PCT)
<imgP> – pointer to source Image structure
<level> – the level to begin procedure
<choice> – Predicate test chosen: (1) pure uniformity; (2) local mean vs. global; (3) local std. deviation vs. global mean; (4) Number of pixels within 2 times standard deviation; (5) Weighted gray level distance test; (6) Texture Homogeneity Test
<parameters> – Cutoff value usage determined by predicate test
<Run_PCT> – Choice to run PCT on color images

Image ***threshold_segment**(Image *inputIMAGE, unsigned int threshval, CVIP_BOOLEAN thresh_inbyte)
<inputIMAGE> – pointer to Image structure
<threshval> – threshold value
<thresh_inbyte> – CVIP_NO (0) apply threshval directly to image data; CVIP_YES (1) threshval is CVIP_BYTE range; remap to image data range before thresholding.

Image ***watershed_segment**(Image *inputIMAGE, float threshold, CVIP_BOOLEAN choice)
<inputIMAGE> – pointer to Image structure
<threshold> – threshold value
<choice> – CVIP_NO (0) do not merge result; CVIP_YES (1) to merge result

12.16 Spatial Filter Library—SpatialFilter.lib

The spatial filter library, *SpatialFilter.lib*, contains all functions relating to spatial filtering. Many of the edge detection functions—Kirsch, Robinson, pyramid, Laplacian, Sobel, Roberts, Prewitt, and Frei–Chen—can all be accessed via the function, *edge_detect_filter,* which has preprocessing and postprocessing functions built in. This library also contains the Hough transform, the unsharp masking algorithm, and the visual acuity/night vision simulation function. Note that some of the spatial filtering functions in CVIPtools are implemented using *convolve_filter* and *get_default_filter.*

SPATIAL FILTER LIBRARY FUNCTION PROTOTYPES

Image ***ace2_filter**(Image *inputIMAGE, int size, float alpha, float beta)
<inputImage> – pointer to an Image
<size> – mask size (3, 5, 7, 9, . . .)
<alpha> – local mean factor
<beta> – local gain factor

Image ***acuity_nightvision_filter**(Image *cvipIMAGE, char reason, int threshold, int choice)
<cvipIMAGE> – pointer to an Image
<reason> – y = nightvision, n = acuity simulation
<threshold> – binary threshold for nightvision simulation (pass −1 if acuity selected)
<choice> – visual acuity value, 20, 30, 40, . . .(pass −1 if nightvision selected)

Image ***adaptive_contrast_filter**(Image *inputIMAGE, float k1, float k2, unsigned int kernel_size, float min_gain, float max_gain)
<inputIMAGE> – pointer to Image structure
<k1> – local gain factor multiplier
<k2> – local mean multiplier
<kernal_size> – size of local window (must be odd)
<min_gain> – local gain factor minimum
<max_gain> – local gain factor maximum

Image ***adapt_median_filter**(Image *inputIMAGE, int wmax)
<inputImage> – pointer to an Image
<wmax> – window maximum size (3, 5, 7, 9, . . .)

Image ***alpha_filter**(Image *imageP, int mask_size, int p)
<imageP> – pointer to an Image.
<mask_size> – size of the filtering window (3-> 3×3)
<p> – number of maximum and minimum pixels to be excluded from the mean calculation.

Image ***boiecox_filter**(Image *inputImage, float var, unsigned int do_thresh, unsigned int do_hyst, unsigned int do_thin, float high_factor, float low_factor, Image *Imagethld, Image *Imagehyst)
<inputImage> – pointer to the input image structure
<var> – variance 0.5 <= var <= 5 of Gaussian filter
<do_thresh> – 0 or 1
<do_hyst> – 0 or 1
<do_thin> – 0 or 1
<high_factor> – high threshold scale factor for the hysterisys threshold
<low_factor> – low threshold scale for the hysterisys threshold or threshold scale factor for normal thresholding
<Imagethld> – pointer to an intermediate image structure
<Imagehyst> – pointer to an intermediate image structure

Image ***canny_filter** (float low, float high, float var, Image *inputImage, Image *nonmax_mag, Image *nonmax_dir)
<low> – Low threshold scale factor for the hysteresis threshold value estimated from the image
<high> – High threshold scale factor for the hysteresis threshold value estimated from the image
<var> – Variance, range: 0.5 <= var <= 5
<inputImage> – pointer to the input image structure
<nonmax_mag> – pointer to an intermediate image structure
<nonmax_dir> – pointer to an intermediate image structure

Image ***contra_filter**(Image *imageP, int mask_size, int p)
<imageP> – pointer to an Image
<mask_size> – size of the filtering window (3-> 3×3)
<p> – filter order

Image ***convolve_filter**(Image *imageP, Matrix *filP)
<imageP> – pointer to an Image
<filP> – pointer to a Matrix containing the kernel to be convolved with <imageP>

Image ***edge_detect_filter**(Image *imageP, int program, int mask_choice, int mask_size, int keep_dc, int threshold, int threshold1, int thresh, int thr)
<imageP> – pointer to an Image
<program> – desired edge detector: EDGE_KIRSCH, EDGE_ROBINSON, EDGE_PYRA MID, EDGE_LAPLACIAN, EDGE_SOBEL, EDGE_ROBERTS, EDGE_PREWITT, EDGE_ FREI
<mask_choice> – type of smoothing filter: 1 = Gaussian blur, 2 = generic lowpass 1, 3 = generic lowpass 2, 4 = neighborhood average

<mask_size> – Laplacian/Roberts (1,2); Sobel/Prewitt (3, 5, 7);
<keep_dc> – 0 (no) or 1 (yes)
<threshold> – value for post-processing binary threshold
<threshold1> – Frei–Chen projection method: 1 = Project onto edge subspace, 2 = Project onto line subspace, 3 = Show complete projection
<thresh> – Frei–Chen projection threshold: 1 = Set threshold on edge projection, 22 = Set threshold on line projection, 3 = Smallest angle between the above
<thr> – if <thresh> = 1 or 2, set threshold for angle (in radians) for Frei-Chen

Image ***edge_link_filter**(IMAGE *cvipIMAGE, int connection)
<cvipIMAGE> – pointer to an Image
<connection> – maximum connect distance

Image ***exp_ace_filter**(Image *inputIMAGE, int size, float beta, float alpha)
<inputImage> – pointer to an Image
<size> – mask size (3, 5, 7, 9, . . .)
<alpha> – local gain factor
<beta> – local mean factor

Image ***geometric_filter**(Image *imageP, int mask_size)
<imageP> – pointer to an Image
<mask_size> – size of the filtering window (3-> 3 × 3)

Matrix ***get_default_filter**(PROGRAMS type, int dimension,int direction)
<type> – type of filter needed: BLUR_SPATIAL, DIFFERENCE_SPATIAL, LOWPASS_SPATIAL, LAPLACIAN_SPATIAL, HIGHPASS_SPATIAL
<dimension> – size of blur filter needed
<direction> – direction for difference filter; 0 = horizontal, 1 = vertical

Image ***harmonic_filter**(Image *imageP, int mask_size)
<imageP> – pointer to an Image
<mask_size> – size of the filtering window (3-> 3 × 3)

Image ***hough_filter**(Image *cvipIMAGE, char *name, char *degree_string, int threshold, int connection, int interactive)
<inputIMAGE> – pointer to a binary Image structure
<name> – name of the input image
<degree_string> – a string indicating angles of interest
<threshold> – minimum number of pixels to define a line
<connection> – maximum distance to link on a line
<interactive> – 0 = use above parameters; 1 = read degree_string, thresold, and connection from standard input

void **image_sharp**(Image *inputImage)
<inputImage> – pointer to an Image structure

Image ***log_ace_filter**(Image *inputIMAGE, int size, float alpha, float beta)
<inputImage> – pointer to an Image
<size> – mask size (3, 5, 7, 9, . . .)
<alpha> – local mean factor
<beta> – local gain factor

Image ***maximum_filter**(Image *imageP, int mask_size)
<imageP> – pointer to an Image
<mask_size> – size of the filtering window (3-> 3 × 3)

Image ***mean_filter**(Image *imageP, int mask_size)
<imageP> – pointer to an Image
<mask_size> – size of the filtering window (3-> 3 × 3)

Image ***median_filter**(Image *inputIMAGE, int size)
<inputIMAGE> – pointer to an Image
<size> – mask size (3,5,7,9,. . .)

Image ***midpoint_filter**(Image *imageP, int mask_size)
<imageP> – pointer to an Image.
<mask_size> – size of filtering window (3-> 3 × 3)

Image ***minimum_filter**(Image *imageP, int mask_size)
<imageP> – pointer to an image.
<mask_size> – size of filtering window (3-> 3 × 3)

Image ***mmse_filter**(Image *inputIMAGE, float noise_var, unsigned int kernel_size)
<inputIMAGE> – pointer to an Image
<noise_var> – noise variance of input image
<kernel_size> – kernel size (an odd number)

float **pratt_merit** (Image *inputImage1, Image *inputImage2, float a)
<inputImage1> – pointer to the ideal edge image structure
<inputImage2> – pointer to the output edge image structure from edge detection operation
<a> – Scaling constant that can be adjusted to adjust the penalty for offset edges

Image ***raster_deblur_filter**(Image *cvip_image)
<cvip_image> – pointer to an Image structure

Image ***shen_castan_filter** (Image *inImage, Image *zeroInter, float b, int window_size, float low_thresh, float high_thresh, int thinFactor)
<inImage> – pointer to the input image structure
<zeroInter> – pointer to an intermediate image structure
<b> – smoothing factor for the ISEF function ($0 < b < 1$).
<window_size> – size of window under consideration
<low_thresh> – Low threshold scale factor for the hysteresis threshold value estimated from the image
<high_thresh> – High threshold scale factor for the hysteresis threshold value estimated from the image
<thinFactor> – distance between final line points

Image ***single_filter**(Image *orig_image, float s_c, float s_r, int r_cen, int c_cen, float rot, float beta, int N, float *h, int choice)
<orig_image> – pointer to an Image
<s_c> – horizontal sizing factor, 1 for no change
<s_r> – vertical sizing factor, 1 for no change
<r_cen> – row coordinate for new center, 0 for no change

<c_cen> – column coordinate for new center, 0 for no change
<rot> – angle of rotation, 0 for no change
<beta> – value for beta, typically 0.3 – 0.8
<N> – kernel size (3, 5, 7, . . .)
<h> – kernel array (of size N * N)
<choice> – operation of filter: 1 = (– –); 2 = (+ +); 3 = (+ –); 4 = (– +)

Image ***smooth_filter**(IMAGE *inputIMAGE, int kernel)
<inputIMAGE> – pointer to an Image
<kernel> – kernel size, from 2 to 10

Matrix ***specify_filter**(int row, int col, float **temp)
<row> – number of rows of mask
<col> – number of columns of mask
<temp> – mask value array

Image ***unsharp_filter**(Image *inputIMAGE, int lower, int upper, float low_clip, float high_clip)
<inputIMAGE> – pointer to an Image structure
<lower> – lower limit for histogram shrink (0–254)
<upper> – upper limit for histogram shrink (1–255)
<low_clip> – percentage of low values to clip during hist_stretch
<high_clip> – percentage of high values to clip during hist_stretch

Image ***Ypmean_filter**(Image *imageP, int mask_size, int p)
<imageP> – pointer to an Image structure
<mask_size> – size of the filtering window (3-> 3 × 3)
<p> – filter order

12.17 Transform Library—Transform.lib

The transform library, *Transform.lib,* contains all functions relating to frequency or sequency domain transforms. The transforms have all been implemented with fast algorithms, so they require the input images to have dimensions that are powers of two. If the input images have non-power-of-two dimensions, the images will be automatically padded with zeros to conform to this criterion. If zero-padding is required, using a small block size will minimize it. These functions all return Image structures.

TRANSFORM LIBRARY FUNCTION PROTOTYPES

Image ***fft_transform**(Image *input_IMAGE, int block_size)
<input_IMAGE> – pointer to an Image
<block_size> – size of the subimages on which to perform the transform (e.g., 8 for 8 × 8 blocks)

Image ***fft_phase**(Image *fftIMAGE, int remap_norm, float k)
<fftIMAGE> – pointer to a complex image structure

<remap_norm> – 0 = remaps the phase data and returns a CVIP_BYTE image; 1 = normalizes the magnitude, using value of k, returns a complex image
<k> – constant to normalize the magnitude

Image ***dct_transform**(Image *inputImage, int blocksize)
<inputIMAGE> – pointer to an Image
<blocksize> – size of the subimages on which to perform the transform (e.g., 8 for 8 × 8 blocks)

Image ***haar_transform**(Image *in_IMAGE, int ibit, int block_size)
<in_IMAGE> – pointer to an Image
<ibit> – 1 (forward transform) or 0 (inverse transform)
<block_size> – block size (4, 8, 16,.largest_dimension/2)

Image ***idct_transform**(Image *inputImage, int blocksize)
<inputIMAGE> – pointer to an Image
<blocksize> – block size used for forward transform

Image ***ifft_transform**(Image *in_IMAGE, int block_size)
<in_IMAGE> – pointer to an Image
<block_size> – block size used for forward transform

Image ***wavdaub4_transform**(Image *image, int isign, int lowband)
<image> – pointer to an Image
<isign> – 1 (forward transform) or 2 (inverse transform)
<lowband> – # of rows/(2^([(# bands desired − 1)/3] − 1))

Image ***wavhaar_transform**(Image *image, int isign, int lowband)
<image> – pointer to an Image
<isign> – 1 (forward transform) or 2 (inverse transform)
<lowband> – # of rows/(2^([(# bands desired − 1)/3] − 1))

Image ***walhad_transform**(Image *in_IMAGE, int ibit, int block_size)
<in_IMAGE> – pointer to an Image
<ibit> – 0 = inverse Walsh transform, 1 = Walsh transform, 2 = inverse Hadamard transform, 3 = Hadamard transform
<block_size> – block size (4,8,16,...,largest_dimension/2)

12.18 Transform Filter Library—TransformFilter.lib

The transform filter library, *TransformFilter.lib*, contains all functions relating to transform domain filtering. Lowpass, highpass, bandpass, bandreject filters, are available in both ideal and butterworth filter types. A high-frequency emphasis filter is also included, which uses a Butterworth highpass filter. These filters take an Image structure as input, which is assumed to be the output from a transform function, and they output the filtered transform data as an Image structure. In order to get the filtered image back, the corresponding inverse transform must be applied to the output from these functions. These standard filters assume FFT symmetry; consequently, use with a non-FFT transform requires the use of the function *nonfft_xformfilter*. These filters all use a circular filter shape.

Frequency domain restoration filters, such as Wiener filters and inverse filters are contained in this library. The restoration filters will accept either the original images or the transformed images as inputs. To obtain the restored image, the inverse FFT (*ifft_transform*) must be applied to the restoration filter output. A utility function to create various point spread function (PSF) images for these filters is called *h_image*. The homomorphic filter, typically used in image enhancement to equalize uneven contrast, is also contained in this library.

TRANSFORM FILTER LIBRARY FUNCTION PROTOTYPES

Image ***Butterworth_Band_Pass**(Image *in_IMAGE, int block_size, int dc, int inner, int outer, int order)
<in_IMAGE> – pointer to an Image
<block_size> – desired block size
<dc> – drop(0) or retain(1) dc component
<inner> – inner cutoff frequency
<outer> – outer cutoff frequency
<order> – filter order

Image ***Butterworth_Band_Reject**(Image *in_IMAGE, int block_size, int dc, int inner, int outer, int order)
<in_IMAGE> – pointer to an image
<block_size> – desired block size
<dc> – drop(0) or retain(1) dc component
<inner> – inner cutoff frequency
<outer> – outer cutoff frequency
<order> – filter order

Image ***Butterworth_High**(Image *in_IMAGE, int block_size, int dc, int cutoff, int order)
<in_IMAGE> – pointer to an image
<block_size> – desired block size
<dc> – drop(0) or retain(1) dc component
<cutoff> – cutoff frequency
<order> – filter order

Image ***Butterworth_Low**(Image *in_IMAGE, int block_size, int dc, int cutoff, int order)
<in_IMAGE> – pointer to an image
<block_size> – desired block size
<dc> – drop(0) or retain(1) dc component
<cutoff> – cutoff frequency
<order> – filter order

Image ***geometric_mean**(Image *degr, Image *degr_fn, Image *p_noise, Image *p_orig, float gamma, float alpha, int choice, int cutoff)
<degr> – pointer to the degraded image
<degr_fn> – pointer to the degradation function
<p_noise> – pointer to the noise power spectral density
<p_orig> – pointer to the original image power spectral density
<gamma> – 'gamma' in the generalized restoration equation.

<alpha> – 'alpha' in the generalized restoration equation.
<choice> – 1 to let R(u,v) = 1, or 2 to let R(u,v) = 0; (R(u,v) is restoration filter, this is used when denominator = 0)
<cutoff> – cutoff frequency for filtering

Image ***High_Freq_Emphasis**(Image *in_IMAGE, int block_size, int dc, int Cutoff, float alfa, int order)
<in_IMAGE> – pointer to an Image
<block_size> – desired block size
<dc> – drop(0) or retain(1) dc component
<Cutoff> – cutoff frequency
<alfa> – a constant (typically 1.0 to 2.0)
<order> – filter order

Image ***h_image**(int type, unsigned int height, unsigned int width)
<type> – mask type: 1-Constant, 2-Center weighted, 3-Gaussian
<height> – height of the mask image
<width> – width of the mask image

Image ***homomorphic**(Image *cvipIMAGE, float upper, float lower, int cutoff)
<cvipIMAGE> – pointer to an Image
<upper> – upper limit, > 1
<lower> – lower limit, < 1
<cutoff> – cutoff frequency

Image ***Ideal_Band_Pass**(Image *in_IMAGE, int block_size, int dc, int inner, int outer)
<in_IMAGE> – pointer to an image
<block_size> – desired block size
<dc> – drop(0) or retain(1) dc component
<inner> – inner cutoff frequency
<outer> – outer cutoff frequency

Image ***Ideal_Band_Reject**(Image *in_IMAGE, int block_size, int dc, int inner, int outer)
<in_IMAGE> – pointer to an image
<block_size> – desired block size
<dc> – drop(0) or retain(1) dc component
<inner> – inner cutoff frequency
<outer> – outer cutoff frequency

Image ***Ideal_High**(Image *in_IMAGE, int block_size, int dc, int cutoff)
<in_IMAGE> – pointer to an image
<block_size> – desired block size
<dc> – drop(0) or retain(1) dc component
<cutoff> – cutoff frequency

Image ***Ideal_Low**(Image *in_IMAGE, int block_size, int dc, int cutoff)
<in_IMAGE> – pointer to an image
<block_size> – desired block size
<dc> – drop(0) or retain(1) dc component
<cutoff> – cutoff frequency

Image ***inverse_xformfilter**(Image *numP, Image *denP, int choice, float cutoff)
<numP> – pointer to the numerator, the degraded image
<denP> – pointer to the denominator, the inverse filter (PSF)
<choice> – 1 to let R(u,v) = 1, or 2 to let R(u,v) = 0; (R(u,v) is restoration filter, this is used when denominator = 0)
<cutoff> – cutoff frequency

Image ***least_squares**(Image *degr, Image *degr_fn, Image *snr_approx, float gamma, int choice, int cutoff)
<degr> – pointer to the degraded image
<numP> – pointer to the degradation function
<denP> – pointer to an approximation of Pn/Pf
<gamma> – gamma in least_squares equation
<cutoff> – cutoff frequency
<choice> – 1 to let R(u,v) = 1, or 2 to let R(u,v) = 0; (R(u,v) is restoration filter, this is used when denominator = 0)

Image ***nonfft_xformfilter**(Image *imgP, int block_size, int dc, int filtertype, int p1, int p2, int order)
<imgP> – pointer to an Image structure
<block_size> – size of blocks used in the transform
<dc> – retain the DC term (1) or not (0)
<filtertype> – one of: IDEAL_LOW, BUTTER_LOW, IDEAL_HIGH, BUTTER_HIGH, IDEAL_BAND, BUTTER_BAND, IDEAL_REJECT, BUTTER_REJECT, HIGH_FREQ_EMPHASIS
<p1> – cutoff frequency for lowpass and highpass, lower cutoff for bandpass
<p2> – upper cutoff for bandpass filters
<order> – filter order, if Butterworth filter selected

Image ***notch**(Image *cvipIMAGE, char *name, NOTCH_ZONE *zone, int number, CVIP_BOOLEAN interactive)
<cvipIMAGE> – input image data
<name> – the name of the image
<zone> – a data structure containing information about which part of the image to remove
<number> – number of notches to perform (ignored if interactive = CVIP_YES)
<interactive> – ask for input from keyboard (CVIP_YES or CVIP_NO)
NOTCH_ZONE data structure:

```
typedef struct {
   int x;
   int y;
   int radius;
} NOTCH_ZONE
```

Image ***parametric_wiener**(Image *degr, Image *degr_fn, Image *p_noise, Image *p_orig, float gamma, int choice, int cutoff)
<degr> – pointer to the degraded image
<degr_fn> – pointer to the degradation function
<p_noise> – pointer to the noise power spectral density
<p_orig> – pointer to the original image power spectral density
<gamma> – 'gamma' in the parametric wiener filter equation

<choice> – 1 to let R(u,v) = 1, or 2 to let R(u,v) = 0; (R(u,v) is restoration filter, this is used when denominator = 0)
<cutoff> – cutoff frequency for filtering

Image ***power_spect_eq**(Image *degr, Image *degr_fn, Image *p_noise, Image *p_orig, int choice, int cutoff)
<degr> – pointer to the degraded image
<degr_fn> – pointer to the degradation function
<p_noise> – pointer to the noise power spectral density
<p_orig> – pointer to the original image power spectral density
<choice> – 1 to let R(u,v) = 1, or 2 to let R(u,v) = 0; (R(u,v) is the restoration filter, this is used when denominator = 0)
<cutoff> – cutoff frequency for filtering

Image ***simple_wiener**(Image *degr, Image *degr_fn, Image *denP, float k)
<degr> – pointer to the degraded image
<denP> – pointer to the degradation function
<k> – a constant

Image ***wiener**(Image *degr, Image *degr_fn, Image *p_noise, Image *p_orig, int choice, int cutoff)
<degr> – pointer to the degraded image
<degr_fn> – pointer to the degradation function
<p_noise> – pointer to the noise power spectral density
<p_orig> – pointer to the original image power spectral density
<choice> – 1 to let R(u,v) = 1, or 2 to let R(u,v) = 0; (R(u,v) is restoration filter, this is used when denominator = 0)
<cutoff> – cutoff frequency for filtering

Section V

Appendices

Appendix A

The CVIPtools CD-ROM

The CVIPtools CD-ROM contains all the necessary files and information to setup and maintain a CVIPtools environment. This includes:

- CVIPtools C source code
- CVIPtools COM source code
- CVIPtools GUI source code
- CVIPtools COM dynamically linked library for Windows, cviptools.dll
- CVIPtools executable for Windows, CVIPtools.exe
- CVIPlab source code
- CVIPtools C code libraries for Windows, *.lib
- CVIPtools header include files for C code
- CVIPlab executable for Windows
- CVIPtools environment installation program
- Help pages for C functions
- Help pages for libraries
- Help pages for COM functions
- Help pages for using CVIPtools
- Images from the book figures
- Book Figures

ORGANIZATION OF THE CD

The CD contains directories of

- ***'CVIPtools'***
  Contains source code, help pages, images, binaries, libraries and setup program for CVIPtools and CVIPlab.
- ***'Book Figures'***
  Contains linked HTML files and Word documents of all book figures. The linked html files are very useful for instructors to use during lectures.
- ***'Networking'***
  Contains a setup program for instructors whose students will use CVIPlab in a networked environment where the students do not have write access to local drives.

and the file

- *'autorun.ini'*
  The file will automatically run the setup program for CVIPtools after you insert the CD into computer: this will install CVIPtools. If it fails to do so, please go to the directory 'CVIPtools' manually and click 'SETUP.EXE' to begin the installation.

Appendix B

Installing and Updating CVIPtools

The entire CVIPtools environment can be setup on any computer running a Windows operating system. This includes the CVIPtools files, the CVIPlab files, the Help files, and the images. The CVIPtools environment can be installed from the CD, and updates can be obtained via the Internet.

CVIPtools ENVIRONMENT

The CVIPtools environment variable *$CVIPtoolsHOME* is specified during installation. During installation, *$CVIPtoolsHOME* is initialized to the directory in which you install CVIPtools. Typically it may be *CVIPtools* or *CVIPtools4.3*.

$CVIPtoolsHOME is set to the top of the CVIPtools source tree. It is used by CVIPtools to locate its various modules. It is also used by the accompanying CVIPlab module to locate libraries and include files.

The Windows version of CVIPtools uses its own image viewer. The related C source code is located in the *$CVIPtoolsHOME*\CVIPC\Display directory.

INSTALLATION OF CVIPtools

- The SETUP program will automatically run when the CVIPtools CD is inserted into the computer. If this fails, you can manually install CVIPtools by running the installation program "SETUP.EXE" by double clicking on the file 'SETUP.EXE' located in directory 'CDROM Drive:\CVIPtools\'.
- Follow the installation instructions in the install window.
- After a successful installation, CVIPtools can be executed by clicking the CVIPtools icon on the desktop.

Note: Please uninstall the old versions of CVIPtools before you install a new version of CVIPtools. However, versions previous to 4.x can remain installed.

ABOUT THE INSTALL PROGRAM

- You can specify the destination directory (CVIPtoolsHOME)
- You can specify the installation method, which has three choices

  ➢ Compact installation (Binary installation)
  This is the easiest way, which will install the CVIPtools executables and other run-time support modules, including the full documentation. It installs the \bin, \HELP and \images directories. With the binary installation, you will

be able to run CVIPtools with the graphical user interface. This type of installation requires about 60 megabytes of disk space.

➢ Typical installation (Library installation)
In addition to the binary installation files, you will also get the libraries which can be linked to create your own applications. You will also get CVIPlab, which is a skeleton program showing how to create stand-alone applications using the libraries. It installs the \bin, \CVIPlab, \HELP, \images, \include and \lib directories. This type of installation requires about 68.2 megabytes of disk space.

➢ Custom installation (Source installation)
With default Custom installation, all the directories under *$CVIPtoolsHOME* are installed. This type of installation requires about 80 megabytes of disk space.

CVIPtools directory organization after installation

- *$CVIPtoolsHOME*\bin
Contains the CVIPtools binary executables 'CVIPtools.exe' and DLL 'cviptools.dll'.

  NOTE: Normally 'cviptools.dll' is registered automatically during the installation. If you would like to unregister it, use command 'regsvr32/u *$CVIPtoolsHOME*\-bin\cviptools.dll' is used to manually register the DLL in DOS shell if necessary.

- *$CVIPtoolsHOME*\CVIPC
Contains the C functions for CVIPtools, only with Source Installation
- *$CVIPtoolsHOME*\CVIPCOM
Contains the COM interface functions for linking the GUI and the underlying C function together, only with Source Installation.
- *$CVIPtoolsHOME*\CVIPGUI
Contains the Visual Basic source code for the CVIPtools GUI, only with Source Installation.
- *$CVIPtoolsHOME*\HELP
Contains the compiled help file 'CVIPtools.chm'.
- *$CVIPtoolsHOME*\images
A default path used by CVIPtools to locate images, which contains all the images in the book figures.
- *$CVIPtoolsHOME*\include
Contains the header files used for C functions of CVIPtools, for use with CVIPlab.
- *$CVIPtoolsHOME*\lib
Contains the libraries for the CVIPtools C functions, for use with CVIPlab.

TO REMOVE (UNINSTALL) CVIPtools

Please follow the steps below to remove old versions of CVIPtools before you install a new version of CVIPtools. Note: Versions previous to 4.x can remain installed.

- Click on 'Start -> Settings -> Control Panel -> Add or Remove Programs'
- Select 'CVIPtools'
- Click 'Change/Remove' button.

TO GET UPDATED VERSIONS OF CVIPtools VIA THE INTERNET

- Access the CVIPtools Homepage at http://www.ee.siue.edu/CVIPtools
- Follow the directions for downloading CVIPtools
- Run the installation program and update CVIPtools as desired

Appendix C

CVIPtools C Functions

This document contains the most recent listing of all C functions available to CVIPtools developers. The functions are grouped by class and library. There are two classes of libraries within CVIPtools—Toolkit and Toolbox.

The Toolkit libraries contain low-level functions, such as data handling and memory management. The Toolbox libraries contain the functions that are typically used by CVIPlab programmers, such as transforms or segmentation routines. This organization is a hierarchical grouping of libraries in which each class successively builds upon the previous class by using the lower-level functions to create higher-level functions. For more detailed information on a particular function see the associated Help pages in CVIPtools.

C.1 Toolkit Libraries

Band.lib—data handling of multi-spectral imagery

assemble_bands	—assembles multiband image from single band images
bandcast	—cast image data to greater precision
bandcopy	—copy band data
band_minmax	—find the min and max values of each band
extract_band	—extracts one band from a multiband image
matalloc	—allocate an array of matrices
matfree	—free memory allocated by matalloc
vecalloc	—allocate an array of vectors
vecfree	—free memory allocated by vecalloc

Image.lib—basic image class methods for type conversion, memory management, etc. (*Note:* see $CVIPtoolsHOME\include\CVIPimage.h for the get and set macros, such as getData_Image)

cast_Image	—cast an image
delete_Image	—Image class destructor
dump_Image	—print image information
duplicate_Image	—create a new instance of an existing image
getBand_Image	—reference a band of matrix data
getBandVector_Image	—unload image bands into a vector
getColorSpace_Image	—get color space of image (e.g. RGB, GRAY, etc.)
getDataFormat_Image	—get data format (i.e. REAL or COMPLEX)
getDataType_Image	—get data type of image (e.g. CVIP_BYTE, CVIP_FLOAT, etc.)
getData_Image	—returns pointer to data (macro in CVIPimage.h)

getFileFormat_Image	—get file format of image (e.g. PPM, PGM, etc.)
getImagPixel_Image	—read an imaginary pixel sample from the image
getImagRow_Image	—reference an imaginary row of the image
getNoOfBands_Image	—get number of data bands of image
getNoOfCols_Image	—get width of image
getNoOfRows_Image	—get height of image
getPixel_Image	—same as "getRealPixel_Image"
getRealPixel_Image	—read a real pixel sample from the image
getRealRow_Image	—reference a real row of the image
getRow_Image	—same as "getRealRow_Image"
history_add	—add info to image history structure
history_check	—check if an operation has been done on an image
history_copy	—copies information from old_story
history_get	—get info about an operation done on an image
history_show	—setup routine for history print
history_print	—performs output of history structure to h_story
makeComplex_Image	—make real image complex
makeReal_Image	—make complex image real
new_Image	—Image class constructor
setBand_Image	—add a new reference to a band of matrix data
setImagPixel_Image	—write an imaginary pixel sample to the image
setPixel_Image	—same as "setRealPixel_Image"
setRealPixel_Image	—write a real pixel sample to the image data

IO.lib—general purpose Input/Output, memory management routines

allocMatrix3D_CVIP	—allocate memory for volume matrix
allocMatrix_CVIP	—allocate memory for regular matrix
close_CVIP	—close a file for reading or writing
error_CVIP	—print error message to terminal
freeMatrix3D_CVIP	—free memory associated with volume matrix
freeMatrix_CVIP	—free memory associated with regulator matrix
getFloat_CVIP	—get floating point value from the user
getInt_CVIP	—get integer value from user
getString_CVIP	—get character string
getUInt_CVIP	—get unsigned integer value from user
init_CVIP	—parse standard info. from command line
msg_CVIP	—print regular message to terminal
openRead_CVIP	—open a file for reading (handles "stdin")
openWrite_CVIP	—open a file for writing (handles "stdout")
perror_CVIP	—print system error message to terminal
print_CVIP	—same as "msg_CVIP" minus extra argument
quiet_CVIP	—turn off messaging
usage_CVIP	—print usage message
verbose_CVIP	—turn on messaging

ObjectManager.lib—object managers/handlers

addhead_DLL	—add link to head of list
addnext_DLL	—add link following the current link
addtail_DLL	—add link to tail of list
delete_DLL	—double Linked List class destructor
find_DLL	—find a particular object in the list

head_DLL	—set current link to head of list
isempty_DLL	—is the list empty?
ishead_DLL	—is current link pointing to head?
istail_DLL	—is current link pointing to tail?
new_DLL	—double Linked List class constructor
next_DLL	—point to next link
previous_DLL	—point to previous link
print_DLL	—print list
Print_reverse_DLL	—print list in reverse order
promote_DLL	—promote current link to head of list
removecurr_DLL	—remove current link
removehead_DLL	—remove link from head of list
removetail_DLL	—remove link from tail of list
replace_DLL	—replace object pointed to be current link
retrieve_DLL	—retrieve object pointed to by current link
size_DLL	—get size of list (number of links)
tail_DLL	—set current link to tail of list
addhead_LL	—add link to head of list
addnext_LL	—add link following the current link
delete_LL	—linked list class destructor
find_LL	—find a particular object in the list
head_LL	—set current link to head of list
isempty_LL	—is the list empty?
ishead_LL	—is current link pointing to head?
istail_LL	—is current link pointing to tail?
new_LL	—linked list class constructor
next_LL	—point to next link
previous_LL	—point to previous link
print_LL	—print list
promote_LL	—promote current link to head of list
removehead_LL	—remove head link
removenext_LL	—remove next link
replace_LL	—replace object pointed to be current link
retrieve_LL	—retrieve object pointed to by current link
size_LL	—return size of list
tail_LL	—set current link to tail of list
addobject_HT	—add object using separate chaining technique
delete_HT	—hash table class destructor
findobject_HT	—find object
new_HT	—hash table class constructor
setkey_HT	—set the hash table key
isempty_Stack	—determine whether a stack is empty
new_Stack	—create a new instance of an object stack
pop_Stack	—pop on object off of the stack
push_Stack	—push an object onto the stack

Mapping.lib—image data mapping functions

condRemap_Image	—if the range is 0–255 no remap is done, if it exceeds this range it is linearly remapped to 0–255
linearTrans_Image	—perform linear mapping of an image through a transformation matrix

logMap_Image	—map image data logarithmically for better display of FFT-transformed images
remap_Image	—map image data into a specified range
trun_Image	—remap image data, maintain relative size of each data band

Matrix.lib—matrix algebra, manipulation and numerical analysis routines (*Note*: see $CVIPtoolsHOME\include\CVIPmatrix.h for the get and set macros, such as getData_ Matrix)

add_Matrix	—add two matrices
and_Matrix	—perform a bitwise AND on two matrices
cbrt_Matrix	—finds cube root of a matrix (real/complex)
clone_Matrix	—returns a new matrix
conj_Matrix	—find complex conjugate of matrix
copy_Matrix	—copy matrix a to matrix b
covariance_Matrix	—find the covariance estimate of N data bands
crop_Matrix	—create a new matrix from region of original
detele_Matrix	—Matrix class destructor
det_Matrix	—find the determinant of a matrix
duplicate_Matrix	—create new instance of an existing matrix
eigenSystem_Matrix	—find the eigenvectors of a matrix
fastCopy_Matrix	—faster copy if data types are the same
getDataFormat_Matrix	—get data format (i.e. REAL or COMPLEX)
getData_Matrix	—same as "getRealData_Matrix"
getDataType_Matrix	—get data type of matrix (e.g. CVIP_BYTE, CVIP_FLOAT, etc.)
getImagData_Matrix	—reference imaginary data (mapped into rows)
getImagRow_Matrix	—get row of imaginary row
getImagVal_Matrix	—get an "imaginary" matrix element
getNoOfCols_Matrix	—get number of columns in matrix
getNoOfRows_Matrix	—get number of rows in matrix
getRealData_Matrix	—reference real data (mapped into rows)
getRealRow_Matrix	—get row of real data
getRealVal_Matrix	—get a "real" matrix element
getRow_Matrix	—same as "getRealRow_Matrix"
getVal_Matrix	—same as "getRealVal_Matrix"
invert_Matrix	—invert a matrix
mag_Matrix	—find magnitude of a matrix (real/complex)
makeComplex_Matrix	—make real matrix complex
makeReal_Matrix	—make complex matrix real
mult_Matrix	—perform vector multiplication of two matrices
multPWise_Matrix	—perform piece-wise multiplication
new_Matrix	—Matrix class constructor
print_Matrix	—print contents of matrix in row major form
read_Matrix	—read a matrix structure from disk
rect2pol_Matrix	—convert from rectangular to polar coordinates
scale_Matrix	—scale a matrix by some factor
setImagVal_Matrix	—set an "imaginary" matrix element
setRealVal_Matrix	—set a "real" matrix element
setVal_Matrix	—same as "setRealVal_Matrix"
sqrt_Matrix	—find square root of matrix (real/complex)
square_mag_Matrix	—find magnitude squared of a matrix (real/complex)

sub_Matrix —subtract two matrices
transpose_Matrix —find the transpose of a matrix
write_Matrix —write a matrix structure to disk

Object.lib—object analysis and identification routines
build_ChainCode —find the chain-code "contour" of an object
delete_ChainCode —delete an instance of a chain code object
delete_Object —object class destructor
drawBB_Objects —draw a bounding box around all objects
draw_ChainCode —draw the contour of an object onto an image using the object's chain code
getProp_Object —find object moment properties
getProp__Objects —find moment properties of multiple objects
getXY_ChainCode —turn a chain code into a list of X–Y coordinates
label_Objects —sequentially label objects (used by label function in libfeature)
listToVector_Objects —create an object vector from an object list
match_Object —match an object
new_ChainCode —create a new instance of a chain code object
new_Object —object class constructor
print_ChainCode —print the chain code results to a file
printLabel_Objects —print an object list to a file
print_Object —print object statistics to file
printProp_Objects —print a list of object properties to a file
read_ChainCode —read a chain code from a file
readLabel__Objects —read an object list from a file
read_Object —read object statistics from disk
readProp_Objects —read a list of object properties from a file
report_ChainCode —print out the chain code values
trimList_Objects —trim an object list based on properties

ROI.lib—region/area of interest designation, manipulation of an image
asgnFullImage_ROI —assign ROI as full image dimension
asgnImage_ROI —assign a ROI to an image
delete_ROI —ROI class destructor
getDataFormat_ROI —get data format of ROI
getDataType_ROI —get data type of ROI
getHorOffset_ROI —get horizontal offset from pixel (0, 0)
getHorSize_ROI —get height/horizontal size of region
getImagePixel_ROI —get/read imaginary pixel sample from ROI
getImagRow_ROI —reference imaginary row from the ROI
getNoOfBands_ROI —get number of data bands in ROI
getNoOfCols_ROI —same as "getHorSize_ROI"
getNoOfRows_ROI —same as "getVerSize_ROI"
getPixel_ROI —same as "getRealPixel_ROI"
getRealPixel_RO —get/read real pixel sample from ROI
getReal/Row_ROI —reference real row from the ROI
getRow_ROI —same as "getRealRow_ROI"
getVerOffset_ROI —get vertical offset from pixel (0, 0)
getVerSize_ROI —get width/vertical size of region
loadRow_ROI —load data from a buffer into ROI

new_ROI —ROI class constructor
setImagPixel_RO —set/write imaginary pixel sample to ROI
setPixel_ROI —same as "setRealPixel_ROI"
setRealPixel_ROI —set/write real pixel sample to ROI
unloadRow_ROI —unload row of data from ROI into buffer

Vector.lib—vector algebra and manipulation routines
band2pixel_Vector —convert a band vector to a pixel vector
convolve_Vector —convolve two vectors
copy_Vector —copy vector a to vector b
findHisto_Vector —find the histogram of a vector
findMaxVal_Vector —return maximum value in vector
findMinVal_Vector —return minimum value in vector
normalize_Vector —normalize a vector between 0 and 1
pixel2band_Vector —convert a pixel vector to a band vector
printHisto_Vector —print histogram values out to a file
subSample_Vector —sub-sample a list of vector points

C.2 Toolbox Libraries

ArithLogic.lib—arithmetic and logical operations on images
add_Image —add two images
and_Image —perform a logical AND on two images
divide_Image —divide one image by another
multiply_Image —multiply two images
not_Image —perform a logical NOT on an imae
or_Image —perform a logical OR on two images
subtract_Image —subtract one image from another
xor_Image —perform a logical XOR on two images

Color.lib—color map utilities and color transforms
colorxform —performs seven color transforms, and inverse transforms
ipct —performs the inverse principal components transform
luminance_Image —performs color to luminance transform
lum_average —performs color to monchrome using average of all bands
pct —performs the principal components transform in RGB-space
pct_color —handles both forward and inverse PCT
pseudocol_freq —pseudocolor transform using FFT spectrum and filters

Compression.lib—image compression/data reduction routines
bit_compress —decomposes grey level image into eight bit planes. Each bitplane is then run-length encoded and stored in a binary file
bit_decompress —decompresses each binary file (corresponding to a particular bit plane) into corresponding binary images
bit_planeadd —decompresses bitplane files and add any combinations of bitplanes to produce the resultant graylevel image.
btc_compress —compress the image in 4*4 blocks, store it in a binary file

btc_decompress —decompress the image from the encoded binary file
btc2_decompress —decompress multilevel BTC encoded image
btc3_decompress —decompress predictive BTC encoded image
btc2_compress —multilevel block truncation coding (BTC) image compression
btc3_compress —predictive BTC compression
dpc_compress —differential predictive coding compression
dpc_decompress —differential predictive coding decompression
frac_compress —fractal compression
frac_decompress — fractal decompression
glr_compress —performs graylevel runlength coding for any window length specified by the user (window range 1–125)
glr_decompress —perform graylevel runlength decoding from the encoded binary file
huf_compress —perform huffman coding and store the probability table and encoded data into a binary file
huff_decompress —perform huffman decoding from the encoded binary file
jpg_compress —JPEG compression
jpg_decompress —JPEG decompression
rms_error —calculates root-mean-square error between two images
snr —calculates peak signal-to-noise ratio in decibels
vq_compress —vector quantization compression
vq_decompress —vector quantization decompression
xvq_compress —transform/vector-qunatization compression
xvq_decompress —transform/vector_quantization decompression
zon_compress —zonal coding based compression, DC quantize
zon_decompress —zonal coding based decompression, DC quantize
zon2_compress —zonal coding based compression, no DC quantize
zon2_decompress —zonal coding based decompression, no DC quantize
zvl_compress —Ziv-Lempel compression
zvl_decompress —Ziv-lempel decompression

Conversion.lib—image conversion, I/O utilities
bintocvip —convert binary (raw) file format to CVIPtools data structure
bmptocvip —convert BMP (Windows raw data) to CVIPtools data structure
ccctocvip —convert CCC file format to CVIPtools data structure
cviptoiris —convert SGI IRIX file format to CVIPtools data structure
CVIPhalftone —quantizes gray image to binary, dithering options
cviptobin —convert CVIPtools data structure to binary (raw) file format
cviptobmp —convert CVIPtools data structure to BMP file format
cviptoccc —convert CVIPtools data structure to CCC file format
cviptoeps —convert CVIPtools data structure to EPS file format
cviptoitex —convert CVIPtools data structure to ITEX file format
cviptojpg —convert CVIPtools data structure to JPEG file format
cviptoras —convert CVIPtools data structure to Sun RAS file format
cviptopnm —convert CVIPtools data structure to PNM file format
cviptotiff —convert CVIPtools data structure to TIFF file format
cviptovip —convert CVIPtools data structure to VIP file format
epstocvip —convert EPS file format to CVIPtools data structure
giftocvip —convert GIF file format to CVIPtools data structure

gray_binary —converts natural binary code to gray code and gray to binary
iristocvip —convert SGI IRIX file format to CVIPtools data structure
itextocvip —convert ITEX file format to CVIPtools data structure
jpgtocvip —convert JPEG file format to CVIPtools data structure
pnmtocvip —convert PNM file format to CVIPtools data structure
rastocvip —convert Sun RAS file format to CVIPtools data structure
read_Image —read image from disk
tifftocvip —convert TIFF file format to CVIPtools data structure
viptocvip —convert VIP file format to CVIPtools data structure
write_Image —write image to disk

Display.lib—display and view functions
view_Image —general purpose image view function for CVIPlab (note: it is a C/C++ function)

Feature.lib—feature extraction functions
area —find area of binary object (number of pixels)
aspect —find aspect ratio (based on bounding box) of binary object
centroid —find row and column coordinates of a binary object
euler —find Euler number of a binary object
hist_feature —find histogram features: mean, standard deviation, skew, energy, entropy
irregular —find irregularity (1/thinness ratio) of binary object
label —labels connected objects in an image
orientation —finds orientation of a binary object via axis of least second moment
perimeter —find the perimeter length of a binary object
projection —find row and column projections of size normalized object
rst_invariant —finds 7 rotation/scale/translation-invariant moment based on features of binary object
spectral_feature —finds spectral features based on FFT power in rings and sectors
texture —finds 14 texture features for four orientations
thinness —finds thinness ratio of binary object

Geometry.lib—geometry manipulation routines
bilinear_interp —shrinks or enlarges an image using bilinear interpolation to calculate the gray-level value of new pixels
copy_paste —copy a subimage from one image and paste it into another
create_black —create an all-black image
create_cosine —create a cosine wave image of any size and desired frequency
create_checkboard —create a checkerboard image
create_circle —create a binary image of a circle
create_degenerate_circle —create a circle with a linear blur from circle radius to blur radius
create_ellipse —create a binary image of an ellipse
create_line —create a line image
create_rectangle —create a rectangular image
create_sine —create a sine wave image of any size and desired frequency

create_squarewave	—create a square wave image of any size and desired frequency
crop	—crop a subimage from an image
display_mesh	—displays a mesh file as an image, used in warp
enlarge	—enlarges image to a user-specified size
keyboard_to_mesh	—creates a mesh structure from keyboard entry for mesh_warping
mesh_warping	—geometrically distort an image
mesh_to_file	—saves a mesh structure to a file, used in warp
object_crop	—special crop function used in *border mask* and *border image*
rotate	—rotate the given image by an angle specified by the user (Range 1–360 degrees)
shrink	—shrinks the given image by a factor specified by the user (Range 0.1–1)
solve_c	—solves bilinear equation, used with warp
spatial_quant	—quantize an image by one of the following methods: average, median, or decimation
translate	—move the entire image horizontally and/or vertically; also used for cut-and-paste of subimage
zoom	—zoom the given image by a factor specified by the user (Range 1–10)

Histogram.lib—image histogram modification/contrast manipulation routines

define_histogram	—allows user to specify an equation for histogram modification
get_histogram	—generates a histogram array from an image
get_histogram_Image	—generates a histogram Image from an image
gray_linear	—graylevel linear modification
gray_multiply	—remap (if necessary) to byte and multiply, clip at 255
gray_multiply2	—cast image to float, multiplies by constant, outputs float image
histeq	—histogram equalization
histogram_show	—prints ASCII representation of a histogram
hist_spec	—perform histogram manipulation using formula specified in by character string(s) for the equation(s)
histogram_spec	—perform histogram manipulation using formula specified by define_histogram
hist_slide	—histogram slide
hist_stretch	—histogram stretch, specify range and percent clip both ends
local_histeq	—local histogram equalization
make_histogram	—generates an image of a histogram
showMax_histogram	—creates an image of a histogram of an image

Morphological.lib—morphological image processing routines

MorphClose_Image	—perform grayscale morphological closing
MorphClose	—perform grayscale morphological closing
MorphDilate_Image	—perform grayscale morphological dilation
MorphDilate	—perform grayscale morphological dilation
MorphErode_Image	—perform grayscale morphological erosion
morphErode	—perform grayscale morphological erosion
morphIterMod_Image	—perform iterative morphological modification of an image
morpho	—performs iterative modification of an image

MorphOpen_Image —perform grayscale morphological opening
MorphOpen —perform grayscale morphological opening

Noise.lib —noise generating routines
gamma_noise —add gamma noise to an image
gaussian_noise —add gaussian noise
neg_exp_noise —add negative-exponential noise
rayleigh_noise —add rayleigh noise
speckle_noise —add speckle (salt-and-pepper) noise
uniform_noise —add uniform noise

Segmentation.lib—image segmentation routines
fuzzyc_segment —perform Fuzzy c-Means color segmentation
gray_quant_segment —perform gray level quantization
hist_thresh_segment —perform adaptive thresholding segmentation
igs_segment —perform improved gray scale (IGS) quantization
median_cut_segment —perform median cut segmentation
multi_resolution_segment —perform multiresolution segmentation
pct_median_segment —perform PCT/median cut segmentation
sct_split_segment —perform SCT/Center Split segmentation
split_merge_segment —perform split and merge, and multiresolution segmentation
threshold_segment —perform binary threshold on an image
watershed_segment —perform watershed segmentation on an image

SpatialFilter.lib—spatial filtering routines and noise creation
ace2_filter —adaptive contrast filter, adapts to local graylevel statistics
acuity_night_vision_filter —visual acuity and night vision application (various blur levels)
adaptive_contrast_filter —adaptive contrast filter, adapts to local gray level statistics
adaptive_median_filter —median filter algorithm that retains image details
alpha_filter —perform an alpha-trimmed mean filter
boiecox_filter —perform a Boie-Cox edge detection
canny_filter —perform a Canny edge detection
contra_filter —perform a contra-harmonic mean filter
convolve_filter —convolve an image with a matrix
edge_detect_filter —perform edge detection on an image (Frei–Chen, Kirsch, Laplacian, Prewitt, Pyramid, Roberts, Robinson, or Sobel)
edge_link_filter —links edge points into lines
exp_ace_filter —adaptive contrast filter using exponential equation
geometric_filter —performs a geometric mean filter
get_default_filter —gets Matrix for predefined spatial masks, used with convolve_filter
harmonic_filter —performs a harmonic mean filter
hough_filter —performs an hough transform, links specified lines
image_sharp —sharpening algorithm II
log_ace_filter —adaptive contrast filter using log equation
maximum_filter —performs a maximum filter
mean_filter —perform a mean filter
median_filter —performs a fast histogram-method median filter
midpoint_filter —performs a midpoint filter
minimum_filter —performs a minimum filter on an image
mmse_filter —minimum mean squared error restoration filter

pratt_merit	—calculate Pratt's figure of merit
raster_deblur_filter	—raster deblurring filter
shen_castan_filter	—perform a Shen-Castan edge detection
single_filter	—performs geometric manipulation and enhancement with a single spatial filter
smooth_filter	—smooths the given image (kernel size in the range 2–10)
unsharp_filter	—performs unsharp masking algorithm
Ypmean_filter	—performs a Yp mean filter

Transform.lib—two-dimensional transforms

fft_phase	—normalizes magnitude and remaps phase data into a BYTE image
fft_transform	—performs block-wise Fast Fourier Transform
dct_transform	—performs block-wise Discrete Cosine Transform
haar_transform	—performs forward or inverse Haar transform
idct_transform	—performs inverse Discrete Cosine Transform
ifft_transform	—performs inverse Fast Fourier Transform
wavtdaub4_transform	—performs wavelet transform based on Daubechies wavelet
wavthaar_transform	—performs wavelet transform based on Haar wavelet
walhad_transform	—performs Walsh/Hadamard transform (forward or inverse)

TransformFilter.lib—transform filtering routines

Butterworth_Band_Pass	—apply Butterworth bandpass filter in transform domain
Butterworth_Band_Reject	—apply Butterworth bandreject filter in transform domain
Butterworth_High	—apply Butterworth highpass filter in transform domain
Butterworth_Low	—apply Butterworth lowpass filter in transform domain
geometric_mean	—geometric mean restoration filter
High_Freq_Emphasis	—perform a high frequency emphasis (HP Butterworth + offset)
h_image	—creates an image for the degradation function, h(r,c)
homorphic	—perform homomorphic filtering
Ideal_Band_Pass	—apply ideal bandpass filter in transform domain
Ideal_Band_Reject	—apply an ideal bandreject filter in transform domain
Ideal_High	—apply ideal highpass filter in transform domain
Ideal_Low	—apply ideal lowpass filter in transform domain
inverse_xformfilter	—perform inverse restoration filter
least_squares	—perform least squares restoration filter
nonfft_xformfilter	—perform standard filters (lowpass, highpass, etc.) on non-FFT symmetry transforms
notch	—perform a notch filter
parametric_wiener	—parametric wiener restoration filter, variable gamma
power_spect_eq	—power spectrum equalization restoration filter
simple_wiener	—simple wiener restoration filter (K parameter)
wiener	—wiener restoration filter

Appendix D

CVIP Resources

This appendix contains useful resources for those involved with computer imaging education, research and application development. In these days of instant world-wide communication it is not meant to be comprehensive, but can serve as a guide to you in the quest for more information.

D.1 Useful CVIP Software (Free or Shareware)

CVIPtools —*http://www.ee.siue.edu/CVIPtools*

JPEG —JPEG C source code
http://www.jpg.org
http://www.jpeg.org

LIBTIFF —Portable library of routines for TIFF files
http://www.libtiff.org

PBMPLUS —Portable library of routines for pnm image files formats
http://www.acme.com/software/pbmplus

ImageJ —Java-based image processing and analysis from National Institutes of Health
http://rsbweb.nih.gov/ij/

ImageTool —Image processing and analysis program
http://ddsdx.uthscsa.edu/dig/itdesc.html

GIMP —GNU image manipulation program
http://www.gimp.org

AdOculus —Imaging educational software, free student version
http://www.theimagingsource.com/prod/soft/adoculos/adoculos.htm

LaboImage —Image processing and analysis software
http://cuiwww.unige.ch/~vision/LaboImage/labo.html

D.2 Useful Internet Resources

Computer Vision Homepage —Site contains links to software, images databases, etc
www.cs.cmu.edu/~cil/vision.html

PEIPA —Contains many links with focus on computer vision
http://peipa.essex.ac.uk —(Pilot European Image Processing Archive)

Open Directory Project —Allows for organized search of imaging software *http://dmoz.org/Computers/Software* —Search on image analysis, image processing or computer vision

Gonzalez and Woods Image Database Links
http://www.imageprocessingbook.com/DIP2E/image databases/image databases.htm

D.3 Professional Societies

These societies sponsor conferences and publish conference proceedings, magazines and journals of interest to imaging professionals

Institute of Electrical and Electronic Engineers (IEEE)
http://www.ieee.org

International Association of Pattern Recognition
http://www.iapr.org/

Society of Motion Picture and Television Engineers (SMPTE)
http://smpte.org

Society of Photo-Optical Instrumentation Engineers (SPIE)
http://spie.org

Association for Computing Machinery (ACM)
http://www.acm.org

International Association of Science and Technology for Development (IASTED)
http://www.iasted.com/

Society for Imaging Science and Technology (IS&T)
http://www.imaging.org/

D.4 CVIP-Related Standards

Acronyms

NTSC —National Television Standards Committee
SECAM —Sequential Coleur Avec Memoire (Sequential Color with Memory)
PAL —Phase Alternation Line
ITU —International Telecommunications Union (formerly CCITT)
ITU-R —International Telecommunications Union-Radio (formerly CCIR)
ISO —International Standards Organization: JPEG/MPEG
ANSI —American Standards Institute
CIE —Commission International de 1'Eclairage
VESA —Video Electronics Standards Association
FCC —Federal Communication Commission

ANSI Standards

ANSI
Washington, DC Headquarters
1819 L Street, NW, 6th F1.
Washington, DC. 20036
Tel: 202 293 8020
Fax: 202 293 9287 11
http://www.ansi.org

ISO Standards

International Organization for standardization (ISO)
1, rue de Varembé, Case postale 56
CH-1211 Geneva 20, Switzerland
Tel: +41 22 749 01 11; Fax: +41 22 733 34 30
http://www.iso.org/iso/en/ISOOnline.frontpage

ITU Standards

International Telecommunication Union
Central Library
Place des Nations
CH-1211 Geneva 20
Switzerland
Tel: +41 22 730 69 00
Fax: +41 22 730 53 26
E-mail: library@itu.int
http://www.itu.int/library/
http://www.itu.int/home/index.html

D.5. Journals and Trade Magazines

IEEE Transactions on Image Processing
IEEE Transactions on Medical Imaging
IEEE Transactions on Pattern Analysis and Machine Intelligence
Pattern Recognition—The Journal of the Pattern Recognition Society
Journal of Electronic Imaging—SPIE and IS&T
IEEE Signal Processing Magazine
IEEE Computer Graphics and Applications
IEEE Robotics and Automation Magazine
OE Magazine—SPIE
Advanced Imaging, www.advancedimagingmag.com
Vision Systems Design, vsd.pennet.com/home.cfm
Photonics Spectra, www.photonics.com/Spectra
Computer Graphics World, cgw.pennet.com/home.cfm
Computerized Medical Imaging and Graphics, publisher Elsevier
Journal of Visual Communication and Image Representation, publisher Elsevier
Image and Vision Computing, publisher Elsevier

Appendix E

CVIPtools Software Organization

E.1 Overview

CVIPtools for Windows is made of several layers: each layer has its own distinct purpose. Lower layers process the image and deal with the algorithms; middle layers provide transparent access to the lower algorithm layers and offer data security to the higher layers. The higher layer interacts with the user and sends commands to execute the operations.

E.2 The Four Layers

CVIPtools for Windows is implemented in four layers: the algorithms code layer, the COM interface layer, the CVIPImage layer (OOP Layer), and the Graphical User Interface (GUI) (see Figure E.1). The algorithms code layer, which is mainly based on the previous version of CVIPtools, consists of all image and data processing procedures and functions. The COM interface layer implements the COM interface for each higher level CVIPtools function, primarily the previous Toolbox functions with a few Toolkit functions, as in Chapter 12. The CVIPImage layer encapsulates the COM interface functions and provides an Object Oriented Architecute (OOA) for the CVIPtools functions, and the Class

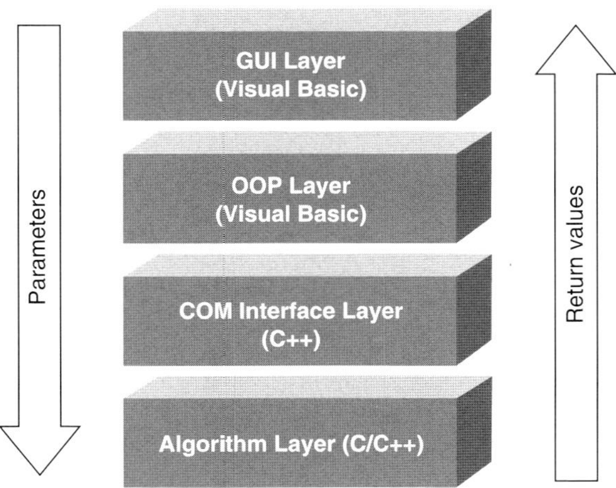

FIGURE E.1
CVIPtools application infrastructure.

CVIPImage helps to consolidate data safety and memory management. The graphical user interface layer, written in Visual Basic, implements the image queue and manages user input and resultant output.

E.3 File and Directory Organization

After CVIPtools installation from the CD, the file and directory organization is as follows. The primary directory at the top level is referred to as *$CVIPtoolsHOME* and is typically called *CVIPtools* or *CVIPtools4.3*, etc. Under this directory are the following directories:

CVIPC contains the C source code for the CVIPtools functions, only with source installation

CVIPCOM contains the C++ COM code and all other associated files including links to the C source code and header files, only with source installation

CVIPGUI contains the Visual Basic code for the GUI, only with source installation

CVIPlab contains the entire CVIPlab source C/C++ code, and all other associated files including links to the libraries and header files (see Chapter 11)

HELP contains the CVIPtools Help executable, *CVIPtoolsHelp.CHM*, compiled with Visual CHM using html files

include contains the header files for CVIPtools C source code

lib contains CVIPtools C/C++ source code libraries for use with CVIPlab

images contains images from the figures in the book

bin contains the executable to run CVIPtools, *CVIPtools.exe*, the library *cviptools.dll* and other associated files and directories used by CVIPtools, including: (1) codebook, (2) compression, (3) feature, (4) mesh, (5) pseudocolor, (6) remap

Appendix F

Common Object Module (COM) Functions—cviptools.dll

The following functions are in the dynamically linked library, *cviptools.dll*. They are all based on Microsoft's common object module interface (COM), and can be used directly by those familiar with this interface. The COM functions are essentially the CVIPtools C functions with a COM wrapper on top. These functions are in the *CVIPTools.cpp* file in the *$CVIPtoolsHOME\CVIPCOM* directory.

Note that the CVIPtools installation program performs system registration for *cviptools.dll*, which is required to run CVIPtools.exe. If any changes are made to *cviptools.dll*, the user needs to perform system registration. This can be done in a DOS shell with this command: *regsvr32 cviptools.dll*.

Ace2_Filter([in] long* image,[in] long wsize,[in]double alpha, [in] double beta,[out,retval] long* Result_Image);

Acuity_Nightvision_Filter([in] long* image,[in] long reason, [in] long threshold, [in] long choice,[out,retval] long* Result_Image);

Adapt_Median_Filter([in] long* image,[in] long mask_size,[out,retval] long* Result_Image);

Adaptive_Contrast_Filter([in] long* image,[in] double k1, [in] double k2,[in] long kernel_size,[in] double min_gain,[in] double max_gain,[out,retval] long* Result_Image);

Add_Image([in] long* input_im1,[in] long* input_im2, [out,retval] long* result_im);

Alpha_Filter([in] long* image,[in] long mask_size,[in] long p,[out,retval] long* Result_Image);

And_Image([in] long* input_im1,[in] long* input_im2, [out,retval] long* result_im);

Area([in] long* image,[in] int r, [in] int c, [out,retval] long* Result);

Aspect([in] long* image,[in] int r, [in] int c, [out,retval] long* Result);

Assemble Bands([in] long* image,[out,retval] long* Result_Image);

Bilinear_Interp([in] long* image,[in] float factor,[out,retval] long* Result_Image);

Bintocvip([in] BSTR*File_Name, [in] long data_bands, [in] long color_order, [in] long interleaved, [in] long height, [in] int width, [in] long verbose, [out, retval] long* Result_Image);

Bit_Compress([in] BSTR *File_Name, [in] long* Input_image, [in]long sect);

Bit_Decompress([in] BSTR*File_Name, [out,retval] long *Result_Image);

Boiecox_Filter([in] long *image, [in] double var, [in] long do_threshold, [in] long do_hyst, [in] long thin, [in] double high_threshold, [in] double low_threshold, [out] long *imageThreshold, [out] long *imageHyst, [out,retval] long* Result_Image);

Btc_Compress([in] BSTR *File_Name, [in] long* Input_image, [in]long blocksize);

Btc_Decompression([in] BSTR *File_Name,[out,retval] long* Result_Image);

Btc2_Compress([in] BSTR *File_Name, [in] long* Input_image, [in]long blocksize);

Btc2_Decompress([in] BSTR *File_Name, [out,retval] long* Result_Image);

Btc3_Compress([in] BSTR *File_Name, [in] long* Input_image, [in]long blocksize);

Btc3_Decompress([in] BSTR *File_Name, [out,retval] long* Result_Image);

Butterworth_Band_Pass([in] long* image,[in] long block_size, [in] long dc,[in] long inner, [in] long outer, [in] long order, [out,retval] long* Result_Image);

Butterworth_Band_Reject([in] long* image,[in] long block_size, [in] long dc, [in] long inner, [in] long outer, [in] long order, [out,retval] long* Result_Image);

Butterworth_High([in] long* image,[in] long block_size, [in] long dc, [in] long cutoff, [in] long order,[out,retval] long* Result_Image);

Butterworth_Low([in] long* image,[in] long block_size, [in] long dc, [in] long cutoff, [in] long order,[out,retval] long* Result_Image);

Canny_Filter([in] long* image, [in] double low, [in] double high, [in] double var, [in] long* nonmax_mag, [in] long *nonmax_dir, [out,retval] long *Result_Image);

Cast_Image([in] long* image,[in] long dtype);

CentroID([in] long* input_iml, [in] long r, [in] long c, [out,retval] VARIANT *result_array);

Check_Bin([in] long* image,[out,retval] long* Result);

Check_xform_history([in] long* image,[in] int filter, [out,retval] long* Result):

Close_Consol( );

Close_Console( );

ColorXform([in] long* image, ([in]long newcspace, [in]double *norm, [in]doubl*ref white,[in] long dir, [out,retval] long* Result_Image);

CondRemap_Image([in] long * image, long dtype, int min, int max,[out,retval] long *Result_Image);

Contra_Filter([in] long* image, [in] long mask_size, [in] long p,[out,retval] long* Result_Image);

Copy_Paste([in] long* src_image, [in] long* dest_image,[in] int start_r, [in] int start_c, [in] int height, [in] int width, [in] int dest_r, [in] int dest_c, [in] long transparent, [out,retval] long* Result_Image);

Create_Black([in] int width, [in] int height, [out,retval] long* Result_Image);

Create_Checkboard([in] int width, [in]int height,[in]int firstx,[in]int firsty,[in]int blockx, [in]int blocky,[out,retval] long* Result_Image);

Create_Circle([in] int width, [in] int height,[in]int centerx,[in]int centery.[in]int radius, out,retval] long* Result_Image);

Create_Cosine([in] int img_size, [in] int frequency, [in] int choice,[out,retval] long* Result_Image);

Create_Degenerate_Circle([in] int width, [in] int height,[in]int centerx,[in]int centery, [in]int radius 1,[in]int radius2,[out,retval] long* Result_Image);

Create_Ellipse([in] int width, [in] int height, [in]int centerx,[in]int centery,[in]int hor_length,[in]int ver_length,[out,retval] long* Result_Image);

Create_Line([in] int width, [in] int height, [in] int tlx, [in] int tly, [in] int brx, [in] int bry,[out,retval] long* Result_Image);

Create_Rectangle([in] int width, [in] int height, [in] int tlx, [in] int tly, [in] int sqwidth, [in] int sqheight, [out,retval] long* Result_Image);

Create_Sine([in] int img_size, [in] int frequency, [in] int choice, [out,retval] long* Result_Image);

Create_Squarewave([in] int img_size, [in] int frequency, [in] int choice, [out,retval] long* Result_Image);

Crop([in] long* image, [in] int row_offset, [in] int col_offset, [in] int rows, [in] int cols,[out,retval][long* Result_Image);

CVIP_OUTPUT([out,retval] BSTR* sMESSAGE);

CVIPhalftone([in] long* image, [in] int halftone, [in] int maxval, [in] float fthreshval, [in] long retain_image, [in] long verbose,[out,retval] long* Result_Image);

Cviptoccc([in] BSTR* File_Name, [in] long maxcolor,[in] long verbose, [in] long dermvis,[in] long* cvip_Image);

Cviptoeps([in] BSTR* File_Name,[in] long* cvip_Image, [in] double scale_x, [in] double scale_y, [in] long band, [in] long verbose);

Cviptogif([in] BSTR *File_Name,[in] long cvip_Image, [in] long interlace, [in] long verbose);

Cviptoiris([in] BSTR *File_Name,[in] long* cvip_Image,[in] long verbose, [in] long prt_type);

Cviptoitex([in] BSTR *File_Name,[in] long* cvip_Image, [in] BSTR *comment,[in] long verbose);

Cviptojpg([in] BSTR *File_Name, [in] *long Input_Image, [in] int quality, [in] long grayscale, [in] long optimize, [in] int smooth, [in] long verbose, [in]BSTR *qtablesFile);

Cviptopnm([in] BSTR *File_Name,[in] long* cvip_Image, [in] long verbose);

Cviptoras([in] BSTR *File_Name,[in] long* cvip_Image, [in] long pr-type, [in] long verbose);

Cviptotiff([in] BSTR *File_Name,[in] long* cvip_Image, [in] long compression [in] long fillorder, [in] long g3options, [in] long predictor, [in] long rowsperstrip, [in] long verbose);

Cviptovip([in] BSTR *File_Name,[in] long* cvip_Image, [in] long verbose,[in] long save_history,[in] long is_compressed);

Date_Range([in] long* input_im1, [out,retval] VARIANT *result_array);

Dct_Transform([in] long* image, [in] long block_size,[out,retval] long* Result_Image);

Delete_Image([in] long* image);

Divide_Image([in] long* input_im1,[in] long* input_im2,[in] longzero2num, [out,retval] long* result_im);

Dpc_Compress([in] BSTR *File_Name, [in] long Input_image, [in]float ratio,[in]int bit_length, [in] int clipping, [in]int direction, [in]int origin, [in]float deviation);

Dpc_Decompress([in] BSTR *File_Name, [out,retval] long* Result_Image);

Draw_mesh([in] long* image,[in] long* pmesh);

Duplicate_Image([in] long* image,[out,retval] long* Result_Image);

DynRLC_Compression([in] BSTR* File_Name, [in] long inputImage, [in] long WindowSize, [out, retval] long *result);

DynRLC_deCompression([in] BSTR* File_Name, [out,retval] long *Result_Image);

Edge_Detect_Filter([in] long* image, [in] long program, [in] long mask_choice, [in] long mask_size, [in] long keep_dc, [in] long threshold,[in] long threshold1, [in] long thresh,[in] long thr.[out,retval] long* Result_Image);

Edge_Link_Filter([in] long* image,[in] long connection,[out,retval] long* Result_Image);

Enlarge([in] long* image, [in] int row, [in] int col,[out,retval] long* Result_Image);

Epstocvip([in] BSTR *File_Name,[in] long verbose);

Euler([in] long* image, [in] long row, [in] long col, [out,retval] long *Result_Image);

Exp_Ace_Filter([in] long* image,[in] long wsize,[in]double alpha, [in] double beta, [out,retval] long* Result_Image);

Extract_Band([in] long* image,[in] int bandno, [out,retval] long* Result_Image);

Fft_Phase([in] long* image,[in] long remap_norm, [in] double k,[out,retval] long* Result_Image);

Fft_Transform([in] long* image, [in] long block_size,[out,retval] long* Result_Image);

File_To_Mesh([in] BSTR* File_Name, [out,retval] long* Result_Image);

Fractal_Compression([in] long *inputImage, [in] BSTR *File_Name, [in] double Tolerate, [in] long min_part, [in] long max_part, [in] long dom_type, [in] long dom_step, [in] long c1, [in] long c2, [in] long s_bits, [in] long o_bits, [out,retval] long *result);

Fractal_deCompression([in] BSTR *File_Name, [out, retval] long *Result_Image);

Fuzzyc_Segment([in] long* image,[in] double variance, [out,retval] long* Result_Image);

Gamma_Noise([in] long* image,[in] double var, [in] int alpha,[out,retval] long* Result_Image);

Gaussian_Noise([in] long* image, [in] double var, [in] double mean,[out,retval] long* Result_Image);

Geometric_Filter([in] long* image,[in] long mask_size, [out,retval] long* Result_Image);

Geometric_Mean([in] long* image,[in] long* image1, [in] long* image2,[in] long* image3,[in] double gamma,[in] double alpha,[in] long choice,[in] long cutoff,[out, retval] long* Result_Image);

get_hist_real([in] long* image,[out,retval] long* Result_Image);

Get_Histogram_Image([in] long* image,[out,retval] long* Result_Image);

Get_max_min_value([in] long* image,[in] double* Result);

GetDataFormat_Image([in] long* image,[out,retval] long* Result);

GetDataType_Image([in] long* image,[out,retval] long* Result);

GetImageInfo([in] long* orig_im, [out,retval] VARIANT *result_array);

Getlast_Hist([in] long* input_im, [in] long* program,[in] int nprogs, [out,retval] long* Result);

GetNoOfBands_Image([in] long* image,[out,retval] long* Result);

GetNoOfCols_Image([in] long* image, [out,retval] long* Results);

GetNoOfRows_Image([in] long* image,[out,retval] long* Result);

giftocvip ([in] BSTR *File_Name,[in] long* cvip_Image, [in] long imageNumber,[in] long showmessage);

Glr_Compress([in] BSTR File_Name, [in] long *Input_image, [in]long win);

Glr_Decompress([in] BSTR *File_Name,[out,retval] long* Result_Image);

Gray_Binary([in] long* image, [in] int direction,[out,retval] long* Result_Image);

Gray_Linear([in] long* image, [in] double start, [in] double end,[in] double s_gray, [in] double slope, [in] int change, [in] int band, [out,retval] long* Result_Image);

Gray_Multiplication([in] long* image,[in] float ratio,[in][long options, [out,retval] long* Result_Image);

Gray_Multiply([in] long* image,[in] float ratio,[out,retval] long* Result_Image);

Gray_Multiply2([in] long* image,[in] float ratio,[out,retval] long* Result_Image);

Gray_Quant_Segment([in] long* image,[in] long num-bits,[out,retval] long* Result_Image);

Graylevel_Quant([in] long* image,[in] long num_bits, [in] long choice, [out,retval] long* Result_Image);

Graylevel_Remap([in] long* inputImage,[in] long bandR, [in] long bandG, [in] long bandB, [out,retval] long *Result_Image);

H_image([in] long type, [in] long height, [in] long width, [out,retval] long* Result_Image);

Haar_Transform([in] long* image,[in] log param1,[in] long param2, [out,retval long* Result_Image);

Harmonic_Filter([in] long* image,[in] long mask_size, [out,retval] long* Result_Image);

High_Freq_Emphasis([in] long* image,[in] long block_size, [in] long dc, [in] long Cutoff, [in] double alfa, [in] long order, [out,retval] long* Result_Image);

Highboost([in] long *inputImage, [in] long MaskSize, [in] long CenterValue, [in] long ifAdd2Origin, [out, retval] long *Result_Image);

HighFreq_Emphasis([in] long * inputImage, [in] long TransformMethod, [in] long CutoffFreq, [in] long FilterOrder, [in] double OffSet, [in] long KeepDC, [out, retval] long * Result_Image);

Highpass_Spatial([in] long *inputImage, [in] long ifAdd2Origin, [out, retval] long * Result_Image);

Hist_Feature([in] long* orig_im, [in] long* labeled_im, [in] long r, [in] long c, [out,retval VARIANT *result_array);

Hist_Slide([in] long* image,[in] int slide,[out,retval] long* Result_Image);

Hist_Spec([in] long * InputImage, [in] BSTR *bandR, [in] BSTR *bandG, [in] BSTR *bandB, [out, retval] long Result_Image);

Hist_Stretch([in] long* image, [in] int low_limit, [in] int high_limt,[in] float low_clip, [in] float high_clip,[out,retval] long* Result_Image);

Hist_Thresh_Segment([in] long* image,[out,retval] long* Result_Image);

Histeq([in] long* image,[in] int mb,[out,retval] long* Result_Image);

Histogram_Spec([in] long* InputImage, [in] double* sped_in,[out, retval] long* Result_Image);

History_Add([in]long* input_im,[in] long* input_history);

History_Check([in] long* image, [in] int program, [out,retval] long* Result);

History_Copy([in] long* input_from,[in] long* input_to);

History_create([in] long prog, [in] long type, [in] float value, [out, retval] long* Result);

History_Get([in] long* image,[in] int program,[out,retval] long* Result);

History_get_data([in] long* history,[in] int program, [out,retval] long* Result);

Homomorphic([in] long* image,[in] double upper, [in] double lower,[in] long cutoff, [in] long ifAdd2Origin, [out,retval] long* Result_Image);

Horizontal_Flip([in] long* image,[out,retval] long* Result_Image);

Hough_Filter([in] long *image, [in] BSTR * name_in, [in] BSTR* degree_string_in,[in] long threshold, [in] long connection, [in] long interactive, [in] long delta_length, [in] long segment_length,[out, retval] long* Result_Image);

Huf_Compress([in] BSTR *File_Name, [in] long* Input_Image);

Huf_Decompress ([in] BSTR *File_Name, [out,retval] long* Result_Image);

Idct_Transform([in] long* image, [in] long block_size, [out,retval] long* Result_Image);

Ideal_Band_Pass([in] long* image,[in] long block_size, [in] long dc, [in] long inner, [in] long outer, [out,retval] long* Result_Image);

Ideal_Band_Reject([in] long* image,[in] long block_size, [in] long dc, [in] long inner, [in] long outer, [out,retval] long* Result_Image);

Ideal_High([in] long* image,[in] long block_size, [in] long dc, [in] long cutoff, [out,retval] long* Result_Image);

Ideal_Low([in] long* image,[in] long block_size, [in] long dc, [in] long cutoff,[out,retval] long* Result_Image);

Ifft_Transform([in] long* image,[in] long block_size,[out,retval] long* Result_Image);

Igs_Segment([in] long* image,[in] long gray_level,[out,retval] long* Result_Image);

Input_Mesh([in] long * mesh_array,[out, retval] long* Result_Image);

Intensity_Slicing([in] long *inputImage,[in] long *lookupTable, [in] long options, [out,retval] long *Result_Image);

Inverse_Xformfilter([in] long* image,[in] long* image1,[in] long choice,[in] double cutoff,[out,retval] long* Result_Image);

Ipct([in] long* image, [in] long is_mask, [out,retval] long* Result_Image);

Irregular([in] long* image,[in] int r, [in] int c, [out,retval] long* Result);

Jpg_Compress([in] BSTR *File_Name, [in] long Input_Image,[in] int quality, [in] long grayscale, [in] long optimize, [in] int smooth, [in] long verbose, [in]BSTR *qtablesFile);

Jpg_Decompress([in] BSTR *File_Name, int colors, [in] long blocksmooth, [in] long grayscale, [in] long nodither, [in] long verbose, [out,retval] long* Result_Image);

Jpgtocvip ([in] BSTR *File_Name, [in] int colors, [in] long blocksmooth, [in] long grayscale [in] long nodither, [in] long verbose,[out, retval] long* Result_Image);

Label([in] long* image,[out,retval] long* Result_Image);

Least_Squares([in] long* image, ([in] long* image1, [in] long* image2,[in] double gamma, [in] long choice,[in] long cutoff,[out,retval] long* Result_Image);

Local_Histeq([in] long* image,[in] int size, [in] int mb,[out,retval] long* Result_Image);

Log_Ace_Filter([in] long* image,[in] long wsize,[in]double alpha, [in] double beta, [out,retval] long* Result_Image);

Log_Remap([in] long* image, [in] long band, [out,retval] long* Result_Image);

LogMap_Image([in] long* image, int band,[out,retval] long* Result_Image);

Lum_Average([in] long* image,[out,retval) long* Result_Image);

Luminance_Image([in] long* image,[out,retval] long* Result_Image);

Make_Histogram([in]double* sped_in[in]long image_format,[in]long color_format,[in] long bands,[out, retval] long* Result_Image);

Maximum_Filter([in] long* image,[in] long mask_size,[out,retval] long* Result_Image);

Mean_Filter([in] long* image,[in] long mask_size,[out,retval] long* Result_Image);

Median_Cut_Segment ([in] long* image,[in] long newcolors, [in] long is_bg, [in] long r_bg, [in] long g_bg, [in] long b_bg, [out,retval] long* Result_Image);

Median_Filter([in] long* image,[in] long mask_size [out,retval] long* Result_Image);

Mesh_To_File([in] BSTR *File_Name [in] long* pmesh);

Mesh_Warping([in] long* image,[in] long* pmesh, [in] long method,[out,retval] long* Result_Image);

Mesh_WarpingRI([in] long* image,[in] long* pmesh, [in] long method, [in] long zero_out,[out,retval] long* Result_Image);

Midpoint_Filter([in] long* image,[in] long mask_size,[out,retval] long* Result_Image);

Minimum_Filter([in] long* image,[in] long mask_size,[out,retval] long* Result_Image);

Mmse_Filter([in] long* image,[in] double noise_var, [in] long kernel_size,[out,retval] long* Result_Image);

MorphClose([in] long* image,int k_type, int ksize, int height,int width,[out,retval] long* Result_Image);

MorphDilate([in] long* image, int k_type, int ksize, int height, int width,[out,retval] long* Result_image);

MorphErode([in] long* image,int k_type, int ksize, int height, int width, [out,retval] long* Result_Image);

Morpho([in] long* input_im1, [in] BSTR *File_Name, [in] long rotate,[in] long boolFUNC, [in] long connectedness, [in] long no_of_iter, [in] long fields, [out,retval] long* Result_Image);

Morpho_com([in] long* image, [in] BSTR *surb_set, [in] long p1, [in] long p2 [in] long p3, [in] long p4, [in] long p5, [out,retval] long* Result_Image);

MorphOpen([in] long* image,int k_type, int ksize, int height, int width,[out,retval] long* Result_Image);

Multi_Resolution_Segment([in] long* image,[in] long choice, [in] double param1, [in] double param2 [in] long run_PCT, [out,retval] long* Result_Image); Multiply_Image ([in] long* input_im 1, [in] long* input_im2, [out,retval] long* result_im);

Neg_Exp_Noise([in] long* image,[in] double var, [out,retval] long* Result_Image);

New_Image (long image_format, long color_space, int bands, int height, int width, long data_type, long data_format,[out,retval] long* Result_Image);

Nonfft_Xformfilter([in] long* image,[in] long block_size, [in] long dc, [in] long filtertype, [in] long p1, [in] double p2, [in] long order, [out,retval] long* Result_Image);

Not_Image([in] long* image,[out,retval] long* Result_Image);

Notch([in] BSTR *File_Name, [in] long x, [in] long y, [in] long radius, [in] long image, [in] long number, [in] long interactive, [out,retval] long* Result_Image);

Object_Crop([in] long* image, [in] long * coordinates, [in] long format, [in] long R, [in] long G,[in] long B,[out,retval] long* Result_Image);

Open_Consol_Redirect_Output( );

Or_Image([in] long* input_im1, [in] long* input_im2, [out,retval] long* result_im);

Orientation([in] long* image, [in] int r, [in] int c, [out,retval] long* Result);

Parametric_Wiener([in] long* image,[in] long* image1,[in] long* image2,[in] long* image3,[in] double gamma,[in] long choice,[in] long cutoff,[out,retval] long* Result_Image);

Pct([in] long* image,[in] long is_mask,[in] double *maskP,[out,retval] long Result_Image);

Pct_Color([in] long* image,[in] long is_mask[in] double *maskP,[in] long choice, [out,retval] long* Result_Image):

Pct_Median_Segment([in] long* image,[in] long colors,[out,retval] long* Result_Image);

Perimeter([in] long* image,[in] int r, [in] int c, [out,retval] long* Result);

Power_Spect_Eq([in] long* image,[in] long* image1,[in] long* image2,[in] long* image3,[in] long choice,[in] long cutoff,[out,retval] long* Result_Image);

Pratt_Merit([in] long *image1, [in] long *image2, [in] double scale_factor, [out,retval] double *Result)

Projection([in] long* input_im1, [in] int r, [in] int c, [in] int height, [in] int width,[out,retval] VARIANT *result_array);

Pseudo_Remap([in] long* image,[in] long lookupTable,[out,retval] long* Result_Image);

Pseudocol_Freq([in] long* image,[in] int inner, [in] int outer, [in] int blow,[in] int bband, [in] int bhigh, [out,retval] long* Result_Image);

Raster_Deblur_Filter([in] long* image, [out,retval] long* Result_Image);

Rayleigh_Noise([in] long* image,[in] double var, [out,retval] long* Result_Image);

Read Image([in] BSTR File_Name, [out, retval] long* Result_Image);

Remap_Image([in] long* image,long dtype, long dmin, long dmax,[out,retval] long* Result_Image);

REMAPP([in] long* image,long dtype, long dmin, long dmax,[out,retval] long* Result_Image);

Rms_Error([in] long* input_im1,[in] long* input_im2, [out,retval] VARIANT *result_array);

Rotate([in] long* image,[in] float degrees,[out,retval] long* Result_Image);

Rst_invariant([in] long* input_im1, [in] long r, [in] long c,[out,retval] VARIANT *result_array);

Save_Compressed Data([in] long* input_im, [in] BSTR *File_Name);

Sct_Split_Segment([in] long* image,[in] long A_split, [in] long B_split,[out,retval] long* Result_Image);

Set_Console([in] long Handle);

Sharpen_I([in] long *inputImage, [in] long ifRemap, [in] long MaskChoice, [in] long MaskSize, [in] float LowClip, [in] float HighClip, [in] long ifAdd2Origin, [out,retval] long *Result_Image);

Sharpen_II([in] long *inputImage, [in} long ifAdd2Origin, [out, retval] long * Result_Image);

Shen_Castan_Filter([in] long *Image,[in] double b, [in] long window_size, [in] double low_threshold, [in double high_threshold, [in] long thin_factor, [in] long *ZeroInter-Image, [out,retval] long *Result_Image);

Show_Image([in} int dc, [in] int x, [in] int y, [in] long* image);

Show_Image_Ex([in] int dc,[in] long* mem_dc, [in] int x, [in]int y, [in] long* image);

Shrink([in] long* image,[in] float factor,[out,retval] long* Result_Image);

Simple_Wiener([in] long* image,[in] long* image1,[in] double k,[out,retval] long* Result_Image);

Single_Filter([in] long* image, [in] double sx, [in] double sy, [in] long xcen, [in] long ycen,[in] double rot, [in] double beta, [in] long N, [in] double *h, [in] long choice, [out,retval] long* Result_Image);

Smooth_Filter([in] long* image,[in] long kernel,[out,retval] long* Result_Image);

Snr([in] long* input_iml,[in] long* input_im2, [out,retval] VARIANT *result_array);

Spatial_Quant([in] long* image,[in] int row,[in] int col,[in] int method,[out,retval] long* Result_Image);

Spec_Hist_Image([in] long *InputImage, [in] BSTR * bandR, [in] BSTR * bandG, [in] BSTR * bandB, [out,retval] long *Result_Image);

Specify_Filter([in] long* image, [in] long mask_height,[in] long mask_width,[in] double *maskP, [in] int normalization, [out,retval] long* Result_Image);

Speckle_Noise([in] long* image,[in] double psalt, [in] double ppepper,[out,retval] long* Result_Image);

Spectral_Feature(long *input_iml, long *input_im2, [in] long no_of_bands, [in] long no_of_sectors,[in] long r,[in] long c,[out,retval] VARIANT *result_array);

Split_Merge_Segment([in] long* image,[in] long level, [in] long choice, [in] double parameter0, [in] double parameter,[in] long Run_PCT, [out,retval] long* Result_Image);

Subtract_Image([in] long* input_im1,[in] long* input_im2, [out,retval] long* result_im);

TextureFeature([in] long* orig_im, [in] long* labeled_im, [in] long r, [in] long c, [in] long distance, [in] long hex_equiv,[out,retval] VARIANT *result_array);

Thinness([in] long* image,[in] int r, [in] int c, [out,retval] double* Result);

Threshold_Segment([in] long* image,[in] long threshval, [in] long thresh_inbyte, [out,retval] long* Result_Image);

Tifftocvip([in] BSTR *File_Name, [in] int verbose, [out, retval] long* Result_Image);

Tile_by_name([in] BSTR *File_Name);

Transform_Compression([in] long *image, [in] BSTR * filename, [in] long color_space, [in] long xform, [in] long WaveletBasis, [in] long subimage_size, [in] long quant, [in] long JPEG_Q_Table, [in] long coding, [in] long data_type, [in] long remap_type, [in] long KeepDC, [out,retval] long *Result_Image);

Transform Sharpen([in] long * inputImage, [in] long TransformMethod, [in] long Cutoff-Freq, [in] long FilterOrder, [in] long Offset, [in] long KeepDC, [in] long Add2Origin, [out, retval] long * Result_Image);

Transform_Smoothing([in] long * inputImage, [in] long TransformMethod, [in] long CutoffFreq, [in] long FilterOrder, [in] long Offset, [in] long KeepDC, [out, retval] long * Result_Image);

Translate([in] long* image, [in] long do_wrap, [in] int y_off, [in] int x_off,[in] int y_mount,[in] int x_mount,[in] int y_slide, [in] int x_zlide, [in] float fill_out,[out,retval] long* Result_Image);

Uniform_Noise([in] long* image, [in] double var, [in] double mean,[out,retval] long* Result_Image);

Unsharp_Filter([in] long* image, [in] long lower, [in] long upper, [in] double low_clip, [in] double high_clip,[out,retval] long* Result_Image);

Vq_Compress([in] long *inputImage,[in] BSTR *File_Name,[in] long cdbook_in_file, [in] long fixed_codebook, [in] double in_error_thres,[in] BSTR* cdbook_file,[in] long in_no_of_entries,[in] long in_row_vector,[in] long in_col_vector, [out, retval] long * result);

Vq_Decompress([in] BSTR *File_Name, [out,retval] long *Result_Image);

Walhad_Transform([in] long* image,[in] long param1,[in] long param2, [out,retval] long* Result_Image);

Watershed_Segment([in] long* inputImage, [in] int choice, [in] float threshold, [out,retval] long * Result_Image);

Wavdaub4_Transform([in] long* image,[in] long param1,[in] long param2, [out,retval] long* Result_Image);

Wavhaar_Transform([in] long* image,[in] long param1,[in] long param2, [out,retval] long* Result_Image);

Wiener([in] long* image,[in] long* image1,[in] long* image2,[in] long* image3,[in] long choice,[in] long cutoff,[out,retval] long* Result_Image);

Write_Image([in] long* input_im,[in]BSTR*File_Name, [in] long retain_image, [in] long set_up, [in]long new_format, [in] long showmessages);

Xor_Image([in] long* input_im1,[in] long* input_im2, [out,retval] long* result_im);

Xvq_Compress([in] long* image, [in] long xform, [in] long scheme, [in] BSTR *filename, [in] long file_type, [in] long remap_type, [in] long dc, [out,retval] long* Result_Image);

Xvq_Decompress([in] BSTR *filename, [out, retval] long *Result_Image);

Ypmean_Filter([in] long* image,[in] long mask_size, [in] long p,[out,retval] long* Result_Image);

Zon_Compress([in] BSTR *File_Name, [in] long* Input_image, [in]int block_size,[in]int choice, [in]int mask_type, [in]float compress_ratio);

Zon_Decompress([in] BSTR *File_Name, [out,retval] long* Result_Image);

Zon2_Compress([in] BSTR *File_Name, [in] long* Input_image, [in]int block_size,[in]int choice, [in]int mask_type, [in]float compress_ratio);

Zon2_Decompress([in] BSTR *File_Name, [out,retval] long* Result_Image);

Zoom([in] long* image, [in] int quadrant, [in] int X, [in] int Y, [in] int dx, [in] int dy, [in] float temp_factor, [out,retval] long* Result_Image);

Zvl_Compress([in] BSTR *File_Name, [in] long* Input_image);

Zvl_Decompress([in] BSTR *File_Name, [out,retval] long* Result_Image);

Index

D

J

K

Q

R

S

T